SELECTED MATERIAL FROM

VOLUME ONE

CHEMISTRY, CELL BIOLOGY AND GENETICS

SELECTED MATERIAL FROM

BIOLOGY

VOLUME ONE

CHEMISTRY, CELL BIOLOGY AND GENETICS

ROBERT J. BROOKER
UNIVERSITY OF MINNESOTA

ERIC P. WIDMAIER
BOSTON UNIVERSITY

LINDA GRAHAM
UNIVERSITY OF WISCONSIN

PETER STILING
UNIVERSITY OF SOUTH FLORIDA

 **Learning Solutions**

Boston   Burr Ridge, IL   Dubuque, IA   New York   San Francisco   St. Louis
Bangkok   Bogotá   Caracas   Lisbon   London   Madrid
Mexico City   Milan   New Delhi   Seoul   Singapore   Sydney   Taipei   Toronto

SELECTED MATERIAL FROM BIOLOGY
VOLUME ONE: *CHEMISTRY, CELL BIOLOGY AND GENETICS*

3 4 5 6 7 8 9 0 DOW DOW 0 9

ISBN 13: 978-0-07-335332-6
ISBN 10: 0-07-335332-9

Editor: Shirley Grall
Production Editor: Jessica Portz
Printer/Binder: RR Donnelley

Robert J. Brooker

Robert J. Brooker (Ph.D., Yale University) received his B.A. in biology at Wittenberg University in 1978. At Harvard, he studied the lactose permease, the product of the *lacY* gene of the *lac* operon. He continues working on transporters at the University of Minnesota, where he is a Professor in the Department of Genetics, Cell Biology, and Development and has an active research laboratory. At the University of Minnesota, Dr. Brooker teaches undergraduate courses in biology, genetics, and cell biology. In addition to many other publications, he has written two editions of the undergraduate genetics text *Genetics: Analysis & Principles,* McGraw-Hill, copyright 2005.

Eric P. Widmaier

Eric P. Widmaier received his Ph.D. in 1984 in endocrinology from the University of California at San Francisco. His research is focused on the control of body mass and metabolism in mammals, the mechanisms of hormone action, and the postnatal development of adrenal gland function. Dr. Widmaier is currently Professor of Biology at Boston University. Among other publications, he is a co-author of *Vander's Human Physiology: The Mechanisms of Body Function*, 10th edition, published by McGraw-Hill, copyright 2006.

Linda E. Graham

Linda E. Graham received her Ph.D. in botany from the University of Michigan, Ann Arbor. Her research explores the evolutionary origin of land-adapted plants, focusing on their cell and molecular biology as well as ecological interactions. Dr. Graham is now Professor of Botany at the University of Wisconsin-Madison. She teaches undergraduate courses in biology and plant biology. She is the co-author of, among other publications, *Algae,* copyright 2000, a major's textbook on algal biology, and *Plant Biology,* copyright 2006, both published by Prentice Hall/Pearson.

Peter D. Stiling

Peter Stiling obtained his Ph.D. from University College, Cardiff, Wales in 1979. Subsequently, he became a Postdoc at Florida State University and later spent two years as a lecturer at the University of the West Indies, Trinidad. During this time, he began photographing and writing about butterflies and other insects, which led to publication of several books on local insects. Dr. Stiling is currently a Professor of Biology at the University of South Florida at Tampa. He teaches graduate and undergraduate courses in ecology and environmental science. He has published many scientific papers and is the author of *Ecology: Theories and Applications*, 4th edition by Prentice Hall/Pearson, copyright 2002. Dr. Stiling's research interests include plant-insect relationships, parasite-host relationships, biological control, restoration ecology, and the effects of elevated carbon dioxide levels on plant herbivore interactions.

The authors are grateful for the help, support, and patience of their families, friends, and students,

Deb, Dan, Nate, and Sarah Brooker,

Maria, Carrie, and Rick Widmaier,

Jim, Michael, and Melissa Graham, and

Jacqui, Zoe, Leah, and Jenna Stiling.

Left to right: Eric Widmaier, Linda Graham, Peter Stiling, and Rob Brooker

BRIEF CONTENTS

A New Biology Book with a Modern Perspective

In addition to being active researchers and experienced writers, our author team has taught majors biology for years. We have taught with the same books that you have. Our goal in creating something new is to offer something better—a comprehensive, modern textbook featuring an evolutionary focus with an emphasis on scientific inquiry.

Through our classroom experiences and research work, we became inspired by the prospect that a new Biology text could move biology education forward. In listening to educators and students, it became clear that we needed to concentrate our efforts on seven crucial areas. These are described briefly below. We will return to each in more detail in the pages that follow.

1. **Experimentation** During the 1970s and 1980s, biological information began to expand at an exponential rate. Biology textbooks grew and to some extent the content suffered as the scientific process was squeezed out by the avalanche of new details. We are committed to striking a better balance between general concepts and experimentation by showing the connection between scientific inquiry and the principles that are learned from such experimentation, especially through the Feature Investigation sections in every chapter.

2. **Modern content** Science is a moving target. Although the content and organization of our Biology textbook is not a dramatic departure from other books, we have added modern content that will better prepare students for future careers in biology. Toward this end, we have received content reviews from over three hundred faculty members from around the world. We are convinced they have helped us produce a book with the most up-to-date content possible.

Striking examples where we feel the content demonstrates a modern approach with an emphasis on recent experimentation include the following:

- Chapter 6 of our Cell Biology Unit explores cell biology at the level of "systems biology" in which the cell is viewed as a group of interacting parts. This allows students to understand how the parts of a cell work together.

- Our Genetics Unit takes a "molecular first" approach so that students will first understand what a gene is, and then consider how genes affect the traits of organisms.

- The Evolution Unit often takes a molecular perspective, and highlights cladistic methods to generate evolutionary relationships. This approach connects evolution at the molecular level and at the level of organisms in their native environments.

- The Diversity Unit has incorporated the newest information regarding evolutionary relationships among modern species. The connection between evolutionary innovation and reproductive success allows students to appreciate why organisms have certain types of traits.

- In our Plant Biology Unit, a much more modern connection has been made between plant structure, function and genetics. Recent information from *Arabidopsis* is often discussed.

- Each chapter in our Animal Biology Unit ends with a section on the modern Impact on Public Health, including the molecular basis of many diseases. In addition, Neuroscience is covered as a mini-Unit of its own, with three complete chapters incorporating some of the most recent information in this exciting area of biology.

- The Ecology Unit also incorporates an evolutionary theme and has an expanded discussion of species interactions. This approach provides students with a deeper understanding of evolutionary adaptations that organisms have.

3. **Evolutionary Perspective** A study of the processes and outcomes of evolution serves to unify the field of biology and the units of our text. Whether describing evolutionary mechanisms at the molecular level or surveying the diversity of life through a view of modern systematics, an understanding of evolution serves to connect and integrate the disciplines of biology.

4. **A Visual Outline** We were determined to create a new art program using both graphics and photography to serve as a "visual outline." We have worked with a large team of scientific illustrators, photographers, educators and students to build an accurate, up-to-date and visually appealing new illustration program that is easy to follow, realistic and instructive.

5. **Ensuring Accuracy** We chose to work as a team of authors to create this new Biology text because the information in our discipline is increasing so rapidly. Each member of our team has experience researching and writing in our respective areas, allowing us to combine efforts to stay abreast of the field. Likewise, we have worked with a much larger team of reviewers, advisors, editors, and accuracy checkers to ensure that this text is as current and accurate as humanly possible.

6. **A Learning System** Starting with a simple outline at the beginning of each chapter, we have focused on the crucial topics in a clear and easy to follow manner. We emphasize

critical thinking and active learning by constantly returning to how science is done, and through several pedagogical devices including our Biological Inquiry Questions which appear often in figure legends throughout the text. We end each chapter with a thorough review section which returns to our outline and emphasizes higher level learning through multiple question types.

7. **Media—Teaching and Learning with Technology**
Our new book is accompanied by a vast array of electronic teaching and learning tools. We have focused on creating new student content that is built upon learning outcomes and assessing student performance. We have also created an unprecedented array of presentation and course management tools to enable instructors to enhance their lectures and manage their classrooms more effectively. Finally, we are committed to offering several electronic book and customized print options to best fit your needs.

Experimentation in Biology Reveals General Principles

Biology is the study of life. The primary way that biologists study life is through experimentation. In this textbook, we have maintained a parallel focus on the general principles of biology and experimentation.

Each chapter is divided into a few sections focusing on general principles in biology. These sections begin with an overview of why the topic is important. We then describe the features of the topic that we and our reviewers have felt are the most important and sometimes the most difficult to grasp.

In describing the principles of biology, we have woven experimentation into each chapter. To be prepared for a career in biology, students need to understand the techniques that are used in biology and bolster their critical thinking skills. Each chapter has a Feature Investigation that shows the steps in the scientific process (for example, see Figure 10.13, p. 200). These investigations include a description of the methods and end with an analysis of the data. This deeper approach allows students to appreciate how the general principles of biology were derived from experiments.

In addition, many scientists are mentioned throughout each chapter with a brief description of how their work contributed to the general principles of biology, reinforcing a sense that biology is an enterprise carried out by scientists around the world. This will prepare students for their next step into the scientific literature in future courses.

An Emphasis on Evolution Provides a Modern Perspective

Evolution is the unifying theme that connects the various areas of biology. We have chosen to explain this theme from a modern perspective by relating the information in each chapter to the genetic material, namely the genomes of organisms. Likewise, because most genes encode proteins, a logical extension is to also relate information to proteomes—the collection of proteins that a cell or organism can make.

We use our Genomes and Proteomes subsections as one way to integrate the various disciplines of biology. For example, let's consider our Cell Biology Unit. In this unit, we emphasize how gene regulation is responsible for the differences between a nerve and muscle cell and how descent with modification occurs at the protein level to produce families of proteins with related cellular functions. Likewise, let's consider our Ecology Unit. In this unit we explain how the modification of particular genes has enabled organisms to compete effectively in their environment. The Ecology Unit also considers how the genomes of organisms have evolved in response to environmental changes over many generations.

As a team, our authors are committed to the idea of integrating the various disciplines in Biology, not separating them. We feel it important, fun, engaging, and actually easy to highlight the evolutionary theme of biology by relating information in each chapter to Genomes and Proteomes. They are intended to provide "perspective". We keep returning to the idea that everything in biology stems from the evolution of genomes, and that genomes primarily encode proteins that ultimately provide organisms with their traits. We feel strongly that this approach provides a modern perspective that will serve our readers well in their future careers. The Genomes and Proteomes subsections are just one way that we integrate the various fields of biology. Genes, proteins, and the molecular mechanisms of life are discussed throughout the entire book.

Textbook Illustrations Are a Key to Learning

In discussions with many of our students, we have come to realize that many, probably most students, are visual learners. They read the textbook for information, but when it comes time to study, their main emphasis is on the figures. Likewise, instructors often rely heavily on good illustrations for their lectures. Therefore, a top priority in the development of our Biology textbook has been the conceptualization and rendering of the illustrations.

As you will see when you scan through this book, the illustrations are very easy to follow, particularly those that have multiple steps. We have taken the attitude that students should be able to look at the figures and understand what is going on, without having to glance back and forth between the text and art. Many figures contain text boxes that explain what the illustration is showing. In those figures with multiple steps, the boxes are numbered so that students will understand that the steps occur in a particular order. In some cases, the numbering was critical when the illustration involved features that could not be presented in a linear manner. For example, a description of hearing in Figure 45.7 is much easier to follow because we have guided the student through the process by using numbered text boxes.

Likewise, technology can help us to engage, to educate and even to inspire our students. As you will see when you skim through the pages, the drawings in this textbook are technologically advanced. They are primarily intended to educate the student and we have maintained a commitment to simplicity. Even so, the illustrations in this textbook are also aimed at being interesting and inspiring. Art elements are drawn with a strong sense of realism and three-dimensionality. The illustrations come to life, which, after all, is important to students who are interested in the study of life. We expect students to occasionally look at a figure and think, "Wow, that's cool!" We invite you to skim through the pages and see for yourself.

Accuracy Is a Top Priority

Inaccuracies in a science textbook come from two primary sources. The first is human error. Authors, editors, and illustrators occasionally make mistakes. Fortunately, such mistakes are rare. Each chapter of our Biology book has been read by dozens of people, including several accuracy checkers whose sole job was to find mistakes. Using our 360-degree developmental process (p. viii), we have worked to ensure an unprecedented level of content accuracy in our textbook.

A second and more common source of error is out-of-date material. Biology is continually changing. New information arrives on a daily basis. Some of this information makes us realize that past information was incorrect. Therefore, textbooks that fail to maintain a current perspective become progressively more inaccurate and misleading. Having an experienced author team with extensive research credentials helps to build a textbook with the most current and accurate information. In addition, over three-hundred faculty members have reviewed our Biology textbook for its content.

We are confident that our book has the most modern content that the industry has to offer. This modern perspective pushes the accuracy of our book to the highest standard possible. In future editions, we will continue to strive for cutting-edge content that maintains a high level of accuracy. In future editions, we will continue to employ the help of many reviewers and accuracy checkers to maintain our commitment to modern content and accuracy.

Our Review Process Ensures a Textbook with the Right Content

If the best writer in the world wrote a textbook single-handedly without input from others, the book would not turn out well. Extensive and open-minded reviews are essential to producing a book that is superior. As we developed our book, we took the attitude that we must always return to a previous draft, analyze it critically, and then revise it accordingly.

From an author's perspective, the review process can be pretty daunting. We turn in chapters that we think are letter perfect, and then receive back reviews that make us painfully aware that writing a textbook is harder than it looks. At the start, we created a process that would make it easier for the editors and reviewers to critically evaluate our work. The first-draft chapters were sent to many outside reviewers who are faculty that either teach General Biology, are experts in the topics found in the chapter, or both. Each first-draft chapter was reviewed by up to 15 different people.

The reviews were collected and provided to the author of a given chapter and the editorial staff. Our editors read the chapters and the reviews, and then gave each author advice on how to make the next draft better. A second important type of input also occurred at the first-draft stage. For each unit, we conducted Focus Groups in which faculty members who had reviewed the chapters of an entire unit came together for a two-day meeting to discuss the chapters with the authors and editors. While written feedback is great, face-to-face discussion often brings out bigger-picture issues that may not be found in written reviews.

At the second-draft stage, we decided upon an innovation that profoundly enhanced the quality of our Biology book. Although the illustrations in a Biology textbook are very expensive to make, we realized that a strong connection between the text and illustrations is critical to produce a superior textbook. Instead of making the illustrations at the final-draft stage, which is typical of textbook publishing, we made them very early in the process. This helped us in two ways. First, reviewers could see the art as the book developed, and make critical changes to it. Second, it allowed the authors and editors to develop a keen sense of consistency between the text and art.

Also at the second-draft stage, we assembled the text and illustrations into a format that looked like a chapter from an actual textbook. We had to keep reminding the reviewers that "These are not finalized chapters. These are early drafts that we want you to critically evaluate." This format allowed the reviewers, authors, and editors to understand how the pieces of the book would fit together. Although it was an exhaustive and rigorous process, adopting this step early in the writing process allowed us to produce a book with a sharp consistency between the text and figures.

We Are Committed to Serving Teachers and Learners

Writing a new textbook is a daunting task. To accurately and thoroughly cover a course as wide ranging as biology, we felt it was essential that our team reflect the diversity of the field. We saw an opportunity to reach students at an early stage in their education and provide their biology training with a solid and up-to-date foundation. We have worked to balance coverage of classic research with recent discoveries that extend biological concepts in surprising new directions or that forge new concepts. Some new discoveries were selected because they highlight scientific controversies, showing students that we don't have all the answers yet. There is still a lot of work for new generations of biologists. With this in mind, we've also spotlighted discoveries made by diverse people doing research in different countries to illustrate the global nature of modern biological science.

As active teachers and writers, one of the great joys of this process for us is that we have been able to meet many more educators and students during the creation of this textbook. It is humbling to see the level of dedication our peers bring to their teaching. Likewise, it is encouraging to see the energy and enthusiasm so many students bring to their studies. We hope this book and its media package will serve to aid both faculty and students in meeting the challenges of this dynamic and exciting course. For us, this remains a work in progress and we encourage you to let us know what you think of our efforts and what we can do to serve you better.

Rob Brooker   brook005@umn.edu
Eric Widmaier   widmaier@bu.edu
Linda Graham   lkgraham@wisc.edu
Peter Stiling   pstiling@cas.usf.edu

The Next Step in Textbook Development

- 10 developmental editors
- 7 developmental focus groups
- 7 art focus groups
- 25 + accuracy checkers
- more than 1,200 reviews by over 350 reviewers across the world
- 11 multiple-day symposia with over 215 majors biology educators participating
- an art development team who worked closely with the authors
- media board of consultants
- 3 photo consultants

The following groups of individuals have been instrumental in ensuring the highest standard of content and accuracy in this textbook. We are deeply indebted to them for their tireless efforts.

Developmental Focus Groups

Cell Unit

Russell Borski,
North Carolina State University
Peter Fajer,
Florida State University
Brad Mehrtens,
University of Illinois–Urbana/Champaign
Randall Walikonis,
University of Connecticut
Sue Simon Westendorf,
Ohio University
Mark Staves,
Grand Valley State University

Genetics Unit

Karl Aufderheide,
Texas A&M University
John Doctor,
Duquesne University
Arlene Larson,
University of Colorado–Denver
Subhash Minocha,
University of New Hampshire
John Osterman,
University of Nebraska–Lincoln
Jill Reid,
Virginia Commonwealth University

Evolution Unit

Mark Decker,
University of Minnesota–Minneapolis
Robert Dill,
Bergen Community College
Jennifer Regan,
University of Southern Mississippi
Michelle Shuster,
New Mexico State University
Fred Wasserman,
Boston University

Diversity Unit

Ernest DuBrul,
University of Toledo
Roland Dute,
Auburn University
Florence Gleason,
University of Minnesota–St. Paul
Ann Rushing,
Baylor University
Randall Yoder,
Lamar University

Plants Unit

Fred Essig,
University of South Florida
Steve Herbert,
University of Wyoming
Mike Muller,
University of Illinois–Chicago
Stuart Reichler,
University of Texas–Austin
Scott Russell,
University of Oklahoma
Rani Vajravelu,
University of Central Florida

Animals Unit

Linda Collins,
University of Tennessee–Chattanooga
William Collins,
Stony Brook University
David Kurjiaka,
Ohio University
Phil Stephens,
Villanova University
David Tam,
University of North Texas
Charles Walcott,
Cornell University

Ecology Unit

James Adams,
Dalton State College
Stanley Faeth,
Arizona State University
Barbara Frase,
Bradley University
Daniel Moon,
University of North Florida
Dan Tinker,
University of Wyoming

Media Focus Group

Russell Borski,
North Carolina State University
Mark Decker,
University of Minnesota
Jon Glase,
Cornell University
John Merrill,
Michigan State University
Melissa Michael,
University of Illinois–Urbana/Champaign
Randall Phillis,
University of Massachusetts–Amherst
Mitch Price,
Pennsylvania State University

Accuracy Checkers

David Asch,
Youngstown State University
Karl Aufderheide,
Texas A&M University
Deborah Brooker
Linda Collins,
*University of Tennessee–
Chattanooga*
Mark Decker,
University of Minnesota
Laura DiCaprio,
Ohio University
Marjorie Doyle,
*University of Wisconsin–
Madison*
Peter Fajer,
Florida State University
Pete Franco,
University of Minnesota
Barbara Frase,
Bradley University

John Graham,
Bowling Green State University
Eunsoo Kim,
*University of Wisconsin–
Madison*
Arlene Larson,
*University of Colorado–
Denver*
David Pennock,
Miami University
Anthony M. Rossi,
University of North Florida
Martin Silberberg,
McGraw-Hill chemistry author
Kevin Strang,
*University of Wisconsin–
Madison*
Fred Wasserman,
Boston University
Jane E. Wissinger,
University of Minnesota

Class Testers

We would like to thank the students and faculty at Ohio University, UCLA, and Calvin College for class testing our book.

End-of-Chapter Questions

Robert Dill,
Bergen Community College
Arlene Larson,
*University of Colorado–
Denver*

Jennifer Regan,
*University of Southern
Mississippi*

Photo Consultants

John Osterman,
*University of Nebraska–
Lincoln*
Sue Simon Westendorf,
Ohio University

Kevin Strang,
*University of Wisconsin–
Madison*

General Biology Symposia

Every year McGraw-Hill conducts several General Biology Symposia, which are attended by instructors from across the country. These events are an opportunity for editors from McGraw-Hill to gather information about the needs and challenges of instructors teaching the major's biology course. It also offers a forum for the attendees to exchange ideas and experiences with colleagues they might not have otherwise met. The feedback we have received has been invaluable, and has contributed to the development of Biology and its supplements.

2006

Michael Bell, *Richland College*
Scott Bowling, *Auburn University*
Peter Busher, *Boston University*
Allison Cleveland,
*University of South Florida–
Tampa*
Sehoya Cotner,
University of Minnesota
Kathryn Dickson,
*California State College–
Fullerton*
Cathy Donald-Whitney,
*Collin County Community
College*
Stanley Faeth,
Arizona State University
Karen Gerhart,
University of California–Davis
William Glider,
University of Nebraska– Lincoln
Stan Guffey,
The University of Tennessee
Bernard Hauser,
*University of Florida–
Gainesville*
Mark Hens,
*University of North Carolina–
Greensboro*
James Hickey,
*Miami University of Ohio–
Oxford*
Sherry Krayesky,
*University of Louisiana–
Lafayette*
Brenda Leady,
University of Toledo
Michael Meighan,
*University of California–
Berkeley*
Comer Patterson,
Texas A&M University
Debra Pires,
*University of California–
Los Angeles*
Robert Simons,
*University of California–
Los Angeles*
Steven D. Skopik,
University of Delaware
Ashok Upadhyaya,
*University of South Florida–
Tampa*
Anthony Uzwiak,
Rutgers University
Dave Williams,
*Valencia Community College–
East Campus*
Jay Zimmerman,
St. John's University

2005

Donald Buckley,
Quinnipiac University
Arthur Buikema,
Virginia Polytechnic Institute
Anne Bullerjahn,
Owens Community College
Garry Davies,
*University of Alaska–
Anchorage*
Marilyn Hart,
Minnesota State University
Daniel Flisser,
Camden County College
Elizabeth Godrick,
Boston University
Miriam Golbert,
College of the Canyons
Sherry Harrel,
Eastern Kentucky University
William Hoese,
*California State University–
Fullerton*
Margaret Horton,
*University of North Carolina
at Greensboro*
Carol Hurney,
James Madison University
James Luken,
Coastal Carolina University
Mark Lyford,
University of Wyoming
Gail McKenzie,
Jefferson State Junior College
Melissa Michael,
*University of Illinois at
Urbana-Champaign*
Subhash C. Minocha,
University of New Hampshire
Leonore Neary,
Joliet Junior College
K. Sata Sathasivan,
University of Texas at Austin
David Senseman,
*University of Texas–
San Antonio*
Sukanya Subramanian,
*Collin County Community
College*
Randall Terry,
Lamar University
Sharon Thoma,
*University of Wisconsin–
Madison*
William Tyler,
*Indian River Community
College*

2004

Jonathan Akin,
Northwestern State University of Louisiana

David Asch,
Youngstown State University

Diane Bassham,
Iowa State University

Donald Buckley,
Quinnipiac University

Ruth Buskirk,
University of Texas, Austin

Charles Creutz,
University of Toledo

Lydia Daniels,
University of Pittsburgh

Laura DiCaprio,
Ohio University

Michael Dini,
Texas Tech University

John Doctor,
Duquesne University

Ernest DuBrul,
University of Toledo

John Elam,
Florida State University

Samuel Hammer,
Boston University

Marilyn Hart,
Minnesota State University

Marc Hirrel,
University of Central Arkansas

Carol Johnson,
Texas A&M University

Dan Krane,
Wright State University

Karin Krieger,
University of Wisconsin–Green Bay

Josephine Kurdziel,
University of Michigan

Martha Lundell,
University of Texas, San Antonio

Roberta Maxwell,
University of North Carolina–Greensboro

John Merrill,
Michigan State University

Melissa Michael,
University of Illinois at Urbana-Champaign

Peter Niewarowski,
University of Akron

Ronald Patterson,
Michigan State University

Peggy Pollak,
Northern Arizona University

Uwe Pott,
University of Wisconsin, Green Bay

Mitch Price,
Pennsylvania State University

Steven Runge,
University of Central Arkansas

Thomas Shafer,
University of North Carolina, Wilmington

Richard Showman,
University of South Carolina

Michèle Shuster,
New Mexico State University

Dessie Underwood,
California State University–Long Beach

Mike Wade,
Indiana University

Elizabeth Willott,
University of Arizona

Carl Wolfe,
University of North Carolina, Charlotte

Reviewers

James K. Adams,
Dalton State College

Sylvester Allred,
Northern Arizona University

Jonathan W. Armbruster,
Auburn University

Joseph E. Armstrong,
Illinois State University

David K. Asch,
Youngstown State University

Amir M. Assadi-Rad,
Delta College

Karl J. Aufderheide,
Texas A&M University

Anita Davelos Baines,
University of Texas–Pan American

Lisa M. Baird,
University of San Diego

Diane Bassham,
Iowa State University

Donald Baud,
University of Memphis

Vernon W. Bauer,
Francis Marion University

Ruth E. Beattie,
University of Kentucky

Michael C. Bell,
Richland College

Steve Berg,
Winona State University

Arlene G. Billock,
University of Louisiana at Lafayette

Kristopher A. Blee,
California State University, Chico

Heidi B. Borgeas,
University of Tampa

Russell Borski,
North Carolina State University

Scott A. Bowling,
Auburn University

Robert Boyd,
Auburn University

Eldon J. Braun,
University of Arizona

Michael Breed,
University of Colorado, Boulder

Randy Brewton,
University of Tennessee, Knoxville

Peggy Brickman,
University of Georgia

Cheryl Briggs,
University of California, Berkeley

Peter S. Brown,
Mesa Community College

Mark Browning,
Purdue University

Cedric O. Buckley,
Jackson State University

Don Buckley,
Quinnipiac University

Arthur L. Buikema, Jr.,
Virginia Tech University

Anne Bullerjahn,
Owens Community College

Ray D. Burkett,
Southeast Tennessee Community College

Stephen P. Bush,
Coastal Carolina University

Peter E. Busher,
Boston University

Jeff Carmichael,
University of North Dakota

Clint E. Carter,
Vanderbilt University

Patrick A. Carter,
Washington State University

Merri Lynn Casem,
California State University, Fullerton

Domenic Castignetti,
Loyola University of Chicago

Maria V. Cattell

David T. Champlin,
University of Southern Maine

Jung H. Choi,
Georgia Institute of Technology

Curtis Clark,
Cal Poly Pomona

Allison Cleveland,
University of South Florida

Janice J. Clymer,
San Diego Mesa College

Linda T. Collins,
University of Tennessee at Chattanooga

Jay L. Comeaux,
Louisiana State University

Bob Connor II,
Owens Community College

Daniel Costa,
University of California at Santa Cruz

Sehoya Cotner,
University of Minnesota

Mack E. Crayton III,
Xavier University of Louisiana

Louis Crescitelli,
Bergen Community College

Charles Creutz,
University of Toledo

Karen A. Curto,
University of Pittsburgh

Mark A. Davis,
Macalester College

Mark D. Decker,
University of Minnesota

Jeffery P. Demuth,
Indiana University

Phil Denette,
Delgado Community College

Donald W. Deters,
Bowling Green State University

Hudson R. DeYoe,
University of Texas–Pan American

Laura DiCaprio,
Ohio University

Randy DiDomenico,
University of Colorado, Boulder

Robert S. Dill,
Bergen Community College

Kevin Dixon,
University of Illinois–Urbana/Champaign

John S. Doctor,
Duquesne University

Michael Meighan,
University of California, Berkeley

Douglas Meikle,
Miami University

Allen F. Mensinger,
University of Minnesota, Duluth

John Merrill,
Michigan State University

Richard Merritt,
Houston Community College

Brian T. Miller,
Middle Tennessee State University

Hugh A. Miller III,
East Tennessee State University

Thomas E. Miller,
Florida State University

Sarah L. Milton,
Florida Atlantic University

Dennis J. Minchella,
Purdue University

Subhash C. Minocha,
University of New Hampshire

Patricia Mire,
University of Louisiana at Lafayette

Daniela S. Monk,
Washington State University

Daniel C. Moon,
University of North Florida

Janice Moore,
Colorado State University

Mathew D. Moran,
Hendrix College

Jorge A. Moreno,
University of Colorado, Boulder

Roderick M. Morgan,
Grand Valley State University

James V. Moroney,
Louisiana State University

Molly R. Morris,
Ohio University

Michael Muller,
University of Illinois at Chicago

Michelle Mynlieff,
Marquette University

Allan D. Nelson,
Tarleton State University

Raymond L. Neubauer,
University of Texas at Austin

Jacalyn S. Newman,
University of Pittsburgh

Colleen J. Nolan,
St. Mary's University

Shawn E. Nordell,
St. Louis University

Margaret Nsofor,
Southern Illinois University, Carbondale

Dennis W. Nyberg,
University of Illinois at Chicago

Nicole S. Obert,
University of Illinois, Urbana-Champaign

David G. Oppenheimer,
University of Florida

John C. Osterman,
University of Nebraska– Lincoln

Brian Palestis,
Wagner College

Julie M. Palmer,
University of Texas at Austin

C. O. Patterson,
Texas A&M University

Ronald J. Patterson,
Michigan State University

Linda M. Peck,
University of Findlay

David Pennock,
Miami University

Shelley W. Penrod,
North Harris College

Beverly J. Perry,
Houston Community College System

Chris Petersen,
College of the Atlantic

Jay Phelan,
UCLA

Eric R. Pianka,
The University of Texas at Austin

Thomas Pitzer,
Florida International University

Peggy E. Pollak,
Northern Arizona University

Richard B. Primack,
Boston University

Lynda Randa,
College of Dupage

Marceau Ratard,
Delgado Community College

Robert S. Rawding,
Gannon University

Jennifer Regan,
University of Southern Mississippi

Stuart Reichler,
University of Texas at Austin

Jill D. Reid,
Virginia Commonwealth University

Anne E. Reilly,
Florida Atlantic University

Linda R. Richardson,
Blinn College

Laurel Roberts,
University of Pittsburgh

Kenneth R. Robinson,
Purdue University

Chris Ross,
Kansas State University

Anthony M. Rossi,
University of North Florida

Kenneth H. Roux,
Florida State University

Ann E. Rushing,
Baylor University

Scott Russell,
University of Oklahoma

Christina T. Russin,
Northwestern University

Charles L. Rutherford,
Virginia Tech University

Margaret Saha,
College of William and Mary

Kanagasabapathi Sathasivan,
The University of Texas at Austin

Stephen G. Saupe,
College of St. Benedict

Jon B. Scales,
Midwestern State University

Daniel C. Scheirer,
Northeastern University

H. Jochen Schenk,
California State University, Fullerton

John Schiefelbein,
University of Michigan

Deemah N. Schirf,
University of Texas at San Antonio

Mark Schlueter,
College of Saint Mary

Scott Schuette,
Southern Illinois University, Carbondale

Dean D. Schwartz,
Auburn University

Timothy E. Shannon,
Francis Marion University

Richard M. Showman,
University of South Carolina

Michele Shuster,
New Mexico State University

Robert Simons,
UCLA

J. Henry Slone,
Francis Marion University

Phillip Snider, Jr.,
Gadsden State Community College

Nancy G. Solomon,
Miami University

Lekha Sreedhar,
University of Missouri– Kansas City

Bruce Stallsmith,
University of Alabama, Huntsville

Susan J. Stamler,
College of Dupage

Mark P. Staves,
Grand Valley State University

William Stein,
Binghamton University

Philip J. Stephens,
Villanova University

Antony Stretton,
University of Wisconsin– Madison

Gregory W. Stunz,
Texas A&M University– Corpus Christi

Julie Sutherland,
College of Dupage

David Tam,
University of North Texas

Roy A. Tassava,
Ohio State University

Sharon Thoma,
University of Wisconsin– Madison

Shawn A. Thomas,
College of St. Benedict/ St. John's University

Daniel B. Tinker,
University of Wyoming

Marty Tracey,
Florida International University

Marsha Turell,
Houston Community College

J. M. Turbeville,
Virginia Commonwealth University

Rani Vajravelu,
University of Central Florida

Neal J. Voelz,
St. Cloud State University

Samuel E. Wages,
South Plains College

Jyoti R. Wagle,
Houston Community College System–Central

Charles Walcott,
Cornell University

Randall Walikonis,
University of Connecticut

Jeffrey A. Walker,
University of Southern Maine

Delon E. Washo-Krupps,
Arizona State University
Frederick Wasserman,
Boston University
Steven A. Wasserman,
*University of California,
San Diego*
R. Douglas Watson,
*University of Alabama
at Birmingham*
Cindy Martinez Wedig,
*University of Texas–
Pan American*

Arthur E. Weis,
University of California–Irvine
Sue Simon Westendorf,
Ohio University
Howard Whiteman,
Murray State University
Susan Whittemore,
Keene State College
David L. Wilson,
University of Miami
Robert Winning,
*Eastern Michigan
University*

Michelle D. Withers,
*Louisiana State
University*
Clarence C. Wolfe,
*Northern Virginia Community
College*
Gene K. Wong,
Quinnipiac University
Richard P. Wunderlin,
University of South Florida

Joanna Wysocka-Diller,
Auburn University
H. Randall Yoder,
Lamar University
Marilyn Yoder,
*University of Missouri–
Kansas City*
Scott D. Zimmerman,
*Southwest Missouri State
University*

International Reviewers

Heather Addy,
University of Calgary
Mari L. Acevedo,
*University of Puerto Rico
at Arecibo*
Heather E. Allison,
University of Liverpool, UK
David Backhouse,
University of New England
Andrew Bendall,
University of Guelph
Marinda Bloom,
*Stellenbosch University,
South Africa*
Tony Bradshaw,
Oxford-Brookes University, UK
Alison Campbell,
University of Waikato
Bruce Campbell,
Okanagan College

Clara E. Carrasco, Ph.D.,
*University of Puerto Rico–
Ponce Campus*
Keith Charnley,
University of Bath, UK
Ian Cock,
Griffith University
Margaret Cooley,
University of NSW
R. S. Currah,
University of Alberta
Logan Donaldson,
York University
Theo Elzenga,
*Rijks Universiteit Groningen,
Netherlands*
Neil C. Haave,
University of Alberta

Tom Haffie,
University of Western Ontario
Louise M. Hafner,
*Queensland University
of Technology*
Annika F. M. Haywood,
*Memorial University
of Newfoundland*
William Huddleston,
University of Calgary
Shin-Sung Kang,
KyungBuk University
Wendy J. Keenleyside,
University of Guelph
Christopher J. Kennedy,
Simon Fraser University
Bob Lauder,
Lancaster University

Richard C. Leegood,
Sheffield University, UK
Thomas H. MacRae,
Dalhousie University
R. Ian Menz,
Flinders University
Kirsten Poling,
University of Windsor
Jim Provan,
Queens University, Belfast, UK
Richard Roy,
McGill University
Han A.B. Wösten,
*Utrecht University,
Netherlands*

ACKNOWLEDGMENTS

The lives of most science-textbook authors do not revolve around an analysis of writing techniques. Instead, we are people who understand science and are inspired by it, and we want to communicate that information to other people. Simply put, we need a lot of help to get it right.

Editors are a key component that help the authors modify the content of their book so it is logical, easy to read, and inspiring. The editorial team for this Biology textbook has been a catalyst that kept this project rolling. The members played various roles in the editorial process. Lisa Bruflodt (Senior Developmental Editor) has been the master organizer. Frankly, this is a ridiculously hard job. Coordinating the efforts of dozens of people and keeping them on schedule is not always fun. Lisa's success at keeping us on schedule has been truly amazing. We are also grateful to Kris Tibbetts (Director of Development) who was involved in the early steps of the book, and kept the focus groups on track.

Our Biology book also has had 6 additional developmental editors who scrutinized each draft of their respective chapters with an emphasis on improving content, clarity, and readability. These developmental editors analyzed educational materials and reviewers' comments, and gave the authors advice on how to improve succeeding drafts. They also provided a list of the general principles that most instructors want in their Biology textbook. These general principles have been a cornerstone for the organization of our chapters.

Suzanne Olivier (Lead Freelance Developmental Editor) did an outstanding job of coordinating the staff of developmental editors. She also played an important role in editing chapters in the Genetics and Plant Biology units. Her early editing of the Genetics Unit, in particular, set the tone for many of the pedagogical features that became established throughout the entire textbook. Other developmental editors focused on particular units. Alice Fugate was involved with the Chemistry, Cell Biology, and Animal Biology Units. Her knack for getting the level of the writing appropriate for majors biology was invaluable, as was her attention to detail. Joni Fraser focused on the Cell Biology, Diversity, and Ecology Units. Somehow she successfully juggled the tasks of addressing all the reviewers' concerns while maintaining the necessary chapter length. Patricia Longoria played an important role in the early editing of the Diversity and Plant Biology Units. Patricia contributed many useful ideas for content and expression, and her unfailing enthusiasm smoothed the way over rocky parts of the process. Robin Fox edited three key chapters in the Genetics Unit. Her attention to detail and the explanation of mathematical principles were invaluable. And finally, Alan Titche was also involved with Animal Biology Unit, and played a major role in developing some of the most challenging chapters in that Unit. We would also like to thank Dr. Jim Deshler and Dr. Mary Erskine for their valuable contributions to the writing of several chapters in the Animal Biology Unit.

Deborah Brooker (Art/Text Coordinating Editor) analyzed all of the chapters in the textbook with one primary question in mind. Do the written text and figures tell a parallel story? With excruciating care, she made sure that the text and figures are consistent, and that the figures, by themselves, are accurate and easy to follow.

Imagineering Media Services Inc., of Ontario, Canada, did a fantastic job of illustrating our Biology book. They were involved early in the process by first making rough sketches based on the material in the first drafts, and then later progressed to drawings with finer detail. Their ability to make realistic, three-dimensional drawings is second to none. We're particularly grateful to Kierstan Hong, who provided a critical line of communication between the publisher, authors, and illustrators throughout most of this process, and also to Mark Mykytiuk, who also played a lead role in overseeing the art development. We would also like to gratefully acknowledge our photography researchers at Pronk & Associates of Ontario, Canada, and particularly to Fiona D'souza for keeping us on schedule. Likewise, we are grateful to John Leland, Photo Research Coordinator, at McGraw-Hill for his coordination of the photo selection process.

We would also like to thank our advisors and contributors:

Media Board of Advisors

Mark Decker,
University of Minnesota–Minneapolis

Naomi Friedman, *Developmental Editor*

Jon Glase,
Cornell University

John Merrill,
Michigan State University

Melissa Michael,
University of Illinois–Urbana/Champaign

Randall Phyllis,
University of Massachusetts

Mitch Price,
Pennsylvania State University

Tutorial Questions

Scott Bowling,
Auburn University

Don Buckley,
Quinnipiac University

Ernest DuBrul,
University of Toledo

Frederick B. Essig,
University of Florida

Jon Glase,
Cornell University

Norman A. Johnson
University of Massachusetts–Amherst

Kari Beth Krieger,
University of Wisconsin–Green Bay

Patricia Mire,
University of Louisiana–Lafayette

Allan Smits,
Quinnipiac University

Test Questions

Russell Borski,
  North Carolina State University
Robert Dunn,
  North Carolina State University
John Godwin,
  North Carolina State University
Mary Beth Hawkins,
  North Carolina State University
Harold Heatwole,
  North Carolina State University
James Mickle,
  North Carolina State University
Gerald Van Dyke,
  North Carolina State University ARIS

ARIS

Brad Mehrtens,
  University of Illinois–Urbana/Champaign

Instructor's Manual

Daniel Moon,
  University of North Florida

Student Study Guide

Michelle Shuster,
  New Mexico State University
Amy Marion,
  New Mexico State University

Lecture Outlines

Brenda Leady,
  University of Toledo

Animations

Kevin Dixon,
  University of Illinois–Urbana/Champaign

Another important aspect of the editorial process is the actual design, presentation, and layout of materials. It's confusing if the text and art aren't on the same page, or if a figure is too large or two small. We are indebted to the tireless efforts of Joyce Berendes (Lead Project Manager) and Wayne Harms (Design Manager) of McGraw-Hill. Their artistic talents, ability to size and arrange figures, and attention to the consistency of the figures have been remarkable. We also wish to thank John Joran (Designer) who cleverly crafted both the interior and exterior designs.

We would like to acknowledge the ongoing efforts of the superb marketing staff at McGraw-Hill. Kent Peterson (Vice President, Director of Marketing) oversees a talented staff of people who work tirelessly to promote our book. Special thanks to Chad Grall (Marketing Director), Wayne Vincent (Internet Marketing Manager), Debra Hash (Senior Marketing Manager) and Heather Wagner (Systems and Promotions Marketing Manager) for their ideas and enthusiasm for this book. The proposal of making a video website for the book *www.brookerbiology.com* was scary for the authors but actually turned out to be fun.

Finally, other staff members at McGraw-Hill Higher Education have ensured that the authors and editors were provided with adequate resources to achieve the goal of producing a superior textbook. These include Kurt Strand (President, Science, Engineering, and Math), Marty Lange (Vice President, Editor-in-Chief), Michael Lange (Vice President, New Product Launches), Janice Roerig-Blong (Publisher) and Patrick Reidy (Executive Editor). In particular, Michael and Patrick communicated with the authors on a regular basis regarding the progress of this project. They attended most of the focus groups and author meetings, and even provided occasional input regarding the content of the book. The bottom line is that the author team is grateful that you have believed in this project, and have provided us with the resources to make it happen.

Student Supplements

Designed to help students maximize their learning experience in biology—we offer the following options to students:

ARIS (Assessment, Review, and Instruction System) is an electronic study system that offers students a digital portal of knowledge. Students can readily access a variety of **digital learning objects** which include:

- chapter level quizzing
- pretests
- animations
- videos
- flashcards
- answers to Biological Inquiry Questions
- answers to all end-of-chapter questions
- MP3 and MP4 downloads of selected content
- learning outcomes and assessment capability woven around key content

Student Study Guide
ISBN: 0-07-299588-2

Helping students focus their time and energy on important concepts, the study guide offers students a variety of tools:

1. Practice Questions—approximately 10–12 multiple choice questions
2. Active Learning Questions—approximately 5–8 open-ended questions that ask the student to explore

something and delve into content a little deeper, reinforcing content through experiential learning.

3. Outline/Summary of Fundamental Concepts—efficient listing of key concepts.
4. Key Terms
5. Strategies for Difficult Concepts

Content Delivery Flexibility

Brooker et al., **Biology** is available in many formats in addition to the traditional textbook to give instructors and students more choices when deciding on the format of their biology text. Choices include:

Volumes

The complete text has been split into three natural segments to allow instructors more flexibility and students more purchasing options.

Volume 1—Units 1 (Chemistry), 2 (Cell), and 3 (Genetics)
ISBN 0-07-335332-9
Volume 2—Units 6 (Plants) and 7 (Animals)
ISBN 0-07-335331-0
Volume 3—Units 4 (Evolution), 5 (Diversity), and 8 (Ecology)
ISBN 0-07-335333-7

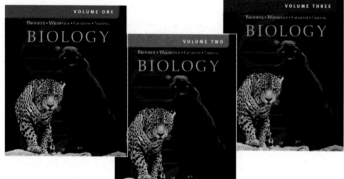

Color Custom by Chapter

For even more flexibility, we offer the Brooker: **Biology** text in a full-color, custom version that allows instructors to pick the chapters they want included. Students pay for only what the instructor chooses.

eBook

The entire text is available electronically through the ARIS website. This electronic text offers not only the text in a digital format but includes embedded links to figures, tables, animations, and videos to make full use of the digital tools available and further enhance student understanding.

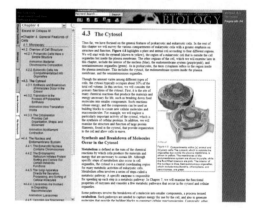

Instructor Supplements

Dedicated to providing high quality and effective supplements for instructors, the following Instructor supplements were developed for **Biology:**

ARIS with Presentation Center

Assessment, Review, and Instruction System, also known as ARIS, is an electronic homework and course management system designed for greater flexibility, power, and ease of use than any other system. Whether you are looking for a preplanned course or one you can customize to fit your course needs, ARIS is your solution.

In addition to having access to all student digital learning objects, ARIS allows instructors to:

Build Assignments

- Choose from pre-built assignments or create your own custom content by importing your own content or editing an existing assignment from the pre-built assignment.

- Assignments can include quiz questions, animations, and videos . . . anything found on the website.
- Create announcements and utilize full course or individual student communication tools
- Assign **unique multi-level tutorial questions** developed by content experts that provide intelligent feedback through a series of questions to help students truly understand a concept; not just repeat an answer.

Track Student Progress

- Assignments are automatically graded
- Gradebook functionality allows full course management including:
  - Dropping the lowest grades
  - Weighting grades / manually adjusting grades
  - Exporting your gradebook to Excel, WebCT or BlackBoard

- Manipulating data allowing you to track student progress through multiple reports

Offer More Flexibility

- **Sharing Course Materials with Colleagues** — Instructors can create and share course materials and assignments with colleagues with a few clicks of the mouse allowing for multiple section courses with many instructors (and TAs) to continually be in synch if desired.
- **Integration with BlackBoard or WebCT**—once a student is registered in the course, all student activity within McGraw-Hill's ARIS is automatically recorded and available to the instructor through a fully integrated grade book that can be downloaded to Excel, WebCT, or Blackboard.

Presentation Center

Build instructional materials wherever, whenever, and however you want!

ARIS Presentation Center is an online digital library containing assets such as photos, artwork, animations, PowerPoints, and other media types that can be used to create customized lectures, visually enhanced tests and quizzes, compelling course websites, or attractive printed support materials.

Access to your book, access to all books!

The Presentation Center library includes thousands of assets from many McGraw-Hill titles. This ever-growing resource gives instructors the power to utilize assets specific to an adopted textbook as well as content from all other books in the library.

Nothing could be easier!

Accessed from the instructor side of your textbook's ARIS website, Presentation Center's dynamic search engine allows you to explore by discipline, course, textbook chapter, asset type, or keyword. Simply browse, select, and download the files you need to build engaging course materials. All assets are copyright McGraw-Hill Higher Education but can be used by instructors for classroom purposes.

Instructor's Testing and Resource CD-ROM

ISBN: 0-07-295658-5
This cross-platform CD-ROM provides these resources for instructors:

- **Instructor's Manual**—This manual contains instructional strategies and activities, student misconceptions,

etymology of key terms, "Beyond the Book" interesting facts, and sources for additional web resources.

- **Test Bank**—The test bank offers multiple-choice and true/false questions that can be used for homework assignments or the preparation of exams.
- **Computerized Test Bank**—This software can be utilized to quickly create customized exams. The user-friendly program allows instructors to sort questions by format or level of difficulty; edit existing questions or add new ones; and scramble questions and answer keys for multiple versions of the same test.

Student Response System

Wireless technology brings interactivity into the classroom or lecture hall. Instructors and students receive immediate feedback through wireless response pads that are easy to use and engage students. This system can be used by instructors to:

- Take attendance
- Administer quizzes and tests
- Create a lecture with intermittent questions
- Manage lectures and student comprehension through the use of the gradebook
- Integrate interactivity into their PowerPoint presentations

Transparencies
ISBN: 0-07-295657-7
This boxed set of overhead transparencies includes every piece of line art in the textbook. The images have been modified to ensure maximum readability in both small and large classroom settings.

BIOLOGY LABORATORY MANUAL
Darrell S. Vodopich, *Baylor University*
Randy Moore, *University of Minnesota*
ISBN: 0-07-332398-5

This laboratory manual is designed to accompany Brooker et al: *Biology*. The experiments and procedures are simple, safe, easy to perform, and especially appropriate for large classes. Few experiments require a second class-meeting to complete the procedure. Each exercise includes many photographs, traditional topics, and experiments that help students learn about life. Procedures within each exercise are numerous and discrete so that an exercise can be tailored to the needs of the students, the style of the instructor, and the facilities available.

BIOLOGICAL INVESTIGATIONS LAB MANUAL
Warren D. Dolphin, *Iowa State University*
ISBN: 0-07-332399-3

Developed to accompany Brooker et al: *Biology*, this lab manual focuses on labs that are investigative and ask students to use more critical thinking and hands-on learning. The author emphasizes investigative, quantitative, and comparative approaches to studying the life sciences.

A VISUAL JOURNEY

Our art program was painstakingly designed in conjunction with the text development to ensure 1) each concept is accurately portrayed, 2) consistency is maintained between the text and art, and 3) it's appropriately placed on the page. The art serves as a visual outline for students, often offering textboxes that explain difficult concepts. For multistep processes, these textboxes are numbered so that the student can easily follow the process from beginning to end.

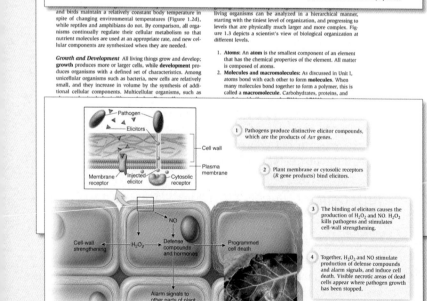

(a) Stages of transcription

1 Initiation: The promoter functions as a recognition site for sigma factor. RNA polymerase is bound to sigma factor, which causes it to recognize the promoter. Following binding, the DNA is unwound into a bubble known as the open complex.

2 Elongation/synthesis of the RNA transcript: Sigma factor is released and RNA polymerase slides along the DNA in an open complex to synthesize RNA.

3 Termination: When RNA polymerase reaches the terminator, it and the RNA transcript dissociate from the DNA.

Figure 12.6 Stages of transcription. (a) Transcription can be divided into initiation, elongation, and termination. The inset emphasizes the direction of RNA synthesis and base pairing between the DNA template strand and RNA. **(b)** Three-dimensional structure of a bacterial RNA polymerase.

(b) Structure of a bacterial RNA polymerase

and birds maintain a relatively constant body temperature in spite of changing environmental temperatures (Figure 1.2d), while reptiles and amphibians do not. By comparison, all organisms continually regulate their cellular metabolism so that nutrient molecules are used at an appropriate rate, and new cellular components are synthesized when they are needed.

Growth and Development All living things grow and develop; **growth** produces more or larger cells, while **development** produces organisms with a defined set of characteristics. Among unicellular organisms such as bacteria, new cells are relatively small, and they increase in volume by the synthesis of additional cellular components. Multicellular organisms, such as

living organisms can be analyzed in a hierarchical manner, starting with the tiniest level of organization, and progressing to levels that are physically much larger and more complex. Figure 1.3 depicts a scientist's view of biological organization at different levels.

1. **Atoms:** An **atom** is the smallest component of an element that has the chemical properties of the element. All matter is composed of atoms.
2. **Molecules and macromolecules:** As discussed in Unit I, atoms bond with each other to form **molecules.** When many molecules bond together to form a polymer, this is called a **macromolecule.** Carbohydrates, proteins, and

1 Pathogens produce distinctive elicitor compounds, which are the products of *Avr* genes.

2 Plant membrane or cytosolic receptors (*R* gene products) bind elicitors.

3 The binding of elicitors causes the production of H_2O_2 and NO. H_2O_2 kills pathogens and stimulates cell-wall strengthening.

4 Together, H_2O_2 and NO stimulate production of defense compounds and alarm signals, and induce cell death. Visible necrotic areas of dead cells appear where pathogen growth has been stopped.

The members of the same species are closely related genetically. In Units VI and VII, we will examine plants and animals at the level of cells, tissues, organs, and complete organisms.

7. **Population:** A group of organisms of the same species that occupy the same environment is called a **population.**
8. **Community:** A biological **community** is an assemblage of populations of different species. The types of species that are found in a community are determined by the environment and by the interactions of species with each other.
9. **Ecosystem:** Researchers may extend their work beyond living organisms and also study the environment. Ecologists analyze **ecosystems,** which are formed by

interactions of a community of organisms with their physical environment. Unit VIII considers biology from populations to ecosystems.

10. **Biosphere:** The **biosphere** includes all of the places on the Earth where living organisms exist, encompassing the air, water, and land.

Modern Forms of Life Are Connected by an Evolutionary History

Life began on Earth as primitive cells about 3.5 to 4 billion years ago. Since that time, those primitive cells underwent evolutionary changes that ultimately gave rise to the species we see today.

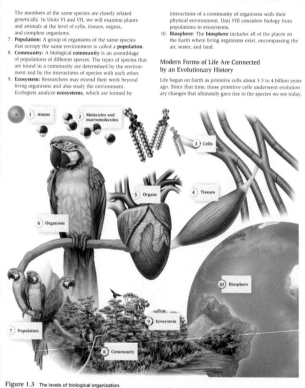

Figure 1.3 The levels of biological organization.

BIOLOGICAL INQUIRY QUESTIONS

These questions are designed to help students delve more deeply into a concept or experimental approach described in the art. These questions challenge a student to analyze the content of the figure they are looking at.

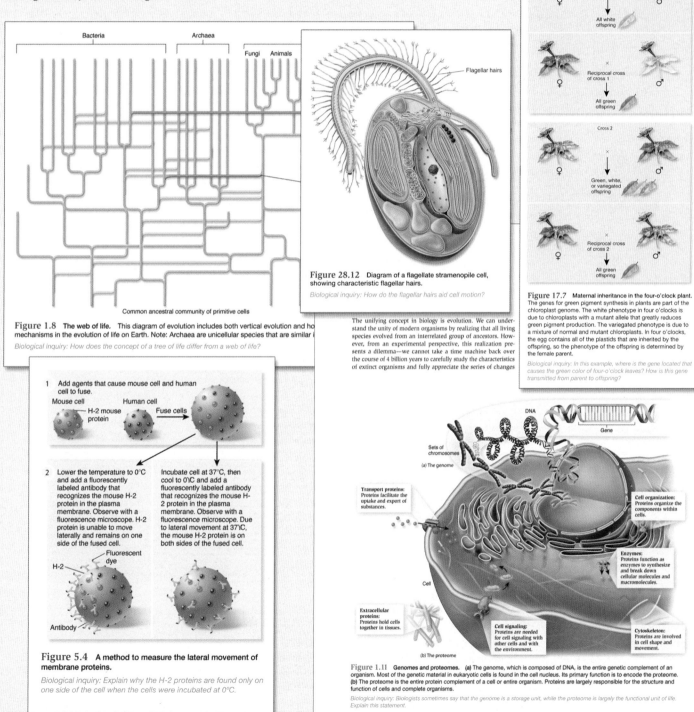

Figure 1.8 **The web of life.** This diagram of evolution includes both vertical evolution and ho... mechanisms in the evolution of life on Earth. Note: Archaea are unicellular species that are similar i...

Biological inquiry: How does the concept of a tree of life differ from a web of life?

Figure 28.12 Diagram of a flagellate stramenopile cell, showing characteristic flagellar hairs.

Biological inquiry: How do the flagellar hairs aid cell motion?

Figure 17.7 Maternal inheritance in the four-o'clock plant. The genes for green pigment synthesis in plants are part of the chloroplast genome. The white phenotype in four o'clocks is due to chloroplasts with a mutant allele that greatly reduces green pigment production. The variegated phenotype is due to a mixture of normal and mutant chloroplasts. In four o'clocks, the egg contains all of the plastids that are inherited by the offspring, so the phenotype of the offspring is determined by the female parent.

Biological inquiry: In this example, where is the gene located that causes the green color of four-o'clock leaves? How is this gene transmitted from parent to offspring?

The unifying concept in biology is evolution. We can understand the unity of modern organisms by realizing that all living species evolved from an interrelated group of ancestors. However, from an experimental perspective, this realization presents a dilemma—we cannot take a time machine back over the course of 4 billion years to carefully study the characteristics of extinct organisms and fully appreciate the series of changes

Figure 5.4 **A method to measure the lateral movement of membrane proteins.**

Biological inquiry: Explain why the H-2 proteins are found only on one side of the cell when the cells were incubated at 0°C.

Figure 1.11 **Genomes and proteomes.** **(a)** The genome, which is composed of DNA, is the entire genetic complement of an organism. Most of the genetic material in eukaryotic cells is found in the cell nucleus. Its primary function is to encode the proteome. **(b)** The proteome is the entire protein complement of a cell or entire organism. Proteins are largely responsible for the structure and function of cells and complete organisms.

Biological inquiry: Biologists sometimes say that the genome is a storage unit, while the proteome is largely the functional unit of life. Explain this statement.

EXPERIMENTAL & MODERN CONTENT

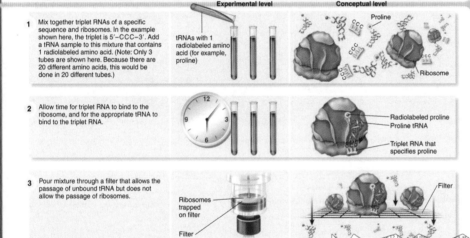

FEATURE INVESTIGATION

Nirenberg and Leder Found That RNA Triplets Can Promote the Binding of tRNA to Ribosomes

In 1964, Nirenberg and Leder discovered that RNA molecules containing any three nucleotides (that is, any triplet) can stim- later in the chapter, but for now just keep in mind that tRNAs interact with mRNA on a ribosome during the synthesis of polypeptides. This sample was divided into 20 tubes. To each tube, they next added a mixture of cellular tRNAs that already had amino acids attached to them. However, each mixture of

Figure 12.14 Nirenberg and Leder's use of triplet binding assays to decipher the genetic code.

Overview

HYPOTHESIS A triplet RNA can bind to a ribosome and promote the binding of the tRNA that carries the amino acid that the triplet RNA specifies.

STARTING MATERIALS Components of an *in vitro* translation system, including ribosomes and tRNAs. Preparations containing all of the different tRNA molecules were given 1 radiolabeled amino acid; the other 19 amino acids were nonlabeled. For example, in 1 sample, radiolabeled glycine was added and the other 19 amino acids were nonlabeled. In a different sample, radiolabeled proline was added and the other 19 amino acids were nonlabeled. The tRNA preparation also contained the enzymes that attach amino acids to tRNAs.

Steps

1 Mix together triplet RNAs of a specific sequence and ribosomes. In the example shown here, the triplet is 5′–CCC–3′. Add a tRNA sample to this mixture that contains 1 radiolabeled amino acid. (Note: Only 3 tubes are shown here. Because there are 20 different amino acids, this would be done in 20 different tubes.)

2 Allow time for triplet RNA to bind to the ribosome, and for the appropriate tRNA to bind to the triplet RNA.

3 Pour mixture through a filter that allows the passage of unbound tRNA but does not allow the passage of ribosomes.

Data & Analysis

5 THE DATA

Triplet	Radiolabeled amino acid trapped on the filter	Triplet
5′ – AAA – 3′	Lysine	5′ – GAC – 3′
5′ – ACA – 3′, 5′ – ACC – 3′	Threonine	5′ – GCC – 3′
5′ – AGA – 3′	Arginine	5′ – GGU – 3′, 5′ – GGC – 3′
5′ – AUA – 3′, 5′ – AUU – 3′	Isoleucine	5′ – GUU – 3′
5′ – CCC – 3′	Proline	5′ – UAU – 3′
5′ – CGC – 3′	Arginine	5′ – UGU – 3′
5′ – GAA – 3′	Glutamic acid	5′ – UUG – 3′

FEATURE INVESTIGATION

Focusing on hypothesis testing and discovery-based science, the Feature Investigations describe a key experiment, including 1) an overview of the hypothesis or goal of the experiment, 2) the steps of the experiment, and 3) ending with an analysis of data. This encourages an appreciation of the scientific process.

GENOMES & PROTEOMES

Comparisons of Small Subunit rRNAs Among Different Species Provide a Basis for Establishing Evolutionary Relationships

Translation is a fundamental process that is vital for the existence of all living species. The components that are needed for translation arose very early in the evolution of life on our planet. In fact, they arose in an ancestor that gave rise to all known living species. For this reason, all organisms have translational components that are evolutionarily related to each other. For example, the rRNA found in the small subunit of ribosomes is similar in all forms of life, though it is slightly larger in eukaryotic species (18S) than in bacterial species (16S). In other words, the gene for the small subunit rRNA (SSU rRNA) is found in the genomes of all organisms.

One way that geneticists explore evolutionary relationships is to compare the sequences of evolutionarily related genes. At the molecular level, gene evolution involves changes in DNA sequences. After two different species have diverged from each other during evolution, the genes of each species have an opportunity to accumulate changes, or mutations, that alter the sequences of those genes. After many generations, evolutionarily related species contain genes that are similar but not identical to each other, because each species will accumulate different

...in the primordial gene that gave rise to modern species and, because these sequences may have some critical function, have not been able to change over evolutionary time. Those sequences shaded in green are identical in all three mammals, but differ compared to one or more bacterial species. Actually, if you scan the mammalian species, you may notice that all three sequences are identical to each other in this region. The sequences shaded in red are identical in two or three bacterial species, but differ compared to the mammalian small subunit rRNA genes. The sequences from *E. coli* and *Serratia marcescens* are more similar to each other than the sequence from *Bacillus subtilis* is to either of them. This is consistent with the idea that *E. coli* and *S. marcescens* are more closely related evolutionarily than either of them is to *B. subtilis*.

12.6 The Stages of Translation

Like transcription, the process of translation occurs in three stages called initiation, elongation, and termination. **Figure 12.20** provides an overview of the process. During initiation, mRNA, the first tRNA, and ribosomal subunits assemble into a complex. Next, in the elongation stage, the ribosome moves from the start codon in the mRNA toward the stop codon, synthesizing a polypeptide according to the sequence of codons in the mRNA. Finally, the process is terminated when the ribosome reaches a stop codon and the complex disassembles, releasing the completed polypeptide. In this section, we will examine the steps in this process as they occur in living cells.

GATTAAGAGGGACGGCCGGGGGCATTCGTATTGCGCCGCTAGAGGTGAAATTC — Human
GATTAAGAGGGACGGCCGGGGGCATTCGTATTGCGCCGCTAGAGGTGAAATTC — Mouse
GATTAAGAGGGACGGCCGGGGGCATTCGTATTGCGCCGCTAGAGGTGAAATTC — Rat
CAAGCTTGAGTCTCGTAGAGGGGGGTAGAATTCCAGGTGTAGCGGTGAAATGC — E. coli
CAAGCTTGAGTCTCGTAGAGGGGGGTAGAATTCCAGGTGTAGCGGTGAAATGC — S. marcescens
GAGACTTGAGTACAGAAGAGGAGAGTGGAATTCCACGTGTAGCGGTGAAATGC — B. subtilis

Figure 12.19 Comparison of small subunit rRNA gene sequences from three eukaryotes and three bacterial species. Note the many similarities (yellow) and differences (green and red) among the sequences.

GENOMES AND PROTEOMES

Providing an evolutionary foundation for our understanding of biology, each Genomes and Proteomes subsection describes modern information regarding the genomic composition of organisms and how this relates to proteomes (their protein composition) and evolution.

END-OF-CHAPTER MATERIALS

The end-of-chapter materials offer students many different opportunities to focus in on key concepts and help them work at improving their knowledge:

Chapter Summary
The Chapter Summary provides the student with an overview of the biological principles and experimental approaches that have been described in the chapter. The summary is organized according to the sections of each chapter, and presents a bulleted list of key concepts.

Test Yourself
These multiple-choice questions are designed to provide students with the sense of how well they understand the material in the chapter. Answers are provided on the ARIS website.

Conceptual Questions
The aim of conceptual questions is to test a student's knowledge of biological principles, such as how a biological mechanism works or the features of a biological process or structure.

Experimental Questions
These questions challenge the student to consider the experiments found in a chapter and to critically evaluate technical procedures and analyze biological data.

Collaborative Questions
Broad in nature, students may benefit by discussing these questions with their peers.

CHAPTER SUMMARY

3.1   The Carbon Atom and the Study of Organic Molecules

- Organic chemistry is the science of studying carbon-containing molecules, which are found in living organisms. Wöhler's work with urea marked the birth of organic chemistry. (Figure 3.1)
- One property of the carbon atom that makes life possible is its ability to form four covalent bonds with other atoms. Carbon can form both polar and nonpolar bonds. The combination of different elements and different types of bonds allows a vast number of organic compounds to be formed from only a few chemical elements. (Figures 3.2, 3.3)
- Organic molecules may occur in various shapes. The structures of molecules determine their functions.

3.2   Classes of Organic Molecules and Macromolecules

- The four major classes of organic molecules are carbohydrates, lipids, proteins, and nucleic acids. Macromolecules are large organic molecules that are composed of many thousands of atoms. Some macromolecules are polymers because they are formed by linking together many smaller molecules called monomers.
- Carbohydrates are composed of carbon, hydrogen, and oxygen atoms. Most cells can break down carbohydrates, releasing energy and storing it in newly created bonds in ATP.
- Carbohydrates include monosaccharides (the simplest sugars), disaccharides, and polysaccharides. The polysaccharides starch (in plant cells) and glycogen (in animal cells) provide an efficient means of storing energy. The plant polysaccharide cellulose serves a support or structural function. (Figures 3.6, 3.7, 3.8)

TEST YOURSELF

1. Molecules that contain the element _____ are considered organic molecules.
   a. hydrogen   d. nitrogen
   b. carbon   e. calcium
   c. oxygen

2. _____ was the first scientist to synthesize an organic molecule. The organic molecule synthesized was _____.
   a. Kolbe, urea   d. Kolbe, acetic acid
   b. Wöhler, urea   e. Wöhler, glucose
   c. Wöhler, acetic acid

3. The versatility of carbon to serve as the backbone for a variety of different molecules is due to
   a. the ability of carbon atoms to form four covalent bonds.
   b. the fact that carbon usually forms ionic bonds with many different atoms.
   c. the abundance of carbon in the environment.
   d. the ability of carbon to form covalent bonds with many different types of atoms.
   e. both a and d.

CONCEPTUAL QUESTIONS

1. Define isomers.
2. List the four classes of organic molecules and give a function of each.
3. Explain the difference between saturated and unsaturated fatty acids.
4. List the seven characteristics of life and explain a little about each.
5. Give the levels of organization from the simplest to most complex.
6. Discuss the difference between discovery-based science and hypothesis testing.
7. What are the steps in the scientific method, also called hypothesis testing?
8. When conducting an experiment, explain how a control sample and an experimental sample differ from each other.

EXPERIMENTAL QUESTIONS

1. Before the experiments conducted by Anfinsen, what were the common beliefs among scientists about protein folding?
2. Explain the hypothesis tested by Anfinsen.
3. Why did Anfinsen use urea and β-mercaptoethanol in his experiments? Explain the result that was crucial to the discovery that the tertiary structure of a protein is dependent on the primary structure.
4. List the seven characteristics of life and explain a little about each.
5. Give the levels of organization from the simplest to most complex.
6. List the taxonomic groups from most inclusive to least inclusive.
7. Explain how actin filaments are involved in movement.
8. Explain the function of the Golgi apparatus.

COLLABORATIVE QUESTIONS

1. Discuss several types of carbohydrates.
2. Discuss some of the roles that proteins play in organisms. Discuss several differences between plant and animal cells.
3. Discuss the relationship between the nucleus, the rough endoplasmic reticulum, and the Golgi apparatus.
4. Discuss the two categories of transport proteins found in plasma membranes.

www.brookerbiology.com
This website includes answers to the Biological Inquiry questions found in the figure legends and all end-of-chapter questions.

CONTENTS

UNIT I Chemistry

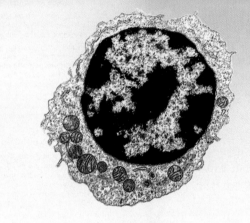

UNIT II Cell

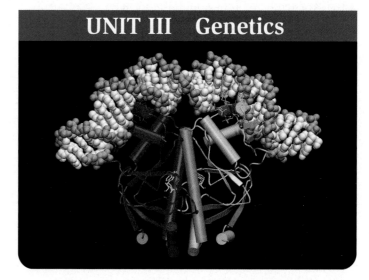

UNIT III   Genetics

1

An Introduction to Biology

Chapter Outline

Spotted and black jaguars.

Biology is the study of life. The diverse forms of life found on Earth provide biologists with an amazing array of organisms to study. In many cases, the investigation of living things leads to unforeseen discoveries that no one would have imagined. For example, researchers determined that the venom from certain poisonous snakes contains a chemical that lowers blood pressure in humans (**Figure 1.1a**). By analyzing that chemical, drugs were later developed to treat high blood pressure. Biologists also found that nine-banded armadillos usually give birth to identical quadruplets (**Figure 1.1b**). Because of this unique way of producing young, nine-banded armadillos are studied to learn more about multiple births and other reproductive issues. The ancient Romans discovered that the bark of the white willow tree can be used to fight fever (**Figure 1.1c**). Modern chemists determined that willow bark contains a substance called salicylic acid, which led to the development of the related compound acetylsalicylic acid, more commonly known as aspirin. In the last century, biologists studied soil bacteria that naturally produce "chemical weapons" to kill competing bacteria in their native environment (**Figure 1.1d**). These chemicals have been characterized and used to develop antibiotics such as streptomycin to treat bacterial infections. As you may have seen, jellyfish naturally produce a greenish glow (**Figure 1.1e**), which is due to a molecule they make called green fluorescent protein (GFP). Scientists have been able to transfer GFP to other organisms and use it as a research tool to study the functions of cells. GFP transferred to mice makes them glow in the dark! Finally, for many decades, biologists have known that the Pacific yew tree produces a toxin in its bark and needles that kills insects (**Figure 1.1f**). Since the 1990s, this toxin, known by the drug name Taxol®, has been used to treat patients with ovarian and breast cancer. These are but a few of the many discoveries that make biology an intriguing discipline. The study of life not only reveals the fascinating characteristics of living species but also leads to the development of drugs and research tools that benefit the lives of people.

To make new discoveries, biologists view life from many different perspectives. What is life made of? How is it organized? How do organisms reproduce? Sometimes, the questions posed by biologists are fundamental and even philosophical in nature. Where did we come from? Can we live forever? What is the physical basis for memory? Can we save endangered species? Biologists are scientific explorers looking for answers to some of the world's most enduring mysteries. Unraveling these mysteries presents an exciting challenge to the best and brightest minds. Our society has been substantially impacted by discoveries in biology, and future biologists will continue to make important advances. The rewards of a career in biology include the excitement of forging into uncharted territory, the thrill of making discoveries that affect the health and lives of people, and the impact of biology on the preservation of the environment and endangered species. For these and many other compelling reasons, students seeking challenging and rewarding careers may wish to choose biology as a lifelong pursuit.

In this chapter, we will begin our survey of biology by examining the basic features that are common to all living organisms. We will consider how evolution has led to the development of modern genomes—the entire genetic compositions of organisms—which can explain the unity and diversity that we observe among modern species. In the second section, we will explore the general approaches that scientists follow when making new discoveries.

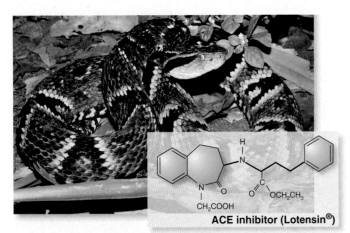

ACE inhibitor (Lotensin®)

(a) A chemical in the venom of the Brazilian arrowhead viper lowers blood pressure. Derivatives of this chemical, called angiotensin-converting enzyme (ACE) inhibitors, are now commonly used to treat high blood pressure in people.

(b) The nine-banded armadillo usually gives birth to identical quadruplets. Armadillos are studied by researchers to learn more about the mechanisms that cause multiple births.

Aspirin

(c) The bark of the white willow contains a chemical that is closely related to aspirin. Modern aspirin, acetylsalicylic acid, was developed after analyzing this chemical in willow trees.

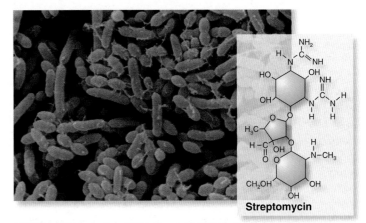

Streptomycin

(d) This soil bacterium (*Streptomycin griseus*) naturally produces a molecule called streptomycin, which it uses to kill competing bacteria in the soil. Doctors administer streptomycin to people as an antibiotic to treat bacterial infections.

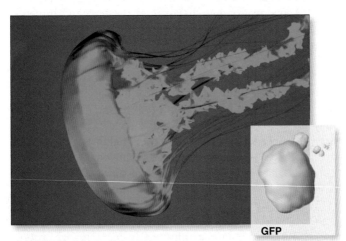

GFP

(e) Jellyfish naturally produce a green glow caused by green fluorescent protein (GFP). GFP can be transferred to other organisms, such as mice, and used as a research tool to study the functions of cells.

Taxol®

(f) The bark and needles of the Pacific yew tree produce a toxin that kills insects. This toxin, called Taxol, is effective in the treatment of ovarian and breast cancer.

Figure 1.1 Amazing discoveries in biology.

1.1 The Properties of Life: Past and Present

Unity and *diversity* are two words that often are used to describe the living world. Unity can be viewed from both modern and prehistorical perspectives. In this section, we first examine how all modern forms of life display a common set of characteristics that distinguish them from nonliving objects. This unity of common traits is rooted in the phenomenon of biological evolution. Life on Earth is united by an evolutionary past in which modern organisms have evolved from pre-existing organisms.

This evolutionary unity does not mean that organisms are exactly alike. Rather, the Earth has many different types of environments, ranging from tropical rain forests to salty oceans, hot and dry deserts, and cold mountaintops. Diverse forms of life have evolved to prosper in the myriad of environments that the Earth has to offer. In this section, we will also begin to examine the diversity that exists within the biological world.

A Set of Characteristics Is Common to All Forms of Modern Life

A fitting way to begin a biology textbook is to distinguish living organisms from nonliving objects. At first, the distinction might seem intuitively obvious. A person is alive, but a rock is not. However, the distinction between living and nonliving may seem less obvious when we consider microscopic entities. Is a bacterium alive? Is a virus alive? Is a chromosome alive? Biologists have wrestled with such questions and have determined that all living organisms display seven characteristics that set them apart from nonliving things.

Cells and Organization The concept of organization is so fundamental to biology that the term **organism** (which comes from the same Latin root, *organum*) can be applied to all living things. Organisms maintain an internal order that is separated from the environment (**Figure 1.2a**). The simplest unit of such organization is the **cell**, which we will examine in Unit II. The **cell theory** states that all organisms are made of cells. Unicellular organisms are composed of one cell, while multicellular organisms such as plants and animals contain many cells. In plants and animals, each cell has internal order, and the cells within the body have specific arrangements and functions.

Energy Use and Metabolism The maintenance of organization requires energy. Therefore, all living organisms acquire energy from the environment and use that energy to maintain their internal order. Cells use energy by catalyzing a variety of chemical reactions that are responsible for the breakdown of nutrients and the synthesis of the components that make up individual cells and living organisms. These chemical reactions are collectively known as **metabolism**. Plants, algae, and certain bacteria can directly harness light energy to produce their own nutrients in the process of **photosynthesis** (**Figure 1.2b**).

(a) Cells and organization: Organisms maintain an internal order. The simplest unit of organization is the cell.

(b) Energy use and metabolism: To maintain their internal order, energy is needed by organisms. Energy is utilized in chemical reactions collectively known as metabolism.

(c) Response to environmental changes: Organisms react to environmental changes to promote their survival.

(d) Regulation and homeostasis: Organisms regulate their cells and bodies to maintain relatively stable internal conditions, a process called homeostasis.

(e) Growth and development: Growth produces more or larger cells, while development produces organisms with a defined set of characteristics.

(f) Reproduction: To sustain life over many generations, organisms must reproduce. Due to genetic material, offspring tend to have traits like their parents.

(g) Biological evolution: Populations of organisms change over the course of many generations. Evolution results in traits that promote survival and reproductive success.

Figure 1.2 Seven characteristics that are common to life.

They are primary producers of food on Earth. In contrast, some organisms, such as animals and fungi, are consumers—they must eat other organisms as food to obtain energy.

Response to Environmental Changes To survive, living organisms must be able to respond to environmental changes. For example, bacterial cells have mechanisms to sense that certain nutrients in the environment are in short supply while others are readily available. Also, plants can respond to changes in the angle of the sun. If you place a plant in a window, it will grow toward the light (**Figure 1.2c**). In the winter, many species of mammals develop a thicker coat of fur to protect them from the cold temperatures. Responses to environmental changes are examples of **adaptations**—processes and structures by which organisms adjust to short-term or long-term changes in their environment.

Regulation and Homeostasis As we have just seen, one way that organisms can respond to environmental variation is to change themselves. The growth of thick fur in the wintertime is an example. A common reason for certain adaptations, including this example, is to maintain homeostasis (from the Greek meaning "to stay the same"). Although life is a dynamic process, living cells and organisms regulate their cells and bodies to maintain relatively stable internal conditions, a process called **homeostasis**. The degree to which homeostasis is achieved varies among different organisms. For example, most mammals and birds maintain a relatively constant body temperature in spite of changing environmental temperatures (**Figure 1.2d**), while reptiles and amphibians do not. By comparison, all organisms continually regulate their cellular metabolism so that nutrient molecules are used at an appropriate rate, and new cellular components are synthesized when they are needed.

Growth and Development All living things grow and develop; **growth** produces more or larger cells, while **development** produces organisms with a defined set of characteristics. Among unicellular organisms such as bacteria, new cells are relatively small, and they increase in volume by the synthesis of additional cellular components. Multicellular organisms, such as plants and animals, begin life at a single-cell stage (for example, a fertilized egg) and then undergo multiple cell divisions to develop into a complete organism with many cells (**Figure 1.2e**).

Reproduction All living organisms have a finite life span and will eventually die. To sustain life over many generations, organisms must **reproduce** (**Figure 1.2f**). A key feature of reproduction is that offspring tend to have characteristics that greatly resemble those of their parent(s). The reason for this is that all living organisms contain genetic material composed of **DNA (deoxyribonucleic acid)**, which provides a blueprint for the organization, development, and function of living things. As discussed in Unit III, DNA harbors **genes**, which contribute to the characteristics or traits of organisms. During reproduction, a copy of this blueprint is transmitted from parents to offspring. The central dogma of genetics is that most genes are transcribed into a type of RNA (ribonucleic acid) molecule called messenger RNA (mRNA) that is then translated into a polypeptide with a specific amino acid sequence. This process is called molecular gene expression. Polypeptides are the structural units of functional proteins. The functioning of proteins is largely responsible for the traits of living organisms.

Biological Evolution The first six characteristics of life, which we have just considered, apply to individual organisms over the short run. Over the long run, another universal characteristic of life is **biological evolution**, which refers to the phenomenon that populations of organisms change over the course of many generations. As a result of evolution, some organisms become more successful at survival and reproduction. Populations become better adapted to the environment in which they live. For example, the long snout of an anteater is an adaptation that enhances its ability to obtain food, namely ants (**Figure 1.2g**). The long snout occurred via biological evolution in which modern anteaters evolved from organisms that did not have such long snouts. Unit IV is devoted to the topic of evolution, while Unit V surveys the evolutionary diversity among different forms of life.

Living Organisms Can Be Viewed at Different Levels of Organization

As we have just learned, life exhibits a set of characteristics, beginning with the concept of organization. The organization of living organisms can be analyzed in a hierarchical manner, starting with the tiniest level of organization, and progressing to levels that are physically much larger and more complex. **Figure 1.3** depicts a scientist's view of biological organization at different levels.

1. **Atoms:** An **atom** is the smallest component of an element that has the chemical properties of the element. All matter is composed of atoms.
2. **Molecules and macromolecules:** As discussed in Unit I, atoms bond with each other to form **molecules**. When many molecules bond together to form a polymer, this is called a **macromolecule**. Carbohydrates, proteins, and nucleic acids (for example, DNA and RNA) are important macromolecules found in living organisms.
3. **Cells:** Molecules and macromolecules associate with each other to form larger structures such as membranes. A cell is formed from the association of these larger structures.
4. **Tissues:** In the case of multicellular organisms such as plants and animals, many cells of the same type associate with each other to form **tissues**. An example is muscle tissue.
5. **Organs:** An **organ** is composed of two or more types of tissue. For example, the heart of a parrot is composed of several types of tissues, including muscle, nervous, and connective tissue.
6. **Organism:** All living things can be called organisms. A single organism possesses the set of characteristics that define life. Biologists classify organisms as belonging to a particular **species**, which is a related group of organisms that share a distinctive form and set of attributes in nature.

The members of the same species are closely related genetically. In Units VI and VII, we will examine plants and animals at the level of cells, tissues, organs, and complete organisms.

7. **Population:** A group of organisms of the same species that occupy the same environment is called a **population**.

8. **Community:** A biological **community** is an assemblage of populations of different species. The types of species that are found in a community are determined by the environment and by the interactions of species with each other.

9. **Ecosystem:** Researchers may extend their work beyond living organisms and also study the environment. Ecologists analyze **ecosystems**, which are formed by interactions of a community of organisms with their physical environment. Unit VIII considers biology from populations to ecosystems.

10. **Biosphere:** The **biosphere** includes all of the places on the Earth where living organisms exist, encompassing the air, water, and land.

Modern Forms of Life Are Connected by an Evolutionary History

Life began on Earth as primitive cells about 3.5 to 4 billion years ago. Since that time, those primitive cells underwent evolutionary changes that ultimately gave rise to the species we see today.

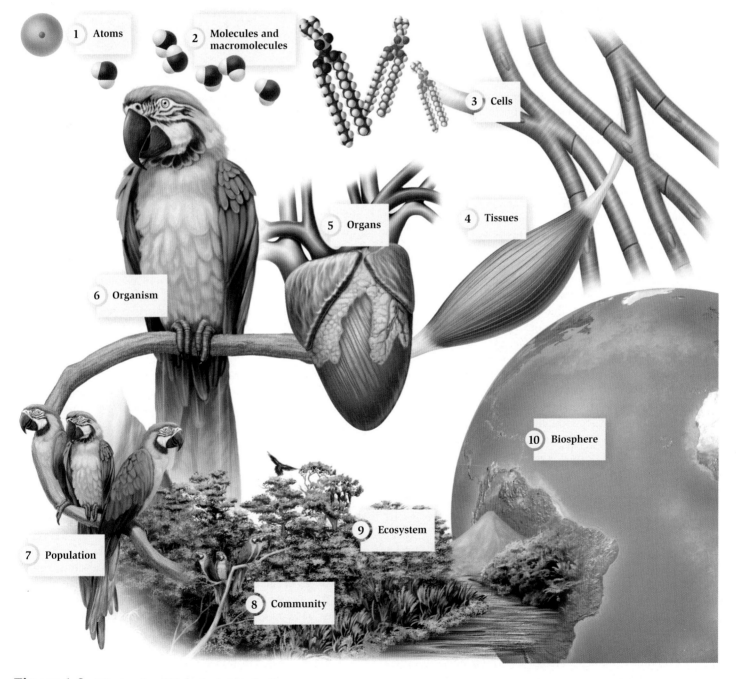

Figure 1.3   The levels of biological organization.

Figure 1.4 An example of modification of a structure for a new function.   The bird shown in the photograph has used a modified milk carton in which to build its nest. By analogy, evolution also involves the modification of pre-existing structures for a new function.

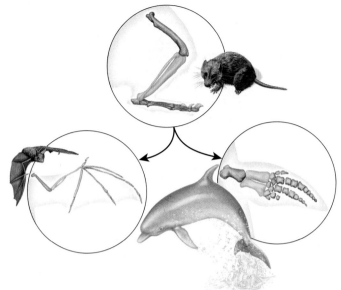

Figure 1.5 An example showing a modification that has occurred as a result of biological evolution.   The wing of a bat and the flipper of a dolphin were modified from a limb that was used for walking in a pre-existing ancestor.

Biological inquiry: Among mammals, give two examples of how the tail has been modified for different purposes.

Understanding the evolutionary history of species often provides key insights regarding the structure and function of an organism's body. As a metaphor to help you appreciate this idea, **Figure 1.4** shows a photograph of a bird that is using a milk carton in which to build a nest. If we did not know that the milk carton had served an earlier purpose, namely to contain milk, we might wonder why the bird had made a nesting site that resembled a milk carton. Obviously, we do not worry about this because we immediately grasp that the milk carton had a previous history, and that it has been modified by a person to serve a new purpose—a nesting site for a bird. Understanding history allows us to make sense out of this nest.

Likewise, evolutionary change involves modifications of characteristics in pre-existing populations. Over long periods of time, populations may change such that structures with a particular function may become modified to serve a new function. For example, the wing of a bat is used for flying, while the flipper of a dolphin is used for swimming (**Figure 1.5**). Both structures were modified from a limb that was used for walking in a pre-existing ancestor.

Evolutionary change occurs by two mechanisms, vertical descent with mutation and horizontal gene transfer. Let's take a brief look at each of these mechanisms.

Vertical Descent with Mutation The traditional way to view evolution involves a progression of changes in a series of ancestors. Such a series is called a **lineage**. **Figure 1.6** shows a portion of the lineage that gave rise to modern horses. This type of evolution is called **vertical evolution** because biologists have traditionally depicted such evolutionary change in a vertical diagram like the one shown in Figure 1.6. In this mechanism of evolution, new species evolve from pre-existing species by the accumulation of **mutations**, which are changes in the genetic material of organisms. But why would some mutations accumulate in a population and eventually change the characteristics of an entire species? One reason is that a mutation may alter the traits of organisms in a way that increases their chances of survival or reproduction. When a mutation causes such a beneficial change, the mutation is more likely to increase in a population over the course of many generations, a process called **natural selection**, which is discussed in Units IV and V. Evolution also involves the accumulation of neutral changes that do not benefit a species, and even rare changes that may be harmful.

Horizontal Gene Transfer In addition to vertical evolution, which produces a lineage, species also evolve by another process that involves genetic exchanges between different species. Sexually reproducing species usually mate with members of their own species. Similarly, asexual species such as bacteria can occasionally transfer genetic material between cells, but again, that tends to occur most readily between members of the same bacterial species. However, on relatively rare occasions, genetic exchanges occur between different species. For example, you may have heard in the news media that resistance to antibiotics among bacteria is a growing medical problem. Genes that confer antibiotic resistance are sometimes transferred between different bacterial species (**Figure 1.7**).

When genes are transferred from one species to another, this event is called **horizontal gene transfer**. In a lineage in

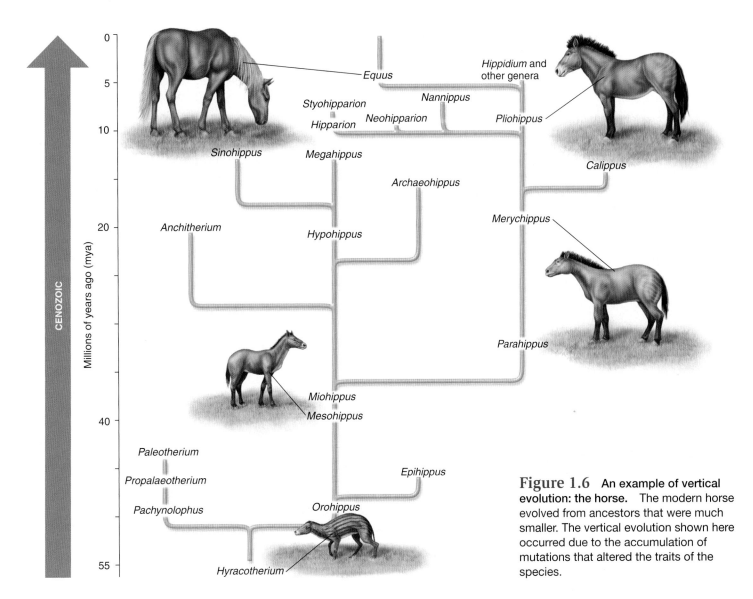

Figure 1.6 **An example of vertical evolution: the horse.** The modern horse evolved from ancestors that were much smaller. The vertical evolution shown here occurred due to the accumulation of mutations that altered the traits of the species.

which the timescale is depicted on a vertical axis, horizontal gene transfer is shown as a horizontal line between two different species (**Figure 1.8**). Genes that are transferred horizontally may be acted upon by natural selection to eventually promote changes in an entire species. This has been an important mechanism of evolutionary change, particularly among bacterial species. In addition, during the early stages of evolution, which occurred a few billion years ago, horizontal gene transfer was an important part of the process that gave rise to all modern species.

Traditionally, biologists have described evolution using diagrams that depict the vertical evolution of species on a long timescale. This is the type of evolutionary tree that was shown in Figure 1.6. For many decades, the simplistic view held that all living organisms evolved from a common ancestor, resulting in a "tree of life," which could describe the vertical evolution that gave rise to all modern species. Now that we understand the great importance of horizontal gene transfer in the evolution of life on Earth, biologists have needed to re-evaluate the concept of evolution as it occurs over time. Rather than a tree of

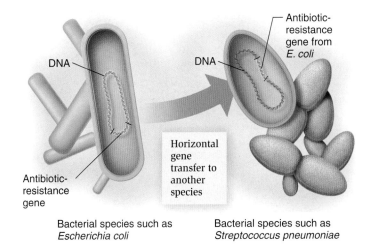

Figure 1.7 **An example of horizontal gene transfer: antibiotic resistance.** One bacterial species may transfer a gene to a different bacterial species, such as a gene that confers resistance to an antibiotic.

life, a more appropriate way to view the unity of living organisms is to describe it as a "web of life," which accounts for both vertical evolution and horizontal gene transfer. Figure 1.8 illustrates such a diagram.

The Classification of Living Organisms Allows Biologists to Appreciate the Unity and Diversity of Life

As biologists discover new species, they try to place them in groups based on their evolutionary history. This is an arduous task because researchers estimate that the Earth has between 10 and 100 million different species! The rationale for categorization is usually based on vertical descent. Species with a recent common ancestor are grouped together, while species whose common ancestor is in the very distant past are placed into different groups. The grouping of species is termed **taxonomy**.

Let's first consider taxonomy on a broad scale. You may have noticed that Figure 1.8 showed three main groups of organisms.

All forms of life can be placed into three large categories or domains called **Bacteria**, **Archaea**, and **Eukarya** (**Figure 1.9**). Bacteria and Archaea are microorganisms that are also termed **prokaryotic** because their cell structure is relatively simple. At the molecular level, bacterial and archaeal cells show significant differences in their lipid composition, metabolic pathways, and mechanisms of gene expression. By comparison, organisms in domain Eukarya are **eukaryotic** and have larger cells with internal compartments that serve various functions. A defining distinction between prokaryotic and eukaryotic cells is that eukaryotic cells have a **cell nucleus** in which the genetic material is surrounded by a membrane. The organisms in domain Eukarya have been further subdivided into four major categories or kingdoms called Animalia (animals), Plantae (plants), Protista (protists), and Fungi. However, as discussed in Chapter 26 and Unit V, the traditional view of four eukaryotic kingdoms is now under revision as biologists have gathered new information regarding the evolutionary relationships of these organisms.

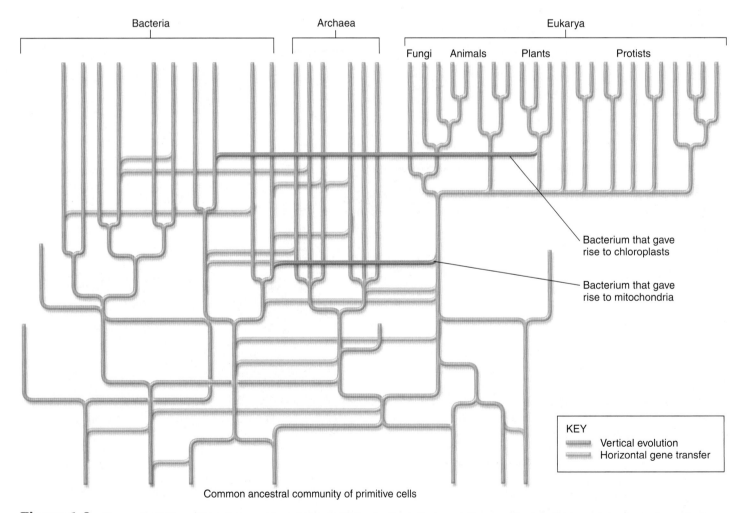

Common ancestral community of primitive cells

Figure 1.8   The web of life.   This diagram of evolution includes both vertical evolution and horizontal gene transfer as important mechanisms in the evolution of life on Earth. Note: Archaea are unicellular species that are similar in structure to bacteria.

Biological inquiry: How does the concept of a tree of life differ from a web of life?

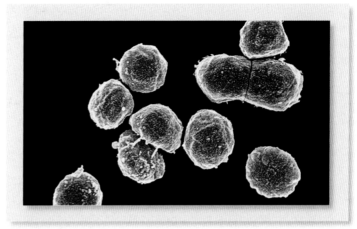

(a) Domain Bacteria: Unicellular prokaryotes that inhabit many diverse environments on Earth.

(b) Domain Archaea: Unicellular prokaryotes that are less common than bacteria. Some live in extreme environments such as hot springs.

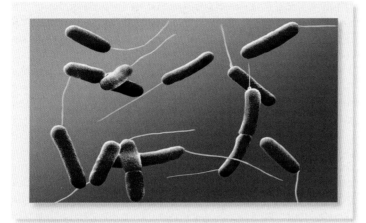

Kingdom Animalia: Multicellular organisms that usually have a nervous system and are capable of locomotion. They must eat other organisms to live.

Kingdom Plantae: Multicellular organisms that can carry out photosynthesis.

Kingdom Protista: Unicellular and small multicellular organisms that are now subdivided into several different kingdoms based on their evolutionary relationships.

Kingdom Fungi: Unicellular and multicellular organisms that have a cell wall but cannot carry out photosynthesis. Fungi usually survive on decaying organic material.

(c) Domain Eukarya

Figure 1.9 **The three domains of life.** Two of these domains, **(a)** Bacteria and **(b)** Archaea, are prokaryotes, while the third domain, **(c)** Eukarya, comprises species that are eukaryotes.

Taxonomy involves multiple levels in which particular species are placed into progressively smaller and smaller groups of organisms that are more closely related to each other evolutionarily (**Figure 1.10**). Such an approach emphasizes the unity and diversity of different species. As an example, let's consider the jaguars, shown on the cover of your textbook. The broadest grouping for the jaguar is the domain, namely Eukarya, followed by progressively smaller divisions, from kingdom (Animalia) to species. In the animal kingdom, jaguars are part of a phylum, Chordata, which is subdivided into classes. Jaguars are in a class called Mammalia, which includes all mammals. The common ancestor that gave rise to mammals arose over 200 million years ago. Mammalia is subdivided into several smaller orders. The jaguar is in the order Carnivora. The order is in turn divided into families; the jaguar and all other cats belong to the family Felidae. The genus *Panthera* is the smallest group of different species that contains the jaguar. As you can see in Figure 1.10, the genus contains only four modern species, the jaguar and other types of large cats. Therefore, the genus has species that are very similar to each other in form, and have evolved from a common (extinct) ancestor that lived relatively recently on an evolutionary timescale, approximately 5 million years ago.

Biologists use a two-part description, called a **binomial**, to provide each species with a unique scientific name. The scientific name of the jaguar is *Panthera onca*. The first part is the genus and the second part is the specific epithet or species descriptor. By convention, the genus name is capitalized, while the specific epithet is not. Both names are italicized. All scientific names are Latinized.

Taxonomic group	Jaguar is found in:	Approximate time when the common ancestor for this group arose	Approximate number of modern species in this group	
Domain	Eukarya	2,000 mya	> 5,000,000	
Kingdom	Animalia	600 mya	> 1,000,000	
Phylum	Chordata	525 mya	50,000	
Class	Mammalia	200 mya	5,000	
Order	Carnivora	60 mya	270	
Family	Felidae	40 mya	38	
Genus	*Panthera*	5 mya	4	
Species	*onca*	1.5 mya	1	

Figure 1.10 Taxonomic and evolutionary groupings leading to the jaguar.

GENOMES & PROTEOMES

The Study of Genomes and Proteomes Provides an Evolutionary Foundation for Our Understanding of Biology

The unifying concept in biology is evolution. We can understand the unity of modern organisms by realizing that all living species evolved from an interrelated group of ancestors. However, from an experimental perspective, this realization presents a dilemma—we cannot take a time machine back over the course of 4 billion years to carefully study the characteristics of extinct organisms and fully appreciate the series of changes that have led to modern species. Fortunately though, evolution has given experimental biologists a wonderful puzzle to study, namely the genomes of modern species. The term **genome** refers to *the complete genetic makeup of an organism* (**Figure 1.11a**). The genome is critical to life because it performs these functions:

- *Acts as a stable informational unit:* The genome of every organism stores information that provides a blueprint to create their characteristics.
- *Provides continuity from generation to generation:* The genome is copied and transmitted from generation to generation.

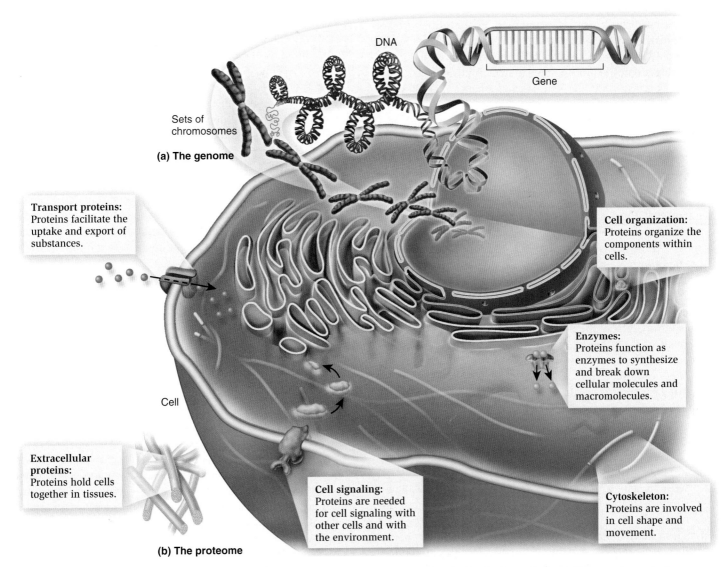

(a) The genome

Transport proteins: Proteins facilitate the uptake and export of substances.

Cell organization: Proteins organize the components within cells.

Enzymes: Proteins function as enzymes to synthesize and break down cellular molecules and macromolecules.

Extracellular proteins: Proteins hold cells together in tissues.

Cell signaling: Proteins are needed for cell signaling with other cells and with the environment.

Cytoskeleton: Proteins are involved in cell shape and movement.

(b) The proteome

Figure 1.11   **Genomes and proteomes.**   (a) The genome, which is composed of DNA, is the entire genetic complement of an organism. Most of the genetic material in eukaryotic cells is found in the cell nucleus. Its primary function is to encode the proteome. (b) The proteome is the entire protein complement of a cell or entire organism. Proteins are largely responsible for the structure and function of cells and complete organisms.

Biological inquiry: Biologists sometimes say that the genome is a storage unit, while the proteome is largely the functional unit of life. Explain this statement.

- *Acts as an instrument of evolutionary change:* Every now and then, the genome undergoes a mutation that may alter the characteristics of an organism. In addition, a genome may acquire new genes by horizontal gene transfer. The accumulation of such changes over the course of many generations produces the evolutionary changes that alter species and produce new species.

The evolutionary history and relatedness of all living organisms can be illuminated by genome analysis. The genome of every organism carries the results and the evidence of millions of years of evolution. The genomes of prokaryotes usually contain a few thousand genes, while those of eukaryotes may contain tens of thousands. An exciting advance in biology over the past couple of decades has been the ability to analyze the DNA sequence of genomes, a technology called **genomics**. For instance, we can compare the genomes of a frog, a giraffe, and a petunia and discover intriguing similarities and differences. These comparisons help us to understand how new traits evolved. For example, all three types of organisms have the same kinds of genes that are needed for the breakdown of nutrients such as sugars. In contrast, only the petunia carries genes that allow it to carry out photosynthesis.

An extension of genome analysis is the study of **proteomes**, which refers to *the complete complement of proteins that a cell or organism can make*. The function of most genes is to encode polypeptides that become units in proteins. As shown in **Figure 1.11b**, these include transport proteins; extracellular proteins; proteins that function in cell organization, in cell signaling, and as enzymes; and proteins that form a cytoskeleton. Proteins are the key participants in maintaining cell structure and carrying out most cell functions. Therefore, the genome of each species carries the information to make its proteome, the hundreds or thousands of proteins that each cell of that species makes. Proteins are largely responsible for the structures and functions of cells and organisms. The technical approach called **proteomics** involves the analysis of the proteome of a single species and the comparison of the proteomes of different species. Proteomics helps us to understand how the various levels of biology are related to one another, from the molecular level—at the level of protein molecules—to the higher levels, such as how the functioning of proteins produces the characteristics of cells and organisms, and the ability of populations of organisms to survive in their natural environments.

As a concrete way to understand the unifying theme of evolution in biology, a recurring theme in the chapters that follow is a brief topic called "Genomes & Proteomes" that will allow you to appreciate how evolution produced the characteristics of modern species. These topics explore how the genomes of different species are similar to each other, and how they are different. You will learn how genome changes affect the proteome and thereby control the traits of modern species. Ultimately, these concepts provide you with a way to relate information at the molecular level to the traits of organisms and their survival within ecosystems.

Along these lines, the cover of your textbook provides food for thought. The cats on the cover are jaguars. A black jaguar is sometimes called a panther, but it is still the same species, *Panthera onca*, as a spotted jaguar. How are the genomes of the spotted and black jaguars different? How are their proteomes different? Can this information tell us anything about the ecosystem in which these animals live? Does this have anything to do with evolutionary change? If we analyzed the genomes of spotted and black jaguars, we would discover that they are overwhelmingly similar to each other (**Figure 1.12a**). Of the 20,000 or so genes, the majority would be identical to each other or nearly so. However, based on their differences in appearance, we would expect the DNA sequence within one particular gene to be different, namely a gene that plays a role in producing pigment in the fur. At the level of the proteome, this slight genome difference causes the spotted jaguar to make proteins (enzymes) that synthesize a background coat pigment that is tan, while a black jaguar makes a background coat pigment that is black. Under bright light, you can see that a black jaguar actually has spots, but the dark background pigment greatly masks their appearance (**Figure 1.12b**).

Do spotted versus black coats have any impact on the life of these animals? The answer is yes. The spotted animals are thought to be better hunters in lighter grassland environments. The black animals are more likely to survive in darker forests, where they are less likely to be seen by their prey. In nature, mixed populations of spotted and black jaguars are often observed on the edges of forests, where both light and dark environments exist. Therefore, biologists have speculated that this is an example of evolutionary change. Genetic mutations have occurred that promote the survival and reproductive success of these animals, which vary in light and dark environments.

1.2 Biology as a Scientific Discipline

What is science? Surprisingly, the definition of science is not easy to state. Most people have an idea of what science is, but actually articulating that idea proves difficult. In biology, we might define **science** as *the observation, identification, experimental investigation, and theoretical explanation of natural phenomena*.

Science is conducted in different ways and at different levels. Some biologists study the molecules that compose life, while others try to understand how organisms survive in their natural environments. In some cases, experiments are designed to test the validity of ideas that are suggested by researchers. In this section, we will examine how biologists follow a standard approach, called the **scientific method**, to test their ideas. We will learn that scientific insight is not based on intuition. Instead, scientific knowledge makes predictions that can be experimentally tested.

Even so, not all discoveries are the result of researchers following the scientific method. Some discoveries are simply made

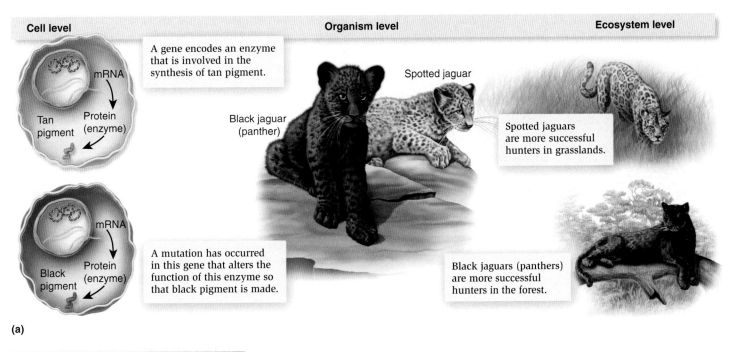

Cell level | **Organism level** | **Ecosystem level**

A gene encodes an enzyme that is involved in the synthesis of tan pigment.

mRNA

Tan pigment → Protein (enzyme)

A mutation has occurred in this gene that alters the function of this enzyme so that black pigment is made.

mRNA

Black pigment → Protein (enzyme)

Black jaguar (panther)

Spotted jaguar

Spotted jaguars are more successful hunters in grasslands.

Black jaguars (panthers) are more successful hunters in the forest.

(a)

(b)

Figure 1.12 How the study of genomes and proteomes can provide us with connections to different biological levels. (a) Spotted jaguars differ from panthers because they make an enzyme that produces a tan pigment, while a mutation in the same gene in panthers results in black pigment. With regard to hunting success, the spotted jaguars are better in grasslands, while the panthers are better in the forest. (b) A close-up view of a panther, showing its spots.

by gathering new information. As described earlier in Figure 1.1, the characterization of many plants and animals has led to the development of many important medicines and research tools. In this section, we will also consider how researchers often set out on "fact-finding missions" that are aimed at uncovering new information that may eventually lead to new discoveries in biology.

Biologists Investigate Life at Different Levels of Organization

Earlier in Figure 1.3, we examined the various levels of biological organization. The study of these different levels depends not only on the scientific interests of biologists but also on the tools that are available to them. Prior to the development of high-quality microscopes, biologists primarily focused their attention on characteristics they could observe with their unaided eyes. They studied the activities of organisms in their natural environments, a branch of biology called **ecology** (**Figure 1.13a**). In addition, researchers have examined the structures and functions of plants and animals, which are disciplines called **anatomy** and **physiology** (**Figure 1.13b**). As microscope technology improved, many researchers shifted their attention to the study of

cells. **Cell biology**, which is the study of cells, became an important branch of biology in the early 1900s and remains so today (**Figure 1.13c**). In the 1970s, genetic tools became available to study single genes and the proteins they encode. This genetic technology enabled researchers to study individual molecules, such as proteins, in living cells. Genetic technology spawned the field of **molecular biology**. Together with the efforts of chemists and biochemists, molecular biologists focus their efforts on the structure and function of the molecules of life (**Figure 1.13d**). Such researchers want to understand how biology works at the molecular and even atomic levels. Overall, the 20th century saw a progressive increase in the number of biologists who used a reductionist approach to understand biology. **Reductionism** involves reducing complex systems to simpler components as a way to understand how the system works. In biology, reductionists study the parts of a cell or organism as individual units.

In the 1980s, the pendulum began to swing in the other direction. Scientists have invented new tools that allow us to study groups of genes (genomic techniques) and groups of proteins (proteomic techniques). Biologists now use the term **systems biology** to describe research that is aimed at understanding how the properties of life arise by complex interactions.

Ecologists study species in their native environments.

(a) Ecology—population/ community/ecosystem levels

Anatomists and physiologists study how the structure of organisms are related to their functions.

(b) Anatomy and physiology— tissue/organ/organism levels

Cell biologists often use the microscope to learn how cells function.

(c) Cell biology—cellular levels

Molecular biologists and biochemists study the molecules and macromolecules that make up cells.

(d) Molecular biology— molecular/atomic levels

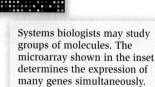

Systems biologists may study groups of molecules. The microarray shown in the inset determines the expression of many genes simultaneously.

(e) Systems biology—all levels, shown here at the molecular level

Figure 1.13   Biological investigation at different levels.

This term is often applied to the study of cells. In this context, systems biology may involve the investigation of groups of proteins with a common goal (**Figure 1.13e**). For example, a systems biologist may conduct experiments that try to characterize an entire cellular process, which is driven by dozens of different proteins. Systems biology is not new. Animal and plant physiologists have been studying the functions of complex organ systems for centuries. Likewise, ecologists have been characterizing ecosystems for a very long time. The novelty and excitement of systems biology in recent years has been the result of new experimental tools that allow us to study complex interactions at the molecular level. As described throughout this textbook, the investigation of genomes and proteomes has provided important insights regarding many interesting topics in systems biology. For example, as discussed in Chapter 6, systems biology has enabled researchers to understand how the various parts of a cell work together as an integrated system.

A Hypothesis Is a Proposed Idea, While a Theory Is a Broad Explanation Backed by Extensive Evidence

Let's now consider the process of science. In biology, a **hypothesis** is a proposed explanation for a natural phenomenon. It is a proposition based on previous observations or experimental studies. For example, with knowledge of seasonal changes, you might hypothesize that maple trees drop their leaves in the autumn because of the shortened amount of daylight. An alternative hypothesis might be that the trees drop their leaves because of colder temperatures. In biology, a hypothesis requires more work by researchers to either accept or reject it.

A useful hypothesis must make predictions that can be shown to be correct or incorrect. The validity of the predictions is usually determined by additional observations or experimentation. If the predictions do not agree with new data, the hypothesis is rejected. Alternatively, a hypothesis may be correct so that further work will not disprove it. Even so, a hypothesis is never really proven but rather always remains provisional. Researchers accept the possibility that perhaps they have not yet conceived of the correct hypothesis. However, after many experiments, biologists may say that they accept a hypothesis, but they should never say that the hypothesis is proven.

By comparison, the term **theory**, as it is used in biology, is a broad explanation of some aspect of the natural world that is substantiated by a large body of evidence. Biological theories incorporate observations, hypothesis testing, and the laws of other disciplines such as chemistry and physics. The power of theories is that they allow us to make many predictions regarding the properties of living organisms. As an example, let's consider the theory that DNA is the genetic material, and that it is organized into units called genes. An overwhelming body of evidence has substantiated this theory. Thousands of living species have been analyzed, and all of them have been found to use DNA as their genetic material, and to express genes that produce the proteins that lead to their characteristics. This theory makes many valid predictions. For example, certain types of mutations in genes are expected to affect the traits of organisms. This prediction has been confirmed experimentally. Similarly, genetic material is copied and transmitted from parents to offspring. By comparing the DNA of parents and offspring, this prediction has also been confirmed. Furthermore, the theory explains the observation that offspring resemble their parents. Overall, two key attributes of a scientific theory are consistency with a vast amount of known data, and the ability to make many correct predictions. Two other important biological theories that we have touched on in this chapter are the cell theory and the theory of evolution by natural selection.

The meaning of the term *theory* is sometimes muddled because it is used in different situations. In everyday language, a "theory" is often viewed as little more than a guess or a hypothesis. For example, a person might say, "My theory is that Professor Simpson did not come to class today because he went to the beach." However, in biology, a theory is much more than a guess. A theory is an established set of ideas that explains a vast amount of data and offers valid predictions that can be tested. Like a hypothesis, a theory can never be proven to be true. Scientists acknowledge that they do not know everything. Even so, biologists would say that theories are extremely likely to be true, based on all known information. In this regard, theories are viewed as **knowledge**, which is the awareness and understanding of information.

Discovery-Based Science and Hypothesis Testing Are Scientific Approaches That Help Us Understand Biology

The path that leads to an important discovery is rarely a straight line. Rather, scientists ask questions, make observations, ask modified questions, and may eventually conduct experiments to test their hypotheses. The first attempts at experimentation may fail, and new experimental approaches may be needed. To suggest that scientists follow a rigid scientific method is an oversimplification of the process of science. Scientific advances often occur as scientists dig deeper and deeper into a topic that interests them. Curiosity is the key phenomenon that sparks scientific inquiry. As discussed next, researchers typically follow two general types of approaches—discovery-based science and hypothesis testing.

Discovery-Based Science The collection and analysis of data, without the need for a preconceived hypothesis is called **discovery-based science** or simply **discovery science**. The information gained from discovery-based science may have practical applications that benefit people. Drug companies, for example, may test hundreds or even thousands of drugs to determine if any of them are useful in the treatment of disease (**Figure 1.14a**). Once a drug has been discovered that is effective in disease treatment, researchers may dig deeper and try to understand how the drug exerts its effects. In this way, discovery-based science may help us learn about basic concepts in medicine

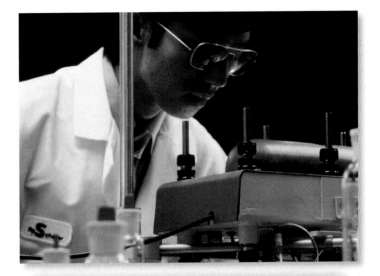

Drug companies may screen hundreds or thousands of different compounds trying to discover ones that may prove effective in the treatment of a particular disease.

(a) Drug discovery

Genetic researchers search through the genomes of humans and other species trying to discover new genes. Such discoveries may help us understand molecular biology and provide insight into the causes of inherited diseases in people.

(b) Discovery of genes

Figure 1.14   Discovery-based science.

and biology. Another example involves the study of genomes (**Figure 1.14b**). Over the past few decades, researchers have identified and begun to investigate newly discovered genes within the human genome without already knowing the function of the gene they are studying. The goal is to gather additional clues that may eventually allow them to propose a hypothesis that explains the gene's function. Discovery-based science often leads to hypothesis testing.

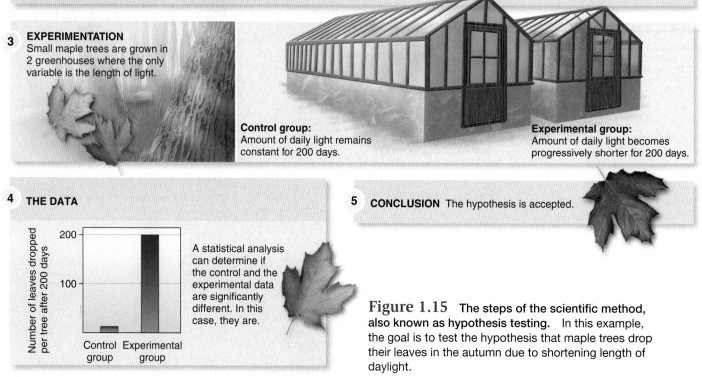

1 **OBSERVATIONS** The leaves on maple trees fall in autumn when the days get colder and shorter.

2 **HYPOTHESIS** The shorter amount of daylight causes the leaves to fall.

3 **EXPERIMENTATION**
Small maple trees are grown in 2 greenhouses where the only variable is the length of light.

Control group:
Amount of daily light remains constant for 200 days.

Experimental group:
Amount of daily light becomes progressively shorter for 200 days.

4 **THE DATA**

(Bar graph: y-axis labeled "Number of leaves dropped per tree after 200 days" with values 100 and 200; x-axis with "Control group" and "Experimental group")

A statistical analysis can determine if the control and the experimental data are significantly different. In this case, they are.

5 **CONCLUSION** The hypothesis is accepted.

Figure 1.15 **The steps of the scientific method, also known as hypothesis testing.** In this example, the goal is to test the hypothesis that maple trees drop their leaves in the autumn due to shortening length of daylight.

Hypothesis Testing In biological science, the scientific method, also known as **hypothesis testing**, is often followed to test the validity of a hypothesis. This strategy may be described as a five-stage process:

1. Observations are made regarding natural phenomena.
2. These observations lead to a hypothesis that tries to explain the phenomena. As mentioned, a useful hypothesis is one that is testable because it makes specific predictions.
3. Experimentation is conducted to determine if the predictions are correct.
4. The data from the experiment are analyzed.
5. The hypothesis is accepted or rejected.

The scientific method is intended to be an objective way to gather knowledge.

As an example, let's return to our scenario of maple trees dropping their leaves in autumn. By observing the length of daylight throughout the year, and comparing that data with the time of the year when leaves fall, one hypothesis might be that shorter daylight causes the leaves to fall (**Figure 1.15**). This hypothesis makes a prediction—exposure of maple trees to shorter daylight will cause their leaves to fall. To test this prediction, researchers would design and conduct an experiment.

Although hypothesis testing may follow many paths, certain experimental features are common to this approach. First, data are often collected in two parallel manners. One set of experi-ments is done on the **control sample**, while another set is conducted on the **experimental sample**. In an ideal experiment, the control and experimental samples differ by only one factor. For example, an experiment could be conducted in which two groups of trees would be observed and the only difference between their environments would be the length of light each day. To conduct such an experiment, researchers would grow small trees in a greenhouse where they could keep factors such as temperature and water the same between the control and exper-imental samples, while providing them with different amounts of daylight. In the control group, the number of hours of light pro-vided by light bulbs would be kept constant each day, while in the experimental group, the amount of light each day would be-come progressively shorter to mimic seasonal light changes. The researchers would then record the amount of leaves that were dropped by the two groups of trees over a certain period of time.

Another key feature of hypothesis testing is data analysis. The result of experimentation is a set of data from which a biol-ogist tries to draw conclusions. Biology is a quantitative science. As such, data often come in the form of numbers that may or may not have important meaning. When experimentation in-volves a control and experimental sample, a common form of analysis is to determine if the data collected from the two sam-ples are significantly different from each other. In this regard, the word *significant* means statistically significant. Biologists apply statistical analyses to their data to determine if the con-trol and experimental samples are likely to be different from each other because of the single variable that is different between the two samples. When they are statistically significant, this means that the differences between the control and experimental data are not likely to have occurred as a matter of random chance. In our tree example shown in Figure 1.15, the trees in the control sample dropped far fewer leaves than did those in the experimental sample. A statistical analysis could determine if the data collected from the two greenhouses are significantly different from each other. If the two sets of data are found not to be significantly different, we would reject our hypothesis. Alternatively, if the differences between the two sets of data are significant, as shown in Figure 1.15, we would accept our hypothesis, though it is not proven.

As described next, discovery-based science and hypothesis testing are often used together to learn more about a particular scientific topic. As an example, let's look at how both approaches have led to successes in the study of the disease called cystic fibrosis.

The Study of Cystic Fibrosis Provides Examples of Both Discovery-Based Science and Hypothesis Testing

Let's consider how biologists made discoveries related to cystic fibrosis (CF), which affects about 1 in every 3,500 Americans. Persons with CF produce abnormally thick and sticky mucus that obstructs the lungs and causes life-threatening lung infections. The thick mucus also blocks the pancreas, which prevents the digestive enzymes that this organ produces from reaching the intestine. For this reason, CF patients tend to have excessive appetites but poor weight gain. Persons with this disease may also experience liver damage because the thick mucus can obstruct the liver. The average life span for people with CF is currently in their mid- to late 30s. Fortunately, as more advances have been made in treatment, this number has steadily increased.

Because of its medical significance, many scientists are interested in this disorder and have conducted studies aimed at gaining greater information regarding its underlying cause. The hope is that a better understanding of the disorder may lead to improved treatment options, and perhaps even a cure. As described next, discovery-based science and hypothesis testing have been critical to gaining a better understanding of this disorder.

The* CF *Gene and Discovery-Based Science In 1945, Dorothy Anderson determined that cystic fibrosis is a genetic disorder. Persons with CF have inherited two faulty *CF* genes, one from each parent. Over 40 years later, researchers used discovery-based science to identify the *CF* gene. Their search for the *CF* gene did not require any preconceived hypothesis regarding the function of the gene. Rather, they used genetic strategies that are similar to those described in Chapter 20. In 1989, research groups headed by Lap-Chi Tsui, Francis Collins, and John Riordan identified the *CF* gene.

The discovery of the gene made it possible to devise diagnostic testing methods to determine if a person carries a faulty *CF* gene. In addition, the identification of the *CF* gene may provide a potential treatment option for people with this disorder.

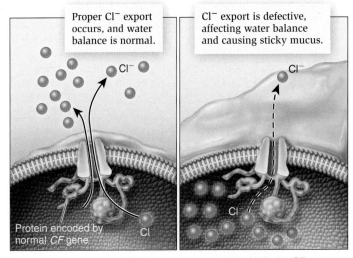

Lung cell with normal *CF* gene   Lung cell with faulty *CF* gene

Figure 1.16 A hypothesis that suggests an explanation of the function of the gene that is defective in patients with cystic fibrosis. The normal *CF* gene, which does not carry a mutation, encodes a transporter protein that transports chloride ions (Cl⁻) across the plasma membrane to the outside of the cell. In persons with CF, this transporter is defective due to a mutation in the *CF* gene.

Biological inquiry: Explain how discovery-based science helped researchers to hypothesize that the CF gene encodes a transporter protein.

As discussed in Chapter 20, gene therapy is a technology in which persons with a faulty gene are given treatments that involve the introduction of the normal gene into their bodies. Clinical trials are underway to determine if the *CF* gene from normal individuals can be used to overcome the symptoms of the faulty gene in CF patients.

The characterization of the *CF* gene provided important clues regarding its function. Researchers observed striking similarities between the *CF* gene and other genes that were already known to encode proteins called transporters that function in the transport of substances across membranes. Based on this observation, as well as other kinds of data, the researchers hypothesized that the function of the normal *CF* gene is to encode a transporter. In this way, the identification of the *CF* gene led researchers to conduct experiments that were aimed at testing a hypothesis of its function.

The* CF *Gene and Hypothesis Testing Based on the characterization of the *CF* gene and on other studies showing that patients with the disorder have an abnormal regulation of salt balance across their plasma membranes, researchers hypothesized that the normal *CF* gene encodes a protein that transports chloride ions (Cl⁻), a component of common table salt (NaCl), across the membrane of cells (**Figure 1.16**). This hypothesis led to experimentation in which researchers tested normal cells and cells from CF patients for their ability to transport Cl⁻. The CF cells were found to be defective in chloride transport. In 1990, scientists successfully transferred the normal gene to CF cells in the laboratory.

The introduction of the normal *CF* gene into the cells from CF patients corrected the defect in chloride transport. Overall, the results showed that the *CF* gene encodes a protein that transports Cl⁻ across the plasma membrane. A mutation in this gene causes it to encode a defective transporter protein, leading to a salt imbalance that affects water levels outside the cell, which explains the thick and sticky mucus in CF patients. In this example, hypothesis testing has provided a way to accept or reject an idea regarding how a disease is caused by a genetic change.

FEATURE INVESTIGATION

Observation and Experimentation Form the Core of Biology

Because biology is the study of life, a biology textbook that focuses only on a description of living organisms would miss the main point. Biology is largely about the process of discovery. Therefore, a recurring theme of this textbook is discovery-based science and hypothesis testing. While each chapter contains many examples of data collection and experiments, a consistent element is a "Feature Investigation"—an actual study by current or past researchers. Some of these involve discovery-based science in which biologists collect and analyze data in an attempt to make discoveries that are not hypothesis driven. Alternatively, most Feature Investigations involve hypothesis testing in which a hypothesis is stated, and the experiment and resulting data are presented.

The Feature Investigations allow you to appreciate the connection between science and scientific theories. We hope you will find this a more interesting and rewarding way to learn about biology. As you read a Feature Investigation, you may find yourself thinking about different approaches and alternative hypotheses. Different people can view the same data and arrive at very different conclusions. As you progress through the experiments in this textbook, you will enjoy biology far more if you try to develop your own skills at formulating hypotheses, designing experiments, and interpreting data.

Finally, it is worthwhile to point out that science is a social discipline. After performing observations and experiments, scientists report their conclusions to the scientific community (**Figure 1.17**). They comment on each other's ideas and work, eventually shaping together the information that builds into scientific theories over many years. As you develop your skills at scrutinizing experiments, it is satisfying to discuss your ideas with other people, including fellow students and faculty members. Importantly, you do not need to "know all the answers" before you enter into a scientific discussion. Instead, a more rewarding way to view science is as an ongoing and never-ending series of questions.

Figure 1.17   **The social aspects of science.**   At scientific meetings, researchers gather together to discuss new data and discoveries. Research that is conducted by professors, students, lab technicians, and industrial participants is sometimes hotly debated.

CHAPTER SUMMARY

- Biology is the study of life. Discoveries in biology help us understand how life exists, and they also have many practical applications such as the development of drugs to treat human diseases. (Figure 1.1)

1.1 The Properties of Life: Past and Present

- Seven characteristics are common to all forms of life. All living things (1) are composed of cells; (2) use energy; (3) respond to environmental changes; (4) regulate their internal conditions (homeostasis); (5) grow and develop; (6) reproduce; and (7) evolve over the course of many generations. (Figure 1.2)

- Living organisms can be viewed at different levels of complexity: atoms, molecules and macromolecules, cells, tissues, organs, organisms, populations, communities, ecosystems, and the biosphere. (Figure 1.3)

- Changes in species often occur as a result of modification of pre-existing structures. (Figures 1.4, 1.5)

- Vertical evolution involves mutations in a lineage that alter the characteristics of species over many generations. During this process, natural selection results in the survival of individuals with greater reproductive success. Over the long run, this process alters species and may produce new species. (Figure 1.6)

- Horizontal gene transfer is the transfer of genes between different species. Along with vertical evolution, it is also an important force in biological evolution. (Figures 1.7, 1.8)

- Taxonomy involves the grouping of species according to their evolutionary relatedness to other species. Going from broad to narrow, each species is placed into a domain, kingdom, phylum, class, order, family, and genus. (Figures 1.9, 1.10)

- The genome is the genetic composition of a species. It provides a blueprint for the traits of an organism, is transmitted from parents to offspring, and acts as an instrument for evolutionary change. The proteome is the collection of proteins that a cell or organism can make. Each chapter in this textbook has a brief discussion of "Genomes & Proteomes" for you to understand this fundamental concept in biology. (Figure 1.11)

- An understanding of genomes and proteomes helps us to understand the characteristics of individuals and how they survive in their native environments. (Figure 1.12, and book cover)

1.2   Biology as a Scientific Discipline

- Biological science involves the observation, identification, experimental investigation, and theoretical explanation of natural phenomena.

- Biologists study life at different levels, ranging from ecosystems to molecular components in cells. (Figure 1.13)

- A hypothesis is a proposal to explain a natural phenomenon. A biological theory is a broad explanation that makes many valid predictions. A theory is based on vast amounts of data.

- Discovery-based science is an approach in which researchers conduct experiments without a preconceived hypothesis. It is a fact-finding mission. (Figure 1.14)

- The scientific method, also called hypothesis testing, is a series of steps to test the validity of a hypothesis. The experimentation often involves a comparison between control and experimental samples. (Figure 1.15)

- The study of cystic fibrosis is an interesting example in which both discovery-based science and hypothesis testing have provided key insights regarding the nature of the disease. (Figure 1.16)

- Each chapter in this textbook has a "Feature Investigation" to help you appreciate how science has led to key discoveries in biology.

- Advances in science often occur when scientists gather together and discuss their data. (Figure 1.17)

TEST YOURSELF

1. The process where living organisms maintain a relatively stable internal condition is
   a. adaptation.
   b. evolution.
   c. metabolism.
   d. homeostasis.
   e. development.

2. Populations of organisms change over the course of many generations. Many of these changes result in increased survival and reproduction. This phenomenon is
   a. evolution.
   b. homeostasis.
   c. development.
   d. genetics.
   e. metabolism.

3. All of the places on Earth where living organisms are found is
   a. the ecosystem.
   b. a community.
   c. the biosphere.
   d. a viable land mass.
   e. a population.

4. Horizontal evolution is the result of
   a. accumulation of mutations over many generations.
   b. transfer of genetic material between individuals of different species.
   c. transfer of genetic material from parent to offspring.
   d. all of the above.
   e. a and c only.

5. The scientific name for humans is *Homo sapiens*. The name *Homo* is the _____ to which humans are classified.
   a. kingdom
   b. phylum
   c. order
   d. genus
   e. species

6. The complete genetic makeup of an organism is called
   a. the genus.
   b. the genome.
   c. the proteome.
   d. the genotype.
   e. the phenotype.

7. A proposed explanation for a natural phenomenon is
   a. a theory.
   b. a law.
   c. a prediction.
   d. a hypothesis.
   e. an assay.

8. In science, a theory should
   a. be equated with knowledge.
   b. be supported by a substantial body of evidence.
   c. provide the ability to make many correct predictions.
   d. all of the above.
   e. b and c only.

9. Conducting research without a preconceived hypothesis is called
   a. discovery-based science.
   b. the scientific method.
   c. hypothesis testing.
   d. a control experiment.
   e. none of the above.

10. What is the purpose of using a control in scientific experiments?
    a. A control allows the researcher to practice the experiment first before actually conducting it.
    b. A researcher can compare the results in the experimental group and control group to determine if a single variable is causing a particular outcome in the experimental group.

c. A control provides the framework for the entire experiment so the researcher can recall the procedures that should be conducted.

d. A control allows the researcher to conduct other experimental changes without disturbing the original experiment.

e. All of the above.

CONCEPTUAL QUESTIONS

1. List the seven characteristics of life and explain a little about each.

2. Give the levels of organization from the simplest to most complex.

3. List the taxonomic groups from most inclusive to least inclusive.

EXPERIMENTAL QUESTIONS

1. Discuss the difference between discovery-based science and hypothesis testing.

2. What are the steps in the scientific method, also called hypothesis testing?

3. When conducting an experiment, explain how a control sample and an experimental sample differ from each other.

COLLABORATIVE QUESTIONS

1. Discuss the terms genomes and proteomes.

2. Discuss the levels of organization of life.

www.brookerbiology.com

This website includes answers to the Biological Inquiry questions found in the figure legends and all end-of-chapter questions.

2

THE CHEMICAL BASIS OF LIFE I: ATOMS, MOLECULES, AND WATER

CHAPTER OUTLINE

Crystals of sodium chloride (NaCl).

Biology—the study of life—is founded on the principles of chemistry and physics. All living organisms are a collection of atoms and molecules bound together and interacting with each other through the forces of nature. Throughout this textbook, we will see how chemistry can be applied to living organisms as we discuss the components of cells, the functions of proteins, the flow of nutrients in plants and animals, and the evolution of new genes. This chapter lays the groundwork for understanding these interactions. We begin with an overview of **inorganic chemistry**—that is, the nature of atoms and molecules with the exception of those that contain rings or chains of carbon. Such carbon-containing molecules form the basis of **organic chemistry** and are covered in Chapter 3.

2.1 Atoms

All life is composed of atoms, which in turn are composed of smaller, subatomic particles. A major role of the physicist is to uncover the properties of subatomic particles. Chemists, by contrast, are interested in the properties of atoms and molecules. Chemistry and physics merge when one attempts to understand the mechanisms by which atoms and molecules interact. When atoms and molecules are studied in the context of a living organism, the science of biochemistry emerges. No living creature is immortal, but atoms never "die." Instead, they exist *ad infinitum* as solitary atoms, or as components of a single molecule, or they shuttle between countless molecules over vast eons of time. In this section, we explore the physical properties of atoms so we can understand how atoms combine to form molecules of biological importance.

Atoms Are the Smallest Functional Units in Living Organisms

Atoms are the smallest functional units of matter that form all chemical substances and that cannot be further broken down into other substances by ordinary chemical or physical means. Many types of atoms are known. The simplest atom, hydrogen, is approximately 1 angstrom (10^{-10} meters) in diameter, roughly one-millionth the diameter of a human hair. Each specific type of atom—nitrogen, hydrogen, oxygen, and so on—occurs as a **chemical element**.

Three subatomic particles—**protons**, **neutrons**, and **electrons**—are found within atoms. The protons and neutrons are confined to a very small volume at the center of an atom, the **atomic nucleus**, whereas the electrons are found in regions at various distances from the nucleus. The numbers of protons and electrons in a given type of atom are identical, but the number of neutrons may vary. Each of the subatomic particles has a different electric charge. Protons have one unit of positive charge, electrons have one unit of negative charge, and neutrons are electrically neutral (Table 2.1).

Table 2.1	Characteristics of Major Subatomic Particles		
Particle	**Location**	**Charge**	**Mass relative to electron**
Electron	Around the nucleus	−1	1
Proton	Nucleus	+1	1,836
Neutron	Nucleus	0	1,839

Because the protons are located in the atomic nucleus, the nucleus has a net positive charge equal to the number of protons it contains. The entire atom has no net electric charge, however, because the number of negatively charged electrons around the nucleus is equal to the number of positively charged protons in the nucleus. As shown in Table 2.1, the masses of protons and neutrons are similar to each other and much greater than the mass of electrons.

FEATURE INVESTIGATION

Rutherford Determined the Modern Model of the Atom

Nobel laureate Ernest Rutherford was born in 1871 in New Zealand, but he did his greatest work at McGill University in Montreal, Canada, and later at the University of Manchester in England. At that time, scientists knew that atoms contained charged particles but had no idea how those particles were arranged. Neutrons had not yet been discovered, and many scientists believed that the positive charge and the mass of an atom were evenly dispersed throughout the atom.

In a now-classic experiment, Rutherford aimed a fine beam of positively charged alpha particles (helium nuclei) at an extremely thin sheet of gold foil only 400 atoms thick (**Figure 2.1**). Surrounding the gold foil was a zinc sulfide screen that registered any alpha particles passing through or bouncing off the foil, much like film in a camera detects light. Rutherford hypothesized that if the positive charges of the gold atoms were uniformly distributed, most of the alpha particles would be slightly deflected as they passed through the foil, because one of the most important features of electric charge is that like charges repel each other. Due to their much smaller mass, he did not expect electrons to have any impact on the ability of an alpha particle to move through the metal foil.

Although some of the alpha particles were indeed deflected as they passed through the foil, more than 98% of them passed right through as if the foil was not there, and a few bounced nearly straight back! To explain the 98% that passed right through, Rutherford concluded that most of the volume of an atom is empty space. To explain the few alpha particles that bounced back, he postulated that most of the atom's positive charge was localized in a highly compact area. The existence of this small, dense region of highly concentrated positive charge—which today we call the atomic nucleus—explains how some alpha particles could be so strongly deflected by the gold foil. Alpha particles would bounce back if they directly collided with

Figure 2.1 Rutherford's gold foil experiment demonstrating that most of the volume of an atom is empty space.

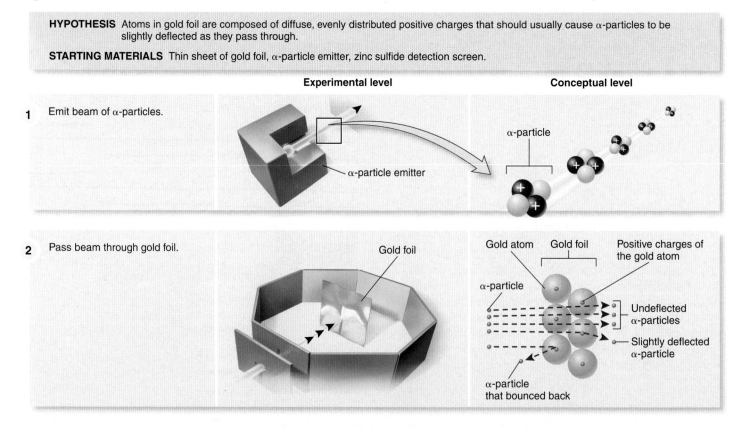

HYPOTHESIS Atoms in gold foil are composed of diffuse, evenly distributed positive charges that should usually cause α-particles to be slightly deflected as they pass through.

STARTING MATERIALS Thin sheet of gold foil, α-particle emitter, zinc sulfide detection screen.

Experimental level   Conceptual level

1   Emit beam of α-particles.

α-particle emitter

α-particle

2   Pass beam through gold foil.

Gold foil

Gold atom   Gold foil   Positive charges of the gold atom

α-particle

Undeflected α-particles

Slightly deflected α-particle

α-particle that bounced back

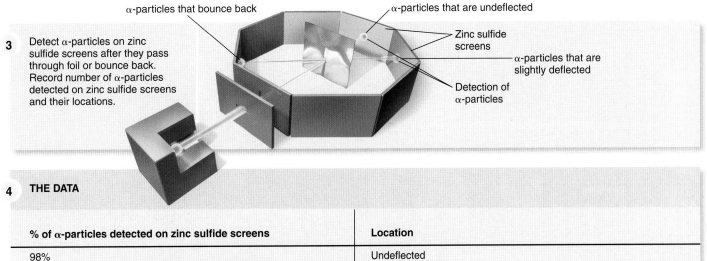

3 Detect α-particles on zinc sulfide screens after they pass through foil or bounce back. Record number of α-particles detected on zinc sulfide screens and their locations.

α-particles that bounce back

α-particles that are undeflected

Zinc sulfide screens

α-particles that are slightly deflected

Detection of α-particles

4 **THE DATA**

% of α-particles detected on zinc sulfide screens	Location
98%	Undeflected
<2%	Slightly deflected
0.01%	Bounced back

an atomic nucleus. Therefore, based on these results, Rutherford rejected his original hypothesis that atoms are composed of diffuse, evenly distributed positive charges.

From a single experiment Rutherford proposed our modern model of an atom, with its small, positively charged nucleus surrounded at relatively great distances by negatively charged electrons. Today we know that more than 99.99% of an atom's volume is outside the nucleus. Indeed, the nucleus accounts for only about 1/10,000 of an atom's diameter—most of an atom is empty space!

Electrons Occupy Orbitals Around an Atom's Nucleus

After Rutherford's experiments, scientists initially visualized an atom as a mini–solar system, with the nucleus being the sun and the electrons traveling in clearly defined orbits around it. Electrons move at terrific speeds. Some estimates suggest that the electron in a typical hydrogen atom could circle the Earth in less than 20 seconds! **Figure 2.2** shows a diagram of the two simplest atoms, hydrogen and helium, which have the smallest numbers of protons. This model of the atom is now considered an oversimplification, because electrons do not actually orbit the nucleus in a defined path. However, this depiction of an atom remains a convenient way to diagram atoms in two dimensions.

At any given moment, it is impossible to precisely predict where a given electron will be located. Electrons travel within regions surrounding the nucleus in which the probability is high of finding that electron. These areas are called **orbitals**. Thus, another way of depicting atoms is a central nucleus surrounded by cloudlike orbitals. The cloud represents the region in which a given electron is most likely to be found. Some orbitals are spherical, called *s* orbitals, while others assume a shape that is often described as similar to a propeller or dumbbell and are called *p* orbitals (**Figure 2.3**). An orbital can contain a maximum of two electrons. Consequently, any atom with more than two electrons must contain additional orbitals.

Orbitals occupy so-called energy shells or energy levels. Atoms with progressively more electrons have orbitals within energy shells that are at greater and greater distances from the nucleus. These shells are numbered so that shell number 1 is closest to the nucleus. Different energy shells may contain one or more orbitals, each orbital with up to two electrons. The innermost shell of all atoms has room for only two electrons, which spin in opposite directions within a spherical orbital (1*s*). The second shell is composed of one spherical orbital (2*s*) and three dumbbell-shaped orbitals (2*p*). Thus, the second shell can hold up to four pairs of electrons (Figure 2.3).

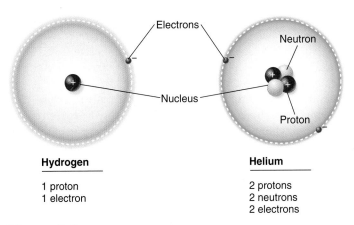

Electrons

Neutron

Nucleus

Proton

Hydrogen

1 proton
1 electron

Helium

2 protons
2 neutrons
2 electrons

Figure 2.2 **The nuclei of two simple atoms and their electrons.** Early depictions of the atom envisioned a nucleus surrounded by electrons in discrete, measurable orbits, much like planets around the sun. This is a model of the two simplest atoms, hydrogen and helium. Note: In all figures of atoms, the sizes and distances are not to scale.

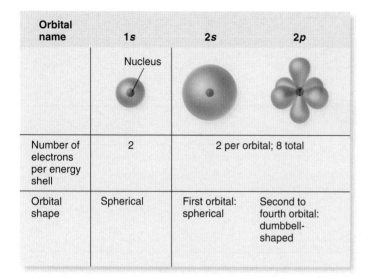

Orbital name	1*s*	2*s*	2*p*
	Nucleus		
Number of electrons per energy shell	2	2 per orbital; 8 total	
Orbital shape	Spherical	First orbital: spherical	Second to fourth orbital: dumbbell-shaped

Figure 2.3   Diagrams of individual electron orbitals. Electrons are found outside the nucleus in orbitals that may resemble spherical or dumbbell-shaped clouds. The orbital cloud represents a region of high probability of locating a particular electron.

Electrons vary in the amount of energy they have. The shell closest to the nucleus fills up with the lowest energy electrons first, and then each subsequent shell fills with higher and higher energy electrons, one shell at a time. Within a given shell, the energy of electrons can also vary among different orbitals. In the second shell, for example, the spherical orbital has lower energy, while the three dumbbell-shaped orbitals have slightly higher and roughly equal energies. In that case, therefore, two electrons fill the spherical orbital first. Any additional electrons fill the dumbbell-shaped orbitals one electron at a time.

Although electrons are actually found in orbitals of varying shapes, as shown in Figure 2.3, chemists often use more simplified diagrams when depicting the energy shells of electrons. **Figure 2.4a** illustrates an example involving nitrogen. An atom of this element has seven protons and seven electrons. Two electrons fill the first shell, and five electrons are found in the outer shell. Two of these fill the 2*s* orbital and are shown as a pair of electrons in the second shell. The other three electrons in the second shell are found singly in each of the three *p* orbitals. The diagram in Figure 2.4a makes it easy to see whether electrons are paired within the same orbital, and whether the outer shell is full. **Figure 2.4b** shows a more realistic depiction of a nitrogen atom, showing how the electrons actually occupy orbitals with different shapes.

Most atoms have outer shells that are not completely filled with electrons. Nitrogen, as we just saw, has a first shell filled with two electrons and a second shell with five electrons (Figure 2.4a). Because the second shell can actually hold eight electrons, the outer shell of a nitrogen atom is not full. As discussed later in this chapter, atoms that have unfilled energy shells tend to share, release, or obtain electrons to fill their outer shell. Those electrons in the outer shell that are available to combine

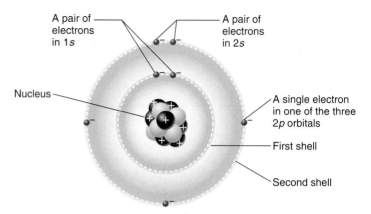

(a) Simplified depiction of a nitrogen atom (7 electrons; 2 electrons in first shell, 5 in second shell)

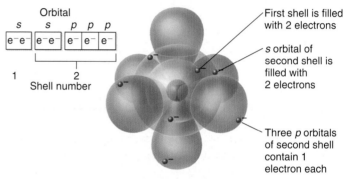

(b) Nitrogen atom showing electrons in orbitals

Figure 2.4   Diagrams showing the multiple shells and orbitals of a nitrogen atom.   The nitrogen atom shown **(a)** simplified and **(b)** with all of its orbitals and shells. An atom's shells fill up one by one. In shells containing more than one orbital, the orbital with lowest energy fills first. Subsequent orbitals gain one electron at a time, shown schematically in boxes, where *e* represents an electron.

Biological inquiry: What is the difference between an energy shell and an orbital?

with other atoms are called the **valence electrons**. Such electrons allow atoms to form chemical bonds with each other.

Each Element Has a Unique Number of Protons

Each chemical element has a specific and unique number of protons that distinguishes one element from another. The number of protons in an atom is its **atomic number**. For example, hydrogen, the simplest atom, has an atomic number of 1, corresponding to its single proton. Magnesium has an atomic number of 12, corresponding to its 12 protons. The term atom refers to a particle that is electrically neutral. Therefore, the atomic number is also equal to the number of electrons in the atom, resulting in a net charge of zero.

Atomic number and electron shells are a useful means of organizing the chemical elements. **Figure 2.5** shows the first three rows of the periodic table of the elements. A one- or two-letter symbol is used as an abbreviation for each element. (Three-letter

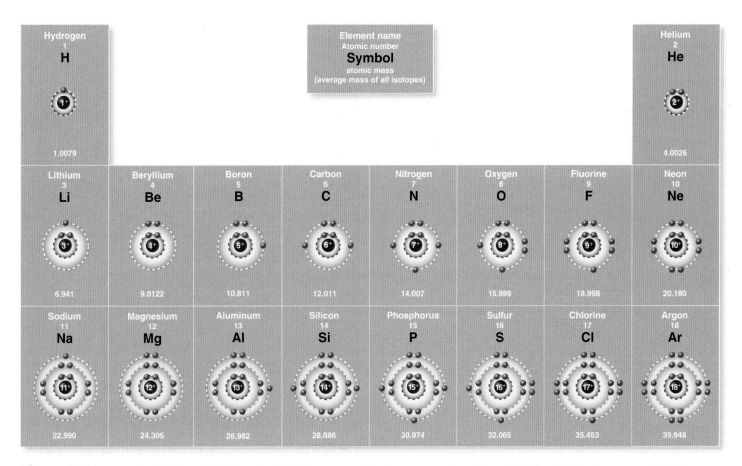

Figure 2.5 **A portion of the periodic table of the elements.** The atoms are shown in models that depict the energy shells in different colors. The occupancy of orbitals is that of the elements in their pure state.

symbols are temporary until a two-letter symbol is chosen officially.) The elements are listed in the order of their atomic numbers. The rows indicate the number of energy shells. Hydrogen (H) has one shell, lithium (Li) has two shells, and sodium (Na) has three shells. The columns, from left to right, indicate the numbers of electrons in the outer shell. The outer shell of lithium (Li) has one electron, beryllium (Be) has two, boron (B) has three, and so forth. This organization of the periodic table tends to arrange elements based on similar chemical properties. For example, helium (He), neon (Ne), and argon (Ar) are inert gases and don't participate in chemical reactions. The inert gases have completely full outer shells. By comparison, beryllium (Be) and magnesium (Mg) are metals that have two electrons in their outer shell. The similarities of elements within a column occur because they have the same number of electrons in their outer shells, and therefore they have similar chemical bonding properties. These properties will be discussed later in this chapter.

Atoms Have a Small but Measurable Mass

Atoms are extremely minute and therefore have very little mass. A single hydrogen atom, for example, has a mass of only 1.67×10^{-24} g (grams). Protons and neutrons are nearly equal in mass, and both are more than 1,800 times the mass of an electron. Despite the great difference in mass of protons and electrons,

however, they nonetheless possess equal but opposite amounts of electric charge (see Table 2.1).

The **atomic mass** scale indicates an atom's mass relative to the mass of other atoms. By convention, the most common type of carbon atom, which has six protons and six neutrons, is assigned an atomic mass of exactly 12. On this scale, a hydrogen atom has an atomic mass of 1, indicating that it has 1/12 the mass of a carbon atom. A magnesium atom, with an atomic mass of 24, has twice the mass of a carbon atom. The term mass is sometimes confused with weight, but these two terms refer to different features of matter. Weight is derived from the gravitational pull on a given mass. If a man who weighs 154 pounds on Earth were standing on the moon, for example, he would only weigh about 25 pounds, but he would weigh 21 trillion pounds if he could stand on a neutron star. However, his mass is the same in all locations. Because we are discussing mass on Earth only, we can assume that the gravitational tug on all matter is roughly equivalent and thus the terms become essentially interchangeable for our purpose.

Atomic mass is measured in units called daltons, after the English chemist John Dalton, who postulated that matter is composed of minute, indivisible units he called atoms. One **dalton (Da)** equals 1/12 the mass of a carbon atom, or about the mass of a proton or a hydrogen atom. Thus, the most common type of carbon atom has an atomic mass of 12 daltons.

Because atoms such as hydrogen have a small mass, while atoms such as carbon have a larger mass, 1 g of hydrogen would have more atoms than 1 g of carbon. A **mole** of any substance contains the same number of particles as there are atoms in exactly 12 g of carbon. Twelve grams of carbon equals 1 mole, while 1 g of hydrogen equals 1 mole. As first described by Italian physicist Amedeo Avogadro, 1 mole of any element contains the same number of atoms—6.022×10^{23}. For example, 12 g of carbon contain 6.022×10^{23} atoms, and 1 g of hydrogen, whose atoms have 1/12 the mass of a carbon atom, also has 6.022×10^{23} atoms. This number, which is known today as **Avogadro's number**, is large enough to be somewhat mind-boggling, and thus gives us an idea of just how small atoms really are. To visualize the enormity of this number, imagine that people could move through a turnstile at a rate of 1 million people per second. It would require almost 20 billion years for 6.022×10^{23} people to move through that turnstile!

Isotopes Vary in Their Number of Neutrons

Although the number of neutrons in an atom is often equal to the number of protons, many elements can exist in multiple forms, called **isotopes**, that differ in the number of neutrons they contain. For example, the most abundant form of the carbon atom, ^{12}C, contains six protons and six neutrons, and thus has an atomic number of 6 and an atomic mass of 12 daltons, as described previously. The superscript placed to the left of ^{12}C is the sum of the protons and neutrons. The rare carbon isotope ^{14}C, however, contains six protons and eight neutrons, giving it an atomic number of 6 but an atomic mass of 14 Da. Nearly 99% of the carbon in living organisms is ^{12}C. Thus, the average atomic mass of carbon is very close to, but actually slightly greater than, 12 Da because of the existence of a small amount of heavier isotopes. This explains why the atomic masses given in the periodic table do not add up exactly to the predicted masses based on the atomic number and the number of neutrons of a given atom (Figure 2.5).

Many isotopes found in nature are inherently unstable and do not exist for long periods of time. Such isotopes are called **radioisotopes**, and they lose energy by emitting subatomic particles and/or radiation. At the very low amounts found in nature, radioisotopes usually pose no serious threat to life, but exposure of living organisms to high amounts of radioactivity can result in the disruption of cellular function and even death.

Modern medicine makes use of the high energy level of radioactive compounds in many ways. For example, solutions containing radioactive isotopes of iodine can be given to a person with an overactive thyroid gland. The thyroid is a gland in the neck that controls many important body functions. It is the only structure in the human body that uses iodine in large quantities, and so the isotope becomes concentrated in the gland. This localizes the radiation of the isotope to the thyroid, killing the hyperactive regions of the gland without harming other parts of the body. Another application makes use of the fact that radiation is easily detectable using various imaging techniques.

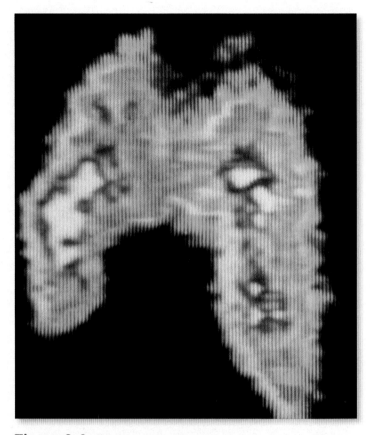

Figure 2.6   Diagnostic image of the human body using radioisotopes.   A procedure called positron-emission tomography (PET) scanning highlights a region of the neck that contains the thyroid gland. A radioactive isotope of iodine was administered to the patient to reveal the size and activity of the gland. Radioactivity in this image shows up as a color.

These techniques, such as the PET scan shown in **Figure 2.6**, can detect the activities of body cells following injection of a compound that contains a radioactive isotope such as ^{131}I.

Researchers also use isotopes to study biological processes, such as tracking the movement of cellular compounds as they are shuttled between different cellular structures. For example, to determine which structures within a cell utilize cholesterol as part of their structure, researchers can incubate cells with cholesterol containing a radioactive isotope of hydrogen called tritium, ^{3}H, in place of one of its ordinary hydrogens. Cells are then examined at various times to determine where the tritium is located. For instance, much of the radioactivity will appear in the cell's plasma membrane, indicating that cholesterol is a component of the outer covering of cells.

Four Elements Constitute the Vast Majority of Living Organisms

Just four elements—oxygen, carbon, hydrogen, and nitrogen—account for the vast majority of atoms in living organisms (**Table 2.2**). These elements typically make up about 95% of the mass

Table 2.2	Chemical Elements Essential for Life in Most Organisms*	
Element	**Symbol**	**% Human body mass**
Most abundant in living organisms (approximately 95% of total mass)		
Oxygen	O	65
Carbon	C	18
Hydrogen	H	9
Nitrogen	N	3
Mineral elements (less than 1% of total mass)		
Calcium	Ca	
Chlorine	Cl	
Magnesium	Mg	
Phosphorus	P	
Potassium	K	
Sodium	Na	
Sulfur	S	
Trace elements (less than 0.01% of total mass)		
Chromium	Cr	
Cobalt	Co	
Copper	Cu	
Fluorine	F	
Iodine	I	
Iron	Fe	
Manganese	Mn	
Molybdenum	Mo	
Selenium	Se	
Silicon	Si	
Tin	Sn	
Vanadium	V	
Zinc	Zn	

* While these are the most common elements in living organisms, many other trace and mineral elements have reported functions. For example, aluminum is believed to be a cofactor for certain chemical reactions in animals, but it is generally toxic to plants.

of living organisms. Much of the hydrogen and oxygen occur in the form of water, which accounts for approximately 60% of the mass of most animals and up to 95% or more in some plants. Nitrogen is a vital element in all proteins, and carbon is a major building block of all living matter. Other vital elements in living organisms include the mineral elements, such as calcium and phosphorus, which are important constituents of the skeletons and shells of animals. Minerals like sodium and potassium are key regulators of water movement and electrical currents that occur across the surfaces of many cells.

In addition, all living organisms require trace elements. These atoms are present in extremely small quantities but still are essential for normal growth and function (Table 2.2). For example, iron plays an important role in how vertebrates store oxygen in their blood, and copper serves a similar role in some invertebrates.

2.2 Chemical Bonds and Molecules

The linkage of atoms with other atoms serves as the basis for life, and also gives life its great diversity. Two or more atoms bonded together make up a **molecule**. Atoms can combine with each other in several ways. For example, two oxygen atoms can combine to form one oxygen molecule, represented as O_2. This representation is called a **molecular formula**, and it consists of the chemical symbols for all of the atoms present (here, O for oxygen) and a subscript that tells you how many of those atoms are present (in this case, two). The term **compound** refers to a molecule composed of two or more different elements. Examples include water (H_2O), with two hydrogen atoms and one oxygen atom, and the sugar glucose ($C_6H_{12}O_6$), which has 6 carbon atoms, 12 hydrogen atoms, and 6 oxygen atoms.

One of the most important features of compounds is their emerging physical properties. This means that the properties of a compound differ greatly from the elements that combined to form it. Let's consider sodium as an example. Pure sodium (Na), also called elemental sodium, is a soft, silvery white metal that you can cut with a knife. When sodium forms a compound with chlorine, table salt (NaCl) is made. NaCl is a white, relatively hard crystal that dissolves in water. Thus the properties of sodium in a compound can be dramatically different from its properties as an element.

The atoms in molecules are held together by chemical bonds. Important types of chemical bonds include covalent and ionic bonds. In this section, we will examine how these types of bonds form, and how they determine the structures of molecules.

Covalent Bonds Join Atoms Through the Sharing of Electrons

Covalent bonds, in which atoms share a pair of electrons, can occur between atoms whose outer shells are not full. A fundamental principle of chemistry is that *atoms tend to be most stable when their outer shells are full of electrons*. **Figure 2.7** shows this principle as it applies to the formation of hydrogen fluoride. The outer shell of a hydrogen atom is full when it contains two electrons, though a hydrogen atom has only one electron. The outer shell of a fluorine atom has seven electrons, though its outer shell would be full if it contained eight electrons. When hydrogen fluoride (HF) is made, the two atoms share a pair of electrons. This allows both of their outer shells to be full. Covalent bonds are strong chemical bonds, because the shared electrons behave as if they belong to each atom.

When the structure of a molecule is diagrammed, each covalent bond is represented by a line indicating a pair of shared electrons. For example, hydrogen fluoride is diagrammed as

H—F

A molecule of water can be diagrammed as

H—O—H

The structural formula of water indicates that the oxygen atom is covalently bound to two hydrogen atoms. Alternatively, as mentioned previously, water can be written by its shorthand molecular formula, H_2O.

Each atom forms a characteristic number of covalent bonds, which depends on the number of electrons that is needed to fill the outer shell. The atoms of some elements important for life, notably carbon, form more than one covalent bond and become linked simultaneously to two or more other atoms. **Figure 2.8** shows the number of covalent bonds that are formed by several atoms that are commonly found in the molecules of living cells.

Learning the **octet rule** may help you remember that atoms are stable when their outer shell is full. For many, but not all, types of atoms, their outer shell is full when they contain eight electrons, an octet. This rule applies to many types of atoms that are found in living organisms including carbon, nitrogen, oxygen, phosphorus, and sulfur. These atoms form a character-

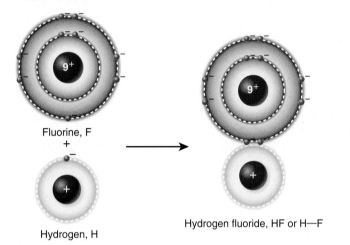

Figure 2.7 **The formation of covalent bonds.** In covalent bonds, electrons from the outer shell of two atoms are shared with each other, in order to complete the outer shells of both atoms. This simplified illustration shows hydrogen forming a covalent bond with fluorine.

istic number of covalent bonds to make an octet in their outer shell (Figure 2.8). However, the octet rule does not always apply. For example, hydrogen has an outer shell that can contain only two electrons, not eight.

In some molecules, a **double bond** occurs when atoms share two pairs of electrons rather than one pair. As shown in **Figure 2.9**, this is the case for an oxygen molecule (O_2), which can be diagrammed as

$$O=O$$

Another common example occurs when two carbon atoms form bonds in compounds. They may share one pair of electrons (single bond) or two pairs (double bond), depending on how many other covalent bonds each carbon forms with other atoms. In rare cases, carbon can even form triple bonds, where three pairs of electrons are shared between two atoms.

Electrons Are Not Always Evenly Shared Between Atoms

Some atoms attract shared electrons more readily than do other atoms. The **electronegativity** of an atom is a measure of its ability to attract electrons in a bond from another atom. When two atoms with different electronegativities form a covalent bond, the shared electrons are more likely to be closer to the atom of higher electronegativity rather than the atom of lower electronegativity. Such bonds are called **polar covalent bonds**, because the distribution of electrons around the atoms creates a polarity, or difference in electric charge, across the molecule. Water is the classic example of a molecule containing polar covalent bonds. The shared electrons at any moment tend to be closer to the oxygen atom rather than either of the hydrogens. This unequal sharing of electrons gives the molecule a region of partial negative charge and two regions of partial positive charge (**Figure 2.10**).

Atoms with high electronegativity, such as oxygen and nitrogen, have a relatively strong attraction for electrons. These atoms

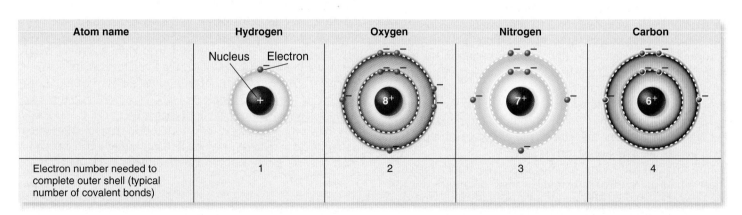

Atom name	Hydrogen	Oxygen	Nitrogen	Carbon
	Nucleus Electron			
Electron number needed to complete outer shell (typical number of covalent bonds)	1	2	3	4

Figure 2.8 **The most abundant elements found in living organisms.** These elements form different numbers of covalent bonds due to the electron configurations in their outer shells.

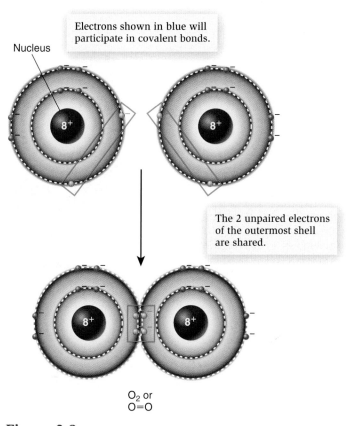

Figure 2.9 A double bond between two oxygen atoms.

Biological inquiry: Explain how an oxygen molecule obeys the octet rule.

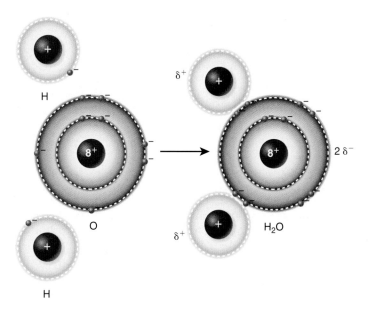

Figure 2.10 Polar covalent bonds in water molecules. In a water molecule, two hydrogen atoms share electrons with an oxygen atom. Because oxygen has a higher electronegativity, the shared electrons spend more time closer to oxygen. This gives oxygen a partial negative charge, designated $2\,\delta^-$, and each hydrogen a partial positive charge, designated δ^+.

form polar covalent bonds with hydrogen atoms, which have low electronegativity. Examples of polar bonds include O—H and N—H. In contrast, bonds between carbon atoms (C—C) and between carbon and hydrogen atoms (C—H) are electrically neutral or nonpolar. Molecules containing significant numbers of polar bonds are known as **polar molecules**, whereas molecules composed predominantly of nonpolar bonds are called **nonpolar molecules**. A single molecule may have different regions with nonpolar bonds and polar bonds. For example, the detergent molecules found in soap have polar and nonpolar ends. The nonpolar ends dissolve in the oil of your skin, and the polar ends help the detergent rinse off in water. As you will see later, the physical characteristics of polar and nonpolar molecules, especially their solubility in water, are quite different.

Hydrogen Bonds Allow Interactions Within and Between Molecules

An important result of certain polar covalent bonds is the ability of one molecule to loosely associate with another molecule through a weak interaction called a **hydrogen bond**. A hydrogen bond forms when a hydrogen atom from one polar molecule becomes electrically attracted to an electronegative atom, such as an oxygen or nitrogen atom, in another polar molecule. Hydrogen bonds, like those between water molecules, are represented in diagrams by dashed or dotted lines to distinguish them from covalent bonds (**Figure 2.11a**). A single hydrogen bond is very weak. The strength of a hydrogen bond is only a few percent of the polar covalent bonds linking the hydrogen and oxygen within a water molecule.

Hydrogen bonds can also occur within a single large molecule. Many large molecules may have dozens, hundreds, or more hydrogen bonds within their structure. Collectively, many hydrogen bonds may add up to a strong force that helps maintain the three-dimensional structure of a molecule. This is particularly true in DNA—the molecule that makes up the genetic material of living organisms. DNA exists as two long, twisting strands of many thousands of atoms and molecules. The two strands are held together all along their length by hydrogen bonds (**Figure 2.11b**). Due to the large number of hydrogen bonds, considerable energy is needed to separate the two strands of DNA.

In contrast to the cumulative strength of many hydrogen bonds, the weakness of individual bonds is also important. When an interaction between two molecules involves relatively few hydrogen bonds, such interactions tend to be short-lived.

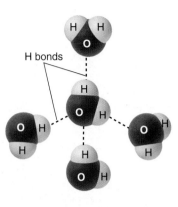

The hydrogen bond (H bond) is a weak attraction between a partially positive hydrogen and a partially negative atom such as oxygen.

H bonds

(a)

A DNA molecule consists of 2 twisted strands held together along its entire length by millions of hydrogen bonds.

(b)

H bonds

Figure 2.11   **Examples of hydrogen bonds.**   Hydrogen bonds are important because they allow for interactions between atoms within a molecule or between atoms from different molecules. **(a)** This example depicts hydrogen bonds (shown as dashed lines) between water molecules. In this diagram, the atoms are depicted as solid spheres, which represent the outer shell. This is called a space-filling model for an atom. **(b)** A DNA molecule is composed of two twisting strands connected to each other by hydrogen bonds (dashed lines). Although each individual bond is weak, the sum of all the hydrogen bonds in a large molecule like DNA imparts considerable stability to the molecule.

Biological inquiry: As discussed in Chapter 11, DNA strands must separate for DNA to be replicated. Do you think the process of strand separation requires energy, or do you think the strands can separate spontaneously?

The reversible nature of hydrogen bonds allows molecules to interact and then to become separated again. For example, as discussed in Chapter 7, small molecules may bind to proteins called enzymes via hydrogen bonds. The small molecules are later released after the enzymes have changed their structure.

Ionic Bonds Involve an Attraction Between Positive and Negative Ions

Atoms are electrically neutral because they contain equal numbers of negative electrons and positive protons. If an atom or molecule gains or loses one or more electrons, it acquires a net electric charge and becomes an **ion**. For example, when a sodium atom (Na), which has 11 electrons, loses one electron, it becomes a sodium ion (Na^+) with a net positive charge. A sodium ion still has 11 protons, but only 10 electrons. Ions such as Na^+ are depicted with a superscript that indicates the net charge of the ion. Some atoms can gain or lose more than one electron. For instance, a calcium atom loses two electrons to become a calcium ion, depicted as Ca^{2+}. On the other hand, a chlorine atom (Cl), which has 17 electrons, can gain an electron and become a chloride ion (Cl^-) with a net negative charge—it has 18 electrons but only 17 protons. Hydrogen atoms and most mineral and trace element atoms readily form ions. **Table 2.3** lists the ionic forms of several elements. The ions listed in this table are relatively stable because the outer electron shells of the ions are full. For example, a sodium atom has one electron in its third (outer) shell. If it loses this electron to become Na^+, it no longer has a third shell, and the second shell, which is full, becomes its outer shell. Alternatively, a Cl

atom has seven electrons in its third (outer) shell. If it gains an electron to become a chloride ion (Cl^-), its outer shell becomes full with eight electrons.

Ions that have a net positive charge are called **cations**, while those that have a net negative charge are **anions**. Ionization, the process of ion formation, can occur in single atoms or in atoms that are covalently linked in molecules.

An **ionic bond** occurs when a cation binds to an anion. **Figure 2.12a** shows an ionic bond between Na^+ and Cl^- to form NaCl. The general name, salt, is given to compounds that are formed from an attraction between a positively charged ion (a cation) and negatively charged ion (an anion). Examples of salts include NaCl, KCl, and $CaCl_2$. Salts may form crystals in which the cations and anions form a regular array. **Figure 2.12b** shows a NaCl crystal in which the sodium and chloride ions are held together by ionic bonds.

Table 2.3	Ionic Forms of Some Common Elements			
Atom	**Chemical symbol**	**Ion**	**Ion symbol**	**Electrons gained or lost**
Calcium	Ca	Calcium ion	Ca^{2+}	2 lost
Chlorine	Cl	Chloride ion	Cl^-	1 gained
Hydrogen	H	Hydrogen ion	H^+	1 lost
Magnesium	Mg	Magnesium ion	Mg^{2+}	2 lost
Potassium	K	Potassium ion	K^+	1 lost
Sodium	Na	Sodium ion	Na^+	1 lost

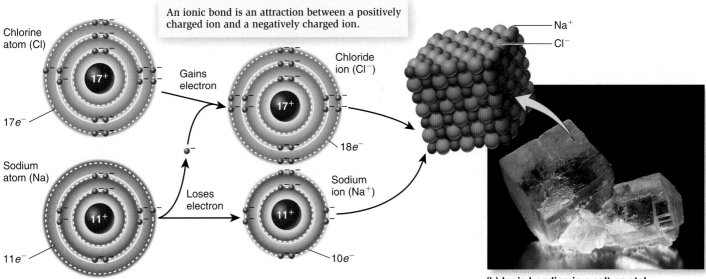

An ionic bond is an attraction between a positively charged ion and a negatively charged ion.

(a) Formation of an ionic bond

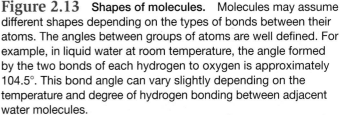

(b) Ionic bonding in a salt crystal

Figure 2.12 Ionic bonding in table salt (NaCl). **(a)** When an electron is transferred from a sodium atom to a chlorine atom, the resulting ions are attracted to each other via an ionic bond. **(b)** In a salt crystal, a lattice is formed in which the positively charged sodium ions (Na^+) are attracted to negatively charged chloride ions (Cl^-).

Molecules May Change Their Shapes

When atoms combine, they can form molecules with various three-dimensional shapes, depending on the arrangements and numbers of bonds between their atoms. As an example, let's consider the arrangements of covalent bonds in a few simple molecules (**Figure 2.13**). These molecules form new orbitals that cause the atoms to have defined angles relative to each other. This gives groups of atoms very specific shapes as shown in the three examples of Figure 2.13.

Molecules containing covalent bonds are not rigid, inflexible structures. Think of a covalent bond, for example, as an axle around which the joined atoms can rotate. Within certain limits, the shape of a molecule can change without breaking its covalent bonds. As illustrated in **Figure 2.14a**, a molecule of six carbon atoms bonded together can assume a number of shapes as a result of rotations around various covalent bonds. The three-dimensional, flexible shape of molecules contributes to their biological properties. As shown in **Figure 2.14b**, the binding of one molecule to another may affect the shape of one of the molecules. An animal can taste food, for instance, because food molecules interact with receptors on its tongue. When a food molecule encounters a receptor, the two molecules recognize each other by their unique shapes, much like a key fitting into a lock. As atoms in the receptor are attracted by hydrogen bonds to atoms in the food, the shape of the receptor changes. As we will see when we look at how the brain receives information from other parts of the body, the altered shape of the receptor initiates a signal that communicates to the animal's brain that the food tastes good (see Chapter 44).

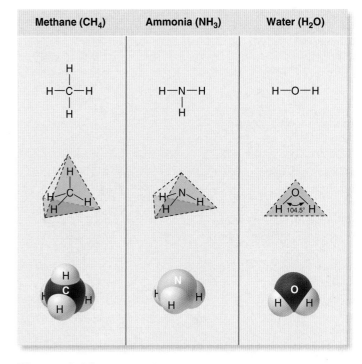

Figure 2.13 Shapes of molecules. Molecules may assume different shapes depending on the types of bonds between their atoms. The angles between groups of atoms are well defined. For example, in liquid water at room temperature, the angle formed by the two bonds of each hydrogen to oxygen is approximately 104.5°. This bond angle can vary slightly depending on the temperature and degree of hydrogen bonding between adjacent water molecules.

Free Radicals Are a Special Class of Highly Reactive Molecules

Recall that an atom or ion is most stable when each of its orbitals is occupied by its full complement of electrons. A molecule containing an atom with a single, unpaired electron in its outer shell is known as a **free radical**. Free radicals can react with other molecules to "steal" an electron from one of their atoms,

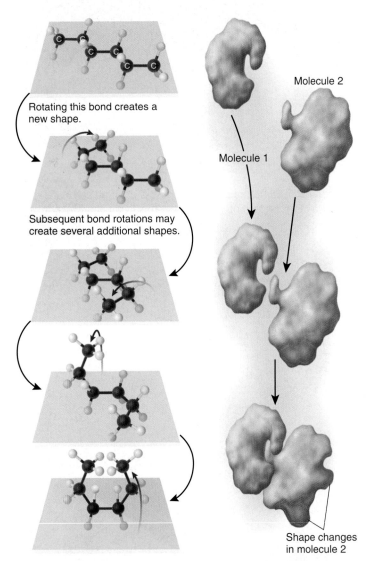

Rotating this bond creates a new shape.

Subsequent bond rotations may create several additional shapes.

Molecule 2

Molecule 1

Shape changes in molecule 2

(a) Bond rotation in a small molecule   **(b) Noncovalent interactions that may alter the shape of molecules**

Figure 2.14   Shape changes in molecules.   A single molecule may assume different three-dimensional shapes without breaking any of the covalent bonds between its atoms, as shown in **(a)** for a six-carbon molecule. Hydrogen atoms above the blue plane are shown in white; those below the blue plane are blue. **(b)** Two molecules are shown schematically as having complementary shapes that permit them to interact. Upon interacting, the flexible nature of the molecules causes molecule 2 to twist sufficiently to assume a new shape. This change in shape is often an important mechanism by which one molecule influences the activity of another.

thereby filling the orbital in the free radical. In the process, this may create a new free radical in the donor molecule, setting off a chain reaction.

Free radicals can be formed in several ways, including exposure of cells to radiation and toxins. Free radicals can do considerable harm to living cells—for example, by causing a cell's membrane to rupture or damaging the genetic material. Surprisingly, the lethal effect of free radicals is sometimes put to good use. Some cells in animals' bodies create free radicals and use them to kill invading cells such as bacteria. Likewise, people use hydrogen peroxide to kill bacteria, as in a dirty skin wound. Hydrogen peroxide can break down to create free radicals, which can then attack bacteria in the wound.

Despite the exceptional case of fighting off bacteria, though, most free radicals that arise in an organism need to be inactivated so that they do not kill healthy cells. Protection from free radicals is afforded by molecules that can donate electrons to the free radicals without becoming highly reactive themselves. Examples of such protective compounds are certain vitamins found in fruits and vegetables, and the numerous plant compounds known as flavonoids. This is one reason why a diet rich in fruits and vegetables is beneficial to our health.

Free radicals are diagrammed with a dot next to the atomic symbol. Examples of biologically important free radicals are superoxide anion, $O_2 \bullet^-$; hydroxyl radical, $OH\bullet$; and nitric oxide, $NO\bullet$. Note that free radicals can be either charged or neutral.

Chemical Reactions Change Elements or Compounds into Different Compounds

A **chemical reaction** occurs when one or more substances are changed into other substances. This can happen when two or more elements or compounds combine with each other to form a compound, when one compound breaks down into two or more molecules, or when electrons are added to or taken away from an atom. Chemical reactions share many similar properties. First, they all require a source of energy for molecules to encounter each other. Such energy is provided partly by heat. In the complete absence of any heat (a temperature called absolute zero), atoms and molecules would be totally stationary and unable to interact. Heat energy causes atoms and molecules to vibrate and move, a phenomenon known as Brownian motion. Second, chemical reactions that occur in living organisms often require more than just Brownian motion to proceed at a reasonable rate. Such reactions need to be catalyzed. As discussed in Chapter 7, a catalyst is a substance that speeds up a chemical reaction. All cells contain many kinds of catalysts called enzymes. Third, chemical reactions tend to proceed in a particular direction, but will eventually reach a state of equilibrium. As an example, let's consider a chemical reaction between methane, a component found in natural gas, and oxygen. These molecules react with each other to produce carbon dioxide and water.

$$CH_4 + 2\,O_2 \rightleftharpoons CO_2 + 2\,H_2O$$

(methane) (oxygen)   (carbon dioxide) (water)

As it is written here, methane and oxygen are the **reactants**, while carbon dioxide and water are the **products**. Whether a chemical reaction is likely to proceed in a forward or reverse direction depends on changes in free energy, as described in Chapter 7. If we began with only methane and oxygen, the forward reaction is very favorable. The reaction would produce a large amount of carbon dioxide and water, as well as heat. This is why natural gas is used as a fuel to heat homes. However, all chemical reactions will eventually reach **equilibrium** in which the rate of the forward reaction is balanced by the rate of the reverse reaction. In the case of the reaction described above, this equilibrium would occur when nearly all of the reactants had been converted to products.

A final feature common to chemical reactions in living organisms is that they occur in water environments. Just as a crystal of sodium chloride will not dissolve in air but will dissolve in water, so too do the chemical reactions in organisms require water. Next, we will examine the properties of this amazing liquid.

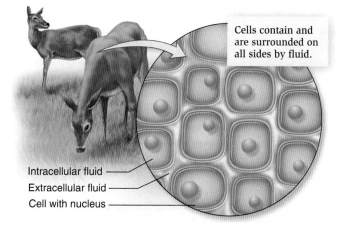

Figure 2.15   **Fluids inside and outside of cells.**   Aqueous solutions exist in the intracellular fluid and in extracellular fluid. Chemical reactions are always ongoing in both fluids.

2.3 Properties of Water

It would be difficult to imagine life without water. People can survive for a month or more without food but usually die in less than a week without water. The bodies of organisms are composed largely of water. Up to 95% of the weight of certain plants comes from water. In people, typically 60–70% of body weight is from water. The brain is roughly 70% water, blood is about 80% water, and the lungs are nearly 90% water. Even our bones are about 20% water! In addition, water is an important liquid in the environments of living organisms. For example, many species are aquatic organisms that survive in a watery environment.

Thus far in this chapter we have considered the features of atoms and molecules, and the nature of chemical reactions between atoms and molecules. In this section, we will turn our attention to issues related to the liquid properties of living organisms, and the environment in which they live. Most of the chemical reactions that occur in nature involve molecules that are dissolved in water, including those reactions that happen inside cells and in the spaces that surround cells of living organisms (**Figure 2.15**). However, not all molecules dissolve in water. In this section, we will examine the properties of chemicals that influence whether they dissolve in water, and consider how biologists measure the amounts of dissolved substances. In addition, we examine some of the other special properties of water that make it a vital component of living organisms and their environments.

Polar Molecules and Ions Readily Dissolve in Water

Substances dissolved in a liquid are known as **solutes**, and the liquid in which they are dissolved is the **solvent**. Solutes dissolve in a solvent to form a **solution**. In all living organisms, the solvent for chemical reactions is water, which is the most abundant solvent in nature. Solutions made with water are called **aqueous solutions**. To understand why a substance dissolves in

Figure 2.16   **NaCl crystals dissolving in water.**   The ability of water to dissolve sodium chloride crystals depends on the electrical attraction between the polar water molecules and the charged sodium and chloride ions. Water molecules surround each ion as it becomes dissolved.

water, we need to consider the chemical bonds in the solute molecule and those in water. As discussed earlier, the covalent bonds linking the two hydrogen atoms to the oxygen atom in a water molecule are polar. Therefore, the oxygen in water has a slight negative charge, and each hydrogen has a slight positive charge. To dissolve in water, a substance must be polar and electrically attracted to water molecules. For example, table salt (NaCl) is a solid crystalline substance because of the strong ionic bonds between positive sodium ions (Na^+) and negative chloride ions (Cl^-). When a crystal of sodium chloride is placed in water, the polar water molecules are attracted to the charged Na^+ and Cl^- (**Figure 2.16**). The ions become surrounded by clusters of water molecules, allowing the Na^+ and Cl^- to separate from each other and enter the water—that is, to dissolve.

Generally, molecules that contain polar covalent bonds will dissolve in water. Such molecules are said to be **hydrophilic**, which literally means "water-loving." In contrast, molecules composed predominantly of carbon and hydrogen are relatively insoluble in water, because carbon-carbon and carbon-hydrogen bonds are nonpolar. These molecules do not have partial positive and negative charges, and therefore are not attracted to water molecules. Such molecules are **hydrophobic**, or "water-fearing." Oils are a familiar example of hydrophobic molecules. Try mixing vegetable oil with water and observe the result. The two liquids separate into an oil phase and water phase. Very little oil dissolves in the water.

Although hydrophobic molecules dissolve poorly in water, they normally dissolve readily in nonpolar solvents. For example, cholesterol is a compound found in the blood and cells of animals. It is a hydrophobic molecule that is barely soluble in water but that easily dissolves in nonpolar solvents used in chemical laboratories, such as ether. Biological membranes like those that encase cells contain about 50% nonpolar compounds. Because of this, cholesterol also inserts into biological membranes, where it helps to maintain the membrane structure.

Molecules that have both polar or ionized regions at one or more sites and nonpolar regions at other sites are called **amphipathic**—consisting of two parts. When mixed with water, long amphipathic molecules may form spheres called **micelles**, with their polar (hydrophilic) regions at the surface of the micelle, where they are attracted to the surrounding water molecules. The nonpolar (hydrophobic) ends are oriented toward the interior of the micelle (**Figure 2.17**). Such an arrangement minimizes the interaction between water molecules and the nonpolar ends of the amphipathic molecules. Nonpolar molecules can dissolve in the central nonpolar regions of these clusters and thus exist in an aqueous environment in far higher amounts than would otherwise be possible based on their low solubility in water. We already considered one familiar example of amphipathic molecules, soap, which can form micelles that help to dissolve oils and nonpolar molecules found in dirt.

Instead of micelles, other amphipathic molecules form structures called bilayers. As you will learn in Chapter 5, lipid bilayers play a key role in membrane structure.

The Amount of a Dissolved Solute Is Its Concentration

Solute **concentration** is defined as the amount of a solute dissolved in a unit volume of solution. For example, if 1 gram (g) of NaCl were dissolved in enough water to make 1 liter of solution, we would say that its solute concentration is 1 g/L.

A comparison of the concentrations of two different substances on the basis of the number of grams per liter of solution does not directly indicate how many molecules of each substance are present. For example, let's compare 10 g each of glucose ($C_6H_{12}O_6$) and sodium chloride (NaCl). Because the individual molecules of glucose have more mass than those of NaCl,

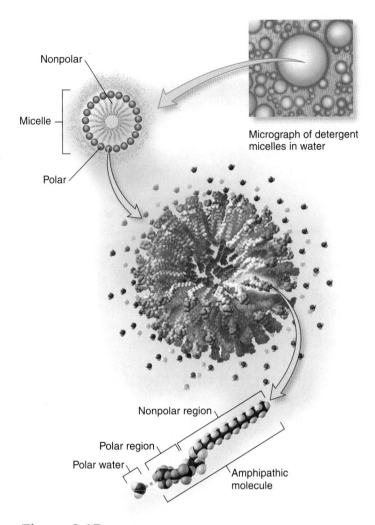

Nonpolar

Micelle

Polar

Micrograph of detergent micelles in water

Nonpolar region

Polar region

Polar water

Amphipathic molecule

Figure 2.17   **The formation of micelles by amphipathic molecules.**   In water, amphipathic molecules tend to arrange themselves so their nonpolar regions are directed away from water molecules, and the polar regions are directed toward the water and can form hydrogen bonds with it.

10 g of glucose will contain fewer molecules than 10 g of NaCl. Therefore, another way to describe solute concentration is according to the moles of dissolved solute per volume of solution. To make this calculation, we must know three things: the amount of dissolved solute, the molecular mass of the dissolved solute, and the volume of the solution.

The **molecular mass** of a molecule is equal to the sum of the atomic masses of all the atoms in the molecule. For example, glucose ($C_6H_{12}O_6$) has a molecular mass of 180 ([6 × 12] + [12 × 1] + [6 × 16] = 180). As mentioned earlier, 1 mole (abbreviated mol) of a substance is the amount of the substance in grams equal to its atomic or molecular mass. The **molarity** of a solution is defined as the number of moles of a solute dissolved in 1 L of solution. A solution containing 180 g of glucose (1 mol) dissolved in enough water to make 1 L is a 1 **molar** solution of glucose (1 mol/L). By convention, a 1 mol/L solution is usually written as 1 M, where the capital M stands for molar and is defined as mol/L. If 90 g of glucose (half its molecular mass) were dissolved in enough water to make 1 L, the solution would have a concentration of 0.5 mol/L, or 0.5 M.

The concentrations of solutes dissolved in the fluids of living organisms are usually much less than 1 mol/L. Many have concentrations in the range of millimoles per liter (1 mM = 0.001 M = 10^{-3} M), while others are present in even smaller concentrations—micromoles per liter (1 μM = 0.000001 M = 10^{-6} M) or nanomoles per liter (1 nM = 0.000000001 M = 10^{-9} M).

H$_2$O Exists as Ice, Water, and Water Vapor

Let's now consider some general features of water and how dissolved solutes affect its properties. H$_2$O is an abundant compound on Earth that exists in all three states of matter—solid (ice), liquid (water), and gas (water vapor). At the temperatures found over most regions of the planet, H$_2$O is found primarily as a liquid in which the weak hydrogen bonds between water molecules are continuously being formed, broken, and formed again. If the temperature rises, the rate at which hydrogen bonds break increases, and molecules of water escape into the gaseous state, becoming water vapor. If the temperature falls, hydrogen bonds are broken less frequently so that larger and larger clusters of water molecules are formed, until at 0°C water freezes into a crystalline matrix—ice. The H$_2$O molecules in ice tend to lie in a more "open" arrangement, which makes ice less dense than water. This is why ice floats on water (**Figure 2.18**). Compared to water, ice is also less likely to participate in most types of chemical reactions.

Changes in state, such as changes between the solid, liquid, and gaseous states of H$_2$O, involve an input or release of energy. For example, when energy is supplied to make water boil, it changes from the liquid to the gaseous state. This is called vaporization. The heat required to vaporize 1 mole of any substance at its boiling point under standard pressure is called the substance's **heat of vaporization**. For water, this value is very high. It takes more than five times as much heat to vaporize water than it does to raise the temperature of water from 0°C to 100°C. In contrast, energy is released when water freezes to form ice. Water has a high **heat of fusion**, which is the amount of heat energy that must be withdrawn or released from a substance to cause it to change from the liquid to the solid state. These two features, the high heats of vaporization and fusion, mean that water is extremely stable as a liquid. Not surprisingly, therefore, living organisms have evolved to function best within a range of temperatures consistent with the liquid phase of H$_2$O.

The temperature at which a solution freezes or vaporizes is influenced by the amounts of dissolved solutes. These are examples of **colligative properties**, which depend strictly on the concentration of dissolved solute particles, not on the specific type of particle. Pure water freezes at 0°C and vaporizes at 100°C. Addition of solutes to water lowers its freezing point below 0°C

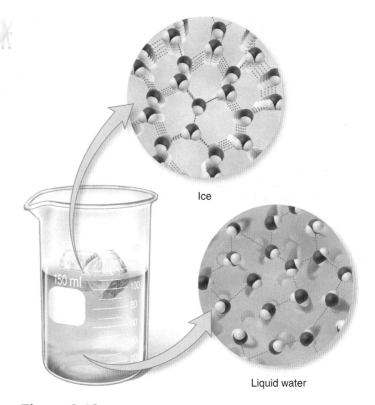

Ice

Liquid water

Figure 2.18   Structure of water and ice.   In its liquid form, the hydrogen bonds between water molecules continually form, break, and re-form, resulting in a changing arrangement of molecules from instant to instant. At temperatures at or below its freezing point, water forms a crystalline matrix called ice. In this solid form, hydrogen bonds are more stable. Ice has a hexagonally shaped crystal structure. The greater space between H$_2$O molecules in this crystal structure causes ice to have a lower density compared to water. For this reason, ice floats on water.

and raises its boiling point above 100°C. Adding a small amount of the compound ethylene glycol—antifreeze—to the water in your car's radiator, for instance, prevents the water from freezing in cold weather. Similarly, the presence of large amounts of solutes partly explains why the oceans do not freeze when the temperature falls below 0°C. Likewise, the colligative properties of water also account for the remarkable ability of certain ectothermic animals, which are unable to maintain warm body temperatures in cold environments, to nonetheless escape becoming frozen solid. Such "cold-blooded" animals produce antifreeze molecules that dissolve in their body fluids, thereby lowering the freezing point of the fluids and preventing their blood and cells from freezing in the extreme cold. The emerald rockcod (*Trematomus bernacchii*), found in the waters of Antarctica, for example, manages to live in ocean waters that are at or below 0°C (**Figure 2.19a**). Similarly, many insects such as the larvae of the parasitic wasp (*Brachon cephi*) also make use of natural antifreeze to stay alive in extreme conditions (**Figure 2.19b**).

(a) Emerald rockcod in the waters of Antarctica

(b) Wasp larvae, which can withstand freezing temperatures

Figure 2.19 Antifreeze in living organisms. Many animals, such as **(a)** the emerald rockcod (*Trematomus bernacchii*) and **(b)** the larvae of the parasitic wasp (*Brachon cephi*), can withstand extremely cold temperatures thanks to natural antifreeze molecules in their body fluids.

Water Performs Many Other Important Tasks in Living Organisms

As discussed earlier, water is the primary solvent in the fluids of all living organisms, from unicellular bacteria to the largest Sequoia tree. Water permits atoms and molecules to interact in ways that would be impossible in their nondissolved states. In Unit II we will consider a myriad of ions and molecules that are solutes in living cells. Even so, it is important to recognize that in addition to acting as a solvent, water serves many other remarkable functions that are critical for the survival of living organisms. For example, water molecules themselves take part in many chemical reactions of this general type:

$$R1—R2 + H—O—H \quad \rightarrow \quad R1—OH + H—R2$$

R is a general symbol to represent a group of atoms. In this equation, *R1* and *R2* are distinct groups of atoms. On the left side, *R1—R2* is a compound in which the groups of atoms are connected by a covalent bond. To be converted to products, a covalent bond is broken in each reactant, R1—R2 and H—O—H, and OH and H (from water) form covalent bonds with R1 and R2, respectively. Reactions of this type are known as hydrolytic reactions (*hydro*, water; *lysis*, break apart), because water is used to break apart another molecule. This process is also called **hydrolysis** (**Figure 2.20a**). As discussed in later chapters, many large molecules are broken down into smaller units by hydrolysis. Alternatively, other chemical reactions in living organisms involve the removal of a water molecule so that a covalent bond can be formed between two separate molecules. For example, let's consider a chemical reaction that is the reverse of our previous hydrolytic reaction:

$$R1—OH + H—R2 \quad \rightarrow \quad R1—R2 + H—O—H$$

Such a reaction involves the removal of a water molecule, and the formation of a covalent bond between two separate molecules. This is termed a **dehydration** or **condensation reaction**. As discussed in later chapters, this is a common reaction that is used to build larger molecules in living organisms.

Another feature of water is that it is incompressible—its volume does not significantly decrease when subjected to high pressure. This has biological importance for many organisms that use water to provide force or support (**Figure 2.20b**). For example, water forms the so-called hydrostatic skeleton of worms and some other invertebrates, and it provides turgidity (stiffness) and support for plants.

Water is also the means by which unneeded and potentially toxic waste compounds are eliminated from an animal's body (**Figure 2.20c**). In mammals, for example, the kidneys filter out soluble waste products derived from the breakdown of proteins and other compounds. The filtered products remain in solution in the watery fluid, which eventually becomes urine and is excreted.

Recall from our discussion of water's properties that it takes considerable energy in the form of heat to convert water from a liquid to a gas. This feature has great biological significance. Although everyone is familiar with the fact that boiling water is converted to water vapor, water can vaporize into the gaseous state even at ordinary temperatures. This process is known as **evaporation**. The simplest way to understand this is to imagine that in any volume of water at any temperature, some water molecules will have higher energy than others. Not every molecule is vibrating identically with the same energy. Those with highest energy escape into the gaseous state. The important point, however, is that even at ordinary temperatures it still requires the same energy to change water from liquid to gas. Thus, the evaporation of sweat from an animal's skin requires considerable energy in the form of body heat, which is then lost to the environment. Evaporation is an important mechanism by which many animals cool themselves on hot days (**Figure 2.20d**).

Another important feature for living organisms is that water has a very high heat capacity, which means that it takes a lot of heat to raise its temperature. This accounts in part for the relatively stable temperatures of large bodies of water compared to inland temperatures. Large bodies of water tend to have a moderating effect on the temperature of nearby land masses.

The hydrogen-bonding properties of water affect its ability to form droplets and to adhere to surfaces. When the molecules within a substance tend to noncovalently attract each other, this phenomenon is called cohesion. Water exhibits strong cohesion due to hydrogen bonding. Cohesion aids in the movement of water through the vessels of plants (**Figure 2.20e**). A property that is similar to cohesion is adhesion, which refers to the ability of water to adhere to another surface. Water tends to cling to surfaces to which it can hydrogen-bond. For this reason, water can coat the surfaces of the digestive tract of animals and act as a lubricant for the passage of food (**Figure 2.20f**). Surface tension is a measure of how difficult it is to break the interface between a liquid and air. In the case of water, the attractive force between hydrogen-bonded water molecules at the interface between water and air is what causes water to form droplets. The surface water molecules attract each other into a configuration (a sphere) that reduces the number of water molecules in contact with air. Likewise, surface tension allows certain insects, such as water striders, to walk on the surface of a pond without sinking (**Figure 2.20g**).

Hydrogen Ion Concentrations Are Changed by Acids and Bases

Pure water has the ability to ionize to a very small extent into hydrogen ions (H^+) and **hydroxide ions** (OH^-). In pure water, the concentrations of H^+ and OH^- are both 10^{-7} mol/L, or 10^{-7} M. An inherent property of water is that the product of the concentrations of H^+ and OH^- is always 10^{-14} M at 25°C. Therefore, in pure water, $[H^+][OH^-] = [10^{-7} \text{ M}][10^{-7} \text{ M}] = 10^{-14}$ M. (The brackets around the symbols for the hydrogen and hydroxide ions indicate concentration.)

When certain substances are dissolved in water, they may release or absorb H^+ or OH^-, thereby altering the relative concentrations of these ions. Molecules that release hydrogen ions in solution are called **acids**. Two examples are:

$$\text{HCl} \rightarrow \text{H}^+ + \text{Cl}^-$$
(hydrochloric acid)   (chloride)

$$\text{H}_2\text{CO}_3 \rightleftharpoons \text{H}^+ + \text{HCO}_3^-$$
(carbonic acid)   (bicarbonate)

Hydrochloric acid is called a **strong acid** because it completely dissociates into H^+ and Cl^- when added to water. By comparison, carbonic acid is a **weak acid** because some of it will remain in the H_2CO_3 state when dissolved in water.

Compared to an acid, a **base** has the opposite effect when dissolved in water—it lowers the H^+ concentration. This can occur in different ways. Some bases, such as sodium hydroxide ($NaOH$), release OH^- when dissolved in water.

$$\text{NaOH} \rightarrow \text{Na}^+ + \text{OH}^-$$

Recall that the product of $[H^+]$ and $[OH^-]$ is always 10^{-14} M. When a base such as NaOH raises the OH^- concentration, some of the hydrogen ions bind to these hydroxide ions to form water. Therefore, increasing the OH^- concentration lowers the H^+ concentration. Alternatively, other bases, such as ammonia, react with water.

$$\text{NH}_3 + \text{H}_2\text{O} \rightleftharpoons \text{NH}_4^+ + \text{OH}^-$$
(ammonia)

—H_2O

Hydrolysis

+ → +

(a) Water participates in chemical reactions.

(b) Water provides support.

Blood enters and is purified by kidney cells.

Waste products are carried away in the urine.

(c) Water is used to eliminate soluble wastes.

(d) Evaporation helps animals dissipate body heat.

(e) The cohesive force of water molecules aids in the movement of fluid through vessels in plants.

(f) Water serves as a lubricant during feeding.

(g) The surface tension of water explains why this water strider doesn't sink.

Figure 2.20   Some amazing roles of water in biology. In addition to acting as a solvent, water serves many crucial functions in nature.

Both NaOH and ammonia have the same effect—they lower the concentration of H$^+$. NaOH achieves this by directly increasing the OH$^-$ concentration, while NH$_3$ reacts with water to produce OH$^-$.

The addition of acids and bases to water can greatly change the H$^+$ and OH$^-$ concentrations over a very broad range. Therefore, chemists and biologists use a log scale to describe the concentrations of these ions. The H$^+$ concentration is expressed as the solution's **pH**, which is defined as the negative logarithm to the base 10 of the H$^+$ concentration.

$$pH = -\log_{10} [H^+]$$

To understand what this equation means, let's consider a few examples. A solution with a H$^+$ concentration of 10^{-7} M has a pH of 7. A concentration of 10^{-7} M is the same as 0.1 μM. A solution in which $[H^+] = 10^{-6}$ M has a pH of 6. 10^{-6} M is the same as 1.0 μM. A solution at pH 6 is said to be more **acidic**, because the H$^+$ concentration is 10-fold higher than a solution at pH 7. Note that as the acidity increases, the pH decreases. A solution where the pH is 7 is said to be neutral because $[H^+]$ and $[OH^-]$ are equal. An acidic solution has a pH that is below 7, while an **alkaline** solution has a pH above 7. **Figure 2.21** considers the pH values of some familiar fluids.

Why is pH of importance to biologists? The answer lies in the observation that H$^+$ and OH$^-$ can readily bind to many kinds of ions and molecules. For this reason, the pH of a solution can affect

- the shapes and functions of molecules;
- the rates of many chemical reactions;
- the ability of two molecules to bind to each other;
- the ability of ions or molecules to dissolve in water.

Due to the various effects of pH, many biological processes function best within very narrow ranges of pH, and even small shifts can have a negative effect. In living cells, the pH ranges from 6.5 to 7.8 and is carefully regulated to avoid major shifts in pH. The blood of the human body has a normal range of about pH 7.35 to 7.45 and is thus slightly alkaline. Certain diseases can reduce or increase blood pH by a few tenths of a unit. When this happens, the enzymes in the body needed for normal metabolism are rendered less functional, leading to illness and even death. As described next, living organisms have molecules called buffers to prevent such changes in pH.

Buffers Minimize Fluctuations in the pH of Fluids

What factors might alter the pH of an organism's fluids? External factors such as acid rain and other forms of pollution can reduce the pH of water entering the roots of plants. In animals, exercise generates lactic acid, and certain diseases can raise or lower the pH of blood.

Organisms have several ways to cope with changes in pH. Complex animals such as mammals, for example, can use struc-

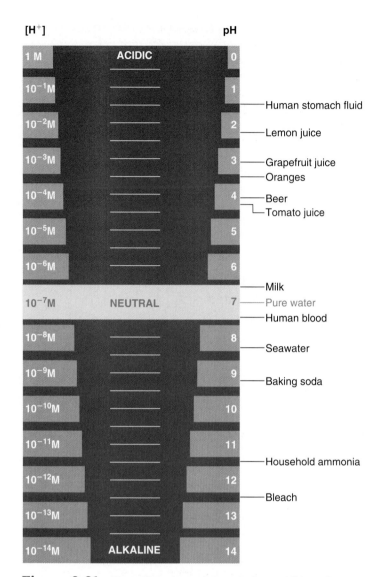

Figure 2.21 The pH scale and the relative acidities of common substances.

Biological inquiry: What is the OH$^-$ concentration at pH 8?

tures like the kidney to secrete acidic or alkaline compounds into the bloodstream when the blood pH becomes imbalanced. Similarly, the kidneys can transfer hydrogen ions from the body into the urine and adjust the body's pH in that way. Another mechanism by which pH balance is regulated in diverse organisms involves the actions of acid-base buffers. An acid-base **buffer** is composed of a weak acid and its related base. One such buffer is the bicarbonate pathway.

$$CO_2 + H_2O \rightleftharpoons H_2CO_3 \rightleftharpoons H^+ + HCO_3^-$$
$$\text{(carbonic acid)} \qquad \text{(bicarbonate)}$$

This buffer system can work in both directions. If the pH of an animal's blood were to increase (that is, the H$^+$ concentration decreased), the bicarbonate pathway would proceed from left to right. CO_2 would combine with water to make carbonic acid, and then the carbonic acid would dissociate into H^+ and bicarbonate. This would raise the H^+ concentration and thereby lower the pH. Alternatively, when the pH of an animal's blood decreases, this pathway runs in reverse. Bicarbonate combines with H^+ to make carbonic acid, which then dissociates to carbon dioxide and water. This process removes H^+ from the blood, restoring it to its normal pH, and the CO_2 is exhaled from the lungs. Many buffers, including this example, exist in nature. Buffers found in living organisms are adapted to function most efficiently at the normal range of pH values seen in that organism.

CHAPTER SUMMARY

2.1 Atoms

- Atoms are the smallest functional units of matter that form all chemical elements and that cannot be further broken down into other substances by ordinary chemical or physical means. Atoms are composed of protons (positive charge), electrons (negative charge), and neutrons (electrically neutral). Electrons are found in orbitals around the nucleus. (Table 2.1, Figures 2.1, 2.2, 2.3, 2.4)

- Each element contains a unique number of protons, its atomic number. The periodic table organizes all known elements by atomic number and energy shells. (Figure 2.5)

- Each atom has a small but measurable mass, measured in daltons. The atomic mass scale indicates an atom's mass relative to the mass of other atoms.

- Many atoms exist as isotopes, which differ in the number of neutrons they contain. Some isotopes are unstable radioisotopes and emit radiation. (Figure 2.6)

- Four elements—oxygen, carbon, hydrogen, and nitrogen—account for the vast majority of atoms in living organisms. In addition, living organisms require mineral and trace elements that are essential for growth and function. (Table 2.2)

2.2 Chemical Bonds and Molecules

- A molecule consists of two or more atoms bonded together. The properties of a molecule are different from the properties of the atoms that combined to form it. A compound is composed of two or more different elements.

- Atoms tend to form bonds that fill their outer shell with electrons.

- Covalent bonds, in which atoms share electrons, are strong chemical bonds. Atoms form two covalent bonds—a double bond—when they share two pairs of electrons. (Figures 2.7, 2.8, 2.9)

- The electronegativity of an atom is a measure of its ability to attract bonded electrons. When two atoms with different electronegativities combine, the atoms form a polar covalent bond because the distribution of electrons around the atoms creates polarity, or difference in electric charge, across the molecule. Polar molecules, such as water, are largely composed of polar bonds, while most nonpolar molecules are composed predominantly of nonpolar bonds. (Figure 2.10)

- An important result of polar covalent bonds is the ability of one molecule to loosely associate with another molecule through weak interactions called hydrogen bonds. (Figure 2.11)

- If an atom or molecule gains or loses one or more electrons, it acquires a net electric charge and becomes an ion. The strong attraction between two oppositely charged ions forms an ionic bond. (Table 2.3, Figure 2.12)

- The three-dimensional, flexible shape of molecules allows them to interact and contributes to their biological properties. (Figures 2.13, 2.14)

- A free radical is an unstable molecule that interacts with other molecules by "stealing" electrons from their atoms.

2.3 Properties of Water

- Chemical reactions change compounds or elements into different compounds. All chemical reactions require energy. In living organisms, chemical reactions take place in a liquid environment, and many are readily reversible. (Figure 2.15)

- Solutes dissolve in a solvent to form a solution. Solute concentration refers to the amount of a solute dissolved in a unit volume of solution. The molarity of a solution is defined as the number of moles of a solute dissolved in 1 L of solution. (Figure 2.16)

- Polar molecules are hydrophilic, while nonpolar molecules composed predominantly of carbon and hydrogen are hydrophobic. Amphipathic molecules, such as detergents, have polar and nonpolar regions. (Figure 2.17)

- Water is the solvent for chemical reactions in all living organisms, which allows atoms and molecules to interact in ways that would be impossible in their nondissolved states.

- H_2O exists as ice, water, and water vapor. (Figure 2.18)

- The colligative properties of water allow it to function as an antifreeze in certain organisms. (Figure 2.19)

- Water's high heat of vaporization and high heat of fusion make it very stable in liquid form.

- Water molecules participate in many chemical reactions in living organisms. Hydrolysis breaks down large molecules into smaller units, and dehydration reactions combine two smaller molecules into one larger one. In living organisms, water provides support, is used to eliminate wastes, dissipates body heat, aids in the movement of liquid through vessels, and serves as a lubricant. Surface tension allows insects to walk on water. (Figure 2.20)

- The pH of a solution refers to its hydrogen ion concentration. The pH of pure water is 7 (a neutral solution). Alkaline solutions have a pH higher than 7, and acidic solutions have a pH lower than 7. (Figure 2.21)

- Buffers are compounds that act to minimize pH fluctuations in the fluids of living organisms. Buffer systems can raise or lower pH as needed.

TEST YOURSELF

1. _____ make up the nucleus of an atom.
   - a. Protons and electrons
   - b. Protons and neutrons
   - c. DNA and RNA
   - d. Neutrons and electrons
   - e. DNA only

2. Living organisms are composed mainly of
   - a. calcium, hydrogen, nitrogen, and oxygen.
   - b. carbon, hydrogen, nitrogen, and oxygen.
   - c. hydrogen, nitrogen, oxygen, and helium.
   - d. carbon, helium, nitrogen, and oxygen.
   - e. carbon, calcium, hydrogen, and oxygen.

3. The ability of an atom to attract bonded electrons is
   - a. polarity.
   - b. electronegativity.
   - c. solubility.
   - d. valence.
   - e. both a and b.

4. Hydrogen bonds differ from covalent bonds in that
   - a. covalent bonds can form between any type of atom and hydrogen bonds only form between H and O.
   - b. covalent bonds involve sharing of electrons and hydrogen bonds involve the complete transfer of electrons.
   - c. covalent bonds result from equal sharing of electrons but hydrogen bonds involve unequal sharing of electrons.
   - d. covalent bonds involve sharing of electrons between atoms but hydrogen bonds are the result of weak attractions between a hydrogen atom of a polar molecule and an electronegative atom of another polar molecule.
   - e. covalent bonds are weak bonds that break easily but hydrogen bonds are strong links between atoms that are not easily broken.

5. A free radical
   - a. is a positively charged ion.
   - b. is an atom with one unpaired electron in its outer shell.
   - c. is a stable atom that is not bonded to another atom.
   - d. can cause considerable cellular damage.
   - e. both b and d.

6. Chemical reactions in living organisms
   - a. require energy to begin.
   - b. usually require a catalyst to initiate the process.
   - c. are usually reversible.
   - d. occur in liquid environments, such as water.
   - e. all of the above.

7. Solutes that easily dissolve in water are said to be
   - a. hydrophobic.
   - b. hydrophilic.
   - c. polar molecules.
   - d. all of the above.
   - e. b and c only.

8. The sum of the atomic masses of all the atoms of a molecule is its
   - a. atomic weight.
   - b. molarity.
   - c. molecular mass.
   - d. concentration.
   - e. polarity.

9. Reactions that involve water in the breaking apart of other molecules are known as _____ reactions.
   - a. hydrophilic
   - b. hydrophobic
   - c. dehydration
   - d. anabolic
   - e. hydrolytic

10. A difference between a strong acid and a weak acid is
    - a. strong acids have a higher molecular mass than weak acids.
    - b. strong acids completely ionize in solution, but weak acids do not completely ionize in solution.
    - c. strong acids give off two hydrogen ions per molecule, but weak acids only give off one hydrogen ion per molecule.
    - d. strong acids are water-soluble, but weak acids are not.
    - e. strong acids give off hydrogen ions and weak acids give off hydroxyl groups.

CONCEPTUAL QUESTIONS

1. What are the types of bonds commonly found in biological molecules?

2. Distinguish between the terms hydrophobic and hydrophilic.

3. List the special properties of water that are ideally suited to life.

EXPERIMENTAL QUESTIONS

1. Before the experiment conducted by Ernest Rutherford, how did many scientists envision the structure of an atom?

2. What was the hypothesis tested by Rutherford?

3. What were the results of the experiment? How did Rutherford interpret the results?

COLLABORATIVE QUESTIONS

1. Discuss the three basic subatomic particles.

2. Discuss several properties of water that make it possible for life to exist.

---

www.brookerbiology.com

This website includes answers to the Biological Inquiry questions found in the figure legends and all end-of-chapter questions.

3

THE CHEMICAL BASIS OF LIFE II: ORGANIC MOLECULES

CHAPTER OUTLINE

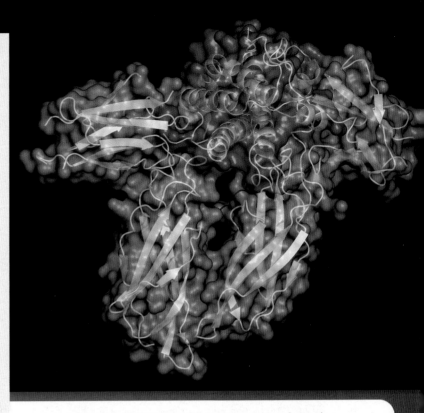

A model showing the structure of a protein—a type of organic macromolecule.

I n Chapter 2, we learned that all life is composed of sub-atomic particles that form atoms, which in turn combine to form molecules. Molecules may be simple in atomic composition, as in water (H_2O) or hydrogen gas (H_2), or may bind with other molecules to form larger molecules. Of the countless possible molecules that can be produced from the known elements in nature, certain types contain carbon and are found in all forms of life. These carbon-containing molecules are collectively referred to as **organic molecules**, so named because they were first discovered in living organisms. Among these are lipids and large, complex compounds called **macromolecules**, which include carbohydrates, proteins, and nucleic acids. In this chapter, we will survey the structures of these molecules and examine their chief functions. We begin by examining the element whose chemical properties are fundamental to the formation of biologically important molecules: carbon. This element provides the atomic scaffold upon which life is built.

3.1 The Carbon Atom and the Study of Organic Molecules

The science of studying carbon-containing molecules is known as organic chemistry. This is a relatively young area of chemical exploration, considering that inorganic chemistry has been studied for hundreds of years, albeit in a rather primitive fashion in the beginning. In this section, we will examine the bonding properties of carbon that create groups of atoms with distinct functions and shapes.

Interestingly, the study of organic molecules was long considered a fruitless endeavor because of a concept called vitalism that persisted into the 19th century. Vitalism held that organic molecules were created by, and therefore imparted with, a vital life force contained within a plant or an animal's body. Supporters of vitalism argued there was no point in trying to synthesize an organic compound, because such molecules could arise only through the intervention of mysterious qualities associated with life. As described next, this would all change due to the pioneering experiments of Friedrich Wöhler in 1828.

Wöhler's Synthesis of an Organic Compound Transformed Misconceptions About Life's Molecules

Friedrich Wöhler (**Figure 3.1a**) was a German physician and chemist who was interested in the properties of inorganic and

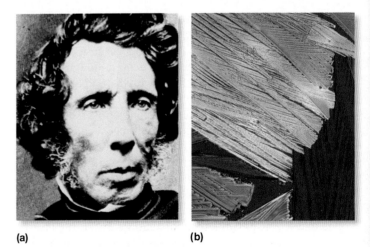

(a)   (b)

Figure 3.1   Wöhler and his crystals of urea.   (a) Friedrich Wöhler. (b) Crystals of urea.

Biological inquiry: How did prior knowledge of urea allow Wöhler to realize he had made urea outside of the body?

organic compounds. He spent some time studying urea, which is found in urine. Urea is a natural organic product formed from the breakdown of proteins in an animal's body. In mammals, urea accumulates in the urine, which is formed by the kidneys, and then is excreted from the body. During the course of his studies, Wöhler purified urea from the urine of mammals. He noted the color, size, shape, and other characteristics of the crystals that formed when urea was isolated. This experience would serve him well in later years when he quite accidentally helped to put the concept of vitalism to rest.

In 1828, while exploring the reactive properties of ammonia and cyanic acid, Wöhler attempted to synthesize an inorganic molecule, ammonium cyanate (NH_4OCN). Instead, Wöhler discovered, to his surprise, that ammonia and cyanic acid reacted to produce a third compound, which, when heated formed familiar-looking crystals (**Figure 3.1b**). After careful analysis, he concluded that these crystals were in fact urea. He announced to the scientific community that he had synthesized urea, an organic compound, "without the use of kidneys, either man or dog." In other words, no mysterious life force was required to create this organic molecule. Subsequently, other scientists, such as Adolph Kolbe in 1845, would demonstrate that organic compounds such as acetic acid could be synthesized directly from their respective elements. These studies were a major breakthrough in the way in which scientists viewed life, and so began the field of science now called organic chemistry. From that time to the present, the fields of chemistry and biology have been understood to be intricately related.

Central to Wöhler's and Kolbe's reactions was the carbon atom. Urea and acetic acid, like all organic compounds, contain carbon atoms bound to other atoms. Let's now consider the chemical features that make carbon such an important element in living organisms.

Carbon Forms Four Covalent Bonds with Other Atoms

One of the properties of the carbon atom that makes life possible is its ability to form four covalent bonds with other atoms, including other carbon atoms. This occurs because carbon has four electrons in its outer shell, and it needs four additional electrons for its outer shell to be full (**Figure 3.2**). In living organisms, carbon atoms most commonly form covalent bonds with other carbons and with hydrogen, oxygen, nitrogen, and sulfur atoms. Bonds between two carbon atoms, between carbon and oxygen, or between carbon and nitrogen can be single or double. The combination of carbon with itself and with different elements and different types of bonds allows a vast number of organic compounds to be formed from only a few chemical elements. This is made all the more impressive because carbon bonds may occur in configurations that are linear, ringlike, or highly branched. Such molecular shapes can produce molecules with a variety of functions.

Because carbon and hydrogen have similar electronegativities, carbon-carbon and carbon-hydrogen bonds are nonpolar. As a consequence, molecules with predominantly hydrogen-

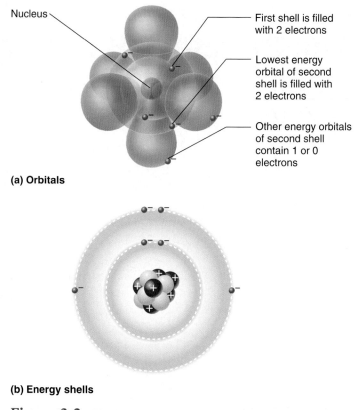

(a) Orbitals

Nucleus

First shell is filled with 2 electrons

Lowest energy orbital of second shell is filled with 2 electrons

Other energy orbitals of second shell contain 1 or 0 electrons

(b) Energy shells

Figure 3.2   Models for the electron orbitals and energy shells of carbon.   Carbon atoms have only four electrons in their outer (second) shell, which allows carbon to form four covalent bonds. When carbon forms four covalent bonds, the result is four hybrid orbitals of equal energy called sp^3 orbitals.

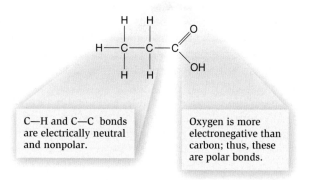

C—H and C—C bonds are electrically neutral and nonpolar.

Oxygen is more electronegative than carbon; thus, these are polar bonds.

Figure 3.3   Nonpolar and polar bonds in an organic molecule.   Carbon can form both nonpolar and polar bonds, and single and double bonds.

carbon bonds, called **hydrocarbons**, tend to be poorly soluble in water. In contrast, when carbon forms polar covalent bonds with oxygen or nitrogen, for example, the molecule is much more soluble in water due to the electrical attraction of polar water molecules. The ability of carbon to form both polar and nonpolar bonds contributes to its ability to serve as the backbone for an astonishing variety of molecules (**Figure 3.3**).

Figure 3.4 The stability of carbon bonds makes life possible even in extreme environments. **(a)** In the Russian Arctic, moss grows beneath the ice. **(b)** Deep-sea vents may reach temperatures greater than 650°F. Organisms can live near these vents, in very warm water.

(a)

(b)

One last feature of carbon that is important to biology is that carbon bonds are stable at the different temperatures associated with life. This property arises in part because the carbon atom is very small compared to most other atoms, and therefore the distance between carbon atoms forming a carbon-carbon bond is quite short. Shorter bonds tend to be stronger and more stable than longer bonds between two large atoms. Thus, carbon bonds are compatible with what we observe about life today, namely that living organisms can inhabit environments ranging from the Earth's icy poles to deep-sea vents (**Figure 3.4**).

Carbon Atoms Can Bond to Several Biologically Important Functional Groups

Aside from the simplest hydrocarbons, most organic molecules and macromolecules contain **functional groups**—groups of atoms with special chemical features that are functionally important. Each type of functional group exhibits the same properties in all molecules in which it occurs. For example, the amino group (NH_2) acts like a base. At the pH found in living organisms, amino groups readily bind H^+ to become NH_3^+, thereby removing H^+ from an aqueous solution and raising the pH. As discussed later in this chapter, amino groups are widely found in proteins and also in other types of organic molecules. **Table 3.1** describes examples of functional groups that are found in many different types of organic molecules. We will discuss each of these groups at numerous points throughout this textbook.

Table 3.1	Some Biologically Important Functional Groups That Bond to Carbon	
Functional group	**Formula**	**Examples of where found***
Amino	$R-N\overset{H}{\underset{H}{}}$	Amino acids (proteins)
Carbonyl**		
Ketone	$R-\overset{\overset{O}{\|}}{C}-R'$	Steroids, eicosanoids, waxes, and proteins
Aldehyde	$R-\overset{\overset{O}{\|}}{C}-H$	
Carboxyl	$R-C\overset{O}{\underset{OH}{}}$	Amino acids, fatty acids
Hydroxyl	$R-OH$	Steroids, alcohol, carbohydrates, some amino acids
Methyl	$R-\overset{\overset{H}{\|}}{\underset{\underset{H}{\|}}{C}}-H$	May be attached to DNA, proteins, and carbohydrates
Phosphate	$R-O-\overset{\overset{O}{\|}}{\underset{\underset{O^-}{\|}}{P}}-O^-$	Nucleic acids, ATP, attached to amino acids
Sulfate	$R-O-\overset{\overset{O}{\|}}{\underset{\underset{O}{\|}}{S}}-O^-$	May be attached to carbohydrates, proteins, and lipids
Sulfhydryl	$R-SH$	Proteins that contain the amino acid cysteine

* This list contains many of the functional groups that are important in biology. However, many more functional groups have been identified by biochemists.
** A carbonyl group is C=O. When the carbon is linked to a free hydrogen atom, this is an aldehyde. In a ketone, the carbon forms covalent bonds with two other carbon atoms.

Carbon-Containing Molecules May Exist in Multiple Forms Called Isomers

When Wöhler did his famous experiment, he was surprised to discover that urea and ammonium cyanate apparently contained the exact same ratio of carbons, nitrogens, hydrogens, and oxygens, yet they were different molecules with distinct chemical and biological properties. Two structures with an identical molecular formula but different structures and characteristics are called **isomers**.

Figure 3.5 depicts three ways in which isomers may occur. **Structural isomers** contain the same atoms but in different

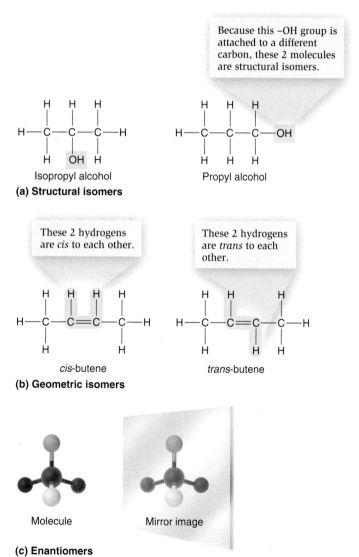

(a) **Structural isomers**

Because this –OH group is attached to a different carbon, these 2 molecules are structural isomers.

Isopropyl alcohol

Propyl alcohol

These 2 hydrogens are *cis* to each other.

These 2 hydrogens are *trans* to each other.

cis-butene

trans-butene

(b) **Geometric isomers**

Molecule

Mirror image

(c) **Enantiomers**

Figure 3.5 Types of isomers. Isomers are compounds with the same molecular formula but different structures. The differences in structure, though small, are sufficient to result in very different biological properties.

bonding relationships. Wöhler's compounds fall into this category. **Stereoisomers** have identical bonding relationships, but the spatial positioning of the atoms differs in the two isomers. Two types of stereoisomers are geometric isomers and enantiomers. In **geometric isomers** like those shown in Figure 3.5b, the two hydrogen atoms linked to the two carbons of a C=C double bond may be on the same side of the carbons, in which case the C=C bond is called a *cis* double bond. If the hydrogens are on opposite sides, it is a *trans* double bond. *Cis-trans* stereoisomers may have very different chemical properties from each other, most notably their stability and sensitivity to heat and light. For instance, the light-sensitive region of your eye contains a molecule called retinal, which may exist in either a *cis* or *trans* form because of a pair of double-bonded carbons in its string of carbon atoms. In darkness, the *cis* form predominates. The energy of sunlight, however, causes retinal to isomerize to the *trans* form. The *trans*-retinal activates the light-capturing cells in the eye.

A second type of stereoisomer, called an **enantiomer**, exists as a pair of molecules that are mirror images. Four different atoms can bind to a single carbon atom in two possible ways, designated a left-handed and a right-handed structure. If the resulting structures are not identical, but instead are mirror images of each other, the molecules are enantiomers (Figure 3.5c). A given pair of enantiomers share identical chemical properties, such as solubility and melting point. However, due to the different orientation of atoms in space, their ability to noncovalently bind to other molecules can be strikingly different. For example, the right-handed form of glucose, called D-glucose, binds very well to certain enzymes in living cells, while the left-handed form, L-glucose, binds poorly. Most enzymes recognize only one type of enantiomer but not both.

3.2 Classes of Organic Molecules and Macromolecules

As we have seen, organic molecules have various shapes due to the bonding properties of carbon. During the past two centuries, biochemists have studied many organic molecules found in living organisms and determined their structures at the molecular level. Many of these compounds are relatively small molecules, containing a few or a few dozen atoms. However, some organic molecules are extremely large macromolecules, being composed of thousands or even millions of atoms. Such large molecules are formed by linking together many smaller molecules called monomers and are thus also known as **polymers** (meaning many small parts). The structure of macromolecules depends on the structure of their monomers, the number of monomers linked together, and the three-dimensional way in which the monomers are linked.

By analyzing the cells of many different species, researchers have determined that all forms of life have organic molecules that fall into four broad categories, based on their chemical and biological properties. The four major types of organic molecules and macromolecules found in all living organisms are carbohydrates, lipids, proteins, and nucleic acids. In this section, we will survey the structures of these organic compounds and begin to examine their biological functions.

Carbohydrates Exist as Sugars and Longer Polymers of Sugars

Carbohydrates are composed of carbon, hydrogen, and oxygen atoms in the proportions represented by the general formula $C_n(H_2O)_n$, where n is a whole number. This formula gives carbohydrates their name—carbon-containing compounds that are hydrated (that is, contain water). Most of the carbon atoms in a carbohydrate are linked to a hydrogen atom and a hydroxyl group. However, other functional groups, such as amino and carboxyl groups, are also found in certain carbohydrates. As discussed next, sugars are relatively small carbohydrates, while polysaccharides are large macromolecules.

Sugars Sugars are small carbohydrates that taste sweet. The simplest sugars are the **monosaccharides** (from the Greek, meaning single sugars). The most common types are molecules with five carbons, called pentoses, and six carbons, called hexoses. Important pentoses are ribose ($C_5H_{10}O_5$) and the closely related deoxyribose ($C_5H_{10}O_4$), which are part of RNA and DNA molecules, respectively. The most common hexose is glucose ($C_6H_{12}O_6$). Like other monosaccharides, glucose is very water-soluble and thus circulates in the blood of animals and the fluids of plants, where it can be transported across plasma membranes. Once inside a cell, glucose is broken down by enzymes. The energy released in this process is used to make many molecules of ATP (adenosine triphosphate), which powers a variety of cellular processes. In this way, sugar is often used as a source of energy by living organisms.

Figure 3.6a illustrates two traditional ways of depicting the bonds between atoms in a monosaccharide. The ring structure is a better approximation of the true shape of the molecule as it mostly exists in solution, with the carbon atoms numbered by convention as shown. The ring is made from the linear structure by an oxygen atom, which forms a bond that bridges two carbons. The hydrogen atoms and the hydroxyl groups may lie above or below the plane of the carbon ring structure.

Figure 3.6b compares different types of isomers of glucose. As mentioned earlier, glucose can exist as D- and L-glucose, which are mirror images of each other. These are enantiomers. Alternatively, other types of isomers are formed by changing the relative positions of the hydrogens and hydroxyl groups along the sugar ring. For example, glucose exists in two interconvertible forms, with the hydroxyl group attached to the number 1 carbon atom lying either above (the β form of glucose, Figure 3.6b) or below (the α form, Figure 3.6a) the plane of the ring. As discussed later, these different isoforms of glucose

have different biological properties. In another example, if the hydroxyl group on carbon atom number 4 of glucose is switched from below to above the plane of the ring, the sugar called galactose is created (Figure 3.6b).

Monosaccharides can join together to form larger carbohydrates. **Disaccharides** (meaning two sugars) are carbohydrates composed of two monosaccharides. A familiar disaccharide is sucrose, or table sugar, which is composed of glucose and fructose (**Figure 3.7**). The linking together of most monosaccharides involves the removal of a hydroxyl group from one monosaccharide and a hydrogen atom from the other, giving rise to a molecule of water and bonding the two sugars together through an oxygen atom. This occurs by a dehydration reaction, also known as a condensation reaction. The bond formed between two sugar molecules is a glycosidic bond. Conversely, hydrolysis of a glycosidic bond in a disaccharide breaks this linkage by adding back the water and thus uncouples the two monosaccharides. Other disaccharides frequently found in nature are maltose, formed in animals during the digestion of large carbohydrates in the intestinal tract, and lactose, present in the milk of mammals. Maltose is α-D-glucose linked to α-D-glucose, while lactose is β-D-galactose linked to β-D-glucose.

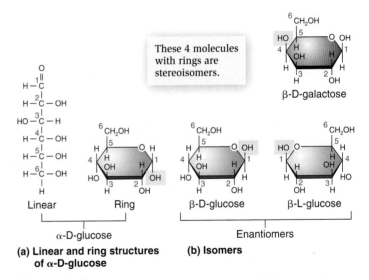

(a) Linear and ring structures of α-D-glucose

(b) Isomers

Figure 3.6 Monosaccharide structure. (a) A comparison of the linear and ring structures of glucose. In solution, such as the fluids of organisms, nearly all glucose is in the ring form. (b) Isomers of glucose. Glucose exists as stereoisomers designated α- and β-glucose, which differ in the position of the —OH group attached to carbon atom number 1. Glucose and galactose differ in the position of the —OH group attached to carbon atom number 4. Enantiomers of glucose, called D-glucose and L-glucose, are mirror images of each other. D-glucose is the form that is used by living cells. Note: The letters *D* and *L* are derived from dextrorotatory (rotating to the right) and levorotatory (rotating to the left).

Biological inquiry: With regard to their binding to enzymes, why do enantiomers such as D- and L-glucose have different biological properties?

Figure 3.7   **Formation of a disaccharide.**   Two monosaccharides can bond to each other to form a disaccharide, such as sucrose, by a dehydration reaction.

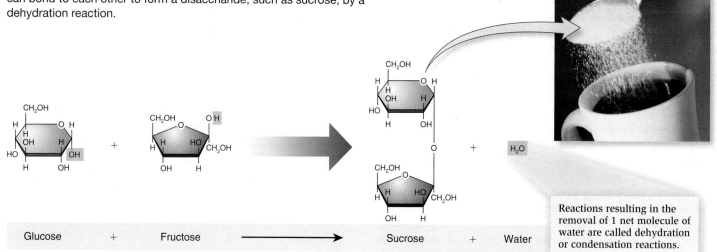

Glucose   +   Fructose $\longrightarrow$ Sucrose   +   Water

Reactions resulting in the removal of 1 net molecule of water are called dehydration or condensation reactions.

Polysaccharides   When many monosaccharides are linked together to form long polymers, **polysaccharides** (meaning many sugars) are made. **Starch**, found in plant cells, and **glycogen**, present in animal cells and sometimes called animal starch, are examples of polysaccharides (**Figure 3.8**). Both of these polysaccharides are composed of thousands of α-D-glucose molecules linked together in long, branched chains, differing only in the extent of branching along the chain. As you can see from the numbering system of the carbon atoms, the bonds that form in polysaccharides are not random but instead form between specific carbon atoms of each molecule. The higher degree of branching in glycogen contributes to its solubility in animal tissues, such as muscle. Starch, because it is less branched, is less soluble and contributes to the properties of plant structures (think of a tough, insoluble kernel of corn).

Certain polysaccharides, such as starch and glycogen, are used to store energy in cells. Like disaccharides, polysaccharides can be hydrolyzed in the presence of water to yield monosaccharides, which are broken down to make ATP. Starch and glycogen, the polymers of α-glucose, provide efficient means of storing energy for those times when a plant or animal cannot obtain sufficient energy for its needs from its environment or diet.

Other polysaccharides provide a structural role, rather than storing energy. The plant polysaccharide **cellulose** is a polymer of β-glucose, with a linear arrangement of carbon-carbon bonds (Figure 3.8). The bond orientations in β-glucose prevent cellulose from being hydrolyzed for ATP production in most organisms. Instead, cellulose forms part of the rigid cell-wall structure characteristic of plants. Cellulose accounts for up to half of all the carbon contained within a typical plant, making it the most common organic compound on Earth. A simple change in bond orientation of glucose molecules, namely α versus β, dramatically alters the biological properties of the resultant polymers, in one case yielding a form of stored fuel and in the other case providing a rigid, protective feature of plant cells.

Some bacteria present in the gastrointestinal tracts of grass and wood eaters, such as cows and termites, can digest cellulose into usable monosaccharides, because they contain an enzyme that can hydrolyze the β-glucose bonds. Humans lack this enzyme, and therefore we eliminate most of the cellulose ingested in our diet as fiber.

Other polysaccharides play structural roles. **Chitin**, a tough, structural polysaccharide, forms the external skeleton of many insects and the cell walls of fungi. The sugar monomers within chitin have nitrogen-containing groups attached to them. **Glycosaminoglycans** are large polysaccharides that play a structural role in animals. For example, glycosaminoglycans are abundantly found in cartilage. These polysaccharides tend to have sugar monomers with carboxyl and sulfate groups.

Lipids Store Energy and Form Membranes and Hormones

Lipids are molecules composed predominantly of hydrogen and carbon atoms. The defining feature of lipids is that they are nonpolar and therefore very insoluble in water. Lipids account for about 40% of the organic matter in the average human body and include fats, phospholipids, and steroids.

Fats   Fats are a mixture of **triglycerides**, also known as triacylglycerols. Fats are formed by bonding glycerol to three fatty acids (**Figure 3.9**). A fatty acid is a chain of carbon and hydrogen atoms with a carboxyl group at the end. Because the carboxyl group (−COOH) releases an H^+ in water to become $−COO^-$, these molecules are called fatty acids. Each of the three hydroxyl groups (−OH) in glycerol is linked to the carboxyl group of a fatty acid by the removal of a molecule of water by a dehydration reaction. The resulting bond is called an ester bond.

The fatty acids found in fats and other lipids may differ with regard to their lengths and the presence of double bonds. Fatty acids are synthesized by the linking together of two-carbon fragments. Therefore, most fatty acids in nature have an even number of carbon atoms, with 16- and 18-carbon fatty acids being the most common in the cells of plants and animals. Fatty acids also differ with regard to the presence of double bonds.

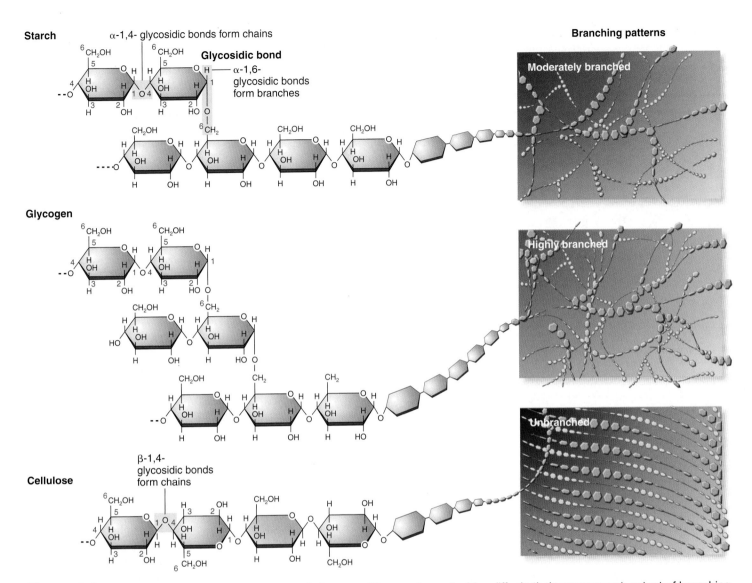

Figure 3.8 **Polysaccharides that are polymers of glucose.** These polysaccharides differ in their arrangement, extent of branching, and type of glucose isomer.

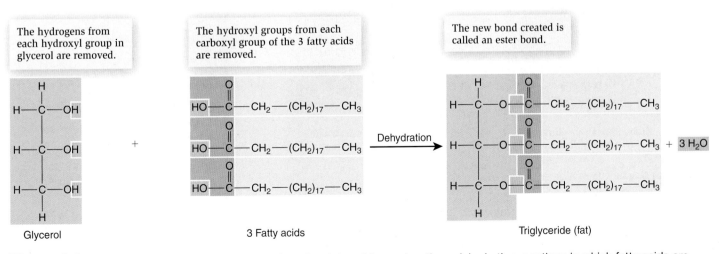

Figure 3.9 **The formation of a fat.** The formation of a triglyceride requires three dehydration reactions in which fatty acids are bonded to glycerol.

When all the carbons in a fatty acid are linked by single covalent bonds, the fatty acid is said to be a **saturated fatty acid**, because all the carbons are saturated with covalently bound hydrogen. Alternatively, some fatty acids contain one or more C=C double bonds and are known as **unsaturated fatty acids**. A fatty acid with one C=C bond is a monounsaturated fatty

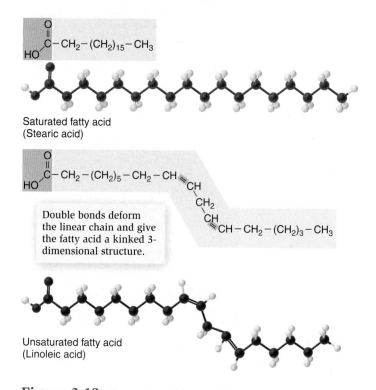

Saturated fatty acid
(Stearic acid)

Double bonds deform the linear chain and give the fatty acid a kinked 3-dimensional structure.

Unsaturated fatty acid
(Linoleic acid)

Figure 3.10 **Examples of fatty acids.** Fatty acids are hydrocarbon chains with a carboxyl functional group at one end and either no double-bonded carbons (saturated) or one or more double bonds (unsaturated). Stearic acid, for example, is an abundant saturated fatty acid in animals, while linoleic acid is an unsaturated fatty acid found in plants.

acid, while two or more C=C bonds constitutes a polyunsaturated fatty acid (**Figure 3.10**). In organisms such as mammals, some fatty acids are necessary for good health but cannot be synthesized by the body. Such fatty acids are called essential fatty acids, because they must be obtained in the diet.

Fats that contain high amounts of saturated fatty acids have high melting points, and therefore they tend to be solid at room temperature. Fats high in unsaturated fatty acids, on the other hand, usually have low melting points and thus are liquids at lower temperatures. Such fats are called oils. Animal fats generally contain a high proportion of saturated fatty acids, whereas vegetable oils contain more unsaturated fatty acids. When you heat a hamburger on the stove, the saturated animal fats melt, and liquid grease appears in the frying pan (**Figure 3.11**). When allowed to cool, however, the oily grease returns to its solid form. By comparison, oils derived from a plant, like olive oil, are liquid even at room temperature.

Like starch and glycogen, fats are important for storing energy. The hydrolysis of triglycerides releases the fatty acids from glycerol, and these products can then be metabolized to provide energy to make ATP. Certain organisms, most notably mammals, have the ability to store large amounts of energy by accumulating fats. One gram of fat stores twice as much energy as 1 g of glycogen or starch. For mobile animals, carrying around less weight is an advantage. In contrast, nonmobile organisms, such as plants, tend to store their energy in the form of polysaccharides. In animals, fats can also play a structural role by forming cushions that support organs. In addition, fats provide insulation under the skin that helps protect terrestrial animals during cold weather and marine mammals in cold water.

Phospholipids Another class of lipids, **phospholipids**, are similar in structure to triglycerides but with one important difference. The third hydroxyl group of glycerol is linked to a phosphate group instead of a fatty acid. In most phospholipids, a small polar or charged nitrogen-containing molecule is attached to this

High temperature converts solid, saturated fats to oil.

After cooling, saturated fats return to their solid form.

Unsaturated fats are oils at room temperature and below.

(a) Animal fats at high and low temperatures

(b) Vegetable fats at low temperature

Figure 3.11 **Fats at different temperatures.** Saturated fats found in animals tend to have high melting points compared to unsaturated fats found in plants.

Biological inquiry: Certain types of fats that are used in baking are called shortenings. They are solid at room temperature. Shortenings are often made from vegetable oils by a process called hydrogenation. What do you think happens to the structure of an oil when it is hydrogenated?

phosphate (**Figure 3.12**). These groups constitute a polar hydrophilic region at one end of the phospholipid, whereas the fatty acid chains provide a nonpolar hydrophobic region at the opposite end. Therefore, phospholipids are amphipathic.

In water, phospholipids become organized into bilayers, with their polar ends attracted to the water molecules and their nonpolar ends facing each other. As you will learn in Chapter 5, this bilayer arrangement of phospholipids is critical for determining the structure of plasma membranes.

Steroids **Steroids** have a distinctly different chemical structure from that of the other types of lipid molecules discussed thus far. Four fused rings of carbon atoms form the skeleton of all steroids (**Figure 3.13a**). A few polar hydroxyl groups may be attached to this ring structure, but they are not numerous enough to make a steroid highly water soluble. For example, steroids with a hydroxyl group are known as sterols—the most well known being cholesterol. Cholesterol is found in the blood of animals, where it can contribute to the formation of clots in major blood vessels.

In steroids, tiny differences in chemical structure can lead to profoundly different biological properties. For example, estrogen is a steroid found in high amounts in female vertebrates (**Figure 3.13b**). Estrogen differs from testosterone (**Figure 13.3c**), a steroid found largely in males, by having one less methyl group,

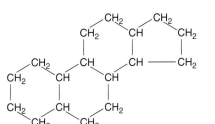

(a) The steroid ring structure

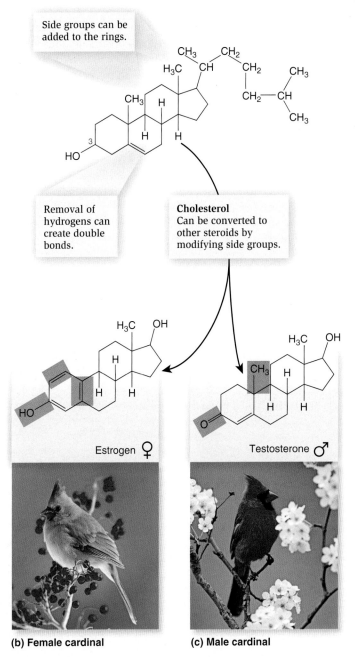

Side groups can be added to the rings.

Removal of hydrogens can create double bonds.

Cholesterol
Can be converted to other steroids by modifying side groups.

Estrogen ♀

Testosterone ♂

(b) Female cardinal   **(c) Male cardinal**

Figure 3.13 **Structure of cholesterol and steroid hormones derived from cholesterol.** **(a)** The structure of a steroid has four rings. Steroids include cholesterol and molecules derived from cholesterol, such as steroid hormones. These include the reproductive hormones **(b)** estradiol, a type of estrogen and **(c)** testosterone.

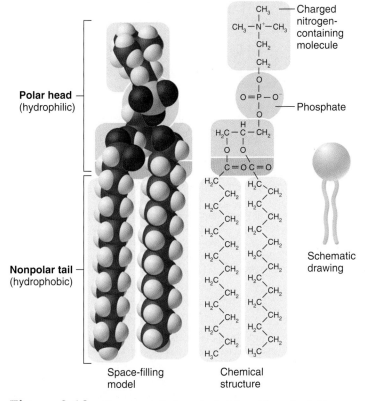

Polar head (hydrophilic)

Charged nitrogen-containing molecule

Phosphate

Nonpolar tail (hydrophobic)

Schematic drawing

Space-filling model

Chemical structure

Figure 3.12 **Structure of phospholipids.** Phospholipids contain both polar and nonpolar regions, making them amphipathic. The fatty acyl tails, formed from fatty acids, are the nonpolar region. The rest of the molecule is polar.

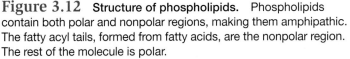

a hydroxyl group instead of a ketone group, and additional double bonds in one of its rings. However, these seemingly small differences are sufficient to make these two molecules largely responsible for whether an animal exhibits male or female characteristics, and whether it is fertile.

Waxes Many plants and animals produce lipids called waxes that are typically secreted onto their surface, such as the leaves of plants and the cuticles of insects. Although any wax may contain hundreds of different compounds, all waxes contain one or more hydrocarbons, and long structures that resemble a fatty acid attached by its carboxyl group to another long hydrocarbon chain. Most waxes are very nonpolar and thus repel water, providing a barrier to water loss. They may also be used as structural elements in colonies like those of bees, where beeswax forms the honeycomb of the hive.

Proteins Are Composed of Amino Acids and Are Involved in Nearly All Life Processes

The word **protein** comes from the Greek *proteios* (meaning of the first rank), which aptly describes their importance. Proteins account for about 50% of the organic material in a typical animal's body, and they play critical roles in almost all life processes (**Table 3.2**).

Proteins are composed of carbon, hydrogen, oxygen, nitrogen, and small amounts of other elements, notably sulfur. The building blocks of proteins are **amino acids**. In other words, proteins are polymers of amino acids. Amino acids have a common structure in which a carbon atom, called the α-carbon, is

linked to an amino group (NH_2) and a carboxyl group (COOH). The α-carbon also is linked to a hydrogen atom and a side chain, which is given a general designation R.

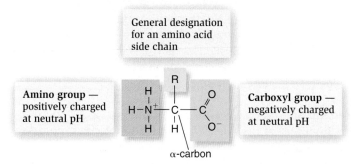

When dissolved in water at neutral pH, the amino group accepts a hydrogen ion and is positively charged, while the carboxyl group loses a hydrogen ion and is negatively charged. The term amino acid is the name given to such molecules because they have an amino group, and also a carboxyl group that behaves like an acid.

The 20 amino acids found in proteins differ with regard to the structures of their side chains (**Figure 3.14**). The amino acids are categorized as those that are nonpolar, polar and uncharged, and polar and charged. The varying structures of the side chains are critical features of protein structure and function. The arrangement and chemical features of the side chains cause proteins to fold and adopt their three-dimensional shapes. In addition, certain amino acids may be critical in protein function. For example, amino acid side chains found within the active sites of enzymes are important in catalyzing chemical reactions.

Table 3.2	Major Categories and Functions of Proteins	
Category	**Functions**	**Examples**
Proteins involved in gene expression and regulation	Make mRNA from a DNA template; synthesize polypeptides from mRNA; regulate genes	RNA polymerase transcribes genes; ribosomal proteins are needed for translation; transcription factor proteins are involved in gene regulation
Motor proteins	Initiate movement	Myosin is a motor protein that provides the contractile force of muscles; kinesin is a key protein that helps cells to sort their chromosomes
Defense proteins	Protect organisms against disease	Antibodies ward off infection due to bacteria or viruses
Metabolic enzymes	Increase rates of chemical reactions	Hexokinase is an enzyme involved in sugar metabolism
Cell signaling proteins	Enable cells to communicate with each other and with the environment	Taste receptors in the tongue allow animals to taste molecules in food
Structural proteins	Support and strengthen structures	Actin provides shape to the cytoplasm of cells such as plant and animal cells; collagen gives strength to tendons
Transporters	Promote movement of solutes across plasma membranes	Ion channels allow movement of charged molecules across plasma membranes; glucose transporters move glucose from outside cells to inside cells, where it can be used for energy

Amino acids are joined together by linking the carboxyl group of one amino acid to the amino group of another. A molecule of water is formed each time two amino acids are joined by a dehydration reaction (**Figure 3.15a**). The covalent bond formed between a carboxyl and amino group is called a **peptide bond**. When many amino acids are joined by peptide bonds, the resulting molecule is called a **polypeptide** (**Figure 3.15b**). The backbone of the polypeptide is highlighted in yellow. The amino acid side chains project from the backbone. When two or more amino acids are linked together, one end of the resulting molecule has a free amino group. This is the amino terminus, or N-terminus. The other end of the polypeptide, called the carboxyl terminus, or C-terminus, has a free carboxyl group. As shown in **Figure 3.15c**, amino acids within a polypeptide are numbered from the amino to the carboxyl terminus.

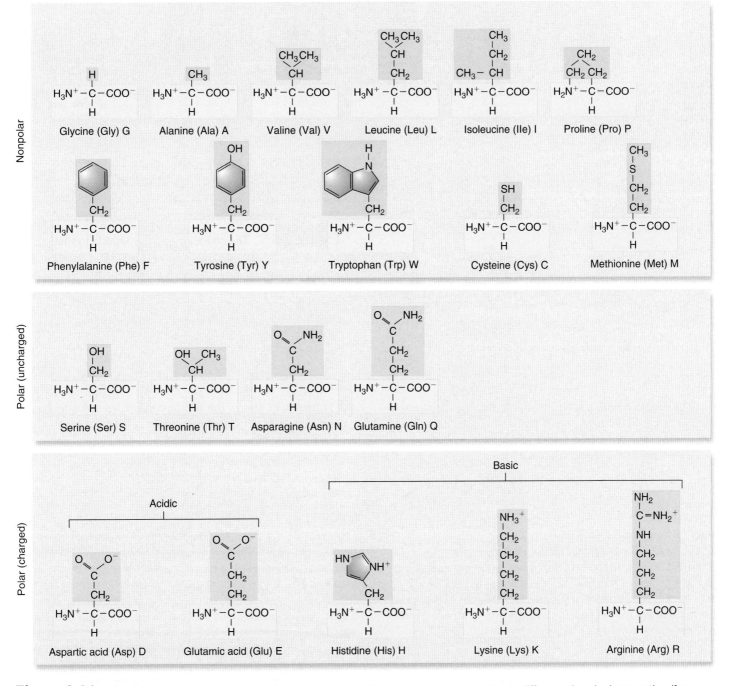

Figure 3.14 **The 20 amino acids found in living organisms.** The various amino acids have different chemical properties (for example, polar versus nonpolar). These properties contribute to the differences in the three-dimensional shapes of proteins, which in turn influence their biological functions.

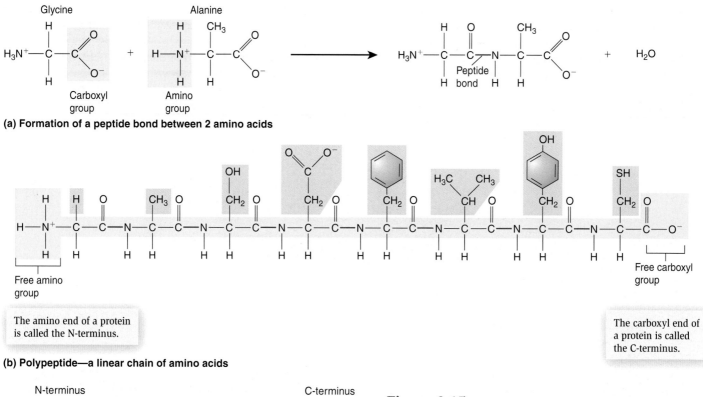

(a) Formation of a peptide bond between 2 amino acids

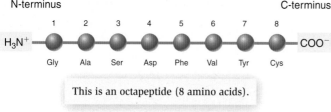

The amino end of a protein is called the N-terminus.

The carboxyl end of a protein is called the C-terminus.

(b) Polypeptide—a linear chain of amino acids

This is an octapeptide (8 amino acids).

(c) Numbering system of amino acids in a polypeptide

Figure 3.15   **The chemistry of polypeptide formation.** Polypeptides are polymers of amino acids. They are formed by linking amino acids together via dehydration reactions to make peptide bonds. Every polypeptide has an amino end, or N-terminus, and a carboxyl end, or C-terminus.

Biological inquiry: How many water molecules would be produced in making a polypeptide that is 72 amino acids long by dehydration reactions?

The term polypeptide refers to a structural unit composed of amino acids. In contrast, a protein is a functional unit composed of one or more polypeptides that have been folded and twisted into precise three-dimensional shapes that carry out a particular function.

Proteins Have a Hierarchy of Structure

Scientists view protein structure at four progressive levels: primary, secondary, tertiary, and quaternary, shown schematically in **Figure 3.16**. These levels of structure are dependent on each other. If one level changes, the other levels may change as a consequence. For example, if the primary structure is changed, this would affect the secondary, tertiary, and quaternary structures. Let's now consider each level separately.

Primary Structure   The **primary structure** of a polypeptide is its amino acid sequence, from beginning to end. The primary structures of polypeptides are determined by genes. As discussed in Chapter 12, genes carry the information to make polypeptides with a defined amino acid sequence.

Figure 3.17 shows the primary structure of ribonuclease, which functions as an enzyme to degrade RNA molecules after they are no longer needed by a cell. Ribonuclease is composed of a relatively short polypeptide with 124 amino acids. An average polypeptide is about 300 amino acids in length, and some genes encode polypeptides that are a few thousand amino acids long.

Secondary Structure   The amino acid sequence of a polypeptide, together with the laws of chemistry and physics, cause a polypeptide to fold into a more compact structure. Amino acids can rotate around bonds within a protein. Consequently, proteins are flexible and can fold into a number of shapes, just as a string of beads can be twisted into many configurations. Folding can be irregular or certain regions can have a repeating folding pattern. Such repeating patterns are called **secondary structure**. The two types are the α helix and β sheet.

In an α helix, the polypeptide backbone forms a repeating helical structure that is stabilized by hydrogen bonds. As shown in Figure 3.16, the hydrogen linked to a nitrogen atom forms a hydrogen bond with an oxygen atom that is double-bonded to a

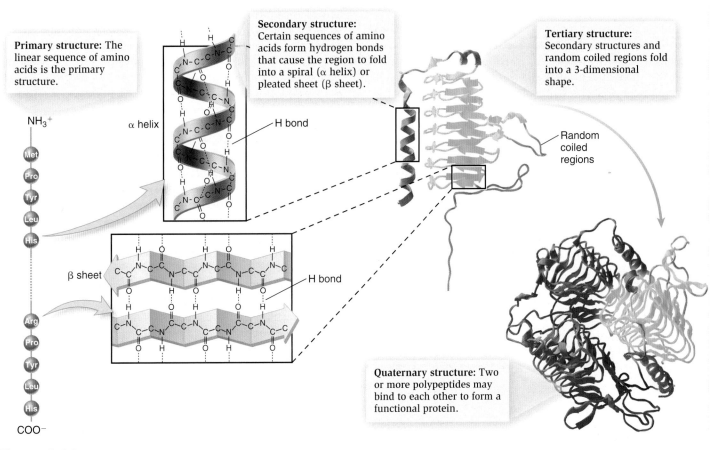

Primary structure: The linear sequence of amino acids is the primary structure.

Secondary structure: Certain sequences of amino acids form hydrogen bonds that cause the region to fold into a spiral (α helix) or pleated sheet (β sheet).

Tertiary structure: Secondary structures and random coiled regions fold into a 3-dimensional shape.

α helix

H bond

β sheet

H bond

Random coiled regions

Quaternary structure: Two or more polypeptides may bind to each other to form a functional protein.

Figure 3.16   The hierarchy of protein structure.

carbon. These hydrogen bonds occur at regular intervals and cause the polypeptide backbone to form a helix. In a β sheet, regions of the polypeptide backbone come to lie parallel to each other. When these parallel regions form hydrogen bonds, again between the hydrogen linked to a nitrogen atom and a double-bonded oxygen, the polypeptide backbone adopts a repeating zigzag shape called a β pleated sheet.

α helices and β sheets are key determinants of a protein's characteristics. For example, α helices in certain proteins are composed primarily of nonpolar amino acids. Proteins containing many such regions with an α helix structure tend to anchor themselves into a lipid-rich environment, such as a cell's plasma membrane. In this way, a protein whose function is needed in a specific location such as a plasma membrane can be retained there. Secondary structure also contributes to the great strength of certain proteins, including the keratins found in hair and hooves, the proteins that make up the silk webs of spiders, and collagen, the chief component of cartilage in mammals.

Some regions along a polypeptide chain do not assume an α helix or β sheet conformation. In other words, they do not have secondary structure. These regions are sometimes called random coiled regions. However, this term is somewhat misleading because the shapes of random coiled regions are usually very specific and are important to the function of a protein.

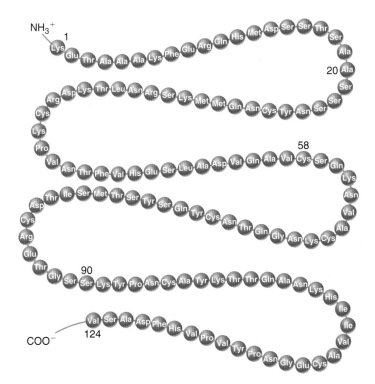

Figure 3.17   **The primary structure of ribonuclease.** The example shown here is ribonuclease from cows, which was studied by Anfinsen as described later in Figure 3.20.

Tertiary Structure As the secondary structure becomes established due to the particular primary structure, a polypeptide folds and refolds upon itself to assume a complex three-dimensional shape—its **tertiary structure** (see Figure 3.16). The tertiary structure is the three-dimensional shape of a single polypeptide. For some proteins, such as ribonuclease, the tertiary structure is the final structure of a functional protein. However, as described next, other proteins are composed of two or more polypeptides and adopt a quaternary structure.

Quaternary Structure Most functional proteins are composed of two or more polypeptides that each adopt a tertiary structure and then assemble with each other (see Figure 3.16). The individual polypeptides are called **protein subunits**. Subunits may be identical polypeptides or they may be different. When proteins consist of more than one polypeptide chain, they are said to have **quaternary structure** and are also known as **multimeric proteins** (meaning many parts).

Factors That Influence Protein Structure Several factors determine the way that polypeptides adopt their secondary, tertiary, and quaternary structures. The amino acid sequences of polypeptides are the defining features that distinguish the struc-

ture of one protein from another. As polypeptides are synthesized in a cell, they fold into secondary and tertiary structures, which assemble into quaternary structures for most proteins. As mentioned, the laws of chemistry and physics, together with the amino acid sequence, govern this process. As shown in **Figure 3.18**, five factors are critical for protein folding and stability:

1. *Hydrogen bonds*—The large number of hydrogen bonds within a polypeptide and between polypeptides adds up to a strong force that promotes protein folding and stability. As we have already learned, hydrogen bonding is a critical determinant of protein secondary structure, and also is important in tertiary and quaternary structure.

2. *Ionic bonds and other polar interactions*—Some amino acid side chains are positively charged while others are negatively charged. Positively charged side chains may bind to negatively charged side chains via ionic bonds. Similarly, uncharged polar side chains in a protein may interact with ionic amino acids. These ionic and polar interactions promote protein folding and stability.

3. *Hydrophobic effect*—Some amino acid side chains are nonpolar. These amino acids tend to avoid water. As a protein folds, the hydrophobic amino acids are likely to be

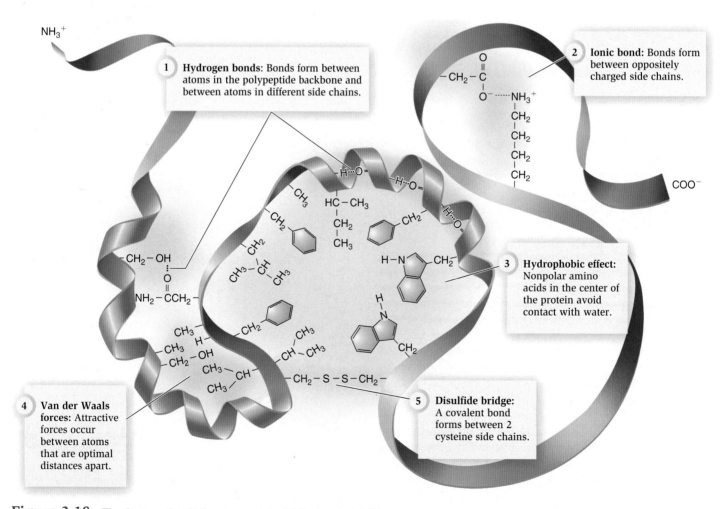

Figure 3.18   The factors that influence protein folding and stability.

found in the center of the protein to avoid contact with water. As mentioned, some proteins have stretches of nonpolar amino acids that anchor them in the hydrophobic portion of membranes.

4. *Van der Waals forces*—Atoms within molecules have weak attractions for each other if they are an optimal distance apart. This optimal distance is called the van der Waals radius, and the weak attraction is the van der Waals force. If two atoms are too close together, their electron clouds will repel each other. If they are too far apart, the van der Waals force will diminish. Similar to hydrogen bonds, many van der Waals forces can contribute to protein folding and stability.

5. *Disulfide bridges*—The side chain of the amino acid cysteine ($-CH_2-SH$) contains a sulfhydryl group, which can react with a sulfhydryl group in another cysteine side chain to produce a disulfide bridge or bond, which links the two amino acid side chains together ($-CH_2-S-S-CH_2-$). Disulfide bonds are covalent bonds that can occur within a single polypeptide or between different polypeptides. Though other forces are usually more important in protein folding, the covalent nature of disulfide bonds can help to stabilize the structure of a protein.

The first four factors just described are also important in the ability of different proteins to interact with each other. As discussed throughout Unit II and other parts of this textbook, many cellular processes involve steps in which two or more different proteins interact with each other. For this to occur, the surface of one protein must bind to the surface of the other. Such binding is usually very specific. The surface of one protein precisely fits into the surface of another (**Figure 3.19**). Such **protein-protein interactions** are critically important so that cellular processes can occur in a series of defined steps. In addition, protein-protein interactions are also important in building large cellular structures that provide shape and organization to cells.

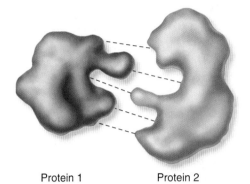

Protein 1   Protein 2

Figure 3.19   **Protein-protein interaction.**   Two different proteins may interact with each other due to hydrogen bonding, ionic bonding, the hydrophobic effect, and van der Waals forces.

FEATURE INVESTIGATION

Anfinsen Showed That the Primary Structure of Ribonuclease Determines Its Three-Dimensional Structure

Prior to the 1960s, the mechanisms by which proteins assume their three-dimensional structures were not understood. Scientists believed that correct folding required unknown cellular factors, or that ribosomes, the site where polypeptides are synthesized, somehow shaped proteins as they were being made. Christian Anfinsen, however, postulated that proteins contain all the information necessary to fold into their proper conformation without the need for organelles or cellular factors. He hypothesized that proteins spontaneously assume their most stable conformation based on the laws of chemistry and physics (**Figure 3.20**).

To test this hypothesis, Anfinsen studied ribonuclease, which we discussed earlier (see Figure 3.17). Biochemists had already determined that ribonuclease has four disulfide bonds between eight cysteine amino acids. Anfinsen began with purified ribonuclease—this is called an *in vitro* experiment, meaning under glass, as in a test tube. The key point is that other cellular components were not present, only the purified protein. He exposed ribonuclease to a chemical called urea that disrupted the hydrogen and ionic bonds, and to another chemical called β-mercaptoethanol that broke the S—S bonds. Following this treatment, he measured the ability of the treated enzyme to degrade RNA. The enzyme had lost nearly all of its ability to degrade RNA. Therefore, Anfinsen concluded that when ribonuclease was unfolded or denatured, it was no longer functional.

The key step in this experiment came when Anfinsen removed the urea and β-mercaptoethanol from the solution. Because these molecules are much smaller than ribonuclease, removing them from the solution was accomplished with a technique called dialysis. In dialysis, solutions are placed in a synthetic bag with microscopic pores that permit small molecules to pass through and leave the bag but retain large macromolecules such as ribonuclease. Anfinsen placed the entire bag in a large beaker of water, into which the urea and β-mercaptoethanol diffused. Then he retested the ribonuclease. The result revolutionized our understanding of proteins. The activity of the ribonuclease was almost completely restored! This meant that even in the complete absence of any cellular factors or organelles, an unfolded protein can refold into its functional structure. This was later confirmed by chemical analyses that demonstrated that the disulfide bonds had re-formed at the proper locations.

Since Anfinsen's time, we have also learned that his experiments with ribonuclease are not representative of all proteins. Some proteins do require certain enzymes and other proteins to assist them in their proper folding. Nonetheless, Anfinsen's experiments provided compelling evidence that the primary structure of a polypeptide is the key determinant of a protein's tertiary structure, a correct observation that earned him a Nobel Prize in 1972.

Figure 3.20   Anfinsen's experiments with ribonuclease demonstrating that the primary structure of a polypeptide plays a key role in protein folding.

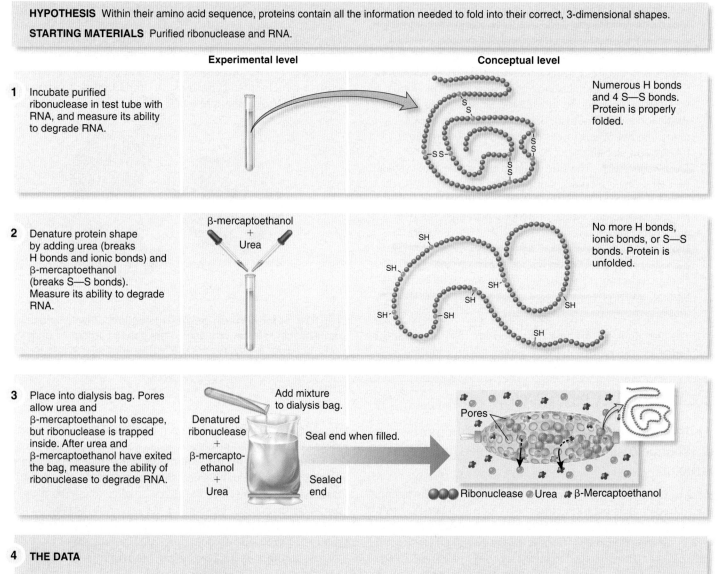

HYPOTHESIS Within their amino acid sequence, proteins contain all the information needed to fold into their correct, 3-dimensional shapes.

STARTING MATERIALS Purified ribonuclease and RNA.

Experimental level | Conceptual level

1 Incubate purified ribonuclease in test tube with RNA, and measure its ability to degrade RNA.

Numerous H bonds and 4 S—S bonds. Protein is properly folded.

2 Denature protein shape by adding urea (breaks H bonds and ionic bonds) and β-mercaptoethanol (breaks S—S bonds). Measure its ability to degrade RNA.

β-mercaptoethanol + Urea

No more H bonds, ionic bonds, or S—S bonds. Protein is unfolded.

3 Place into dialysis bag. Pores allow urea and β-mercaptoethanol to escape, but ribonuclease is trapped inside. After urea and β-mercaptoethanol have exited the bag, measure the ability of ribonuclease to degrade RNA.

Add mixture to dialysis bag.
Denatured ribonuclease + β-mercapto-ethanol + Urea
Seal end when filled.
Sealed end

Pores

Ribonuclease ● Urea ✿ β-Mercaptoethanol

4 THE DATA

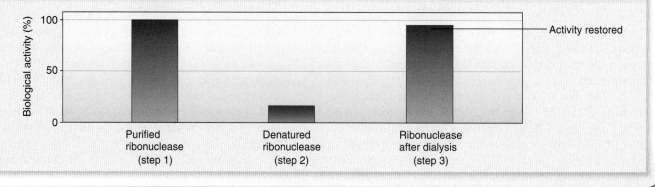

Biological activity (%)

Activity restored

Purified ribonuclease (step 1) | Denatured ribonuclease (step 2) | Ribonuclease after dialysis (step 3)

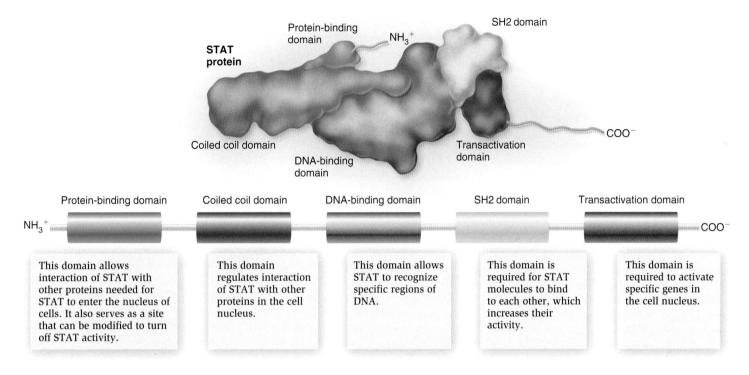

Figure 3.21 **The domain structure of a STAT protein.** The boxes represent different domains, connected by chains of amino acids (the connecting lines in this figure).

GENOMES & PROTEOMES

Proteins Contain Functional Domains Within Their Structures

Modern research into the functions of proteins has revealed that many proteins have a modular design. This means that portions within proteins, called modules or **domains**, have distinct structures and functions. These units of amino acid sequences have been duplicated during evolution so that the same kind of domain may be found in several different proteins. When the same domain is found in different proteins, the domain has the same three-dimensional shape and performs a function that is characteristic of that domain.

As an example, **Figure 3.21** shows a protein that is known to play a critical role in regulating how certain genes are turned on and off in living cells. This protein bears the cumbersome name of *signal transducer and activator of transcription* (STAT) protein. In the primary structure, each domain of the STAT protein is shown as a separate box, linked to the next domain by a straight line. These boxes and lines represent strings of amino acids, which have secondary and tertiary structures.

Each domain of this protein is involved in a distinct biological function, a common occurrence in proteins with multiple domains. For example, one of the domains is labeled the SH2 domain (Figure 3.21). In many different proteins, this domain allows proteins to recognize other proteins in a very specific way. The function of SH2 domains is to bind to tyrosines that are phosphorylated by cellular enzymes. Many proteins contain SH2 domains, and as might be predicted, they all bind to phosphorylated tyrosines in the proteins they recognize.

As a second example, a STAT protein has another domain called a DNA-binding domain. This portion of the protein has a structure that specifically binds to DNA. Overall, the domain structure of proteins enables them to have multiple regions, each with its own structure and purpose in the functioning of the protein.

Nucleic Acids Are the Source of Genetic Information

Nucleic acids account for only about 2% of the weight of animals like ourselves, yet these molecules are extremely important because they are responsible for the storage, expression, and transmission of genetic information. The expression of genetic information in the form of specific proteins determines whether one is a human, a mouse, an onion, or a bacterium. Likewise, genetic information determines whether a cell is part of a muscle or a leaf.

The two classes of nucleic acids are **deoxyribonucleic acid (DNA)** and **ribonucleic acid (RNA)**. DNA molecules store genetic information coded in the sequence of their monomer building blocks. RNA molecules are involved in decoding this information into instructions for linking together a specific sequence of amino acids to form a polypeptide chain.

Like other macromolecules, both types of nucleic acids are polymers and consist of linear sequences of repeating monomers.

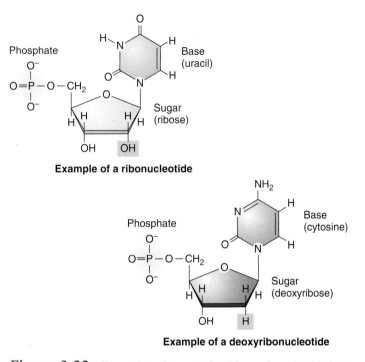

Example of a ribonucleotide

Example of a deoxyribonucleotide

Figure 3.22   Examples of two nucleotides.   A nucleotide has a phosphate group, a five-carbon sugar, and a nitrogenous base.

Each monomer, known as a **nucleotide**, has three components: a phosphate group, a five-carbon sugar (either ribose or deoxyribose), and a single or double ring of carbon and nitrogen atoms known as a **base** (**Figure 3.22**). The phosphate group of one nucleotide is linked to the sugar of the adjacent nucleotide to form a polynucleotide strand with the bases protruding from the side of the phosphate-sugar backbone (**Figure 3.23**). This is a phosphoester bond.

DNA   The nucleotides in DNA contain the five-carbon sugar **deoxyribose**. Four different nucleotides are present in DNA, corresponding to the four different bases that can be linked to deoxyribose. The **purine** bases, **adenine (A)** and **guanine (G)**, have double (fused) rings of nitrogen and carbon atoms, and the **pyrimidine** bases, **cytosine (C)** and **thymine (T)**, have only a single ring (see Figure 3.23).

A DNA molecule consists of two strands of nucleotides coiled around each other to form a double helix (**Figure 3.24**). The two strands are held together by hydrogen bonds between a purine base in one strand and a pyrimidine base in the opposite strand. The ring structure of each base lies in a flat plane perpendicular to the sugar-phosphate backbone, somewhat like steps on a spiral staircase. This base pairing maintains a constant distance between the sugar-phosphate backbones of the two strands as they coil around each other.

As we will see in Chapter 11, only certain bases can pair with others, due to the location of the hydrogen-bonding groups in the four bases (see Figure 3.24). Two hydrogen bonds can be formed between adenine and thymine (A-T pairing), while three hydrogen bonds are formed between guanine and cytosine (G-C pairing). In a DNA molecule, A is always paired with T,

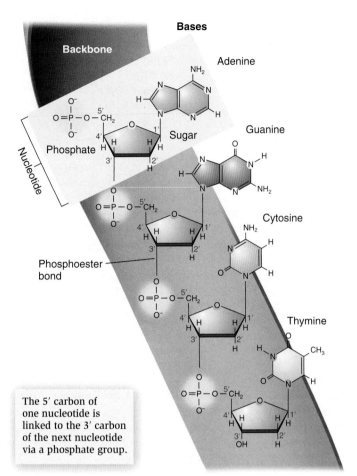

The 5′ carbon of one nucleotide is linked to the 3′ carbon of the next nucleotide via a phosphate group.

Figure 3.23   Structure of a DNA strand.   Nucleotides are linked to each other to form a strand of DNA. The four bases found in DNA are shown. A strand of RNA would be similar except the sugar would be ribose and uracil would be substituted for thymine.

and G with C. If we know the amount of one type of base in a DNA molecule, we can predict the relative amounts of each of the other three bases. For example, if a DNA molecule were composed of 20% A bases, then there must also be 20% T bases. That leaves 60% of the bases that must be C and G combined. Because the amounts of C and G must be equal, this particular DNA molecule must be composed of 30% each of C and G. This specificity provides the mechanism for duplicating and transferring genetic information.

RNA   RNA molecules differ in only a few respects from DNA. RNA consists of a single rather than double strand of nucleotides. In RNA, the sugar in each nucleotide is **ribose** rather than deoxyribose. Also, the pyrimidine base thymine in DNA is replaced in RNA with the pyrimidine base **uracil (U)** (see Figure 3.22). The other three bases—adenine, guanine, and cytosine—are found in both DNA and RNA. Certain forms of RNA called messenger RNA (mRNA) and transfer RNA (tRNA) are responsible for converting the information contained in DNA into the formation of a new polypeptide. This topic will be discussed in Chapter 12.

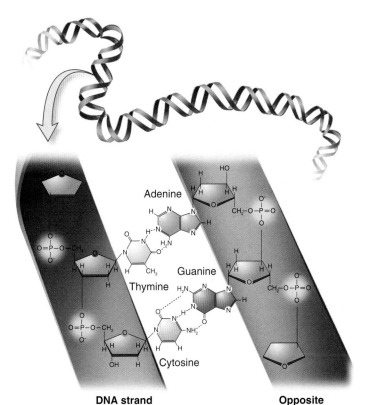

Figure 3.24 The double-stranded structure of DNA. DNA consists of two strands coiled together into a double helix. The bases form hydrogen bonds in which A pairs with T, and G pairs with C.

CHAPTER SUMMARY

3.1 The Carbon Atom and the Study of Organic Molecules

- Organic chemistry is the science of studying carbon-containing molecules, which are found in living organisms. Wöhler's work with urea marked the birth of organic chemistry. (Figure 3.1)

- One property of the carbon atom that makes life possible is its ability to form four covalent bonds with other atoms. Carbon can form both polar and nonpolar bonds. The combination of different elements and different types of bonds allows a vast number of organic compounds to be formed from only a few chemical elements. (Figures 3.2, 3.3)

- Organic molecules may occur in various shapes. The structures of molecules determine their functions.

- Carbon bonds are stable at the different temperatures associated with life. (Figure 3.4)

- Organic compounds may contain functional groups. (Table 3.1)

- Carbon-containing molecules can exist as isomers, which have identical molecular composition but different structures and characteristics. Structural isomers contain the same atoms but in different bonding relationships. Stereoisomers have identical bonding relationships but different spatial positioning of their atoms. Enantiomers exist as mirror images of each other. (Figure 3.5)

3.2 Classes of Organic Molecules and Macromolecules

- The four major classes of organic molecules are carbohydrates, lipids, proteins, and nucleic acids. Macromolecules are large organic molecules that are composed of many thousands of atoms. Some macromolecules are polymers because they are formed by linking together many smaller molecules called monomers.

- Carbohydrates are composed of carbon, hydrogen, and oxygen atoms. Most cells can break down carbohydrates, releasing energy and storing it in newly created bonds in ATP.

- Carbohydrates include monosaccharides (the simplest sugars), disaccharides, and polysaccharides. The polysaccharides starch (in plant cells) and glycogen (in animal cells) provide an efficient means of storing energy. The plant polysaccharide cellulose serves a support or structural function. (Figures 3.6, 3.7, 3.8)

- Lipids, composed predominantly of hydrogen and carbon atoms, are nonpolar and very insoluble in water. Major classes of lipids include fats, phospholipids, steroids, and waxes.

- Fats, a mixture of triglycerides, are formed by bonding glycerol with three fatty acids. In a saturated fatty acid, all the carbons are linked by single covalent bonds. Unsaturated fatty acids contain one or more C=C double bonds. Animal fats generally contain a high proportion of saturated fatty acids, and vegetable fats contain more unsaturated fatty acids. (Figures 3.9, 3.10, 3.11)

- Phospholipids are similar in structure to triglycerides, except they are amphipathic because one fatty acid is replaced with a charged polar group. (Figure 3.12)

- Steroids are constructed of four fused rings of carbon atoms. Small differences in steroid structure can lead to profoundly different biological properties, such as the differences between estrogen and testosterone. (Figure 3.13)

- Waxes, another class of lipids, are nonpolar and repel water, and they are often found as protective coatings on the leaves of plants and the outer surfaces of animals' bodies.

- Proteins are composed of carbon, hydrogen, oxygen, nitrogen, and small amounts of other elements, such as sulfur. Proteins are macromolecules that play critical roles in almost all life processes. The proteins of living organisms are composed of the same set of 20 amino acids, corresponding to 20 different side chains. (Figure 3.14, Table 3.2)

- Amino acids are joined together by linking the carboxyl group of one amino acid to the amino group of another, forming a peptide bond. A polypeptide is a structural unit composed of amino acids, while a protein is a functional unit composed of one or more polypeptides that have been folded and twisted into precise three-dimensional shapes. (Figure 3.15)

- The four levels of protein structure are primary (its amino acid sequence), secondary (bending or twisting into helices or β sheets), tertiary (folding and refolding to assume a three-dimensional shape), and quaternary (multimeric proteins that consist of more than one polypeptide chain). If the primary structure of a protein changes, the other levels would change as a consequence. The three-dimensional structure of a protein determines its function, for example, by creating binding sites for other molecules. (Figures 3.16, 3.17, 3.18, 3.19, 3.20, 3.21)

- Nucleic acids are responsible for the storage, expression, and transmission of genetic information. The two types of nucleic acids are deoxyribonucleic acid (DNA) and ribonucleic acid (RNA). (Figures 3.22, 3.23)

- DNA molecules store genetic information coded in the sequence of their monomers. A DNA molecule consists of two strands of nucleotides coiled around each other to form a double helix, held together by hydrogen bonds between a purine base on one strand and a pyrimidine base on the opposite strand. (Figure 3.24)

- RNA molecules are involved in decoding this information into instructions for linking a specific sequence of amino acids to form a specific polypeptide chain. RNA consists of a single strand of nucleotides. The sugar in each nucleotide is ribose rather than deoxyribose, and the base uracil replaces thymine.

TEST YOURSELF

1. Molecules that contain the element _____ are considered organic molecules.
   a. hydrogen
   b. carbon
   c. oxygen
   d. nitrogen
   e. calcium

2. _____ was the first scientist to synthesize an organic molecule. The organic molecule synthesized was _____.
   a. Kolbe, urea
   b. Wöhler, urea
   c. Wöhler, acetic acid
   d. Kolbe, acetic acid
   e. Wöhler, glucose

3. The versatility of carbon to serve as the backbone for a variety of different molecules is due to
   a. the ability of carbon atoms to form four covalent bonds.
   b. the fact that carbon usually forms ionic bonds with many different atoms.
   c. the abundance of carbon in the environment.
   d. the ability of carbon to form covalent bonds with many different types of atoms.
   e. both a and d.

4. _____ are molecules that have the same molecular composition but differ in structure and/or bonding association.
   a. Isotopes
   b. Isomers
   c. Free radicals
   d. Analogues
   e. Ions

5. _____ is a storage polysaccharide commonly found in the cells of animals.
   a. Glucose
   b. Sucrose
   c. Glycogen
   d. Starch
   e. Cellulose

6. In contrast to other fatty acids, essential fatty acids
   a. are always saturated fats.
   b. cannot be synthesized by the organism and are necessary for survival.
   c. can act as building blocks for large, more complex macromolecules.
   d. are the simplest form of lipids found in plant cells.
   e. are structural components of plasma membranes.

7. Phospholipids are said to be amphipathic, which means these molecules
   a. are partially hydrolyzed during cellular metabolism.
   b. are composed of a hydrophilic portion and hydrophobic portion.
   c. may be poisonous to organisms if in combination with certain other molecules.
   d. are molecules composed of lipids and proteins.
   e. all of the above.

8. The monomers of proteins are _____ and these are linked by polar covalent bonds commonly referred to as _____ bonds.
   a. nucleotides, peptide
   b. amino acids, ester
   c. hydroxyl groups, phosphodiester
   d. amino acids, peptide
   e. monosaccharides, glycosidic

9. The _____ of a nucleotide determines whether it is a component of DNA or a component of RNA.
   a. phosphate group
   b. five-carbon sugar
   c. side chain
   d. fatty acid
   e. both b and d

10. A _____ is a portion of protein with a particular structure and function.
    a. peptide bond
    b. domain
    c. phospholipid
    d. wax
    e. monosaccharide

CONCEPTUAL QUESTIONS

1. Define isomers.
2. List the four classes of organic molecules and give a function of each.
3. Explain the difference between saturated and unsaturated fatty acids.

EXPERIMENTAL QUESTIONS

1. Before the experiments conducted by Anfinsen, what were the common beliefs among scientists about protein folding?
2. Explain the hypothesis tested by Anfinsen.
3. Why did Anfinsen use urea and β-mercaptoethanol in his experiments? Explain the result that was crucial to the discovery that the tertiary structure of a protein is dependent on the primary structure.

COLLABORATIVE QUESTIONS

1. Discuss several types of carbohydrates.
2. Discuss some of the roles that proteins play in organisms.

www.brookerbiology.com
This website includes answers to the Biological Inquiry questions found in the figure legends and all end-of-chapter questions.

4

GENERAL FEATURES OF CELLS

CHAPTER OUTLINE

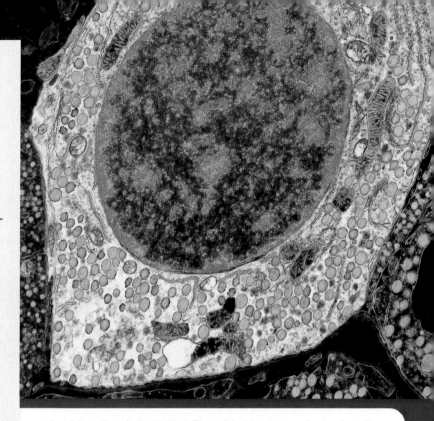

A cell from the pituitary gland. The cell in this micrograph was viewed by a technique called transmission electron microscopy, which is described in this chapter. The micrograph is artificially colored using a computer to enhance the visualization of certain cell structures.

When German botanist Matthias Schleiden studied plant material under the microscope, he was struck by certain consistent features. In particular, he noted the presence of many similar-looking compartments, each of which contained a dark area. (Today we call those compartments cells and the dark area is the nucleus.) In 1838, Schleiden speculated that cells are living entities and plants are aggregates of cells arranged according to definite laws.

Schleiden was a good friend of the German physiologist Theodor Schwann. Over dinner one evening, their conversation turned to the nuclei of plant cells, and Schwann remembered having seen similar structures in animal tissue. Schwann conducted additional studies that showed large numbers of nuclei in animal tissue, at regular intervals, and also located in cell-like compartments. In 1839, Schwann extended Schleiden's hypothesis to animals. In 1855, German biologist Rudolf Virchow proposed that *omnis cellula e cellula* ("every cell originates from another cell"). This idea arose from his research, which showed that diseased cells divide to produce more diseased cells.

According to the **cell theory** or **cell doctrine**, which is credited to both Schleiden and Schwann with contributions from Virchow:

1. all living things are composed of one or more cells;
2. cells are the smallest units of living organisms;
3. new cells come only from pre-existing cells by cell division.

Cell biology is the study of individual cells and their interactions with each other.

Most cells are so small that they cannot be seen with the naked eye. However, as cell biologists have begun to unravel cell structure and function at the molecular level, the cell has emerged as a unit of wonderful complexity and adaptability. In this chapter, we will begin our examination of cells with an overview of their structures and functions. Later chapters in this unit will explore certain aspects of cell biology in greater detail. But first, let's look at the tools and techniques that allow us to observe these tiny entities.

4.1 Microscopy

The **microscope** is a magnification tool that enables researchers to study the structure and function of cells. The first compound microscope—a microscope with more than one lens—was invented in 1595 by Zacharias Jansen of Holland. In 1663 an English biologist, Robert Hooke, studied cork under a primitive compound microscope he had made. (He actually observed cell walls because cork cells are dead and have lost their internal components.) Hooke coined the word *cell* (from the Latin word *cellula*, meaning small compartment) to describe the structures he observed. Ten years later, the Dutch merchant Anton van Leeuwenhoek refined techniques of making lenses and was able to observe single-celled microorganisms such as bacteria. Among his many accomplishments, he discovered blood cells and was the first to see living sperm cells of animals.

Three important parameters in microscopy are magnification, resolution, and contrast. **Magnification** is the ratio between the size of an image produced by a microscope and its actual size. For example, if the image size is 100 times larger than its actual size, the magnification is designated 100×.

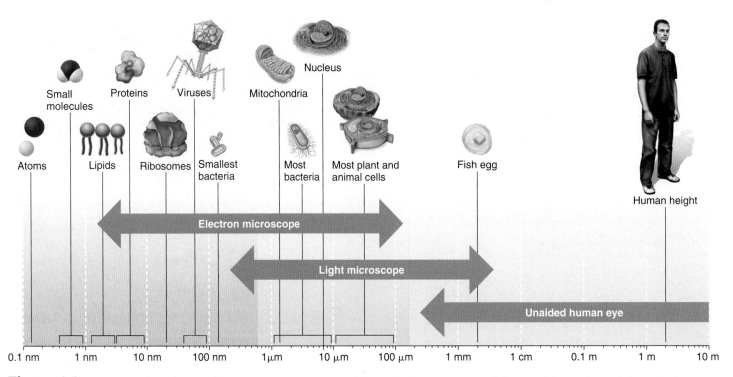

Figure 4.1   A comparison of the sizes of various chemical and biological structures, and the resolving power of the naked eye, light microscope, and electron microscope.   The scale shown at the bottom is a logarithmic scale to accommodate the wide range of sizes shown in this drawing.

Depending on the quality of the lens and illumination source, every microscope has an optimal range of magnification before objects appear too blurry to be readily observed. **Resolution**, a measure of the clarity of an image, is the ability to observe two adjacent objects as distinct from one another. For example, a microscope with good resolution enables a researcher to distinguish two adjacent chromosomes as separate objects, which would appear as a single object under a microscope with poor resolution. The third important parameter in microscopy is **contrast**. The ability to visualize a particular cell structure may depend on how different it looks from an adjacent structure. If the object of interest, such as a particular protein, can be specifically stained with a colored dye, this makes viewing much easier. The application of stains, which selectively label individual components of the cell, greatly improves contrast. However, staining should not be confused with colorization. Many of the micrographs shown in this textbook are colorized to emphasize certain cellular structures (see the chapter opener, for example). In colorization, particular colors are added to micrographs with the aid of a computer. This is done for educational purposes. For example, colorization can help to emphasize different parts of a cell.

Microscopes are categorized into two groups based on the source of illumination. A **light microscope** utilizes light for illumination, while an **electron microscope** uses an electron beam. Very good light microscopes can resolve structures that are 0.2 μm (micron, or micrometer) apart or greater. The resolving power of a microscope depends on several factors including the wavelength of the source of illumination. Resolution is improved when the illumination source has a shorter wavelength.

A major advance in microscopy occurred in 1931 when Max Knoll and Ernst Ruska invented the first electron microscope. Because the wavelength of an electron beam is much shorter than visible light, the resolution of the electron microscope is far better than any light microscope. The limit is typically around 2 nm (nanometers), which is about 100 times better than the light microscope. **Figure 4.1** shows the range of resolving powers of the light and electron microscopes, and compares them to various cells and cell structures.

Over the past several decades, enormous technological advances have made light microscopy a powerful research tool. Improvements in lens technology, microscope organization, sample preparation, sample illumination, and computerized image processing have enabled researchers to create different types of light microscopes, each with its own advantages and disadvantages (**Figure 4.2**).

Similarly, improvements in electron microscopy occurred during the 1930s and 1940s, and by the 1950s the electron microscope was playing a major role in advancing our understanding of cell biology. During **transmission electron microscopy (TEM)**, a beam of electrons is transmitted through a biological sample. In preparation for TEM, a sample is treated with a chemical that binds to cellular molecules and fixes them in place. The sample is placed in a liquid resin and then the resin polymerizes to form a hardened block. The sample embedded within the block is sliced into very thin sections, typically 0.1–1.0 μm in thickness. To provide contrast, the sample is stained with a heavy metal. During staining, the metal binds to certain cellular structures such as membranes. The thin sections of the sample that have been stained with heavy metal are then adhered to a

Micrographs of nerve cells	Type of microscope	Micrographs of nerve cells	Type of microscope

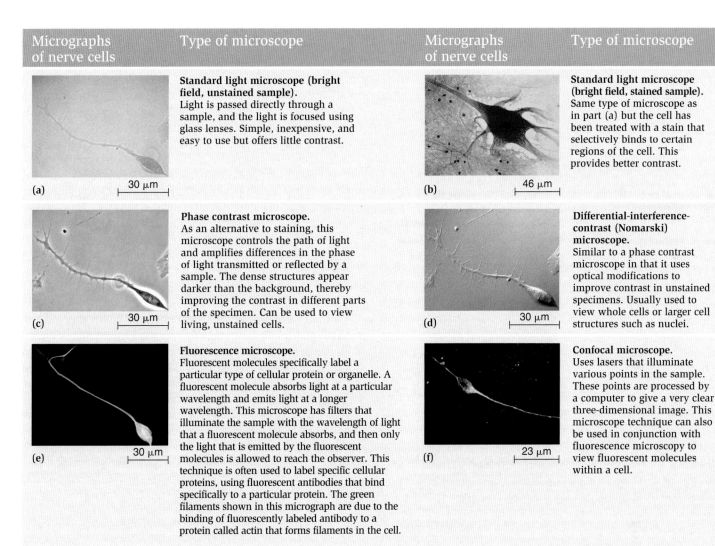

(a) 30 μm

Standard light microscope (bright field, unstained sample).
Light is passed directly through a sample, and the light is focused using glass lenses. Simple, inexpensive, and easy to use but offers little contrast.

(b) 46 μm

Standard light microscope (bright field, stained sample).
Same type of microscope as in part (a) but the cell has been treated with a stain that selectively binds to certain regions of the cell. This provides better contrast.

(c) 30 μm

Phase contrast microscope.
As an alternative to staining, this microscope controls the path of light and amplifies differences in the phase of light transmitted or reflected by a sample. The dense structures appear darker than the background, thereby improving the contrast in different parts of the specimen. Can be used to view living, unstained cells.

(d) 30 μm

Differential-interference-contrast (Nomarski) microscope.
Similar to a phase contrast microscope in that it uses optical modifications to improve contrast in unstained specimens. Usually used to view whole cells or larger cell structures such as nuclei.

(e) 30 μm

Fluorescence microscope.
Fluorescent molecules specifically label a particular type of cellular protein or organelle. A fluorescent molecule absorbs light at a particular wavelength and emits light at a longer wavelength. This microscope has filters that illuminate the sample with the wavelength of light that a fluorescent molecule absorbs, and then only the light that is emitted by the fluorescent molecules is allowed to reach the observer. This technique is often used to label specific cellular proteins, using fluorescent antibodies that bind specifically to a particular protein. The green filaments shown in this micrograph are due to the binding of fluorescently labeled antibody to a protein called actin that forms filaments in the cell.

(f) 23 μm

Confocal microscope.
Uses lasers that illuminate various points in the sample. These points are processed by a computer to give a very clear three-dimensional image. This microscope technique can also be used in conjunction with fluorescence microscopy to view fluorescent molecules within a cell.

Figure 4.2 Micrographs of nerve cells using different types of light microscopes.

copper grid and placed in a transmission electron microscope. When the beam of electrons strikes the sample, some of them hit the heavy metal and are scattered, while those that pass through without being scattered are focused to form an image on a photographic plate or screen (**Figure 4.3a**). Because the scattered electrons are lost from the beam, the metal-stained regions of the sample that scatter electrons appear as areas of reduced electron penetration. These areas are darker than those that allow electrons to pass through them. Compared with other forms of microscopy, TEM has the best resolution of any microscope. TEM provides the greatest resolution of organelles and other cellular structures. However, such microscopes are expensive and are not commonly used to view living cells.

Scanning electron microscopy (SEM) is another type of electron microscopy that utilizes an electron beam to produce an image of the three-dimensional surface of biological samples (**Figure 4.3b**). A biological sample is coated with a thin layer of heavy metal such as gold or palladium and then exposed to an electron beam that scans the surface of the specimen.

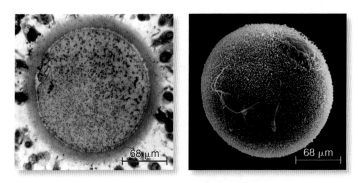

(a) Transmission electron micrograph   68 μm

(b) Scanning electron micrograph   68 μm

Figure 4.3 A comparison of transmission and scanning electron microscopy. (a) A developing human egg cell, observed by TEM, shortly before it was released from an ovary. (b) An egg cell, with a few attached sperm, was coated with heavy metal and observed via SEM. This SEM is colorized.

Biological inquiry: What is the primary advantage of SEM?

The electrons that are scattered from the surface of the sample are detected and create an image on a computer screen. SEM provides a three-dimensional image of the surface of a sample.

And now, let's turn our attention to what microscopes show us about cells.

4.2 Overview of Cell Structure

Cell structure relies on four critical phenomena: (1) matter, (2) energy, (3) organization, and (4) information. In Chapters 2 and 3, we considered the first factor. The matter that is found in living organisms is composed of atoms, molecules, and macromolecules. We will discuss the second factor, energy, throughout this unit, particularly in Chapters 7 and 8. Energy is needed to create molecules and macromolecules, and to carry out many cellular functions. Cells also need energy to become organized, the third phenomenon that underlies cell structure. As discussed throughout this chapter and Unit II, a cell is not a haphazard bag of components. The molecules and macromolecules that constitute cells have specific sites where they are found. For instance, if we compare the structure of a nerve cell in two different humans, or two nerve cells within the same individual, we would see striking similarities in their overall structures. Thus, a key attribute of all living cells is the ability to maintain a particular type of internal structure and organization.

Finally, a fourth critical issue is information. Cell organization requires instructions. These instructions are found in the blueprint of life, namely the genetic material, which is discussed in Unit III. Each living organism has a **genome**, which is defined as the entire complement of its genetic material. Every living cell has a copy of the genome; the **genes** within each species' genome contain the information to create cells with particular structures and functions. This information is passed from cell to cell and from parent to offspring to yield new generations of cells and new generations of life. In this section, we will explore the general structure of cells and examine how the genome contributes to cell structure and function.

Prokaryotic Cells Have a Simple Structure

All forms of life can be placed into two categories based on cell structure—prokaryotes and eukaryotes. We will first consider the **prokaryotes**, which have a relatively simple structure. The term comes from the Greek *pro* and *karyon*, which means "before a kernel"—a reference to the kernel-like appearance of what would later be named the cell nucleus. Prokaryotic cells lack a membrane-enclosed nucleus.

From an evolutionary perspective, the two categories of prokaryotes are **bacteria** and **archaea**. Both types are microorganisms that are relatively small. Bacteria are abundant throughout the world, being found in soil, water, and even our digestive tracts. Most bacterial species are not harmful to humans, but some species are pathogenic—they cause disease. Examples of pathogenic bacteria include *Vibrio cholerae*, the source of cholera, and *Bacillus anthracis*, which causes anthrax. Archaea are less common than bacteria and often occupy extreme environments such as hot springs and deep-sea vents. In this chapter, we will discuss the structure of bacterial cells. We will examine the genetics of bacteria in Chapter 18 and the evolutionary origins of bacteria and archaea in Chapter 22.

Figure 4.4 shows a typical bacterial cell. The **plasma membrane**—a double layer of phospholipids and embedded proteins—forms an important barrier between the cell and its external environment. The **cytoplasm** is the region of the cell that is contained within the plasma membrane. Certain structures in the bacterial cytoplasm are visible via microscopy. These include the **nucleoid**, which is the region of the cell where its genetic material (DNA) is located, and **ribosomes**, which are involved in polypeptide synthesis. We will examine the functions of ribosomes later in this chapter.

Many bacterial structures are located outside the plasma membrane. Nearly all species of prokaryotes have a relatively rigid **cell wall** that supports and protects the plasma membrane and cytoplasm. The cell wall is porous, so it does not prevent most nutrients in the environment from reaching the plasma membrane. Many bacteria also secrete a **glycocalyx**, an outer, viscous covering surrounding the bacterium. The glycocalyx traps water and helps protect bacteria from drying out. Certain strains of bacteria that invade animals' bodies produce a very thick, gelatinous glycocalyx called a **capsule** that may help them avoid being destroyed by the animal's immune (defense) system. Finally, many prokaryotes have appendages such as **pili** and **flagella**. Pili allow prokaryotes to attach to surfaces and to each other. Flagella provide a way for prokaryotes to swim.

For many decades, the cytoplasm of bacterial cells was thought to be a fluid-filled space with relatively little defined organization, other than the nucleoid. The cell wall was believed to be solely involved with forming cell shape. During the past decade, however, these notions have been challenged by a molecular analysis of bacterial cells. Researchers have discovered that bacteria possess an architecture inside their cytoplasm that bears striking similarities to the cytoskeleton found in eukaryotic cells (described later in this chapter). **Figure 4.5** illustrates the functions of three proteins in the organization of bacterial cells. FtsZ, which is important in cell division, is found at the site where a cell divides into two cells. MreB plays a role in cell polarity. In many bacteria such as *Caulobacter crescentus*—an aquatic bacterium that survives in nutrient-poor water—cellular components are asymmetrically distributed throughout the cell. As suggested in Figure 4.5, the concentrations of components at one pole are different compared to the other pole. MreB forms a spiral structure in the cell, and evidence suggests it is necessary for this polarity. Finally, the localization of CreS in certain regions of the cell is critical for cell shape. Interestingly, the bacterial proteins described in Figure 4.5 are evolutionarily related to eukaryotic proteins described later in this chapter. In particular, FtsZ, MreB, and CreS are related to tubulin, actin, and intermediate filament proteins, respectively.

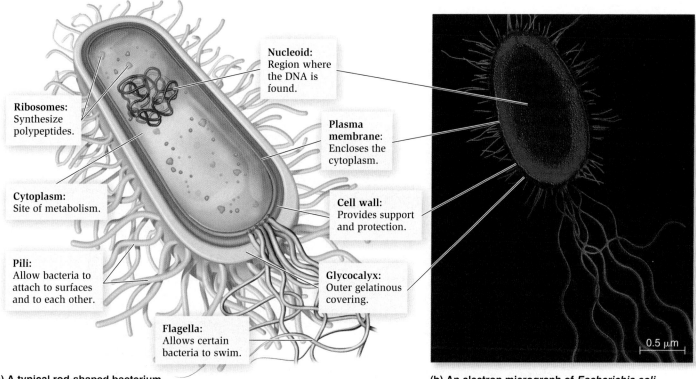

(a) A typical rod-shaped bacterium

Nucleoid:
Region where
the DNA is
found.

Ribosomes:
Synthesize
polypeptides.

Plasma
membrane:
Encloses the
cytoplasm.

Cytoplasm:
Site of metabolism.

Cell wall:
Provides support
and protection.

Pili:
Allow bacteria to
attach to surfaces
and to each other.

Glycocalyx:
Outer gelatinous
covering.

Flagella:
Allows certain
bacteria to swim.

(b) An electron micrograph of *Escherichia coli*.

0.5 μm

Figure 4.4 Structure of a typical prokaryotic cell. Prokaryotic cells, which include bacteria and archaea, lack internal compartmentalization.

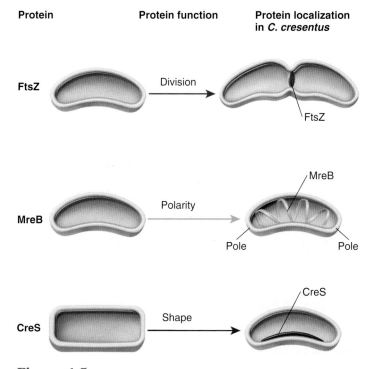

Protein	Protein function	Protein localization in *C. cresentus*
FtsZ	Division	FtsZ
MreB	Polarity	MreB / Pole / Pole
CreS	Shape	CreS

Figure 4.5 Proteins that play a role in the architecture of bacterial cells. FtsZ plays a role in cell division, MreB is involved in cell polarity, and CreS influences cell shape.

Eukaryotic Cells Are Compartmentalized into Organelles

Aside from prokaryotes, all other species are **eukaryotes** (meaning "true nucleus"), which include protists, fungi, plants, and animals. Paramecia and algae are types of protists, while yeasts and molds are types of fungi. The DNA of all eukaryotic cells is housed in a distinct compartment called a **nucleus**, which is an example of an **organelle**—a subcellular structure or membrane-bounded compartment with its own unique structure and function. In contrast to prokaryotes, eukaryotic cells exhibit **compartmentalization**, which means that they have many organelles that separate the cell into different regions. Cellular compartmentalization allows a cell to carry out specialized chemical reactions in different places. For example, protein synthesis and protein breakdown occur in different compartments in the cell.

Figures 4.6 and **4.7** describe the morphology (form and structure) of typical animal and plant cells. Some general features of cell organization, such as a nucleus, are found in nearly all eukaryotic cells. Even so, be aware that the shape, size, and organization of cells vary considerably among different species and even among different cell types of the same species. Micrographs of a human skin cell and a human nerve cell show that, although these cells contain the same types of organelles, their overall morphologies are quite different (see Figure 4.6b).

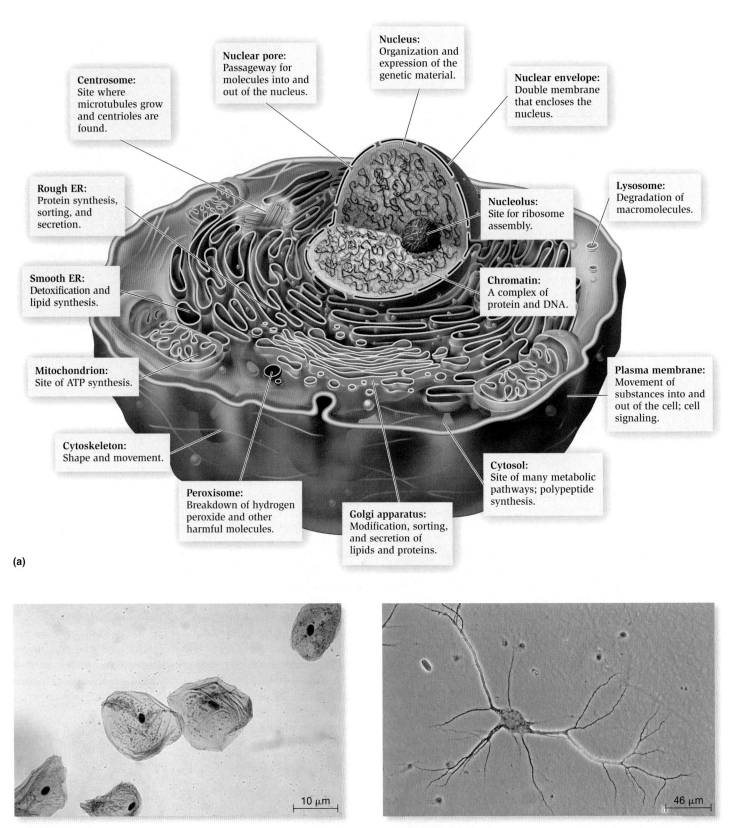

Centrosome:
Site where microtubules grow and centrioles are found.

Nuclear pore:
Passageway for molecules into and out of the nucleus.

Nucleus:
Organization and expression of the genetic material.

Nuclear envelope:
Double membrane that encloses the nucleus.

Rough ER:
Protein synthesis, sorting, and secretion.

Nucleolus:
Site for ribosome assembly.

Lysosome:
Degradation of macromolecules.

Smooth ER:
Detoxification and lipid synthesis.

Chromatin:
A complex of protein and DNA.

Mitochondrion:
Site of ATP synthesis.

Plasma membrane:
Movement of substances into and out of the cell; cell signaling.

Cytoskeleton:
Shape and movement.

Cytosol:
Site of many metabolic pathways; polypeptide synthesis.

Peroxisome:
Breakdown of hydrogen peroxide and other harmful molecules.

Golgi apparatus:
Modification, sorting, and secretion of lipids and proteins.

(a)

10 μm

46 μm

Human skin cell

Human nerve cell

(b)

Figure 4.6   **General structure of an animal cell.**   (a) A schematic drawing of a typical animal cell. (b) Micrographs of a human skin cell (*left*) and a human nerve cell (*right*). Although these cells have the same types of organelles, note that their general morphologies are quite different.

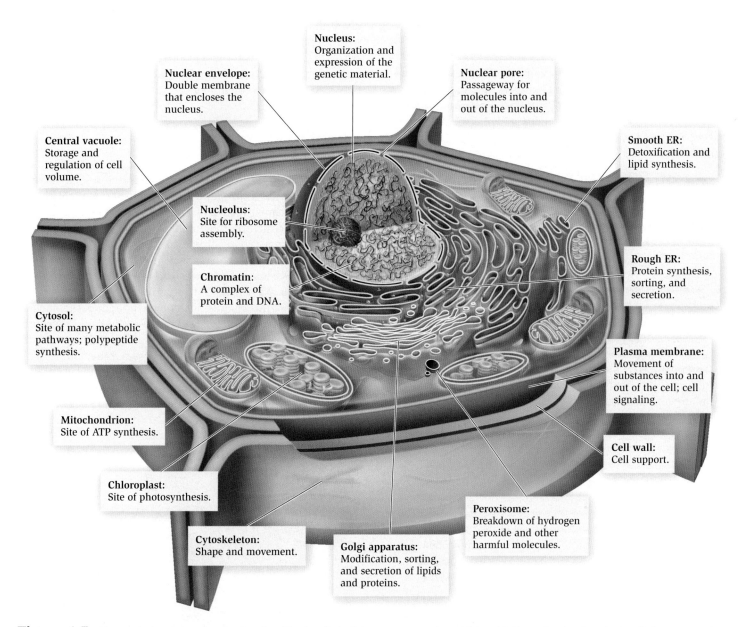

Figure 4.7 **General structure of a plant cell.** Plant cells lack lysosomes and centrioles. Unlike animal cells, plant cells have an outer cell wall; a large central vacuole that functions in storage, digestion, and cell volume; and chloroplasts, which carry out photosynthesis.

GENOMES & PROTEOMES

The Proteome Determines the Characteristics of a Cell

Many organisms, such as plants and animals, are multicellular, meaning that the body of a single organism is composed of many cells. However, the cells of a multicellular organism are not all identical. For example, your body contains nerve cells, muscle cells, skin cells, etcetera. An intriguing question, therefore, is, How does a single organism produce different types of cells?

To answer this question, we need to consider the distinction between genomes and proteomes. Recall that the genome constitutes all types of genetic material that an organism has. Most genes encode the production of polypeptides, which assemble into functional proteins. An emerging theme discussed in this unit is that the structures and functions of proteins are primarily responsible for the structures and functions of cells. A typical eukaryotic cell produces thousands of different types of proteins. For example, researchers estimate that one skeletal muscle cell in your arm produces about 15,000 different types of proteins; the total number of proteins in one cell is far more than 15,000 because many copies of some proteins are made. The **proteome** is defined as all of the types and relative amounts of proteins that are made in a particular cell at a particular time and under specific conditions.

As an example, let's consider muscle cells and nerve cells—two cell types that have dramatically different organization and structure. Actually, the genes in a human muscle cell of a particular individual are identical to those in a human nerve cell. However, their proteomes are different. An important principle in cell biology is that *the proteome of a cell determines its structure and function.* Several phenomena underlie the differences that are observed in the proteomes of different cell types. If we compare muscle and nerve cells, these phenomena include:

1. *Certain proteins that are found in muscle cells are not found in nerve cells, and vice versa.* For a protein to be expressed in a particular cell, the gene that corresponds to that protein must be "turned on," that is, actively expressing that protein. Due to gene regulation, which is described in Chapter 13, certain genes that are turned on in muscle cells are not turned on in nerve cells and vice versa.

2. *The relative amounts of certain proteins are different in muscle and nerve cells.* The amount of a given protein depends on many factors including how strongly the corresponding gene is turned on, how efficiently the protein is synthesized, and how long the protein lasts within a cell. Some proteins are found in both nerve and muscle cells, but in greatly different amounts. For example, a protein called actin, which is discussed later in this chapter, occurs abundantly in muscle cells—where it plays a key role in muscle contraction—but to a much lesser extent in nerve cells.

3. *The amino acid sequences of particular proteins can vary in different cell types.* As discussed in Chapter 13, the mRNA from a single gene can produce two or more polypeptides with slightly different amino acid sequences via a process called alternative splicing. For example, tropomyosin is a protein that regulates cell movement. The form of tropomyosin in muscle cells has a slightly different amino acid sequence from the type in nerve cells.

4. *Nerve and muscle cells may alter their proteins in different ways.* After a protein is made, its structure may be changed in a variety of ways. These include the covalent attachment of molecules such as phosphate and carbohydrate, and the cleavage of a protein to a smaller size.

These four phenomena enable nerve and muscle cells to produce different proteomes, and thus different structures and functions. Likewise, the proteomes of muscle and nerve cells differ from those of other cell types—skin, liver, etcetera.

During the last decade or so, researchers have also discovered an association between proteome changes and disease. For example, the proteomes of healthy cells are different from the proteomes of cancer cells. Furthermore, the proteomes of cancer cells change as the disease progresses. One reason for studying cancer-cell proteomes is to improve the early detection of cancer by identifying proteins that are made in the early stages, when the disease is most treatable. In addition, information about the ways that the proteomes of cancer cells change may help researchers uncover new treatment options. A key challenge for biologists is to understand the synthesis and function of proteomes in different cell types, and how proteome changes may lead to disease conditions.

4.3 The Cytosol

Thus far, we have focused on the general features of prokaryotic and eukaryotic cells. In the rest of this chapter we will survey the various compartments of eukaryotic cells with a greater emphasis on structure and function. **Figure 4.8** highlights a plant and animal cell according to four different regions. We will start with the **cytosol** (shown in yellow), the region of a eukaryotic cell that is outside the cell organelles but inside the plasma membrane. The other regions of the cell, which we will examine later in this chapter, include the interior of the nucleus (blue), the endomembrane system (purple/pink), and semiautonomous organelles (green). As in prokaryotes, the term cytoplasm refers to the region inside the plasma membrane. This includes the cytosol, the endomembrane system inside the plasma membrane, and the semiautonomous organelles.

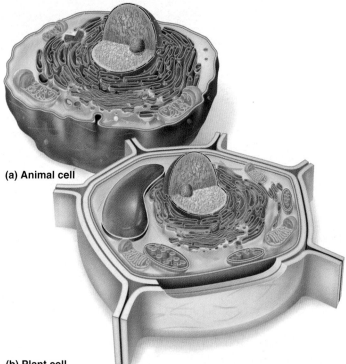

(a) Animal cell

(b) Plant cell

Figure 4.8   Compartments within (a) animal and (b) plant cells.   The cytosol, which is outside the organelles but inside the plasma membrane, is shown in yellow. The membranes of the endomembrane system are shown in purple, while the fluid filled interiors are pink. The interior of the nucleus is blue. Semiautonomous organelles, which include mitochondria, chloroplasts, and peroxisomes, are green.

Though the amount varies among different types of cells, the cytosol typically occupies about 50% of the total cell volume. In this section, we will consider the primary functions of the cytosol. First, it is the site of many chemical reactions that produce the materials and energy necessary for life, such as breaking down food molecules into smaller components. Such reactions release energy, and the components can be used as building blocks to create new cellular molecules and macromolecules. For example, we will explore a particularly important activity of the cytosol, which is the synthesis of cellular proteins. In addition, we will examine the structure and function of large protein filaments, found in the cytosol, that provide organization to the cell and allow cells to move.

Synthesis and Breakdown of Molecules Occur in the Cytosol

Metabolism is defined as the sum of the chemical reactions by which cells produce the materials and energy that are necessary to sustain life. Although specific steps of metabolism also occur in cell organelles, the cytosol is a central coordinating region for many metabolic activities of eukaryotic cells. Metabolism often involves a series of steps called a metabolic pathway. A specific **enzyme** is responsible for speeding up each step in a metabolic pathway. In Chapter 7, we will examine the functional properties of enzymes and consider a few metabolic pathways that occur in the cytosol and cellular organelles.

Some pathways involve the breakdown of a molecule into smaller components, a process termed **catabolism**. Such pathways are needed to capture energy for use by the cell, and also to generate molecules that provide the building blocks to construct cellular macromolecules. Conversely, other metabolic pathways are involved in **anabolism**, the synthesis of cellular molecules and macromolecules. For example, polysaccharides are made by linking together sugar molecules. To create proteins, amino acids are covalently connected to form a polypeptide. This process is described next.

Translation Is the Process of Polypeptide Synthesis

A **polypeptide** is composed of a linear sequence of amino acids. The term polypeptide describes a unit of structure. By comparison, the term **protein** is a unit of function. One or more polypeptides assemble into a three-dimensional protein that performs a particular function. A critical activity of all cells is the synthesis of polypeptides. This process is called **translation** because the information within a gene is ultimately translated into the sequence of amino acids in a polypeptide. Chapter 12 describes the details of the process. In this chapter, we will briefly consider the general features of translation to help us appreciate how cells make proteins, which are vital to cell structure and function.

Translation requires many cellular components including a ribosome and two types of RNA molecules (**Figure 4.9**). The ribosome is the site where polypeptide synthesis occurs. Messenger RNA (mRNA) is produced from a gene and provides the information to make a polypeptide. Transfer RNA (tRNA) molecules, which carry amino acids, bind to the mRNA so that a polypeptide can be made, one amino acid at a time. The role of the ribosome is to facilitate the binding between the mRNA and tRNA molecules, and to catalyze the formation of covalent bonds between adjacent amino acids. Once the entire polypeptide is made, it is released from the ribosome.

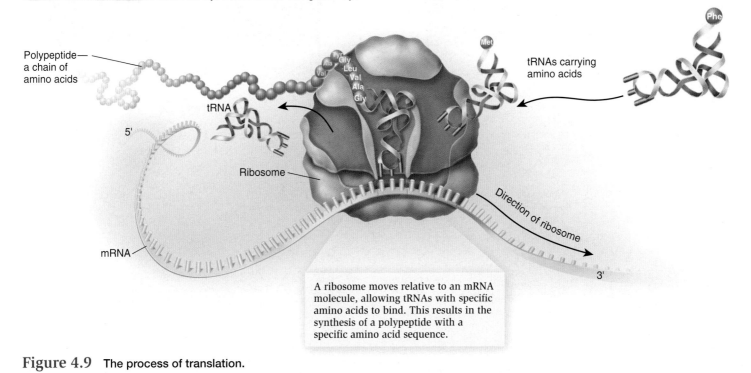

A ribosome moves relative to an mRNA molecule, allowing tRNAs with specific amino acids to bind. This results in the synthesis of a polypeptide with a specific amino acid sequence.

Figure 4.9 The process of translation.

The Cytoskeleton Provides Cell Organization, Shape, and Movement

The **cytoskeleton** is a network of three different types of protein filaments: **microtubules, intermediate filaments**, and **actin filaments** (Table 4.1). Let's first consider the structure of these protein filaments and their roles in the construction and organization of cells. Later, we will examine how they are involved in cell movement.

Microtubules Microtubules are long, hollow, cylindrical structures about 25 nm in diameter composed of the protein tubulin. The assembly of tubulin to form a microtubule results in a polar structure with a plus end and a minus end (Table 4.1). Growth of microtubules occurs at the plus end, while shortening of microtubules can occur at either the plus or minus end. A single microtubule can oscillate between growing and shortening phases, a phenomenon termed **dynamic instability**. Dynamic instability is important in many cellular activities including the sorting of chromosomes during cell division.

The sites where microtubules form within a cell can vary among different types of organisms. Animal cells that are not preparing to divide contain a single structure near their nucleus called the **centrosome** or **microtubule-organizing center** (Table 4.1). Within the centrosome are the **centrioles**, a conspicuous pair of structures arranged perpendicular to each other. In animal cells, microtubule growth starts at the centrosome such that the minus end is anchored there. In contrast, most plant cells and many protists lack centrosomes and centrioles. Microtubules are created at many sites that are scattered throughout a plant cell.

Microtubules are important for cell shape and organization. Organelles such as the Golgi apparatus often are attached to microtubules. In addition, microtubules are involved in the organization of chromosomes during mitosis and in the orientation of cells during cell division. We will examine these events in Chapter 15.

Intermediate Filaments Intermediate filaments are another class of cytoskeletal filament found in the cells of many but not all animal species. Their name is derived from the observation

Table 4.1	Properties of Cytoskeletal Filaments		
Characteristic	**Microtubules**	**Intermediate filaments**	**Actin filaments**
Diameter	25 nm	10 nm	7 nm
Structure	Hollow tubule	Twisted filament	Spiral filament

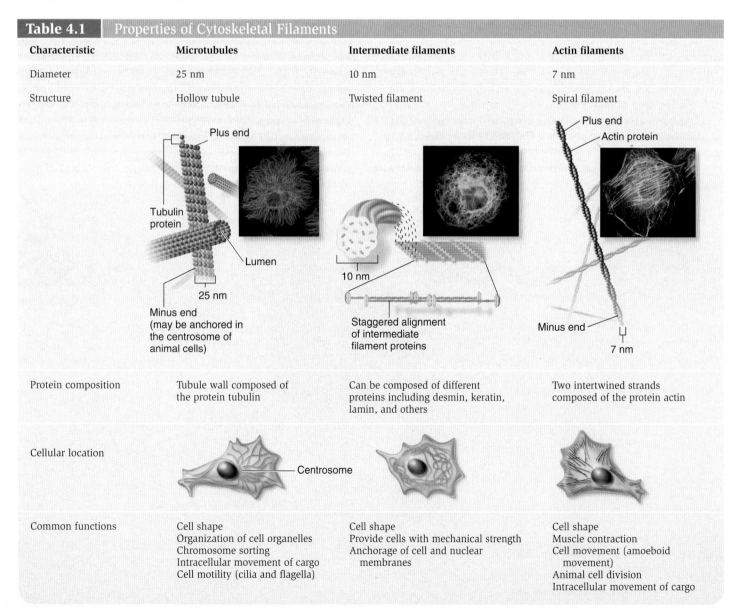

Protein composition	Tubule wall composed of the protein tubulin	Can be composed of different proteins including desmin, keratin, lamin, and others	Two intertwined strands composed of the protein actin
Cellular location			

Common functions	Cell shape
	Organization of cell organelles
	Chromosome sorting
	Intracellular movement of cargo
	Cell motility (cilia and flagella)
		Cell shape
		Provide cells with mechanical strength
		Anchorage of cell and nuclear
		  membranes
			Cell shape
			Muscle contraction
			Cell movement (amoeboid
			  movement)
			Animal cell division
			Intracellular movement of cargo

that they are intermediate in diameter between actin filaments and myosin filaments. (Myosin filaments are described in Chapter 46.) Intermediate filament proteins bind to each other in a staggered array to form a twisted, ropelike structure with a diameter of approximately 10 nm (Table 4.1). Intermediate filaments tend to be more stable than microtubules and actin filaments, which readily polymerize and depolymerize. They function as tension-bearing fibers that help maintain cell shape and rigidity.

Several types of related proteins can assemble into intermediate filaments. Desmins form intermediate filaments in muscle cells and provide mechanical strength. Keratins form intermediate filaments in skin, intestinal, and kidney cells, where they are important for mechanical strength and cell shape. They are also a major constituent of hair and nails. In addition, intermediate filaments are found inside the cell nucleus. As discussed later in this chapter, nuclear lamins form a network of intermediate filaments that line the inner nuclear membrane and provide anchorage points for the nuclear pores.

Actin Filaments Actin filaments—also known as **microfilaments** because they are the thinnest cytoskeletal filaments—are long, thin fibers approximately 7 nm in diameter (Table 4.1).

Each fiber is composed of two strands of actin monomers that spiral around each other. Like microtubules, actin filaments have plus and minus ends, and they are very dynamic structures that grow at the plus end.

Despite their thinness, actin filaments play a key role in cell strength and shape. Although actin filaments are dispersed throughout the cytosol, they tend to be highly concentrated near the plasma membrane. In many types of cells, actin filaments support the plasma membrane and provide strength and shape to the cell. The minus ends of actin filaments are usually anchored at the plasma membrane, which explains why actin filaments are typically found there.

Motor Proteins Interact with Microtubules or Actin Filaments to Promote Cellular Movements

Motor proteins are a category of cellular proteins that use ATP as a source of energy to promote movement. As shown in **Figure 4.10a**, a motor protein consists of three domains called the head, hinge, and tail. The head is the site where ATP binds and is hydrolyzed to ADP and P_i. ATP binding and hydrolysis cause a bend in the hinge, which results in movement. The tail region is attached to other proteins, or to other kinds of cellular molecules.

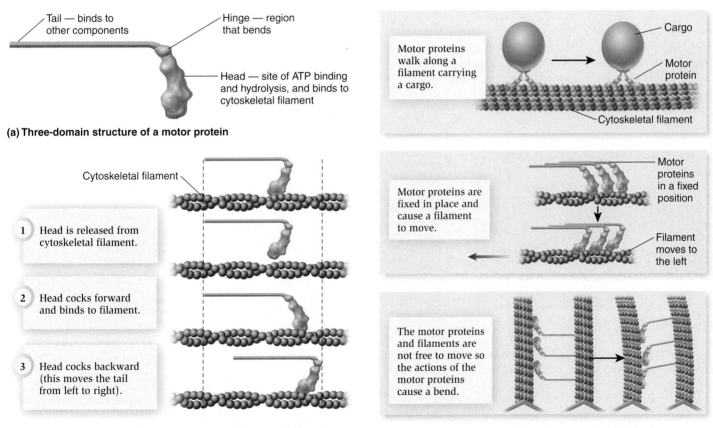

(a) Three-domain structure of a motor protein

Tail — binds to other components
Hinge — region that bends
Head — site of ATP binding and hydrolysis, and binds to cytoskeletal filament

(b) Movement of a motor protein along a cytoskeletal filament

Cytoskeletal filament

1. Head is released from cytoskeletal filament.
2. Head cocks forward and binds to filament.
3. Head cocks backward (this moves the tail from left to right).

(c) Three types of movements facilitated by motor proteins and cytoskeletal filaments

Motor proteins walk along a filament carrying a cargo.
Cargo
Motor protein
Cytoskeletal filament

Motor proteins are fixed in place and cause a filament to move.
Motor proteins in a fixed position
Filament moves to the left

The motor proteins and filaments are not free to move so the actions of the motor proteins cause a bend.

Figure 4.10 **Motor proteins and their interactions with cytoskeletal filaments.** (a) Three-domain structure of a motor protein. Note: The protein subunits of motor proteins often associate with each other along their tails, such that the motor has two tails, two hinges, and two heads. (b) Conformational changes in a motor protein that allow it to walk along a cytoskeletal filament. (c) Three ways that motor proteins and cytoskeletal filaments can cause movement.

To promote cellular movement, the head region of a motor protein interacts with a cytoskeletal filament (**Figure 4.10b**). When ATP binds and is hydrolyzed, the motor protein attempts to "walk" along the filament. The head of the motor protein is initially attached to a filament. To move forward, the head detaches from the filament, cocks forward, binds to the filament, and cocks backward. To imagine how this works, consider the act of walking in which the ground is a cytoskeletal filament, your leg is the head of the motor protein, and your hip is the hinge. To walk, you lift your leg up, you move it forward, you place it on the ground, and then you cock it backward (which propels you forward). This series of events is analogous to how a motor protein walks along a cytoskeletal filament.

Interestingly, cells have utilized the actions of motor proteins to promote three different kinds of movements. In the first example shown in **Figure 4.10c** (top), the tail region is attached to a cargo, so that the motor protein moves the cargo from one location to another. Alternatively, a motor protein can remain in place and cause the filament to move (Figure 4.10c, middle). As discussed in Chapter 46, this occurs during muscle contraction. A third possibility is that both the motor protein and filament are restricted in their movement. In this case, when the motor protein attempts to walk, it exerts a force that causes the filament to bend (Figure 4.10c, bottom). As described next, this occurs during the bending of cilia and flagella.

Let's now consider some specific examples where motor proteins and cytoskeletal filaments interact and result in movement. In certain kinds of cells, microtubules and motor proteins facilitate movement involving cell appendages called **cilia** and **flagella** (singular, *cilium* and *flagellum*). Flagella are usually longer than cilia and present singly or in pairs. A single flagellum may propel a cell such as a sperm cell with a whiplike motion (**Figure 4.11a**). Alternatively, a pair of flagella may move in a synchronized manner to pull a microorganism through the water (think of a human swimmer doing the breaststroke). Certain unicellular algae swim in this manner (**Figure 4.11b**). By comparison, cilia are often shorter than flagella and tend to cover all or part of the surface of a cell. Protists such as paramecia may have hundreds of adjacent cilia that beat in a coordinated fashion to propel the organism through water (**Figure 4.11c**).

Despite their differences in length, cilia and flagella share the same internal structure (**Figure 4.12**). This arrangement containing microtubules, the motor protein dynein, and linking proteins is called an **axoneme**. In the cilia and flagella of most eukaryotic organisms, the microtubules form an arrangement called a 9 + 2 array. Each of the two central microtubules consists of a single microtubule, while the outer nine are doublet microtubules, which are two merged microtubules. The microtubules in cilia and flagella emanate from **basal bodies**, which are anchored to the cytoplasmic side of the plasma membrane. Much like the centrosome of animal cells, the basal bodies provide a site for microtubules to grow. In addition to microtubules, the core structure of a cilium and flagellum also has motor proteins, namely dynein, and linking proteins such as nexin, that hold the axoneme together.

The movement of both cilia and flagella involves the propagation of a bend, which begins at the base of the structure and proceeds toward the tip (see Figure 4.11a). The bending occurs because dynein is activated to walk toward the basal body of the microtubules. ATP hydrolysis is required for this process.

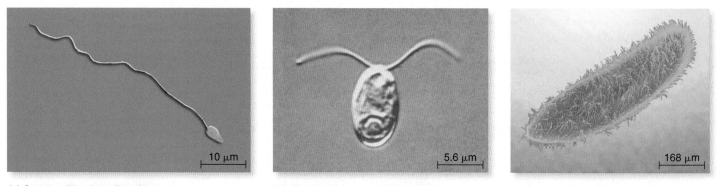

(a) Sperm with a long flagellum   10 μm

(b) *Chlamydomonas* with 2 flagella   5.6 μm

(c) Paramecium with many cilia   168 μm

Figure 4.11   Cellular movements due to the actions of flagella and cilia.   Both flagella and cilia cause movement by a bending motion. In flagella, movement occurs by a whiplike motion that is due to the propagation of a bend from the base to the tip. In addition, the nature of swimming depends on the length of the appendage and whether it involves coordination among multiple flagella or cilia. **(a)** Sperm swim by means of a single, long flagellum that moves in a whiplike motion. **(b)** The swimming of *Chlamydomonas reinhardtii* also involves a whiplike motion at the base, but the motion is precisely coordinated between two flagella. This results in swimming behavior that resembles a breaststroke. **(c)** Ciliated protozoa swim via many shorter cilia. The bending motion is coordinated among multiple adjacent cilia.

Biological inquiry: During the movement of a cilium or flagellum, describe the type of movements that are occurring in dynein and microtubules.

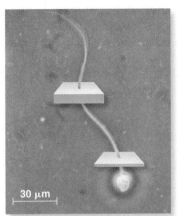

(a) Human sperm cell

30 μm

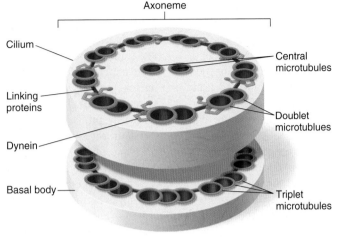

Axoneme

Cilium

Central microtubules

Linking proteins

Dynein

Doublet microtubules

Basal body

Triplet microtubules

(b) Molecular structure of a flagellum

Figure 4.12 **Structure of a eukaryotic flagellum.** **(a)** SEM of a human sperm cell. **(b)** Drawing of the axoneme. The core structure consists of a 9 + 2 arrangement of nine outer doublet microtubules and two central microtubules. This structure is anchored to the basal body, which has nine triplet microtubules, in which three microtubules are fused together. Note: The structure of the basal body is very similar to centrioles in animal cells.

However, the microtubules and dynein are not free to move relative to each other because of linking proteins. Therefore, instead of dyneins freely walking along the microtubules, they exert a force that bends the microtubules (see Figure 4.10c, bottom). The dyneins at the base of the structure are activated first, followed by dyneins that are progressively closer to the tip of the appendage. The resulting movement propels the organism.

As a second example, let's consider how motor proteins interact with actin. As discussed in Chapter 46, actin and a motor protein called myosin are responsible for the movement observed in muscle cells. In protists such as the amoeba, movement occurs via the dynamic rearrangement of the actin cytoskeleton (**Figure 4.13**). Actin filaments are formed near the leading edge to create a projection called a lamellipodium. The cell is pulled toward the leading edge using motor proteins such as myosin, which tugs on actin filaments and promotes cellular movement. Actin filaments and myosin motors are also involved in other types of movement, as we will see next in our Feature Investigation.

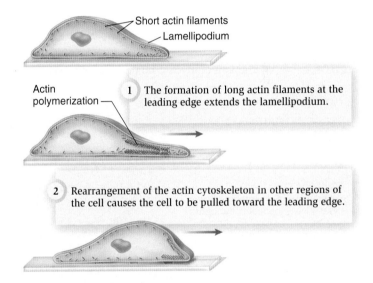

Short actin filaments

Lamellipodium

Actin polymerization

1 The formation of long actin filaments at the leading edge extends the lamellipodium.

2 Rearrangement of the actin cytoskeleton in other regions of the cell causes the cell to be pulled toward the leading edge.

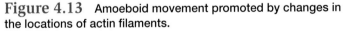

Figure 4.13 Amoeboid movement promoted by changes in the locations of actin filaments.

FEATURE INVESTIGATION

Sheetz and Spudich Showed That Myosin Walks Along Actin Filaments

Cell biologists often gain clues to protein function from studying living cells. In the early 1950s, Hugh Huxley, Andrew F. Huxley, and their colleagues studied muscle contraction via microscopy and other techniques. They proposed the sliding-filament model for muscle contraction, in which actin filaments slide past thicker filaments composed of myosin. This model was based on observations involving intact muscle cells. Research at that time and during subsequent decades indicated that actin and myosin were necessary for muscle movement.

Biologists want to understand how a process, such as muscle contraction, works at the molecular level. However, studying living cells has its drawbacks. Recall that the proteome of every living cell is composed of thousands of different proteins, making it difficult to establish the function of any one protein. As an alternative to studying a process in living cells, another approach is to isolate and purify cellular components and study their functions outside the cell. This is an ***in vitro*** approach (literally "in glass," such as in a glass test tube). By comparison, studying living cells is termed ***in vivo***, meaning "in life."

In 1983, approximately 30 years after the sliding-filament model was proposed, Michael Sheetz and James Spudich devised a clever approach to study myosin function *in vitro*.

Prior to their work, researchers had learned how to purify myosin protein from muscle cells. A fragment of myosin could be attached to a fluorescently labeled bead, making it possible to follow myosin movement using a fluorescence microscope. However, to study movement, Sheetz and Spudich also needed actin filaments. Although actin filaments can be purified from cells, they become a tangled mess during the purification process. As an alternative, the researchers were aware that the alga *Nitella axillaris* has arrays of actin filaments that lie inside the cell and parallel to the plasma membrane. These parallel arrays function in the phenomenon known as **cytoplasmic streaming**, in which the cytoplasm circulates throughout the cell to distribute resources efficiently in large cells. Along these lines, another advantage of *Nitella* for this experiment was that the cells are relatively large—several hundred microns in diameter and several centimeters in length.

Figure 4.14 illustrates the *in vitro* approach of Sheetz and Spudich. Their procedure involved cutting open *Nitella* cells and pinning down the plasma membrane to expose the actin filaments. Except for chloroplasts, which are found between the plasma membrane and the actin filaments, the rest of the cellular contents were washed away. Next, a solution containing purified myosin attached to fluorescent beads was added and observed via fluorescence microscopy. Sheetz and Spudich conducted their experiments with and without ATP, which is needed for muscle cell movement. In addition, the researchers tested the effects of N-ethylmaleimide (NEM), a chemical that was already known to bind to myosin and inhibit its function. As shown in the data, myosin was able to move along actin filaments in the presence of ATP. Furthermore, when myosin was treated with NEM, the movement was inhibited.

Taken together, these *in vitro* experiments confirmed that myosin is a motor protein that uses ATP to walk along actin filaments. Furthermore, this purified system provided evidence that all that is needed for movement are actin, myosin, and ATP. In addition, the results are consistent with the idea that different types of movement, such as cytoplasmic streaming in algae and muscle contraction in animals, use the same underlying molecular mechanism—in this case, an interaction between the motor protein myosin and actin filaments.

Figure 4.14   Movement of myosin-coated beads along actin filaments.

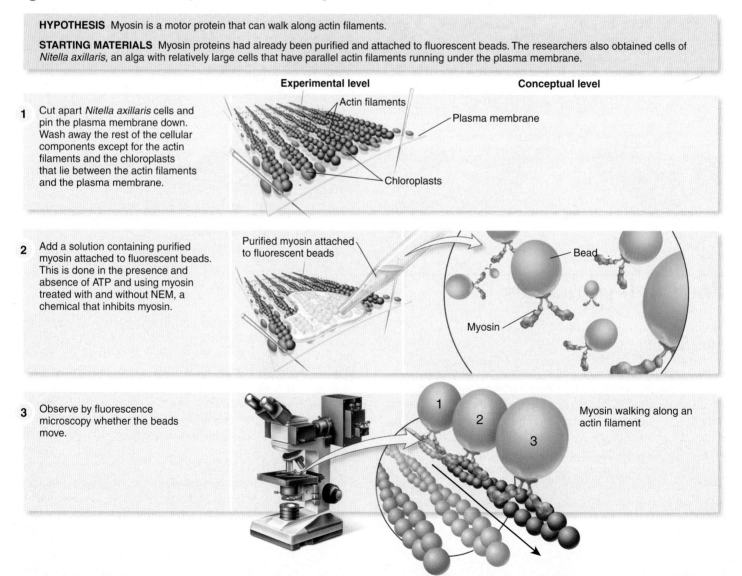

HYPOTHESIS Myosin is a motor protein that can walk along actin filaments.

STARTING MATERIALS Myosin proteins had already been purified and attached to fluorescent beads. The researchers also obtained cells of *Nitella axillaris*, an alga with relatively large cells that have parallel actin filaments running under the plasma membrane.

Experimental level   **Conceptual level**

1 Cut apart *Nitella axillaris* cells and pin the plasma membrane down. Wash away the rest of the cellular components except for the actin filaments and the chloroplasts that lie between the actin filaments and the plasma membrane.

Actin filaments
Plasma membrane
Chloroplasts

2 Add a solution containing purified myosin attached to fluorescent beads. This is done in the presence and absence of ATP and using myosin treated with and without NEM, a chemical that inhibits myosin.

Purified myosin attached to fluorescent beads
Bead
Myosin

3 Observe by fluorescence microscopy whether the beads move.

Myosin walking along an actin filament

4 THE DATA

Results from step 3:

Conditions	Movement observed?
Plus ATP, plus NEM	No
No ATP, no NEM	No
Plus ATP, no NEM	Yes

This is a time-lapse micrograph in which photos were taken every few seconds. It shows a single bead (yellow) moving from left to right. The red objects are stained chloroplasts, which are below the actin filaments. The actin filaments are not visible in this micrograph because they are not fluorescently labeled.

4.4 The Nucleus and Endomembrane System

In Chapter 2, we learned that the nucleus of an atom contains protons and neutrons. In cell biology, the term **nucleus** has a different meaning. It is an organelle found in eukaryotic cells that contains most of the cell's genetic material. A small amount of genetic material is also found in mitochondria and chloroplasts.

The membranes that enclose the nucleus are part of a larger network of membranes called the **endomembrane system**. These include not only the nuclear envelope, which encloses the nucleus, but also the endoplasmic reticulum, Golgi apparatus, lysosomes, and vacuoles. The prefix *endo*—meaning "inside"—originally referred only to these organelles and internal membranes. However, we now know that the plasma membrane is also part of this integrated membrane system (**Figure 4.15**). Some of these membranes, such as the nuclear envelope and the membrane of the endoplasmic reticulum, have direct connections to one another. Other organelles of the endomembrane system pass materials to each other via **vesicles**—small membrane-enclosed spheres. The movement of vesicles occurs in both directions. For example, vesicles that are formed from the endoplasmic reticulum can fuse with the Golgi apparatus and vesicles from the Golgi apparatus can fuse with the endoplasmic reticulum. Therefore, the endomembrane system forms a dynamic, integrated network of membranes that requires constant sorting to maintain the functional properties of each organelle.

In this section, we will survey the structures and functions of the organelles of the endomembrane system. We will also examine the plasma membrane, which is formed from the endomembrane system.

The Eukaryotic Nucleus Contains Chromosomes

The nucleus is the internal compartment that is enclosed by a double-membrane structure termed the **nuclear envelope** (**Figure 4.16**). In most cells, the nucleus is a relatively large organelle

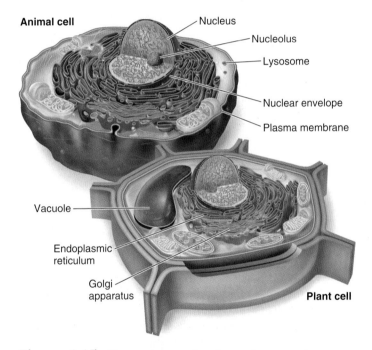

Figure 4.15   **The nucleus and endomembrane system.** This figure highlights the internal compartment of the nucleus (blue), the membranes of the endomembrane system (purple), and the fluid-filled interiors of the endomembrane system (pink). The nuclear envelope is considered part of the endomembrane system, but the interior of the nucleus is not.

that typically occupies 10–20% of the total cell volume. The outer membrane of the nuclear envelope is continuous with the endoplasmic reticulum membrane. **Nuclear pores** are formed where the inner and outer nuclear membranes make contact with each other. The pores provide a passageway for the movement of molecules and macromolecules into and out of the nucleus. Although cell biologists view the nuclear envelope as part of the endomembrane system, the materials within the nucleus are not (see Figure 4.15).

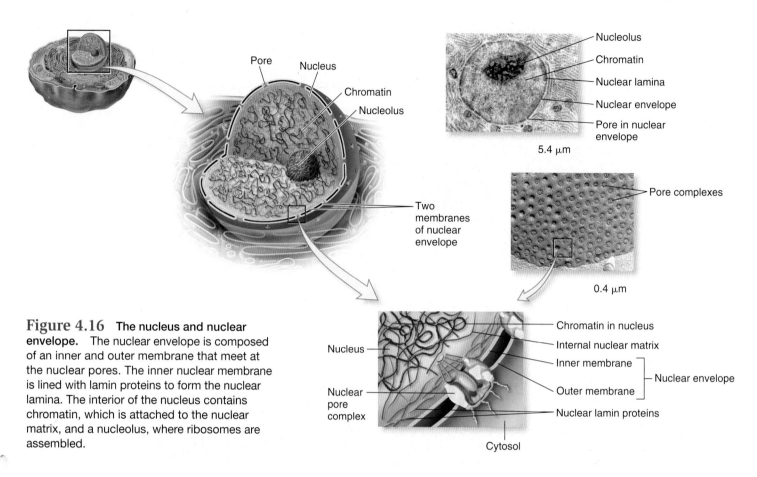

Pore

Nucleus

Chromatin

Nucleolus

Two membranes of nuclear envelope

Nucleolus

Chromatin

Nuclear lamina

Nuclear envelope

Pore in nuclear envelope

5.4 μm

Pore complexes

0.4 μm

Nucleus

Nuclear pore complex

Chromatin in nucleus

Internal nuclear matrix

Inner membrane

Outer membrane

Nuclear envelope

Nuclear lamin proteins

Cytosol

Figure 4.16 **The nucleus and nuclear envelope.** The nuclear envelope is composed of an inner and outer membrane that meet at the nuclear pores. The inner nuclear membrane is lined with lamin proteins to form the nuclear lamina. The interior of the nucleus contains chromatin, which is attached to the nuclear matrix, and a nucleolus, where ribosomes are assembled.

Inside the nucleus are the **chromosomes** and a filamentous network of proteins called the **nuclear matrix**. Each chromosome is composed of genetic material, namely DNA, and many types of proteins that help to compact the chromosome to fit inside the nucleus. The complex formed by DNA and such proteins is termed **chromatin**. The nuclear matrix consists of two parts: the nuclear lamina, which is composed of intermediate filaments that line the inner nuclear membrane, and an internal nuclear matrix, which is connected to the lamina and fills the interior of the nucleus. The nuclear matrix serves to organize the chromosomes within the nucleus. Each chromosome is located in a distinct, nonoverlapping **chromosome territory,** which is visible when cells are exposed to dyes that label each type of chromosome (**Figure 4.17**).

The primary function of the nucleus involves the protection, organization, and expression of the genetic material. Another important function is the assembly of ribosomes. Ribosome assembly occurs in the **nucleolus** (plural, *nucleoli*), a prominent region in the nucleus of nondividing cells. A ribosome is composed of two subunits, one small and one large; each subunit contains one or more RNA molecules and several types of proteins. The RNA molecules that are components of ribosomes are made at the nucleolus because the genes that encode these RNA molecules are located there. The ribosomal

proteins are produced in the cytosol, imported into the nucleus through the nuclear pores, and assembled with the RNA molecules to form the ribosomal subunits. The subunits then exit through the nuclear pores into the cytosol, where they are needed for protein synthesis.

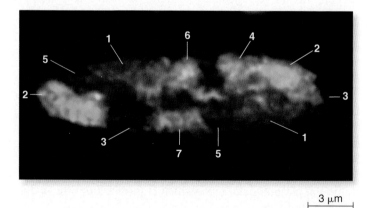

3 μm

Figure 4.17 **Chromosome territories in the cell nucleus.** Chromosomes from a chicken were labeled with chromosome-specific probes. Each of the seven types of chicken chromosomes is colored with a different dye. Each chromosome occupies its own distinct, nonoverlapping territory within the cell nucleus.

The Endoplasmic Reticulum Initiates Protein Sorting and Carries Out Certain Metabolic Functions

The **endoplasmic reticulum (ER)** is a convoluted network of membranes that form flattened, fluid-filled tubules or **cisternae** (**Figure 4.18**). The terms endoplasmic (which means "in the cytoplasm") and reticulum ("little net") refer to the location and shape of this organelle when viewed under a microscope. The term **lumen** describes the internal space of an organelle. The ER membrane encloses a single compartment called the **ER lumen**. In some cells, the ER membrane makes up more than half of the total membrane in the cell. The rough ER has its outer surface studded with ribosomes, giving it a bumpy appearance. Once bound to the ER membrane, the ribosomes actively synthesize proteins through the ER membrane. The smooth ER lacks ribosomes.

Rough ER The **rough endoplasmic reticulum (rough ER)** plays a key role in the initial synthesis and sorting of proteins that are destined for the ER, Golgi apparatus, lysosomes, vacuoles, plasma membrane, or outside of the cell. To reach any of these locations, a protein must first be directed to the ER membrane. In conjunction with protein sorting, a second function of the rough ER is the insertion of certain newly made proteins into the membrane. A third important function of the rough ER is the attachment of carbohydrate to proteins and lipids. This process is called **glycosylation**. The topics of protein sorting, membrane protein insertion, and protein glycosylation will be discussed in Chapter 6, because they are important in the maintenance of cell organization.

Smooth ER The **smooth endoplasmic reticulum (smooth ER)**, which is continuous with the rough ER, functions in diverse metabolic processes. The extensive network of smooth ER membranes allows increased surface area for key enzymes that play important metabolic roles. In liver cells, enzymes in the smooth ER detoxify many potentially harmful organic molecules including barbiturate drugs and ethanol. These enzymes convert hydrophobic toxic molecules into more hydrophilic molecules, which are easily excreted from the body. Chronic alcohol consumption, as in alcoholics, leads to a greater amount of smooth ER in liver cells, which increases the rate of alcohol breakdown. This explains why people who consume alcohol regularly must ingest more alcohol to experience its effects.

The smooth ER of liver cells also plays a role in carbohydrate metabolism. The liver cells of animals store energy in the form of glycogen, which is a polymer of glucose. Glycogen granules, which are in the cytosol, sit very close to the smooth ER membrane. When chemical energy is needed, enzymes are activated that break down the glycogen to glucose-6-phosphate. Then, an enzyme in the smooth ER called glucose-6-phosphatase removes the phosphate group, and glucose is released into the bloodstream.

Another important function of the smooth ER in all eukaryotes is the accumulation of calcium ions. The smooth ER contains calcium pumps that transport Ca^{2+} into the ER lumen. The regulated release of Ca^{2+} into the cytosol is involved in many vital cellular processes, including muscle contraction in animals.

Finally, enzymes in the smooth ER are critical in the synthesis and modification of lipids. For example, steroid hormones such as estrogen and testosterone are derived from the lipid

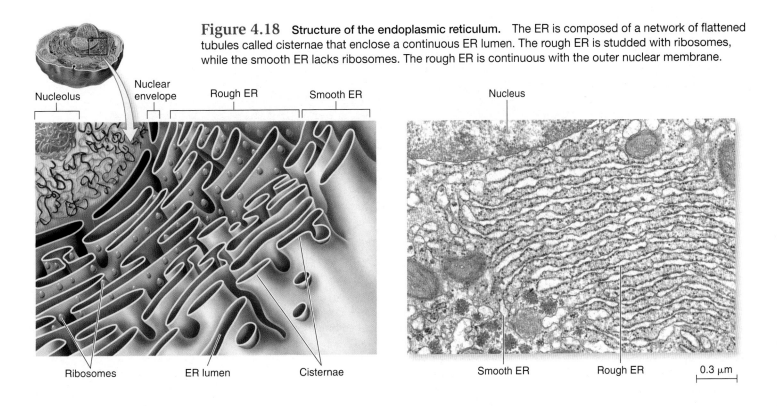

Figure 4.18 **Structure of the endoplasmic reticulum.** The ER is composed of a network of flattened tubules called cisternae that enclose a continuous ER lumen. The rough ER is studded with ribosomes, while the smooth ER lacks ribosomes. The rough ER is continuous with the outer nuclear membrane.

cholesterol. Enzymes in the smooth ER are necessary for certain modifications that are needed to produce these hormones. In addition, the smooth ER is the primary site for the synthesis of phospholipids, which are the main lipid component of eukaryotic cell membranes.

The Golgi Apparatus Directs the Secretion, Processing, and Sorting of Cellular Molecules

The **Golgi apparatus** (also called the Golgi body, Golgi complex, or simply Golgi) was discovered by the Italian microscopist Camillo Golgi in 1898. It consists of a stack of flattened, membranes; each flattened membrane encloses a single compartment (**Figure 4.19**). The Golgi stacks are named according to their orientation in the cell. The *cis* Golgi is close to the ER membrane, the *trans* Golgi is near the plasma membrane, and the *medial* Golgi is found in the middle. Materials are transported between the Golgi stacks via membrane vesicles that bud from one compartment in the Golgi (for example, the *cis* Golgi) and fuse with another compartment (for example, the *medial* Golgi).

The Golgi apparatus performs three overlapping functions: (1) secretion, (2) processing, and (3) protein sorting. The Golgi apparatus packages different types of materials into **secretory vesicles** that later fuse with the plasma membrane, thereby releasing their contents outside the plasma membrane (Figure 4.19). Proteins that are destined for secretion are synthesized into the ER, travel to the Golgi, and then to the plasma membrane for secretion. This route, called the secretory pathway, is described in Chapter 6.

Enzymes in the Golgi apparatus process, or modify, certain proteins and lipids. As mentioned earlier, carbohydrates can be attached to proteins and lipids in the endoplasmic reticulum. Glycosylation continues in the Golgi. For this to occur, a protein or lipid is transported via vesicles from the ER to the *cis* Golgi. Most of the glycosylation occurs in the *medial* Golgi. A second type of processing event is **proteolysis**—enzymes called **proteases** cut proteins into smaller polypeptides. For example, the hormone insulin is first made as a large precursor protein termed proinsulin. In the Golgi apparatus, proinsulin is packaged with proteases into vesicles. Prior to secretion, the proteases cut out a portion of the proinsulin to create a smaller insulin molecule that is a functional hormone.

The third function of the Golgi is protein sorting. After a protein enters the Golgi from the ER, it will be directed to one of six locations. Either it will stay in the Golgi, or it will be transported via vesicles to the ER, a lysosome (in animal cells), a vacuole (in plant cells), the plasma membrane, or the exterior of the cell.

Lysosomes Are Involved in Degrading Macromolecules

Lysosomes are small organelles found in animal cells that are able to lyse macromolecules, hence their name. Lysosomes contain many **acid hydrolases**, which are hydrolytic enzymes that use a molecule of water to break a covalent bond. This type of chemical reaction is called hydrolysis. The hydrolases found in a lysosome function optimally at an acidic pH. The fluid-filled

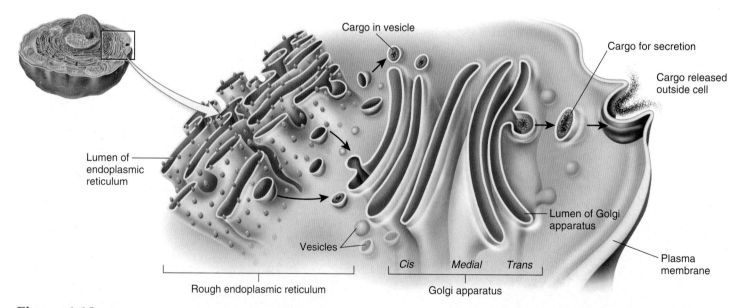

Cargo in vesicle

Cargo for secretion

Cargo released outside cell

Lumen of endoplasmic reticulum

Lumen of Golgi apparatus

Vesicles

Cis   *Medial*   *Trans*

Plasma membrane

Rough endoplasmic reticulum

Golgi apparatus

Figure 4.19   The Golgi apparatus and secretory pathway.   The Golgi is composed of stacks of membranes that enclose separate compartments. Transport to and from the Golgi compartments occurs via membrane vesicles. Vesicles can bud from the ER and go to the Golgi, and vesicles from the Golgi can fuse with the plasma membrane to release cargo to the outside. The pathway from the ER to the Golgi to the plasma membrane is termed the secretory pathway.

Biological inquiry: If we consider the Golgi apparatus as three compartments (cis, medial, and trans), describe the compartments that a protein will travel through to be secreted.

interior of a lysosome has a pH of approximately 4.8. If a lysosomal membrane breaks, releasing acid hydrolases into the cytosol, the enzymes are not very active because the cytosolic pH is neutral (approximately pH 7.0). This prevents significant damage to the cell from accidental leakage.

Lysosomes contain many different types of acid hydrolases that can break down proteins, carbohydrates, nucleic acids, and lipids. This enzymatic function enables lysosomes to break down complex materials. One function of lysosomes involves the digestion of substances that are taken up from outside the cell. This process, called endocytosis, is described in Chapter 5. In addition, lysosomes help digest intracellular materials. In a process known as **autophagy** (meaning the eating of one's self), cellular material, such as a worn-out organelle, becomes enclosed in a double membrane (**Figure 4.20**). This **autophagosome** then fuses with a lysosome, and the material inside the autophagosome is digested. The small molecules that are released from this digestion are recycled back into the cytosol.

Vacuoles Are Specialized Compartments That Function in Storage, the Regulation of Cell Volume, and Degradation

The term **vacuole** (literally, "empty space") came from early microscopic observations of these compartments. We now know that vacuoles are not empty but instead contain fluid and sometimes even solid substances. Most vacuoles are made from the fusion of many smaller membrane vesicles. Vacuoles are prominent organelles in plant cells, fungal cells, and certain protists. In animal cells, vacuoles tend to be smaller and are more commonly used to temporarily store materials or transport substances. In animals, such vacuoles are sometimes called storage vesicles.

The functions of vacuoles are extremely varied, and they differ among cell types and even environmental conditions. The best way to appreciate vacuole function is to consider a few examples. Mature plant cells often have a large **central vacuole** that occupies 80% or more of the cell volume (**Figure 4.21a**).

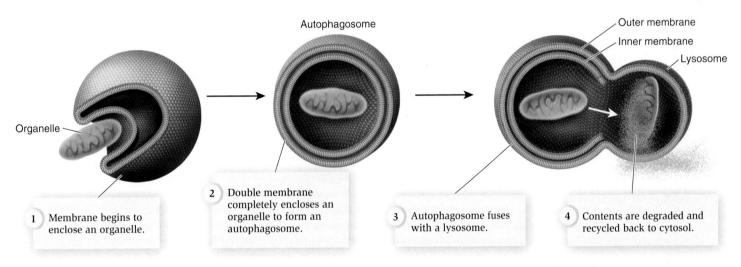

Autophagosome

Organelle

Outer membrane

Inner membrane

Lysosome

1 Membrane begins to enclose an organelle.

2 Double membrane completely encloses an organelle to form an autophagosome.

3 Autophagosome fuses with a lysosome.

4 Contents are degraded and recycled back to cytosol.

Figure 4.20 **Autophagy.** In the example shown here, a double membrane surrounds a mitochondrion to form an autophagosome, which fuses with a lysosome. The contents of the autophagosome are degraded and released back into the cytosol.

Biological inquiry: Why do you think autophagy is useful to a cell?

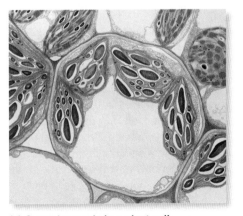

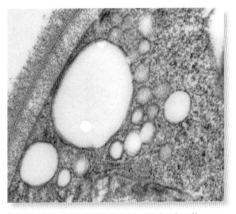

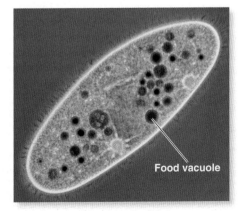

(a) Central vacuole in a plant cell

(b) Contractile vacuoles in an algal cell

(c) Food vacuoles in a paramecium

Food vacuole

Figure 4.21 **Examples of vacuoles.**

The membrane of this vacuole is called the **tonoplast**. The central vacuole serves two important purposes. First, it stores a large amount of water, enzymes, and inorganic ions such as calcium; it also stores other materials including proteins and pigments. Second, it performs a space-filling function. The large size of the vacuole exerts a pressure on the cell wall called turgor pressure. If a plant becomes dehydrated and this pressure is lost, a plant will wilt. Turgor pressure is important in maintaining the structure of plant cells and the plant itself, and it helps to drive the expansion of the cell wall, which is necessary for growth.

Certain species of protists also use vacuoles to maintain cell volume. Freshwater organisms such as the alga *Chlamydomonas reinhardtii* have small, water-filled **contractile vacuoles** that expand as water enters the cell (**Figure 4.21b**). Once they reach a certain size, the vacuoles suddenly contract, expelling their contents to the exterior of the cell. This mechanism is necessary to remove the excess water that continually enters the cell by diffusing across the plasma membrane.

Another function of vacuoles is degradation. Some protists engulf their food into large **phagocytic vacuoles** or **food vacuoles** (**Figure 4.21c**). As in the lysosomes of animal cells, food vacuoles contain digestive enzymes to break down the macromolecules within the food. Macrophages, a type of cell found in animals' immune systems, engulf bacterial cells into phagocytic vacuoles, where the bacteria are destroyed.

The Plasma Membrane Is the Interface Between a Cell and Its Environment

The cytoplasm of all cells is surrounded by a plasma membrane, which is part of the endomembrane system and provides a boundary between a cell and the extracellular environment. Proteins in the plasma membrane play many important roles that affect the activities inside the cell. First, many plasma membrane proteins are involved in **membrane transport** (**Figure 4.22**). Some of these proteins function to transport essential nutrients or ions into the cell, while others are involved in the export of substances. Due to the functioning of these transporters, the plasma membrane is selectively permeable; it allows only certain substances in and out. We will examine the structures and functions of a variety of transporters in Chapter 5.

A second vital function of the plasma membrane is **cell signaling**. To survive and adapt to changing conditions, cells must be able to sense changes in their environment. In addition, the cells of a multicellular organism need to communicate with each other to coordinate their activities. The plasma membrane of all cells contains receptors that recognize signaling molecules—either environmental agents or molecules secreted by other cells. Once signaling molecules bind to a receptor, this elicits a series of steps known as a signal cascade that causes the cell to respond to the signal (Figure 4.22). For example, when you eat a meal, the hormone insulin is secreted into your bloodstream. This hormone binds to receptors in the plasma membrane of your cells, which results in a cellular response that allows your cells to take up the glucose from the food into the cytosol.

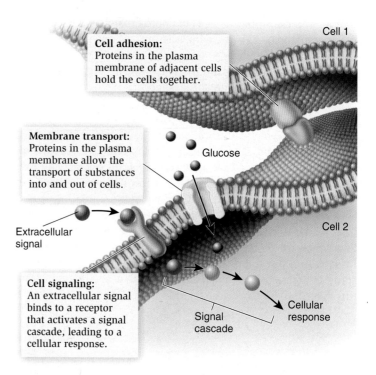

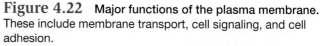

Cell adhesion: Proteins in the plasma membrane of adjacent cells hold the cells together.

Membrane transport: Proteins in the plasma membrane allow the transport of substances into and out of cells.

Cell signaling: An extracellular signal binds to a receptor that activates a signal cascade, leading to a cellular response.

Cell 1

Glucose

Extracellular signal

Cell 2

Signal cascade

Cellular response

Figure 4.22 Major functions of the plasma membrane. These include membrane transport, cell signaling, and cell adhesion.

We will explore the details of receptors and signal cascades in Chapter 9.

A third important role of the plasma membrane in animal cells is **cell adhesion**. Proteins in the plasma membranes of adjacent cells bind to each other and promote cell-to-cell adhesion (Figure 4.22). This phenomenon is critical for animal cells to properly interact to form a multicellular organism and allows cells to recognize each other. The structures and functions of proteins that are involved in cell adhesion will be examined in Chapter 10.

4.5 Semiautonomous Organelles

In the rest of this chapter, we will examine those organelles in eukaryotic cells that are considered semiautonomous: mitochondria, chloroplasts, and peroxisomes. These organelles can grow and divide to reproduce themselves, but they are not completely autonomous because they depend on other parts of the cell for their internal components (**Figure 4.23**). For example, most of the proteins that are found in mitochondria are imported from the cytosol. In this section, we will survey the structures and functions of the semiautonomous organelles in eukaryotic cells. In Chapter 6 we will consider the evolutionary origins of these organelles, and in Chapters 7 and 8 we will explore the functions of mitochondria and chloroplasts in greater depth.

Animal cell

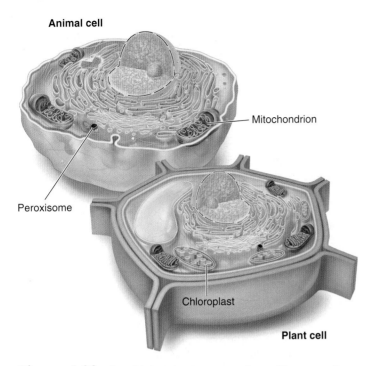

Mitochondrion

Peroxisome

Chloroplast

Plant cell

Figure 4.23 Semiautonomous organelles. These are the mitochondria, chloroplasts, and peroxisomes.

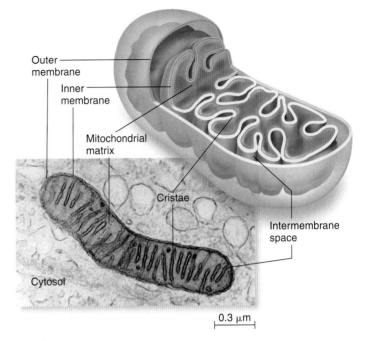

Outer membrane

Inner membrane

Mitochondrial matrix

Cristae

Intermembrane space

Cytosol

0.3 μm

Figure 4.24 Structure of a mitochondrion. This organelle is enclosed in two membranes. Numerous cristae in the inner membrane increase the membrane's surface area. The mitochondrial matrix lies inside the inner membrane. The micrograph is a colorized TEM.

Mitochondria Supply Cells with Most of Their ATP

Mitochondrion (plural, *mitochondria*) literally means "thread granule," which is what mitochondria look like under a light microscope. They are similar in size to bacteria. Depending on its function, a cell may contain a few hundred to a few thousand mitochondria—cells with particularly heavy energy demands, such as muscle cells, have more mitochondria than other cells. Indeed, research has shown that regular exercise increases the number and size of mitochondria in human muscle cells to meet the expanded demand for energy.

A mitochondrion has an outer membrane and an inner membrane separated by a region called the intermembrane space (**Figure 4.24**). The inner membrane is highly invaginated (folded) to form projections called **cristae**. These invaginations greatly increase the surface area of the inner membrane, which is the site where ATP is made. The compartment inside the inner membrane is the **mitochondrial matrix**.

The primary role of mitochondria is to make ATP. Even though mitochondria produce most of a cell's ATP, it is incorrect to think that mitochondria create energy. Rather, their primary function is to convert chemical energy that is stored within the covalent bonds in organic molecules into a form that can be readily used by cells. Covalent bonds in sugars, fats, and amino acids store a large amount of energy. The breakdown of these molecules into simpler molecules releases energy that is used to make ATP. Many proteins in living cells utilize ATP to carry out their functions, such as muscle contraction, uptake of nutrients, cell division, and many other cellular processes.

Mitochondria perform other functions as well. They are involved in the synthesis, modification, and breakdown of several types of cellular molecules. For example, the synthesis of certain hormones requires enzymes that are found in mitochondria. Therefore, if mitochondria do not function properly, this affects not only ATP synthesis but the synthesis of other products as well. Another interesting role of mitochondria is to generate heat in specialized fat cells known as brown fat cells. Groups of brown fat cells serve as "heating pads" that help to revive hibernating animals and protect sensitive areas of young animals from the cold.

Chloroplasts Carry Out Photosynthesis

Chloroplasts are organelles that can capture light energy and use some of that energy to synthesize organic molecules such as glucose. This process, called **photosynthesis**, is described in Chapter 8. Chloroplasts are found in nearly all species of plants and algae. **Figure 4.25** shows the structure of a typical chloroplast. Like the mitochondrion, a chloroplast contains an outer and inner membrane. An intermembrane space lies between these two membranes. A third membrane, the **thylakoid membrane**, forms many flattened, fluid-filled tubules that enclose a single, convoluted compartment. These tubules tend to stack on top of each other to form a structure called a **granum** (plural, *grana*). The **stroma** is the compartment of the chloroplast that is inside the inner membrane but outside the thylakoid membrane.

The **thylakoid lumen** is enclosed by the thylakoid membrane. This organization facilitates the process of photosynthesis.

Chloroplasts are a specialized version of plant organelles that are more generally known as **plastids**. All plastids are derived from unspecialized **proplastids**. The various types of plastids are distinguished by their synthetic abilities and the types of pigments they contain. Chloroplasts, which carry out photosynthesis, contain the green pigment chlorophyll. The abundant number of chloroplasts in the leaves of plants gives them their green color (**Figure 4.26a**). Chromoplasts, a second type of plastid, function in synthesizing and storing yellow, orange, and red pigments. Chromoplasts give many fruits and flowers their colors (**Figure 4.26b**). In autumn, the chromoplasts also give many leaves their yellow, orange, and red colors. A third type of

plastid, leucoplasts, typically lacks pigment molecules. An amyloplast is a leucoplast that synthesizes and stores starch. Amyloplasts are common in underground structures such as roots and tubers (**Figure 4.26c**).

Peroxisomes Catalyze Detoxifying Reactions

Peroxisomes, which were discovered by Christian de Duve in 1965, are relatively small organelles found in all eukaryotic cells. Peroxisomes consist of a single membrane that encloses a fluid-filled lumen (**Figure 4.27**). A typical eukaryotic cell contains several hundred of them. Peroxisomes are viewed as semi-autonomous because peroxisomal proteins are imported into the peroxisome in a manner that is very similar to the targeting of proteins to the mitochondria and chloroplasts. Another similarity is that new peroxisomes are produced by the division of pre-existing peroxisomes. However, the origin of peroxisomes remains controversial. As discussed in Chapter 6, mitochondria and chloroplasts have DNA because they are evolutionarily derived from bacteria that took up residence in a primordial eukaryotic cell. Likewise, some researchers speculate that peroxisomes could also have originated in this manner and lost their DNA during evolution. Alternatively, other scientists postulate that peroxisomes are derived from an invagination of cellular membranes around cytosolic enzymes. Further research is needed to resolve this controversy.

The general function of peroxisomes is to catalyze certain chemical reactions, typically those that break down molecules by removing hydrogen or adding oxygen. In mammals, for example, large numbers of peroxisomes can be found in the cells of the liver, where toxic molecules accumulate and are broken down. A by-product of this type of chemical reaction is hydrogen peroxide, H_2O_2, which is also broken down in this organelle (hence the name peroxisome). Hydrogen peroxide has the potential to be highly toxic. In the presence of metals such as iron (Fe^{2+}) that are found naturally in living cells, one way that hydrogen peroxide can be broken down causes the formation of a hydroxide ion (OH^-) and a molecule called a hydroxide free radical ($OH\cdot$).

$$Fe^{2+} + H_2O_2 \rightarrow Fe^{3+} + OH^- + OH\cdot \text{ (hydroxide free radical)}$$

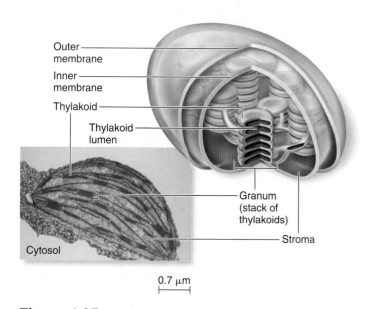

Outer membrane
Inner membrane
Thylakoid
Thylakoid lumen
Granum (stack of thylakoids)
Stroma
Cytosol

0.7 μm

Figure 4.25   **Structure of a chloroplast.**   Like a mitochondrion, a chloroplast is enclosed in a double membrane. In addition, it has an internal thylakoid membrane system that forms flattened compartments. These compartments stack on each other to form grana. The stroma is located inside the inner membrane but outside the thylakoid membrane. The micrograph is a colorized TEM.

(a) Leaves, which contain chloroplasts

(b) Fruit, which contain chromoplasts

(c) Roots, which contain amyloplasts

Figure 4.26   **Types of plastids.**   (a) Chloroplasts are involved in photosynthesis and give plants their green color. (b) Chromoplasts store yellow, orange, and red pigments, typically found in fruit and flowers. (c) Amyloplasts are colorless plastids that store starch in roots.

The hydroxide free radical is highly reactive and can damage proteins, lipids, and DNA. Therefore, it is beneficial for cells to break down hydrogen peroxide in an alternative manner that does not form a hydroxide free radical. Peroxisomes contain an enzyme called **catalase** that breaks down hydrogen peroxide to make water and oxygen gas.

$$2\ H_2O_2 \xrightarrow{\text{Catalase}} 2\ H_2O + O_2$$

Depending on the cell type, peroxisomes also contain enzymes that detoxify other molecules. These reactions may generate hydrogen peroxide, which is broken down via catalase.

Aside from detoxification, peroxisomes usually contain enzymes involved in the metabolism of fats and amino acids. For instance, plant seeds contain specialized organelles called **glyoxysomes** that are similar to peroxisomes. Glyoxysomes contain enzymes that are needed to convert fats to sugars. These enzymes become active when a seed germinates and the seedling begins to grow.

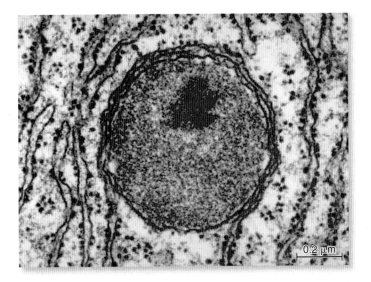

Figure 4.27 **Structure of a peroxisome.** This micrograph is a TEM.

CHAPTER SUMMARY

4.1 Microscopy

- Three important parameters in microscopy are magnification, resolution, and contrast. A light microscope utilizes light for illumination, while an electron microscope uses an electron beam.
- Transmission electron microscopy (TEM) provides the best resolution of any form of microscopy, and scanning electron microscopy (SEM) produces an image of a three-dimensional surface. (Figures 4.1, 4.2, 4.3)

4.2 Overview of Cell Structure

- Cell structure relies on four phenomena: matter, energy, organization, and information. Every living organism has a genome. The genes within the genome contain the information to create cells with particular structures and functions.
- We can classify all forms of life into two categories based on cell structure: prokaryotes and eukaryotes.
- The prokaryotes have a relatively simple structure and lack a membrane-enclosed nucleus. The two categories of prokaryotes are bacteria and archaea. Structures in prokaryotic cells include the plasma membrane, cytoplasm, nucleoid, and ribosomes. Many prokaryotes also have a cell wall and a glycocalyx. (Figures 4.4, 4.5)
- Eukaryotic cells are compartmentalized into organelles and contain a nucleus that houses most of their DNA. (Figures 4.6, 4.7, 4.8)
- The proteome of a cell determines its structure and function.

4.3 The Cytosol

- The cytosol is a central coordinating region for many metabolic activities of eukaryotic cells. A critical activity of all cells is translation, the synthesis of polypeptides. (Figure 4.9)
- The cytoskeleton is a network of three different types of protein filaments: microtubules, intermediate filaments, and actin filaments. Microtubules are important for cell organization, shape, and movement.

Intermediate filaments help maintain cell shape and rigidity. Actin filaments support the plasma membrane and play a key role in cell strength, shape, and movement. (Table 4.1, Figures 4.10, 4.11, 4.12, 4.13, 4.14)

4.4 The Nucleus and Endomembrane System

- The primary function of the nucleus involves the organization and expression of the cell's genetic material. A second important function is the assembly of ribosomes. (Figures 4.15, 4.16, 4.17)
- The endomembrane system includes the nuclear envelope, endoplasmic reticulum, Golgi apparatus, lysosomes, vacuoles, and plasma membrane. The rough endoplasmic reticulum (rough ER) plays a key role in the initial sorting of proteins. The smooth endoplasmic reticulum (smooth ER) functions in metabolic processes such as detoxification, carbohydrate metabolism, accumulation of calcium ions, and synthesis and modification of lipids. The Golgi apparatus performs three overlapping functions: secretion, processing, and protein sorting. Lysosomes degrade macromolecules and help digest substances taken up from outside the cell (endocytosis) and inside the cell (autophagy). (Figures 4.18, 4.19, 4.20)
- Types and functions of vacuoles include central vacuoles, contractile vacuoles, and phagocytic or food vacuoles. (Figure 4.21)
- Proteins in the plasma membrane perform many important roles that affect activities inside the cell, including membrane transport, cell signaling, and cell adhesion. (Figure 4.22)

4.5 Semiautonomous Organelles

- Mitochondria, chloroplasts, and peroxisomes are considered semiautonomous because they can grow and divide, but they still depend on other parts of the cell for their internal components. (Figure 4.23)
- Mitochondria produce most of a cell's ATP, which is utilized by many proteins to carry out their functions. Other mitochondrial functions include the synthesis, modification, and breakdown of cellular molecules and the generation of heat in specialized fat cells. (Figure 4.24)

- Chloroplasts, which are found in nearly all species of plants and algae, carry out photosynthesis. (Figure 4.25)
- Plastids, such as chloroplasts, chromoplasts, and amyloplasts differ in their function and the pigments they store. (Figure 4.26)
- Peroxisomes catalyze certain chemical reactions, typically those that break down molecules by removing hydrogen or adding oxygen. Peroxisomes usually contain enzymes involved in the metabolism of fats and amino acids. (Figure 4.27)

TEST YOURSELF

1. The cell doctrine states that
   a. any living organism is composed of one or more cells.
   b. new cells are derived from organic molecules from the environment.
   c. the smallest units of living organisms are atoms.
   d. the function of cells depends on the shape of the cell.
   e. all of the above.

2. When using microscopes, the resolution refers to
   a. the ratio between the size of the image produced by the microscope and the actual size of the object.
   b. the degree to which a particular structure looks different from other structures around it.
   c. how well a structure takes up certain dyes.
   d. the clarity of an image.
   e. the degree to which the image is magnified.

3. A bacterial cell may possess a _____, which may protect it from the immune system of other multicellular organisms that it may infect.
   a. cell wall
   b. flagellum
   c. pili
   d. nucleoid
   e. capsule

4. Different cells of the same multicellular individual have different proteomes due to all of the following except
   a. differences in the types of proteins made in different cell types.
   b. differences in the genomes of the different cell types.
   c. the abundance of certain proteins may not be the same in different cell types.
   d. the amino acid sequences of proteins may be different in different cell types.
   e. different cell types may alter the proteins in different ways.

5. The process of polypeptide synthesis is called
   a. metabolism.
   b. transcription.
   c. translation.
   d. hydrolysis.
   e. both c and d.

6. Each of the following is part of the endomembrane system except
   a. the nuclear envelope.
   b. the endoplasmic reticulum.
   c. the Golgi apparatus.
   d. lysosomes.
   e. peroxisomes.

7. Molecules move into and out of the nucleus by
   a. diffusing through the nuclear membrane.
   b. transport proteins.
   c. moving through nuclear pores.
   d. attaching to the nucleolus.
   e. all of the above.

8. Functions of the smooth endoplasmic reticulum include
   a. detoxification of harmful organic molecules.
   b. metabolism of carbohydrates.
   c. protein sorting.
   d. all of the above.
   e. a and b only.

9. The central vacuole in many plant cells is important for
   a. storage.
   b. photosynthesis.
   c. structural support.
   d. all of the above.
   e. a and c only.

10. Peroxisomes
    a. are vesicles similar to lysosomes that break down different classes of macromolecules.
    b. play an important role in the synthesis of ATP.
    c. are vesicles that contain enzymes necessary for manufacturing complex sugars.
    d. are the organelles primarily involved in protein synthesis.
    e. contain the enzyme catalase, which breaks down hydrogen peroxide to water and oxygen.

CONCEPTUAL QUESTIONS

1. Define organelle.
2. Explain how actin filaments are involved in movement.
3. Explain the function of the Golgi apparatus.

EXPERIMENTAL QUESTIONS

1. What hypothesis was tested in the experiment of Figure 4.14? What observations led to the proposal of this hypothesis?
2. What is the benefit of purifying cellular components and studying them *in vitro* instead of in intact cells? What was the benefit of using *Nitella axillaris* to determine the function of myosin? What was the purpose of using the fluorescent beads in the experiment? What was the purpose of NEM in the experiment?
3. Explain the results of the experiment of Figure 4.14.

COLLABORATIVE QUESTIONS

1. Discuss several differences between plant and animal cells.
2. Discuss the relationship between the nucleus, the rough endoplasmic reticulum, and the Golgi apparatus.

www.brookerbiology.com

This website includes answers to the Biological Inquiry questions found in the figure legends and all end-of-chapter questions.

5

MEMBRANE STRUCTURE AND TRANSPORT

CHAPTER OUTLINE

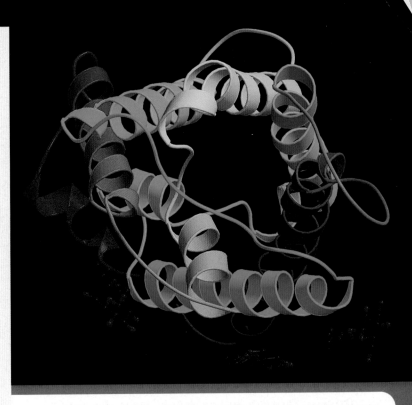

A model for the structure of aquaporin. This protein, found in the plasma membrane of many cell types, such as red blood cells and plant cells, allows the rapid movement of water molecules across the membrane.

Cellular membranes, also known as biological membranes or biomembranes, are an essential characteristic of all living cells. The **plasma membrane** is the biomembrane that separates the internal contents of a cell from its external environment. With such a role, you might imagine that the plasma membrane would be thick and rigid. Remarkably, the opposite is true. Cellular membranes are thin, typically 5–10 nm thick, and somewhat fluid. It would take 5,000 to 10,000 membranes stacked on top of each other to equal the thickness of the page you are reading. Despite their thinness, cellular membranes are impressively dynamic structures that effectively maintain the separation between a cell and its surroundings, and they provide an interface to carry out many vital cellular activities. Table 5.1 lists some of their key roles.

In this chapter, we will begin by considering the components that provide the structure of membranes. Then, we examine one of a membrane's primary functions, **membrane transport**.

Biomembranes regulate the traffic of substances into and out of the cell and its organelles. In the second part of this chapter, we will focus on the various ways to transport ions, small molecules, and large macromolecules across membranes. Chapters 6 through 10 will examine other functions of membranes.

5.1 Membrane Structure

As we progress through this textbook, a theme that will emerge is "structure determines function." This paradigm is particularly interesting when we consider how the structure of cellular membranes enables them to compartmentalize the cell while selectively importing and exporting vital substances. The two primary components are lipids, which form the basic matrix of a membrane, and proteins, which are embedded in the membrane or loosely attached to its surface. A third component is carbohydrate, which may be attached to membrane lipids and proteins. In this section, we are mainly concerned with the organization of these components to form a biological membrane, and how they are important in the overall function of membranes. We also consider several interesting experiments that provided insight into the dynamic properties of membranes.

Biological Membranes Are a Mosaic of Lipids, Proteins, and Carbohydrates

Figure 5.1 shows the biochemical organization of cellular membranes, which are similar in composition among all living organisms. The framework of the membrane is the **phospholipid bilayer**, which consists of two layers of lipids. The most abundant lipids found in membranes are the phospholipids. Recall

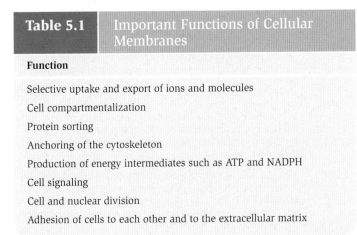

Table 5.1	Important Functions of Cellular Membranes
Function	
Selective uptake and export of ions and molecules	
Cell compartmentalization	
Protein sorting	
Anchoring of the cytoskeleton	
Production of energy intermediates such as ATP and NADPH	
Cell signaling	
Cell and nuclear division	
Adhesion of cells to each other and to the extracellular matrix	

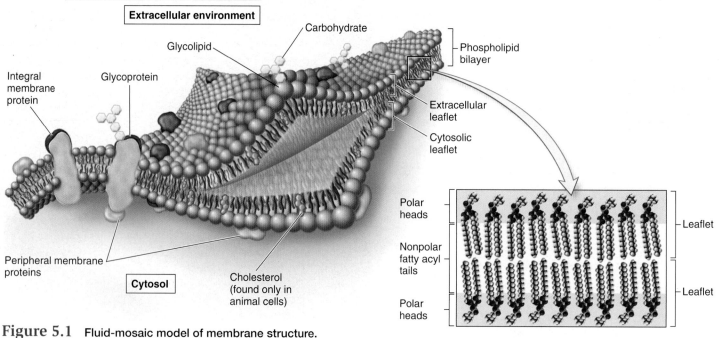

Figure 5.1   Fluid-mosaic model of membrane structure. The basic framework of a plasma membrane is a phospholipid bilayer. Proteins are inserted into the membrane and may be bound on the surface to other proteins or to lipids. Proteins and lipids, which have covalently bound carbohydrate, are called glycoproteins and glycolipids, respectively.

from Chapter 3 that phospholipids are **amphipathic** molecules, meaning they have a hydrophobic (water-fearing) region and a hydrophilic (water-loving) region. The hydrophobic tails of the lipids, referred to as fatty acyl tails, form the interior of the membrane and the hydrophilic head groups are on the surface. Experimentally, artificial membranes can be made from lipid alone.

Cellular membranes contain proteins, and most membranes also have carbohydrates attached to lipids and proteins. The relative amounts of lipids, proteins, and carbohydrates vary among different membranes. Some membranes, such as the inner mitochondrial membrane, have no carbohydrate while the plasma membrane of eukaryotic cells can have a large amount. A typical membrane found in cell organelles contains 50% protein by mass; the remainder is mostly lipids. However, the smaller lipid molecules outnumber the proteins by about 50 to 1 because the mass of one lipid molecule is much less than the mass of a protein.

Overall, the membrane is considered a mosaic of lipid, protein, and carbohydrate molecules. The membrane structure illustrated in Figure 5.1 is referred to as the **fluid-mosaic model**, originally proposed by Jonathan Singer and Garth Nicolson in 1972. As discussed later, the membrane exhibits properties that resemble a fluid because lipids and proteins can move relative to each other within the membrane. **Table 5.2** summarizes some of the historical experiments that led to the formulation of the fluid-mosaic model.

Table 5.2	Historical Developments That Led to the Formulation of the Fluid-Mosaic Model
Date	**Description**
1917	Irving Langmuir made artificial membranes experimentally by creating a monolayer of lipids on the surface of water. The polar heads interacted with water, and nonpolar tails projected into the air.
1925	Evert Gorter and F. Grendel proposed that lipids form bilayers around cells. This was based on careful measurements of lipid content enclosing red blood cells that showed there was just enough lipid to surround the cell with two layers.
1935	Because proteins were also found in membranes, Hugh Davson and James Danielli proposed incorrectly that a phospholipid bilayer was sandwiched between two layers of protein.
1950s	Electron microscopy revealed that membranes look like a train track—two dark lines separated by a light space. Initially, these results were misinterpreted. Researchers thought the two dark lines were layers of proteins and the light area was the phospholipids bilayer. Later, it was correctly determined that the dark lines in these experiments are the phospholipid heads, which were heavily stained, while the light region between them is their phospholipid tails.
1966	Using freeze fraction electron microscopy (described later in this chapter), Daniel Branton concluded that membranes are bilayers, because the freeze fracture procedure splits membranes in half, thus revealing proteins in the two membrane leaflets.
1972	Jonathan Singer and Garth Nicolson proposed the fluid-mosaic model described in Figure 5.1. Their model was consistent with the observation that membrane proteins are globular, and some are known to span the phospholipid bilayer and project from both sides.

Half of a phospholipid bilayer is termed a **leaflet**. Each leaflet faces a different region. For example, the plasma membrane contains a cytosolic leaflet and an extracellular leaflet (Figure 5.1). With regard to lipid composition, the two leaflets of cellular membranes are highly asymmetrical. The most striking asymmetry occurs with glycolipids—lipids with carbohydrate attached. These are found primarily in the extracellular leaflet so that the carbohydrate portion of a glycolipid protrudes into the extracellular medium.

Membrane Proteins Are Attached to or Embedded in the Phospholipid Bilayer

Although the phospholipid bilayer forms the basic foundation of cellular membranes, the protein component carries out most other functions (see Table 5.1). Proteins can bind to membranes in three different ways (**Figure 5.2**). **Transmembrane proteins** have one or more regions that are physically embedded in the hydrophobic region of the phospholipid bilayer. These regions, the **transmembrane segments**, are stretches of hydrophobic amino acids that span or traverse the membrane from one leaflet to the other. In most transmembrane proteins, the transmembrane segment is folded into an α-helix structure that is stabilized by hydrogen bonds. Such a segment is stable in a membrane because the nonpolar amino acids can interact favorably with the hydrophobic fatty acyl tails of the lipid molecules.

A second way for proteins to associate with the membrane is via **lipid anchors**. A lipid anchor involves the covalent attachment of a lipid to an amino acid side chain within a protein. The fatty acyl tails keep the protein firmly bound to the membrane. Both transmembrane proteins and lipid-anchored proteins are classified as **integral membrane proteins**, also called intrinsic membrane proteins, because they cannot be released from the membrane unless the membranes are dissolved with an organic solvent or detergent—in other words, you would have to disrupt the integrity of the membrane to remove them.

Peripheral membrane proteins, also called extrinsic proteins, are a third class of membrane proteins. These proteins do not interact with the hydrophobic interior of the phospholipid bilayer. Instead, they are noncovalently bound to regions of integral membrane proteins that project out from the membrane, or they are bound to the polar head groups of phospholipids. Peripheral membrane proteins are typically bound to the membrane by hydrogen and/or ionic bonds. For this reason, they usually can be removed from the membrane experimentally by exposing the membrane to high salt concentrations. Researchers can use this treatment to distinguish between peripheral and integral membrane proteins.

GENOMES & PROTEOMES

Approximately 25% of All Genes Encode Membrane Proteins

Membrane proteins participate in some of the most important and interesting cellular processes. These include transport, energy transduction, cell signaling, secretion, cell recognition, and cell-to-cell contact. Research studies have revealed that cells devote a sizeable fraction of their energy and metabolic machinery to the synthesis of membrane proteins. These proteins are particularly important in human medicine—approximately 70% of all medications exert their effects by binding to membrane proteins. Examples include the drugs aspirin, ibuprofen, and acetaminophen, which are widely used to relieve pain and inflammatory conditions such as arthritis. These drugs bind to cyclooxygenase, a protein in the ER membrane that is necessary for the synthesis of chemicals that play a role in inflammation and pain sensation.

Because membrane proteins are so important biologically and medically, researchers have analyzed the genomes of many species and asked the question, What percentage of genes encodes membrane proteins? They have developed tools to predict the likelihood that a gene encodes a protein that is a membrane protein. For example, the occurrence of transmembrane α helices can be predicted from the amino acid sequence of a protein. All 20 amino acids can be ranked according to their preference for a hydrophobic or hydrophilic environment. With these values, the amino acid sequence of a protein can be analyzed using computer software to determine the average hydrophobicity of short amino acid sequences within the protein. A stretch of 18 to 20 amino acids in an α helix is long enough to span the membrane.

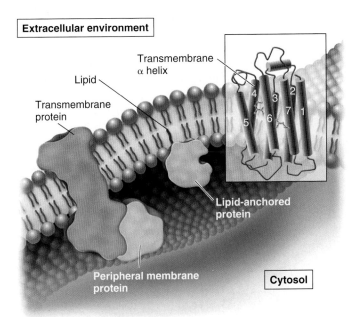

Figure 5.2   Three types of membrane proteins.   Integral membrane proteins may have transmembrane segments that traverse the phospholipid bilayer or they may contain lipid anchors. Peripheral membrane proteins are noncovalently bound to the hydrophilic regions of integral membrane proteins or to the polar head groups of lipids. Inset: The protein shown here contains seven transmembrane segments in an α-helix structure. The transmembrane α helices are depicted as cylinders. This particular protein, bacteriorhodopsin, functions as an ion pump in halophilic (salt-loving) archaea.

If such a stretch contains a high percentage of hydrophobic amino acids, it is predicted to be a transmembrane α helix. However, such computer predictions must eventually be verified by experimentation.

Using a computer approach, many research groups have attempted to calculate the percentage of genes that encode membrane proteins in various species. Table 5.3 shows the results of one such study. Note that the estimated percentage

Table 5.3	Estimated Percentage of Genes That Encode Membrane Proteins*
Organism	**Percentage of genes that encode membrane proteins**
Archaea	
Archaeoglobus fulgidus	24.2
Methanococcus jannaschii	20.4
Pyrococcus horikoshii	29.9
Bacteria	
Escherichia coli	29.9
Bacillus subtilis	29.2
Haemophilus influenzae	25.3
Eukaryotes	
Homo sapiens	29.7
Drosophila melanogaster	24.9
Arabidopsis thaliana	30.5
Saccharomyces cerevisiae	28.2

* Data from Stevens and Arkin (2000) *Proteins: Structure, Function, and Genetics* 39: 417–420. While the numbers may vary due to different computer programs and estimation techniques, the same general trends have been observed in other similar studies.

of membrane proteins is substantial: 20–30% of all genes may encode membrane proteins. This trend is found throughout all domains of life including archaea, bacteria, and eukaryotes. For example, in *E. coli*, approximately 30% of its genes encode membrane proteins. The genome of this bacterium contains about 4,300 genes, so roughly 1,290 genes may encode different membrane proteins. The human genome also has about 30% of genes that encode membrane proteins. However, the human genome is larger, containing 20,000 to 25,000 different genes, so the total number of genes that encode different membrane proteins is estimated at 6,000 to 7,500. The functions of many of them have yet to be determined. Identifying their functions will help researchers gain a better understanding of human biology. Likewise, medical researchers and pharmaceutical companies are interested in the identification of new membrane proteins that could be targets for effective new medications.

Membranes Are Semifluid

Let's now turn our attention to the dynamic properties of membranes. Although a membrane provides a critical interface between a cell and its environment, it is not a solid, rigid structure. Rather, biomembranes exhibit properties of **fluidity**, which means that individual molecules remain in close association yet have the ability to readily move within the membrane. Though membranes are often described as fluid, it is more appropriate to say they are **semifluid**. In a fluid substance, molecules can move in three dimensions. By comparison, most lipids can rotate freely around their long axes and move laterally within the membrane leaflet (**Figure 5.3**). This type of motion is considered two-dimensional, which means that it occurs within the

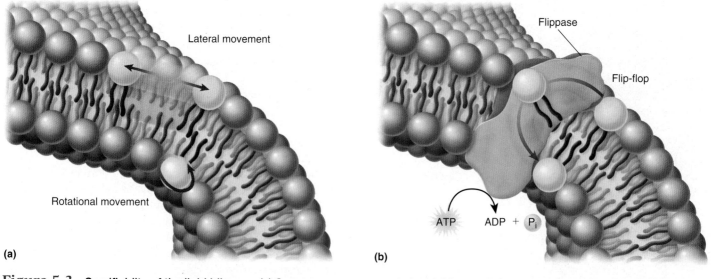

(a) Lateral movement / Rotational movement

(b) Flippase / Flip-flop / ATP → ADP + P$_i$

Figure 5.3   Semifluidity of the lipid bilayer.   (a) Spontaneous movements in the bilayer. Lipids can rotate (that is, move 360°) and move laterally (for example, from left to right in the plane of the bilayer). (b) Flip-flop does not happen spontaneously because the polar head group would pass through the hydrophobic region of the bilayer. Instead, the enzyme flippase uses ATP to flip phospholipids from one leaflet to the other.

plane of the membrane. Because rotational and lateral movements keep the fatty acyl tails within the hydrophobic interior, such movements are energetically favorable. In fact, the movements of lipids within cellular membranes are quite pronounced. At 37°C, a typical lipid molecule exchanges places with its neighbors about 10^7 times per second, and it can move several micrometers per second. At this rate, a lipid could traverse the length of a bacterial cell (approximately 1 μm) in only 1 second and the length of a typical animal cell in 10–20 seconds.

In contrast to rotational and lateral movements, the "flip-flop" of lipids from one leaflet to the opposite leaflet does not occur spontaneously (Figure 5.3). Energetically, such movements are extremely unfavorable because the polar head of a phospholipid would have to be transported through the hydrophobic interior of the membrane. For this reason, the transport of lipids from one leaflet to another requires the action of the enzyme flippase, which uses energy from the hydrolysis of ATP to flip a lipid from one leaflet to the other.

The biochemical properties of phospholipids have a profound effect on the fluidity of the phospholipid bilayer. One key factor is the length of their fatty acyl tails, which range from 14 to 24 carbon atoms with 18 to 20 carbons being the most common. Shorter acyl tails are less likely to interact, which makes the membrane more fluid. A second important factor is the presence of double bonds in the acyl tails. When a double bond is found, the lipid is said to be **unsaturated** with respect to the number of hydrogens that can be bound to the carbon atoms. A double bond creates a kink in the fatty acyl tail, making it more difficult for neighboring tails to interact and making the bilayer more fluid.

A third factor is the presence of cholesterol, which is a short and fairly rigid molecule that is produced by animal cells (refer back to Figure 3.13a). Cholesterol tends to stabilize membranes; its effects depend on temperature. At higher temperatures, such as those observed in mammals that maintain a constant body temperature, cholesterol makes the membrane less fluid. At lower temperatures, such as in icy water, cholesterol has the opposite effect. It makes the membrane more fluid and prevents it from freezing.

An optimal level of bilayer fluidity is essential for normal cell function, growth, and division. If a membrane is too fluid, which may occur at higher temperatures, it can become leaky. On the other hand, if a membrane becomes too solid, which may occur at lower temperatures, the functioning of membrane proteins will be inhibited. Cells adapt to changes in temperature by altering the lipid composition of their membranes. For example, when the water temperature drops, certain fish incorporate more cholesterol in their membranes. If a plant cell is exposed to high temperatures for many hours or days, it alters its lipid composition to have longer fatty acyl tails and fewer double bonds.

Like lipids, many integral membrane proteins may rotate and laterally move throughout the plane of a membrane. Because membrane proteins are larger than lipids, they move through the membrane at a much slower rate. Flip-flop of integral membrane proteins does not occur because the proteins also contain hydrophilic regions that project out from the phospholipid bilayer. It would be energetically unfavorable for the hydrophilic regions of membrane proteins to pass through the hydrophobic portion of the phospholipid bilayer.

Researchers can examine the lateral movements of lipids and integral membrane proteins by a variety of methods. In 1970, Larry Frye and Michael Edidin conducted an experiment that verified the lateral movement of membrane proteins (**Figure 5.4**). Mouse and human cells were mixed together and exposed to agents that caused them to fuse with each other. Some cells were cooled to 0°C while others were incubated at 37°C before being cooled. Both sets of cells were then exposed to fluorescently labeled antibodies that became specifically bound to a mouse membrane protein called H-2. The fluorescent label was observed with a fluorescence microscope. If the cells were maintained at 0°C, a temperature that greatly inhibits lateral movement, the fluorescence was seen on only one side of the fused cell. However, if the cells were incubated for several hours at 37°C

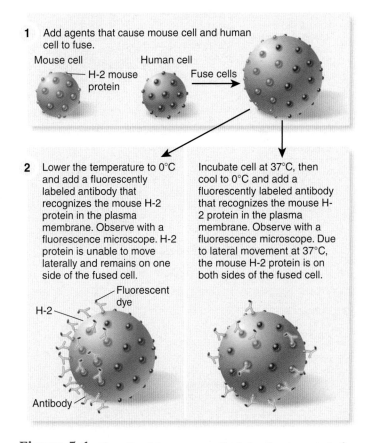

1 Add agents that cause mouse cell and human cell to fuse.

Mouse cell   Human cell

— H-2 mouse   Fuse cells
protein

2 Lower the temperature to 0°C and add a fluorescently labeled antibody that recognizes the mouse H-2 protein in the plasma membrane. Observe with a fluorescence microscope. H-2 protein is unable to move laterally and remains on one side of the fused cell.

Incubate cell at 37°C, then cool to 0°C and add a fluorescently labeled antibody that recognizes the mouse H-2 protein in the plasma membrane. Observe with a fluorescence microscope. Due to lateral movement at 37°C, the mouse H-2 protein is on both sides of the fused cell.

Fluorescent dye

H-2

Antibody

Figure 5.4 A method to measure the lateral movement of membrane proteins.

Biological inquiry: Explain why the H-2 proteins are found only on one side of the cell when the cells were incubated at 0°C.

and then cooled to 0°C, the fluorescence was distributed throughout the fused cell. This occurred because the higher temperature allowed the lateral movement of the H-2 protein throughout the plasma membrane.

A second approach to studying lateral movement, fluorescence recovery after photobleaching (FRAP), was developed in 1976 by Watt Webb and colleagues (**Figure 5.5**). In the experiment shown here, proteins on the surface of a cell were covalently labeled with a fluorescent molecule. In this example, the molecule emitted a red color so the entire surface of the cell appeared red. A laser beam was then focused on a small region of the cell surface. The energy of the laser beam altered the structure of the fluorescent molecules and eliminated the red color, a phenomenon called photobleaching. Immediately after photobleaching, a small region of the cell surface appeared white. Over time, bleached molecules within the white spot spread outward, and the white region filled in with red fluorescent molecules. These results indicate that proteins can laterally move in the membrane.

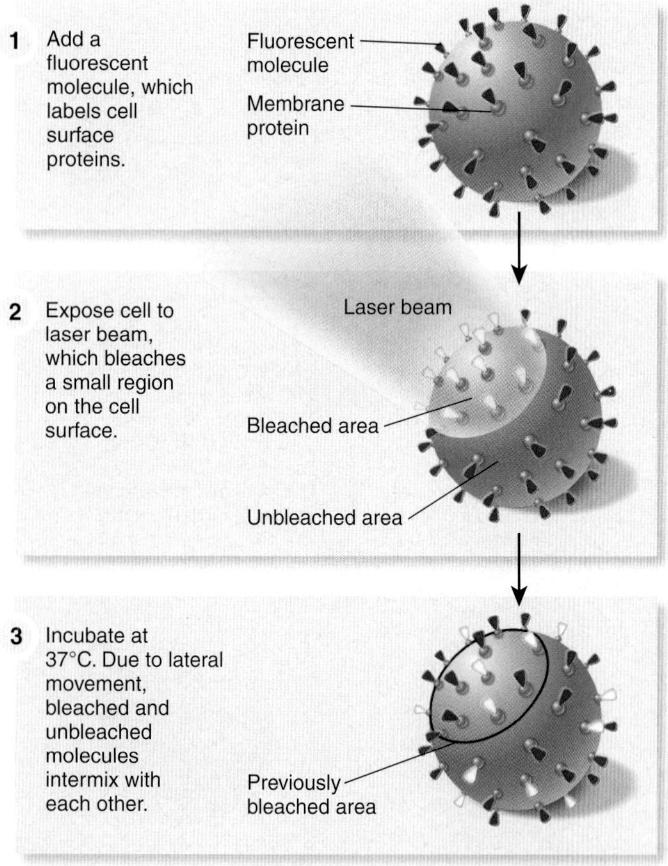

Unlike the examples shown in Figures 5.4 and 5.5, not all integral membrane proteins are capable of rotational and lateral movement. Depending on the cell type, 10–70% of membrane proteins may be restricted in their movement. Integral membrane proteins may be bound to components of the cytoskeleton, which restricts the proteins from moving laterally (**Figure 5.6**). Also, membrane proteins may be attached to molecules that are outside the cell, such as the interconnected network of proteins that forms the extracellular matrix of animal cells.

Glycosylation of Lipids and Proteins Serves a Variety of Cellular Functions

As mentioned earlier, the third constituent of biomembranes is carbohydrate. **Glycosylation** refers to the process of covalently attaching a carbohydrate to a protein or lipid. When a carbohydrate is attached to a lipid, this creates a **glycolipid**, while attachment to a protein produces a **glycoprotein**. We will discuss the mechanism of protein glycosylation in Chapter 6.

Though the roles of carbohydrate in cell structure and function are not entirely understood, some functional consequences of glycosylation are beginning to emerge. The carbohydrates that are attached to proteins and lipids have well-defined structures that serve in some cases as recognition signals for other cellular proteins. For example, proteins that are destined for the lysosome are glycosylated and have a sugar—mannose-6-phosphate—that is recognized by other proteins in the cell that target the glycosylated protein to the lysosome. Similarly, membrane glycolipids and glycoproteins often play a role in cell surface recognition.

1 Add a fluorescent molecule, which labels cell surface proteins.

Fluorescent molecule

Membrane protein

2 Expose cell to laser beam, which bleaches a small region on the cell surface.

Laser beam

Bleached area

Unbleached area

3 Incubate at 37°C. Due to lateral movement, bleached and unbleached molecules intermix with each other.

Previously bleached area

Figure 5.5 **Fluorescence recovery after photobleaching (FRAP).** In this method, the surface of cells was coated with fluorescent molecules. A laser beam bleached a small section shown in white. Over time, bleached molecules left the area and unbleached molecules invaded the bleached area by lateral movement.

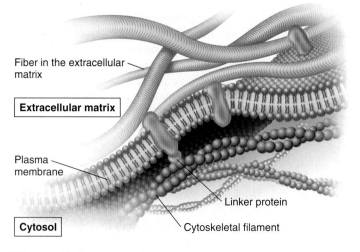

Fiber in the extracellular matrix

Extracellular matrix

Plasma membrane

Linker protein

Cytosol

Cytoskeletal filament

Figure 5.6 **Attachment of transmembrane proteins to the cytoskeleton and extracellular matrix.** Some membrane proteins have regions that project into the cytosol and are anchored to large cytoskeletal filaments (via linker proteins). The binding to these large filaments restricts lateral movement. Similarly, some integral membrane proteins may bind to large, immobile components in the extracellular matrix that also restrict the movement of the proteins.

During embryonic development in animals, significant cell movement occurs. Layers of cells slide over each other to create body structures such as the spinal cord and internal organs. The proper migration of individual cells and cell layers relies on the recognition of cell types via the carbohydrates on their cell surfaces.

In addition to its role as a recognition marker, carbohydrate can have a protective effect. The term **cell coat** or **glycocalyx** is used to describe the carbohydrate-rich zone on the cell surface that shields the cell from mechanical and physical damage (**Figure 5.7**). The carbohydrate portion of glycosylated proteins protects them from the harsh conditions of the extracellular environment and degradation by extracellular proteases, which are enzymes that digest proteins.

Membrane Structure Can Be Viewed with an Electron Microscope

Electron microscopy, discussed in Chapter 4, is a valuable tool to probe membrane structure and function. In transmission electron microscopy (TEM), a biological sample is thin sectioned and stained with heavy-metal dyes such as osmium tetroxide. This compound binds tightly to the polar head groups of phospholipids, but it does not bind well to the fatty acyl tails.

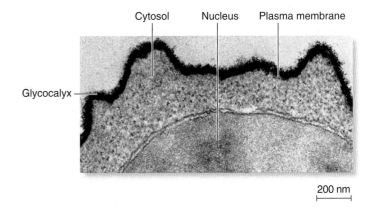

200 nm

Figure 5.7   A micrograph of the cell coat or glycocalyx. This figure shows a lymphocyte—a type of white blood cell—stained with ethidium red, which emphasizes the thick carbohydrate layer that surrounds the cell.

As shown in **Figure 5.8a**, membranes stained with osmium tetroxide resemble a railroad track. Two thin dark lines, which are the stained polar head groups, are separated by a uniform light space about 2 nm thick. This railroad track morphology is seen consistently when cell membranes are subjected to electron microscopy.

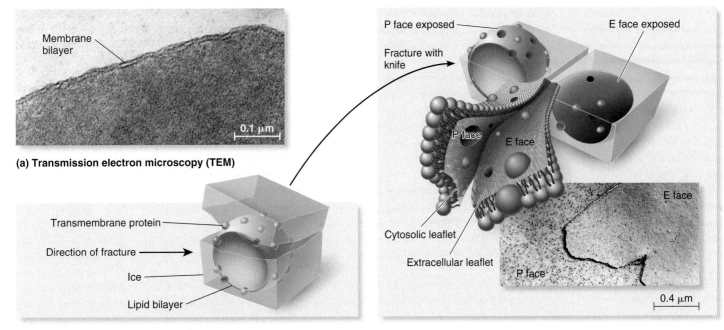

(a) Transmission electron microscopy (TEM)

(b) Freeze fracture electron microscopy (FFEM)

Figure 5.8   Electron micrographs of a cellular membrane.   **(a)** In the standard form of TEM, the membrane appears as two dark parallel lines. These lines are the lipid head groups, which stain darkly with osmium tetroxide. The fatty acyl tails do not stain well and appear as a light region sandwiched between the dark lines. **(b)** In the technique of freeze fracture electron microscopy, a sample is frozen in liquid nitrogen, split, and fractured. The sample is then coated with metal and viewed under the electron microscope. Note: The knife does not actually cut the two leaflets of the membrane apart. It simply separates the sample into two parts, and membranes that happen to be located at the fractured region will split along their bilayers.

Biological inquiry: If a heavy metal labeled the hydrophobic tails rather than the polar head groups (as osmium tetroxide does), do you think you would see a bilayer (that is, a railroad track) under TEM?

Due to the incredibly small size of biological membranes, scientists have not been able to invent instruments small enough to dissect them. However, a specialized form of TEM, freeze fracture electron microscopy (FFEM), can be used to analyze the interiors of phospholipid bilayers. Russell Steere invented this method in 1957. In FFEM, a sample is frozen in liquid nitrogen and split with a knife (**Figure 5.8b**). The knife does not actually cut through the bilayer, but it fractures the frozen sample. Due to the weakness of the central membrane region, the leaflets separate into a P face (the protoplasmic face that was next to the cytosol) and the E face (the extracellular face). Most transmembrane proteins do not break in half. They remain embedded within one of the leaflets, usually in the P face. The samples, which are under a vacuum, are then sprayed with a heavy metal such as platinum that coats the sample and reveals architectural features within each leaflet. When viewed with an electron microscope, membrane proteins are visible as bumps that provide significant three-dimensional detail about their form and shape.

5.2 Membrane Transport

If plasma membranes consisted of only a phospholipid bilayer, they would not permit the uptake of most nutrients and the export of waste products. However, the plasma membrane is a **selectively permeable** barrier between the cell and its external environment. As a protective envelope, its structure ensures that essential molecules such as glucose and amino acids enter the cell, metabolic intermediates remain in the cell, and waste products exit. The selective permeability of the plasma membrane allows the cell to maintain a favorable internal environment.

In this section, we begin with a discussion of the phospholipid bilayer, how it presents a barrier to the movement of ions and molecules across membranes, and the concept of gradients across membranes. We will then focus on **transport proteins**, which are embedded within the phospholipid bilayer. Transport proteins allow membranes to be selectively permeable by providing a passageway for the movement of some but not all substances across the membrane. Because different cells require various mixtures of ions and low-molecular-weight molecules, the plasma membrane of each cell type contains a specific set of transport proteins that allow only certain ions and molecules to cross. We will examine how transport proteins play a key role in the selective uptake and export of materials, and consider a mechanism found in eukaryotic cells for the transport of substances across membranes via membrane vesicles.

The Phospholipid Bilayer Is a Barrier to the Diffusion of Hydrophilic Substances

Because of their hydrophobic interiors, phospholipid bilayers present a formidable barrier to the movement of ions and hydrophilic molecules. **Diffusion** occurs when a solute (that is,

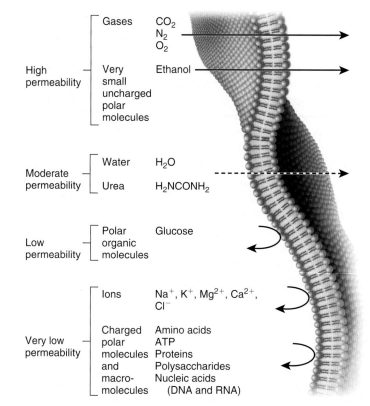

Figure 5.9 Relative permeability of an artificial phospholipid bilayer to a variety of solutes.   Solutes that easily penetrate are shown with a *straight arrow* that passes through the bilayer. The *dotted line* indicates solutes that have moderate permeability. The remaining solutes shown at the bottom are relatively impermeable. Note: Permeability is also related to solute concentration. At high concentrations, solutes are more likely to pass through a membrane.

a dissolved substance) moves from a region of high concentration to a region of lower concentration. When diffusion occurs through a membrane without the aid of a transport protein, it is called **passive diffusion**. The rate of passive diffusion depends on the chemistry of the solute and its concentration. **Figure 5.9** compares the relative permeabilities of various solutes through an artificial phospholipid bilayer that does not contain any proteins or carbohydrates. Gases and a few small, uncharged polar molecules can passively diffuse across the bilayer. However, the rate of diffusion of ions and larger polar molecules such as sugars and amino acids is relatively slow. Similarly, macromolecules, such as proteins and large carbohydrates, do not readily cross a lipid bilayer.

When we consider the steps of passive diffusion among different solutes, the greatest variation occurs in the ability of solutes to enter the hydrophobic interior of the bilayer. As an example, let's compare urea and diethylurea. Diethylurea is much more hydrophobic because it contains two nonpolar ethyl groups ($-CH_2CH_3$) (**Figure 5.10**). For this reason, it can more readily pass through the hydrophobic region of the phospholipid

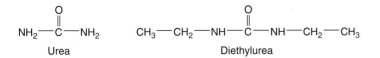

Figure 5.10   Structures of urea and diethylurea.

Biological inquiry: Which molecule would you expect to pass through a phospholipid bilayer more quickly, methanol (CH₃OH) or methane (CH₄)?

bilayer. The rate of passive diffusion of diethylurea through a phospholipid bilayer is about 50 times faster than urea.

Cells Maintain Gradients Across Their Membranes

As we have just seen, phospholipid bilayers are quite impermeable to ions and most hydrophilic molecules. A hallmark of living cells is their ability to maintain a relatively constant internal environment that is distinctively different from their external environment. This involves establishing gradients of solutes across the plasma membrane and organellar membranes. When we speak of a **transmembrane gradient**, we mean that the concentration of a solute is higher on one side of a membrane than the other. For example, immediately after you eat a meal, a higher concentration of glucose is found outside your cells compared to inside. Gradients involving ions have two components—electrical and chemical. An **ion electrochemical gradient** is a dual gradient that has both an electrical gradient and chemical gradient. For example, let's consider a gradient involving Na^+. An electrical gradient could exist in which the amount of net positive charge outside a cell is greater than inside. At the same time, a chemical gradient could exist in which the concentration of Na^+ outside is greater than inside. These two gradients together constitute a Na^+ electrochemical gradient. Transmembrane gradients of ions and other solutes are a universal feature of all living cells.

One way to view the transport of solutes across membranes is to consider how the transport process affects the pre-existing gradients across membranes. **Passive transport** refers to the diffusion of a solute across a membrane in a process that is energetically favorable and does not require an input of energy. Passive transport tends to dissipate a pre-existing gradient. (Note: The adjective *passive* can have different meanings in biology. It can mean without an input of energy, as in passive transport, or without the aid of a transport protein, as in passive diffusion.)

Passive transport can occur in two ways, via passive diffusion or facilitated diffusion. As mentioned earlier, passive diffusion is the diffusion of a solute directly through the phospholipid bilayer to move across the membrane (**Figure 5.11a**). This is not a common way for most solutes to move. The second pathway for passive transport is **facilitated diffusion**, which involves the aid of transport proteins (**Figure 5.11b**). Transport proteins facilitate the movement of various nutrients and—in some types of cells—water across the membrane.

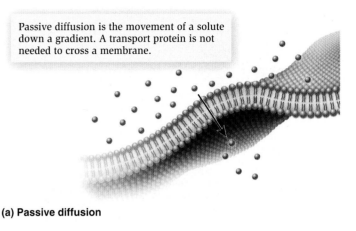

Passive diffusion is the movement of a solute down a gradient. A transport protein is not needed to cross a membrane.

(a) Passive diffusion

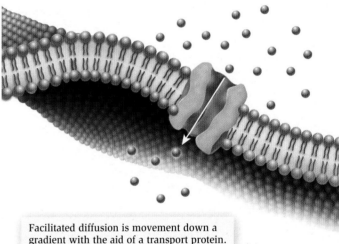

Facilitated diffusion is movement down a gradient with the aid of a transport protein.

(b) Facilitated diffusion

Figure 5.11   Types of passive transport. (a) Passive diffusion. (b) Facilitated diffusion.

Osmosis Is the Movement of Water Across Membranes to Balance Solute Concentrations

Lipid bilayers are relatively impermeable to many solutes, yet somewhat permeable to water. When the solute concentrations on both sides of the plasma membrane are equal, the two solutions are said to be **isotonic.** However, we have also seen that transmembrane gradients commonly exist across membranes. When the solute concentration inside the cell is higher, it is said to be **hypertonic** relative to the outside of the cell, which is **hypotonic**. If the solutes cannot move across the membrane, water will move and tend to balance the solute concentrations. In this process, called **osmosis**, water diffuses through a membrane from the hypotonic compartment into the hypertonic compartment.

Cells generally have a high internal concentration of a variety of solutes including ions, sugars, amino acids, etc. Animal cells,

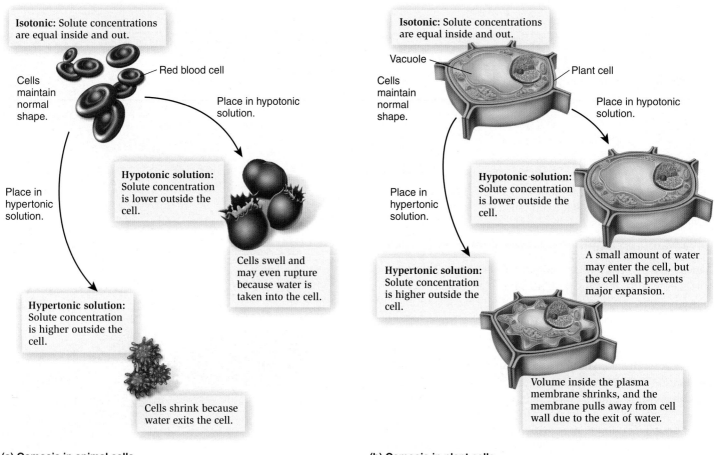

(a) Osmosis in animal cells

(b) Osmosis in plant cells

Figure 5.12   The phenomenon of osmosis.   (a) In animal cells that lack a cell wall, osmosis may promote cell shrinkage (crenation) or swelling. **(b)** In plant cells that have a rigid cell wall, a hypertonic medium causes the plasma membrane to pull away from the cell wall, while a hypotonic medium causes only a minor amount of expansion.

Biological inquiry: The inside of a cell has 100 mM KCl, while the outside has 10 mM KCl. If the membrane is impermeable to KCl, which direction will water move?

which are not surrounded by a rigid cell wall, must maintain a balance between the extracellular and intracellular solute concentrations. As discussed later, animal cells contain a variety of transport proteins that can sense changes in cell volume and allow the necessary movements of solutes across the membrane to prevent osmotic changes and maintain normal cell shape. However, if animal cells are placed in a hypertonic medium, water will exit the cells via osmosis and equalize solute concentrations on both sides of the membrane, causing them to shrink in a process called **crenation** (**Figure 5.12a**). Alternatively, if cells are placed in a hypotonic medium, water will diffuse into them to equalize solute concentrations on both sides of the membrane. In extreme cases, a cell may take up so much water that it bursts, a phenomenon called osmotic lysis.

Osmosis can also affect cells with a rigid cell wall, such as bacteria, fungi, algae, and plant cells. If the extracellular fluid surrounding a plant cell is hypertonic, water will exit the cell and the plasma membrane will pull away from the cell wall (**Figure 5.12b**). Alternatively, if the extracellular fluid is hypotonic, a plant cell will take up a small amount of water, but the cell wall prevents major changes in cell size. The tendency

of water to move into any cell creates an **osmotic pressure**, which is defined as the hydrostatic pressure required to stop the net flow of water across a membrane due to osmosis. In plant cells, osmotic pressure is also called **turgor pressure** or simply cell turgor. The turgor pressure pushes the plasma membrane against the rigid cell wall. An appropriate level of turgor is needed for cells to maintain their proper structure (**Figure 5.13**). If a plant has insufficient water, the extracellular fluid surrounding plant cells becomes hypertonic, the plasma membrane pulls away from the cell wall, and the osmotic pressure drops. The result is plasmolysis, which is associated with wilting.

Some freshwater organisms, such as amoebae and paramecia, can exist in extremely hypotonic environments where the external solute concentration is always much lower than the concentration of solutes in their cytosol. Because of the great tendency for water to move into the cell by osmosis, some organisms contain a contractile vacuole to prevent osmotic lysis. A contractile vacuole takes up water from the cytosol and periodically discharges it by fusing the vacuole with the plasma membrane (**Figure 5.14**).

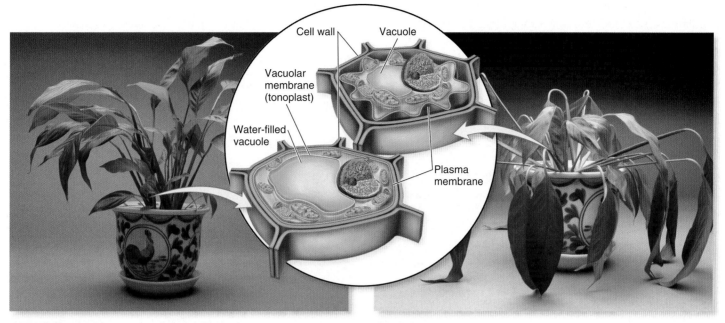

(a) Sufficient water   **(b) Wilting**

Figure 5.13   **Wilting in plants.**   **(a)** When a plant has plenty of water, the slightly hypotonic surroundings cause the vacuole to store water. The increased size of the vacuole influences the volume of the cytosol, thereby exerting a turgor pressure against the cell wall. **(b)** Under dry conditions, water is released from the cytosol into the extracellular medium. The vacuole also shrinks, because it loses water to the cytosol. Turgor pressure is lost. This causes the plant to wilt.

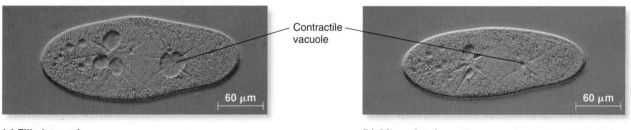

(a) Filled vacuole   **(b) After releasing water**

Figure 5.14   **The contractile vacuole in *Paramecium caudatum*.**   **(a)** A contractile vacuole is filled with water from radiating canals that collect fluid from the cytosol. **(b)** The contractile vacuole has recently fused with the plasma membrane (which would be above the plane of this page) and released the water from the cell.

FEATURE INVESTIGATION

Agre Discovered That Osmosis Occurs More Quickly in Cells with Transport Proteins That Allow the Facilitated Diffusion of Water

In living cells, the flow of water may occur by passive diffusion through the phospholipid bilayer. However, in the 1980s, researchers also discovered that certain cell types allow water to move across the plasma membrane at a much faster rate than would be predicted by passive diffusion. For example, water moves very quickly across the membrane of red blood cells, which causes them to shrink and swell in response to changes in extracellular

solute concentrations. Likewise, bladder and kidney cells, which play a key role in regulating water balance in the bodies of vertebrates, allow the rapid movement of water across their membranes. Based on these observations, researchers speculated that certain cell types might have proteins in their plasma membranes that permit the rapid movement of water.

One approach that is used to characterize a new protein is to first identify a protein based on its relative abundance in a particular cell type, and then attempt to determine the protein's function. This rationale was applied to the discovery of proteins that allow the rapid movement of water across membranes.

Peter Agre and his colleagues first identified a protein that was abundant in red blood cells and kidney cells, but not found in many other cell types. Though they initially did not know the function of the protein, its physical structure was similar to other proteins that were already known to function as transport proteins. They named this protein CHIP28, which stands for <u>ch</u>annel-forming <u>i</u>ntegral membrane <u>p</u>rotein with a molecular mass of <u>28</u>,000 Da. During the course of their studies, they also identified and isolated the gene that encodes CHIP28.

In 1992, Agre and his colleagues conducted experiments to determine if CHIP28 functions in the transport of water across membranes (**Figure 5.15**). Because they already had isolated the gene that encodes CHIP28, they could make many copies of this gene in a test tube (*in vitro*) using gene cloning techniques

Figure 5.15   The discovery of water channels.

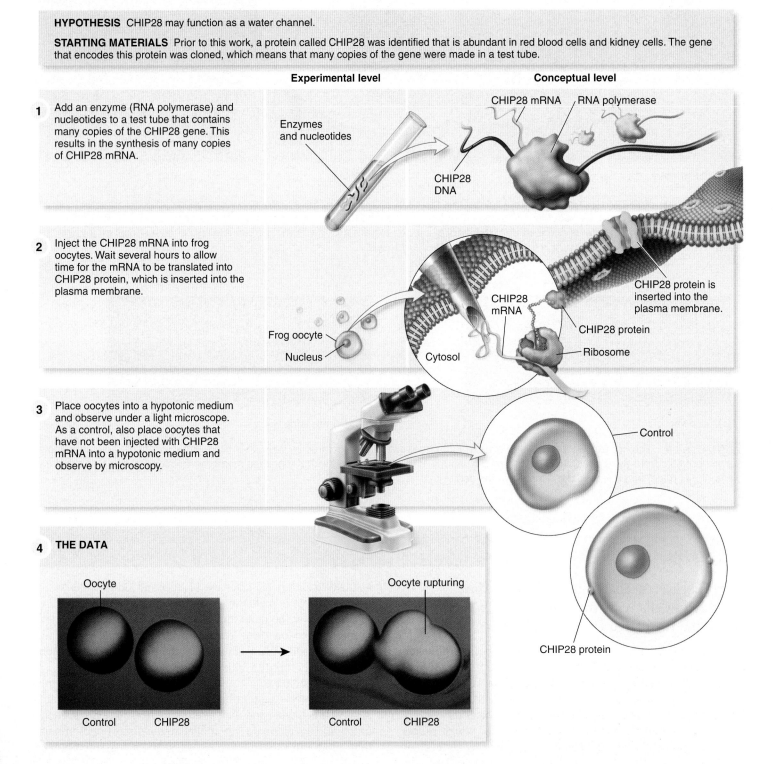

(see Chapter 20). Starting with many copies of the gene *in vitro*, they added an enzyme to transcribe the gene into mRNA that encodes the CHIP28 protein. This mRNA was then injected into frog oocytes, chosen because frog oocytes are large, easy to inject, and lack pre-existing proteins in their plasma membranes that allow the rapid movement of water. After injection, the mRNA was expected to be translated into CHIP28 proteins that should be inserted into the plasma membrane of the oocytes. After allowing sufficient time for this to occur, the oocytes were placed in a hypotonic medium. As a control, oocytes that had not been injected with CHIP28 mRNA were also exposed to a hypotonic medium.

As you can see in the data, a striking difference was observed between oocytes that expressed CHIP28 versus the control. Within minutes, oocytes that contained the CHIP28 protein were seen to swell due to the rapid uptake of water. Three to 5 minutes after being placed in a hypotonic medium, they actually burst! By comparison, the control oocytes did not swell as rapidly, and they did not rupture even after 1 hour. Taken together, these results are consistent with the hypothesis that CHIP28 functions as a transport protein that allows the facilitated diffusion of water across the membrane. Many subsequent studies confirmed this observation. Later, CHIP28 was renamed **aquaporin** to indicate its newly identified function of allowing water to diffuse through a pore in the membrane (**Figure 5.16**). More recently, the three-dimensional structure of aquaporin was determined (see inset to Figure 5.16). Agre was awarded the Nobel Prize in 2003 for this work.

Aquaporin is an example of a transport protein called a channel. Next, we will discuss the characteristics of channels and other types of transport proteins.

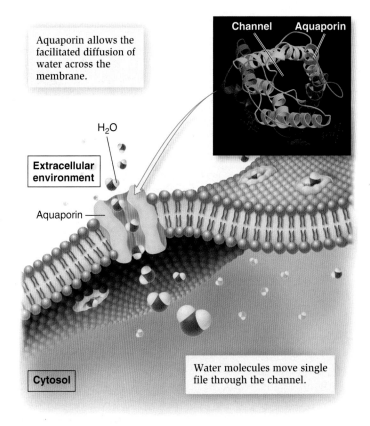

Aquaporin allows the facilitated diffusion of water across the membrane.

Channel   Aquaporin

H_2O

Extracellular environment

Aquaporin

Cytosol

Water molecules move single file through the channel.

Figure 5.16   Function and structure of aquaporin. Aquaporin is found in the membrane of certain cell types and allows the rapid diffusion of water across the membrane. The inset shows the structure of aquaporin, which was determined by X-ray crystallography.

Transport Proteins Cause Biological Membranes to Be Selectively Permeable

Because the phospholipid bilayer is a physical barrier to the passive diffusion of ions and most hydrophilic molecules, cells are able to separate their internal contents from their external environment. However, this barrier also poses a severe problem because cells must take up nutrients from the environment and export waste products. To overcome this dilemma, species have evolved a multitude of transport proteins—transmembrane proteins that provide a passageway for the movement of ions and hydrophilic molecules across membranes. Transport proteins enable biological membranes to be selectively permeable. We can categorize transport proteins into two classes, channels and transporters, based on the manner in which they move solutes across the membrane.

Channels **Channels** are transmembrane proteins that form an open passageway for the facilitated diffusion of ions or molecules across the membrane (**Figure 5.17**). In other words, solutes move directly through a channel to get to the other side. Aquaporin, discussed in the Feature Investigation, is a channel that allows the movement of water across the membrane. When a channel is open, the transmembrane movement of solutes can be extremely rapid, up to 100 million ions or molecules per second!

Most channels are **gated**, which means they can open to allow the diffusion of solutes and close to prohibit diffusion. The phenomenon of gating allows cells to regulate the movement of solutes. Researchers have discovered a variety of gating mechanisms (**Figure 15.18**). Gating sometimes involves the direct binding of a molecule to the channel protein itself. **Ligand-gated channels** are controlled by the noncovalent binding of small molecules—called ligands—such as hormones or neurotransmitters. These ligands are often important in the transmission of signals between nerve and muscle cells or between two nerve cells. Alternatively, intracellular proteins may bind noncovalently to channels and control their ability to open and close. For example, certain types of calcium channels that are found in nerve cells are controlled by regulatory proteins that bind to the channel and cause it to open. Another gating mechanism involves the covalent binding of a small molecule, such as a phosphate group, to the channel protein. Regulatory proteins involved in cell-signaling pathways may covalently attach phosphate to proteins as a way to regulate their function. In some cases, the targets of these regulatory proteins are channels.

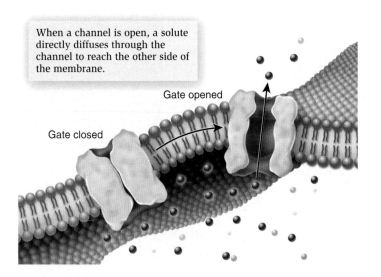

When a channel is open, a solute directly diffuses through the channel to reach the other side of the membrane.

Gate opened

Gate closed

Figure 5.17   Mechanism of transport by a channel protein.

As an example, chloride channels in the human lung are opened by phosphorylation.

Channel gating can also occur by mechanisms that do not involve the direct binding of a molecule to the channel protein. For example, some channels are **voltage-gated**, meaning that the channel opens and closes in response to changes in the amount of electric charge across the membrane. Sodium and potassium channels in nerve cells are voltage-gated. Another interesting gating mechanism involves **mechanosensitive channels**, which are sensitive to changes in membrane tension. For example, our ability to hear depends, in part, on the functioning of mechanosensitive channels in the cells of the inner ear that are sensitive to different frequencies of sound. The opening of these ion channels in response to sound transmits signals to the brain that a particular frequency has been detected.

Transporters   **Transporters**, also known as **carriers**, bind their solutes in a hydrophilic pocket and undergo a conformational change that switches the exposure of the pocket to the other side of the membrane (**Figure 5.19**). Transporters tend to be much slower than channels. Their rate of transport is typically 100 to 1,000 ions or molecules per second. Transporters provide the principal pathway for the uptake of organic molecules, such as sugars, amino acids, and nucleotides. In animals, they also allow cells to take up certain hormones and neurotransmitters. In addition, many transporters play a key role in export. Waste-products of cellular metabolism must be released from cells before they reach toxic levels. For example, a transporter removes lactic acid, a by-product of muscle cells during exercise. Excessive lactic acid buildup is partly responsible for the burning sensation you feel during a strenuous workout. Other transporters, which are involved with ion transport, play an important role in cellular processes such as regulating internal pH and controlling cell volume.

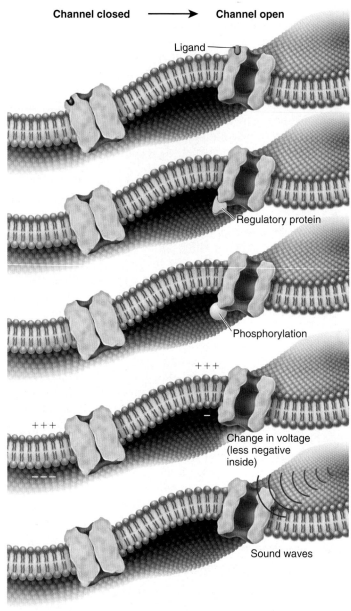

Channel closed ⟶ Channel open

Ligand

Regulatory protein

Phosphorylation

+ + +

− − −

+ + +

−

Change in voltage (less negative inside)

Sound waves

Figure 5.18   Different ways that channels are gated.

Transporters are subdivided according to the number of solutes they bind and the direction of transport (**Figure 5.20**). **Uniporters** bind a single molecule or ion and transport it across the membrane. **Symporters** or **cotransporters** bind two or more ions or molecules and transport them in the same direction. **Antiporters** bind two or more ions or molecules and transport them in opposite directions.

A **pump** is a transporter that directly couples its conformational changes to an energy source, such as ATP hydrolysis (**Figure 5.21**). The conformational changes of pumps are energetically driven. A common category of pumps found in all living cells are **ATP-driven pumps**, which have a binding site for ATP.

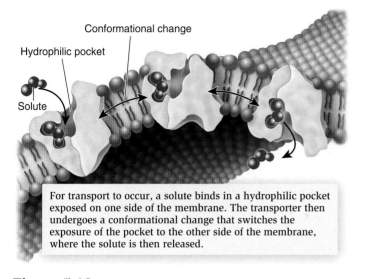

For transport to occur, a solute binds in a hydrophilic pocket exposed on one side of the membrane. The transporter then undergoes a conformational change that switches the exposure of the pocket to the other side of the membrane, where the solute is then released.

Figure 5.19 Mechanism of transport by a transporter, also called a carrier.

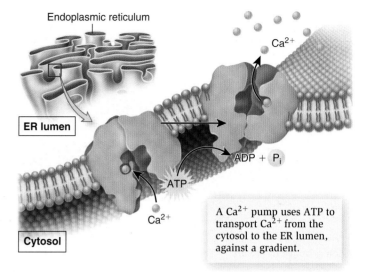

A Ca^{2+} pump uses ATP to transport Ca^{2+} from the cytosol to the ER lumen, against a gradient.

Figure 5.21 Transporters that function as pumps.

The hydrolysis of ATP provides energy that controls the sequence of conformational changes. The energy obtained from ATP hydrolysis can be used to pump solutes against a gradient. Pumps can be uniporters, symporters, or antiporters. As discussed next, pumps use energy to achieve active transport.

Active Transport Is the Movement of Substances Against a Gradient

Active transport is the movement of a solute across a membrane against its gradient—that is, from a region of low concentration to higher concentration. Active transport is energetically unfavorable and requires the input of energy. **Primary active transport** involves the functioning of pumps that directly use energy to transport a solute against a gradient. **Figure 5.22** shows a pump that uses ATP to transport H$^+$ against a gradient. Such a pump can establish a large H$^+$ electrochemical gradient across a membrane.

Secondary active transport involves the utilization of a pre-existing gradient to drive the active transport of a solute (Figure 5.22). For example, an H$^+$/sucrose symporter can utilize an H$^+$ electrochemical gradient, established by an ion pump, to move sucrose against its concentration gradient. In this regard, only sucrose is actively transported. Hydrogen ions move down (with) their electrochemical gradient. H$^+$/solute symporters are more common in bacteria, fungi, algae, and plant cells, because H$^+$ pumps are found in their plasma membranes. In animal cells, a pump that exports Na$^+$ maintains the Na$^+$ gradient across the plasma membrane. Na$^+$/solute symporters are prevalent in animal cells.

Symport enables cells to actively import nutrients against a gradient. Symport proteins use the energy stored in the electrochemical gradient of Na$^+$ or H$^+$ to power the uphill movement of organic solutes such as sugars, amino acids, and other needed

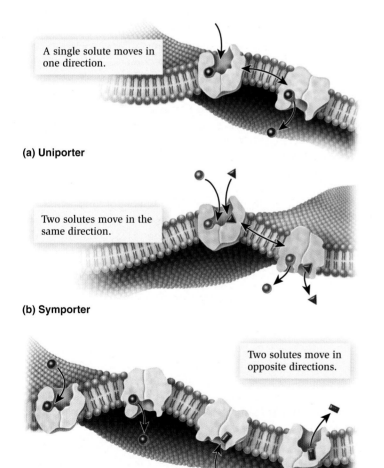

(a) Uniporter

A single solute moves in one direction.

(b) Symporter

Two solutes move in the same direction.

(c) Antiporter

Two solutes move in opposite directions.

Figure 5.20 Types of transporters based on the direction of transport. **(a)** Uniporter. **(b)** Symporter. **(c)** Antiporter.

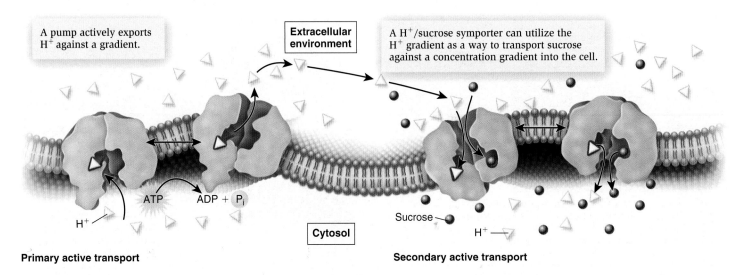

A pump actively exports H⁺ against a gradient.

Extracellular environment

A H⁺/sucrose symporter can utilize the H⁺ gradient as a way to transport sucrose against a concentration gradient into the cell.

ATP   ADP + P_i

H⁺

Cytosol

Sucrose

H⁺

Primary active transport

Secondary active transport

Figure 5.22 Types of active transport.   During primary active transport, a pump directly uses energy, in this case from ATP, and generates a gradient. The pump shown here uses ATP to establish an H⁺ electrochemical gradient. Secondary active transport via symport involves the utilization of this gradient to drive the active transport of a solute, such as sucrose.

solutes. Therefore, with symporters in their plasma membrane, cells can scavenge nutrients from the extracellular environment and accumulate them to high levels within the cytoplasm.

ATP-Driven Ion Pumps Generate Ion Electrochemical Gradients

The concept of active transport was discovered in the 1940s based on the study of ion movements using radioisotopes of Na^+ and K^+. After analyzing the movement of these ions across the plasma membrane of muscle cells, nerve cells, and red blood cells, researchers determined that the export of sodium ions (Na^+) is coupled to the import of potassium ions (K^+) (**Figure 5.23a**). In the late 1950s, Danish biochemist Jens Skou proposed that a single transporter is responsible for this phenomenon. By studying the membranes of nerve cells from crabs, he was the first person to describe an ATP-driven ion pump, which was later named the Na^+/K^+-ATPase. This pump can actively transport Na^+ and K^+ against their gradients by using the energy from ATP hydrolysis.

Interestingly, Skou initially had trouble characterizing this pump. He focused his work on the large nerve cells found in the shore crab (*Carcinus maenas*). In membranes from these cells, he was able to identify a transporter that could hydrolyze ATP, but the rate of hydrolysis was too low to account for the known levels of ATP hydrolysis associated with Na^+ and K^+ pumping. When he added Na^+ to his membranes, the ATP hydrolysis rate was not greatly affected. Then he tried adding K^+, but the ATP hydrolysis rate still did not increase. This was a frustrating period for Skou. Eventually, he did the critical experiment in which he added both Na^+ and K^+ to his membranes. With both ions present, ATP hydrolysis soared dramatically. This observation led to the identification and purification of the Na^+/K^+-ATPase. Jens Skou was awarded the Nobel Prize in 1997, over 40 years after his original work.

Let's take a closer look at the Na^+/K^+-ATPase that Skou discovered. Every time one ATP is hydrolyzed, the Na^+/K^+-ATPase functions as an antiporter that pumps three Na^+ into the extracellular environment and two K^+ into the cytosol. Because one cycle of pumping results in the net export of one positive charge, the Na^+/K^+-ATPase also produces an electrical gradient across the membrane. For this reason, it is considered an **electrogenic pump**, meaning that it generates an electrical gradient. The plasma membrane of a typical animal cell contains thousands of Na^+/K^+-ATPase pumps.

By studying the interactions of Na^+, K^+, and ATP with the Na^+/K^+-ATPase, researchers have pieced together a molecular roadmap of the steps that direct the pumping of ions across the membrane. These steps are termed the **reaction mechanism** (**Figure 5.23b**). A central precept of the reaction mechanism is that the Na^+/K^+-ATPase can alternate between two conformations, designated E1 and E2. In E1, the ion-binding sites are accessible from the cytosol—sodium ions bind tightly to this conformation while potassium ions have a low affinity. In E2, the ion-binding sites are accessible from the extracellular environment—sodium ions have a low affinity and potassium ions bind tightly.

To examine the reaction mechanism, let's begin with the E1 conformation. Three sodium ions bind to the Na^+/K^+-ATPase from the cytosol (Figure 5.23b). When this occurs, ATP is hydrolyzed to ADP and phosphate. Temporarily, the phosphate is covalently bound to the pump, an event called phosphorylation. The pump then switches to the E2 conformation. The sodium ions are released into the extracellular environment because they have a lower affinity for the E2 conformation, and then two potassium ions bind from the outside. The binding of K^+ causes the release of phosphate, which in turn causes a switch to E1. Because the E1 conformation has a low affinity for K^+, the potassium ions are released into the cytosol. The Na^+/K^+-ATPase is now ready for another round of pumping.

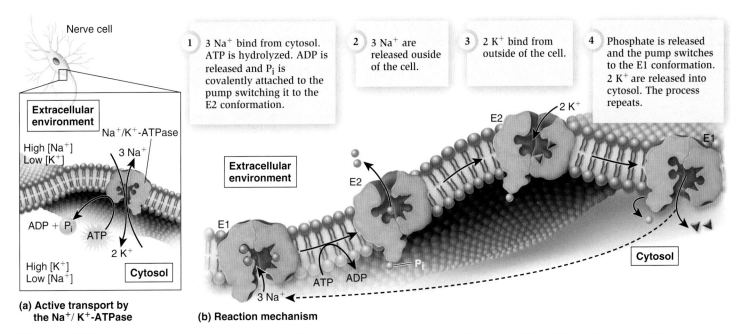

Figure 5.23 Structure and function of the Na⁺/K⁺-ATPase. (a) Active transport by the Na⁺/K⁺-ATPase. Each time this protein hydrolyzes one ATP molecule, it pumps out three Na⁺ and pumps in two K⁺. (b) Reaction mechanism. The figure illustrates the protein conformational changes between E1 and E2. As this occurs, ATP is hydrolyzed to ADP and phosphate. During the reaction mechanism, phosphate is covalently attached to the protein but is released after potassium ions bind.

Biological inquiry: If a cell had ATP and sodium ions, but potassium ions were missing, how far through the reaction mechanism could the Na⁺/K⁺-ATPase proceed?

The Na⁺/K⁺-ATPase is a key cellular enzyme in animal cells because it functions as an ion pump that maintains Na⁺ and K⁺ gradients across the plasma membrane. Many other types of ion pumps are also found in the plasma membrane and in organellar membranes. Ion pumps play the primary role in the formation and maintenance of ion gradients that drive many important cellular processes (**Table 5.4**). Biologists have come to understand that the transport of ions against their gradients is a never-ending activity of all living cells. ATP is commonly the source of energy to drive ion pumps, and cells typically use a substantial portion of their cellular ATP to keep them working. For example, nerve cells use up to 70% of their ATP just to operate ion pumps!

Macromolecules and Large Particles Are Transported via Endocytosis and Exocytosis

We have seen that most small substances are transported via membrane proteins such as pumps, transporters, and channels, which provide a passageway for the movement of substances across the membrane. Eukaryotic cells have two other mechanisms, exocytosis and endocytosis, to transport larger molecules such as proteins and polysaccharides, and even very large particles. Both mechanisms involve the packaging of the transported substance, sometimes called the cargo, into a membrane vesicle or vacuole. Table 5.5 describes important examples of exocytosis and endocytosis.

Table 5.4	Important Functions of Ion Electrochemical Gradients
Function	**Description**
Transport of ions and molecules	Symporters and antiporters use H⁺ and Na⁺ gradients to take up nutrients and export waste products.
Production of energy intermediates	In the mitochondrion and chloroplast, H⁺ gradients are used to synthesize ATP.
Regulation of cystolic pH	Transporters sense pH changes and regulate the internal pH.
Osmotic regulation	Animal cells control their internal volume by regulating ion gradients between the cytosol and extracellular fluid.
Nerve signaling	Na⁺ and K⁺ gradients are involved in conducting action potentials, the signals transmitted by nerve cells.
Muscle contraction	Ca²⁺ gradients regulate the ability of muscle fibers to contract.
Bacterial swimming	H⁺ gradients drive the rotation of bacterial flagella.

Exocytosis **Exocytosis** is a process in which material inside the cell, which is packaged into vesicles, is excreted into the extracellular environment (**Figure 5.24**). These vesicles are usually derived from the Golgi apparatus. As the vesicles form, a specific cargo is loaded into their interior. For example, in animal cells, large polysaccharides are made within the lumen of the Golgi and packaged within vesicles that bud from the Golgi.

Table 5.5	Examples of Exocytosis and Endocytosis

Exocytosis	Description
Hormones	Certain hormones, such as insulin, are composed of polypeptides. To exert its effect, insulin is secreted via exocytosis into the bloodstream from B cells of the pancreas.
Digestive enzymes	Digestive enzymes that function in the lumen of the small intestine are secreted via exocytosis from cells of the pancreas.
Extracellular matrix	Most of the components of the extracellular matrix that surrounds animal cells are secreted via exocytosis.

Endocytosis	Description
Uptake of vital nutrients	Many important nutrients are very insoluble in the bloodstream. Therefore, they are bound to proteins in the blood and then taken into cells via endocytosis. Examples include the uptake of lipids (bound to low-density lipoprotein) and iron (bound to transferrin protein).
Root nodules	Nitrogen-fixing root nodules found in certain species of plants such as legumes are formed by the endocytosis of bacteria. After endocytosis, the bacterial cells are contained within a membrane-enclosed compartment in the nitrogen-fixing tissue of functional nodules.
Immune system	Cells of the immune system, known as macrophages, engulf and destroy bacteria via endocytosis.

The budding process involves the formation of a protein coat around the emerging vesicle. The assembly of coat proteins on the surface of the membrane causes the bud to form. Eventually, the bud separates from the membrane to form a vesicle. After the vesicle is released, the coat is shed. Finally, the vesicle fuses with the plasma membrane and releases its contents into the extracellular medium.

The process of exocytosis illustrates the dynamic and fluid nature of membranes. They readily bud and fuse with other membranes. The same is true for endocytosis, which is described next.

Endocytosis During **endocytosis**, the plasma membrane invaginates, or folds inward, to form a vesicle that brings substances into the cell (**Figure 5.25**). A common form of endocytosis is **receptor-mediated endocytosis**, in which a receptor is specific for a given cargo. When a receptor binds to that cargo, this stimulates the binding of coat proteins to the membrane, which initiates the formation of a vesicle. Many receptors aggregate together as a protein coat forms around the vesicle. Once inside the cell, the vesicle sheds its coat. In most cases, the vesicle fuses with an internal membrane organelle, such as a lysosome, and the receptor releases its cargo. Depending on the cargo, the lysosome may release it directly into the cytosol or digest it into simpler building blocks before releasing it.

Other specialized forms of endocytosis occur in certain types of cells. **Pinocytosis** (meaning "cell drinking") involves the formation of membrane vesicles from the plasma membrane as a way for cells to internalize the extracellular fluid. This allows cells to sample the extracellular solutes. Pinocytosis is particularly important in cells that are actively involved in nutrient absorption, such as cells that line the intestine in animals.

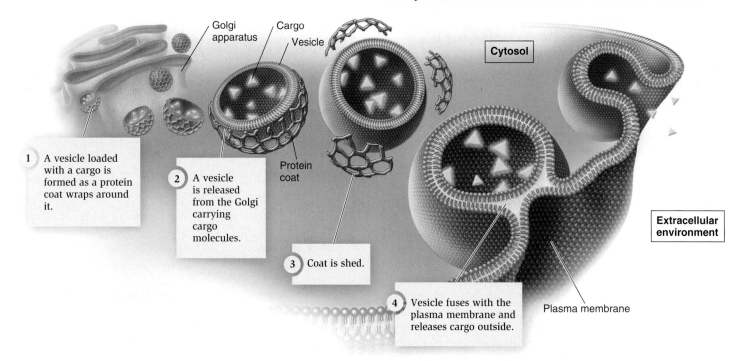

Figure 5.24 **Exocytosis.** Membrane vesicles bud from an organelle inside the cell, such as the Golgi apparatus, and are loaded with a specific cargo. The budding process involves the formation of a protein coat around the emerging vesicle, which is later shed. When the vesicle fuses with the plasma membrane, it releases its contents into the extracellular environment.

Phagocytosis ("cell eating") is an extreme form of endocytosis. It involves the formation of an enormous membrane vesicle called a phagosome, or phagocytic vacuole, that engulfs a large particle such as a bacterium. Only certain kinds of cells can carry out phagocytosis. For example, macrophages, which

are cells of the immune system in mammals, kill bacteria via phagocytosis. Once inside the cell, the phagosome fuses with lysosomes and the digestive enzymes within the lysosomes destroy the bacterium.

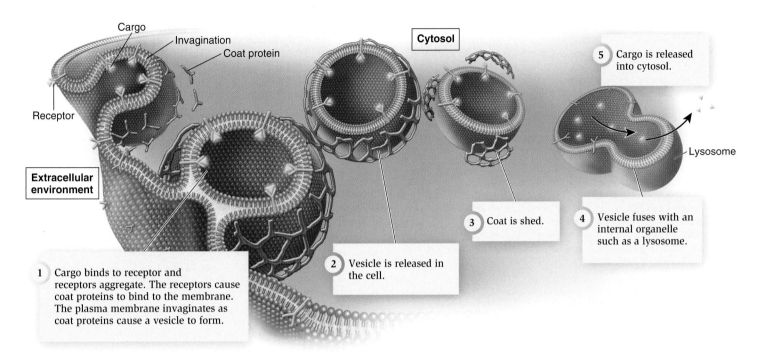

Figure 5.25 **Receptor-mediated endocytosis.** A cargo binds to a receptor, which stimulates the aggregation of many receptors and the formation of a vesicle with a protein coat. Once inside the cell, the vesicle sheds its coat. In most cases, the vesicle fuses with an internal membrane organelle and releases its cargo into the cytosol.

Chapter Summary

5.1 Membrane Structure

- Cellular membranes are dynamic structures that separate a cell from its surroundings yet provide an interface to carry out vital cellular activities. One of their crucial functions is membrane transport. (Table 5.1)

- The accepted model of the plasma membrane is the fluid-mosaic model, and its basic framework is the phospholipid bilayer. Cellular membranes also contain proteins, and most membranes have attached carbohydrates. (Figure 5.1, Table 5.2)

- Proteins can bind to membranes in three different ways: as transmembrane proteins, lipid-anchored proteins, or peripheral membrane proteins. Transmembrane proteins and lipid-anchored proteins are classified as integral membrane proteins. Researchers are working to identify new membrane proteins and their functions because these proteins are so important biologically and medically. (Figure 5.2, Table 5.3)

- Bilayer semifluidity is essential for normal cell function, growth, and division. The chemical properties of phospholipids, such as tail length, the presence of double bonds, and the presence of cholesterol, have a profound effect on the fluidity of the phospholipid bilayer. (Figures 5.3, 5.4, 5.5, 5.6)

- Glycosylation, which produces glycolipids or glycoproteins, has a variety of cellular functions. Carbohydrate can serve as a recognition marker or a protective cell coat. (Figure 5.7)

- Electron microscopy is a valuable tool for studying membrane structure and function. Freeze fracture electron microscopy (FFEM) can be used to analyze the interiors of phospholipid bilayers. (Figure 5.8)

5.2 Membrane Transport

- Living cells maintain a constant internal environment that is separated from their external environment. This involves establishing transmembrane gradients across the plasma membrane and organellar membranes. (Figure 5.9, Table 5.4)

- Diffusion occurs when a solute moves from a region of high concentration to a region of lower concentration. Passive diffusion occurs through a membrane without the aid of a transport protein. (Figure 5.10)

- Passive transport refers to the diffusion of a solute across a membrane in a process that does not require an input of energy. Passive transport can occur via passive diffusion or facilitated diffusion. (Figure 5.11)

- In the process of osmosis, water diffuses through a membrane from a solution that is hypotonic (lower solute concentration) into a solution that is hypertonic (higher solute concentration). Solutions with identical solute concentrations are isotonic. The tendency of water to move into any cell creates an osmotic (turgor) pressure. (Figures 5.12, 5.13, 5.14)

- Transport proteins enable biological membranes to be selectively permeable. The two classes of transport proteins are channels and transporters. Channels form an open passageway for the direct diffusion of solutes across the membrane; one example is aquaporin, which allows the movement of water. Most channels are gated, which allows cells to regulate the movement of solutes. Transporters (carriers), which tend to be slower than channels, bind their solutes in a hydrophilic pocket and undergo a conformational change that switches the exposure of the pocket to the other side of the membrane. A pump is a transporter that directly couples its conformational changes to an energy source. All living cells contain ATP-driven pumps. (Figures 5.15, 5.16, 5.17, 5.18, 5.19, 5.20, 5.21)

- Active transport is the movement of a solute across a membrane against its gradient. Primary active transport involves pumps that directly use energy and generate a solute gradient. Secondary active transport involves the utilization of a pre-existing gradient. (Figure 5.22)

- The Na^+/K^+-ATPase is an electrogenic ATP-driven pump. The reaction mechanism refers to the steps that direct the pumping of ions across the membrane. (Figure 5.23)

- Eukaryotic cells have two other mechanisms, exocytosis and endocytosis, to transport large molecules and particles. Exocytosis is a process in which material inside the cell is packaged into vesicles and excreted into the extracellular environment. During endocytosis, the plasma membrane folds inward to form a vesicle that brings substances into the cell. Forms of endocytosis include pinocytosis, phagocytosis, and receptor-mediated endocytosis. (Figures 5.24, 5.25, Table 5.5)

TEST YOURSELF

1. Which of the following statements best describes the chemical composition of biomembranes?
   a. Biomembranes are bilayers of proteins with associated lipids and carbohydrates.
   b. Biomembranes are composed of two layers, one layer of phospholipids and one layer of proteins.
   c. Biomembranes are bilayers of phospholipids with associated proteins and carbohydrates.
   d. Biomembranes are composed of equal numbers of phospholipids, proteins, and carbohydrates.
   e. Biomembranes are composed of lipids with proteins attached to the outer surface.

2. _____ is a lipid that helps stabilize membranes of animal cells by regulating fluidity as temperature changes.
   a. Cholesterol   d. ATP
   b. Prostaglandin   e. Acetone
   c. Glycerol

3. The presence of double bonds in the fatty acyl tail will
   a. decrease fluidity because of the attraction between the unsaturated tails.
   b. increase fluidity due to the difficulty of the kinked acyl tail to interact.
   c. decrease fluidity by increasing the space between the phospholipids.
   d. increase fluidity by allowing more room for cholesterol to move into the membrane.
   e. decrease fluidity by decreasing the amount of proteins in the membrane.

4. Carbohydrates of the plasma membrane
   a. are associated with a protein or lipid.
   b. are located on the outer surface of the membrane.
   c. can function as cell markers for recognition by other cells.
   d. all of the above.
   e. a and c only.

5. Which of the following can easily diffuse through a lipid bilayer?
   a. sodium ions   d. oxygen
   b. amino acids   e. DNA
   c. glucose

6. The tendency for Na^+ to move into the cell is due to
   a. the higher numbers of Na^+ outside the cell resulting in a chemical concentration gradient.
   b. the net negative charge inside the cell attracting the positively charged Na^+.
   c. the attractive force of K^+ inside the cell pulling Na^+ into the cell.
   d. all of the above.
   e. a and b only.

7. The hydrostatic pressure required to stop osmosis is
   a. an electrochemical gradient.   d. osmotic pressure.
   b. filtration.   e. partial pressure.
   c. tonicity.

8. The selectively permeable characteristic of plasma membranes is mainly due to the presence of
   a. phospholipids.
   b. transport proteins.
   c. glycolipids on the outer surface of the membrane.
   d. concentration gradients across the membrane.
   e. cholesterol.

9. During _____ , materials are moved across the plasma membrane against their concentration gradient.
   a. facilitated diffusion   d. filtration
   b. osmosis   e. simple diffusion
   c. active transport

10. Large particles or large volumes of fluid can be brought into the cell by
    a. facilitated diffusion.   d. exocytosis.
    b. active transport.   e. all of the above.
    c. endocytosis.

CONCEPTUAL QUESTIONS

1. Explain the fluid-mosaic model of membrane structure.

2. What are three types of membrane proteins?

3. What is the difference between passive diffusion and passive transport?

EXPERIMENTAL QUESTIONS

1. What observations led to the experiment of Figure 5.15 to identify proteins that may increase water movement into the cell?

2. How did Agre and associates choose a candidate protein that may function as a water channel in plasma membranes? What was the hypothesis tested by the researchers? Briefly explain how they were able to test their hypothesis.

3. What were the results of the experiment of Figure 5.15? Do they support the proposed hypothesis?

COLLABORATIVE QUESTIONS

1. Discuss the concept of solute concentrations in cells.

2. Discuss the two categories of transport proteins found in plasma membranes.

www.brookerbiology.com

This website includes answers to the Biological Inquiry questions found in the figure legends and all end-of-chapter questions.

6

SYSTEMS BIOLOGY OF CELL ORGANIZATION

CHAPTER OUTLINE

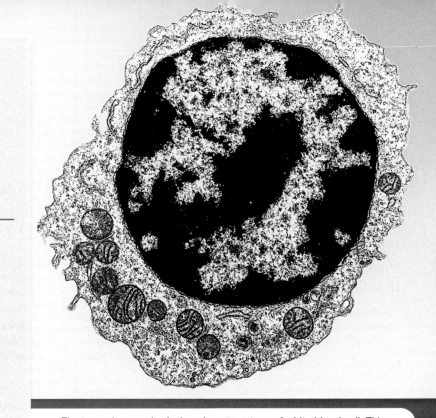

Electron micrograph of a lymphocyte, a type of white blood cell. This cell has a large nucleus and several mitochondria (colored blue). A cell can be viewed as a system of four interacting parts: the nucleus, endomembrane system, cytosol, and semiautonomous organelles.

The first few chapters of this textbook laid the foundation for understanding cell structure and function. We learned that life depends on organic molecules, which form the building blocks for macromolecules such as proteins, nucleic acids, and carbohydrates. In addition, we considered cell organization at a higher level. Cells contain complex structures such as membranes, chromosomes, ribosomes, and a cytoskeleton. Eukaryotic cells have organelles that provide specialized compartments to carry out various cellular functions.

Thus far, we have surveyed the structures and functions of cells with an emphasis on describing the various parts of cells. We have taken a reductionist approach, which involves reducing a complex system to simpler components to understand how the system works. In cell biology, reductionists study the parts of a cell as individual units. In this chapter, however, we will view the cell from a broader perspective. In **systems biology**, researchers study living organisms in terms of their underlying network structure—groups of structural and functional connections—rather than their individual molecular components. A "system" can be anything from a metabolic pathway to a cell, an organ, or even an entire organism. In this chapter, we focus on the cell as a system. Our goal is to understand how the organization of a cell arises by complex interactions between its various components and compartments. In the first section, we will examine the general principles of cell organization. In the remaining two sections we focus on dynamic interactions among the different compartments of eukaryotic cells.

6.1 Principles of Cell Organization

Now that we are familiar with the general properties of cells and membranes, we can ask, How are the various parts of a cell formed and maintained? We usually think of a cell as having form and structure, and at any given instant, it does. However, a cell is also profoundly dynamic, which we can better appreciate when we look at its organization from the perspective of systems biology. Let's consider a football analogy. A living cell is less like a football field—with goalposts, lines, and turf—and more like the players who follow rules and have various formations and movements. Like the players in a football game, the components of a cell are dynamic entities. Like a successful team, these components work together as a system to maintain cell organization and carry out vital functions.

We begin this section by considering the phenomena that underlie the existence of cells. These include information, namely, a genome, functional molecules, and pre-existing organization. We will then examine how the genome allows cells to synthesize proteins at the correct time and in the correct cell type. Likewise, the genome provides proteins with sorting signals that are recognized by a sorting system, which directs them to the proper cellular location. This system, which is critical for cell organization, will be described in later sections.

Another factor that underlies cell organization is the ability of certain proteins to interact with each other to form larger complexes, sometimes called molecular machines. We will examine how molecular machines assemble from their individual components, and how they carry out complex functions and provide cell organization. Finally, the organization of living cells is a dynamic process that involves the turnover of cellular molecules and macromolecules. We will also explore how macromolecules such as proteins are broken down to their building blocks so that a cell can continually replenish its macromolecules.

Information, Functional Molecules, and Organization Must Already Exist for Cells to Maintain Their Structure

As discussed in Chapter 22, the first primitive cells arose from nonliving materials between 3.5 and 4.0 billion years ago. According to the cell doctrine, *all modern cells come from pre-existing cells by division.* Modern cells, which are much more complex than the first primitive cells, can grow, adapt, and modify their organization, but they do not arise spontaneously from nonliving materials. At least three key factors are responsible for this principle. First, all cells must possess a genome, which provides the information to make RNA and proteins. The genomes of modern species are the products of 3.5–4.0 billion years of evolution and contain the information to make the thousands of proteins necessary to maintain cell structure and function.

Even so, if a researcher were able to synthesize the genome of a living organism and put it in a test tube full of water, nothing would happen. Living cells require a second ingredient— pre-existing molecules such as enzymes and organic molecules— to make things happen. For instance, RNA polymerase, a protein that functions as an enzyme, is needed to make new RNA molecules, using nucleotides (organic molecules) as building blocks and DNA as a template. As another example, RNA molecules and proteins are needed to construct a ribosome, which is a critical component for synthesizing polypeptides. Cell vitality requires a genome to provide information and also active molecules such as RNA, proteins, and small molecules to access that information.

So let's suppose that a researcher could somehow synthesize the genome of a species and produce all of its RNA molecules and proteins. That would be an amazing feat! If the genome, RNAs, and proteins were combined in a test tube with other small molecules that are needed for energy and as building blocks, do you think a cell would arise? The answer is no. Some enzymatic activity would probably occur, but a living cell would not result. In addition to a genome and functional molecules, cells require a third key factor: pre-existing organization. For example, cells rely on a plasma membrane, which contains multiple transport proteins that allow the selective uptake and excretion of ions and molecules. In addition, organelles compartmentalize cells, and most organelles must be made from pre-existing organelles.

Thus, the formation of cells relies on a genome, functional molecules, and organization, all of which come from pre-existing cells. In the rest of this chapter, we will examine the interplay that occurs among these three phenomena to maintain cellular organization and promote the formation of cells.

Cell Organization Relies on Genetic Information, Which Produces a Proteome

The genome of every organism contains the information necessary to produce a system of RNA and protein molecules that provides the foundation for cell structure, function, and organization. An important paradigm in biology is that "structure determines function." The information in most genes is used to make mRNA molecules that encode the amino acid sequences of proteins. The amino acid sequence of proteins, together with chemical principles, governs their three-dimensional structure. The three-dimensional structure of proteins, in turn, determines their function.

Although the genome contains the information to make proteins with specific structures and functions, the study of individual proteins does not allow us to completely understand how the parts of cells are made, and how they maintain cellular organization. To appreciate the dynamic form and function of living cells, we need to take a broader, more integrative look at their molecular components.

1. *The proteome—the entire collection of proteins that a cell makes—is largely responsible for the structures and functions of living cells* (**Figure 6.1a**). Some proteins play a role in cell shape and organization. Other proteins and a few RNA molecules function as enzymes to synthesize or break down other cellular components. For example, as discussed later in this chapter, phospholipids are made by enzymes in the cell. The synthesis of most components that form the foundation for cell structure relies on enzymes. Some exceptions are water and inorganic ions, which are taken up from the environment. Likewise, transport proteins are important in the uptake and export of ions and molecules across cell membranes. Finally, the activities of cells are controlled by many proteins that are generally known as regulatory proteins. As described next, some of these proteins regulate genes in the cell nucleus, while others regulate the functions of other proteins.

2. *Gene and protein regulation cause the proteome to be dynamic.* In multicellular organisms, the regulation of genes causes the amounts of proteins to vary in different cell types. This allows a multicellular organism to have cells with specialized structures and functions (**Figure 6.1b**). Nerve and muscle cells have a different organization because their proteomes are different. In addition, the regulation of genes and proteins allows a cell to respond in a dynamic way to changes in the environment (**Figure 6.1c**). For example, when a cell is exposed to a higher temperature, it will produce proteins that help it cope with the increased heat.

Enzymes are involved in the synthesis and breakdown of cellular molecules and macromolecules.

Regulatory proteins control the expression of genes and the functions of other proteins.

Proteins play a structural role that influences cell shape, organization, and movement.

Transport proteins allow the movement of substances across membranes.

(a) The proteome is a diverse collection of proteins that carry out cell functions and promote cell organization.

Muscle fiber cell

Nerve cell

(b) Differences in cell morphology among specialized cells can be attributed to differences in the proteome.

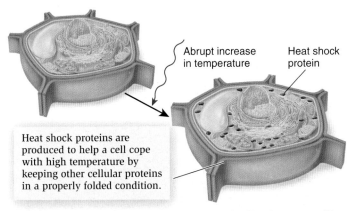

Abrupt increase in temperature

Heat shock protein

Heat shock proteins are produced to help a cell cope with high temperature by keeping other cellular proteins in a properly folded condition.

(c) Cells adapt to environmental changes by altering the composition of their proteomes.

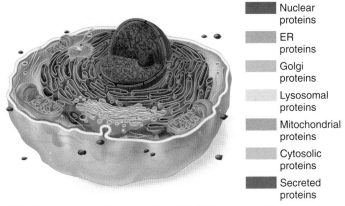

Nuclear proteins

ER proteins

Golgi proteins

Lysosomal proteins

Mitochondrial proteins

Cytosolic proteins

Secreted proteins

(d) Proteins have sorting signals within their amino acid sequences that direct them to the correct cellular compartment.

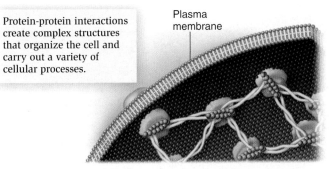

Protein-protein interactions create complex structures that organize the cell and carry out a variety of cellular processes.

Plasma membrane

(e) Proteins may have surfaces that cause them to interlock via protein-protein interactions to form larger cellular structures.

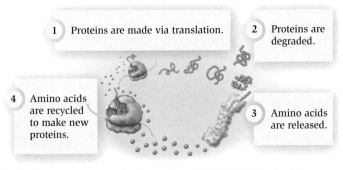

1 Proteins are made via translation.

2 Proteins are degraded.

4 Amino acids are recycled to make new proteins.

3 Amino acids are released.

(f) Cellular macromolecules are made and later broken down to recycle their building blocks.

Figure 6.1 Genomes and proteomes and their relationship to cell structure, function, and organization.

3. *Proteins have **sorting signals**—short amino acid sequences in their structure—that direct them to their correct location* (**Figure 6.1d**). For example, one sorting signal directs certain proteins to the mitochondria while a different sorting signal sends other proteins to the endoplasmic reticulum. The last two sections of this chapter will explore the molecular mechanisms that facilitate protein sorting, a key event that maintains cell organization.

4. *Because their structures may bind to each other and interlock, proteins often undergo **protein-protein interactions**.* These interactions are very specific and can build larger structures that provide organization to the cell (**Figure 6.1e**). An example is the cytoskeleton that forms filaments in the cell. In addition, protein-protein interactions can produce molecular machines that carry out complex cellular functions.

5. *Because most molecules and macromolecules in cells are short-lived, cells must continually synthesize new components and break down unwanted components.* Proteins, for example, are made during the process of translation and typically exist for just several minutes or a few hours (**Figure 6.1f**). As discussed later, cells have different mechanisms to degrade proteins to individual amino acids so that the amino acids can be recycled—that is, used to make new proteins.

Interactions Among Proteins and RNA Molecules Make Complex Molecular Machines

A machine is an object that has moving parts and does useful work. If the size of the machine is measured in nanometers, it is appropriately called a **molecular machine**. Even a single protein could be considered a molecular machine if it undergoes conformational changes as part of its function. The association of proteins with each other and also with RNA molecules may form larger molecular machines. These machines provide structure and organization to cells and enable them to carry out complicated processes.

Large molecular machines are formed by an assembly process. In some cases, interactions between the protein components of a machine may occur spontaneously, without requiring an input of energy. Sometimes additional proteins are needed for the machine to form but are not retained in its final structure. Some machines may require an input of energy, such as from ATP hydrolysis, to promote the assembly process.

Figure 6.2 describes the assembly of ATP synthase, a machine that makes ATP. Because all living organisms use ATP for energy, ATP synthase is found in both prokaryotes and eukaryotes. ATP synthase is composed of eight different protein subunits called a, b, c, alpha (α), beta (β), gamma (γ), delta (δ),

and epsilon (ϵ). The a, b, and c subunits are transmembrane proteins, found either in the bacterial plasma membrane or the mitochondrial and thylakoid membranes of eukaryotes. Nine to twelve of the c subunits assemble together to form a ring in the membrane, and one a subunit and two b subunits bind to this ring. The α, β, γ, δ, and ϵ subunits associate with each other to form a complex of three α subunits, three β subunits, and one subunit each of γ, δ, and ϵ. This complex then binds to the membrane components to complete the assembly process.

Why does ATP synthase assemble in this manner? The reason is **molecular recognition**—surfaces on the various protein subunits recognize each other in a very specific way, causing them to bind to each other and promote the assembly process. Said another way, the amino acid sequences of these proteins produce surfaces that fold in a way that causes the proteins to interlock. Thus, the amino acid sequences of proteins, which are stored in the genome, contain the information for protein-protein interactions.

Molecular machines carry out complex cellular functions. The ribosome (**Figure 6.3a**), a molecular machine composed of

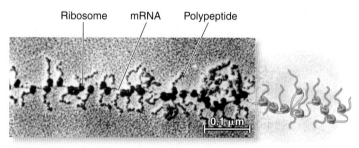

Ribosome   mRNA   Polypeptide

0.1 µm

(a) Ribosomes in the act of making polypeptides

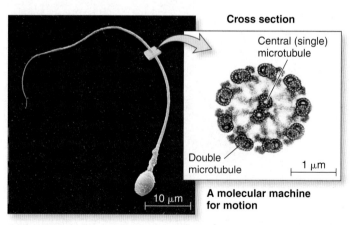

Cross section

Central (single) microtubule

Double microtubule

1 µm

10 µm

A molecular machine for motion

(b) The flagella of sperm contain a molecular machine for locomotion

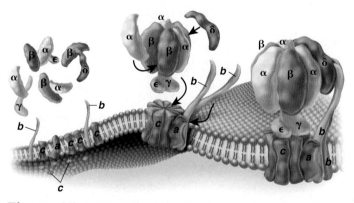

Figure 6.2 **Assembly of a molecular machine, the ATP synthase found in bacteria, mitochondria, and chloroplasts.** Nine to twelve of the c subunits form a ring in the membrane, and one a subunit and two b subunits bind to the ring. The α, β, γ, δ, and ε subunits form a complex of three α subunits, three β subunits, and one subunit each of γ, δ, and ε. This complex binds to the membrane components.

Biological inquiry: Explain why the ATP synthase assembles the way it does.

Figure 6.3 Other examples of molecular machines. (a) The ribosome, which plays a role in protein synthesis, is composed of many types of proteins and several large RNA molecules. In the electron micrograph shown here, many ribosomes are gliding along an mRNA molecule, synthesizing polypeptides as they go. (b) The flagellum of a sperm contains a molecular machine composed of microtubules and accessory proteins. The cross section shows the 9 + 2 arrangement of microtubules in which nine double microtubules form a ring with two central microtubules that are single microtubules.

many types of proteins and several large RNA molecules, functions as a molecular arena for the synthesis of new polypeptides. The flagellum (**Figure 6.3b**) is a molecular machine composed of microtubules, motor proteins, and many other proteins that enables eukaryotic cells, such as sperm cells, to move.

Molecular machines are also vital in promoting cell organization. Perhaps the most important example is the cytoskeleton, which we consider next.

The Cytoskeleton Is a Molecular Machine That Organizes the Cell and Directs Cellular Movements

Recall from Chapter 4 that the cytoskeleton plays a key role in cell organization and many of the dynamic processes that maintain it. For example, in many cell types, actin filaments form a band just inside the plasma membrane that provides mechanical strength and plays a role in cell shape. Protein-protein interactions occur between actin and many other cellular proteins. Actin filaments are sometimes linked to proteins that are embedded in the plasma membrane, an important factor in the shape of many cells such as the biconcave-disk appearance of human red blood cells (**Figure 6.4**).

The cytoskeleton also organizes and directs intracellular and cellular movements. For example, various types of intracellular cargo, including chromosomes and even organelles such as mitochondria and chloroplasts, are transported within the cell by moving along microtubules and actin filaments. For instance, chloroplasts are sometimes moved to the side of a plant cell that receives more light. **Figure 6.5** illustrates how a **vesicle**—a small, membrane-enclosed sac—moves along microtubules in a nerve cell. In this example, accessory proteins connect a motor protein to the vesicle. The motor protein uses the energy from ATP hydrolysis to "walk" along a microtubule.

Molecules Are Broken Down and Their Building Blocks Are Recycled

Thus far, we have considered how the interactions of cellular components provide organization and complex functions to living cells. The maintenance of cell organization is a dynamic process. Except for DNA, which is stably maintained and inherited from cell to cell, other large molecules such as RNA, proteins, lipids, and polysaccharides have finite lifetimes. Biologists often speak of the **half-life** of molecules, which is the time it takes for 50% of the molecules to be broken down and recycled. For example, a population of mRNA molecules in prokaryotes has an average half-life of about 5 minutes, while mRNAs in eukaryotes tend to exist for longer periods of time, on the order of 30 minutes to 24 hours or even several days.

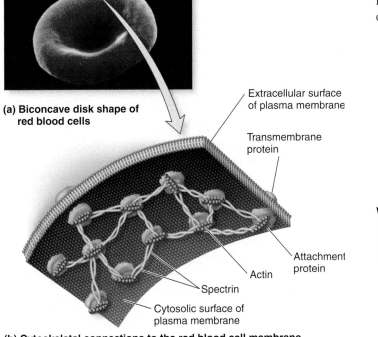

(a) Biconcave disk shape of red blood cells

(b) Cytoskeletal connections to the red blood cell membrane

Figure 6.4 Role of the cytoskeleton in promoting cell shape. Many different proteins are involved in forming the shape of a cell. (a) A micrograph of a red blood cell, which looks like a biconcave disk. (b) To create this shape, proteins within the red blood cell membrane are anchored to an intricate group of cytoskeletal proteins including actin and spectrin.

Biological inquiry: Describe the types of protein-protein interactions that produce the biconcave-disk shape of red blood cells.

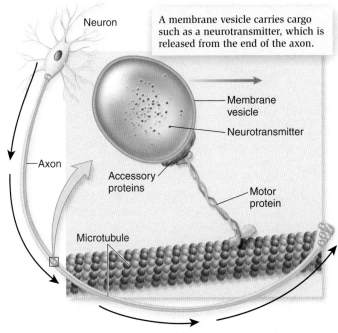

Figure 6.5 Movement of a membrane vesicle along a microtubule. In this example, a motor protein walks along a microtubule in the axon of a nerve cell. The vesicle carries a neurotransmitter, which is released at the end of the axon. Accessory proteins link the motor protein to the vesicle and also control the activity of the motor protein.

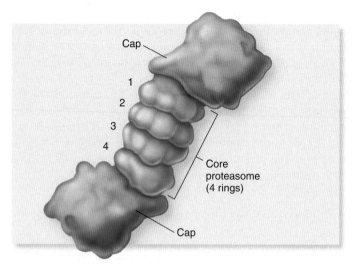

(a) Structure of the eukaryotic proteasome

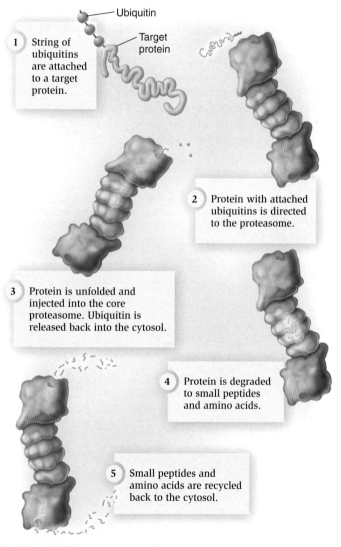

1 String of ubiquitins are attached to a target protein.

2 Protein with attached ubiquitins is directed to the proteasome.

3 Protein is unfolded and injected into the core proteasome. Ubiquitin is released back into the cytosol.

4 Protein is degraded to small peptides and amino acids.

5 Small peptides and amino acids are recycled back to the cytosol.

(b) Steps of protein degradation in eukaryotic cells

Figure 6.6   Protein degradation via the proteasome.

The breakdown of large, cellular molecules and the recycling of their building blocks occur by a variety of mechanisms. As discussed in Chapter 4, lysosomes can degrade materials in the cytosol by a process called autophagy (refer back to Figure 4.20). The components of a mitochondrion, for example, can be degraded in this manner, and building blocks are returned to the cytosol, where they can be used to make new macromolecules. Lysosomes also degrade proteins that are imported into the cell via endocytosis. In addition, cytosolic enzymes break down RNA, polysaccharides, and lipids into smaller building blocks. For example, ribonucleases are enzymes that degrade mRNA molecules.

To survive and respond to environmental changes, cells must continually degrade proteins that are faulty or nonfunctional and synthesize new ones. To be degraded, proteins are recognized by **proteases**—enzymes that cleave the bonds between adjacent amino acids. Although lysosomes in eukaryotic cells are involved in protein breakdown via autophagy and endocytosis, the primary pathway for protein degradation in archaea and eukaryotic cells is via a molecular machine called a **proteasome**. The core of the proteasome is formed from four stacked rings, each composed of seven protein subunits (**Figure 6.6a**). The proteasomes of eukaryotic cells also contain cap structures at each end that control the entry of proteins into the proteasome.

In eukaryotic cells, unwanted proteins are directed to a proteasome by the covalent attachment of a small protein called **ubiquitin**. **Figure 6.6b** describes the steps of protein degradation via eukaryotic proteasomes. First, a string of ubiquitin proteins are attached to the target protein. This event directs the protein to a proteasome cap, which has binding sites for ubiquitin. The cap also has enzymes that unfold the protein and inject it into the internal cavity of the proteasome core. The ubiquitin proteins are removed during entry and return to the cytosol. Inside the proteasome, proteases degrade the protein into small peptides and amino acids. The process is completed when the peptides and amino acids are recycled back into the cytosol. The amino acids can be used to make new proteins.

Ubiquitin-targeting has two advantages. First, the enzymes that attach ubiquitin to its target recognize improperly folded proteins, allowing cells to identify and degrade nonfunctional components. Second, changes in cellular conditions may warrant the rapid breakdown of particular proteins. For example, cell division requires a series of stages called the cell cycle, which depends on the degradation of specific proteins. Ubiquitin targeting directs these proteins to the proteasome for degradation.

Eukaryotic Cells Are a System of Four Interacting Parts That Work Together

We can view a eukaryotic cell as a system of four interacting parts: the interior of the nucleus, the cytosol, the endomembrane system, and the semiautonomous organelles (**Figure 6.7**). These four regions play a role in their own structure and organization, as well as the structure and organization of the entire cell.

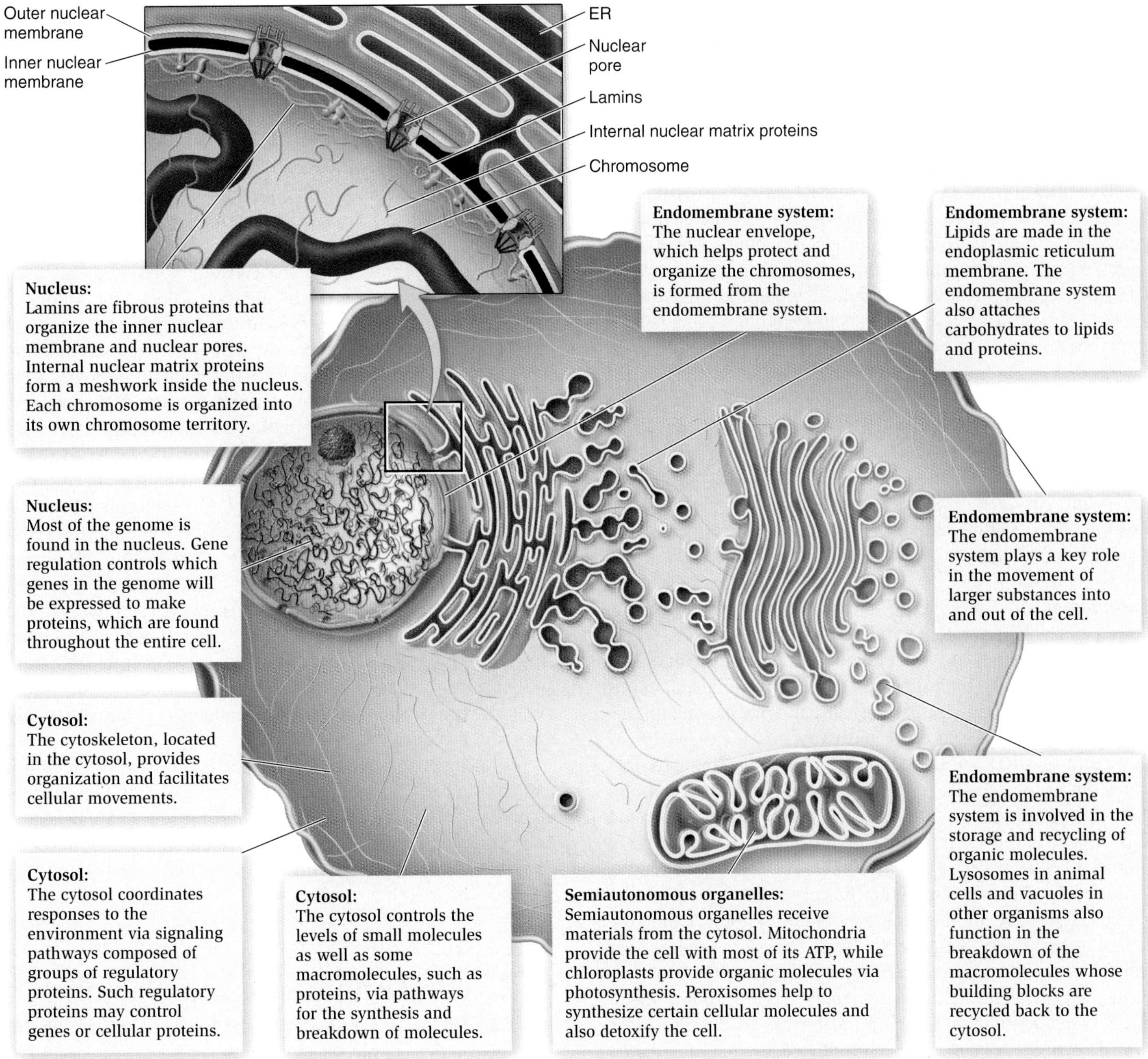

Outer nuclear membrane

Inner nuclear membrane

ER

Nuclear pore

Lamins

Internal nuclear matrix proteins

Chromosome

Endomembrane system:
The nuclear envelope, which helps protect and organize the chromosomes, is formed from the endomembrane system.

Endomembrane system:
Lipids are made in the endoplasmic reticulum membrane. The endomembrane system also attaches carbohydrates to lipids and proteins.

Nucleus:
Lamins are fibrous proteins that organize the inner nuclear membrane and nuclear pores. Internal nuclear matrix proteins form a meshwork inside the nucleus. Each chromosome is organized into its own chromosome territory.

Nucleus:
Most of the genome is found in the nucleus. Gene regulation controls which genes in the genome will be expressed to make proteins, which are found throughout the entire cell.

Endomembrane system:
The endomembrane system plays a key role in the movement of larger substances into and out of the cell.

Cytosol:
The cytoskeleton, located in the cytosol, provides organization and facilitates cellular movements.

Endomembrane system:
The endomembrane system is involved in the storage and recycling of organic molecules. Lysosomes in animal cells and vacuoles in other organisms also function in the breakdown of the macromolecules whose building blocks are recycled back to the cytosol.

Cytosol:
The cytosol coordinates responses to the environment via signaling pathways composed of groups of regulatory proteins. Such regulatory proteins may control genes or cellular proteins.

Cytosol:
The cytosol controls the levels of small molecules as well as some macromolecules, such as proteins, via pathways for the synthesis and breakdown of molecules.

Semiautonomous organelles:
Semiautonomous organelles receive materials from the cytosol. Mitochondria provide the cell with most of its ATP, while chloroplasts provide organic molecules via photosynthesis. Peroxisomes help to synthesize certain cellular molecules and also detoxify the cell.

Figure 6.7 **The four interacting parts of eukaryotic cells.** These include the nucleus, cytosol, endomembrane system, and semiautonomous organelles.

Nucleus The nucleus houses the genome. Earlier in this chapter, we learned how the genome plays a key role in producing the proteome. The collection of proteins that a cell makes is largely responsible for the structure and function of the entire cell. Gene regulation, which occurs in the cell nucleus, is very important in creating specific cell types and enabling cells to respond to environmental changes. Chapters 12 and 13 will examine the topics of gene expression and regulation in greater detail.

The nucleus itself is organized by a collection of filamentous proteins called the nuclear matrix (see inset to Figure 6.7). These proteins form extensive protein-protein interactions that perform two roles. First, proteins known as lamins lie along the nuclear envelope and help to organize the inner nuclear membrane and nuclear pores. Second, internal nuclear matrix proteins form a meshwork inside the nucleus. Each chromosome is organized into its own chromosome territory (refer back to Figure 4.17).

Cytosol The cytosol is the region that is outside of the organelles and inside the plasma membrane. It is an important coordination center for cell function and organization. Along with the plasma membrane, the cytosol coordinates responses to the environment. Factors in the environment may stimulate signaling pathways in the cytosol that affect the functions of cellular proteins and change the expression of genes in the cell nucleus.

The cytosol also has a large impact on cell structure because it is the compartment where many small molecules are metabolized in the cell. The region receives molecules that are taken up from the environment. In addition, many pathways for the synthesis and breakdown of cellular molecules are found in the cytosol, and pathways in organelles are often regulated by events there.

A particularly important component of cell organization is the cytoskeleton, which is found in the cytosol. The formation and function of the cytoskeleton is caused by an amazing series of protein-protein interactions. As discussed earlier, the cytoskeleton provides organization to the cell and facilitates cellular movements. In most cells, the cytoskeleton is a dynamic structure, enabling its composition to respond to environmental and developmental changes.

Endomembrane System The endomembrane system can be viewed as a smaller system within the confines of a cell, which is a larger system. The endomembrane system includes the nuclear envelope, endoplasmic reticulum (ER), Golgi apparatus, lysosomes, vacuoles, secretory vesicles, and plasma membrane. This system forms a **secretory pathway** that is crucial in the movement of larger substances, such as proteins and carbohydrates out of the cell. The export of proteins and carbohydrates plays a key role in the organization of materials that surround cells. The structure of this extracellular matrix is discussed in Chapter 10. This pathway can also run in reverse, the **endocytic pathway**, to take substances into the cell.

A key feature of the endomembrane system is that the membranes are very dynamic and their structures change over time. In some cases, the change can be dramatic. For example, during cell division, the nuclear envelope breaks up into membrane vesicles, thereby allowing the release of chromosomes into the cytosol so they can be sorted to daughter cells. Later, vesicles coalesce to re-form a nuclear envelope in each daughter cell. The ability of the nuclear envelope to break up into vesicles and later re-form a nuclear envelope is a vital feature of cell division in eukaryotic cells.

The endomembrane system also contributes to the overall structure and organization of eukaryotic cells in other ways. Most of a cell's lipids are made in the endoplasmic reticulum membrane and distributed to other parts of the cell. In addition, the endomembrane system attaches carbohydrates to lipids and proteins. This process plays a role in protein sorting, and also helps to organize the materials that are found outside of cells.

Finally, another important function of the endomembrane system that serves the needs of the entire cell is the storage and recycling of organic molecules. Vacuoles often play a role in the storage of organic molecules such as proteins, carbohydrates, and fats. When needed, lysosomes in animal cells and vacuoles in the cells of other organisms also assist in breaking down these macromolecules. The building blocks are then recycled back to the cytosol and used to construct new macromolecules.

Semiautonomous Organelles The semiautonomous organelles include the mitochondria, chloroplasts, and peroxisomes. Regarding organization, these organelles tend to be rather independent. They exist in the cytosol much like a bacterium would grow in a laboratory medium. While a bacterium would take up essential nutrients from the growth medium, the semiautonomous organelles take up molecules from the cytosol. The organelles use these molecules to carry out their functions and maintain their organization. Like bacteria, the semiautonomous organelles divide by binary fission to produce more of themselves.

Although the semiautonomous organelles rely on the rest of the cell for many of their key components, they also give back to the cell in ways that are vital to maintaining cell organization. Mitochondria take up organic molecules from the cytosol and give back ATP, which is used throughout the cell to drive processes that are energetically unfavorable. This energy is crucial for cell organization. The chloroplasts capture light energy and synthesize organic molecules. These organic molecules also store energy and can be broken down when energy is needed. In addition, organic molecules, such as sugars and amino acids, are used as building blocks to synthesize many different types of cellular molecules, such as carbohydrate polymers and proteins.

Later in this chapter, we will focus on how mitochondria and chloroplasts are made and how they maintain their distinct organization. The functions of these organelles involve many interesting pathways that are described in greater detail in Chapters 7 and 8, respectively.

6.2 The Endomembrane System

Thus far, we have examined some of the general features that account for the maintenance of cell organization. In this section, we will take a closer look at the organization of the endomembrane system. As mentioned earlier, this system is a collection of membranes and cell compartments in eukaryotic cells that includes the nuclear envelope, endoplasmic reticulum (ER), Golgi apparatus, lysosomes, vacuoles, secretory vesicles, and plasma membrane (**Figure 6.8**). The intracellular organelles of the endomembrane system reside in the cytosol. The system is impressively dynamic. Much of its activity is related to the transport of membrane vesicles between the various compartments. Transport of vesicles occurs in both directions. For example, membrane vesicles bud from the ER and fuse with the Golgi, and vesicles from the Golgi can return to the ER. As we learned in Chapter 5, exocytosis involves the fusion of intracellular vesicles with the plasma membrane, while endocytosis

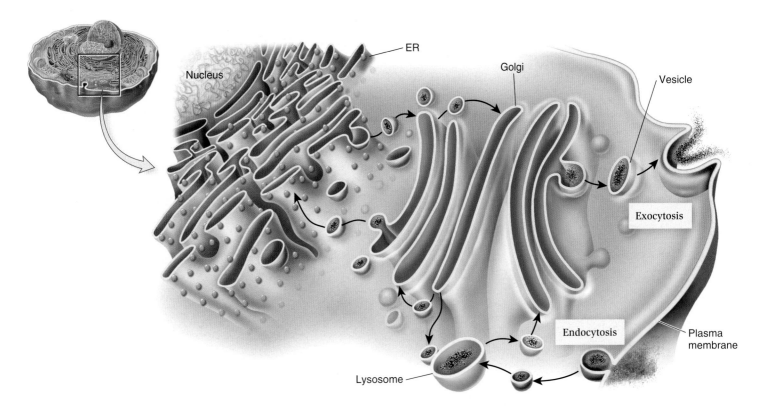

Figure 6.8 **The dynamic nature of the endomembrane system.** Vesicle transport occurs in both directions, allowing materials to be shared among the various membranes and compartments of the endomembrane system.

involves the inward folding of the plasma membrane to form intracellular vesicles.

The endomembrane system is critical for lipid synthesis, protein synthesis and sorting, and the attachment of carbohydrates to lipids and proteins. Its dynamic nature serves to distribute these materials throughout the cell, modify them for use inside the cell, or process them for secretion to be used outside the cell. The past few decades have seen exciting advances in our understanding of how these events occur at the molecular level. In this section, we will examine the roles of the endomembrane system in cell structure, function, and organization.

Lipid Synthesis Occurs at the ER Membrane

In eukaryotic cells, the cytosol and endomembrane system work together to synthesize most lipids. This process occurs at the cytosolic leaflet of the ER membrane. **Figure 6.9** shows a simplified pathway for the synthesis of phospholipids, the main components of cell and organelle membranes. The building blocks for a phospholipid are two fatty acids each with an acyl tail, one glycerol molecule, one phosphate, and a polar group. These building blocks are made via enzymes in the cytosol or they are taken into cells from the diet. To begin the process of phospholipid synthesis, the fatty acids are activated by attachment to an organic molecule called coenzyme A (CoA). This acti-

vation promotes the bonding of the two fatty acids to a glycerol-phosphate molecule, and the resulting molecule is inserted into the cytosolic leaflet of the ER membrane. The phosphate is removed from glycerol, and then a polar molecule that is linked to phosphate is attached to glycerol. In this example, the polar head group contains choline, but many other types are possible. Phospholipids are initially made in the cytosolic leaflet, but flippases in the ER membrane transfer some to the other leaflet.

The lipids that are made in the ER membrane can be transferred to other membranes in the cell by a variety of mechanisms. Phospholipids in the ER can diffuse laterally to the nuclear envelope (**Figure 6.10a**). In addition, lipids can be transported through the cytosol via vesicles to the Golgi, lysosomes, vacuoles, and plasma membrane (**Figure 6.10b**). A third mode of lipid movement involves **lipid exchange proteins**, which extract a lipid from one membrane, diffuse through the cell, and insert the lipid into another membrane (**Figure 6.10c**). Such transfer can occur between any two membranes, even between the endomembrane system and semiautonomous organelles. For example, lipid exchange proteins can transfer lipids between the ER and mitochondria. In addition, chloroplasts and mitochondria can synthesize certain types of lipids that can be transferred from these organelles to other cellular membranes via lipid exchange proteins.

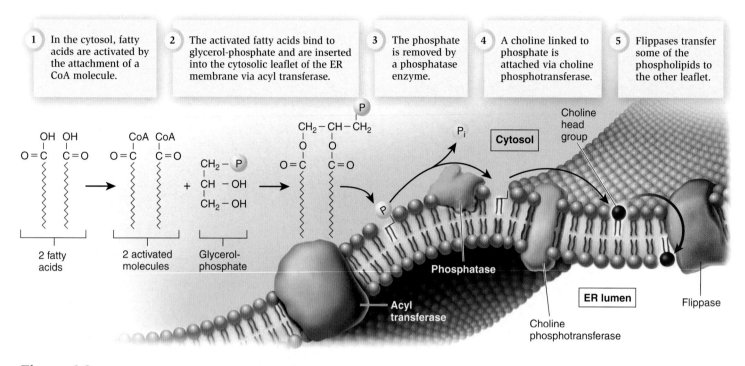

1 In the cytosol, fatty acids are activated by the attachment of a CoA molecule.

2 The activated fatty acids bind to glycerol-phosphate and are inserted into the cytosolic leaflet of the ER membrane via acyl transferase.

3 The phosphate is removed by a phosphatase enzyme.

4 A choline linked to phosphate is attached via choline phosphotransferase.

5 Flippases transfer some of the phospholipids to the other leaflet.

Figure 6.9 A simplified pathway for the synthesis of membrane phospholipids at the ER membrane.

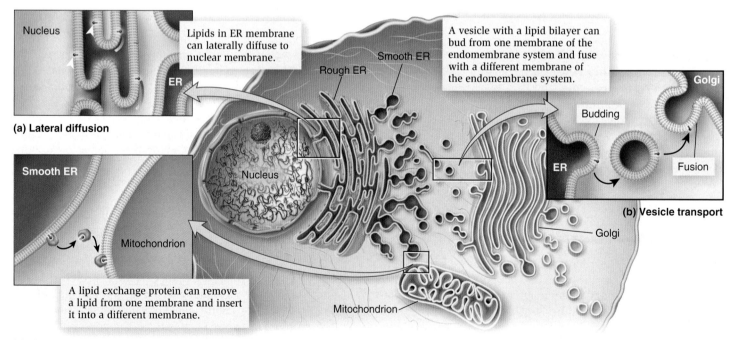

(a) Lateral diffusion

Lipids in ER membrane can laterally diffuse to nuclear membrane.

A vesicle with a lipid bilayer can bud from one membrane of the endomembrane system and fuse with a different membrane of the endomembrane system.

(b) Vesicle transport

(c) Lipid exchange proteins

A lipid exchange protein can remove a lipid from one membrane and insert it into a different membrane.

Figure 6.10 Mechanisms of lipid transfer throughout a eukaryotic cell. As shown in the insets, lipids can be distributed throughout the cell in various ways. **(a)** Lateral diffusion occurs between the ER and nuclear envelope. **(b)** Vesicle transport occurs among the membranes of the endomembrane system. **(c)** Lipid exchange proteins can transfer lipids between any two membranes, such as between the ER and a mitochondrion.

FEATURE INVESTIGATION

Palade Demonstrated That Secreted Proteins Move Sequentially Through Organelles of the Endomembrane System

Eukaryotic cells make thousands of different proteins. In most cases, a protein functions in only one compartment within a cell, or it functions only after it is secreted from the cell. Therefore, proteins must be sorted to the correct locations. One of the first indications that proteins are sorted intracellularly came from studies by George Palade and his collaborators in the 1960s.

Palade's team conducted **pulse-chase experiments**, in which the researchers administered a pulse of radioactive amino acids to cells so that they made radioactive proteins. A few minutes later, the cells were given a large amount of nonradioactive amino acids. This is called a "chase" because it chases away the ability of the cells to make any more radioactive proteins. In this way, radioactive proteins were produced only briefly. Because they were labeled with radioactivity, the fate of these proteins could be monitored over time. The goal of a pulse-chase experiment is to determine where the radioactive proteins are produced and the pathways they take as they travel through a cell.

Palade chose to study the cells of the pancreas. This organ secretes enzymes and protein hormones that play a role in digestion. Therefore, these cells were chosen because their primary activity is protein secretion. To study the pathway for protein secretion, Palade and colleagues injected a radioactive version of the amino acid leucine into the bloodstream of guinea pigs, followed 3 minutes later by an injection of nonradioactive leucine (**Figure 6.11**). At various times after the second injection, samples of pancreatic cells were removed from the animals. The cells were then prepared for transmission electron microscopy (see Chapter 4). The sample was stained with a heavy metal that became bound to membranes and showed the locations of the cell organelles. In addition, the sample was coated with a radiation-sensitive emulsion. When radiation was emitted from radioactive proteins, it interacted with the emulsion in a way that caused the precipitation of silver atoms, which became tightly bound to the sample. In this way, the precipitated silver atoms marked the location of the radiolabeled proteins. Unprecipitated silver chloride in the emulsion was later washed away. Because silver atoms are electron dense, they produce dark spots in a transmission electron micrograph. Therefore, dark spots revealed the locations of radioactive proteins.

Figure 6.11   Palade's use of the pulse-chase method to study protein secretion.

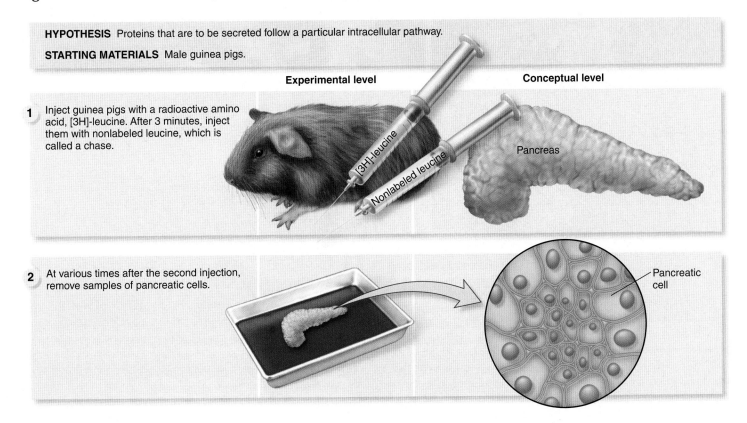

HYPOTHESIS Proteins that are to be secreted follow a particular intracellular pathway.

STARTING MATERIALS Male guinea pigs.

Experimental level   Conceptual level

1 Inject guinea pigs with a radioactive amino acid, [3H]-leucine. After 3 minutes, inject them with nonlabeled leucine, which is called a chase.

[3H]-leucine

Nonlabeled leucine

Pancreas

2 At various times after the second injection, remove samples of pancreatic cells.

Pancreatic cell

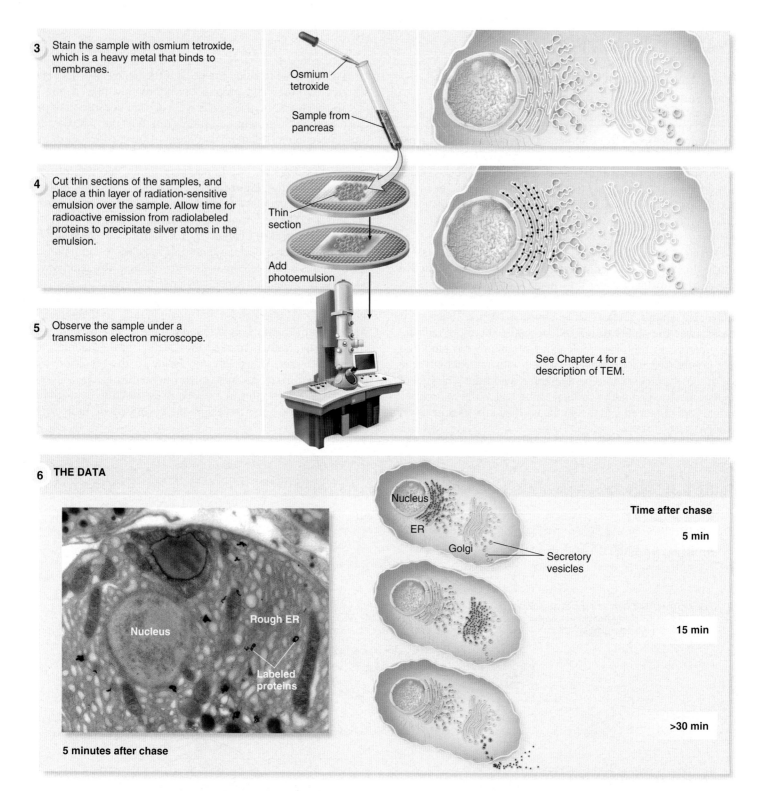

3 Stain the sample with osmium tetroxide, which is a heavy metal that binds to membranes.

Osmium tetroxide

Sample from pancreas

4 Cut thin sections of the samples, and place a thin layer of radiation-sensitive emulsion over the sample. Allow time for radioactive emission from radiolabeled proteins to precipitate silver atoms in the emulsion.

Thin section

Add photoemulsion

5 Observe the sample under a transmisson electron microscope.

See Chapter 4 for a description of TEM.

6 THE DATA

Nucleus

ER

Golgi

Secretory vesicles

Time after chase

5 min

15 min

>30 min

Nucleus

Rough ER

Labeled proteins

5 minutes after chase

The micrograph in the data of Figure 6.11 illustrates the results that were observed 5 minutes after the completion of the pulse-chase injections. Very dark objects, namely radioactive proteins, were observed in the rough ER. As shown schematically to the right of the actual data, later time points indicated that the radioactive proteins moved from the ER to the Golgi, and then to secretory vesicles near the plasma membrane. In this way, Palade followed the intracellular pathway of protein movement. His experiments provided the first evidence that secreted proteins were synthesized into the rough ER and moved through a series of cellular compartments before they were secreted. These findings caused researchers to wonder how proteins are targeted to particular organelles and how they move from one compartment to another. These topics are described next.

Protein Localization Involves Sorting Signals and Vesicle Transport

Since Palade's pioneering studies, scientists have learned a great deal about the localization of proteins. The process of evolution has produced a proteome in every organism in which each type of protein functions in a particular cellular compartment or performs its role after being secreted from the cell. For example, acid hydrolases function in the lysosome to degrade macromolecules. In eukaryotes, most proteins contain short stretches of amino acid sequences that direct them to their correct cellular location. These sequences are called **sorting signals** or **traffic signals**. Each sorting signal is recognized by specific cellular components that facilitate the proper routing of that protein. **Table 6.1** describes the sorting signals found in proteins that are destined for the nucleus, endomembrane system, or secretion. We will consider the signals involved in semi-autonomous organelles later in this chapter.

Most eukaryotic proteins begin their synthesis on ribosomes in the cytosol. The cytosol provides amino acids, which are used as building blocks to make these proteins. The synthesis of a protein, a process called translation, is completed in the cytosol for those proteins destined for the cytosol, nucleus, mitochondria, chloroplasts, or peroxisomes. Cytosolic proteins lack any sorting signal, so they stay there. By comparison, the uptake of proteins into the nucleus, mitochondria, chloroplasts, and peroxisomes occurs after the protein is completely made (that is, completely translated). This is called **post-translational sorting** because sorting does not happen until translation is finished. We will examine the uptake of proteins into the mitochondrion later in the chapter. By comparison, the synthesis of other eukaryotic proteins begins in the cytosol and then halts temporarily until the ribosome has become bound to the ER membrane. After this occurs, translation resumes and the polypeptide is synthesized into the ER lumen or ER membrane. Proteins that are destined for the ER, Golgi, lysosome, vacuole, plasma membrane, or secretion are first directed to the ER. This is **cotranslational sorting** because the first step in the sorting process begins while translation is occurring.

The concept of sorting signals in proteins was first proposed by Günter Blobel in the 1970s. Blobel and colleagues discovered a sorting signal in proteins that sends them to the ER membrane, which is the first step in cotranslational sorting (**Figure 6.12**). To be directed to the rough ER membrane, a polypeptide must contain a sorting signal called an **ER signal sequence**, which is usually located near the amino terminus. As the ribosome is making the polypeptide in the cytosol, the ER signal sequence emerges from the ribosome and is recognized by a protein/RNA complex called a **signal recognition particle (SRP)**. The SRP has two functions. First, it recognizes the ER signal sequence and pauses translation. Second, SRP binds to a receptor in the ER membrane, which docks the ribosome over a channel. At this stage, SRP is released and translation resumes. The growing polypeptide is threaded through the channel to cross the ER membrane. In most cases, the ER signal sequence is removed by signal peptidase. If the protein is not a membrane protein, it will be released into the lumen of the ER. In 1999, Blobel won the Nobel Prize for his discovery of sorting signals in proteins. It is worth noting that the process shown in Figure 6.12 illustrates another important role of protein-protein interactions—a series of interactions causes the steps of a process to occur in a specific order.

Some proteins are meant to function in the ER. Such proteins contain ER retention signals in addition to the ER signal sequence. Alternatively, other proteins that are destined for the Golgi, lysosomes, vacuoles, plasma membrane, or secretion must be sorted to these other locations. Such proteins leave the ER and are transported to their correct location. This transport process occurs via membrane vesicles that are formed from one compartment and then move through the cytosol and fuse with another compartment. Vesicles from the ER may go to the Golgi, and then vesicles from the Golgi may go to the lysosomes, vacuoles, or plasma membrane. Sorting signals within their amino acid sequences are responsible for directing them to their correct location.

Table 6.1	Sorting Signals of Proteins That Are Destined for the Nucleus, Endomembrane System, or Secretion
Type of signal	**Description**
Nuclear-localization signal	Can be located almost anywhere in the polypeptide sequence. The signal sequence is four to eight amino acids in length and contains several positively charged residues and usually one or more prolines.
ER signal	A sequence of about 6 to 12 amino acids near the amino terminus that is composed of mostly nonpolar amino acids.
ER retention signal	A sequence of four amino acids, lysine-aspartic acid-glutamic acid-leucine, located at the carboxyl terminus of the protein.
Golgi retention signal	A sequence of 20 hydrophobic amino acids that form a transmembrane segment flanked by positively charged residues.
Lysosome-sorting signal	A grouping of amino acids in the tertiary structure of of a polypeptide. Positively charged residues within this grouping are thought to play an important role. This grouping causes lysosomal proteins to be covalently modified to contain a mannose-6-phosphate residue (a carbohydrate), which directs the protein to the lysosome.

Destination of a cellular protein	**Type of signal the protein contains within its amino acid sequence**
Nucleus	Nuclear-localization signal.
ER	ER signal and an ER retention signal.
Golgi	ER signal and a Golgi retention signal.
Lysosome	ER signal and a lysosomal sorting signal.
Plasma membrane	ER signal and a hydrophobic transmembrane domain that anchors it in the membrane.
Secretion	ER signal. No additional signal required.

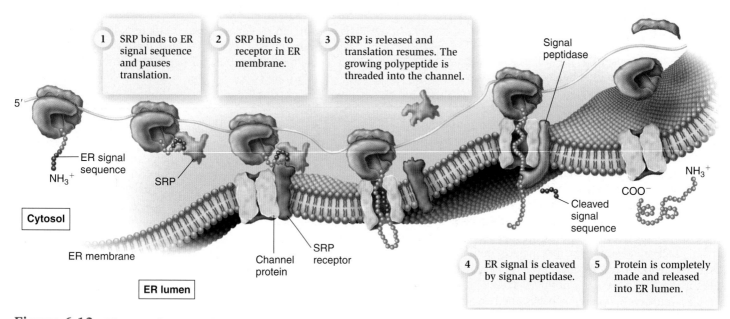

Figure 6.12   First step in cotranslational protein localization: cotranslational sorting.

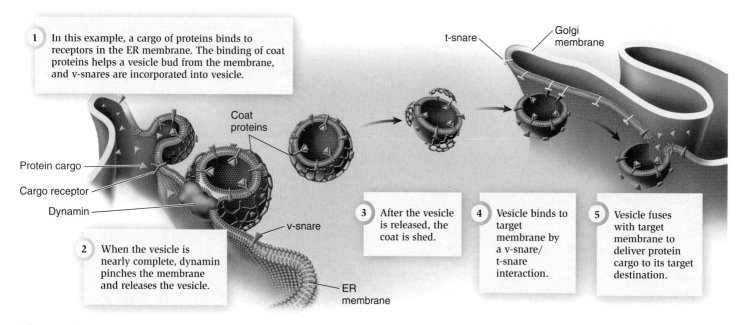

Figure 6.13   Second step in cotranslational protein localization: vesicle transport from the endoplasmic reticulum.

Figure 6.13 describes vesicle transport from the ER to the Golgi. A cargo, such as soluble protein molecules, is loaded into a developing vesicle by binding to cargo receptors in the ER membrane. Vesicle formation is facilitated by **coat proteins**, which help a vesicle to bud from a given membrane. As a vesicle forms, other proteins called **v-snares** are incorporated into the vesicle membrane, hence the name v-snare. Many types of v-snares are known to exist; the particular v-snare that is found in a vesicle membrane depends on the type of cargo that it carries. Finally, a protein called dynamin pinches the membrane and releases the vesicle. After a vesicle is released from one compartment such as the ER, the coat is shed. The vesicle then travels through the cytosol. But how does the vesicle know where to go? The answer is that the v-snares in the vesicle membrane are recognized by **t-snares** in a target membrane. After the v-snare recognizes a t-snare, the vesicle fuses with the membrane containing the t-snare. The recognition between v-snares and t-snares ensures that a vesicle carrying a specific cargo moves to the correct target membrane in the cell. Like the sorting of proteins to the ER membrane, the formation and sorting of vesicles also involves a series of protein-protein interactions that cause the steps to occur in a defined manner.

Most Transmembrane Proteins Are First Inserted into the ER Membrane

Our previous discussion of protein targeting involved a soluble protein (that is, a nonmembrane protein) that had an ER signal sequence. After the signal was removed, the protein was released into the ER lumen. With the exception of proteins destined for semiautonomous organelles, most membrane proteins are recognized by SRPs and synthesized into the ER membrane. From there, they may be transported via vesicles to other membranes of the endomembrane system.

As discussed in Chapter 5, if a sequence within a polypeptide contains a stretch of 20 amino acids that are mostly hydrophobic, this region will become a transmembrane segment. In the example shown in **Figure 6.14**, the polypeptide contains one such sequence. After the ER signal sequence is cleaved or removed, this will create a membrane protein with a single transmembrane segment. Other polypeptides may contain two or more transmembrane segments. Each time a polypeptide sequence contains a stretch of 20 hydrophobic amino acids, an additional transmembrane segment is synthesized into the membrane. For some membrane proteins, the ER signal sequence may not be removed. When it remains, it will usually function as a transmembrane segment.

Glycosylation of Proteins Occurs in the ER and Golgi Apparatus

Glycosylation is the attachment of carbohydrate to a protein, producing a glycoprotein. Carbohydrates may also be attached to lipids by glycosylation, but here we will focus on the glycosylation of proteins. In proteins, glycosylation may aid in protein folding, and it may protect a protein from extracellular factors that could harm its structure. In addition, glycosylation may play a role in protein sorting. For example, proteins destined for the lysosome have attached carbohydrate that serves as a sorting signal (see Table 6.1).

Two forms of protein glycosylation are known to occur in eukaryotes: N-linked and O-linked. N-linked glycosylation, which also occurs in archaea, involves the attachment of a carbohydrate to the amino acid asparagine in a polypeptide chain. It is called N-linked because the carbohydrate attaches to a nitrogen atom of the asparagine side chain. For this to occur, a group of 14 sugar molecules are built onto a lipid called dolichol. This carbohydrate tree is then transferred to an asparagine as a polypeptide is synthesized into the ER lumen (**Figure 6.15**). The carbohydrate tree is attached only to asparagines occurring in the sequence asparagine-X-serine or asparagine-X-threonine, where X could be any amino acid except proline. An enzyme in the ER, oligosaccharide transferase, recognizes this sequence and transfers the carbohydrate tree from dolichol to the asparagine.

Following this initial glycosylation step, the carbohydrate tree is further modified as other enzymes in the ER attach additional sugars or remove sugars. After a glycosylated protein is transferred to the Golgi by vesicle transport, enzymes in the Golgi usually modify the carbohydrate tree as well. N-linked glycosylation commonly occurs on membrane proteins

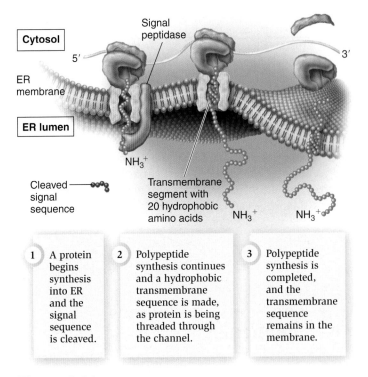

1 A protein begins synthesis into ER and the signal sequence is cleaved.	**2** Polypeptide synthesis continues and a hydrophobic transmembrane sequence is made, as protein is being threaded through the channel.	**3** Polypeptide synthesis is completed, and the transmembrane sequence remains in the membrane.

Figure 6.14   Insertion of membrane proteins into the ER membrane.

Biological inquiry: What structural feature of a protein causes a region to form a transmembrane segment?

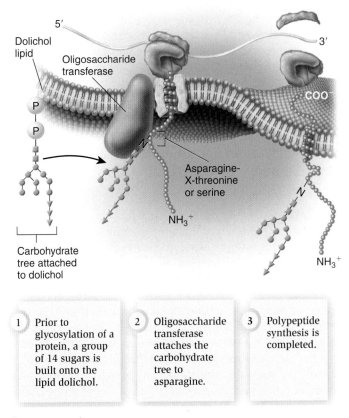

1 Prior to glycosylation of a protein, a group of 14 sugars is built onto the lipid dolichol.	**2** Oligosaccharide transferase attaches the carbohydrate tree to asparagine.	**3** Polypeptide synthesis is completed.

Figure 6.15   N-linked glycosylation in the endoplasmic reticulum.

that are transported to the cell surface. In some cell surface proteins, N-linked glycosylation plays a role in cell-to-cell recognition, a crucial phenomenon in the migration of cells during embryonic development in animals.

The second form of glycosylation, O-linked glycosylation, occurs only in the Golgi apparatus. This form involves the addition of a string of sugars to the oxygen atom of serine or threonine side chains in polypeptides. In animals, O-linked glycosylation is important for the production of proteoglycans, which are highly glycosylated proteins that are secreted from cells and help to organize the extracellular matrix that surrounds cells. Proteoglycans are also a component of mucus, a slimy material that coats many cell surfaces and is secreted into fluids such as saliva. High concentrations of carbohydrates give mucus its slimy texture.

6.3 The Semiautonomous Organelles

The other types of organelles found in the cytosol are the semiautonomous organelles—mitochondria, chloroplasts, and peroxisomes—so named because they divide by fission to produce more of themselves. In Chapter 4, we examined their general structures and functions. Although they are compartments within the cytosol, the mitochondria, chloroplasts, and peroxisomes are somewhat like independent systems. For example, mitochondria and chloroplasts have their own genetic material, synthesize some proteins, and divide independently of when the cell divides. However, they are not entirely autonomous because they rely on the rest of the cell for raw materials and even import most of their proteins from the cytosol. As we will examine in this section, the seemingly mysterious behavior of these organelles can be traced to their evolutionary origin about 2 billion years ago.

Mitochondria and Chloroplasts Contain Their Own Genetic Material and Divide by Binary Fission

To appreciate the structure and organization of mitochondria and chloroplasts, we need to briefly examine their genetic properties. In 1951, Y. Chiba exposed plant cells to Feulgen, a DNA-specific dye, and discovered that the chloroplasts became stained. Based on this observation, he was the first to suggest that chloroplasts contain their own DNA. Researchers in the 1970s and 1980s isolated DNA from both chloroplasts and mitochondria. These studies revealed that the DNA of these organelles resembled smaller versions of bacterial chromosomes.

The chromosomes found in mitochondria and chloroplasts are referred to as the **mitochondrial genome** and **chloroplast genome**, while the chromosomes found in the nucleus of the cell constitute the **nuclear genome**. The genomes of most mito-

chondria and chloroplasts are composed of a single circular, double-stranded chromosome. Compared to the nuclear genome, they are very small. For example, the amount of DNA in the human nuclear genome (about 3 billion base pairs) is about 200,000 times greater than in the mitochondrial genome. In terms of genes, the human genome has approximately 20,000 to 25,000 different genes while the human mitochondrial genome has only several dozen. Chloroplast genomes tend to be larger than mitochondrial genomes, and they have a correspondingly greater number of genes. Depending on the particular species of plant or algae, a chloroplast genome is about 10 times larger than the mitochondrial genome of human cells.

Just as the genomes of mitochondria and chloroplasts resemble bacterial genomes, the production of new mitochondria and chloroplasts bears a striking resemblance to the division of bacterial cells. Like their bacterial counterparts, mitochondria and chloroplasts increase in number via **binary fission**, or splitting in two. **Figure 6.16** illustrates the process for mitochondria. The mitochondrial genome in the nucleoid is duplicated and the organelle divides into two separate organelles. Mitochondrial and chloroplast division are needed to maintain a full

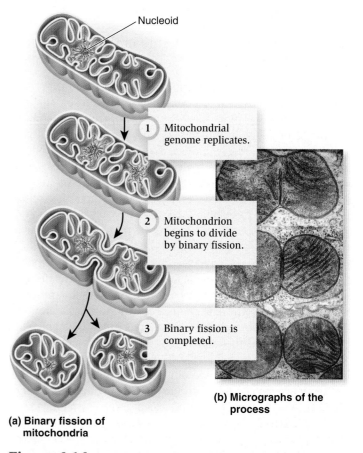

Nucleoid

1 Mitochondrial genome replicates.

2 Mitochondrion begins to divide by binary fission.

3 Binary fission is completed.

(a) Binary fission of mitochondria

(b) Micrographs of the process

Figure 6.16 Division of mitochondria by binary fission. (a) The mitochondrial genome is duplicated and the organelle divides into two separate organelles. (b) Micrographs of the process.

complement of these organelles when cell growth occurs following cell division. In addition, environmental conditions may influence the sizes and numbers of these organelles. For example, when plants are exposed to more sunlight, the number of chloroplasts in leaf cells increases.

GENOMES & PROTEOMES

Mitochondria and Chloroplasts Are Derived from Ancient Symbiotic Relationships

The observation that mitochondria and chloroplasts contain their own genetic material may seem puzzling. Perhaps you might think that it would be simpler for a eukaryotic cell to have all of its genetic material in the nucleus. The distinct genomes of mitochondria and chloroplasts can be traced to their evolutionary origin, which involved an ancient symbiotic association.

A symbiotic relationship occurs when two different species live in direct contact with each other. **Endosymbiosis** describes a symbiotic relationship in which the smaller species—the symbiont—actually lives inside (*endo-*, inside) the larger species. In 1883, Andreas Schimper proposed that chloroplasts were descended from an endosymbiotic relationship between cyanobacteria (a bacterium capable of photosynthesis) and eukaryotic cells. In 1922, Ivan Wallin also hypothesized an endosymbiotic origin for mitochondria.

In spite of these interesting ideas, the question of endosymbiosis was largely ignored until the discovery that mitochondria and chloroplasts contain their own genetic material. In 1970, the issue of endosymbiosis as the origin of mitochondria and chloroplasts was revived by Lynn Margulis in her book *Origin of Eukaryotic Cells*. During the 1970s and 1980s, the advent of molecular genetic techniques allowed researchers to analyze genes from mitochondria, chloroplasts, bacteria, and eukaryotic nuclear genomes. Researchers discovered that genes in mitochondria and chloroplasts are very similar to bacterial genes. Likewise, mitochondria and chloroplasts are strikingly similar in size and shape to certain bacterial species. These observations provided strong support for the **endosymbiosis theory**, which proposes that mitochondria and chloroplasts originated from bacteria that took up residence within a primordial eukaryotic cell (**Figure 6.17**). Over the next 2 billion years, the characteristics of the intracellular bacterial cell gradually changed to those of a mitochondrion or chloroplast. We will return to this topic in Chapter 22.

Symbiosis occurs because the relationship is beneficial to one or both species. According to the endosymbiosis theory, this relationship provided eukaryotic cells with useful cellular characteristics. Chloroplasts, which were derived from cyanobacteria, have the ability to carry out photosynthesis. This benefits plant cells by giving them the ability to use the energy from sunlight. It is less clear how the relationship would have been beneficial to a cyanobacterium. By comparison, mitochondria are

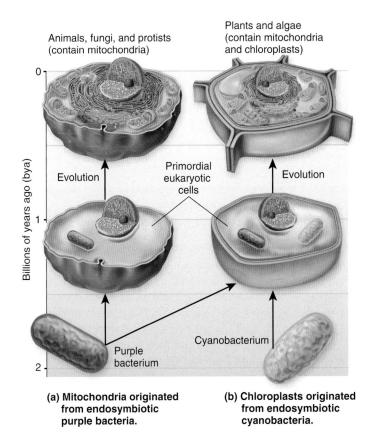

Figure 6.17 **The endosymbiosis theory.** **(a)** According to this concept, modern mitochondria were derived from purple bacteria, also called α-proteobacteria. Over the course of evolution, their characteristics changed into those found in mitochondria today. **(b)** A similar phenomenon occurred for chloroplasts, which were derived from cyanobacteria, a bacterium that is capable of photosynthesis.

Biological inquiry: Discuss the similarities and differences between modern bacteria and mitochondria.

thought to have been derived from a different type of bacteria known as purple bacteria or α-proteobacteria. In this case, the endosymbiotic relationship enabled eukaryotic cells to synthesize greater amounts of ATP.

During the evolution of eukaryotic species, genes that were originally found in the genome of the primordial cyanobacteria and purple bacteria have been transferred from the organelles to the nucleus. This has occurred many times throughout evolution, so that modern mitochondria and chloroplasts have lost most of the genes that still exist in present-day purple bacteria and cyanobacteria. Some biologists have proposed that peroxisomes may also have arisen by an endosymbiotic relationship but have lost all of their genetic material. Alternatively, others suggest that peroxisomes may have their origins in the endomembrane system. The evolutionary origin of peroxisomes remains unclear.

Some researchers speculate that the movement of genes into the nucleus makes it easier for the cell to control the structure, function, and division of mitochondria and chloroplasts. In modern cells, hundreds of different proteins that make up these organelles are encoded by genes that have been transferred to the nucleus. These proteins are made in the cytosol, and then taken up into mitochondria or chloroplasts. We will discuss this topic next.

Most Proteins Are Sorted to Mitochondria, Chloroplasts, and Peroxisomes Post-Translationally

The organization of semiautonomous organelles is largely dependent on the uptake of proteins from the cytosol. Most proteins in mitochondria and chloroplasts, and all proteins in peroxisomes, are synthesized in the cytosol and taken up into their respective organelles. For example, most proteins involved in ATP synthesis are made in the cytosol and taken up into mitochondria after they have been completely synthesized. For this to occur, a protein must have the appropriate targeting sequence

as part of its amino acid sequence. **Table 6.2** summarizes the targeting sequences that direct proteins to mitochondria, chloroplasts, and peroxisomes.

As one example of post-translational sorting, let's consider how a protein is directed to the mitochondrial matrix. As described in **Figure 6.18**, this process involves a series of intricate protein-protein interactions. A protein that is destined for the mitochondrial matrix is first made in the cytosol, where proteins called chaperones keep it in an unfolded state. A receptor protein in the outer mitochondrial membrane recognizes the matrix targeting sequence. The protein is released from the chaperone as it is transferred to a channel in the outer mitochondrial membrane. Because it is in an unfolded state, the mitochondrial protein can be threaded through this channel, and then through another channel in the inner mitochondrial membrane. These channels lie close to each other at contact sites between the outer and inner membranes. As the protein emerges in the matrix, other chaperone proteins that were already in the matrix continue to keep it unfolded. Eventually, the matrix targeting sequence is cleaved and the entire protein is threaded into the matrix. At this stage, the chaperone proteins are released and the protein can adopt its three-dimensional structure.

Table 6.2	Targeting Sequences of Proteins That Are Destined for Semiautonomous Organelles*
Destination	**Mitochondria**
Matrix	Contains a matrix targeting sequence that is usually a short sequence at the amino terminus of a protein with several positively charged residues (that is, lysine and arginine). This sequence folds into an α-helix in which the positive charges are on one face of the helix.
Outer-membrane	Has a matrix targeting sequence followed by a stop transfer sequence that prevents passage through the outer membrane channel.
Intermembrane space	Contains a matrix targeting sequence followed by a cleavage site that causes the release of the protein into the intermembrane space.
Inner-membrane	Contains a matrix targeting sequence and one or more transmembrane regions.
	Chloroplasts
Stroma	Contains a stroma targeting sequence that is an uncharged α-helix with polar amino acids along one side.
Outer-membrane	Not well understood, but may have a stroma targeting sequence and a stop transfer sequence.
Intermembrane space	Not well understood, but may have a stroma targeting sequence and a cleavage site that causes the release of the protein into the intermembrane space.
Inner-membrane	Contains a stroma targeting sequence and one or more transmembrane regions.
Thylakoid membrane	Also has a stroma targeting sequence followed by a thylakoid targeting sequence that is a stretch of hydrophobic amino acids.
	Peroxisomes
Peroxisome membrane and lumen	Most peroxisomal proteins contain a specific sequence of three amino acids, serine-lysine-leucine, that is located near the carboxyl terminus of the protein. A few peroxisomal proteins have a 26 amino acid sorting sequence at the amino terminus.

*These are the most common types of targeting sequences, also called sorting signals. Variations are known to occur in the amino acid sequences that function as the sorting signals.

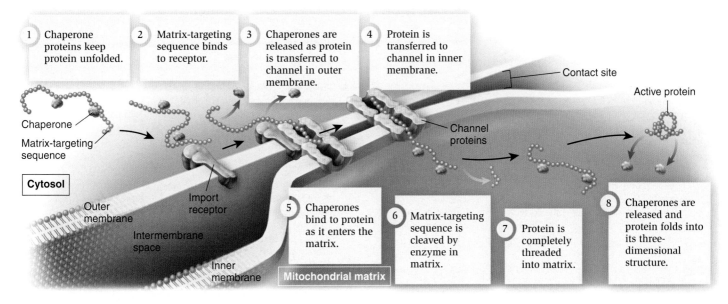

1. Chaperone proteins keep protein unfolded.

2. Matrix-targeting sequence binds to receptor.

3. Chaperones are released as protein is transferred to channel in outer membrane.

4. Protein is transferred to channel in inner membrane.

5. Chaperones bind to protein as it enters the matrix.

6. Matrix-targeting sequence is cleaved by enzyme in matrix.

7. Protein is completely threaded into matrix.

8. Chaperones are released and protein folds into its three-dimensional structure.

Chaperone

Matrix-targeting sequence

Cytosol

Outer membrane

Import receptor

Intermembrane space

Inner membrane

Mitochondrial matrix

Contact site

Active protein

Channel proteins

Figure 6.18 Post-translational sorting of a protein to the mitochondrial matrix.

Biological inquiry: What do you think would happen if chaperone proteins did not bind to a mitochondrial matrix protein before it was imported into the mitochondrion?

CHAPTER SUMMARY

6.1 Principles of Cell Organization

- In systems biology, researchers study living organisms in terms of their underlying network structure—groups of structural and functional connections—rather than their individual molecular components.

- The maintenance of cell structure relies on the genome, functional molecules, and pre-existing organization. Other factors include the ability of certain proteins to interact with each other to form molecular machines, and the dynamic turnover of cellular molecules and macromolecules.

- Modern cells can grow, adapt, and modify their organization, but they do not arise spontaneously from nonliving materials.

- Cell organization depends on genetic information in the cell nucleus, which produces a proteome. (Figure 6.1)

- Interactions among proteins and RNA molecules form molecular machines that carry out complex cellular functions. (Figures 6.2, 6.3)

- The cytoskeleton is a molecular machine that provides cell organization and directs cellular movements. (Figures 6.4, 6.5)

- The maintenance of cell organization is a dynamic process that involves breaking down molecules and recycling their building blocks. (Figure 6.6)

- In eukaryotic cells, four regions work together to produce dynamic organization. The nucleus houses the genome, which plays a key role in producing the proteome. The cytosol is an important coordination center for cell metabolism and organization. The organelles of the endomembrane system perform several important functions in eukaryotic cells. The semiautonomous organelles (mitochondria, chloroplasts, and peroxisomes) perform a variety of crucial functions. (Figure 6.7)

6.2 The Endomembrane System

- In eukaryotic cells, the cytosol and endomembrane system work together to synthesize most lipids. The endomembrane system includes the nuclear envelope, endoplasmic reticulum, Golgi apparatus, lysosomes, vacuoles, and plasma membrane. (Figures 6.8, 6.9, 6.10)

- Palade's pulse-chase experiments demonstrated that secreted proteins move sequentially through the organelles of the endomembrane system. (Figure 6.11)

- Protein localization involves sorting signals and vesicle transport. (Table 6.1, Figures 6.12, 6.13)

- Most transmembrane proteins are first inserted into the ER membrane and then transported via vesicles to other membranes of the endomembrane system. (Figure 6.14)

- Glycosylation of proteins occurs in the ER and Golgi apparatus. Glycosylation may help proteins fold properly, protect them from extracellular factors, and assist in protein sorting. (Figure 6.15)

6.3 The Semiautonomous Organelles

- Mitochondria and chloroplasts contain their own genetic material and divide by binary fission. (Figure 6.16)

- According to the endosymbiosis theory, mitochondria and chloroplasts have evolved from bacteria that took up residence in early eukaryotic cells. (Figure 6.17)

- Most proteins are sorted to mitochondria, chloroplasts, and peroxisomes post-translationally. (Table 6.2, Figure 6.18)

TEST YOURSELF

1. The main structural elements of cells that produce cellular organization are
   a. proteins.
   b. organelles.
   c. membranes.
   d. all of the above.
   e. a and c only.

2. Which of the following statements best supports the requirements to make new cells?
   a. The formation of new cells relies solely on the presence of the genome.
   b. New cell formation requires the correct genetic information and the building blocks necessary to produce the cellular components.
   c. New cell production requires the genome, functional molecules, and pre-existing organization.
   d. The formation of new cells requires the appropriate genetic information and functional molecules for cellular activity only.
   e. All of the above.

3. Large molecular complexes that perform different cellular activities that involve changes in molecular conformation are called molecular
   a. clocks.
   b. motors.
   c. machines.
   d. proteins.
   e. proteomes.

4. Protein conformation is important to protein function. The assembly of many complex proteins relies on _____ , where surfaces of subunits recognize each other and bind together.
   a. protein sorting
   b. traffic signaling
   c. post-transcriptional changes
   d. molecular recognition
   e. proteasome activity

5. In the nucleus, proteins help maintain organization by
   a. forming a meshwork in the nucleus that organizes each chromosome into its own chromosome territory.
   b. organizing the outer nuclear membrane.
   c. organizing the inner nuclear membrane.
   d. all of the above.
   e. a and c only.

6. The cytoskeleton is an important feature of the cytosol that provides organization by
   a. determining cell shape.
   b. determining the structure of the endomembrane system.
   c. transporting proteins to the endoplasmic reticulum for protein sorting.
   d. recycling membrane-bounded proteins.
   e. all of the above.

7. Proteins that function to move lipids from one membrane to another are called
   a. lipases.
   b. membrane-bounded lipoproteins.
   c. proteases.
   d. lipid exchange proteins.
   e. phospholipids.

8. Protein sorting in the cell is possible due to
   a. sorting signals in the amino acid sequences of proteins that determine protein destinations in the cell.
   b. chaperone proteins that function to direct all proteins to the proper location inside or outside the cell.
   c. formation of protein sorting vesicles that carry proteins from the Golgi to lysosomes.
   d. DNA sequences that remain part of the proteins that determine cellular destination.
   e. all of the above.

9. Proteins that remain in the cytosol as opposed to passing through the endomembrane system are sorted
   a. prior to translation by pretranslational sorting.
   b. during translation by cotranslational sorting.
   c. after translation by post-translational sorting.
   d. both b and c.
   e. none of the above.

10. Vesicles move to the appropriate membrane in the cell by the recognition of _____ , proteins in the target membrane that act as binding sites for the vesicle.
    a. v-snares
    b. traffic signals
    c. coat proteins
    d. chaperones
    e. t-snares

CONCEPTUAL QUESTIONS

1. List the components of the endomembrane system and briefly describe the functions of each.

2. Briefly explain how sorting signals function in protein localization.

3. Define glycosylation.

EXPERIMENTAL QUESTIONS

1. Explain the procedure of a pulse-chase experiment as described in Figure 6.11. What was the purpose of the approach?

2. Why were pancreatic cells used for the investigation of Figure 6.11?

3. What were the results of the experiment of Figure 6.11? What did the researchers conclude?

COLLABORATIVE QUESTIONS

1. What roles do the genome and proteome play in the organization of cells?

2. Discuss the theory of how the mitochondria and chloroplast were initially formed.

www.brookerbiology.com

This website includes answers to the Biological Inquiry questions found in the figure legends and all end-of-chapter questions.

7

ENZYMES, METABOLISM, AND CELLULAR RESPIRATION

CHAPTER OUTLINE

Metabolism of food. The food we eat contains a variety of organic molecules that we use for energy.

As discussed in Chapter 2, a **chemical reaction** is a process in which one or more substances are changed into other substances. Such reactions can result in molecules attaching to each other to form larger molecules, molecules breaking apart to form two or more smaller molecules, rearrangements of atoms within molecules, or the transfer of electrons from one atom to another. Every living cell continuously performs thousands of such chemical reactions to sustain life. The term **metabolism** is used to describe the sum total of all chemical reactions that occur within an organism. The term also refers to a specific set of chemical reactions occurring at the cellular level. For example, biologists may speak of sugar metabolism or fat metabolism. Most metabolism involves the breakdown or synthesis of organic molecules. In Chapter 6, we learned that cells maintain their structure by using organic molecules. Such molecules provide the building blocks to construct cells, and the chemical bonds within organic molecules store energy that can be used to drive cellular processes. A key emphasis of this chapter is how, through chemical reactions, cells utilize energy that is stored within the chemical bonds of organic molecules such as the sugar glucose.

When we eat food, we are using much of that food for energy. People often speak of "burning calories." While it is true that metabolism does generate some heat, the chemical reactions that take place in the cells of living organisms are uniquely different from those that occur, say, in a fire. When wood is burned, the reaction produces enormous amounts of heat in a short period of time; the reaction lacks control. In contrast, the metabolism that occurs in living cells is extremely controlled. The food molecules from which we harvest energy give up that energy in a very restrained manner rather than all at once, as in a fire. An underlying theme in metabolism involves the remarkable control that cells possess when coordinating chemical reactions that utilize energy.

In this chapter, we will begin with a general discussion of chemical reactions. We will examine what factors control the direction of a chemical reaction and what determines its rate, paying particular attention to the role of enzymes, a type of biological catalyst. We then consider metabolism at the cellular level. First, we will examine some of the general features of chemical reactions that are vital for the energy needs of living cells. We will also explore the variety of ways in which metabolic processes are regulated and survey a group of chemical reactions that involves the breakdown of carbohydrates, namely, the sugar glucose. As you will learn, cells carry out an intricate series of reactions so that glucose can be "burned" in a very controlled fashion. Finally, the last section explores the synthesis of natural products that play specialized roles in many species.

7.1 Energy, Chemical Reactions, and Enzymes

Two general factors govern the fate of a given chemical reaction in a living cell—its direction and rate. To illustrate this point, let's consider a generalized chemical reaction such as

$$aA + bB \rightleftharpoons cC + dD$$

where A and B are the reactants, C and D are the products, and a, b, c, and d are the number of moles of reactants and products. This reaction is reversible, which means that A + B could be converted to C + D, or C + D could be converted to A + B. The direction of the reaction, whether C + D are made (the forward direction) or A + B are made (the reverse direction), depends on energy and on the concentrations of A, B, C, and D.

In this section, we will begin by examining the interplay of energy and the concentration of reactants as they govern the direction of a chemical reaction. You will learn that cells use energy intermediate molecules, such as ATP, to drive chemical reactions in a desired direction. Many types of chemical reactions, particularly those that involve organic molecules, proceed at a very slow rate unless facilitated by a biological catalyst. In the second part of this section, we will examine how catalysts called enzymes are critical cellular components that speed up the rates of chemical reactions in living organisms.

Energy Exists in Many Forms

To understand why a chemical reaction occurs, we first need to consider **energy**, which we will define as the ability to promote change. Physicists often consider energy in two forms: kinetic energy and potential energy (**Figure 7.1**). **Kinetic energy** is energy associated with movement, such as the movement of a baseball bat from one location to another. By comparison, **potential energy** is the energy that a substance possesses due to its structure or location. The energy contained within covalent bonds in molecules is a type of potential energy called **chemical energy**. The breakage of those bonds is one way that living cells can harness this energy to perform cellular functions. **Table 7.1** summarizes chemical and other forms of energy that are important in biological systems.

An important issue in biology is the ability of energy to be converted from one form to another. The study of energy interconversions is called **thermodynamics**. Physicists have determined that two laws govern energy interconversions:

1. **The first law of thermodynamics**—The first law states that energy cannot be created or destroyed; it is also called the *law of conservation of energy*. However, energy can be transferred from one place to another and can be

(a) Kinetic energy   (b) Potential energy

Covalent bonds in glucose store energy.

Figure 7.1   Examples of energy.   **(a)** Kinetic energy, such as swinging a bat, is energy associated with motion. **(b)** Potential energy is stored energy. Chemical bonds in molecules such as glucose store large amounts of chemical energy.

transformed from one type to another (as when, for example, chemical energy is transformed into heat).

2. **The second law of thermodynamics**—The second law states that the transfer of energy or the transformation of energy from one form to another increases the **entropy**, or degree of disorder of a system (**Figure 7.2**). When energy is converted from one form to another, the increase in entropy causes some energy to become unusable by living organisms.

Next, we will see how these two laws place limits on the ways that living cells can use energy for their own needs.

The Change in Free Energy Determines the Direction of a Chemical Reaction or Any Other Cellular Process

From the perspective of living organisms, energy is a critical component that is necessary for life to exist. Energy powers many cellular processes including chemical reactions, cellular

Table 7.1	Types of Energy That Are Important in Biology	
Energy type	**Description**	**Biological example**
Light	Light is a form of electromagnetic radiation that is visible to the eye. The energy of light is packaged in photons.	During photosynthesis, light energy is captured by pigments in chloroplasts (Chapter 8). Ultimately, this energy is used to reduce carbon, thus producing organic molecules.
Heat	Heat is the transfer of kinetic energy from one object to another, or from an energy source to an object. In biology, heat is often viewed as energy that can be transferred due to a difference in temperature between two objects or locations.	Many organisms, such as humans, maintain their bodies at a constant temperature. This is achieved by chemical reactions that generate heat.
Mechanical	Mechanical energy is the energy that is possessed by an object due to its motion or its position relative to other objects.	In animals, mechanical energy is associated with movements due to muscle contraction such as walking.
Chemical	Chemical energy is energy stored in the chemical bonds of molecules. When the bonds are broken and rearranged, this can release large amounts of energy.	The covalent bonds in organic molecules, such as glucose and ATP, store large amounts of energy. When these bonds are broken, the chemical energy released can be used to drive cellular processes.
Electrical/Ion gradient	The movement of charge or the separation of charge can provide energy. Also, a difference in ion concentration across a membrane constitutes an electrochemical gradient gradient, which is a source of potential energy.	High-energy electrons can release energy (that is, drop down to lower energy levels). The energy that is released can be used to drive cellular processes such as the pumping of H^+ across membranes (as discussed later in this chapter).

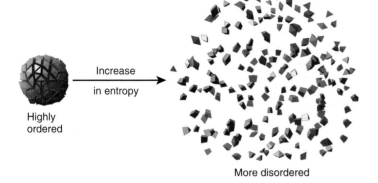

Figure 7.2 Entropy. Entropy is a measure of the disorder of a system. An increase in entropy means an increase in disorder.

Biological inquiry: Which do you think has more entropy, a NaCl crystal at the bottom of a beaker of water or the same beaker of water after the Na+ and Cl− in the crystal have dissolved in the water?

movements such as those occurring in muscle contraction, and the maintenance of cell organization. To understand how living organisms use energy, we need to distinguish between the energy that can be used to do work (usable energy) and the energy that cannot do work (unusable energy).

Total energy = Usable energy + Unusable energy

Why is some energy unusable? The main culprit is entropy. As stated by the second law of thermodynamics, energy transformations involve an increase in entropy, a measure of the disorder that cannot be harnessed to do work. For living organisms, the total energy is termed **enthalpy (H)** and the usable energy—the amount of available energy that can be used to do work—is called the **free energy (G)**. The letter G is in recognition of J. Willard Gibbs, who proposed the concept of free energy in 1878. The unusable energy is the system's entropy (S). Gibbs proposed that these three factors are related to each other in the following way:

$$H = G + TS$$

where T is the absolute temperature in Kelvin (K). Because our focus is on free energy, we can rearrange this equation as

$$G = H - TS$$

A critical issue in biology is whether a process will or will not occur spontaneously. For example, will glucose be broken down into carbon dioxide and water? Another way of framing this question is to ask, Is the breakdown of glucose a spontaneous reaction? A spontaneous reaction or process is one that will occur without an additional input of energy. However, a spontaneous reaction does not necessarily proceed quickly. In some cases, the rate of a spontaneous reaction can be quite slow.

The key way to evaluate if a reaction is spontaneous is to determine the free energy change that occurs as a result of the chemical reaction.

$$\Delta G = \Delta H - T\Delta S$$

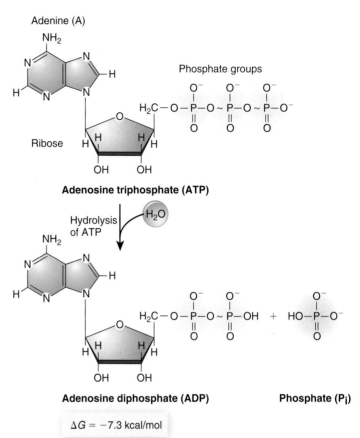

Adenosine triphosphate (ATP)

Hydrolysis of ATP

Adenosine diphosphate (ADP) **Phosphate (Pi)**

$\Delta G = -7.3$ kcal/mol

Figure 7.3 The hydrolysis of ATP to ADP and Pi. As shown in this figure, ATP has a net charge of −4, while ADP and Pi are shown with net charges of −2 each. When these compounds are shown in chemical reactions with other molecules, the net charges will also be shown. Otherwise, these compounds will simply be designated ATP, ADP, and Pi. It should also be noted that at neutral pH, ADP^{2-} will dissociate to ADP^{3-} and H^+. Likewise, P_i^{2-} also dissociates to $P_i^{3-} + H^+$ at neutral pH.

where the Δ sign indicates a change, such as before and after a chemical reaction. If the reaction has a negative free energy change (ΔG < 0), this means that free energy is released. Such a reaction is said to be **exergonic.** Exergonic reactions are spontaneous. Alternatively, if a reaction has a positive free energy change (ΔG > 0), requiring the addition of free energy from the environment, it is termed **endergonic.** An endergonic reaction is not a spontaneous reaction.

If ΔG for a chemical reaction is negative, the reaction favors the formation of products, while a reaction with a positive ΔG favors the formation of reactants. Chemists have determined free energy changes for a variety of chemical reactions. **Adenosine triphosphate (ATP)** is a molecule that is a common energy source for all cells. Let's look at the breakdown of ATP to adenosine diphosphate (ADP) and inorganic phosphate (Pi). Because water is used to remove a phosphate group, chemists refer to this as the hydrolysis of ATP (**Figure 7.3**). In the reaction of converting 1 mole of ATP to 1 mole of ADP and Pi, ΔG equals −7.3 kcal/mole. Because this is a negative value, the reaction strongly favors the formation of products. As discussed later in this chapter, the energy that is liberated by the hydrolysis of ATP is used to drive a variety of cellular processes.

Even when a chemical reaction is associated with a negative free energy change, not all of the reactants are converted to products. The reaction reaches a state of **chemical equilibrium** in which the rate of formation of products equals the rate of formation of reactants. According to the generalized equation:

$$aA + bB \rightleftharpoons cC + dD$$

An equilibrium occurs, such that:

$$K_{eq} = \frac{[C]^c[D]^d}{[A]^a[B]^b}$$

where K_{eq} is the equilibrium constant.

Biologists make two simplifying assumptions when determining values for equilibrium constants. First, the concentration of water does not change during the reaction and the pH remains constant at pH 7. The equilibrium constant under these conditions is designated K_{eq}'. If water is one of the reactants, as in a hydrolysis reaction, it is not included in the chemical equilibrium equation. As an example, let's consider the chemical equilibrium for the hydrolysis of ATP.

$$ATP^{4-} + H_2O \rightleftharpoons ADP^{2-} + P_i^{2-}$$

$$K_{eq}' = \frac{[ADP][P_i]}{[ATP]}$$

Experimentally, the value for K_{eq}' for this reaction has been determined and found to be approximately 1,650,000 M. Such a large value indicates that the equilibrium greatly favors the formation of products—ADP and P_i.

Cells Use ATP to Drive Endergonic Reactions

An important issue that faces living organisms is that many vital processes require the addition of free energy, that is, they are endergonic and will not occur spontaneously. Fortunately, organisms have a strategy to overcome this problem. If an endergonic reaction is coupled to an exergonic reaction, the endergonic reaction will proceed spontaneously if the net free energy change for both processes combined is negative. For example, consider the following reactions:

Glucose + phosphate^{2-} → Glucose-phosphate^{2-} + H$_2$O
$$\Delta G = +3.3 \text{ kcal/mole}$$

ATP^{4-} + H$_2$O → ATP^{2-} + P$_i^{2-}$ $\Delta G = -7.3$ kcal/mole

Coupled reaction:

Glucose + ATP^{4-} → Glucose-phosphate^{2-} + ADP^{2-}
$$\Delta G = -4.0 \text{ kcal/mole}$$

The first reaction, in which phosphate is covalently attached to glucose, is endergonic, while the second, the hydrolysis of ATP, is exergonic. By itself, the first reaction would not be spontaneous. If the two reactions are coupled, however, the net free energy change for both reactions combined is exergonic. In the coupled reaction, a phosphate is directly transferred from ATP to glucose; this coupled reaction proceeds spontaneously because the net free energy change is negative. As discussed later, the transfer of phosphate from ATP to glucose is a first step in the breakdown of glucose to smaller molecules. Exergonic reactions, such as the breakdown of ATP, are commonly coupled to cellular processes that would otherwise be endergonic.

Enzymes Increase the Rates of Chemical Reactions

Thus far we have examined aspects of energy and considered how the laws of physics are related to the direction of chemical reactions. If a chemical reaction has a negative free energy change, the reaction will be spontaneous; it will tend to proceed in the direction of reactants to products. While thermodynamics governs the direction of an energy transformation, it does not control the rate of a chemical reaction. For example, the breakdown of the molecules in gasoline to smaller molecules is a highly exergonic reaction. Even so, we could place gasoline and oxygen in a container and nothing much would happen (provided it wasn't near a flame). If we came back several days later, we would expect to see the gasoline still sitting there. Perhaps if we came back in a few million years, the gasoline would have been broken down. On a timescale of months or a few years, however, the chemical reaction would proceed very slowly.

For most chemical reactions in cells to proceed at a rapid pace, a catalyst is needed. A **catalyst** is an agent that speeds up the rate of a chemical reaction without being consumed during the reaction. In living cells, the most common catalysts are **enzymes**, protein molecules that accelerate chemical reactions. The term was coined in 1876 by a German physiologist, Wilhelm Kühne, who discovered the enzyme trypsin in pancreatic juice. Interestingly, a few biological catalysts are RNA molecules called **ribozymes**. For example, RNA molecules within ribosomes catalyze the formation of bonds that link amino acids to each other.

Why are catalysts necessary to speed up a chemical reaction? When a covalent bond is broken or formed, this process initially involves the straining or contortion of one or more bonds in the starting molecule(s) and it may involve the positioning of two molecules so that they interact with each other properly. Let's consider the reaction in which ATP is used to attach a phosphate to glucose.

Glucose + ATP^{4-} → Glucose-phosphate^{2-} + ADP^{2-}

For a reaction to occur between glucose and ATP, they must collide in the correct orientation and possess enough energy so that chemical bonds can be changed. As glucose and ATP approach each other, their electron clouds cause repulsion. To overcome this repulsion, an initial input of energy, called the **activation energy**, is required (**Figure 7.4**). Activation energy allows the molecules to get close enough to cause a rearrangement of bonds. With the input of activation energy, glucose and ATP can achieve a **transition state** in which the original bonds have stretched to their limit. Once the reactants have reached

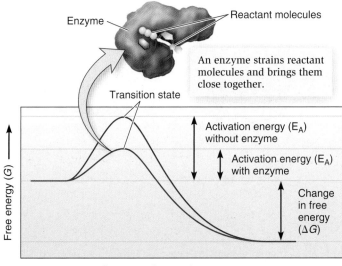

Figure 7.4   Activation energy of a chemical reaction. This figure depicts an exergonic reaction. The activation energy is needed for molecules to achieve a transition state. One way that enzymes lower the activation energy is by straining the reactants so that less energy is required to attain the transition state. A second way is by binding two reactants so they are close to each other and in a favorable orientation.

Biological inquiry: How does lowering the activation energy affect the rate of a chemical reaction?

the transition state, the chemical reaction can readily proceed to the formation of products.

The activation energy required to achieve the transition state is a barrier to the formation of products. This barrier is the reason why the rate of many chemical reactions is very slow. There are two common ways to overcome this barrier and thereby accelerate a chemical reaction. First, the reactants could be exposed to a large amount of heat. For example, as we noted previously, if gasoline is sitting at room temperature, nothing much happens. However, if the gasoline is exposed to a flame or spark, it breaks down rapidly, perhaps at an explosive rate! Alternatively, a second strategy is to lower the activation energy barrier. Enzymes lower the activation energy to a point where a small amount of available heat can push the reactants to a transition state.

How do enzymes work to lower the activation energy barrier of chemical reactions? Enzymes are large proteins that bind relatively small reactants (Figure 7.4). When bound to an enzyme, the bonds in the reactants can be strained (that is, stretched), thereby making it easier for them to achieve the transition state. This is one way that enzymes lower the activation energy. In addition, when a chemical reaction involves two or more reactants, the enzyme provides a site where the reactants are positioned very close to each other and in an orientation that facilitates the formation of new covalent bonds. This also lowers the necessary activation energy for a chemical reaction.

Straining the reactants and bringing them close together are two common ways that enzymes lower the activation energy barrier. In addition, enzymes may facilitate a chemical reaction by changing the local environment of the reactants. For example, amino acids in an enzyme may have charges that affect the chemistry of the reactants. In some cases, enzymes lower the activation energy by directly participating in the chemical reaction. For example, certain enzymes that hydrolyze ATP form a covalent bond between phosphate and an amino acid in the enzyme. However, this is a very temporary condition. The covalent bond between phosphate and the amino acid is quickly broken, releasing phosphate and returning the amino acid back to its original condition. An important example of such an enzyme is Na^+/K^+-ATPase, described in Chapter 5 (refer back to Figure 5.23).

Enzymes Recognize Their Substrates with High Specificity and Undergo Conformational Changes

Thus far, we have considered how enzymes lower the activation energy of a chemical reaction and thereby increase its rate. Let's consider some other features of enzymes that enable them to serve as effective catalysts in chemical reactions. The **active site** is the location in an enzyme where the chemical reaction takes place. The **substrates** for an enzyme are the reactant molecules and/or ions that bind to an enzyme at the active site and participate in the chemical reaction. For example, hexokinase is an enzyme whose substrates are glucose and ATP (**Figure 7.5**). The binding between an enzyme and substrate produces an **enzyme-substrate complex**.

A key feature of nearly all enzymes is that they bind their substrates with a **high affinity** or high degree of specificity. For example, because hexokinase binds glucose very well, we say it has a high affinity for glucose. By comparison, hexokinase has a low affinity for other sugars, such as fructose and galactose, which have similar structures to glucose. In 1894, the German scientist Emil Fischer proposed that the recognition of a substrate by an enzyme resembles the interaction between a lock and key: only the right-sized key (the substrate) will fit into the keyhole (active site) of the lock (the enzyme). Further research revealed that the interaction between an enzyme and its substrates also involves movements or conformational changes in the enzyme itself. As shown in Figure 7.5, these conformational changes cause the substrates to bind more tightly to the enzyme, a phenomenon called **induced fit**. Only after this conformational change takes place does the enzyme catalyze the conversion of reactants to products.

Some enzymes require additional nonprotein molecules or ions to carry out their functions. **Prosthetic groups** are small molecules that are permanently attached to the surface of an enzyme and aid in catalysis. **Cofactors** are usually inorganic ions, such as Fe^{3+} or Zn^{2+}, that temporarily bind to the surface of an enzyme and promote a chemical reaction. Finally, some enzymes use **coenzymes**, organic molecules that participate in the chemical reaction but are left unchanged after the reaction is completed.

The ability of enzymes to increase the rate of a chemical reaction is also affected by the surrounding conditions. In particular, the temperature and pH play an important role in the proper functioning of enzymes. Most enzymes function maximally in a narrow range of temperature and pH. For example, many human

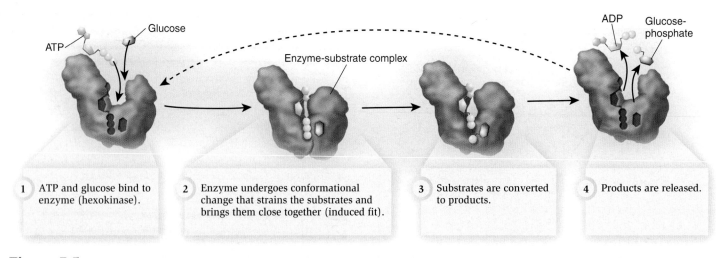

1 ATP and glucose bind to enzyme (hexokinase).

2 Enzyme undergoes conformational change that strains the substrates and brings them close together (induced fit).

3 Substrates are converted to products.

4 Products are released.

Figure 7.5 **The steps of an enzyme-catalyzed reaction.**   The example shown here involves the enzyme hexokinase, which binds glucose and ATP. The products are glucose-phosphate and ADP, which are released from the enzyme.

enzymes work best at 37°C (98.6°F), which is the body's normal temperature. If the temperature was several degrees above or below this value, the function of many cytosolic enzymes would be greatly inhibited. Similarly, enzyme function is sensitive to pH. Many cytosolic enzymes function optimally at pH 7.2, the pH normally found in the cytosol of humans cells. If the pH was significantly above or below this value, enzyme function would be decreased.

7.2 Overview of Metabolism

In the previous section, we examined the underlying factors that govern individual chemical reactions. In living cells, chemical reactions are often coordinated with each other and occur in sequences called **metabolic pathways**, each step of which is catalyzed by a specific enzyme (**Figure 7.6**). These pathways are categorized according to whether the reactions lead to the breakdown or synthesis of substances. **Catabolic reactions** result in the breakdown of molecules into smaller molecules. Such reactions are often exergonic. By comparison, **anabolic reactions** promote the synthesis of larger molecules from smaller precursor molecules. This process usually is endergonic and, in living cells, must be coupled to an exergonic reaction. In this section, we will survey the general features of catabolic and anabolic reactions, and explore the ways in which these metabolic pathways are controlled.

Catabolic Reactions Recycle Organic Building Blocks and Produce Energy Intermediates Such as ATP and NADH

Catabolic reactions involve the breakdown of macromolecules or smaller organic molecules. One reason for the breakdown of these molecules is to recycle the building blocks needed to construct new macromolecules. For example, proteins are composed of a linear sequence of amino acids. When a protein is improp-

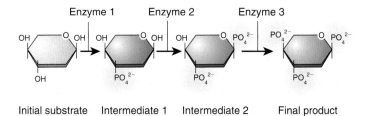

Initial substrate   Intermediate 1   Intermediate 2   Final product

Figure 7.6 **A metabolic pathway.**   In a metabolic pathway, a series of different enzymes catalyze the changes in the structure of a molecule, beginning with a starting substrate and ending with a final product.

erly folded or is no longer needed by a cell, the peptide bonds between amino acids in such a protein are broken by enzymes called proteases. This generates amino acids that can be used in the construction of new proteins.

$$\text{Protein} \xrightarrow{\text{Proteases}} \text{Many individual amino acids}$$

Similarly, the breakdown of RNA by enzymes called nucleases produces nucleotides that can be used in the synthesis of new RNA molecules.

$$\text{RNA} \xrightarrow{\text{Nucleases}} \text{Many individual nucleotides}$$

The breakdown of macromolecules, such as unneeded proteins and RNA molecules, allows a cell to recycle the building blocks that comprise those macromolecules and use them to make new macromolecules.

A second reason for the breakdown of macromolecules and smaller organic molecules is to obtain energy that can be used to drive endergonic processes in the cell. Covalent bonds store a large amount of energy. However, when cells break covalent bonds in organic molecules such as carbohydrates and proteins, they do not directly use the energy that is released in this process.

Instead, the released energy is stored in **energy intermediates**, molecules such as ATP and NADH, that are directly used to drive endergonic reactions in cells.

As an example, let's consider the breakdown of glucose into two molecules of pyruvate. As we'll discuss later in this chapter, the breakdown of glucose to pyruvate involves a catabolic pathway called glycolysis. Some of the energy that is released during the breakage of covalent bonds in glucose is harnessed to synthesize ATP. However, this does not occur in a single step. Rather, glycolysis involves a series of steps in which covalent bonds are broken and rearranged. This process creates molecules that can readily donate a phosphate group to ADP, thereby creating ATP. For example, phosphoenolpyruvate has a phosphate group attached to pyruvate. Due to the arrangement of bonds in phosphoenolpyruvate, this phosphate bond is easily broken. Therefore, the phosphate can be readily transferred to ADP

$$\text{Phosphoenolpyruvate + ADP} \rightarrow \text{Pyruvate + ATP}$$
$$\Delta G = -7.5 \text{ kcal/mole}$$

This is an exergonic reaction and therefore favors the formation of products. In this step of glycolysis, the breakdown of an organic molecule, namely, phosphoenolpyruvate, results in the synthesis of an energy intermediate molecule, ATP, which can then be used by a cell to drive endergonic reactions. This way of synthesizing ATP, termed **substrate-level phosphorylation**, occurs when an enzyme directly transfers a phosphate from one molecule to a different molecule. In this case, a phosphate is transferred from phosphoenolpyruvate to ADP. Another way to make ATP is via **chemiosmosis**. In this process, energy stored in an ion electrochemical gradient is used to make ATP from ADP and P_i. We will consider this mechanism later in the chapter.

An important event that may occur during the breakdown of small organic molecules is **oxidation**, a process that involves the removal of electrons. This event is called oxidation because oxygen is frequently involved in chemical reactions that remove electrons from other molecules. By comparison, **reduction** is the addition of electrons to an atom or molecule. Reduction is so named because the addition of a negatively charged electron reduces the net charge of a molecule.

Electrons do not exist freely in solution. When an atom or molecule is oxidized, the electron that is removed must be transferred to another atom or molecule, which becomes reduced. This type of reaction is termed a **redox reaction**, which is short for a reduction-oxidation reaction. As a generalized equation, an electron may be transferred from molecule A to molecule B as follows:

$$Ae^- + B \rightarrow A + Be^-$$

As shown in the right side of this reaction, A has been oxidized (that is, had an electron removed) and B has been reduced (that is, had an electron added). In general, a substance that has been oxidized has less energy, while a substance that has been reduced has more energy.

During the oxidation of organic molecules such as glucose, the electrons are used to create energy intermediates such as

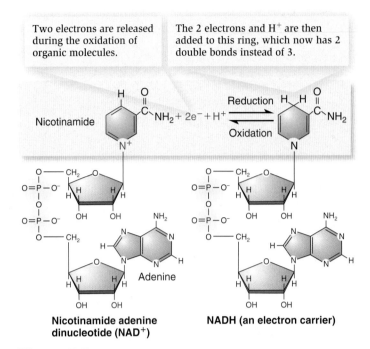

| Two electrons are released during the oxidation of organic molecules. | The 2 electrons and H⁺ are then added to this ring, which now has 2 double bonds instead of 3. |

Figure 7.7   The reduction of NAD⁺ to create NADH. NAD⁺ is composed of two nucleotides, one with an adenine base and one with a nicotinamide base. The oxidation of organic molecules releases electrons that can bind to NAD⁺ (and along with a hydrogen ion) result in the formation of NADH. The two electrons and H⁺ are incorporated into the nicotinamide ring. Note: the actual net charges of NAD⁺ and NADH are minus one and minus two, respectively. They are designated NAD⁺ and NADH to emphasize the net charge of the nicotinamide ring, which is involved in oxidation-reduction reactions.

NADH (**Figure 7.7**). In this process, an organic molecule has been oxidized and **NAD⁺** (**nicotinamide adenine dinucleotide**) has been reduced to NADH. Cells use NADH in two common ways. First, as we will see, the oxidation of NADH is a highly exergonic reaction that can be used to make ATP. Second, NADH can donate electrons to other organic molecules and thereby energize them. Such energized molecules can more readily form covalent bonds. Therefore, as described next, NADH is often needed in anabolic reactions that involve the synthesis of larger molecules through the formation of covalent bonds between smaller molecules.

Anabolic Reactions Require an Input of Energy to Make Larger Molecules

Anabolic reactions are also called **biosynthetic reactions**, because they are necessary to make larger molecules and macromolecules. We will examine the synthesis of macromolecules in several chapters of this textbook. For example, RNA and protein biosynthesis are described in Chapter 12. Cells also need to synthesize small organic molecules, such as amino acids and fats, if they are not readily available from food sources. Such molecules are made by the formation of covalent linkages between precursor molecules. For example, glutamate (an amino acid) is made by the covalent linkage between α-ketoglutarate (a product of sugar metabolism) and ammonium.

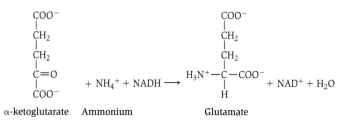

In both reactions, an energy intermediate molecule such as NADH or ATP is needed to drive the reaction forward.

Another amino acid, glutamine, is made from glutamate and ammonium.

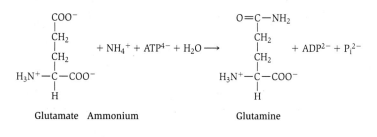

In both reactions, an energy intermediate molecule such as NADH or ATP is needed to drive the reaction forward.

GENOMES & PROTEOMES

Many Proteins Use ATP as a Source of Energy

Over the past several decades, researchers have studied the functions of many types of proteins and discovered numerous examples in which a protein uses ATP to drive a cellular process (**Table 7.2**). In humans, a typical cell uses millions of ATP molecules per second. At the same time, the breakdown of food molecules releases energy that allows us to make more ATP from ADP and P_i. The turnover of ATP occurs at a remarkable pace. An average person hydrolyzes about 100 pounds of ATP per day, yet we do not have 100 pounds of ATP in our bodies. For this to happen, each ATP undergoes about 10,000 cycles of hydrolysis and resynthesis (from ADP and P_i) during an ordinary day (**Figure 7.8**).

By studying the structures of many proteins that use ATP, biochemists have discovered that particular amino acid sequences within proteins function as ATP-binding sites. This information has allowed researchers to predict whether a newly discovered protein uses ATP or not. When an entire genome sequence of a species is experimentally determined, the genes that encode proteins can be analyzed to find out if the encoded proteins have ATP-binding sites in their amino acid sequences. Using this approach, researchers have been able to analyze proteomes—all of the proteins that a given cell can make—and estimate the percentage of proteins that are able to bind ATP. (Most of these proteins are expected to use ATP as a source of energy, though some of them may simply bind ATP without hydrolyzing it to ADP and P_i.) This approach has been applied to the proteomes of bacteria, archaea, and eukaryotes. On average, over 20% of all proteins bind ATP. However, this number

Table 7.2	Examples of Proteins That Use ATP for Energy
Type	**Description**
Metabolic enzymes	Many enzymes use ATP to catalyze endergonic reactions. For example, hexokinase uses ATP to attach phosphate to glucose.
Transporters	Ion pumps, such as the Na^+/K^+-ATPase, use ATP to pump ions against a gradient (see Chapter 5).
Motor proteins	Motor proteins such as myosin use ATP to facilitate cellular movement, as in muscle contraction (see Chapter 46).
Chaperones	Chaperones are proteins that use ATP to aid in the folding and unfolding of cellular proteins (see Chapter 6).
DNA-modifying enzymes	Many proteins such as helicases and topoisomerases use ATP to modify the conformation of DNA (see Chapter 11).
Aminoacyl-tRNA synthetases	These enzymes use ATP to attach amino acids to tRNAs (see Chapter 12).
Protein kinases	Protein kinases are regulatory proteins that use ATP to attach a phosphate to proteins, thereby affecting the function of the phosphorylated protein (see Chapter 9).

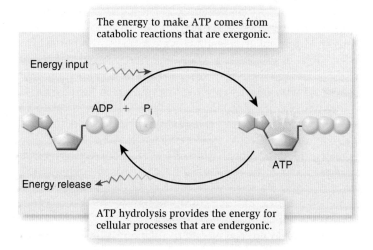

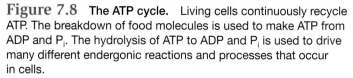

Figure 7.8 **The ATP cycle.** Living cells continuously recycle ATP. The breakdown of food molecules is used to make ATP from ADP and P_i. The hydrolysis of ATP to ADP and P_i is used to drive many different endergonic reactions and processes that occur in cells.

Biological inquiry: If a large amount of ADP was broken down in the cell, how would this affect the ATP cycle?

is likely to be an underestimate of the total percentage of ATP-utilizing proteins because we may not have identified all of the types of ATP-binding sites in proteins. In humans, who have an estimated genome size of 20,000 to 25,000 different genes, a minimum of 4,000 to 5,000 of those genes encode proteins that use ATP. From these numbers, we can see the enormous importance of ATP as a source of energy for living cells.

Metabolic Pathways Are Regulated in Three General Ways

Before we end our general discussion of metabolism, let's consider the various ways in which chemical reactions are regulated in living cells. The regulation of catabolic pathways is important, so that a cell breaks down organic molecules when energy is needed but conserves them when an adequate supply of energy intermediates is available. The control of anabolic pathways is essential so that a cell does not waste energy making too much of the products of such pathways. The regulation of catabolic and anabolic pathways occurs at the genetic, cellular, and biochemical levels.

Gene Regulation Because enzymes in every metabolic pathway are encoded by genes, one way that cells control chemical reactions is via gene regulation. For example, if a bacterial cell is not exposed to a particular sugar in its environment, it will turn off the genes that encode the enzymes that are needed to break down that sugar. Alternatively, if the sugar becomes available, the genes are switched on. Chapter 13 examines the steps of gene regulation in detail.

Cellular Regulation Metabolism is also coordinated at the cellular level. Cells integrate signals from their environment and adjust their chemical reactions to adapt to those signals. As discussed in Chapter 9, cell-signaling pathways often lead to the activation of protein kinases that covalently attach a phosphate group to target proteins. For example, when people are frightened, they secrete a hormone called epinephrine into their bloodstream. This hormone binds to the surface of muscle cells and stimulates an intracellular pathway that leads to the phosphorylation of several intracellular proteins, including enzymes involved in carbohydrate metabolism. These activated enzymes promote the breakdown of carbohydrates, an event that supplies the frightened individual with more energy. Epinephrine is sometimes called the "fight-or-flight" hormone because the added energy prepares an individual to either stay and fight or run away. After a person is no longer frightened, hormone levels drop and other enzymes called phosphatases remove the phosphate groups from enzymes, thereby restoring the original level of carbohydrate metabolism.

Biochemical Regulation A third and very prominent way that metabolic pathways are controlled is at the biochemical level. In this case, the binding of a molecule to an enzyme directly regulates its function. Biochemical regulation is typically categorized according to the site where the regulatory molecule binds. Let's consider two types of regulation that involve regulatory molecules that inhibit enzyme function.

Competitive inhibitors are molecules that bind to the active site of an enzyme and inhibit the ability of the substrate to bind. In other words, such inhibitors "compete" with the substrate for the ability to bind to the enzyme. Competitive inhibitors usually have a structure that mimics the structure of the enzyme's substrate. Competitive inhibition can be overcome by increasing the concentration of the substrate and decreasing the concentration of the inhibitor.

Noncompetitive inhibitors bind to an enzyme at a location that is outside the active site and inhibit the enzyme's function. An example is a form of regulation called **feedback inhibition**, in which the product of a metabolic pathway inhibits an enzyme that acts early in the pathway, thus preventing the overaccumulation of the product (**Figure 7.9**). Many metabolic pathways use feedback inhibition as a form of biochemical regulation. In such cases, the inhibited enzyme has two binding sites. One site is the active site, where the reactants are converted to products. In addition, enzymes that are controlled by feedback inhibition also have an **allosteric site**, where a molecule can bind noncovalently and affect the function of the active site. The binding of a molecule to an allosteric site causes a conformational change in the enzyme that inhibits its catalytic function. Allosteric sites are often found in the enzymes that catalyze the early steps in a metabolic pathway. Such allosteric sites typically bind molecules that are the products of the metabolic pathway. When the products bind to these sites, they inhibit the function of these enzymes and thereby prevent the formation of too much product.

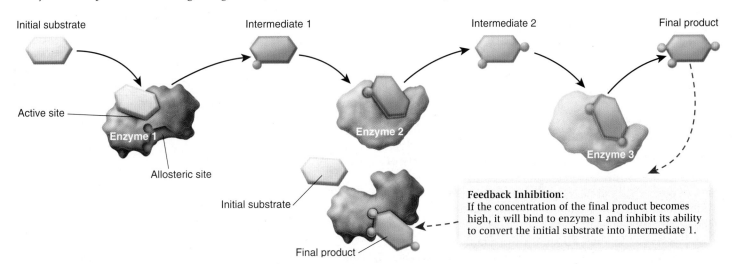

Initial substrate   Intermediate 1   Intermediate 2   Final product

Active site

Allosteric site

Enzyme 1   Enzyme 2   Enzyme 3

Initial substrate

Final product

Feedback Inhibition:
If the concentration of the final product becomes high, it will bind to enzyme 1 and inhibit its ability to convert the initial substrate into intermediate 1.

Figure 7.9 **Feedback inhibition.** In this process, the product of a metabolic pathway inhibits an enzyme that functions in the pathway, thereby preventing the overaccumulation of the product.

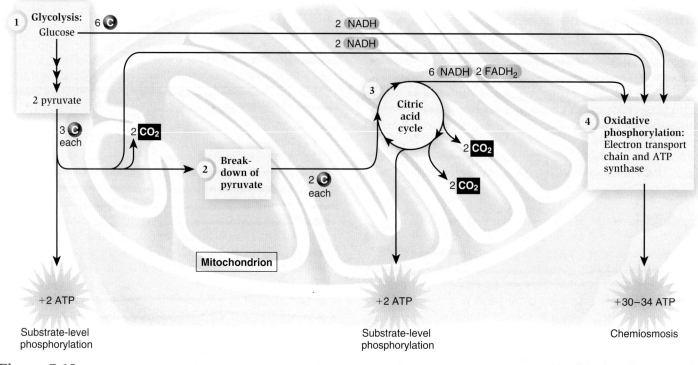

Figure 7.10   **An overview of glucose metabolism.**   Glycolysis occurs in the cytosol and results in the breakdown of glucose into two pyruvates, producing two ATP and two NADH molecules. The two pyruvates enter the mitochondrion, where they are broken down into two acetyl groups (each attached to CoA) and two CO_2 molecules. Two molecules of NADH are made in the process. The two acetyl groups then enter the citric acid cycle, where they are broken down to four CO_2 molecules. Two ATP, six NADH, and two $FADH_2$ are synthesized. The NADH and $FADH_2$ molecules that are made during these various steps are then used during oxidative phosphorylation to synthesize more ATP molecules. The maximum yield of ATP is 34 to 38 molecules for every glucose that is completely broken down.

Cellular and biochemical regulation are important and rapid ways to control chemical reactions in a cell. But when considering a metabolic pathway composed of several enzymes, which enzyme in a pathway should be controlled? In many cases, a metabolic pathway has a **rate-limiting step**, which is the slowest step in a pathway. If the rate-limiting step is inhibited or occurs at a faster rate, such changes will have the greatest impact on the formation of the product of the metabolic pathway. Rather than affecting all of the enzymes in a metabolic pathway, cellular and biochemical regulation are often directed at the enzyme that catalyzes the rate-limiting step. This is an efficient and rapid way to control the amount of product of a pathway.

7.3 Cellular Respiration

Cellular respiration is a process by which living cells obtain energy from organic molecules. A primary aim of cellular respiration is to make the energy intermediates ATP and NADH. When oxygen (O_2) is used, this process is termed **aerobic respiration**. During aerobic respiration, O_2 is consumed and CO_2 is released. When we breathe, we inhale the oxygen that is needed for aerobic respiration and exhale the CO_2 that is a by-product of the process. For this reason, the term respiration has a second meaning, which is the act of breathing.

Different types of organic molecules, such as carbohydrates, proteins, and fats, can be used as energy sources to drive aerobic respiration.

$$\text{Organic molecules} + O_2 \rightarrow CO_2 + H_2O + \text{Energy}$$

In this section, we will largely focus on the use of glucose as an energy source for cellular respiration.

$$C_6H_{12}O_6 + 6 O_2 \rightarrow 6 CO_2 + 6 H_2O + \text{Energy intermediates} + \text{Heat}$$
Glucose

We will examine the metabolic pathways in which glucose is broken down into carbon dioxide and water, thereby releasing a large amount of energy that is used to make many ATP molecules. In so doing, we will focus on four pathways: (1) glycolysis, (2) the breakdown of pyruvate, (3) the citric acid cycle, and (4) oxidative phosphorylation. We will conclude our discussion of cellular respiration with a consideration of the metabolism of other organic molecules, such as proteins and fats, and an examination of anaerobic respiration, a second form of respiration in which cells can oxidize fuel and generate ATP without using oxygen.

Several Metabolic Pathways Are Involved in the Breakdown of Glucose to CO₂

Before we examine the details of cellular respiration, let's take a look at the entire process. We will focus on the breakdown of glucose in a eukaryotic cell in the presence of oxygen. The covalent bonds within glucose contain a large amount of chemical bond energy. When glucose is broken down, ultimately to CO_2 and water, the energy within those bonds is released and used to make three types of energy intermediates: ATP, NADH, and $FADH_2$. The following is an overview of the stages that occur during the breakdown of glucose (**Figure 7.10**):

1. **Glycolysis:** In glycolysis, glucose (6 C, meaning a compound with six carbon atoms) is broken down in the cytosol to two pyruvate molecules (3 C each), producing a net gain of two ATP molecules, via substrate-level phosphorylation, and two NADH molecules.

2. **Breakdown of pyruvate to an acetyl group:** The two pyruvate molecules enter the mitochondrion, where each one is broken down to an acetyl group (2 C) and one CO_2 molecule. For each pyruvate broken down, one NADH molecule is made.

3. **Citric acid cycle:** Each acetyl group (2 C) is broken down to two CO_2 molecules. One ATP, three NADH, and one $FADH_2$ are made in this process. Because there are two acetyl groups, the total yield is four CO_2, two ATP via substrate-level phosphorylation, six NADH, and two $FADH_2$.

4. **Oxidative phosphorylation:** The NADH and $FADH_2$ made in the three previous stages contain high-energy electrons that can be readily transferred in a redox reaction to other molecules. Once removed, these high-energy electrons release some energy and that energy is harnessed to make approximately 30 to 34 ATP molecules via chemiosmosis. As discussed later, oxidative phosphorylation consists of two components, the electron transport chain and the ATP synthase.

Now, let's examine in detail the chemical changes that take place in each of these four stages.

Stage 1: Glycolysis Is a Metabolic Pathway That Breaks Down Glucose to Pyruvate

Glycolysis (from the Greek *glykos*, sweet, and *lysis*, splitting) involves the breakdown of glucose, a simple sugar. This process can occur in the presence or absence of oxygen, that is, under aerobic or anaerobic conditions. Our understanding of glycolysis has a rich history. In 1897, Hans Buchner and Eduard Buchner made an accidental discovery. They were interested in manufacturing cell-free extracts of yeast for possible clinical use. This cell-free extract contained only the internal contents of yeast cells, not the intact cells themselves. To preserve these extracts, they added sucrose, a commonly used preservative in 19th-century chemistry. To their great surprise, they discovered that the cell-free extract converted the sucrose to ethanol. The significance of this finding was extraordinary. The Buchners showed for the first time that metabolism could take place outside of living cells. This observation is considered by many as the birth of **biochemistry**, the study of the chemistry of living organisms.

The Buchners' findings paved the way for the in-depth investigation of the breakdown of glucose. During the 1930s, the efforts of several German biochemists, including Gustav Embden, Otto Meyerhof, and Jacob Parnas, determined that the process involved 10 steps, each one catalyzed by a different enzyme. The elucidation of these steps was a major achievement in the field of biochemistry. Researchers have since discovered that glycolysis is the common pathway for glucose breakdown in bacteria, archaea, and eukaryotes. Remarkably, the steps of glycolysis are virtually identical in nearly all living species, suggesting that glycolysis arose very early in the evolution of life on our planet.

The 10 steps of glycolysis can be grouped into three phases (**Figure 7.11**). The first phase (steps 1–3) involves an energy

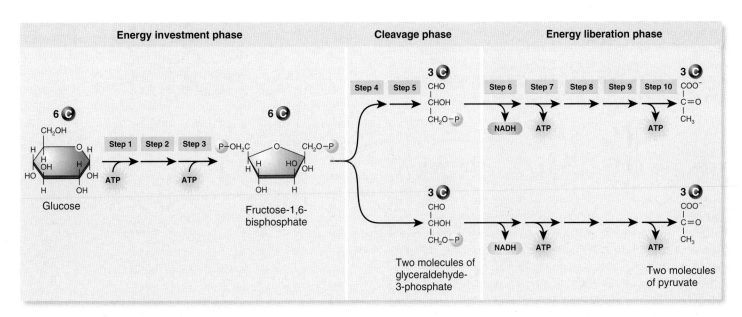

Figure 7.11   Overview of glycolysis.

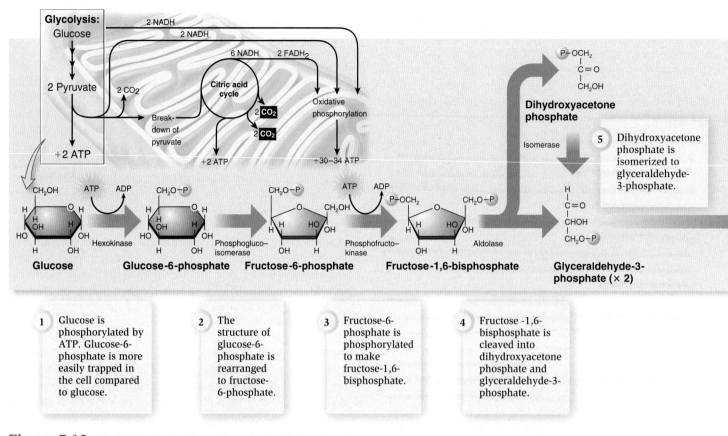

Figure 7.12 An inside look at the steps of glycolysis.

investment. Two ATP molecules are hydrolyzed to create fructose-1,6-bisphosphate. The cleavage phase (steps 4–5) breaks a six-carbon molecule into two molecules of glyceraldehyde-3-phosphate. The third phase (steps 6–10) liberates energy. The two glyceraldehyde-3-phosphate molecules are broken down to two pyruvate molecules. This produces two molecules of NADH and four molecules of ATP. Because two molecules of ATP are used in the energy investment phase, the net yield of ATP is two molecules. **Figure 7.12** describes the details of the 10 reactions of glycolysis. The net reaction of glycolysis is:

$$C_6H_{12}O_6 \; + \; 2 \; NAD^+ \; + \; 2 \; ADP^{2-} \; + \; 2 \; P_i^{2-} \; \rightarrow$$
Glucose

$$2 \; CH_3(C{=}O)COO^- \; + \; 2 \; H^+ \; + \; 2 \; NADH \; + \; 2 \; ATP^{4-} \; + \; 2 \; H_2O$$
Pyruvate

When a cell has a sufficient amount of ATP, feedback inhibition occurs in glycolysis. At high concentration, ATP binds to an allosteric site in phosphofructokinase, which catalyzes the third step in glycolysis and is thought to be the rate-limiting step. When ATP binds to this allosteric site, a conformational change occurs that renders the enzyme functionally inactive. This prevents the further breakdown of glucose and thereby inhibits the overproduction of ATP.

Stage 2: Pyruvate Enters the Mitochondrion and Is Broken Down to an Acetyl Group and CO_2

In eukaryotes, pyruvate is made in the cytosol and then transported into the mitochondrion. Once in the mitochondrial matrix, pyruvate molecules are broken down by an enzyme complex called pyruvate dehydrogenase (**Figure 7.13**). A molecule of CO_2 is removed from each pyruvate and the remaining acetyl group is attached to an organic molecule called coenzyme A (CoA) to create acetyl CoA. During this process, two high-energy electrons are removed from pyruvate and transferred to NAD^+ and together with H^+ create a molecule of NADH. For each pyruvate, the net reaction is:

$$\begin{array}{c} \quad\quad O \;\;\; O \\ \quad\quad \parallel \;\;\; \parallel \\ {}^-O{-}C{-}C{-}CH_3 \; + \; CoA{-}SH \; + \; NAD^+ \rightarrow \\ \text{Pyruvate} \quad\quad\quad CoA \end{array}$$

$$\begin{array}{c} \quad\quad O \\ \quad\quad \parallel \\ CoA{-}S{-}C{-}CH_3 \; + \; CO_2 \; + \; NADH \\ \text{Acetyl CoA} \end{array}$$

The acetyl group is attached to CoA via a covalent bond to a sulfur atom. The hydrolysis of this bond releases a large amount of

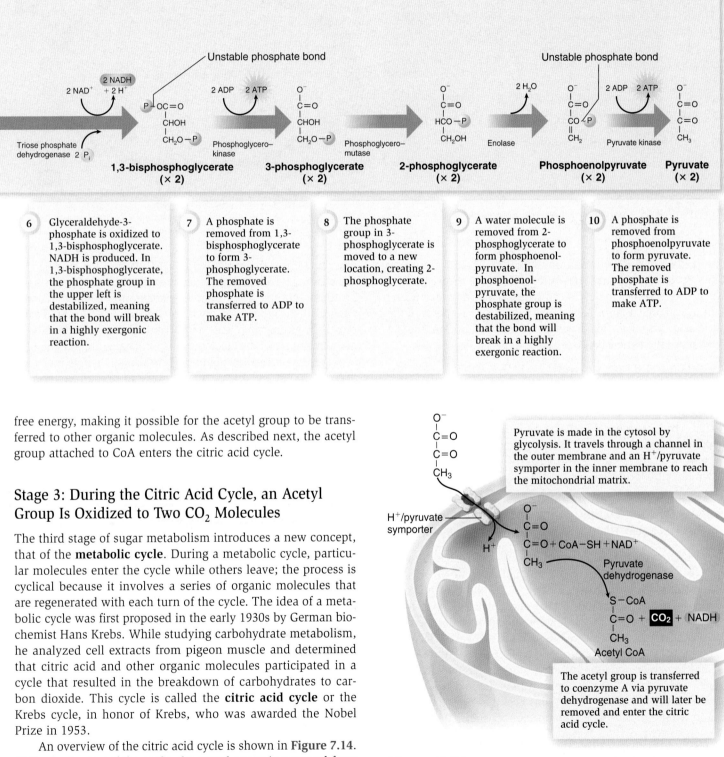

6 | Glyceraldehyde-3-phosphate is oxidized to 1,3-bisphosphoglycerate. NADH is produced. In 1,3-bisphosphoglycerate, the phosphate group in the upper left is destabilized, meaning that the bond will break in a highly exergonic reaction.

7 | A phosphate is removed from 1,3-bisphosphoglycerate to form 3-phosphoglycerate. The removed phosphate is transferred to ADP to make ATP.

8 | The phosphate group in 3-phosphoglycerate is moved to a new location, creating 2-phosphoglycerate.

9 | A water molecule is removed from 2-phosphoglycerate to form phosphoenolpyruvate. In phosphoenolpyruvate, the phosphate group is destabilized, meaning that the bond will break in a highly exergonic reaction.

10 | A phosphate is removed from phosphoenolpyruvate to form pyruvate. The removed phosphate is transferred to ADP to make ATP.

free energy, making it possible for the acetyl group to be transferred to other organic molecules. As described next, the acetyl group attached to CoA enters the citric acid cycle.

Stage 3: During the Citric Acid Cycle, an Acetyl Group Is Oxidized to Two CO_2 Molecules

The third stage of sugar metabolism introduces a new concept, that of the **metabolic cycle**. During a metabolic cycle, particular molecules enter the cycle while others leave; the process is cyclical because it involves a series of organic molecules that are regenerated with each turn of the cycle. The idea of a metabolic cycle was first proposed in the early 1930s by German biochemist Hans Krebs. While studying carbohydrate metabolism, he analyzed cell extracts from pigeon muscle and determined that citric acid and other organic molecules participated in a cycle that resulted in the breakdown of carbohydrates to carbon dioxide. This cycle is called the **citric acid cycle** or the Krebs cycle, in honor of Krebs, who was awarded the Nobel Prize in 1953.

An overview of the citric acid cycle is shown in **Figure 7.14**. At the beginning of the cycle, the acetyl group is removed from acetyl CoA and attached to oxaloacetate (4 C) to form citrate (6 C), also called citric acid. In a series of several steps, two CO_2 molecules are released. As this occurs, three molecules of NADH,

Figure 7.13 **Breakdown of pyruvate and the attachment of an acetyl group to CoA.** Pyruvate enters the mitochondrion by traveling through a channel in the outer membrane and then through an H^+/pyruvate symporter in the inner membrane.

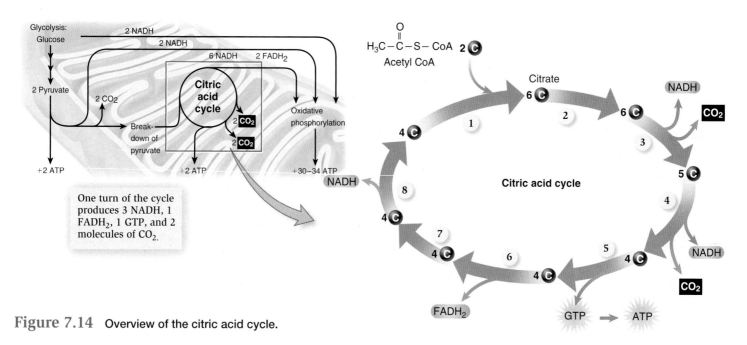

Figure 7.14 Overview of the citric acid cycle.

one molecule of FADH$_2$, and one molecule of GTP are made. The GTP is used to make ATP. After eight steps, oxaloacetate is regenerated so that the cycle can begin again, provided acetyl CoA is available. **Figure 7.15** shows a more detailed view of the citric acid cycle. For each acetyl group attached to CoA, the net reaction of the citric acid cycle is:

$$\text{Acetyl-CoA} + 3\ H_2O + 3\ NAD^+ + FAD + GDP^{2-} + P_i^{2-} \rightarrow$$
$$\text{CoA-SH} + 2\ CO_2 + 3\ NADH + FADH_2 + GTP^{4-}$$

Competitive inhibition is one way that the citric acid cycle is regulated. Oxaloacetate is a competitive inhibitor of succinate dehydrogenase (Figure 7.15). Therefore, when oxaloacetate levels become too high, this inhibits succinate dehydrogenase and slows down the citric acid cycle.

Stage 4: During Oxidative Phosphorylation, NADH and FADH2 Are Used to Make More ATP

Up to this point, the oxidation of glucose has yielded 6 molecules of CO$_2$, 4 molecules of ATP, 10 molecules of NADH, and 2 molecules of FADH$_2$. Let's now consider how high-energy electrons are removed from NADH and FADH$_2$ to make more ATP. This process is called **oxidative phosphorylation** because NADH and FADH$_2$ are oxidized due to the removal of electrons, and ATP is made by the phosphorylation of ADP (**Figure 7.16**). Oxidative phosphorylation usually occurs under aerobic conditions, which means that it typically requires oxygen. As described next, the oxidative process involves the electron transport chain, while the phosphorylation occurs via the ATP synthase.

Electron Transport Chain The **electron transport chain** consists of a group of protein complexes and small organic molecules embedded in the inner mitochondrial membrane. These components are referred to as an electron transport chain because the components can accept and donate electrons to each

other in a linear manner (Figure 7.16). Most of the members of the chain are protein complexes (designated I to IV) that have prosthetic groups. For example, cytochrome oxidase contains two prosthetic groups, each with an iron atom. The iron in each prosthetic group can readily accept and release an electron. One of the members of the electron transport chain, ubiquinone (Q), is not a protein. Rather, ubiquinone is a small organic molecule that can accept and release an electron and can diffuse through the lipid bilayer.

The red line in Figure 7.16 shows the path of electrons as they flow along the electron transport chain. This path is a series of redox reactions in which electrons are transferred to components with increasingly higher electronegativity. At the beginning of the chain, a pair of high-energy electrons from NADH are transferred one at a time to NADH dehydrogenase (complex I), and then to ubiquinone (Q), cytochrome *b-c$_1$* (complex III), cytochrome *c*, cytochrome oxidase (complex IV), and finally O$_2$. At the end of the chain, 2 electrons, 2 H$^+$, and 1/2 O$_2$ combine to form a molecule of water. Similarly, FADH$_2$ transfers electrons to succinate reductase (complex II), then to ubiquinone, and the rest of the chain.

The electron transport chain is also called the **respiratory chain** because the oxygen we breathe is used in this process. One component of the electron transport chain, cytochrome oxidase, is inhibited by carbon monoxide. The deadly effects of carbon monoxide occur because the electron transport chain is shut down, preventing cells from making enough ATP for survival.

As shown in Figure 7.16, the movement of electrons results in the pumping of H$^+$ across the inner mitochondrial membrane to establish a large **H$^+$ electrochemical gradient** in which the concentration of H$^+$ is higher outside of the matrix, and an excess of positive charge exists outside the matrix. Because hydrogen ions consist of protons, the H$^+$ electrochemical gradient is also called the **proton-motive force**. NADH dehydrogenase, cytochrome *b-c$_1$*, and cytochrome oxidase are H$^+$ pumps.

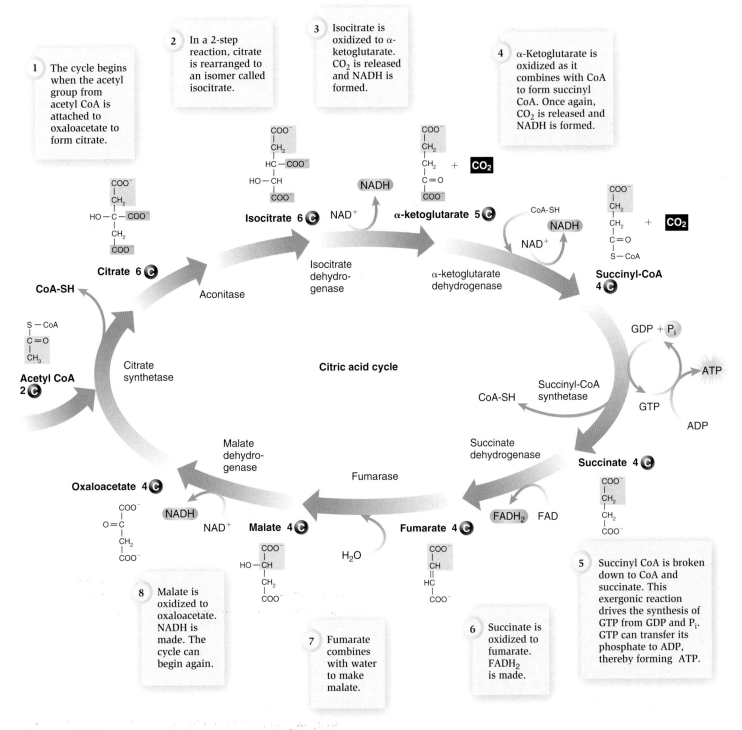

Figure 7.15 A detailed look at the steps of the citric acid cycle.

While traveling along the electron transport chain, electrons release free energy, and some of this energy is captured by these proteins to pump H$^+$ out of the matrix into the intermembrane space. Because the electrons from FADH$_2$ enter the chain at an intermediate step, fewer hydrogen ions are pumped out of the matrix compared to the release of electrons from NADH.

Why do electrons travel from NADH or FADH$_2$ to O$_2$? As you might expect, the answer lies in free energy changes. The electrons found on the energy intermediates have a high amount

of potential energy. As they travel along the electron transport chain, free energy is released (**Figure 7.17**). The movement of one electron from NADH to O$_2$ results in a very negative free energy change of approximately -25 kcal/mole. That is why the process is spontaneous and proceeds in the forward direction. Because it is a highly exergonic reaction, some of the free energy can be harnessed to do cellular work. In this case, some energy is used to pump H$^+$ across the inner mitochondrial membrane and establish an H$^+$ electrochemical gradient.

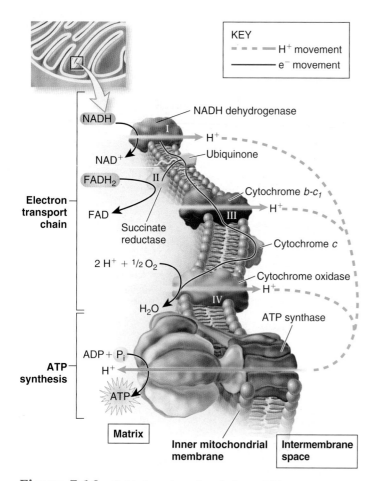

Figure 7.16   Oxidative phosphorylation.   This process consists of two distinct events involving the electron transport chain and the ATP synthase. The electron transport chain removes electrons from NADH or FADH$_2$ and pumps H$^+$ across the inner mitochondrial membrane. The ATP synthase uses the energy in the H$^+$ electrochemical gradient to synthesize ATP.

Biological inquiry: Can you explain the name of cytochrome oxidase? Can you think of another appropriate name?

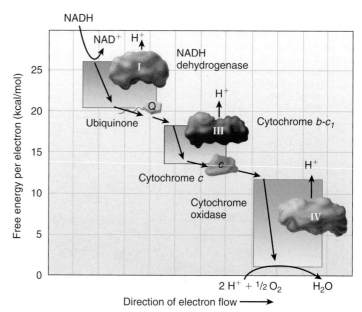

Figure 7.17   The relationship between free energy and electron movement along the electron transport chain. As electrons hop from one site to another along the electron transport chain, they release energy. Some of this energy is harnessed to pump H$^+$ across the inner mitochondrial membrane. The total energy released by a single electron is approximately −25 kcal/mole.

For each molecule of NADH that is oxidized and each molecule of ATP that is made, the two chemical reactions of oxidative phosphorylation can be represented as follows:

$$NADH + H^+ + 1/2\ O_2 \rightarrow NAD^+ + H_2O$$
$$ADP^{2-} + P_i^{2-} \rightarrow ATP^{4-} + H_2O$$

The oxidation of NADH to NAD$^+$ results in an H$^+$ electrochemical gradient in which more hydrogen ions are in the intermembrane space compared to the matrix. The synthesis of one ATP molecule is thought to require the movement of three to four ions into the matrix, down their H$^+$ electrochemical gradient.

When we add up the maximal amount of ATP that can be made by oxidative phosphorylation, most researchers agree that it is in the range of 30 to 34 ATP molecules for each glucose molecule that is broken down to CO$_2$ and water. However, the maximum amount of ATP is rarely achieved for two reasons. First, although 10 NADH and 2 FADH$_2$ are available to create the H$^+$ electrochemical gradient across the inner mitochondrial membrane, a cell may use some of these molecules for anabolic pathways. For example, NADH is used in the synthesis of organic molecules such as glycerol (a component of phospholipids) and lactic acid (which is secreted from muscle cells during strenuous exercise). Second, the mitochondrion may use some of the H$^+$ electrochemical gradient for other purposes. For example, the gradient is used for the uptake of pyruvate into the matrix via an H$^+$/pyruvate symporter. Therefore, the actual amount of ATP synthesis is usually a little less than the maximum number of 30 to 34. Even so, when we compare the amount of

ATP Synthase   The second event of oxidative phosphorylation is the synthesis of ATP by an enzyme called **ATP synthase**. The H$^+$ electrochemical gradient across the inner mitochondrial membrane is a source of potential energy. How is this energy used? The flow of H$^+$ back into the matrix is an exergonic process. The lipid bilayer is relatively impermeable to H$^+$. However, H$^+$ can pass through the membrane-embedded portion of the ATP synthase. This enzyme harnesses some of the free energy that is released as the ions flow through its membrane-embedded region to synthesize ATP from ADP and P$_i$ (see Figure 7.16). This is an example of an energy conversion: energy in the form of an H$^+$ electrochemical gradient, or proton-motive force, is converted to chemical bond energy in ATP. The chemical synthesis of ATP by pushing H$^+$ across a membrane is called chemiosmosis (from the Greek *osmos*, to push), and the theory behind it was proposed by Peter Mitchell, a British chemist who was awarded the Nobel Prize in 1978.

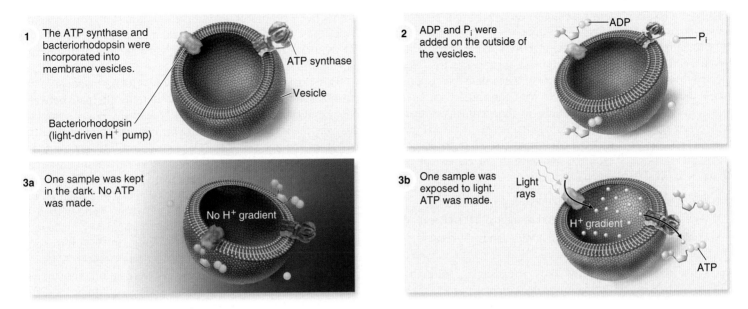

1. The ATP synthase and bacteriorhodopsin were incorporated into membrane vesicles.

ATP synthase

Vesicle

Bacteriorhodopsin (light-driven H+ pump)

2. ADP and P_i were added on the outside of the vesicles.

ADP

P_i

3a. One sample was kept in the dark. No ATP was made.

No H+ gradient

3b. One sample was exposed to light. ATP was made.

Light rays

H+ gradient

ATP

Figure 7.18 The Racker and Stoeckenius experiment showing that an H+ electrochemical gradient drives ATP synthesis via the ATP synthase.

Biological inquiry: Is the functioning of the electron transport chain always needed to make ATP via the ATP synthase?

ATP that can be made by glycolysis (2), the citric acid cycle (2), and oxidative phosphorylation (30–34), we see that oxidative phosphorylation provides a cell with a much greater capacity to make ATP.

Experiments with Purified Proteins in Membrane Vesicles Verified Chemiosmosis

To show experimentally that the ATP synthase actually uses an H+ electrochemical gradient to make ATP, researchers needed to purify the enzyme and study its function *in vitro*. In 1974, Ephraim Racker and Walther Stoeckenius purified the ATP synthase and another protein called bacteriorhodopsin, which is found in certain species of archaea. Previous research had shown that bacteriorhodopsin is a light-driven H+ pump. Racker and Stoeckenius took both purified proteins and inserted them into membrane vesicles (**Figure 7.18**). The ATP synthase was oriented so that its ATP synthesizing region was on the outside of the vesicles. Bacteriorhodopsin was oriented so that it would pump H+ into the vesicles. They added ADP and P_i on the outside of the vesicles. In the dark, no ATP was made. However, when they shone light on the vesicles, a substantial amount of ATP was made. Because bacteriorhodopsin was already known to be a light-driven H+ pump, these results convinced researchers that the ATP synthase uses an H+ electrochemical gradient as an energy source to make ATP.

The ATP Synthase Is a Rotary Machine That Makes ATP as It Spins

The structure and function of the ATP synthase are particularly intriguing and have received much attention over the past few decades (**Figure 7.19**). The ATP synthase is a rotary machine.

The nonmembrane-embedded portion consists of 1 ε, 1 γ, 1 δ, 3 α, and 3 β subunits. Movement of H+ through the *c* subunits causes the γ subunit to rotate. The rotation, in 120° increments, causes the β subunits to progress through a series of 3 conformational changes that lead to the synthesis of ATP from ADP and P_i.

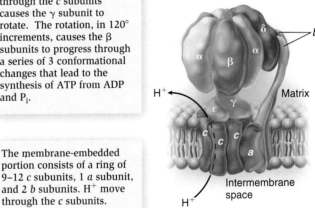

The membrane-embedded portion consists of a ring of 9–12 *c* subunits, 1 *a* subunit, and 2 *b* subunits. H+ move through the *c* subunits.

Figure 7.19 The subunit structure and function of the ATP synthase.

The membrane-embedded region is composed of three types of subunits called *a*, *b*, and *c*. Approximately 9 to 12 *c* subunits form a ring in the membrane. Each *c* subunit is an H+ channel. One *a* subunit is bound to this ring, and two *b* subunits are attached to the *a* subunit and protrude out of the membrane. The nonmembrane-embedded subunits are designated with Greek letters. One ε and one γ subunit bind to the ring of *c* subunits. The γ subunit forms a long stalk that pokes into the center of another ring of three α and three β subunits. The β subunits are the catalytic site where ATP is made.

Finally, the δ subunit forms a connection between the ring of α and β subunits and the two *b* subunits.

When hydrogen ions pass through a *c* subunit, a conformational change causes the γ subunit to turn clockwise (when viewed from the intermembrane space). Each time the γ subunit turns 120° it changes its contacts with the three β subunits, which in turn causes the β subunits to change their conformations. How do these conformational changes promote ATP synthesis? The answer is that the conformational changes occur in a way that favors ATP synthesis and release. The conformational changes in the β subunits happen in the following order:

- Conformation 1: ADP and P_i bind with good affinity.
- Conformation 2: ADP and P_i bind so tightly that ATP is made.
- Conformation 3: ATP (and ADP and P_i) bind very weakly, and ATP is released.

Each time the γ subunit turns 120°, it causes a β subunit to change to the next conformation. After conformation 3, a 120° turn by the γ subunit returns a β subunit back to conformation 1, and the cycle of ATP synthesis can begin again. Because the ATP synthase has three β subunits, each subunit is in a different conformation at any given time.

Paul Boyer proposed the concept of a rotary machine in the late 1970s. In his model, the three β subunits alternate between three conformations, as described previously. Boyer's original idea was met with great skepticism, because the concept that part of an enzyme could spin was very novel, to say the least. John Walker and colleagues were able to determine the three-dimensional structure of the nonmembrane-embedded portion of the ATP synthase. The structure revealed that each of the three β subunits had a different conformation—one with ADP bound, one with ATP bound, and one without any nucleotide bound. This result supported Boyer's model. In 1997, Boyer and Walker shared the Nobel Prize for their work on the ATP synthase. As described in the Feature Investigation, researchers subsequently visualized the rotation of the γ subunit.

FEATURE INVESTIGATION

Yoshida and Kinosita Demonstrated That the γ Subunit of the ATP Synthase Spins

In 1997, Masasuke Yoshida, Kazuhiko Kinosita, and colleagues set out to experimentally visualize the rotary nature of the ATP synthase (**Figure 7.20**). The membrane-embedded region of the ATP synthase can be separated from the rest of the protein by treatment of mitochondrial membranes with a high concentration of salt, releasing the portion of the protein containing the one γ, three α, and three β subunits. The researchers adhered the $γα_3β_3$ complex to a glass slide so that the γ subunit was protruding upwards. Because the γ subunit is too small to be seen with a light microscope, it is not possible to visualize the rotation of the γ subunit directly. To circumvent this problem, the researchers attached a large, fluorescently labeled actin filament to the γ subunit via a linker protein. The fluorescently labeled actin filament is very long compared to the γ subunit and can be readily seen with a fluorescence microscope.

Figure 7.20   Evidence that the ATP synthase is a rotary machine.

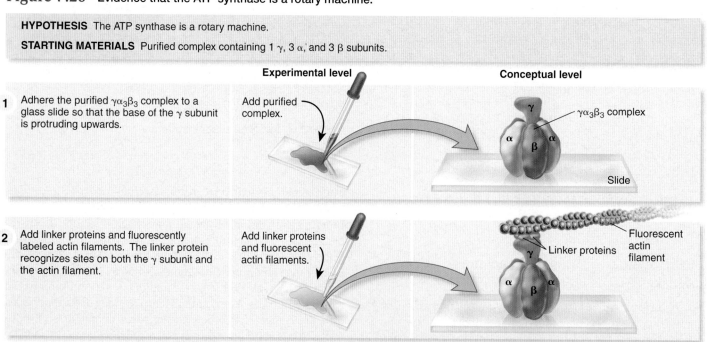

HYPOTHESIS The ATP synthase is a rotary machine.

STARTING MATERIALS Purified complex containing 1 γ, 3 α, and 3 β subunits.

Experimental level | Conceptual level

1 Adhere the purified $γα_3β_3$ complex to a glass slide so that the base of the γ subunit is protruding upwards.

Add purified complex.

γ
$γα_3β_3$ complex
α   α
β
Slide

2 Add linker proteins and fluorescently labeled actin filaments. The linker protein recognizes sites on both the γ subunit and the actin filament.

Add linker proteins and fluorescent actin filaments.

Fluorescent actin filament
γ   Linker proteins
α   α
β

3 Add ATP. As a control, do not add ATP.

4 Observe under a fluorescence microscope. The method of fluorescence microscopy is described in Chapter 4.

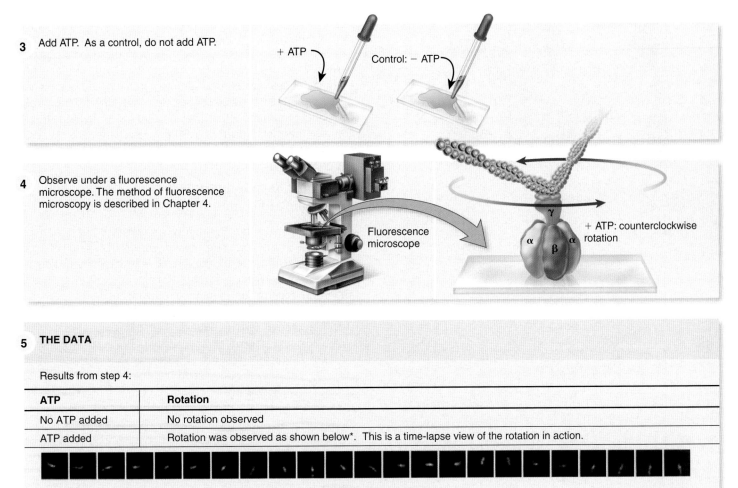

5 THE DATA

Results from step 4:

ATP	Rotation
No ATP added	No rotation observed
ATP added	Rotation was observed as shown below*. This is a time-lapse view of the rotation in action.

*This figure shows a series of micrographs from Noji et al. (1997) *Nature* 386, 299–303.

Because the membrane-embedded portion of the protein is missing, you may be wondering how the researchers could get the γ subunit to rotate. The answer is that they added ATP. Although the normal function of the ATP synthase is to make ATP, it can run backwards. In other words, the ATP synthase can hydrolyze ATP. As shown in the data for Figure 7.20, when the researchers added ATP, the fluorescently labeled actin filament was observed to rotate in a counterclockwise direction, which is opposite to the direction that the γ subunit rotates when ATP is synthesized. In fact, actin filaments were observed to rotate for more than 100 revolutions in the presence of ATP. These results convinced the scientific community that the ATP synthase is indeed a rotary machine.

Metabolic Pathways for Carbohydrate Metabolism Are Interconnected to Pathways for Amino Acid and Fat Metabolism

Thus far, we have focused our attention on the stages of glucose breakdown that result in the release of CO_2 and the production of NADH, $FADH_2$ and ATP. In addition, cells use other organic molecules as a source of energy. When you eat a meal, it is likely to contain not only carbohydrates (including glucose) but also proteins and fats. Proteins and fats are also broken down by the enzymes involved with glucose metabolism.

As shown in **Figure 7.21**, proteins and fats can enter into glycolysis, or the citric acid cycle, at different points. Proteins are first acted upon by enzymes, either in digestive juices or within cells, that cleave the bonds connecting individual amino acids. Because the 20 amino acids differ in their side chains, amino acids and their breakdown products can enter at different points in the pathway. Breakdown products of amino acids can enter glycolysis, or an acetyl group can be removed from certain amino acids and become attached to CoA. Other amino acids can be modified and enter the citric acid cycle. Similarly, fats can be broken down to glycerol and fatty acids.

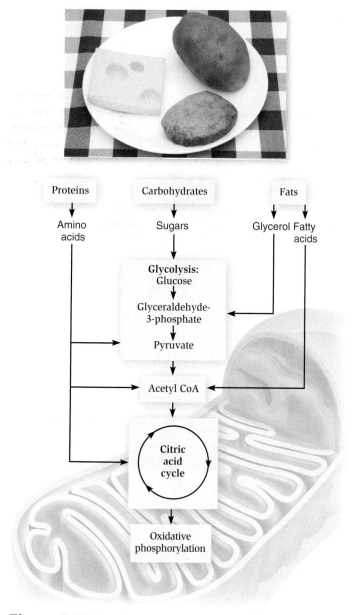

Figure 7.21 Integration of protein, carbohydrate, and fat metabolism. Breakdown products of amino acids and fats can enter the same pathway that is used to break down carbohydrates.

Glycerol can be modified to glyceraldehyde-3-phosphate and enter glycolysis at step 5 (see Figure 7.12). Fatty acyl tails can have two carbon acetyl units removed, which bind to CoA and then enter the citric acid cycle. By utilizing the same pathways for the breakdown of sugars, amino acids, and fats, cells are more efficient because they can use the same enzymes for the breakdown of different starting molecules.

Likewise, carbohydrate metabolism is connected to the metabolism of other cellular components at the anabolic level. Cells may use carbohydrates to manufacture parts of amino acids, fats, and nucleotides. For example, glucose-6-phosphate

of glycolysis is used to construct the sugar and phosphate portion of nucleotides, while oxaloacetate of the citric acid cycle can be used as a precursor for the biosynthesis of purine and pyrimidine bases. Portions of amino acids can be made from products of glycolysis (for example, pyruvate) and components of the citric acid cycle (oxaloacetate). In addition, many other catabolic and anabolic pathways are found in living cells that connect the metabolism of carbohydrates, proteins, fats, and nucleic acids.

Anaerobic Respiration Involves the Breakdown of Organic Molecules in the Absence of Oxygen

In this section, we have primarily surveyed catabolic pathways that result in the complete breakdown of glucose in the presence of oxygen. It is also common for cells to metabolize organic molecules such as glucose without using oxygen, a process called **anaerobic respiration**. Some microorganisms, such as bacteria and fungi, exist in environments that lack oxygen and have to carry out anaerobic respiration to obtain sufficient amounts of energy. Examples include microbes living in your intestinal tract and those living deep in the soil. Similarly, when a person exercises strenuously, the rate of oxygen consumption by muscle cells may greatly exceed the rate of oxygen delivery. Under these conditions, the muscle cells become anaerobic and must use anaerobic respiration to maintain their level of activity.

Organisms have evolved two different strategies to carry out anaerobic respiration. One strategy is to use a substance other than O_2 as the final electron acceptor of the electron transport chain. In the electron transport chain shown in Figure 7.16, cytochrome oxidase recognizes O_2 and catalyzes its reduction to H_2O. The final electron acceptor of the chain is O_2. Many species of bacteria that live under anaerobic conditions have evolved enzymes that function similarly to cytochrome oxidase, but recognize molecules other than O_2 and use them as the final electron acceptor. For example, *Escherichia coli*, which is found in your intestinal tract, has an enzyme called nitrate reductase that can use nitrate (NO_3^-) as the final electron acceptor of the electron transport chain under anaerobic conditions, and thus oxidative phosphorylation can occur.

Other organisms, including yeast and mammals, can use only O_2 as the final electron acceptor of their electron transport chains. When confronted with anaerobic conditions, these organisms must have a different strategy for anaerobic respiration. Such organisms can still carry out glycolysis, which can occur under anaerobic or aerobic conditions. However, a key issue is that glycolysis uses NAD^+ and generates NADH. As discussed earlier in this chapter, NADH is a good electron donor, because it readily gives up its electrons. Under aerobic conditions, oxygen acts as a final electron acceptor, and the high-energy electrons from NADH can be used to make more ATP. However, this cannot occur under anaerobic conditions in these organisms and, as a result, NADH builds up. At high concentrations, NADH will haphazardly donate its electrons to other molecules and

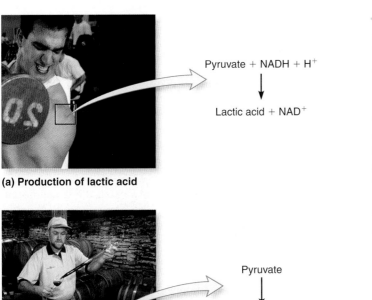

Pyruvate + NADH + H$^+$

↓

Lactic acid + NAD$^+$

(a) Production of lactic acid

Pyruvate

↓

Acetaldehyde + NADH + H$^+$ + CO$_2$

↓

Ethanol + NAD$^+$

(b) Production of ethanol

Figure 7.22 **Examples of anaerobic respiration.** In these examples, NADH is produced by the oxidation of an organic molecule, and then the NADH is used up by donating electrons to a different organic molecule such as pyruvate **(a)** or acetaldehyde **(b)**. Note: at neutral pH, lactic acid releases H$^+$ to form lactate.

promote the formation of free radicals, highly reactive chemicals that can damage DNA and cellular proteins. For this reason, cells of mammals and yeast that are exposed to anaerobic conditions must have a strategy to remove the excess NADH normally generated from the breakdown of organic molecules such as glucose. In addition, cells need to regenerate NAD$^+$ to keep glycolysis running.

In the case of muscle cells working under anaerobic conditions, the pyruvate from glycolysis is converted by NADH to lactate, also known as lactic acid (**Figure 7.22a**). When a person exercises strenuously, the buildup of lactic acid causes a burning sensation or even cramps that often prompt the individual to slow down. Products of anaerobic respiration, such as lactic acid, are eventually secreted from cells. Similarly, during wine making, a yeast cell metabolizes sugar under anaerobic conditions. The pyruvate is broken down to CO$_2$ and a two-carbon molecule called acetaldehyde. The acetaldehyde is then reduced by NADH to make ethanol (**Figure 7.22b**). In these examples, NADH is used up by donating its electrons to pyruvate and acetaldehyde, thereby creating lactic acid and ethanol, respectively. These reactions also regenerate NAD$^+$, which allows cells to metabolize more glucose and keep making ATP.

The term **fermentation** is used to describe the breakdown of organic molecules to produce energy without any net oxida-tion (that is, without any removal of electrons). The breakdown of glucose to lactic acid or ethanol are examples of fermentation. Although electrons are removed from an organic molecule such as glucose to make NADH and pyruvate, the electrons are donated back to an organic molecule in the creation of lactic acid or ethanol. Therefore, there is no net removal of electrons from an organic molecule. Compared with oxidative phosphorylation, fermentation produces far less ATP for two reasons. First, glucose is not oxidized completely to CO$_2$ and water. Second, the NADH that is made during glycolysis cannot be used to make more ATP. Overall, the complete breakdown of glucose in the presence of oxygen yields 34 to 38 ATP molecules. By comparison, the anaerobic breakdown of glucose to lactic acid or ethanol yields only two ATP molecules.

7.4 Secondary Metabolism

Primary metabolism is the synthesis and breakdown of molecules and macromolecules that are found in all forms of life and are essential for cell structure and function. These include compounds such as lipids, sugars, nucleotides, and amino acids and the macromolecules that are derived from them. Cellular respiration, which we considered earlier in this chapter, is an example of primary metabolism. By comparison, **secondary metabolism** involves the synthesis of molecules—**secondary metabolites**—that are not essential for cell structure and growth. Any given secondary metabolite is unique to one species or group of species and is not usually required for survival. Secondary metabolites, also called secondary compounds, are commonly made in plants, bacteria, and fungi, where they play a variety of roles. Many secondary metabolites taste bad. When produced in a plant, for example, such a molecule may prevent an animal from eating the plant. In some cases, secondary metabolites are toxic. Such molecules may act as a chemical weapon that inhibits the growth of nearby organisms. In addition, many secondary metabolites produce a strong smell or bright color that attracts or repels other organisms. For example, the scent from a rose is due to secondary metabolites. The scent attracts insects that aid in pollination. Alternatively, other secondary metabolites may smell bad and prevent an animal from ingesting a plant.

Biologists have discovered thousands of different secondary metabolites, though any given species tends to produce only one or a few types. As you will learn, many of these have been put to practical use by humans, including spices that are used in cooking and antibiotics that inhibit the growth of pathogenic microorganisms and are used to treat diseases. Plants are particularly diverse in the types of secondary metabolites they produce, perhaps because they need clever ways to defend themselves. Bacteria and fungi also produce a large array of these compounds, while animals tend to produce relatively few. In this section, we will survey four categories of secondary metabolites: phenolics, alkaloids, terpenoids, and polyketides.

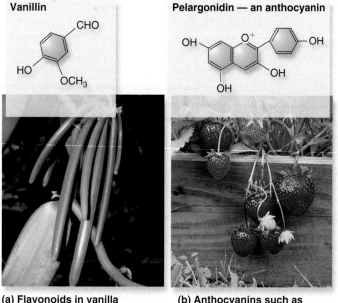

(a) Flavonoids in vanilla provide flavor

(b) Anthocyanins such as pelargonidin give red color

Figure 7.23   Phenolic compounds as secondary metabolites. The two examples shown here are flavonoids, which are a type of phenolic compound. **(a)** The flavor of vanilla is largely produced by flavonoids, an example of which is vanillin. Commercially, vanillin is extracted from the seed capsule of *Vanilla planifolia*. **(b)** Another group of flavonoids that causes red, blue, or purple color are anthocyanins. The red color of strawberries is caused by pelargonidin, an anthocyanin.

Phenolic Compounds Are Antioxidants That Have Intense Flavors and Bright Colors

The **phenolic** compounds all contain a cyclic ring of carbon with three double bonds, known as a benzene ring, within their structure. When a benzene ring is covalently linked to a single hydroxyl group, the compound is known as phenol.

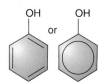

Phenol is the simplest of the phenolic compounds, though free phenol is not significantly accumulated in living organisms. However, more complex molecules that are derived from phenol are made in cells. Such phenolic compounds are synthesized using the side groups of the amino acids phenylalanine (which has a benzene ring) or tyrosine (which has a phenol ring). Common categories of phenolics are the flavonoids, tannins, and lignins. The tannins and lignins are large polymeric molecules composed of many phenolic units.

Flavonoids are produced by many plant species and create a variety of flavors and smells. These can play a role as deterrents to eating a plant or as attractants that promote pollination. The flavors of chocolate and vanilla largely come from a mixture of flavonoid molecules. Vanilla is produced by several spe-

cies of perennial vines of the genus *Vanilla* native to Mexico and tropical America (**Figure 7.23a**). The primary source of commercial vanilla comes from *V. planifolia*. Vanilla extract is obtained from the seed capsules. Another role of flavonoids is pigmentation. Anthocyanins (from the Greek *anthos*, flower, and *kyanos*, blue) produce the red, purple, and blue colors of many flowers, fruits, and vegetables (**Figure 7.23b**).

Biochemists have discovered that flavonoids have remarkable antioxidant properties that prevent the formation of damaging free radicals. In plants, flavonoids are thought to act as powerful antioxidants helping to protect plants from UV damage. In recent times, nutritionists have advocated the consumption of fruits and vegetables that have high amounts of flavonoids, such as broccoli and spinach. Dark chocolate is also rich in these antioxidants!

Tannins are large phenolic polymers, so named because they combine with the protein of animal skins to form leather. This process, known as tanning, also imparts a tan color to animal skins. Tannins are found in many plant species and typically act as a deterrent to animals, either because of a bitter taste or due to toxic effects. They also can inhibit the enzymes found in the digestive tracts of animals, if consumed in large amounts. Tannins are found abundantly in grape skins and play a key role in the flavor of red wine. Aging breaks down tannins, making the wine less bitter.

Lignins are also large phenolic polymers synthesized by plants. Lignin is found in plant cell walls and makes up about one-quarter to one-third of the weight of dry wood. The lignins form polymers that bond with other plant wall components such as cellulose. This strengthens plant cells and enables a plant to better withstand the rigors of environmental stress. To make paper, which is much more malleable than wood, the lignin is removed.

Alkaloids Form a Large Group of Bitter-Tasting Molecules That Also Provide Defense Mechanisms

Alkaloids are a group of structurally related molecules that all contain nitrogen and usually have a cyclic, ringlike structure. More than 12,000 different alkaloids have been discovered. Their name is derived from the observation that they are basic or alkaline molecules. Alkaloids are usually synthesized from amino acids precursors. Alkaloids are commonly made in plant species and occasionally in fungi (mushrooms) and animals (shellfish). Familiar examples include caffeine, nicotine, atropine, morphine, ergot, and quinine.

Like phenolics, many alkaloids serve a defense function in plants. Alkaloids are bitter-tasting molecules and often have an unpleasant odor. These features may prevent an animal from eating a plant or its fruit. For example, an alkaloid in chile peppers called capsaicin elicits a burning sensation. This molecule is so potent that one-millionth of a drop can be detected by the human tongue. Capsaicin may serve to discourage mammals from eating the peppers. Interestingly, however, birds do not experience the burning sensation of capsaicin and serve to disperse the seeds.

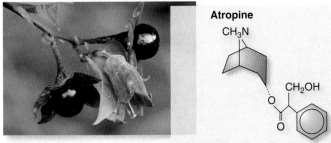

Deadly nightshade
(*Hyoscyamus niger*)

Figure 7.24   Alkaloids as secondary metabolites.   Atropine is an alkaloid produced by the plant called deadly nightshade (*Hyoscyamus niger*). Atropine is toxic because it interferes with nerve transmission. In humans, atropine causes the heart to speed up to dangerous and possibly fatal rates.

Other alkaloids are poisonous, like the alkaloid **atropine**, a potent toxin derived from the deadly nightshade plant (**Figure 7.24**). Animals that eat this plant and consequently ingest atropine become very sick and may die. It is unlikely that an animal that eats deadly nightshade and survives would choose to eat it a second time. Atropine acts by interfering with nerve transmission. In humans, for example, atropine causes the heart to speed up to dangerous rates, because the nerve inputs that normally keep a check on heart rate are blocked by atropine. Other alkaloids are not necessarily toxic but can cause an animal that eats them to become overstimulated (caffeine), understimulated (any of the opium alkaloids like morphine), or simply nauseated because the compound interferes with nerves required for proper functioning of the gastrointestinal system.

Terpenoids Are Molecules with Intense Smells and Color

A third major class of secondary metabolites are the **terpenoids**, of which over 25,000 have been identified, more than any other family of naturally occurring products. Terpenoids are synthesized from five-carbon isoprene units (shown below) and are also called isoprenoids.

Isoprene units are linked to each other to form larger compounds with multiples of five-carbon atoms. In many cases, the isoprene units form cyclic structures.

Terpenoids have a wide array of functions in plants. Notably, because many terpenoids are volatile (they become gases), they are responsible for the odors emitted by many types of plants, such as menthol produced by mint. The odors of terpenoids may attract pollinators or repel animals that eat plants. In addition, terpenoids often impart an intense flavor to plant tissues. Many of the spices used in cooking are rich in different types of terpenoids. Examples include cinnamon, fennel, cloves, cumin, cara-

Flamingo (*Phoenicopterus ruber*)

Figure 7.25   Terpenoids as secondary metabolites. Carotenoids are a type of terpenoid with bright color. The example here is β-carotene, which gives many organisms an orange color. Flamingos (*Phoenicopterus ruber*) receive β-carotene in their diet, primarily from eating shellfish.

way, and tarragon. Terpenoids are found in many traditional herbal remedies and are under medical investigation for potential pharmaceutical effects.

Other terpenoids, like the carotenoids, are responsible for the coloration of many species. An example is β-carotene, which gives carrots their orange color. Carotenoids are also found in leaves, but their color is masked by chlorophyll, which is green. In the autumn, when chlorophyll breaks down, the color of the carotenoids becomes evident. Carotenoids give color to animals such as salmon, goldfish, and flamingos (**Figure 7.25**). Another role of terpenoids is cell signaling. All steroid hormones, which function as signaling molecules in animals and plants (as we will discuss in Chapter 9), are derived from terpenoids.

Polyketides Are Often Used as Chemical Weapons to Kill Competing Organisms

Polyketides are a group of secondary metabolites that are produced by bacteria, fungi, plants, insects, dinoflagellates, mollusks, and sponges. They are synthesized by the polymerization of acetyl (CH_3COOH) and propionyl (CH_3CH_2COOH) groups to create a diverse collection of molecules, often with many ringed structures. Polyketides are usually secreted by the organism that makes them and are often highly toxic to other organisms. For example, the polyketide known as streptomycin is made by the soil bacterium *Streptomyces griseus* (**Figure 7.26**). It is secreted by this bacterium and taken up by other species, where it disrupts protein synthesis and thereby inhibits their growth. In this way, *S. griseus* is able to kill or inhibit the growth of other species in its vicinity. This is an advantage for *S. griseus* because other species may be using limited resources that the bacterium could use for its own growth.

During the past several decades, over 10,000 polyketides have been identified and analyzed. Familiar examples include streptomycin, erythromycin, and tetracycline. The toxic effects of polyketides are often very selective, making them valuable medical tools. For example, streptomycin disrupts protein synthesis in many bacterial species, but it does not adversely affect protein synthesis in mammalian cells. Therefore, it has been used as an antibiotic to treat or prevent bacterial infections in humans and other mammals. Similarly, other polyketides inhibit the growth of fungi, parasites, and insects. More recently, researchers have even discovered that certain polyketides inhibit the growth of cancer cells. The production and sale of polyketides to treat and prevent diseases and as pesticides constitute an enormous industry, with annual sales in the U.S. at over $20 billion.

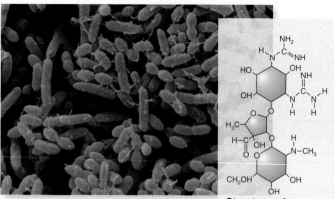

Streptomyces griseus, a soil bacterium   **Streptomycin**

Figure 7.26   Polyketides as secondary metabolites. Streptomycin, whose structure is shown here, is an antibiotic produced by *Streptomyces griseus*, a soil bacterium. The scanning electron micrograph shows *S. grisieus*.

CHAPTER SUMMARY

7.1 Energy, Chemical Reactions, and Enzymes

- The fate of a chemical reaction is determined by its direction and rate.

- Energy, the ability to promote change, exists in many forms. According to the first law of thermodynamics, energy cannot be created or destroyed but it can be converted from one form to another. The second law of thermodynamics states that energy interconversions involve an increase in entropy. (Figures 7.1, 7.2, Table 7.1)

- Free energy is the amount of available energy that can be used to do work. Spontaneous reactions release free energy, which means they have a negative free energy change. (Figure 7.3)

- An exergonic reaction has a negative free energy change, while an endergonic reaction has a positive change. Chemical reactions proceed until they reach a state of chemical equilibrium, where the rate of formation of products equals the rate of formation of reactants.

- Cells use energy intermediates such as ATP to drive endergonic reactions.

- Proteins that speed up the rate of a chemical reaction are called enzymes. They lower the activation energy that is needed to achieve a transition state. (Figure 7.4)

- Enzymes recognize the reactants, also called substrates, with a high specificity. Conformational changes are involved in lowering the activation energy for a chemical reaction. (Figure 7.5)

7.2 Overview of Metabolism

- Metabolism is the sum of the chemical reactions in a living organism. Enzymes often function in pathways that lead to the formation of a particular product. (Figure 7.6)

- Catabolic reactions involve the breakdown of larger molecules into smaller ones. These reactions regenerate small molecules that are used as building blocks to make new molecules. The small molecules are also broken down to make energy intermediates such as ATP and NADH. Such reactions are often redox reactions in which electrons are transferred from one molecule to another. (Figure 7.7)

- Anabolic reactions involve the synthesis of larger molecules and macromolecules.

- Estimates from genome analysis indicate that over 20% of a cell's proteins use ATP. (Table 7.2, Figure 7.8)

- Metabolic pathways are controlled by gene regulation, cell signaling, and feedback inhibition. (Figure 7.9)

7.3 Cellular Respiration

- Cells obtain energy via cellular respiration, which involves the breakdown of molecules.

- The breakdown of glucose occurs in four stages: glycolysis, pyruvate breakdown, citric acid cycle, and oxidative phosphorylation. (Figure 7.10)

- Glycolysis is the breakdown of glucose to two pyruvates, producing two ATP and two NADH. ATP is made by substrate-level phosphorylation. (Figures 7.11, 7.12)

- Pyruvate is broken down to CO_2 and an acetyl group that becomes attached to CoA. NADH is made during this process. (Figure 7.13)

- During the citric acid cycle, the acetyl group attached to CoA is broken down to two CO_2 molecules. Three NADH, one $FADH_2$, and one ATP are made during this process. (Figures 7.14, 7.15)

- Oxidative phosphorylation involves two events. The electron transport chain oxidizes NADH or $FADH_2$ and generates an H^+ electrochemical gradient. This gradient is utilized by the ATP synthase to make ATP via chemiosmosis. (Figures 7.16, 7.17)

- Racker and Stoeckenius showed that the ATP synthase uses an H^+ gradient by reconstituting the ATP synthase with a light-driven H^+ pump. (Figure 7.18)

- The ATP synthase is a rotary machine. The rotation is caused by the movement of H^+ through the c subunits that cause the γ subunit to spin, resulting in conformational changes in the β subunits that promote ATP synthesis. (Figure 7.19)

- Yoshida and Kinosita experimentally demonstrated rotation of the γ subunit by attaching a fluorescently labeled actin filament and watching it spin in the presence of ATP. (Figure 7.20)

- Proteins and fats can enter into glycolysis or the citric acid cycle at different points. (Figure 7.21)

- Anaerobic respiration occurs in the absence of oxygen. Either the final electron acceptor of the electron transport chain is not oxygen, or NAD^+ is regenerated by donating electrons to an organic molecule such as pyruvate (to make lactic acid). (Figure 7.22)

7.4 Secondary Metabolism

- Secondary metabolites are not usually necessary for cell structure and function, but they provide an advantage to an organism that may involve taste, smell, color, or poison. Four categories are phenolic compounds, alkaloids, terpenoids, and polyketides. (Figures 7.23, 7.24, 7.25, 7.26)

TEST YOURSELF

1. According to the second law of thermodynamics
   a. energy cannot be created or destroyed.
   b. each energy transfer decreases the disorder of a system.
   c. energy is constant in the universe.
   d. each energy transfer increases the level of disorder in a system.
   e. chemical energy is a form of potential energy.

2. _____ reactions release free energy.
   a. Exergonic   d. All of the above
   b. Spontaneous   e. Both a and b
   c. Endergonic

3. Enzymes speed up reactions by
   a. providing chemical energy to fuel a reaction.
   b. lowering the activation energy necessary to initiate the reaction.
   c. causing an endergonic reaction to become an exergonic reaction.
   d. substituting for one of the reactants necessary for the reaction.
   e. none of the above.

4. Which of the following factors will alter the function of enzymes?
   a. pH   d. all of the above
   b. temperature   e. b and c only
   c. cofactors

5. In biological systems, ATP functions by
   a. providing the energy necessary for an endergonic reaction by coupling it with an exergonic reaction.
   b. acting as an enzyme and lowering the activation energy of certain reactions.
   c. adjusting the pH of solutions to maintain optimal conditions for enzyme activity.
   d. regulating the speed at which endergonic reactions proceed.
   e. interacting with enzymes as a cofactor to stimulate chemical reactions.

6. During redox reactions, the molecule that donates an electron is said to be
   a. reduced.   d. catabolized.
   b. phosphorylated.   e. methylated.
   c. oxidized.

7. Currently scientists are identifying proteins that use ATP as an energy source by
   a. determining whether those proteins function in anabolic or catabolic reactions.
   b. determining if the protein has a known ATP-binding site.
   c. predicting the free energy necessary for the protein to function.
   d. determining if the protein has an ATP synthase subunit.
   e. all of the above.

8. During glycolysis, ATP is produced by
   a. oxidative phosphorylation.
   b. substrate-level phosphorylation.
   c. redox reactions.
   d. all of the above.
   e. both a and b.

9. The energy necessary to produce ATP during oxidative phosphorylation is provided by
   a. the H^+ concentration gradient produced by the electron transport chain.
   b. GTP produced during the citric acid cycle.
   c. lactic acid metabolism.
   d. the release of CO_2 from the mitochondria.
   e. all of the above.

10. Secondary metabolites
    a. help deter predation of certain organisms by causing the organism to taste bad.
    b. help attract pollinators by producing a pleasant smell.
    c. help organisms compete for resources by acting as a poison to competitors.
    d. provide protection from DNA damage.
    e. all of the above.

CONCEPTUAL QUESTIONS

1. Distinguish between endergonic and exergonic reactions.

2. Define feedback inhibition.

3. The electron transport chain is so named because electrons are transported from one component to another. Describe the purpose of the electron transport chain.

EXPERIMENTAL QUESTIONS

1. The components of the ATP synthase are too small to be visualized by light microscopy. For the experiment of Figure 7.20, how did the researchers observe the movement of the ATP synthase?

2. In the experiment of Figure 7.20, what observation did the researchers make that indicated that the ATP synthase is a rotary machine? What was the control of this experiment? What did it indicate?

3. Were the rotations seen by the researchers in the data of Figure 7.20 in the same direction as expected in the mitochondria during ATP synthesis? Why or why not?

COLLABORATIVE QUESTIONS

1. Discuss several ways in which metabolic pathways are controlled or regulated.

2. Discuss the concept of secondary metabolism and give an example.

www.brookerbiology.com

This website includes answers to the Biological Inquiry questions found in the figure legends and all end-of-chapter questions.

8

PHOTOSYNTHESIS

CHAPTER OUTLINE

A tropical rain forest in Akaka Falls State Park, Hawaii. Plant life in tropical rain forests carries out a large amount of the world's photosynthesis and supplies the atmosphere with a sizeable fraction of its oxygen.

Across the Earth, the most visible color on land is green. We often associate this color with emerging life, as in the growth of plants in the spring. The green color of plants is due to a pigment called chlorophyll. This pigment provides the starting point for the process of **photosynthesis**, in which the energy within light is captured and used to synthesize carbohydrates. Nearly all living organisms ultimately rely on photosynthesis for their nourishment. Photosynthesis is also responsible for producing the oxygen that makes up a large portion of the Earth's atmosphere. Therefore, all aerobic organisms ultimately rely on photosynthesis for respiration.

Organisms can be categorized as heterotrophs and autotrophs. **Heterotrophs** must eat food, organic molecules from their environment, to sustain life. Examples of heterotrophs include most species of bacteria and protists, as well as all species of fungi and animals. In Chapter 7, we learned how cells use food molecules such as glucose for their energy needs and as building blocks to make new cellular molecules and macromolecules. In this chapter, we turn to **autotrophs**, organisms that make organic molecules from inorganic sources, and focus on **photoautotrophs**, those organisms that use light as a source of energy. These include green plants, algae, and some prokaryotic species such as cyanobacteria.

We begin this chapter with an overview of photosynthesis, with an emphasis on how it occurs in green plants. We will then explore the two stages of photosynthesis: the light reactions, in which light energy is captured by the chlorophyll pigments within plants and converted to chemical energy in the form of two compounds, ATP and NADPH; and the Calvin cycle, a series of steps in which these compounds drive the incorporation of CO_2 into carbohydrates. We conclude with a consideration of the variations in photosynthesis that occur in plants existing in hot and dry conditions.

8.1 Overview of Photosynthesis

In the mid-1600s, a Flemish physician, Jan Baptista Van Helmont, conducted an experiment in which he transplanted the shoot of a young willow tree into a bucket of soil of known weight. He watered the tree and allowed it to grow for five years. After this time, the willow tree had added 164 pounds to its original weight, but the soil had lost only 2 ounces. Van Helmont correctly concluded that the willow tree did not get most of its nutrients from the soil. However, he incorrectly concluded that the material that made up the bark, wood, roots, and leaves came from the water he had added over the five years. Although water does contribute to the growth and mass of plants, we now know that CO_2 from the air is also critically important.

In 1771, Joseph Priestley, an English chemist, carried out an experiment in which he placed a burning candle in a closed chamber. The candle burned out very quickly. He conducted another experiment in which he placed a burning candle and a sprig of mint in a second chamber. Similarly, the candle quickly went out. After several days, Priestley was able to relight each candle without opening the two chambers by focusing a beam of sunlight onto the wicks with a mirror. However, only in the chamber with the sprig of mint could the candle burn again. Priestley hypothesized that plants restore to the air whatever burning candles remove. His results occurred because plants release oxygen as a result of photosynthesis.

Shortly thereafter, Jan Ingenhousz, a Dutch physician, immersed green plants underwater and discovered that they released bubbles of oxygen. Moreover, Ingenhousz determined that sunlight was necessary for oxygen production. During this same period, Jean Senebier, a Swiss botanist, found that CO_2 is required for plant growth and Nicolas-Théodore de Saussure, a Swiss chemist, showed that water is also required. With this accumulating information, Julius von Mayer, a German physicist, proposed in 1845 that plants convert light energy from the sun into chemical energy.

For the next several decades, plant biologists studied photosynthesis in prokaryotes, algae, and green plants. Researchers discovered that some photosynthetic bacteria could use hydrogen sulfide (H_2S) instead of water (H_2O) for photosynthesis and that these organisms released sulfur instead of oxygen. In the 1930s, based on this information, Dutch American microbiologist Cornelis van Niel proposed a general equation for photosynthesis that applies to plants, algae, and photosynthetic bacteria alike.

$$CO_2 + 2\ H_2A + \text{Light energy} \rightarrow CH_2O + 2A + H_2O$$

where A is oxygen (O) or sulfur (S) and CH_2O is the general formula for a carbohydrate. This is a redox reaction in which H_2A is oxidized and CO_2 is reduced.

In green plants, A is oxygen and 2A is a molecule of oxygen that is designated O_2. Therefore this equation becomes:

$$CO_2 + 2\ H_2O + \text{Light energy} \rightarrow CH_2O + O_2 + H_2O$$

When the carbohydrate that is produced is glucose, we multiply each side of the equation by six to obtain:

$$6\ CO_2 + 12\ H_2O + \text{Light energy} \rightarrow C_6H_{12}O_6 + 6\ O_2 + 6\ H_2O$$
$$\text{Glucose}$$
$$\Delta G = +685\ \text{kcal/mole}$$

In this redox reaction, CO_2 is reduced during the formation of glucose, and H_2O is oxidized during the formation of O_2. The free energy change required for the production of 1 mole of glucose from carbon dioxide and water is a whopping +685 kcal/mole! As we learned in Chapter 7, endergonic reactions are driven forward by coupling the reaction with an exergonic process, a process that releases free energy. In this case, the energy from light ultimately drives the synthesis of glucose.

In this section, we will survey the general features of photosynthesis as it occurs in green plants. Later sections will examine the various steps in this process.

Photosynthesis Powers the Biosphere

The term **biosphere** describes the regions on the surface of the Earth and in the atmosphere where living organisms exist. Life in the biosphere is largely driven by the photosynthetic power of green plants. For most species to exist, a key energy cycle involves the interplay between organic molecules (such as glucose) and inorganic molecules, namely, CO_2 and O_2. Life on Earth

involves a cycle in which cells use organic molecules for energy and plants replenish those molecules via photosynthesis. As we examined in Chapter 7, cellular respiration involves the breakdown of organic molecules to produce energy intermediates such as ATP. When glucose is broken down to CO_2, the net reaction of cellular respiration in the presence of oxygen can be summarized as:

$$C_6H_{12}O_6 + 6\ O_2 \rightarrow 6\ CO_2 + 6\ H_2O + \textbf{Energy}$$

The net reaction of photosynthesis (in which only six net molecules of H_2O are consumed) can be viewed as the opposite of respiration.

$$6\ CO_2 + 6\ H_2O + \textbf{Energy} \rightarrow C_6H_{12}O_6 + 6\ O_2$$

The breakdown of glucose during cell respiration is an energy-releasing process that drives the synthesis of ATP. By comparison, the energy that is needed to synthesize glucose during photosynthesis ultimately comes from sunlight.

Plants make a large proportion of the Earth's organic molecules via photosynthesis. At the same time, they also produce O_2. These organic molecules are metabolized by the plants themselves as well as by heterotrophs such as animals and fungi. This metabolism generates CO_2, which is released into the atmosphere and can be used by plants to make more organic molecules like glucose. In this way, a cycle exists between photosynthesis and cellular respiration that sustains life on our planet.

In Plants, Photosynthesis Occurs in the Chloroplast

Chloroplasts are organelles found in plant cells and algae that carry out photosynthesis. These organelles contain large quantities of **chlorophyll**, a pigment that gives plants their green color. All green parts of a plant contain chloroplasts and are capable of photosynthesis, although the majority of photosynthesis occurs in the leaves (**Figure 8.1**). The central part of the leaf, called the **mesophyll**, contains cells that carry out the bulk of photosynthesis in plants. For photosynthesis to occur, the mesophyll cells must obtain water and carbon dioxide. The water is taken up by the roots of the plant and is transported to the leaves by small veins. Carbon dioxide gas enters the leaf, and oxygen exits, via pores called **stomata** (singular, *stoma*), from the Greek, meaning mouth. The anatomy of leaves will be examined further in Chapter 35.

Like the mitochondrion, a chloroplast contains an outer and inner membrane, with an intermembrane space lying between the two. A third membrane, called the **thylakoid membrane**, contains pigment molecules, including chlorophyll. The thylakoid membrane forms many flattened, fluid-filled tubules called the **thylakoids**, which enclose a single, convoluted compartment known as the **thylakoid lumen**. Thylakoids stack on top of each other to form a structure called a **granum** (plural, *grana*). The **stroma** is the fluid-filled region of the chloroplast between the thylakoid membrane and the inner membrane.

Photosynthesis Occurs in Two Stages

The process of photosynthesis can be divided into two stages called the **light reactions** and the **Calvin cycle**. The term photosynthesis is derived from the association between these two stages: The prefix <u>photo</u> refers to the light reactions that capture the energy needed for the <u>synthesis</u> of carbohydrates that occurs in the Calvin cycle. Each stage occurs at specific sites in the chloroplast: The light reactions take place at the thylakoid membrane, and the Calvin cycle occurs in the stroma (**Figure 8.2**).

The light reactions involve an amazing series of energy conversions, starting with light energy and ending with chemical energy in the form of covalent bonds. The light reactions produce three chemical products: ATP, NADPH, and O_2. ATP and NADPH are energy intermediates that provide the needed energy and electrons to drive the Calvin cycle. Like NADH, **NADPH (nicotinamide adenine dinucleotide phosphate)** is an electron carrier; its structure differs from NADH by the presence of an additional phosphate group. In the Calvin cycle, atmospheric carbon dioxide is incorporated into organic molecules, some of which are converted to carbohydrates.

O_2 is another important product of the light reactions. As described in Chapter 7, this molecule is vital to the process of aerobic respiration. Nearly all of the O_2 in the atmosphere is produced by photosynthesis from green plants and aquatic microorganisms. A large percentage of atmospheric oxygen is made by regions of the Earth that are rich in plant life, such as tropical rain forests. More than 20% of the world's oxygen is produced in the Amazon rain forest in South America alone.

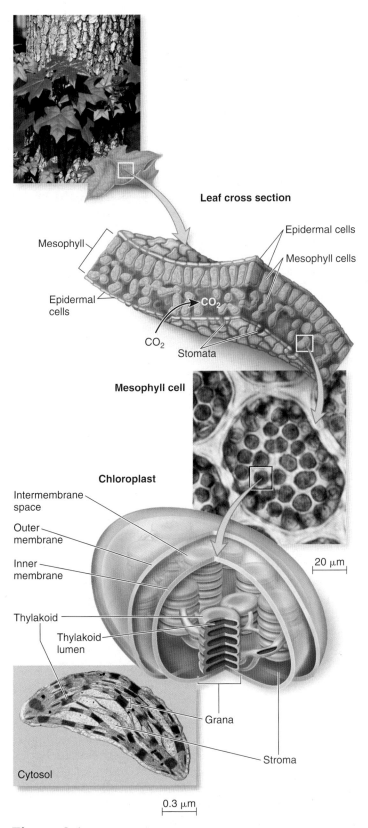

Figure 8.1 **Leaf organization.** Leaves are composed of layers of cells. The epidermal cells are on the outer surface, both top and bottom, with mesophyll cells sandwiched in the middle. The mesophyll cells in most plants are the primary sites of photosynthesis.

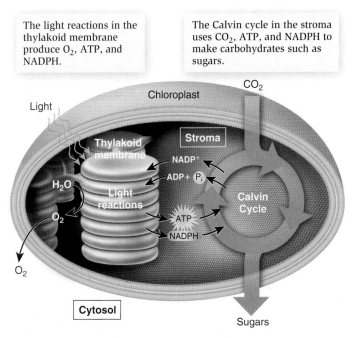

The light reactions in the thylakoid membrane produce O_2, ATP, and NADPH.

The Calvin cycle in the stroma uses CO_2, ATP, and NADPH to make carbohydrates such as sugars.

Figure 8.2 **An overview of the two stages of photosynthesis.** The light reactions, through which O_2, ATP, and NADPH are made, occur at the thylakoid membrane. The Calvin cycle, in which enzymes use ATP and NADPH to incorporate CO_2 into carbohydrate, occurs in the stroma.

Ecologists are alarmed about the rate at which such forests are being destroyed by human activities. Rain forests once covered 14% of the Earth's land surface but now occupy less than 6%. At their current rate of destruction, rain forests may be nearly eliminated in less than 40 years. Such an event may lower the level of oxygen in the atmosphere and thereby have a harmful impact on living organisms on a global scale.

8.2 Reactions That Harness Light Energy

Photosynthesis relies on the first law of thermodynamics. Recall from Chapter 7 that this law states that while energy cannot be created or destroyed, it can be transferred from one place to another and transformed from one type to another. During photosynthesis, energy in the form of light is transferred from the sun, some 92 million miles away, to a pigment molecule in a photosynthetic organism such as a plant. The next thing that happens is an interesting and complex series of energy transformations in which light energy is transformed into electrochemical energy and then into energy stored within chemical bonds.

In this section, we will explore this series of transformations, collectively called the light reactions of photosynthesis. We begin by examining the unique properties of light and then consider the features of chloroplasts that allow them to capture light energy. The remaining sections focus on how the light reactions of photosynthesis create three important products: O_2, ATP, and NADPH.

Light Energy Is a Form of Electromagnetic Radiation

Light is a critical phenomenon that is essential to support life on Earth. Light is a type of electromagnetic radiation, so named because it consists of energy in the form of electric and magnetic fields. Electromagnetic radiation travels as waves caused by the oscillation of the electric and magnetic fields. The **wavelength** is the distance between the peaks in a wave pattern. The **electromagnetic spectrum** encompasses all possible wavelengths of electromagnetic radiation, from relatively short wavelengths (gamma rays) to much longer wavelengths (radio waves) (**Figure 8.3**). Visible light is the range of wavelengths that are detected by the human eye, commonly in the range of 380–740 nm. As discussed later, visible light provides the energy to drive photosynthesis.

Physicists have also discovered that light exhibits behaviors that are characteristic of particles. Albert Einstein formulated the photon theory of light in which he proposed that light is composed of discrete particles called **photons**—massless particles each traveling in a wavelike pattern and moving at the speed of light. Each photon contains a specific amount of energy. An important difference between the various types of electro-

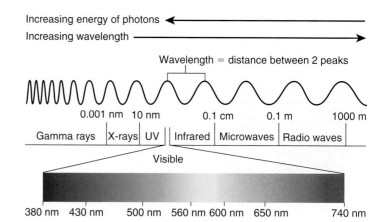

Figure 8.3 **The electromagnetic spectrum.** The bottom portion of this figure emphasizes visible light, the wavelengths of electromagnetic radiation that are visible to the human eye. Light in the visible portion of the electromagnetic spectrum drives photosynthesis.

magnetic radiation described in Figure 8.3 is the amount of energy found in the photons. Shorter wavelength radiation carries more energy per unit of time than longer wavelength radiation. The photons of gamma rays carry more energy than those of radio waves.

The sun radiates the entire spectrum of electromagnetic radiation. The atmosphere prevents much of this radiation from reaching the Earth's surface. For example, the ozone layer forms a thin shield in the upper atmosphere, protecting life on Earth from much of the sun's ultraviolet rays. Even so, a substantial amount of electromagnetic radiation does reach the Earth's surface. The effect of light on living organisms is critically dependent on the energy of the photons. The photons found in gamma rays, X-rays, and UV rays have very high energy. When molecules in cells absorb such energy, the effects can be devastating. Such types of radiation can cause mutations in DNA and even lead to cancer. By comparison, the energy of photons found in visible light is much milder. Molecules can absorb this energy in a way that does not cause permanent harm. Let's now consider how molecules in living cells absorb the energy within visible light.

Photosynthetic Pigments Absorb Light Energy

When light strikes an object, one of three things can happen. First, light may simply pass through an object. Second, the object may change the path of light toward a different direction. A third possibility is that the object may absorb the light. The term **pigment** is used to describe a molecule that can absorb light energy. When light strikes a pigment, some of the wavelengths of light energy are absorbed, while others are reflected. For example, we perceive that leaves are green because they are reflecting radiant energy of the green wavelength. Various pigments in the leaves absorb the other light energy wavelengths.

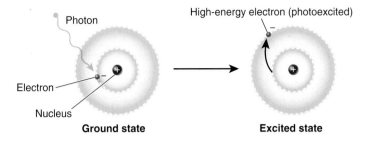

Figure 8.4 **Absorption of light energy by an electron.** When a photon of light of the correct amount of energy strikes an electron, the electron is boosted from the ground (unexcited) state to a higher energy level. When this occurs, the electron occupies an orbital that is farther away from the nucleus of the atom. At this farther distance, the electron is held less firmly and is considered unstable.

Biological inquiry: For an electron to drop down to a lower orbital, describe the three things that could happen.

At the extremes of color reflection are white and black. A white object reflects nearly all of the light energy falling on it, whereas a black object absorbs nearly all of the light energy.

What do we mean when we say that light energy is absorbed? In the visible spectrum, light energy is usually absorbed by boosting electrons to higher energy levels (**Figure 8.4**). Recall from Chapter 2 that electrons are located around the nucleus of an atom. The location that an electron is likely to be found is called its orbital. Electrons in different orbitals possess different amounts of energy. For an electron to absorb light energy and be boosted to an orbital with a higher energy, it must overcome the difference in energy between the orbital it is in and the orbital to which it is going. For this to happen, an electron must absorb a photon that contains precisely that amount of energy. Among different pigment molecules, there exist a variety of electrons that can be shifted to different energy levels. Therefore, the wavelength of light that a pigment absorbs depends on the amount of energy that is needed to boost an electron to a higher orbital.

After an electron absorbs energy, it is said to be in an excited state. Usually, this is an unstable condition. The electron may release the energy in one of two forms. When an excited electron drops back down to a lower energy level, it may release heat. For example, on a sunny day, the sidewalk heats up because it absorbs light energy that is released as heat. A second way that an electron can release energy is in the form of light. Certain organisms, such as jellyfish, possess molecules that make them glow. This glow is due to the release of light when electrons drop down to lower energy levels, a phenomenon called fluorescence.

In the case of photosynthetic pigments, however, a different event happens that is critical for the process of photosynthesis. At a particular step, an excited electron in a photosynthetic pigment is removed from that molecule and transferred to another molecule where the electron is more stable. When this occurs, the energy in that electron is said to be "captured" because the electron does not readily drop down to a lower energy level and release heat or light.

In plants, several different pigment molecules absorb the light energy used to drive photosynthesis. Two types of chlorophyll pigments, termed **chlorophyll *a*** and **chlorophyll *b***, are found in green algae and plants. Their structure was determined in the 1930s by German chemist Hans Fischer (**Figure 8.5a**). In the chloroplast, both chlorophylls *a* and *b* are bound to integral membrane proteins in the thylakoid membrane. The chlorophylls contain a structure called a porphyrin ring that has a bound magnesium ion (Mg^{2+}). In this ring, an electron can follow a path in which it spends some of its time around several different atoms. These electrons, called delocalized electrons because they aren't restricted to a single atom, absorb light energy. Chlorophyll also contains a long hydrocarbon structure called a phytol tail. This tail is hydrophobic and anchors the pigment to the surface of proteins within the thylakoid membrane.

Carotenoids are another type of pigment found in chloroplasts (**Figure 8.5b**). These pigments impart a color that ranges from yellow to orange to red. Carotenoids are often the major pigments in flowers and fruits. In leaves, the more abundant chlorophylls usually mask the colors of carotenoids. In temperate climates where the leaves change colors, the quantity of chlorophyll in the leaf declines during autumn. The carotenoids become readily visible and produce the yellows and oranges of autumn foliage.

An **absorption spectrum** is a diagram that depicts the wavelengths of electromagnetic radiation that are absorbed by a pigment. Each of the pigments shown in **Figure 8.5c** absorbs light in different regions of the visible spectrum. The absorption spectra of chlorophylls *a* and *b* are slightly different, though both chlorophylls absorb light most strongly in the red and violet parts of the visible spectrum and absorb green light poorly. Carotenoids absorb light in the blue and blue-green regions of the visible spectrum.

Having different pigments allows plants to absorb light at many different wavelengths. In this way, plants are more efficient at capturing the energy in sunlight. This phenomenon is highlighted in an **action spectrum**, which describes the rate of photosynthesis plotted as a function of different wavelengths of light (**Figure 8.5d**). The highest rates of photosynthesis correlate with the wavelengths that are strongly absorbed by the chlorophylls and carotenoids. Photosynthesis is poor in the green region of the spectrum, because these pigments do not readily absorb this wavelength of light.

Photosystem II Captures Light Energy and Produces O_2

Photosynthetic organisms have the unique ability not only to absorb light energy but also to capture that energy in a stable way. Many organic molecules can absorb light energy. For example, on a sunny day, molecules in your skin absorb light energy and release the energy as heat. The heat that is released,

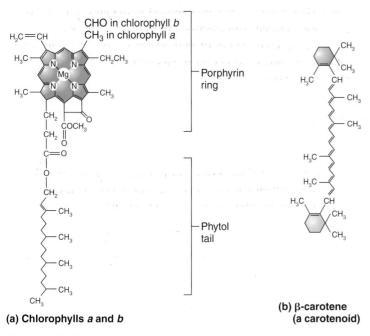

(a) Chlorophylls *a* and *b*

(b) β-carotene (a carotenoid)

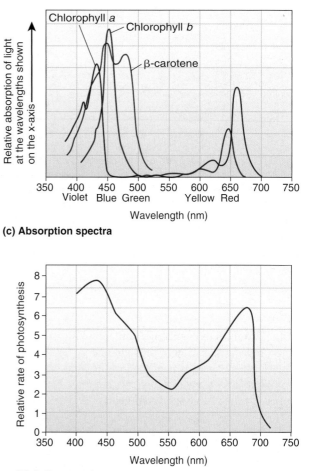

(c) Absorption spectra

(d) Action spectrum

Figure 8.5 Structures and properties of pigment molecules. **(a)** The structure of chlorophylls *a* and *b*. As indicated, chlorophylls *a* and *b* differ only at a single site, at which chlorophyll *a* has a —CH_3 group and chlorophyll *b* has a —CHO group. **(b)** The structure of β-carotene, an example of a carotenoid. The shaded areas are the regions where a delocalized electron can hop from one atom to another. **(c)** Absorption spectra that show the absorption of light by chlorophyll *a*, chlorophyll *b*, and β-carotene. **(d)** An action spectrum of photosynthesis depicting the relative rate of photosynthesis in green plants at different wavelengths of light.

however, cannot be harnessed to do useful work. A key feature of photosynthesis is the ability of pigments to capture light energy and transfer it to other molecules that can hold on to the energy in a stable fashion and ultimately produce energy intermediate molecules that can do cellular work.

Let's now consider how chloroplasts capture light energy. The thylakoid membrane contains two distinct complexes of proteins and pigment molecules called **photosystem I (PSI)** and **photosystem II (PSII)**. Photosystem I was discovered before photosystem II, but because photosystem II is the initial step in photosynthesis, we will examine its function first.

Photosystem II has two main components, a light-harvesting complex and a reaction center (**Figure 8.6**). In 1932, Robert Emerson and an undergraduate student, William Arnold, originally discovered the **light-harvesting complex** in the thylakoid membrane. It is composed of several dozen pigment molecules that are anchored to proteins. The role of the complex is to directly absorb photons of light. When a pigment molecule absorbs a photon, this boosts an electron to a higher energy level. As shown in Figure 8.6, the energy (not the electron itself) can be transferred to adjacent pigment molecules by a process called **resonance energy transfer**. Eventually, the energy may be transferred to a special pigment molecule designated P680, so called

because it is best at absorbing light at a wavelength of 680 nm. When an electron in P680 is excited, it is designated P680*. The P680 pigment molecule is located in the **reaction center** of PSII. The light-harvesting complex is also called the **antenna complex** because it acts like an antenna that absorbs energy from light and funnels that energy to P680 in the reaction center.

A high-energy (photoexcited) electron in a pigment molecule is relatively unstable. It may abruptly release its energy by giving off heat or light. Unlike the pigments in the antenna complex that undergo resonance energy transfer, P680* can actually release its high-energy electron. The role of the reaction center is to quickly remove the high-energy electron from P680* and transfer it to another molecule, where the electron will be more stable. This molecule is called the **primary electron acceptor** (Figure 8.6). The transfer of the electron from P680* to the primary electron acceptor is remarkably fast. It occurs in less than a few picoseconds! (One picosecond equals one-trillionth of a second, also noted as 10^{-12} s.) Because this occurs so quickly, the excited electron does not have much time to release its energy in the form of heat or light.

After the primary electron acceptor has received this high-energy electron, the light energy has been captured and can be used to perform cellular work. As we will discuss shortly, the

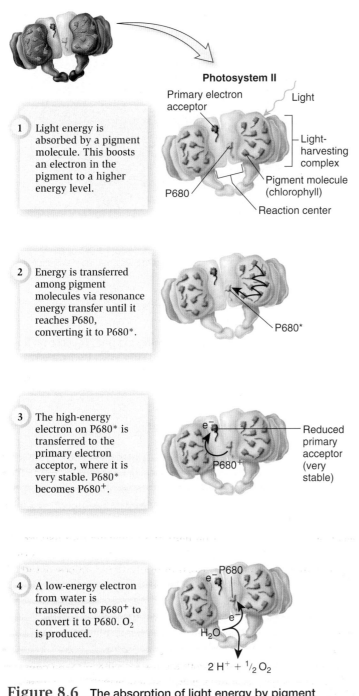

1 Light energy is absorbed by a pigment molecule. This boosts an electron in the pigment to a higher energy level.

Photosystem II

Primary electron acceptor

Light

Light-harvesting complex

Pigment molecule (chlorophyll)

P680

Reaction center

2 Energy is transferred among pigment molecules via resonance energy transfer until it reaches P680, converting it to P680*.

P680*

3 The high-energy electron on P680* is transferred to the primary electron acceptor, where it is very stable. P680* becomes P680$^+$.

e$^-$

P680$^+$

Reduced primary acceptor (very stable)

4 A low-energy electron from water is transferred to P680$^+$ to convert it to P680. O$_2$ is produced.

e$^-$ P680

e$^-$

e$^-$

H$_2$O

2 H$^+$ + $^1/_2$O$_2$

Figure 8.6 The absorption of light energy by pigment molecules in the light-harvesting complex, and the path that leads to the capture of energy by the primary electron acceptor.

work that it first performs is to synthesize the energy intermediates ATP and NADPH. Later, these energy intermediates are used to make carbohydrates.

Before we examine the fate of the high-energy electron that was transferred to the primary electron acceptor, let's consider what happens to the P680 molecule after it has released its high-energy electron. Another function of the reaction center is to

replace the electrons that are removed from pigment molecules. They are replaced with electrons from water molecules (Figure 8.6). The reaction center of photosystem II removes electrons from water and transfers those electrons to oxidized pigment molecules (P680$^+$).

$$H_2O \rightarrow 1/2\ O_2 + 2\ H^+ + 2\ e^-$$

$$2\ P680^+ + 2\ e^- \rightarrow 2\ P680$$

(from water)

The oxidation of water results in the formation of oxygen gas (O$_2$). Photosystem II is the only known protein complex that can oxidize water, resulting in the release of O$_2$ into the atmosphere.

Photosystem II Is an Amazing Redox Machine

Redox reactions are fundamentally important for cells to store and utilize energy and to form covalent bonds in organic molecules. Photosystem II is a particularly remarkable example of a redox machine. As we have learned, this complex of proteins removes electrons from a pigment molecule and transfers them to a primary electron acceptor. Perhaps even more remarkable is that photosystem II can remove electrons from water, a very stable molecule that holds onto its electrons tightly. The removal of electrons is how O$_2$ is made.

Many approaches have been used to study how photosystem II works. In recent years, much effort has been aimed at determining the biochemical composition of the protein complex and the roles of its individual components. The number of protein subunits varies somewhat from species to species and may vary due to environmental changes. Typically, photosystem II contains around 19 different protein subunits. Two subunits, designated D1 and D2, contain the reaction center that carries out the redox reactions (**Figure 8.7a**). Two other subunits, called CP43 and CP47, bind the pigment molecules that form the light-harvesting complex. Many additional subunits regulate the function of photosystem II and provide structural support.

The oxidation of water occurs in a region called the **manganese cluster**. This site is located on the side of D1 that faces the thylakoid lumen. The manganese cluster has four Mn^{2+}, one Ca^{2+}, and one Cl$^-$. Two water molecules bind to this site. D1 catalyzes the removal of four electrons from the two water molecules to create four H$^+$ and O$_2$. The electrons are transferred, one at a time, to a tyrosine (Tyr) in D1 and then to an oxidized pigment molecule (P680$^+$) to produce P680. When the electron on P680 becomes excited, usually by resonance energy transfer, it then moves to the primary electron acceptor, which is an organic molecule called pheophytin (Pp) that is permanently bound to photosystem II. Pheophytin transfers its electron to a plastoquinone molecule, designated Q$_A$, which is also permanently bound to photosystem II. Next, the electron is transferred to another plastoquinone molecule designated Q$_B$, which can accept two high-energy electrons and bind two H$^+$. Q$_B$ can diffuse away from the reaction center.

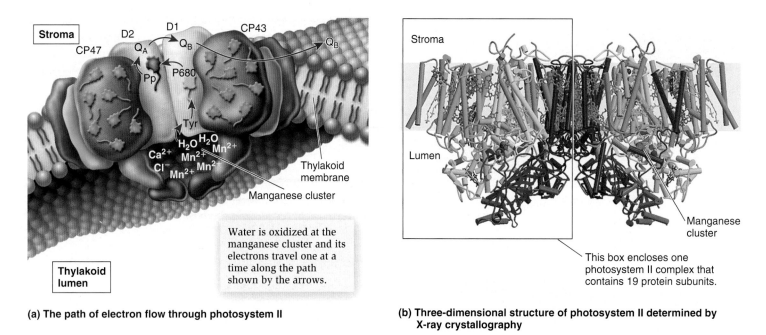

(a) The path of electron flow through photosystem II

(b) Three-dimensional structure of photosystem II determined by X-ray crystallography

Figure 8.7   **A closer look at the structure and function of photosystem II.**   **(a)** Schematic drawing showing the path of electron flow from water to Q_B. The CP47 and CP43 protein subunits wrap around D1 and D2 so that pigments in CP47 and CP43 can transfer energy to P680 by resonance energy transfer. **(b)** The three-dimensional structure of photosystem II as determined by X-ray crystallography. In the crystal structure, the colors are CP47 (red), D2 (orange), D1 (yellow), and CP43 (green).

Biological inquiry: According to this figure, how many redox reactions does photosystem II catalyze?

In 2004, So Iwata, James Barber, and colleagues determined the three-dimensional structure of photosystem II using a technique called **X-ray crystallography**. In this method, researchers must purify a protein or protein complex and expose it to conditions that cause the proteins to associate with each other in an ordered array. In other words, the proteins form a crystal. When a crystal is exposed to X-rays, the resulting pattern can be analyzed mathematically to determine the three-dimensional structure of the crystal's components. Major advances in this technique over the last couple of decades have enabled researchers to determine the structures of relatively large macromolecular machines such as photosystem II and ribosomes. **Figure 8.7b** shows the three-dimensional structure of photosystem II. The structure shown here is a dimer; it has two PSII complexes each with 19 protein subunits. As seen in this figure, the intricacy of the structure of photosystem II rivals the complexity of its function.

Photosystems II and I Work Together to Produce ATP and NADPH

Let's now consider what happens to the high-energy electrons that are transferred to the primary electron acceptor in photosystem II (**Figure 8.8**). After electrons reach Q_B, they enter an electron transport chain located in the thylakoid membrane. The electron transport chain functions similarly to the one found in mitochondria. From Q_B, electrons go to a cytochrome complex, then to plastocyanin (Pc), a small protein, and then

to a pigment molecule in the reaction center of photosystem I. Along the journey from photosystem II to photosystem I, the electron releases some of its energy at particular steps and is transferred to the next component that has a higher electronegativity. The energy that is released is harnessed to move H^+ into the thylakoid lumen. One result of the electron movement is to establish an H^+ electrochemical gradient. Additionally, the splitting of water also adds H^+ into the thylakoid lumen. The synthesis of ATP in chloroplasts is achieved by a chemiosmotic mechanism similar to that used to make ATP in mitochondria.

The key role of photosystem I is to make NADPH. When light strikes the light-harvesting complex of photosystem I, this energy is also transferred to a reaction center, where a high-energy electron is removed from a pigment molecule, designated P700, and transferred to a primary electron acceptor. A protein called ferredoxin (Fd) can accept two high-energy electrons, one at a time, from the primary electron acceptor. Fd then transfers the two electrons to the enzyme $NADP^+$ reductase. This enzyme transfers the two electrons to $NADP^+$ and together with an H^+ creates NADPH. The formation of NADPH consumes one H^+ in the stroma and thereby contributes to the formation of an H^+ electrochemical gradient because it results in fewer H^+ in the stroma.

A key difference between photosystem II and photosystem I is how the oxidized forms of P680 and P700 receive electrons. As discussed earlier, $P680^+$ receives an electron from water. By comparison, $P700^+$—the oxidized form of P700—receives an electron from Pc. Therefore, photosystem I does not need to split water to reduce $P700^+$ and thus does not generate oxygen.

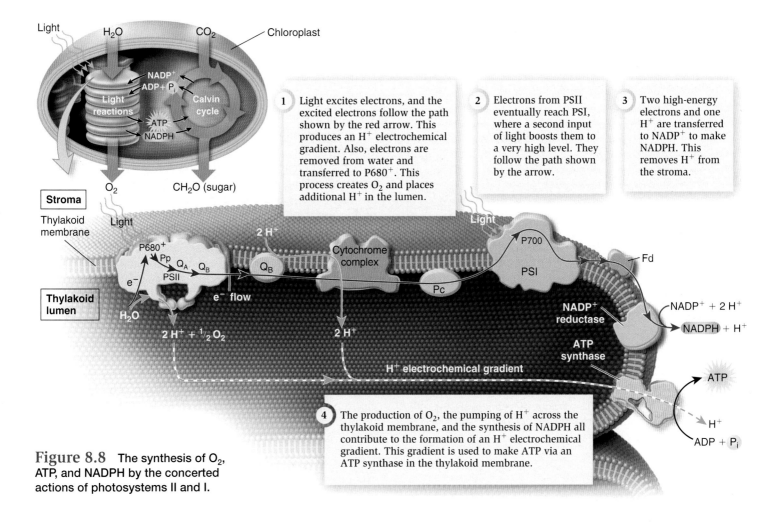

1 Light excites electrons, and the excited electrons follow the path shown by the red arrow. This produces an H^+ electrochemical gradient. Also, electrons are removed from water and transferred to $P680^+$. This process creates O_2 and places additional H^+ in the lumen.

2 Electrons from PSII eventually reach PSI, where a second input of light boosts them to a very high level. They follow the path shown by the arrow.

3 Two high-energy electrons and one H^+ are transferred to $NADP^+$ to make NADPH. This removes H^+ from the stroma.

4 The production of O_2, the pumping of H^+ across the thylakoid membrane, and the synthesis of NADPH all contribute to the formation of an H^+ electrochemical gradient. This gradient is used to make ATP via an ATP synthase in the thylakoid membrane.

Figure 8.8 The synthesis of O_2, ATP, and NADPH by the concerted actions of photosystems II and I.

In summary, the steps of the light reactions of photosynthesis produce three chemical products:

1. O_2 is produced in the thylakoid lumen by the oxidation of water by photosystem II. Two electrons are removed from water, which creates two H^+ and $1/2$ O_2. The two electrons are transferred to $P680^+$ molecules.
2. ATP is produced in the stroma by an H^+ electrochemical gradient. This gradient results from three events: (1) the splitting of water, which places H^+ in the thylakoid lumen, (2) the movement of high-energy electrons from photosystem II to photosystem I, which pumps H^+ into the lumen, and (3) the formation of NADPH, which consumes H^+ in the stroma.
3. NADPH is produced in the stroma from high-energy electrons that start in photosystem II and are boosted a second time in photosystem I. Two high-energy electrons and one H^+ are transferred to $NADP^+$ to create NADPH.

The combined action of photosystem II and photosystem I is termed **noncyclic electron flow** because the electrons move linearly from PSII to PSI and ultimately to $NADP^+$.

The Use of Light Flashes of Specific Wavelengths Provided Experimental Evidence for the Existence of PSII and PSI

The use of light flashes at particular wavelengths has been an important experimental technique for helping researchers understand the light reactions of photosynthesis. In this method, pioneered by Robert Emerson, a photosynthetic organism is exposed to a particular wavelength of light, after which the rate of photosynthesis is measured by the amount of CO_2 consumed or the amount of O_2 produced. In the 1950s, Emerson performed a particularly intriguing experiment that greatly stimulated photosynthesis research (**Figure 8.9**). He subjected algae to light flashes of different wavelengths and obtained a mysterious result. When he exposed algae to a wavelength of 680 nm, he observed a low rate of photosynthesis. A similarly low rate of photosynthesis occurred when he exposed algae to a wavelength of 700 nm. However, when he exposed the algae to both wavelengths of light simultaneously, the rate of photosynthesis was more than double the rates observed at only one wavelength. This phenomenon was termed the **enhancement effect**. We know now that it occurs because 680-nm light can readily activate the pigment (P680) in the reaction center in photosystem II, but is not very efficient at activating pigments in photosystem I. In contrast, light of 700-nm wavelength is optimal at activating photosystem I but not very good at activating photosystem II. When algae are exposed to both wavelengths, however, maximal activation of the pigments in both photosystems is achieved.

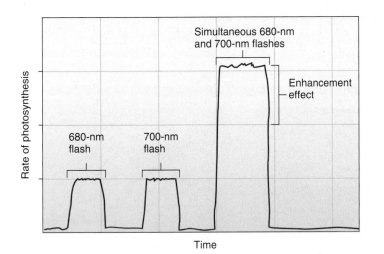

Figure 8.9 **The enhancement effect observed by Emerson.** When photosynthetic organisms such as green plants and algae are exposed to 680-nm and 700-nm light simultaneously, the rate of photosynthesis is much more than double the rate produced by each wavelength individually.

Biological inquiry: Would the enhancement effect be observed if two consecutive flashes of light occurred at 680 nm?

When researchers began to understand that photosynthesis results in the production of both ATP and NADPH, Robin Hill and Fay Bendall also proposed that photosynthesis involves two photoactivation events. According to their model, known as the **Z scheme**, an electron proceeds through a series of energy changes during photosynthesis. The Z refers to the zigzag shape of this energy curve. Based on our modern understanding of photosynthesis, we now know that these events involve increases and decreases in the energy of an electron as it moves from photosystem II to photosystem I (**Figure 8.10**). An electron on a nonexcited pigment molecule in photosystem II has the lowest energy. In photosystem II, light boosts an electron to a much higher energy level. As the electron travels from photosystem II to photosystem I, some of the energy is released. The input of light in photosystem I boosts the electron to an even higher energy than it attained in photosystem II. The electron releases a little energy before it is eventually transferred to $NADP^+$.

Cyclic Electron Flow Produces Only ATP

The mechanism of harvesting light energy, described in Figure 8.8, is called noncyclic electron flow because electrons begin at photosystem II and eventually are transferred to $NADP^+$. The electron flow is a linear process that produces ATP and NADPH in roughly equal amounts. However, as we will see, the Calvin cycle uses more ATP than NADPH. Therefore, in some photosynthetic organisms, excited electrons take an alternative path as a way to make more ATP. In 1959, Daniel Arnon discovered

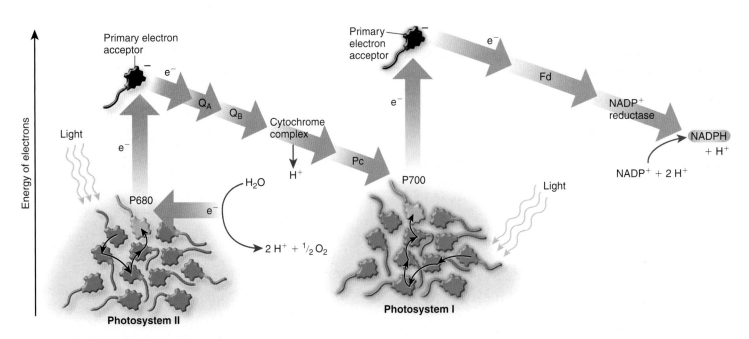

Figure 8.10 **The Z scheme, which depicts the energy of an electron as it moves from photosystem II to $NADP^+$.** During its journey from photosystem II to $NADP^+$, an electron varies in the amount of energy it contains. As seen here, the input of light boosts the energy of the electron two times. At the end of the pathway, 2 electrons are used to make NADPH.

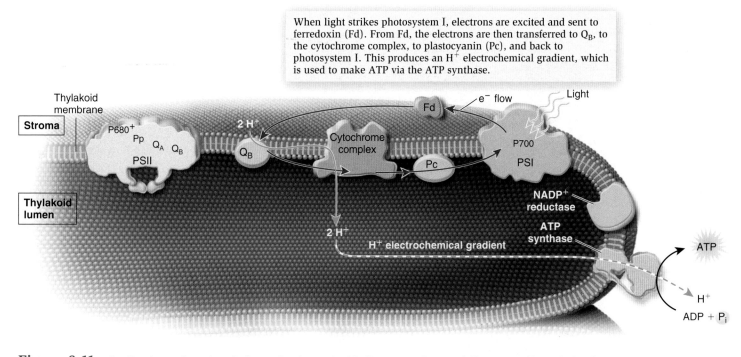

When light strikes photosystem I, electrons are excited and sent to ferredoxin (Fd). From Fd, the electrons are then transferred to Q_B, to the cytochrome complex, to plastocyanin (Pc), and back to photosystem I. This produces an H^+ electrochemical gradient, which is used to make ATP via the ATP synthase.

Figure 8.11 Cyclic photophosphorylation. As shown in this figure, an electron follows a cyclic path that is powered by photosystem I. This contributes to the formation of an H^+ electrochemical gradient, which is used to make ATP.

a pattern of electron flow that is cyclic and generates ATP alone (**Figure 8.11**). Arnon termed the process **cyclic photophosphorylation** because (1) the path of electrons is cyclic, (2) light energizes the electrons, and (3) ATP is made via the phosphorylation of ADP. Due to the path of electrons, the mechanism is also called **cyclic electron flow**.

When light strikes photosystem I, high-energy electrons are sent to the primary electron acceptor and then to ferredoxin (Fd). The key difference in cyclic photophosphorylation is that the high-energy electrons are transferred from ferredoxin to Q_B. From Q_B, the electrons then go the cytochrome complex, to plastocyanin (Pc), and back to photosystem I. As the electrons travel along this cyclic route, they release energy and some of this energy is used to transport H^+ into the thylakoid lumen. The resulting H^+ gradient drives the synthesis of ATP via the ATP synthase.

GENOMES & PROTEOMES

The Cytochrome Complexes of Mitochondria and Chloroplasts Have Evolutionarily Related Proteins in Common

A recurring theme in cell biology is that evolution has resulted in groups of genes that encode proteins that play similar but specialized roles in cells. When two or more genes are similar because they are derived from the same ancestral gene, they are called **homologous genes**. As discussed in Chapter 23, homologous genes encode proteins that have similar amino acid sequences and perform similar functions.

A comparison of the electron transport chains of mitochondria and chloroplasts reveals homologous genes. In particular, let's consider the cytochrome complex found in the thylakoid membrane of plants and algae, called cytochrome b_6-f (**Figure 8.12a**) and cytochrome b-c_1, which is found in the electron transport chain of mitochondria (**Figure 8.12b**; refer back to Figure 7.16). Both cytochrome b_6-f and cytochrome b-c_1 are composed of several protein subunits. One of those proteins is called cytochrome b_6 in cytochrome b_6-f and cytochrome b in cytochrome b-c_1. By analyzing the gene sequences, researchers discovered that cytochrome b_6 and cytochrome b are homologous. These proteins carry out similar functions: both of them accept electrons from a quinone (plastoquinone or ubiquinone) and both donate an electron to another protein within their respective complexes (cytochrome f or cytochrome $c1$). Likewise, both of these proteins function as H^+ pumps that capture some of the energy that is released from electrons to transport H^+ across the membrane. Thus, evolution has produced a family of cytochrome b-type proteins that play similar but specialized roles. Cytochrome b functions as a redox protein and H^+ pump in the electron transport chain of mitochondria, while cytochrome b_6 plays the same role in chloroplasts.

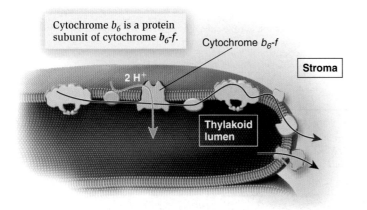

Cytochrome b_6 is a protein subunit of cytochrome b_6-f.

Cytochrome b_6-f

Stroma

2 H$^+$

Thylakoid lumen

(a) Cytochrome b_6 in a chloroplast

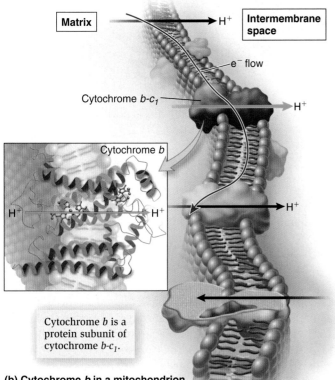

Matrix

H$^+$ Intermembrane space

e$^-$ flow

Cytochrome b-c_1

H$^+$

Cytochrome b

H$^+$ H$^+$

H$^+$

Cytochrome b is a protein subunit of cytochrome b-c_1.

(b) Cytochrome b in a mitochondrion

Figure 8.12   Homologous proteins in the electron transport chains of mitochondria and chloroplasts.   **(a)** Cytochrome b_6, a subunit of the cytochrome b_6-f complex found in chloroplasts, and **(b)** cytochrome b, a subunit of the cytochrome b-c_1 complex found in mitochondria, are homologous proteins that play similar roles in their respective electron transport chains. The inset shows the three-dimensional structure of cytochrome b, which was determined by X-ray crystallography. It is an integral membrane protein with several transmembrane α helices and two heme groups, which are prosthetic groups involved in electron transfer. The structure of cytochrome b_6 has also been determined and found to be very similar. (Note: The orientation of cytochrome b in the inset is such that it would pump H$^+$ out of the matrix. The orientation of cytochrome b_6 is oriented so it would pump H$^+$ out of the stroma.)

Biological inquiry: Explain why the three-dimensional structures of cytochrome b and cytochrome b_6 are very similar.

8.3 Calvin Cycle

In the previous section, we learned how the light reactions of photosynthesis produce O$_2$, ATP, and NADPH. We will now turn our attention to the second phase of photosynthesis, the Calvin cycle, in which ATP and NADPH are used to make carbohydrates. The Calvin cycle consists of a series of steps that occur in a metabolic cycle somewhat similar to the citric acid cycle described in Chapter 7. However, while the citric acid cycle is catabolic, the Calvin cycle is an anabolic pathway leading to the biosynthesis of carbohydrates.

The Calvin cycle takes CO$_2$ from the atmosphere and incorporates the carbon into organic molecules, namely, carbohydrates. As mentioned earlier, carbohydrates are critical for two reasons. First, these organic molecules provide the precursors to make the organic molecules and macromolecules of nearly all living cells. The organic molecules in your body are ultimately derived from the operation of the Calvin cycle in algae and plants. The second key reason why the Calvin cycle is important involves the storage of energy. Recall that molecules such as glucose contain large amounts of chemical energy in their covalent bonds. The Calvin cycle produces carbohydrates, which store energy. These carbohydrates are accumulated inside plant cells. When a plant is in the dark and not carrying out photosynthesis, the stored carbohydrates can be used as a source of energy. Similarly, when an animal consumes a plant, it can use the carbohydrates as an energy source.

In this section, we will examine the three phases of the Calvin cycle and their components. We will also explore the experimental approach of Melvin Calvin and his colleagues that enabled them to elucidate these steps.

The Calvin Cycle Incorporates CO$_2$ into Carbohydrate

The Calvin cycle, also called the Calvin-Benson cycle, was determined by chemists Melvin Calvin and Andrew Adam Benson and their colleagues in the 1940s and 1950s. This cycle requires a massive input of energy. For every 6 carbon dioxide molecules that are incorporated into carbohydrate such as glucose, 18 ATP molecules are hydrolyzed and 12 NADPH molecules are oxidized.

$$6 \; CO_2 + 12 \; H_2O \rightarrow C_6H_{12}O_6 + 6 \; O_2 + 6 \; H_2O$$

Glucose

$$18 \; ATP + 18 \; H_2O \rightarrow 18 \; ADP + 18 \; P_i$$

$$12 \; NADPH \rightarrow 12 \; NADP^+ + 12 \; H^+ + 24 \; e^-$$

Although biologists commonly describe glucose as a product of photosynthesis, glucose is not directly made by the Calvin cycle. Instead, products of the Calvin cycle are used as starting materials for the synthesis of glucose and other molecules, including sucrose. After glucose molecules are made, they may be linked together to form a polymer of glucose called starch, which is

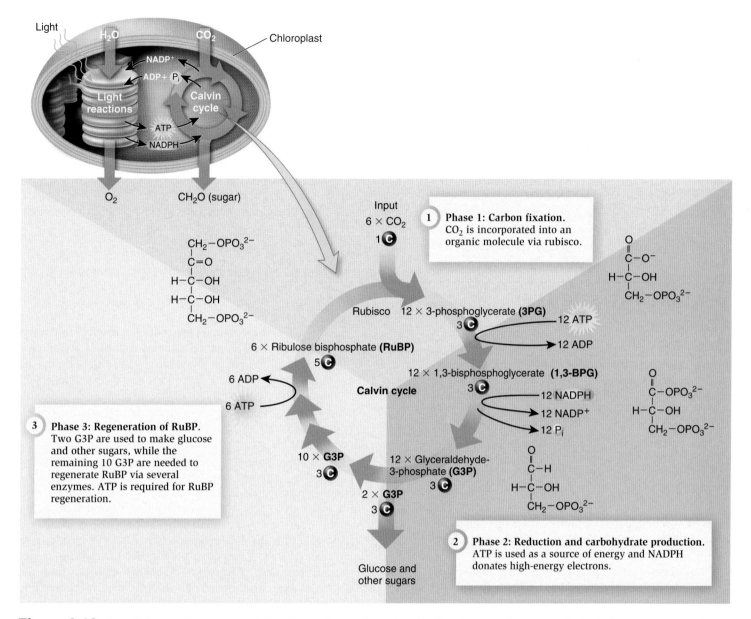

Figure 8.13   The Calvin cycle. This cycle has three phases: (1) carbon fixation, (2) reduction and carbohydrate production, and (3) regeneration of RuBP.

stored in the chloroplast for later use. Alternatively, the disaccharide sucrose may be made and transported out of the leaf to other parts of the plant.

The Calvin cycle can be divided into three phases: carbon fixation, reduction and carbohydrate production, and regeneration of RuBP (**Figure 8.13**).

Carbon Fixation (Phase 1) In **carbon fixation**, CO_2 becomes incorporated into ribulose bisphosphate (RuBP), a five-carbon sugar. The enzyme that catalyzes this step is named RuBP carboxylase/oxygenase, or **rubisco**. This enzyme, which constitutes the most abundant protein in chloroplasts, is perhaps the most abundant protein on Earth. The product of the reaction is a six-carbon intermediate that immediately splits in half to form two molecules of 3-phosphoglycerate (3PG).

Reduction and Carbohydrate Production (Phase 2) In the second phase, ATP is used to convert 3PG to 1,3-bisphosphoglycerate. Next, electrons from NADPH reduce 1,3-bisphosphoglycerate to glyceraldehyde-3-phosphate (G3P). G3P is a carbohydrate with three carbon atoms. The key difference between 3PG and G3P is that G3P has a C—H bond, while the analogous carbon in 3PG forms a C—O bond (Figure 8.13). The C—H bond can occur because the G3P molecule has been reduced by the addition of two electrons from NADPH. Compared to 3PG, the bonds in G3P store more energy and enable G3P to readily form larger organic molecules such as glucose.

Only some of the G3P molecules are used to make glucose or other carbohydrates. Phase 1 began with 6 RuBP molecules and 6 CO_2 molecules. Twelve G3P molecules are made at the end of phase 2. Two of these G3P molecules are used in carbohydrate production. As described next, the other 10 G3P molecules are needed to keep the Calvin cycle turning.

Regeneration of RuBP (Phase 3) In the last phase of the Calvin cycle, a series of enzymatic steps converts the 10 G3P molecules into 6 RuBP molecules, using 6 molecules of ATP. After the RuBP molecules are regenerated, they can serve as acceptors for CO_2, thereby allowing the cycle to continue.

As we have just seen, the Calvin cycle begins by using carbon from an inorganic source, that is, CO_2, and ends with organic molecules that will be used by the plant to make other compounds. You may be wondering why it is not possible to directly link the CO_2 molecules together to form these larger molecules. The answer lies in the number of electrons that orbit carbon atoms. In CO_2, the carbon atom is considered electron poor.

Oxygen is a very electronegative atom that monopolizes the electrons it shares with other atoms. In a covalent bond between carbon and oxygen, the shared electrons are closer to the oxygen atom.

By comparison, in an organic molecule, the carbon atom is electron rich. During the Calvin cycle, ATP provides energy and NADPH donates high-energy electrons so that the carbon originally in CO_2 has been reduced. Put simply, the Calvin cycle places additional electrons onto carbon atoms. Compared to the carbon in CO_2, the carbon in an organic molecule can readily form C—H and C—C bonds, which allows the eventual synthesis of larger essential molecules including glucose, amino acids, and so on. In addition, the covalent bonds within these molecules are capable of storing large amounts of energy.

FEATURE INVESTIGATION

The Calvin Cycle Was Determined by Isotope Labeling Methods

The steps in the Calvin cycle involve the conversion of one type of molecule to another, eventually regenerating the starting material, RuBP. In the 1940s and 1950s, Calvin and his colleagues used ^{14}C, a radioisotope of carbon, as a way to label and trace molecules produced during the cycle (**Figure 8.14**). They injected ^{14}C-labeled CO_2 into cultures of the green algae *Chlorella pyrenoidosa* grown in an apparatus called a "lollipop" (because of its shape). The *Chlorella* cells were given different lengths of time to incorporate the ^{14}C-labeled carbon, ranging from fractions of a second to many minutes. After this incubation period, the cells were abruptly placed into a solution of alcohol to inhibit enzymatic reactions and thereby stop the cycle.

The researchers separated the newly made radiolabeled molecules by a variety of methods. The most commonly used method was two-dimensional paper chromatography. In this approach, a sample containing radiolabeled molecules was spotted onto a corner of the paper at a location called the origin. The edge of the paper was placed in a solvent, such as phenol-water, and the solvent would ascend to the top of the paper. As the solvent rose through the paper, so did the radiolabeled molecules. The rate at which they rose depended on their structures, which determined how strongly they interacted with the paper. This step separated the mixture of molecules spotted onto the paper at the origin. The paper was then dried, turned 90°, and then the edge was placed in a different solvent, such as butanol-propionic acid-water. Again, the solvent would rise through the paper, thereby separating molecules that may not have been adequately separated during the first separation step. After this second separation step, the paper was dried and exposed to X-ray film, a procedure called autoradiography. Radioactive emission from the ^{14}C-labeled molecules caused dark spots to appear on the film.

Figure 8.14 The determination of the Calvin cycle using labeling of CO_2 with ^{14}C.

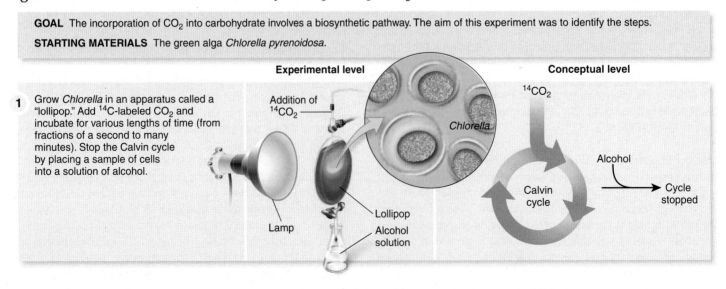

GOAL The incorporation of CO_2 into carbohydrate involves a biosynthetic pathway. The aim of this experiment was to identify the steps.

STARTING MATERIALS The green alga *Chlorella pyrenoidosa*.

Experimental level | Conceptual level

1   Grow *Chlorella* in an apparatus called a "lollipop." Add ^{14}C-labeled CO_2 and incubate for various lengths of time (from fractions of a second to many minutes). Stop the Calvin cycle by placing a sample of cells into a solution of alcohol.

Addition of $^{14}CO_2$

Chlorella

$^{14}CO_2$

Alcohol

Calvin cycle

Cycle stopped

Lamp

Lollipop

Alcohol solution

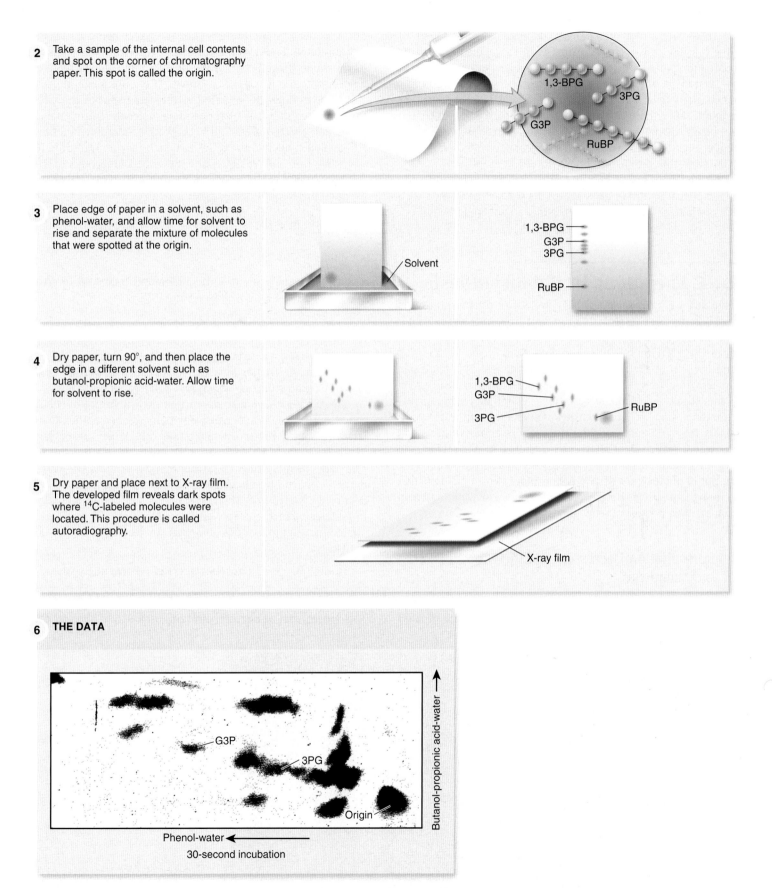

2 Take a sample of the internal cell contents and spot on the corner of chromatography paper. This spot is called the origin.

1,3-BPG

3PG

G3P

RuBP

3 Place edge of paper in a solvent, such as phenol-water, and allow time for solvent to rise and separate the mixture of molecules that were spotted at the origin.

Solvent

1,3-BPG
G3P
3PG

RuBP

4 Dry paper, turn 90°, and then place the edge in a different solvent such as butanol-propionic acid-water. Allow time for solvent to rise.

1,3-BPG
G3P
3PG

RuBP

5 Dry paper and place next to X-ray film. The developed film reveals dark spots where ¹⁴C-labeled molecules were located. This procedure is called autoradiography.

X-ray film

6 THE DATA

Butanol-propionic acid-water →

G3P

3PG

Origin

Phenol-water ←

30-second incubation

The pattern of spots changed, depending on the length of time that the cells were incubated with ^{14}C-labeled CO_2. If the incubation period was short, only molecules that were made in the first steps of the Calvin cycle would be seen, while longer incubations revealed molecules synthesized in later steps. For example, after short incubations, 3-phosphoglycerate (3PG) and 1,3-bisphosphoglycerate (1,3-BPG) would be observed, while longer incubations would show glyceraldehyde-3-phosphate (G3P) and ribulose bisphosphate (RuBP).

A challenge for Calvin and his colleagues was to identify the chemical nature of each spot. This was achieved by a variety of chemical methods. For example, a spot could be cut out of the paper, the molecule within the paper could be washed out, or eluted, and then the eluted molecule could be subjected to the same procedure that included a radiolabeled molecule whose structure was already known. If the unknown molecule and known molecule migrated to the same spot in the paper, this indicated that they were the same molecule. During the late 1940s and 1950s, Calvin and his coworkers identified all of the ^{14}C-labeled spots and the order in which they appeared. In this way, they were able to determine the series of reactions of what we now know as the Calvin cycle. For this work, Calvin was awarded the Nobel Prize in 1961.

8.4 Variations in Photosynthesis

Thus far, we have considered the process of photosynthesis as it occurs in the chloroplasts of green plants and algae. Photosynthesis is a two-stage process in which the light reactions produce O_2, ATP, and NADPH, and the Calvin cycle uses the ATP and NADPH in the synthesis of carbohydrates. This two-stage process is a universal feature of photosynthesis in all green plants, algae, and cyanobacteria. However, certain environmental conditions such as light intensity, temperature, and water availability may influence both the efficiency of photosynthesis and the way in which the Calvin cycle operates. In this section, we begin by examining how hot and dry conditions may reduce the output of photosynthesis. We then explore two adaptations that certain plant species have evolved that conserve water and help to maximize photosynthetic efficiency in such environments.

Photorespiration Decreases the Efficiency of Photosynthesis

In the previous section, we learned that rubisco is a key enzyme of the Calvin cycle. Rubisco functions as a carboxylase because it adds a CO_2 molecule to RuBP, an organic molecule, to create two molecules of 3-phosphoglycerate (3PG).

$$RuBP + CO_2 \rightarrow 2\ 3PG$$

For most species of plants, the incorporation of CO_2 into RuBP is the only way for carbon fixation to occur. Because 3PG is a three-carbon molecule, these plants are called **C_3 plants**. Examples of C_3 plants include wheat and oak trees (**Figure 8.15**).

Researchers have discovered that the active site of rubisco can also function as an oxygenase. As mentioned earlier, that is why it is called rubisco, which stands for RuBP carboxylase/oxygenase. When rubisco adds an O_2 molecule to RuBP, this creates only one molecule of 3-phosphoglycerate and a two-carbon molecule called phosphoglycolate. The phosphoglycolate is then dephosphorylated to glycolate and released from the chloroplast. In a series of several steps, glycolate is eventually oxidized in other organelles to produce an organic molecule plus a molecule of CO_2.

$$RuBP + O_2 \rightarrow \text{3-phosphoglycerate} + \text{Phosphoglycolate}$$

$$\text{Phosphoglycolate} \rightarrow \text{Glycolate} \rightarrow\rightarrow\rightarrow CO_2$$

This process, called **photorespiration**, uses O_2 and liberates CO_2.

(a)   **(b)**

Figure 8.15 Examples of C_3 plants. The structures of **(a)** wheat and **(b)** white oak leaves are similar to that shown in Figure 8.1.

Photorespiration is considered wasteful because it reverses the effects of photosynthesis, thereby reducing the ability of a plant to make carbohydrates, which limits plant growth.

Photorespiration is more likely to occur when plants are exposed to a hot and dry environment. Under these conditions, the stomata of the leaves close, inhibiting the uptake of CO_2 from the air and trapping the O_2 that is produced by photosynthesis. When the level of CO_2 is low and O_2 is high, photorespiration is favored. If C_3 plants are subjected to hot and dry environmental conditions, as much as 25–50% of their photosynthetic work is reversed by the process of photorespiration.

Why do plants carry out photorespiration? The answer is not entirely clear. Photorespiration undoubtedly results in the disadvantage of lowering the efficiency of photosynthesis. One common view is that photorespiration does not offer any advantage and is an evolutionary relic. When rubisco first evolved some 3 billion years ago, the atmospheric oxygen level was low, so that phosphorespiration would not have been a problem. Another view is that photorespiration may have a protective advantage. On hot and dry days, CO_2 levels within a plant may fall and O_2 levels will rise. Under these conditions, highly toxic oxygen-containing molecules such as free radicals may be produced that could cause damage to the plant. Therefore, plant biologists have speculated that the role of photorespiration may be to protect the plant against the harmful effects of such toxic molecules by consuming O_2 and releasing CO_2. In addition, photorespiration may affect the metabolism of other compounds in plants. Recent research suggests that photorespiration may also help plants to assimilate nitrogen into organic molecules.

C_4 Plants Have Evolved a Mechanism to Minimize Photorespiration

Certain species of plants have developed an interesting way to prevent photorespiration. In the early 1960s, Hugo Kortschak discovered that the first product of photosynthesis in sugarcane is not 3-phosphoglycerate but instead is a compound with four carbon atoms. Species such as sugarcane are thus called **C_4 plants** because the first step in carbon fixation produces a four-carbon compound. Later, Marshall Hatch and Roger Slack confirmed this result and identified the compound as oxaloacetate. For this reason, the pathway is sometimes called the Hatch-Slack pathway.

C_4 plants employ an interesting cellular organization to avoid photorespiration (**Figure 8.16**). Unlike C_3 plants, an interior layer in the leaves of many C_4 plants has a two-cell organization composed of mesophyll cells and bundle-sheath cells. CO_2 from the atmosphere enters the mesophyll cells via stomata. Once inside, the enzyme **PEP carboxylase** adds CO_2 to phosphoenolpyruvate (PEP) to produce the four-carbon compound oxaloacetate. PEP carboxylase does not recognize O_2. Therefore, unlike rubisco, PEP carboxylase does not promote photorespiration when CO_2 is low and O_2 is high. Instead, PEP carboxylase continues to fix CO_2.

A key feature of these types of C_4 plants is that a four-carbon compound is transferred between cells. As shown in Figure 8.16, the compound oxaloacetate is converted to the four-carbon compound malate, which is transported into the bundle-sheath cell. Malate is then broken down into pyruvate and CO_2.

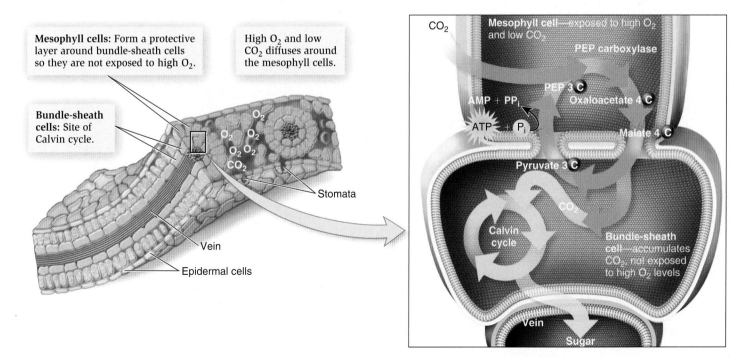

Figure 8.16 **Leaf structure and its relationship to the C_4 cycle.** C_4 plants have mesophyll cells that initially take up CO_2, and bundle-sheath cells, where much of the carbohydrate synthesis occurs. Compare this leaf structure with the structure of C_3 leaves shown in Figure 8.1.

The pyruvate returns to the mesophyll cell, where it is converted to PEP via ATP, and the cycle in the mesophyll cell can begin again. The main outcome of this C_4 cycle is that the mesophyll cell pumps CO_2 into the bundle-sheath cell. The Calvin cycle occurs in the chloroplasts of the bundle-sheath cell. Because the mesophyll cell supplies the bundle-sheath cell with a steady supply of CO_2, the concentration of CO_2 remains high in the bundle-sheath cell. This strategy minimizes photorespiration, which requires low CO_2 and high O_2 levels to proceed.

Which is better—being a C_3 or a C_4 plant? The answer is that it depends on the environment. In warm and dry climates, C_4 plants have an advantage because during the day they can keep their stomata partially closed to reduce water vaporization from the leaf and thereby conserve water. Furthermore, they can avoid photorespiration. C_4 plants are well adapted to habitats with high daytime temperatures and intense sunlight. Some examples of C_4 plants include crabgrass, corn, and sugarcane. In cooler climates, C_3 plants have the edge because it takes less energy for them to fix carbon dioxide. The process of carbon fixation that occurs in C_4 plants uses ATP to regenerate PEP from pyruvate (Figure 8.16), which C_3 plants do not have to expend. Biologists estimate that about 90% of the plant species on Earth are C_3 plants.

CAM Plants Are C_4 Plants That Take Up Carbon Dioxide at Night

We have just learned that certain C_4 plants prevent photorespiration by pumping CO_2 into the bundle-sheath cells, where the Calvin cycle occurs. Another strategy followed by other C_4 plants, called **CAM plants**, is to separate these processes in time. CAM stands for crassulacean acid metabolism, because the process was first studied in members of the plant family Crassulaceae. CAM plants are water-storing succulents such as cacti, bromeliads (including pineapple), and sedums. To avoid water loss, CAM plants keep their stomata closed during the day and open them during the night.

How, then, do CAM plants carry out photosynthesis? **Figure 8.17** compares CAM plants with the other type of C_4 plants we considered in Figure 8.16. During the night, the stomata of CAM plants open, thereby allowing the entry of CO_2 into mesophyll cells, which joins with PEP to form the four-carbon compound oxaloacetate. This is converted to malate, which accumulates during the night in the central vacuoles of the cells. In the morning, the stomata close to conserve moisture. The accumulated malate in the mesophyll cells leaves the vacuole and is broken down to release CO_2, which then drives the Calvin cycle during the daytime.

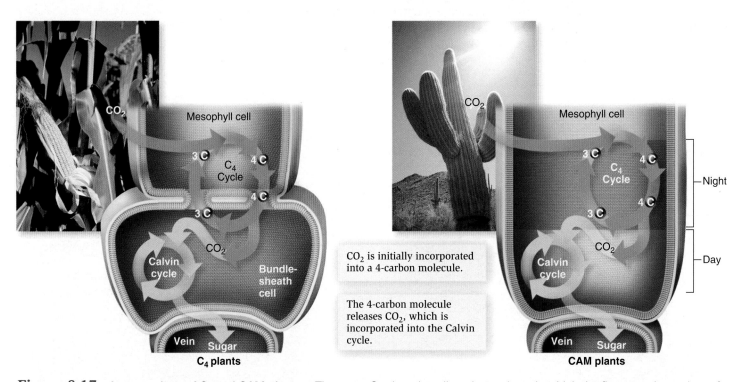

Figure 8.17   A comparison of C_4 and CAM plants.   The name C_4 plant describes those plants in which the first organic product of carbon fixation is a four-carbon compound. Using this definition, CAM plants are a type of C_4 plant because they produce a four-carbon molecule when CO_2 is initially taken up. CAM plants, however, do not separate the functions of making a four-carbon molecule and the Calvin cycle into different types of cells. Instead, they make a four-carbon molecule at night and break down that molecule during the day, so that the CO_2 can be incorporated via the Calvin cycle.

CHAPTER SUMMARY

- Photosynthesis is the process by which plants, algae, and cyanobacteria capture light energy to synthesize carbohydrates.

- Heterotrophs must obtain organic molecules in their food, while autotrophs can make organic molecules from inorganic sources. Photoautotrophs use the energy from light to make organic molecules.

8.1 Overview of Photosynthesis

- The general formula for photosynthesis is that carbon dioxide, water, and energy are needed to make carbohydrates and oxygen.

- In plants and algae, photosynthesis occurs within chloroplasts, which have an outer membrane, inner membrane, and thylakoid membrane. The stroma is found between the thylakoid membrane and inner membrane. In plants, the leaves are the major site of photosynthesis. (Figure 8.1)

- The light reactions of photosynthesis capture light energy to make ATP, NADPH, and O_2. These reactions occur at the thylakoid membrane. Carbohydrate synthesis happens in the stroma and uses ATP and NADPH from the light reactions. (Figure 8.2)

8.2 Reactions That Harness Light Energy

- Light is a form of electromagnetic radiation that travels in waves and is composed of photons with discrete amounts of energy. (Figure 8.3)

- Electrons can absorb light energy and be boosted to a higher energy level. (Figure 8.4)

- Photosynthetic pigments include chlorophylls *a* and *b* and carotenoids. These pigments absorb light energy in the visible spectrum. (Figure 8.5)

- Pigment molecules in photosystem II capture light energy, and that energy is transferred to the reaction center via resonance energy transfer. A high-energy electron from P680* is transferred to a primary electron acceptor. An electron from water is then used to replenish the electron that is lost from P680*. (Figures 8.6, 8.7)

- During noncyclic electron flow, electrons from photosystem II follow a pathway along an electron transport chain in the thylakoid membrane. This pathway generates an H^+ gradient that is used to make ATP. In addition, light energy striking photosystem I boosts an electron to a very-high-energy level that allows the synthesis of NADPH. (Figure 8.8)

- Emerson showed that, compared to single light flashes at 680 nm and 700 nm, light flashes at both wavelengths more than doubled the amount of photosynthesis. This occurred because these wavelengths activate pigments in PSII and PSI, respectively. (Figure 8.9)

- Hill and Bendall proposed the Z scheme for electron activation during photosynthesis. According to this scheme, an electron absorbs light energy twice, and it loses some of that energy as it flows along the electron transport chain in the thylakoid membrane. (Figure 8.10)

- During cyclic photophosphorylation, electrons are activated in PSI and flow from Fd to Q_B to the cytochrome complex to Pc and back to PSI. This cyclic electron route produces an H^+ gradient that is used to make ATP. (Figure 8.11)

- Cytochrome *b* in mitochondria and cytochrome b_6 in chloroplasts are homologous proteins, both of which are involved in electron transport and H^+ pumping. (Figure 8.12)

8.3 Calvin Cycle

- The Calvin cycle can be divided into three phases: carbon fixation, reduction and carbohydrate production, and regeneration of ribulose bisphosphate (RuBP). During this process, ATP is used as a source of energy and NADPH is used as a source of high-energy electrons so that CO_2 can be incorporated into carbohydrate. (Figure 8.13)

- Calvin and Benson determined the steps in the Calvin cycle by isotope labeling methods in which products of the Calvin cycle were separated by chromatography. (Figure 8.14)

8.4 Variations in Photosynthesis

- C_3 plants can only incorporate CO_2 into organic molecules via RuBP to make 3PG, a three-carbon molecule. (Figure 8.15)

- Photorespiration can occur under conditions of high O_2 and low CO_2, which occur under hot and dry conditions. During this process, some CO_2 is liberated and O_2 is used.

- Plants generally known as C_4 plants avoid photorespiration because the CO_2 is first incorporated, via PEP carboxylase, into a four-carbon molecule, which is pumped from mesophyll cells into bundle-sheath cells. This maintains a high concentration of CO_2 in the bundle-sheath cells, where the Calvin cycle occurs. The high CO_2 concentration minimizes photorespiration. (Figure 8.16)

- CAM plants prevent photorespiration by fixing CO_2 into a four-carbon molecule at night and then running the Calvin cycle during the day with their stomata closed. (Figure 8.17)

TEST YOURSELF

1. The water necessary for photosynthesis
   a. is split into H_2 and O_2.
   b. is directly involved in the synthesis of carbohydrate.
   c. provides the electrons to replace lost electrons in photosystem II.
   d. provides H^+ needed to synthesize G3P.
   e. none of the above.

2. The reaction center pigment differs from the other pigment molecules of the light-harvesting complex in that
   a. the reaction center pigment is a carotenoid.
   b. the reaction center pigment absorbs light energy and transfers that energy to other molecules without the transfer of electrons.
   c. the reaction center pigment transfers excited electrons to other molecules.
   d. the other pigments transfer high-energy electrons to other pigment molecules.
   e. the reaction center acts as an ATP synthase to produce ATP.

3. The cyclic electron flow that occurs in photosystem I produces
   a. NADPH.
   b. oxygen.
   c. ATP.
   d. all of the above.
   e. a and c only.

4. During the light reactions, the high-energy electron from P680
   a. eventually moves to $NADP^+$.
   b. becomes incorporated in water molecules.
   c. is pumped into the thylakoid space to drive ATP production.
   d. provides the energy necessary to split water molecules.
   e. falls back to the low-energy state in photosystem II.

5. During the first phase of the Calvin cycle, carbon dioxide is incorporated into ribulose bisphosphate by
   a. oxaloacetate.
   b. rubisco.
   c. RuBP.
   d. quinone.
   e. G3P.

6. The NADPH produced during the light reactions is necessary for
   a. the carbon fixation phase, which incorporates carbon dioxide into an organic molecule of the Calvin cycle.
   b. the reduction phase, which produces carbohydrates in the Calvin cycle.
   c. the regeneration of RuBP of the Calvin cycle.
   d. all of the above.
   e. a and b only.

7. The majority of the G3P produced during the reduction and carbohydrate production phase is used to produce
   a. glucose.
   b. ATP.
   c. RuBP to continue the cycle.
   d. rubisco.
   e. all of the above.

8. Photorespiration
   a. is the process where plants use sunlight to make ATP.
   b. is an inefficient way plants can produce organic molecules and in the process use O_2 and release CO_2.
   c. is a process that plants use to convert light energy to NADPH.
   d. occurs in the thylakoid space.
   e. is the normal process of carbohydrate production in cool, moist environments.

9. Photorespiration is avoided in C_4 plants because
   a. these plants separate the formation of a four-carbon molecule from the rest of the Calvin cycle in different cells.
   b. these plants only carry out anaerobic respiration.
   c. the enzyme PEP functions to maintain high CO_2 concentrations in the bundle-sheath cells.
   d. all of the above.
   e. a and c only.

10. Plants that are commonly found in hot and dry environments that carry out carbon fixation at night are
    a. oak trees.
    b. C_3 plants.
    c. CAM plants.
    d. all of the above.
    e. a and b only.

CONCEPTUAL QUESTIONS

1. Define photosynthesis. Explain the formula for photosynthesis.
2. Explain the function of NADPH.
3. Describe the parts of the chloroplast.

EXPERIMENTAL QUESTIONS

1. What was the purpose of the study conducted by Calvin and his colleagues?
2. In the experiment of Figure 8.14, why did the researchers use ^{14}C? Why did they examine samples at several different time periods? How were the different molecules in the samples identified?
3. What were the results of Calvin's study?

COLLABORATIVE QUESTIONS

1. Discuss the terms heterotroph and photoautotroph.
2. Discuss some of the differences between C_3 and C_4 plants.

---

www.brookerbiology.com

This website includes answers to the Biological Inquiry questions found in the figure legends and all end-of-chapter questions.

9

CELL COMMUNICATION AND REGULATION OF THE CELL CYCLE

CHAPTER OUTLINE

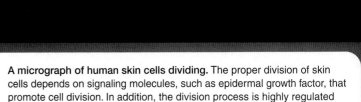

A micrograph of human skin cells dividing. The proper division of skin cells depends on signaling molecules, such as epidermal growth factor, that promote cell division. In addition, the division process is highly regulated to avoid mutations that could harm the resulting daughter cells.

In Chapter 5, we learned that all living cells are surrounded by a plasma membrane, which separates the internal cell contents from the extracellular environment. The plasma membrane enables cells to maintain an internal environment that is well ordered and carries out a variety of cell functions, as described throughout this unit. However, the plasma membrane also tends to isolate a cell from its surroundings. A cell cannot survive if it cannot sense changes in the extracellular environment and respond to them. For example, if the nutrient conditions change, such as a change from carbohydrates to fats, a cell needs to respond in a way that facilitates the uptake and metabolism of fats. If it cannot adapt to environmental changes, a cell is destined to die.

Cell communication is the process through which cells can detect and respond to signals in their extracellular environment. In multicellular organisms, cell communication is also needed to coordinate cellular activities within the whole organism. One of the most important reasons why cells must sense environmental conditions and communicate with each other is to promote cell division. For example, cells should divide only when they have sufficient nutrients to support the division process. The uptake of nutrients acts as a signal to stimulate cell division. In addition, the cells of multicellular organisms must communicate with each other so that cell division in one part of the body is coordinated with other parts. In animals, for example, signaling molecules are released at specific stages of development and promote the division of particular cells.

In this chapter, we will examine how cells respond to environmental signals and also produce signals so they can communicate with other cells. As you will learn, cell communication involves an amazing diversity of cellular proteins, lipids, and steroids that are devoted to this process. We will also consider the cell cycle, which in eukaryotic organisms is a series of events that result in cell division. This chapter focuses on the regulation of the cell cycle, which links cell communication and cell division. Chapter 15 will examine the steps of the cell division process as they pertain to the sorting of chromosomes.

9.1 General Features of Cell Communication

All living cells, including bacteria, fungi, protists, plant cells, and animal cells, are capable of cell communication, also known as cell signaling, a phenomenon that involves both incoming and outgoing signals. A **signal** is an agent that can influence the properties of cells. For example, on a sunny day, cells can sense their exposure to ultraviolet (UV) light and respond accordingly. In humans, UV light acts as an incoming signal to promote the synthesis of melanin, a protective pigment that helps to prevent the harmful effects of UV radiation. In addition, organisms can produce outgoing signals that influence the behavior of neighboring cells. Plant cells, for example, produce hormones that influence the pattern of cell elongation so that the plant grows toward light. Cells of all living organisms both respond to incoming signals and elicit outgoing signals. Cell communication is thus a two-way street.

Communication at the cellular level involves not only sending and receiving signals but also their interpretation. For this to occur, a signal must affect the conformation, or shape, of a cellular protein called a **receptor**. When a signal and receptor interact, a conformational change occurs in the receptor, eventually leading to some type of response in the cell. In this section, we begin by considering why cells need to respond to signals. We will then examine various forms of signaling, which differ in part based on the distance between the cells. Finally, we will examine the main steps that occur when a cell is exposed to a signal and elicits a response to it.

Cells Respond to Signals from Their Environment and from Other Cells

Before getting into the details of cell communication, let's take a general look at why cells need to respond to signals. The first reason is that cells need to respond to a changing environment. Changes in the environment are a persistent feature of life, and living cells are continually faced with alterations in temperature, nutrient availability, and availability of water. A cell may even be exposed to a toxic chemical in its environment. Being able to respond to change, a phenomenon called **adaptation**, is critical for the survival of all living organisms. Adaptation at the cellular level is also referred to as a **cellular response**. As an example, let's consider the response of a yeast cell to glucose in its environment (**Figure 9.1**). Some of the glucose acts as a signaling molecule that causes the cell to respond accordingly. In this case, the cell increases the number of glucose transporters that are needed to take glucose into the cell and also increases the number of enzymes that are required to metabolize glucose once it is inside. The cellular response has therefore allowed the cell to utilize glucose efficiently. We can say that the cell has become adapted to the presence of glucose in its environment.

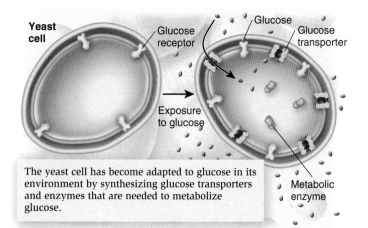

The yeast cell has become adapted to glucose in its environment by synthesizing glucose transporters and enzymes that are needed to metabolize glucose.

Figure 9.1   **Response of a yeast cell to glucose.**   When glucose is absent from the extracellular environment, the cell is not well prepared to take up and metabolize this sugar. However, when glucose is present, some of that glucose binds to a receptor in the membrane, which changes the amounts and properties of intracellular and membrane proteins so that the cell can readily use glucose.

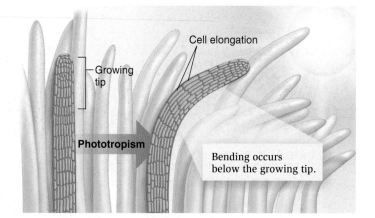

Figure 9.2   **Phototropism in plants.**   This process involves a bend that occurs just beneath the actively growing tip. This bend is caused by cell elongation along the nonilluminated side. The amount of auxin is higher on the nonilluminated side.

Biological inquiry: Does auxin cause cells to elongate or shorten?

A second reason for cell signaling is the need for cells to communicate with each other—a process called cell-to-cell communication. In one of the earliest experiments demonstrating cell-to-cell communication, Charles Darwin and his son Francis Darwin studied phototropism, the phenomenon in which plants grow toward light (**Figure 9.2**). The Darwins observed that the actual bending occurs in a zone below the growing tip. They concluded that the transmission of a signal from the growing tip to cells below the tip had to take place for this to occur. Later research revealed that the signal is a molecule called auxin, which is transmitted from cell to cell. The higher amount of auxin present on the nonilluminated side promotes cell elongation on that side only, thereby causing the plant to bend toward the light source.

Cell-to-Cell Communication Can Occur Between Adjacent Cells and Between Cells That Are Long Distances Apart

Researchers have determined that organisms have evolved a variety of different mechanisms to achieve cell-to-cell communication. The mode of communication depends, in part, on the distance between the cells that need to communicate with each other. Let's first examine the various ways in which signals are transferred between cells. Later in this chapter, we will learn how such signals elicit a cellular response.

One way to categorize cell signaling is by the manner in which the signal is transmitted from one cell to another. Signals are relayed between cells in five common ways, all of which involve a cell that produces a signal and a target cell that receives the signal (**Figure 9.3**).

Direct Intercellular Signaling   In a multicellular organism, cells that are adjacent to each other may have contacts, called

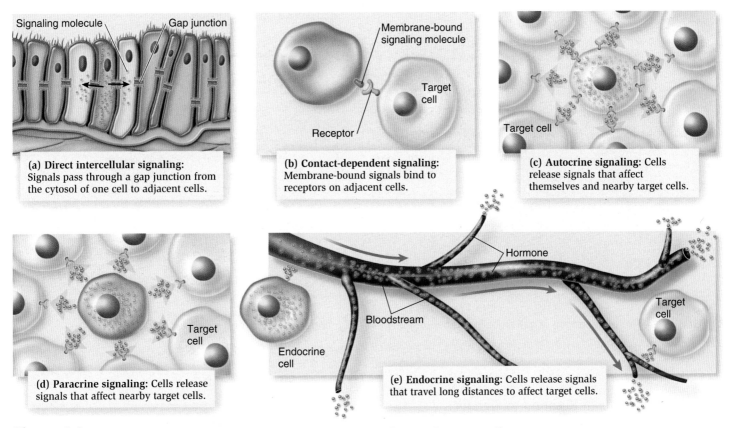

Figure 9.3 Examples of cell-to-cell communication based on the distance between cells.

cell junctions, that enable them to pass signaling molecules and other materials between the cytosol of one cell and the cytosol of another (Figure 9.3a). For example, certain cells that line your lungs have intercellular connections called gap junctions, which we will examine further in Chapter 10. One cell can transmit a signaling molecule to the cytosol of an adjacent cell and thereby affect that cell's behavior. This type of signaling influences the rhythmic movements of cilia on adjacent cells that sweep dust and other unwanted particles out of the lungs.

Contact-Dependent Signaling Not all signaling molecules can readily diffuse from one cell to another. Some molecules are bound to the surface of cells and provide a signal to other cells that make contact with the surface of that cell (Figure 9.3b). In this case, one cell has a membrane-bound signaling molecule that is recognized by a receptor on the surface of another cell. This occurs when portions of nerve cells grow and make contact with other nerve or muscle cells.

Autocrine Signaling In autocrine signaling, a cell secretes signaling molecules that bind to receptors on its own cell surface, stimulating a response (Figure 9.3c). In addition, the signaling molecule can affect neighboring cells of the same cell type. Autocrine signaling is often important for groups of cells to sense cell density. When cell density is high, the concentration of autocrine signals is also high.

Paracrine Signaling In paracrine signaling, a specific cell secretes a signaling molecule that does not affect the cell secreting the signal, but it does influence the behavior of target cells in close proximity (Figure 9.3d). Usually, the signal is broken down too quickly to be carried to other parts of the body and affect distant cells. A specialized form of paracrine signaling called **synaptic signaling** occurs in the nervous system of animals. Neurotransmitters—molecules made in nerve cells that transmit a signal to an adjacent cell—are released at the end of the nerve cell and traverse a narrow space called the synapse. The neurotransmitter then binds to a receptor in a target cell. We will discuss this mechanism of signaling in detail in Chapter 43.

Endocrine Signaling In contrast to the previous mechanisms of cell signaling, endocrine signaling occurs over long distances (Figure 9.3e). In both animals and plants, molecules involved in long-distance signaling are called **hormones**. They are usually longer lasting than signaling molecules involved in autocrine and paracrine signaling. In animals, endocrine signaling involves the secretion of hormones into the bloodstream that may affect virtually all cells of the body, including those that are far from the cells that secrete the signaling molecules. In plants, hormones move through vessels and can also move through adjacent cells. Some signaling molecules are even gases that are secreted into the air. Ethylene is a gas given off by plants that plays a variety of roles, such as the acceleration of fruit ripening.

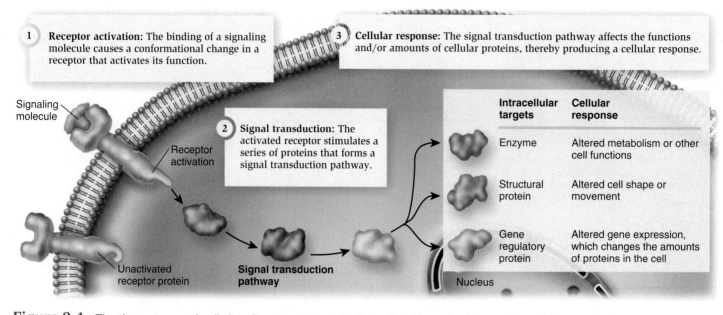

Figure 9.4   The three stages of cell signaling: receptor activation, signal transduction, and a cellular response.

Biological inquiry: For most signaling molecules, explain why a signal transduction pathway is necessary.

The Ability of Cells to Respond to Signals Is Usually a Three-Stage Process

Thus far, we have learned that signals influence the behavior of cells in close proximity or at long distances, usually interacting with receptors to elicit a cellular response. What events occur when a cell encounters a signal? In most cases, the binding of a signal to a receptor causes the receptor to activate a signal transduction pathway, which converts the signal to a cellular response. **Figure 9.4** diagrams the three stages of cell signaling: receptor activation, signal transduction, and a cellular response.

Stage 1: Receptor Activation   In the initial stage, a signaling molecule binds to a receptor, causing receptor activation. The binding of the signal causes a conformational change in the receptor that activates the receptor's function. In most cases, the activated receptor initiates a response by stimulating a sequence of changes in a series of proteins that collectively forms a signal transduction pathway, as described next.

Stage 2: Signal Transduction   During signal transduction, the initial signal is converted—or transduced—to a different signal inside the cell. This process is carried out by a group of proteins that form a **signal transduction pathway**. These proteins undergo a series of changes that may result in the production of another signaling molecule. For example, after a hormone binds to a receptor in the plasma membrane, the activated receptor may interact with intracellular proteins that produce new signals inside the cell.

Stage 3: Cellular Response   Cells can respond to signals in several different ways. Figure 9.4 shows three common cate-

gories of proteins that are controlled by cell signaling. Many signaling molecules exert their effects by altering the activity of one or more enzymes. For example, certain hormones provide a signal that the body needs energy. These hormones activate enzymes that are required for the breakdown of molecules such as carbohydrates.

A second way that cells respond to signals is by altering the functions of structural proteins in the cell. For example, when cells move during embryonic development or when an amoeba moves toward food, signals play a role in the rearrangement of actin filaments, which are components of the cytoskeleton. The coordination of signaling and changes in the cytoskeleton enable cells to move in the correct direction. In addition, changes in structural proteins may cause storage vesicles within a cell to fuse with the plasma membrane. Such vesicles may contain proteins or small molecules that are then secreted by the cell.

Cells may also respond to signals by changing the expression of genes. When exposed to sex hormones, cells can activate genes that change the properties of cells and even the characteristics of entire organisms. As discussed in Chapter 51, estrogens and androgens are responsible for the development of secondary sex characteristics in humans, including breast development in females and beard growth in males.

9.2 Cellular Receptors

As we have learned, signals are needed for cells to respond to environmental changes and to communicate with each other. Communication can occur between adjacent cells or over long distances. In most cases, the binding of a signaling molecule to its receptor activates a signal transduction pathway that ultimately leads to a cellular response. This response may involve effects on enzyme activities, the functions of structural proteins, and/or the transcription of genes.

In this section, we will take a closer look at receptors and how they interact with signaling molecules. We will focus on how these interactions affect the function of the receptor. As you will learn, cells contain many different types of receptors, and their molecular functions are quite diverse. In particular, we will look at differences among receptors based on whether they are located on the cell surface or inside the cell. In this chapter, our focus will be on receptors that respond to chemical signaling molecules. Other receptors discussed in Units VI and VII respond to mechanical motion (mechanoreceptors), temperature changes (thermoreceptors), and light (photoreceptors).

Signals Bind Specifically to Receptors and Alter Their Conformations

The ability of cells to respond to a signal usually requires precise recognition between a receptor and its signal. In many cases, the signal is a molecule—a steroid, peptide, or protein—that binds to the receptor. A signaling molecule binds to a receptor in much the same way that a substrate binds to the active site of an enzyme, as described in Chapter 7. The signaling molecule, which is called a **ligand**, binds noncovalently to the receptor molecule with a high degree of specificity. The binding occurs when the ligand and receptor collide in the correct orientation and with enough energy.

$$\text{[Ligand]} + \text{[Receptor]} \underset{k_{off}}{\overset{k_{on}}{\rightleftarrows}} \text{[Ligand} \cdot \text{Receptor complex]}$$

The value k_{on} is the rate at which binding occurs. After a complex forms between the ligand and its receptor, the noncovalent interaction between a ligand and receptor remains stable for a finite period of time. The term k_{off} is the rate at which the ligand · receptor complex falls apart or dissociates.

In general, the binding and release between a ligand and its receptor are relatively rapid, and therefore an equilibrium is reached when the rate of formation of new ligand · receptor complexes equals the rate at which existing ligand · receptor complexes dissociate:

$$k_{on}\,\text{[Ligand][Receptor]} = k_{off}\,\text{[Ligand} \cdot \text{Receptor complex]}$$

Rearranging,

$$\frac{\text{[Ligand] [Receptor]}}{\text{[Ligand} \cdot \text{Receptor complex]}} = \frac{k_{off}}{k_{on}} = K_d$$

K_d is called the **dissociation constant** between a ligand and its receptor. Let's look carefully at the left side of this equation and consider what it means. At a ligand concentration where half of the receptors are bound to a ligand, the concentration of the ligand · receptor complex equals the concentration of receptor that doesn't have ligand bound. At this ligand concentration, [Receptor] and [Ligand · Receptor complex] cancel out of the

equation because they are equal. Therefore, at a ligand concentration where half of the receptors have bound ligand:

$$K_d = \text{[Ligand]}$$

When the ligand concentration is above the K_d value, most of the receptors are likely to have ligand bound to them. In contrast, if the ligand concentration is substantially below the K_d value, most receptors will not be bound by their ligand. The K_d values for many different ligands and their receptors have been experimentally determined. This information allows researchers to predict when a signaling molecule is likely to cause a cellular response. If the concentration of a signaling molecule is far below the K_d value, a cellular response is not likely because relatively few receptors will form a complex with the signaling molecule.

Unlike enzymes, which convert their substrates into products, receptors do not usually alter the structure of their ligands. Instead, the ligands alter the structure of their receptors, causing a conformational change. This concept is shown in **Figure 9.5**, in which the ligand is a hormone. Once the hormone binds, the receptor undergoes a conformational change. After binding, the hormone · receptor complex is called an activated receptor, because the binding of the hormone to its receptor has changed the receptor in a way that will activate its ability to initiate a cellular response.

Because cell communication is critical for cell survival, the evolution of organisms over the past 3.5 to 4.0 billion years has produced a variety of ways in which cells can respond to signals. Previously, we learned that one way to categorize cell communication is based on the distance that the signal travels. Another way to categorize cell signaling is based on the location and activity of the receptor, as we will explore next.

Cells Contain a Variety of Cell Surface Receptors That Respond to Signals in Different Ways

Most signals, either environmental agents or signals that are secreted from cells, are small hydrophilic molecules or large molecules that do not readily pass through the plasma membrane of cells.

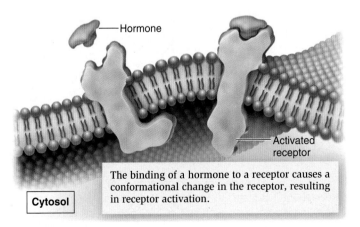

— Hormone

Activated receptor

Cytosol

The binding of a hormone to a receptor causes a conformational change in the receptor, resulting in receptor activation.

Figure 9.5   A conformational change induced by a ligand binding to a receptor.

To respond to such signals, cells possess several different types of **cell surface receptors**—receptors that are found in the plasma membrane. A typical cell is expected to contain dozens or even hundreds of different cell surface receptors that enable the cell to respond to different kinds of signaling molecules.

By analyzing the functions of cell surface receptors from many different organisms, researchers have determined that most fall into one of three categories: enzyme-linked receptors, G-protein-coupled receptors, and ligand-gated ion channels.

Enzyme-Linked Receptors Receptors known as **enzyme-linked receptors** are found in all living species. They typically have two important domains: an extracellular domain, which binds a signaling molecule, and an intracellular domain, which has a catalytic function (**Figure 9.6a**). When a signaling molecule binds to the extracellular domain, a conformational change is transmitted through the membrane-embedded portion of the protein that affects the conformation of the catalytic domain. In most cases, this conformational change causes the catalytic domain to become functionally active.

Most types of enzyme-linked receptors function as **protein kinases**, enzymes that transfer phosphate groups from ATP to a

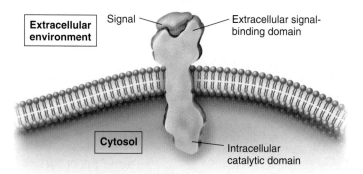

(a) Structure of enzyme-linked receptors

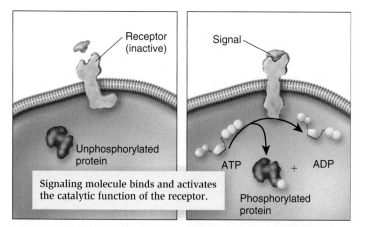

(b) A receptor that functions as a protein kinase

Figure 9.6 Enzyme-linked receptors. **(a)** The enzyme-linked receptor protein has an extracellular domain, which binds a signaling molecule, and an intracellular domain, which has a catalytic function. **(b)** An enzyme-linked receptor can function as a protein kinase, which catalyzes the transfer of a phosphate group from ATP to an intracellular protein.

protein (**Figure 9.6b**). In the absence of a signaling molecule, the catalytic domain remains inactive. However, when a signal binds to the extracellular domain, this transmits a conformational change to the catalytic domain, making it active. Under these conditions, the cell surface receptor may phosphorylate itself or intracellular proteins. Later in this chapter, we will explore in more detail how this event leads to a cellular response.

G-Protein-Coupled Receptors Receptors called **G-protein-coupled receptors (GPCRs)** are commonly found in the cells of eukaryotic species. GPCRs exhibit a typical structure in which the protein contains seven transmembrane segments that wind back and forth through the plasma membrane. The receptors interact with intracellular proteins called **G proteins**, which are so named because of their ability to bind guanosine triphosphate (GTP) and guanosine diphosphate (GDP). GTP is similar in structure to ATP except it has guanine as a base instead of adenine. In the 1970s, the existence of G proteins was first proposed by Martin Rodbell and colleagues, who found that GTP was needed for certain hormone receptors to cause an intracellular response. Later, Alfred Gilman and coworkers used genetic and biochemical techniques to identify and purify a G protein. In 1994, Rodbell and Gilman won the Nobel Prize for their pioneering work.

Figure 9.7 shows how a GPCR and G protein interact. At the cell surface, the binding of a signaling molecule to a GPCR activates the receptor by causing a conformational change. The activated receptor then binds to a membrane-bound G protein, causing the G protein to release GDP and bind GTP instead. GTP binding changes the conformation of the G protein, causing it to dissociate into its α subunit and a β/γ dimer. Eventually, the extracellular signaling molecule will be degraded, thereby lowering its concentration and resulting in the dissociation between the signaling molecule and the GPCR. When this occurs, the GPCR is no longer activated and the cellular response will be reversed, because the α subunit will eventually hydrolyze its bound GTP to GDP and P_i. When this occurs, the α and β/γ subunits reassociate with each other to form an inactive complex. Later in this chapter, we will examine how the α subunit and β/γ dimer interact with other proteins in a signal transduction pathway to elicit a cellular response.

Ligand-Gated Ion Channels Ion channels are proteins that allow the diffusion of ions across cellular membranes. **Ligand-gated ion channels** are a third type of cell surface receptor found in the plasma membrane of animal and plant cells. When a signaling molecule (ligand) binds to this type of receptor, the channel opens and allows the flow of ions through the membrane (**Figure 9.8**).

In animals, ligand-gated ion channels are important in the transmission of signals between nerve and muscle cells and between two nerve cells. In addition, ligand-gated ion channels in the plasma membrane allow the uptake of Ca^{2+} into the cytosol. As discussed later in this chapter, changes in the cytosolic concentration of Ca^{2+} play a role in signal transduction.

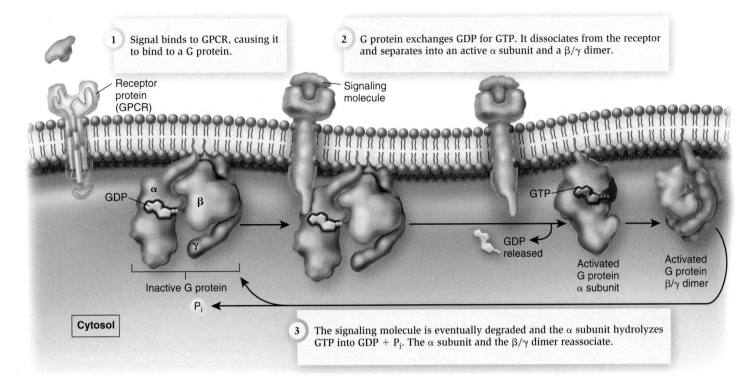

1 Signal binds to GPCR, causing it to bind to a G protein.

2 G protein exchanges GDP for GTP. It dissociates from the receptor and separates into an active α subunit and a β/γ dimer.

Receptor protein (GPCR)

Signaling molecule

GDP

α

β

γ

GTP

GDP released

Inactive G protein

Activated G protein α subunit

Activated G protein β/γ dimer

P_i

Cytosol

3 The signaling molecule is eventually degraded and the α subunit hydrolyzes GTP into GDP + P_i. The α subunit and the β/γ dimer reassociate.

Figure 9.7 The activation of G-protein-coupled receptors.

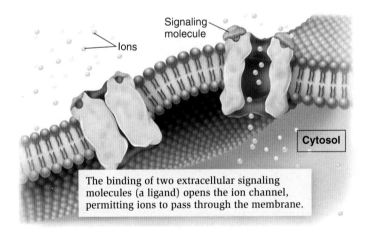

Signaling molecule

Ions

Cytosol

The binding of two extracellular signaling molecules (a ligand) opens the ion channel, permitting ions to pass through the membrane.

Figure 9.8 The function of a ligand-gated ion channel.

Some Signaling Molecules Pass Through the Plasma Membrane and Activate Intracellular Receptors

Although most receptors for signaling molecules are located in the plasma membrane, some are found inside the cell. In these cases, the signaling molecule must pass through the plasma membrane to gain access to its receptor. For example, two types of signaling molecules—steroids and auxins—diffuse into the cell, where they bind to internal receptors. As described in Chapter 51, steroid hormones, such as estrogens and androgens, are secreted into the bloodstream from cells of endocrine glands.

The behavior of estrogen is typical of many steroid hormones (**Figure 9.9a**). Because of its hydrophobic nature, estrogen is capable of passing through the plasma membrane of a target cell and binding to a receptor in the nucleus. Some steroids bind to receptors in the cytosol, which then travel into the nucleus, while other steroid hormones bind to receptors that are in the nucleus; estrogen is one of the latter. After binding, the estrogen • receptor complex undergoes a conformational change that enables it to form a dimer with another estrogen • receptor complex. The dimer then binds to the DNA and activates the transcription of specific genes. The estrogen receptor is an example of a transcription factor, a protein that regulates the transcription of genes.

A second example of an intracellular receptor is the receptor for a group of related plant hormones called auxins, which are important in many plant signaling pathways, including the growth of shoots toward light, the growth of roots into the soil, and the flowering process. In 2005, two research groups led by Ottoline Leyser and Mark Estelle determined that auxin binds to an intracellular receptor called TIR1 (TIR stands for transport inhibitor response, so named because mutations in this receptor alter the response of drugs that inhibit auxin transport in plants). The binding of auxin in the cytosol activates TIR1, and it travels into the nucleus. Like steroid receptors, TIR1 also causes the activation of specific genes. However, TIR1 does not do this directly. After auxin binds, the receptor causes the breakdown of a protein that inhibits several genes (**Figure 9.9b**). When this inhibitory protein is broken down, the gene inhibition is relieved and the genes are transcribed into mRNA.

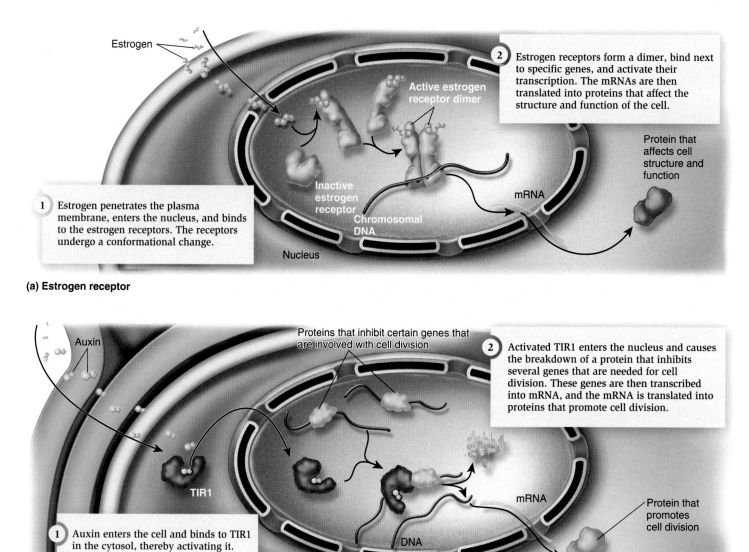

Estrogen

2 Estrogen receptors form a dimer, bind next to specific genes, and activate their transcription. The mRNAs are then translated into proteins that affect the structure and function of the cell.

Active estrogen receptor dimer

Protein that affects cell structure and function

1 Estrogen penetrates the plasma membrane, enters the nucleus, and binds to the estrogen receptors. The receptors undergo a conformational change.

Inactive estrogen receptor

mRNA

Chromosomal DNA

Nucleus

(a) Estrogen receptor

Auxin

Proteins that inhibit certain genes that are involved with cell division

2 Activated TIR1 enters the nucleus and causes the breakdown of a protein that inhibits several genes that are needed for cell division. These genes are then transcribed into mRNA, and the mRNA is translated into proteins that promote cell division.

TIR1

mRNA

Protein that promotes cell division

1 Auxin enters the cell and binds to TIR1 in the cytosol, thereby activating it.

DNA

Nucleus

(b) Auxin receptor

Figure 9.9   Examples of intracellular receptors.

9.3 Signal Transduction and the Cellular Response

In this section, we turn our attention to the intracellular events that enable a cell to respond to a signaling molecule that binds to a cell surface receptor. In most cases, the binding of a signaling molecule to its receptor stimulates a signal transduction pathway. We begin by examining the components of different signal transduction pathways, looking in detail at the two-component regulatory systems common to bacteria and the signaling transduction pathways involving molecules called second messengers. We will use the activities of the hormone epinephrine as a way to explore how signal transduction pathways affect cellular proteins in a way that leads to a cellular response. Lastly, we'll explore the reasons why different types of cells can respond in distinct ways to the same hormone.

Two-Component Regulatory Systems in Bacteria Sense Environmental Changes

The survival of bacteria is largely dependent on their ability to adapt to environmental changes. By studying many different bacterial species, researchers have discovered that a common way for bacteria to adapt to a changing environment is via **two-component regulatory systems**. In such systems, one component, an enzyme-linked receptor called a **sensor kinase**, recognizes a signal found in its environment. The sensor kinase also has the ability to hydrolyze ATP and phosphorylate itself. The phosphate group is then transferred to a second component, a protein called the **response regulator**, which is usually a protein that regulates the expression of many genes.

Figure 9.10 considers a two-component regulatory system, found in many species of bacteria, that senses the presence of

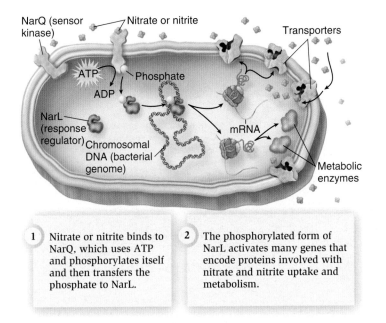

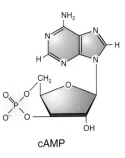

| 1 | Nitrate or nitrite binds to NarQ, which uses ATP and phosphorylates itself and then transfers the phosphate to NarL. | 2 | The phosphorylated form of NarL activates many genes that encode proteins involved with nitrate and nitrite uptake and metabolism. |

Figure 9.10 An example of a two-component regulatory system in bacteria.   The system shown here regulates nitrate and nitrite metabolism. NarQ is a receptor, and NarL is a single component of a signal transduction pathway. When phosphorylated by NarQ, NarL functions as a transcriptional activator.

nitrate (NO_3^-) and nitrite (NO_2^-) in the environment. NarQ is a sensor kinase that recognizes either nitrate or nitrite. When either of these binds to NarQ, NarQ phosphorylates itself via ATP and then transfers a phosphate to the response regulator, NarL. The phosphorylated form of NarL activates the transcription of many genes involved in nitrate and nitrite metabolism and transport. Once these genes are transcribed into mRNA and then translated into specific proteins, the bacterium has become better adapted to the presence of nitrate or nitrite in its environment.

Two-component regulatory systems are also found in fungi and plants, but they do not appear to exist in animals. Why animals lack this form of signaling is not understood.

Second Messengers Are Key Components of Many Signal Transduction Pathways

Cell biologists call signals that bind to a cell surface receptor the first messengers. After first messengers bind to receptors, many signal transduction pathways lead to the production of **second messengers**, small molecules or ions that relay signals inside the cell. The following examples illustrate the roles of second messengers in signal transduction pathways.

Signal Transduction via cAMP As we saw earlier, the binding of a signaling molecule to a G-protein-coupled receptor (GPCR) activates an intracellular G protein by causing it to bind GTP and dissociate into an α subunit and a β/γ dimer (see Figure 9.7). Let's now follow the role of these subunits in signal

transduction pathways. The α subunit and the β/γ dimer can activate several different kinds of proteins that are components of signal transduction pathways.

Mammalian and plant cells make several different types of G protein α subunits. One type binds to **adenylyl cyclase**, an enzyme in the plasma membrane (**Figure 9.11a**). This stimulates adenylyl cyclase to synthesize **cyclic adenosine monophosphate (cAMP)** from ATP. The molecule, cAMP, is an example of a second messenger.

cAMP

One effect of cAMP is to activate protein kinase A (PKA), which is composed of four subunits: two catalytic subunits that phosphorylate specific cellular proteins, and two regulatory subunits that inhibit the catalytic subunits when they are bound to each other. Cyclic AMP binds to the regulatory subunits of PKA. The binding of cAMP separates the regulatory and catalytic subunits, which allows each catalytic subunit to be active.

The catalytic subunit of PKA then phosphorylates specific cellular proteins such as enzymes, structural proteins, and transcription factors. When a protein kinase attaches a phosphate group onto a protein, the shape and function of that protein is altered. Often, phosphorylation activates protein function. In 1955, Edmond Fischer and Edwin Krebs discovered that the phosphorylation of enzymes is important in regulating glycogen breakdown. Since that time, researchers have determined that phosphorylation is a widespread mechanism that regulates protein function in all species. For their key discovery, Fischer and Krebs were awarded the Nobel Prize in 1992.

Let's now consider how the effects of a signaling molecule are reversed. When the signaling molecule is no longer produced and its extracellular levels fall, the molecule dissociates from the receptor, because the binding to its receptor is a reversible process (**Figure 9.11b**). Once the signaling molecule is released, the receptor is no longer activated. Intracellularly, the α subunit hydrolyzes its GTP to GDP, and the α subunit and β/γ dimer reassociate to an inactive G protein. The levels of cAMP decrease due to the action of an enzyme called **phosphodiesterase**, which converts cAMP to AMP. As cAMP levels fall, the regulatory subunits of PKA release cAMP, and the regulatory and catalytic subunits reassociate, thereby inhibiting PKA. Finally, enzymes called **protein phosphatases** are responsible for removing phosphate groups from proteins, thereby reversing the effects of PKA.

In the 1950s, Earl Sutherland determined that many different hormones cause the formation of cAMP in a variety of cell types.

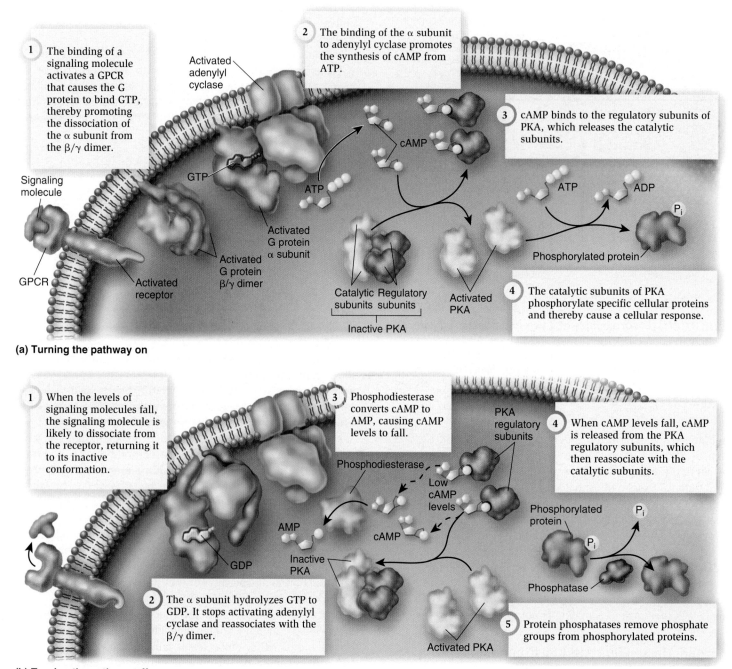

1 The binding of a signaling molecule activates a GPCR that causes the G protein to bind GTP, thereby promoting the dissociation of the α subunit from the β/γ dimer.

2 The binding of the α subunit to adenylyl cyclase promotes the synthesis of cAMP from ATP.

3 cAMP binds to the regulatory subunits of PKA, which releases the catalytic subunits.

4 The catalytic subunits of PKA phosphorylate specific cellular proteins and thereby cause a cellular response.

Activated adenylyl cyclase

Signaling molecule

GTP

ATP

cAMP

ATP   ADP

P$_i$

Phosphorylated protein

GPCR

Activated receptor

Activated G protein α subunit

Activated G protein β/γ dimer

Catalytic subunits   Regulatory subunits

Activated PKA

Inactive PKA

(a) Turning the pathway on

1 When the levels of signaling molecules fall, the signaling molecule is likely to dissociate from the receptor, returning it to its inactive conformation.

3 Phosphodiesterase converts cAMP to AMP, causing cAMP levels to fall.

4 When cAMP levels fall, cAMP is released from the PKA regulatory subunits, which then reassociate with the catalytic subunits.

PKA regulatory subunits

Phosphodiesterase

Low cAMP levels

AMP

cAMP

GDP

Inactive PKA

2 The α subunit hydrolyzes GTP to GDP. It stops activating adenylyl cyclase and reassociates with the β/γ dimer.

Phosphorylated protein

P$_i$

P$_i$

Phosphatase

Activated PKA

5 Protein phosphatases remove phosphate groups from phosphorylated proteins.

(b) Turning the pathway off

Figure 9.11 **The signal transduction pathway involving cAMP.** (a) The pathway leading to the formation of cAMP and subsequent activation of PKA, which is mediated by a G-protein-coupled receptor (GPCR). (b) The steps that shut the pathway off when the signaling molecule is no longer present in the environment.

This observation, for which he won the Nobel prize, stimulated great interest in the study of signal transduction pathways. Since Sutherland's discovery, the production of second messengers such as cAMP has been found to have two important advantages. One benefit is the amplification of the signal; the binding of a signal to a single receptor can cause the synthesis of many cAMP molecules that activate PKA (**Figure 9.12**). Likewise, each molecule of PKA can phosphorylate many proteins in the cell to promote a cellular response. In this way, a signal can have a dramatic effect on the functioning of many proteins in a target cell.

A second advantage of second messengers such as cAMP is speed. Brian Bacskai and colleagues studied the response of nerve cells to a signaling molecule called serotonin, which is a neurotransmitter that binds to a GPCR. In humans, serotonin is believed to play a role in depression, anxiety, and sexual drive. To monitor cAMP levels, nerve cells grown in a laboratory were injected with a fluorescent protein that changes its fluorescence

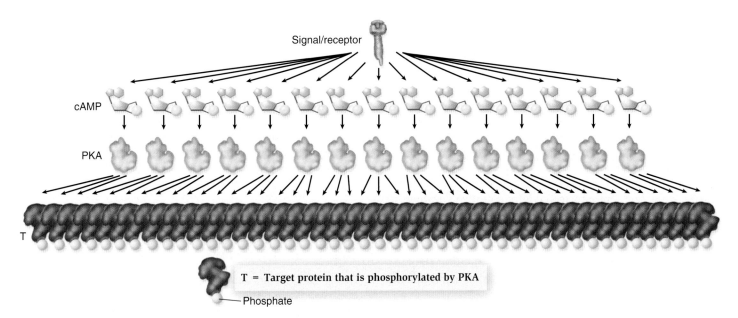

Figure 9.12 Signal amplification. An advantage of a signal transduction pathway is the amplification of a signal. In this case, a single signaling molecule can lead to the phosphorylation of many target proteins.

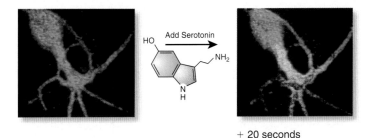

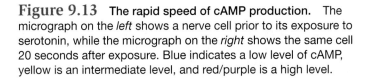

Figure 9.13 The rapid speed of cAMP production. The micrograph on the *left* shows a nerve cell prior to its exposure to serotonin, while the micrograph on the *right* shows the same cell 20 seconds after exposure. Blue indicates a low level of cAMP, yellow is an intermediate level, and red/purple is a high level.

when cAMP is made. As shown in the micrograph on the right in **Figure 9.13**, a substantial amount of cAMP was made within 20 seconds after the addition of serotonin.

Signal Transduction via Ca²⁺ Cells use many different types of second messengers, and more than one type may be used at the same time. Calcium ions are another common second messenger. Cells maintain very large Ca^{2+} gradients (**Figure 9.14**). Two types of calcium pumps are found in the membranes of cells. Ca^{2+}-ATPases use the energy from ATP hydrolysis to pump Ca^{2+} against a gradient, while Na^{+}/Ca^{2+} and H^{+}/Ca^{2+} exchangers or antiporters use the energy within Na^{+} or H^{+} gradients to pump Ca^{2+} across the membrane. These pumps produce a cytosolic Ca^{2+} concentration that is very low, typically 0.1–1.0 μM. By comparison, the Ca^{2+} concentration found outside of cells and inside the endoplasmic reticulum (ER) lumen and mitochondrial matrix is far higher, as much as 1,000- to 10,000-fold higher!

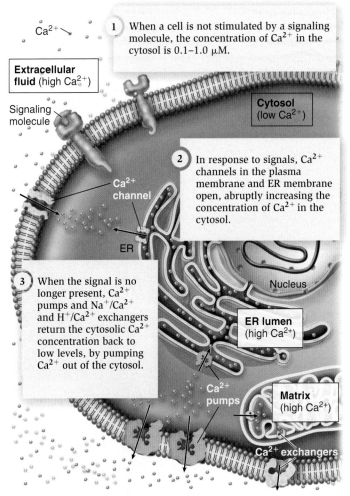

Figure 9.14 Calcium levels inside and outside of eukaryotic cells.

Calcium channels that are found in the plasma membrane and ER membrane play a key role in the ability of calcium to act as a second messenger. When such channels open in response to signaling, the cytosolic concentration of Ca^{2+} can increase dramatically and quickly, in a matter of seconds. The increase in Ca^{2+} concentration in the cytosol acts as a second messenger that elicits a cellular response. In plants, calcium signaling is involved in many processes, including cell elongation in response to light, the opening and closing of stomata, and the ability to sense gravity. In animals, calcium-signaling pathways are also important in many cellular events such as nerve transmission, muscle contraction, and the secretion of digestive enzymes by the pancreas. As we will explore later, such responses are mediated by calcium-binding proteins that are activated when the cytosolic concentration of Ca^{2+} is increased.

The calcium channels that are found in the ER membrane are opened via a second messenger called inositol trisphosphate, a pathway that is described next.

Signal Transduction via Diacylglycerol and Inositol Trisphosphate

We have already examined how a G protein can activate adenylyl cyclase and thereby promote the synthesis of cAMP. Let's now consider a second way that an activated G protein can influence a signal transduction pathway. The α subunit can activate an enzyme called phospholipase C (Figure 9.15). When phospholipase C becomes active, it breaks a covalent bond in a plasma membrane phospholipid, which produces two second messengers, diacylglycerol (DAG) and inositol trisphosphate (IP_3). DAG binds to a membrane-bound enzyme called protein kinase C (PKC). IP_3 is released into the cytosol and binds to a ligand-gated Ca^{2+} channel in the ER membrane. The binding of IP_3 causes the channel to open, releasing Ca^{2+} into the cytosol.

Therefore, this pathway involves another messenger, the calcium ions that we considered previously.

Calcium ions can affect the behavior of cells in a variety of ways, two of which are shown in Figure 9.15. Ca^{2+} can bind to PKC, which, in combination with DAG, causes this enzyme to become activated. Once activated, PKC can phosphorylate specific cellular proteins. In addition, Ca^{2+} can bind to a protein called calmodulin, which is a <u>cal</u>cium-<u>modul</u>ated prote<u>in</u>. The Ca^{2+}-calmodulin complex can then interact with specific cellular proteins and alter their functions. For example, calmodulin regulates proteins involved in carbohydrate breakdown in liver cells.

A Cellular Response Depends on Which Proteins Are Controlled by a Signal Transduction Pathway

As we have seen, signaling molecules exert their effects on eukaryotic cells via signal transduction pathways that control the activities and/or synthesis of specific proteins. In multicellular organisms, one of the amazing effects of signaling molecules such as hormones is their ability to coordinate cellular activities. One example is the hormone epinephrine (also called adrenaline), which is secreted from endocrine cells. Epinephrine is also called the fight-or-flight hormone because it quickly prepares the body for strenuous physical activity. In humans, for example, epinephrine is secreted into the bloodstream when someone is confronted with a stressful event or is exercising vigorously.

Epinephrine has different effects throughout the body (Figure 9.16). In the heart, it stimulates heart muscle cells so that the heart beats faster. In the liver, epinephrine promotes the release of glucose into the bloodstream so that an individual

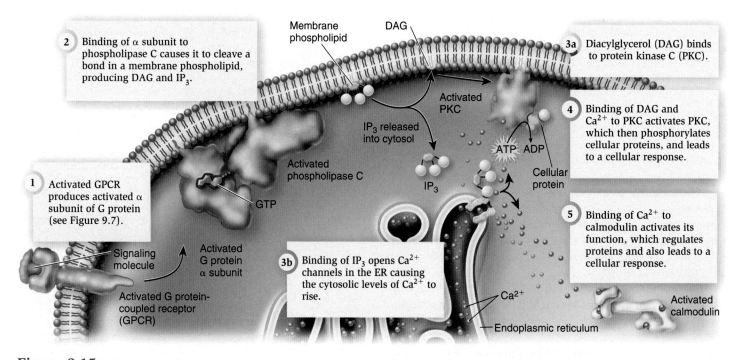

Figure 9.15   Signal transduction pathway involving diacylglycerol (DAG) and inositol trisphosphate (IP_3).

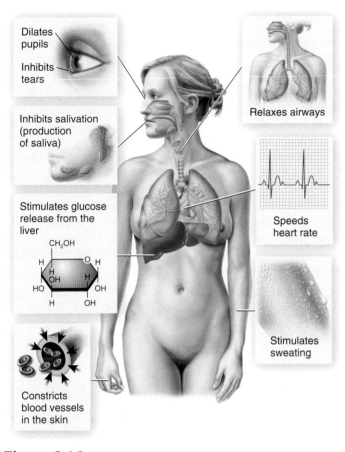

Figure 9.16   The effects of epinephrine in humans.

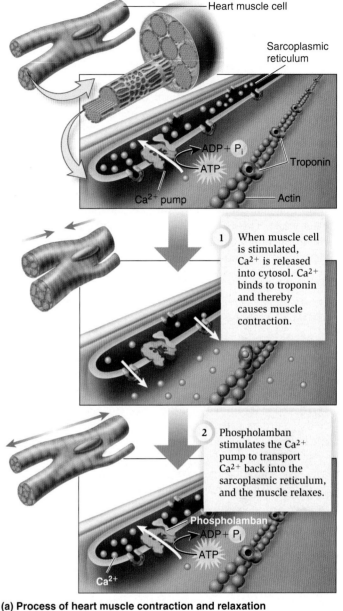

(a) Process of heart muscle contraction and relaxation

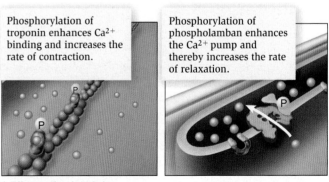

(b) Effects of epinephrine

Figure 9.17   The contraction of heart muscle cells and the effects of epinephrine.   (a) The steps of muscle contraction and relaxation. (b) The effects of epinephrine. This hormone activates PKA, which phosphorylates troponin and phospholamban.

can use the glucose to produce energy (ATP) needed to fight or flee. Let's examine the cell's response to an extracellular signal by considering how epinephrine affects cells in the heart.

Cellular Response of Heart Muscle Cells to Epinephrine
Contraction of heart muscle is controlled by the level of Ca^{2+} in the cytosol. Calcium ions are stored inside an organelle found only in muscle cells called the sarcoplasmic reticulum, which is a specialized form of the endoplasmic reticulum. When a muscle cell is stimulated to contract, calcium ions are released from the sarcoplasmic reticulum into the cytosol, where the ions bind to a protein called troponin (**Figure 9.17a**). The binding of Ca^{2+} causes a conformational change in troponin that subsequently leads to muscle contraction. For the muscle to relax, the ions must be pumped back into the sarcoplasmic reticulum. A protein called phospholamban activates a Ca^{2+} pump, which stimulates the pumping of Ca^{2+} into the sarcoplasmic reticulum. This lowers the concentration of Ca^{2+} in the cytosol and causes the muscle to relax.

How does epinephrine increase the heart rate? In heart muscle cells, the primary effect of epinephrine is to activate adenylyl cyclase, which produces cAMP. This, in turn, activates PKA, which phosphorylates troponin and phospholamban (**Figure 9.17b**). When troponin is phosphorylated, this enhances Ca^{2+} binding and thereby increases the rate of contraction.

When phospholamban is phosphorylated, it stimulates the function of the Ca^{2+} pump. This promotes a faster pumping of Ca^{2+} back into the sarcoplasmic reticulum and thereby enhances the rate of relaxation. Taken together, the phosphorylation of these proteins increases the heart rate by increasing the rates of both contraction and relaxation.

Interestingly, one of the effects of caffeine can be explained by this mechanism. Caffeine inhibits phosphodiesterase, which converts cAMP to AMP (see Figure 9.11b). Phosphodiesterase functions to remove cAMP once a signaling molecule, such as epinephrine, is no longer present. When phosphodiesterase is inhibited by caffeine, cAMP persists for a longer period of time and thereby causes the heart to beat faster. Therefore, even low levels of signaling molecules such as epinephrine will have a greater effect. This is one of the reasons why drinks containing caffeine, including coffee and many energy drinks, provide a feeling of vitality and energy.

GENOMES & PROTEOMES

A Cell's Response to Signaling Molecules Depends on the Proteins It Makes

As Figure 9.16 shows, a hormone such as epinephrine produces diverse responses throughout the body. How do we explain the fact that various cell types can respond so differently to the same hormone? The answer lies in **differential gene regulation**. The transcription of genes is controlled by a variety of processes. As a multicellular organism develops from a fertilized egg, the cells of the body become differentiated into particular types, such as heart and liver cells. Although different cell types such as heart and liver cells contain the same set of genes, they are not expressed in the same pattern. Certain genes that are turned off in heart cells are turned on in liver cells, while some genes that are turned on in heart cells are turned off in liver cells. The pattern of gene regulation in any given cell type is critical in its ability to respond to signaling molecules. The following are examples of how differential gene regulation affects the cellular response:

1. *A cell may or may not express a receptor for a particular signaling molecule.* For example, not all cells of the human body express a receptor for epinephrine. These cells are not affected when epinephrine is released into the bloodstream.
2. *Different cell types have different cell surface receptors that recognize the same signaling molecule.* In humans, a signaling molecule called acetylcholine has two different types of receptors. One acetylcholine receptor is a ligand-gated ion channel that is expressed in skeletal muscle cells. Another acetylcholine receptor is a G-protein-coupled receptor (GPCR) that is expressed in heart muscle cells. Because of this, skeletal and heart muscle cells respond differently to acetylcholine.
3. *Two (or more) receptors may work the same way in different cell types but have different affinities for the same signaling molecule.* Two different GPCRs may recognize the same

hormone, but a receptor expressed in liver cells may have a higher affinity (that is, a lower K_d) for a hormone compared to a receptor expressed in muscle cells. In this case, the liver cells will respond to a lower hormone concentration than the muscle cells will.
4. *The expression of proteins involved in intracellular signal transduction pathways may vary in different cell types.* For example, one cell type may express the proteins that are needed to activate PKA, while another cell type may not.
5. *The expression of proteins that are controlled by signal transduction pathways may vary in different cell types.* In liver cells, the presence of epinephrine leads to the activation of glycogen phosphorylase kinase (GPK), an enzyme involved in glycogen breakdown. However, this enzyme is not expressed in all cells of the body. Glycogen breakdown will only be stimulated by epinephrine if GPK is expressed in that cell.

Taken together, these five examples of differential gene regulation help explain why different cell types respond to signaling molecules in a myriad of ways.

9.4 Regulation of the Cell Cycle

In this chapter, we have examined how cells receive signals and respond to them in meaningful ways. One of the most important ways that a cell may respond to a signal is to divide. Life is a continuum in which new living cells are formed by the division of pre-existing cells. The Latin axiom *Omnis cellula e cellula*, meaning "Every cell from a cell," was first proposed in 1858 by a German pathologist, Rudolf Virchow. From an evolutionary perspective, cell division is a cellular response with a very ancient origin. All living organisms, from unicellular bacteria to multicellular plants and animals, are products of repeated rounds of cell growth and division extending back to the beginnings of life nearly 4 billion years ago.

We now know that cell division is a process that involves remarkable accuracy and precise timing. A cell must be able to sense when conditions are appropriate for cell division to occur and then orchestrate a series of events that will ensure the production of healthy new cells. In Chapter 15, we will examine the mechanics of cell division, with an emphasis on how chromosomes are correctly sorted to new daughter cells. In this section, we will focus on the regulation of the cell cycle. We begin with a description of the phases of the cell cycle. We then explore the factors that affect the cell's decision to divide, and see how the cell cycle is controlled by proteins that carefully monitor the division process to ensure its accuracy.

The Cell Cycle Is a Series of Phases That Lead to Cell Division

Eukaryotic cells that are destined to divide progress through a series of stages known as the **cell cycle** (**Figure 9.18**). The phases consist of G_1 (first gap), **S** (synthesis of DNA, the genetic material), G_2 (second gap), and **M phase** (mitosis and cytokinesis).

The G_1 and G_2 phases were originally described as gap phases, to indicate a pause in activity between DNA synthesis and mitosis. However, we now know that these are critical stages of the cell cycle. In actively dividing cells, the G_1, S, and G_2 phases are collectively known as **interphase**. Alternatively, cells may exit the cell cycle and remain for long periods of time in a stage called G_0. The G_0 stage is a substitute for G_1. A cell in the G_0 phase has postponed making a decision to divide or, in the case of terminally differentiated cells (such as nerve cells in an adult animal), has made a decision to never divide again.

The length of the cell cycle varies considerably among different cells types, ranging from several minutes in quickly growing embryos to several months in slow-growing adult cells. For fast-dividing mammalian cells in adults, the length of the cycle is typically 24 hours. The various stages within the cell cycle also vary in length. G_1 is often the longest and also the most variable phase, while M phase is the shortest. For a cell that divides in 24 hours, the following lengths of time for each phase are typical:

- G_1 phase: 11 hours
- S phase: 8 hours
- G_2 phase: 4 hours
- M phase: 1 hour

The G_1 phase is a period in a cell's life when it may decide to divide. Depending on the environmental conditions and the presence of signaling molecules, a cell in the G_1 phase may accumulate molecular changes that cause it to progress through the rest of the cell cycle. When this occurs, cell biologists say that a cell has reached a special control point called the **restriction point**. Once past the restriction point, a cell is committed to advance to the S phase, during which the chromosomes are replicated. After replication, the two duplicated chromosomes are still joined to each other and referred to as a pair of **sister chromatids** (**Figure 9.19**). When S phase is completed, a cell actually has twice as many chromatids as the number of chromosomes in the G_1 phase. For example, a human cell in the G_1 phase has 46 distinct chromosomes, whereas the same cell in G_2 would have 46 pairs of sister chromatids, for a total of 92 chromatids.

During the G_2 phase, a cell synthesizes proteins that are necessary for chromosome sorting and cell division. It then progresses into the M phase of the cell cycle, when **mitosis** occurs. The primary purpose of mitosis is to divide one cell nucleus into two nuclei, distributing the duplicated chromosomes so that each daughter cell will receive the same complement of chromosomes. For example, a human cell in the G_2 phase has 92 chromatids, which are found in 46 pairs. During mitosis, these pairs of chromatids are separated and sorted so that each daughter cell will receive 46 chromosomes. Mitosis is the name given to this sorting process. In most cases, mitosis is followed by **cytokinesis**, which is the division of the cytoplasm to produce two distinct daughter cells.

The decision to divide is based on external factors, such as environmental conditions and signaling molecules, and internal controls, including cell cycle control molecules and checkpoints, as we will discuss next.

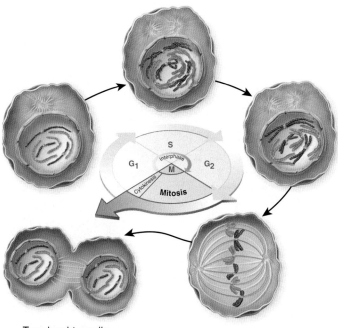

Two daughter cells

Figure 9.18 **The eukaryotic cell cycle.** Dividing cells progress through a series of stages denoted G_1, S, G_2, and M phases. This diagram shows the progression of a cell through the cell cycle to produce two daughter cells. The original diploid cell had three pairs of chromosomes, for a total of six individual chromosomes. During S phase, these have replicated to yield 12 chromatids. After mitosis is complete, two daughter cells each contain six chromosomes.

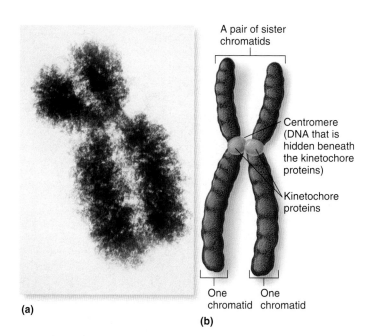

(a)

(b)

Figure 9.19 **Metaphase chromosomes.** (a) Metaphase is a step during mitosis when the chromosomes are highly compacted. This TEM shows a metaphase chromosome that exists in a form called a pair of sister chromatids. (b) A schematic drawing of sister chromatids. This structure has two chromatids that lie side by side. The two chromatids are held together by proteins called kinetochore proteins that bind to each other and to the centromeres on each chromatid.

Environmental Conditions and Signaling Molecules Affect the Decision to Divide

In unicellular organisms, the decision to divide is based largely on environmental conditions. For example, if yeast cells are supplied with a sufficient amount of nutrients and are exposed to the correct temperature and pH, they are likely to divide. In multicellular organisms such as plants and animals, the decision to divide is more complex. Although nutrient availability, temperature, and pH are important to the cells in a multicellular organism, other factors also influence the process of cell division. As development proceeds in a multicellular organism, genetic factors come into play. In mammals, for example, most nerve cells in the adult have lost the capacity to divide, due to a genetic program that begins during embryonic development. This is one reason why it is difficult to recover from certain injuries to the nervous system, such as those that affect the spinal cord.

Multicellular organisms rely on signaling molecules to coordinate cell division throughout the body. In plants, hormones play a key role in promoting cell division. For example, small molecules called **cytokinins**—so named because their presence causes cytokinesis—promote cell division in plants. These compounds have a structure that is similar to adenine. Cytokinins are secreted at the growing tips of plants. In animals, **growth factors**, a group of proteins that stimulate certain cells to grow and divide, are secreted into the bloodstream.

Several signaling pathways that promote cell division have been characterized at the molecular level. **Figure 9.20** describes a simplified signal transduction pathway for epidermal growth factor (EGF). This protein is secreted from endocrine cells and stimulates epidermal cells, such as skin cells, to divide (see chapter-opening photo). The pathway begins when two molecules of EGF bind to two EGF receptor subunits. The binding of EGF causes the subunits to dimerize and phosphorylate each other. The phosphorylated form of the EGF receptor is recognized by an intracellular protein called GRB2. This interaction changes the conformation of GRB2 so that it binds to the protein Sos, which also changes its conformation. The activation of Sos causes another protein called Ras to release GDP and bind GTP. The GTP form of Ras is the active form that binds to Raf-1, which is a protein kinase. This begins a **protein kinase cascade**, the sequential activation of several protein kinases in a row. Recall that protein kinases are enzymes that catalyze the phosphorylation of a target protein. This phosphorylation alters the activity and conformation of the target protein. Raf-1 phosphorylates MEK, which in turn phosphorylates MAPK. MAPK stands for mitogen-activated protein kinase. This kinase was first discovered because it is activated in the presence of mitogens—agents that cause a cell to proceed through mitosis. MAPK enters the nucleus and phosphorylates several proteins, including transcription factors such as Myc, Jun, and Fos, which are proteins that activate the transcription of genes involved in cell division.

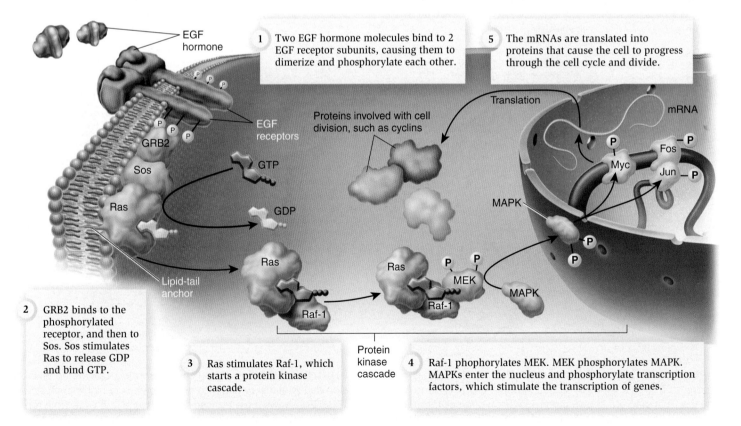

Figure 9.20   The epidermal growth factor (EGF) pathway that promotes cell division.

Biological inquiry: Certain mutations can alter the structure of the Ras protein so that it will not hydrolyze GTP. Such mutations cause cancer. Explain why.

Growth factors such as EGF cause a rapid increase in the expression of many genes in mammals, perhaps as many as 100. As we will discuss in Chapter 14, growth factor signaling pathways are often involved in cancer. Mutations that cause proteins in these pathways to become hyperactive result in cells that divide uncontrollably. These genes encode proteins that are necessary for a cell to divide, including proteins called cyclins, whose function is described next.

The Cell Cycle Is Controlled by Checkpoint Proteins

The progression through the cell cycle is a highly regulated process that ensures that the nuclear genome is intact and that the conditions are appropriate for a cell to divide. Proteins called **cyclins** and **cyclin-dependent kinases (cdks)** are responsible for advancing a cell through the phases of the cell cycle. Cyclins are so named because their amount is cyclical, fluctuating with the cell cycle. To be active, the kinases controlling the cell cycle must bind to (are dependent on) a cyclin. The number of cyclins and cdks varies from species to species.

Figure 9.21 gives a simplified description of how cyclins and cdks work together to advance a cell through G_1 and mitosis. During G_1, the amount of a particular cyclin termed G_1 cyclin increases. The G_1 cyclin binds to cdk to form an activated G_1 cyclin/cdk complex. This complex phosphorylates proteins that are needed to advance the cell to the next stage in the cell cycle. For example, certain proteins involved with DNA synthesis are phosphorylated and activated, thereby allowing the cell to carry on events in S phase. When the cell passes into the S phase, G_1 cyclin is degraded. Similar events advance the cell through other stages of the cell cycle. A different cyclin, called mitotic cyclin, accumulates late in G_2. It binds to cdk to form an activated mitotic cyclin/cdk complex. This complex phosphorylates proteins that are needed to advance into M phase.

Three critical regulatory points called **checkpoints** are found in the cell cycle of eukaryotic cells (Figure 9.21). At these checkpoints, a variety of proteins, referred to as checkpoint proteins, act as sensors to determine if a cell is in the proper condition to divide. The G_1 checkpoint, also called the restriction point, determines if conditions are favorable for cell division. In addition, G_1-checkpoint proteins can sense if the DNA has incurred damage. If so, these checkpoint proteins will prevent the formation of active cyclin/cdk complexes, and thereby stop the progression of the cell cycle. A second checkpoint exists in G_2. This checkpoint also checks the DNA for damage and ensures that all of the DNA has been replicated. In addition, the G_2 checkpoint monitors the levels of proteins that are needed to progress through M phase. A third checkpoint, called the metaphase checkpoint, senses the integrity of the spindle apparatus. As we will see in Chapter 15, the spindle apparatus is involved in chromosome sorting. Metaphase is a step in mitosis during which all of the chromosomes should be attached to the spindle apparatus. If a chromosome is not correctly attached, the metaphase checkpoint will stop the cell cycle. This checkpoint prevents cells from incorrectly sorting their chromosomes during division.

Checkpoint proteins delay the cell cycle until problems are fixed or even prevent cell division when problems cannot be fixed. A primary aim of checkpoint proteins is to prevent the division of a cell that may have incurred DNA damage or harbors abnormalities in chromosome number. When the functions of checkpoint genes are lost due to mutation, cell division may not be directly accelerated. However, as discussed in Chapter 14, the loss of checkpoint protein function increases the likelihood that undesirable genetic changes will occur that can cause mutation and cancerous growth.

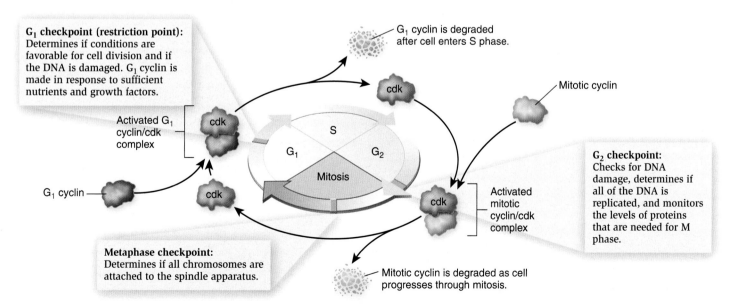

Figure 9.21 **Checkpoints in the cell cycle.** This is a general diagram of the eukaryotic cell cycle. Progression through the cell cycle requires the formation of activated cyclin/cdk complexes. There are different types of cyclin proteins, which are typically degraded after the cell has progressed to the next phase. The formation of activated cyclin/cdk complexes is regulated by checkpoint proteins.

FEATURE INVESTIGATION

Masui and Markert's Study of Oocyte Maturation Led to the Identification of Cyclin and Cyclin-Dependent Kinase

During the 1960s, researchers were intensely searching for the factors that promote cell division. In 1971, Yoshio Masui and Clement Markert developed a way to test whether a substance causes a cell to progress from one phase of the cell cycle to the next. They chose to study frog oocytes—cells produced by female frogs that develop or mature into egg cells. At the time of their work, researchers had already determined that frog oocytes naturally become dormant in the G_2 stage of the cell cycle for up to eight months (**Figure 9.22**). During mating season, female frogs produce a hormone called progesterone. After progesterone binds to receptors in dormant egg cells, they progress from G_2 to the beginning of M phase, where the chromosomes condense and become visible under the microscope. This phenomenon is called maturation. When a sperm fertilizes the egg, M phase is completed, and the zygote continues to undergo cellular divisions.

Because progesterone is a signaling molecule, Masui and Markert speculated that this hormone affects the functions and/or amounts of proteins that trigger the oocyte to undergo maturation. To test this hypothesis, they developed the procedure described in **Figure 9.23**, using the oocytes of the leopard frog (*Rana pipiens*). They began by exposing oocytes to progesterone *in vitro* and then incubating these oocytes for 2 hours or 12 hours. As a control, they also used oocytes that had not been exposed to progesterone. These three types of cells were called the donor oocytes.

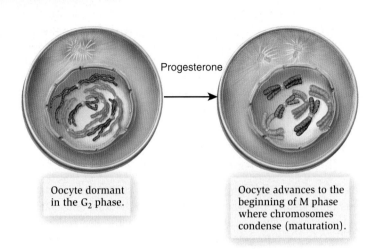

Progesterone

Oocyte dormant in the G_2 phase.

Oocyte advances to the beginning of M phase where chromosomes condense (maturation).

Figure 9.22   Oocyte maturation in certain species of frogs.

Next, they used a micropipette to transfer a small amount of cytosol from the three types of donor oocytes to recipient oocytes that had not been exposed to progesterone. As seen in the data, the recipient oocytes that had been injected with cytosol from the control donor oocytes or from oocytes that had been incubated with progesterone for only 2 hours did not progress to M phase. However, cytosol from donor oocytes that had been incubated with progesterone for 12 hours caused the recipient oocytes to advance to M phase. Masui and Markert concluded that a cytosolic factor, which required more than 2 hours to be synthesized after progesterone treatment, had been

Figure 9.23   The experimental approach used by Masui and Markert to identify cyclin and cyclin-dependent kinase (cdk).

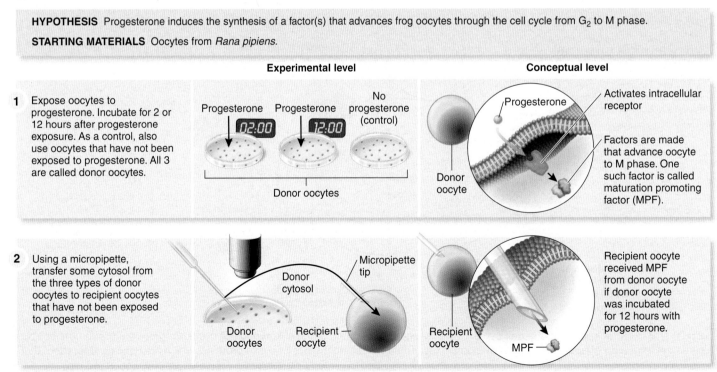

HYPOTHESIS Progesterone induces the synthesis of a factor(s) that advances frog oocytes through the cell cycle from G_2 to M phase.

STARTING MATERIALS Oocytes from *Rana pipiens*.

Experimental level   Conceptual level

1 Expose oocytes to progesterone. Incubate for 2 or 12 hours after progesterone exposure. As a control, also use oocytes that have not been exposed to progesterone. All 3 are called donor oocytes.

Progesterone   Progesterone   No progesterone (control)

02:00   12:00

Donor oocytes

Progesterone

Activates intracellular receptor

Donor oocyte

Factors are made that advance oocyte to M phase. One such factor is called maturation promoting factor (MPF).

2 Using a micropipette, transfer some cytosol from the three types of donor oocytes to recipient oocytes that have not been exposed to progesterone.

Donor cytosol

Micropipette tip

Donor oocytes   Recipient oocyte

Recipient oocyte

MPF

Recipient oocyte received MPF from donor oocyte if donor oocyte was incubated for 12 hours with progesterone.

3 Incubate for several hours and observe the recipient oocytes under the microscope to determine if the recipient oocytes advance to M phase. Advancement to M phase can be determined by the condensation of the chromosomes.

Recipient oocyte that had received cytosol containing MPF from donor oocyte.

Condensed chromosomes

4 THE DATA

Donor oocytes	Recipient oocytes proceeded to M phase?
Control, no progesterone exposure	No
Progesterone exposure, incubation for 2 hours	No
Progesterone exposure, incubation for 12 hours	Yes

transferred to the recipient oocytes and induced maturation. The factor that caused the oocytes to progress (or mature) from G_2 to M phase was originally called the **maturation promoting factor (MPF)**.

After MPF was discovered in frogs, it was found in all eukaryotic species that researchers studied. MPF is important in the division of all types of cells, not just oocytes. It took another 17 years before Manfred Lohka, Marianne Hayes, and James Maller were able to purify the components that make up MPF. This was a difficult undertaking because these components are found in very small amounts in the cytosol, and they are easily degraded during purification procedures. We now know that MPF is a complex made of mitotic cyclin and cyclin-dependent kinase (cdk), as described in Figure 9.21.

CHAPTER SUMMARY

9.1 General Features of Cell Communication

- A signal is an agent that can influence the properties of cells. Cell signaling is needed so that cells can sense environmental changes and communicate with each other.
- When a cell responds to an environmental signal, it has become adapted to its environment. (Figure 9.1)
- Cell-to-cell communication also allows cells to adapt, as when plants grow toward light. (Figure 9.2)
- Cell-to-cell communication can vary in the mechanism and distance that a signal travels. Signals are relayed between cells in five common ways: direct intercellular, contact-dependent, autocrine, paracrine, and endocrine signaling. (Figure 9.3)
- Cell signaling is usually a three-stage process involving receptor activation, signal transduction, and a cellular response. (Figure 9.4)

9.2 Cellular Receptors

- A signaling molecule, also called a ligand, binds to a receptor with an affinity that is measured as a K_d value. The binding of a ligand to a receptor is usually very specific and alters the conformation of the receptor. (Figure 9.5)
- Most receptors involved in cell signaling are found on the cell surface.
- Enzyme-linked receptors have some type of catalytic function. Many of them are protein kinases that can phosphorylate proteins. (Figure 9.6)

- G-protein-coupled receptors (GPCRs) interact with G proteins to initiate a cellular response. (Figure 9.7)
- Some receptors are ligand-gated ion channels that allow the flow of ions across cellular membranes (Figure 9.8)
- Some receptors, such as the estrogen receptor and auxin receptor, are intracellular receptors. (Figure 9.9)

9.3 Signal Transduction and the Cellular Response

- Two-component regulatory systems involve a sensor kinase and a response regulator. An example is NarQ and NarL, which regulate nitrate and nitrite utilization in bacteria. (Figure 9.10)
- Second messengers, such as cAMP, play a key role in signal transduction pathways. These pathways are reversible once the signal is degraded. (Figure 9.11)
- Second messenger pathways amplify the signal and occur with great speed. (Figures 9.12, 9.13)
- Changes in Ca^{2+} concentration play a key role in many signal transduction pathways. (Figure 9.14)
- Inositol trisphosphate (IP_3) and diacylglycerol (DAG) are other examples of second messengers involved in signal transduction. (Figure 9.15)
- Hormones such as epinephrine exert different effects throughout the body. (Figure 9.16)
- In heart muscle cells, epinephrine increases the rate of contraction due to signaling pathways that phosphorylate troponin and phospholamban. (Figure 9.17)

- The way in which any particular cell responds to a signaling molecule depends on the types of proteins that it makes. These include the types of receptors, proteins involved in signaling transduction pathways, and proteins that carry out the cellular response. The amounts of these proteins are controlled by differential gene regulation.

9.4 Regulation of the Cell Cycle

- The eukaryotic cell cycle consists of four phases called G_1 (first gap), S (synthesis of DNA), G_2 (second gap), and M phase (mitosis and cytokinesis). The G_1, S, and G_2 phases are collectively known as interphase. (Figure 9.18)

- Once a cell passes a restriction point in G_1, it is destined to duplicate its DNA and to divide. During S phase, chromosomes are replicated and form pairs of sister chromatids. (Figure 9.19)

- Signaling pathways influence whether or not a cell will divide. An example is the pathway that is stimulated by epidermal growth factor. (Figure 9.20)

- An interaction between cyclin and cyclin-dependent kinase is necessary for cells to progress through the cell cycle. Checkpoint proteins sense the environmental conditions and the integrity of the genome, and they control whether or not the cell progresses through the cell cycle. (Figure 9.21)

- Masui and Markert studied the maturation of frog oocytes to identify a substance that was necessary for oocytes to progress through the cell cycle. This substance was later identified as a complex of mitotic cyclin and cyclin-dependent kinase and called maturation promoting factor (MPF). (Figures 9.22, 9.23)

TEST YOURSELF

1. The ability of a cell to respond to changes in its environment is termed
   a. signaling.
   b. adaptation.
   c. irritability.
   d. cell communication.
   e. stimulation.

2. When a cell secretes a signaling molecule that binds to receptors on neighboring cells as well as the same cell, this is called _____ signaling.
   a. direct intercellular
   b. contact-dependent
   c. autocrine
   d. paracrine
   e. endocrine

3. Which of the following does not describe a typical cellular response to signaling molecules?
   a. activation of enzymes within the cell
   b. change in the function of structural proteins, which determine cell shape
   c. alteration of levels of certain proteins in the cell by changing the level of gene expression
   d. change in a gene sequence that encodes a particular protein
   e. all of the above are examples of cellular responses

4. A cell's ability to respond to a particular signal depends on
   a. whether or not the cell possesses the appropriate receptor for the signal.
   b. the chemical nature of the signal molecule.
   c. whether or not the signal molecule is water soluble.
   d. the concentration of ATP in the cell.
   e. none of the above.

5. _____ bind to receptors inside cells.
   a. Steroid hormones
   b. Ions
   c. Auxins
   d. All of the above
   e. Both a and c

6. Small molecules, such as cAMP, that relay signals within the cell are called
   a. secondary metabolites.
   b. ligands.
   c. G proteins.
   d. second messengers.
   e. transcription factors.

7. The benefit of second messengers in signal transduction pathways is
   a. an increase in the speed of a cellular response.
   b. duplication of the ligands in the system.
   c. amplification of the signal.
   d. all of the above.
   e. a and c only.

8. All cells of a multicellular organism may not respond in the same way to a particular signal that binds to a cell surface receptor. The difference in response may be due to
   a. the concentration of the signal molecule in the cytoplasm.
   b. the functional differences of the receptors.
   c. the structural differences that may occur in ligands.
   d. mutations that occur during development.
   e. none of the above.

9. Whether or not a cell divides depends on
   a. nutrient availability.
   b. environmental conditions.
   c. the presence of signal molecules that regulate cell division.
   d. cellular proteins that regulate cell division.
   e. all of the above.

10. Checkpoints during the cell cycle are important because they
    a. allow the organelle activity to catch up to cellular demands.
    b. ensure the integrity of the cell's DNA.
    c. allow the cell to generate sufficient ATP for cellular division.
    d. are the only time DNA replication can occur.
    e. all of the above.

CONCEPTUAL QUESTIONS

1. What are the two general reasons that cells need to communicate?

2. Explain the three stages of cell signaling.

3. What are protein kinases?

EXPERIMENTAL QUESTIONS

1. At the time of Masui and Markert's study shown in Figure 9.23, what was known about the effects of progesterone on oocytes?

2. What hypothesis did Masui and Markert propose to explain the function of progesterone? Explain the procedure used to test the hypothesis.

3. How did the researchers explain the difference between the results using 2-hour-exposed donor oocytes versus 12-hour-exposed donor oocytes?

COLLABORATIVE QUESTIONS

1. Discuss several different types of cell-to-cell communication.

2. Discuss how differential gene regulation enables various cell types to respond differently to the same signaling molecule.

www.brookerbiology.com

This website includes answers to the Biological Inquiry questions found in the figure legends and all end-of-chapter questions.

10

MULTICELLULARITY

CHAPTER OUTLINE

An oak tree.   The wood of an oak tree is largely composed of material that is outside of cells. This material connects cells to each other and provides structural support for the tree.

In unicellular species, an entire organism consists of a single cell. Prokaryotes, which include bacteria and archaea, are usually unicellular. Likewise, most protists are unicellular. As discussed in Chapter 22, multicellular organisms came into being approximately 1 billion years ago. In a species that is **multicellular**, a single organism is composed of more than one cell. Some species of protists are multicellular, as are most species of fungi. In this chapter we will focus on plants and animals, which are always multicellular organisms.

The main benefit of multicellularity arises from the division of labor between different cells of the body. For example, the intestinal cells of animals and the root cells of plants have become specialized for nutrient uptake. Other cells in a multicellular organism perform different roles, such as reproduction. In animals, most of the cells of the body are devoted to the growth, development, and survival of the organism, while specialized cells called gametes function in reproduction.

Multicellular species usually have much larger genomes than unicellular species. The increase in genome size is associated with an increase in proteome size; multicellular organisms produce a larger array of proteins than do unicellular species. The additional proteins play a role in three general phenomena. First, in a multicellular organism, cell communication is vital for the proper organization and functioning of cells. Many more proteins involved in cell signaling are made in multicellular species. Second, both the arrangement of cells within the body and the attachment of cells to each other require a greater variety of proteins in multicellular species than in unicellular species. Finally, additional proteins play a role in cell specialization because proteins that are needed for the structure and function of one cell type may not be needed in a different cell type, and vice versa.

In this chapter, we consider characteristics specific to the cell biology of multicellular organisms. We will begin by exploring the material that is produced by animal and plant cells and then secreted into the surrounding medium to form an extracellular matrix. This matrix plays many important roles in the organization and functioning of cells within multicellular organisms. We will then turn our attention to cell junctions, specialized structures that enable cells to make physical contact with one another. Cells within multicellular organisms form junctions that help to create a cohesive and well-organized body. Finally, we examine the organization and function of tissues, groups of cells that have a similar structure and function. In this chapter, we will survey the general features of tissues from a cellular perspective. Units VI and VII will explore the characteristics of plant and animal tissues in greater detail.

10.1 Extracellular Matrix

Organisms are not composed solely of cells. A large portion of an animal or plant consists of a network of material that is secreted from cells and forms a complex meshwork outside of cells called the **extracellular matrix (ECM)**. The ECM is a major component of certain parts of animals and plants. For example, bones and cartilage in animals and the woody portions of plants are composed largely of ECM. In fact, while the cells within wood eventually die, the ECM that they have produced provides a rigid structure that can support the plant for years or even centuries.

Over the past few decades, cell biologists have examined the synthesis, composition, and function of the ECM in animals and plants. In this section, we will begin by examining the structure and role of the ECM in animal cells, focusing on the function of the major ECM protein and carbohydrate macromolecules. We will then explore the cell wall, the extracellular component of plant cells, and consider how it differs in structure and function from the ECM of animal cells.

The Extracellular Matrix in Animals Supports and Organizes Cells and Plays a Role in Cell Signaling

Unlike the cells of bacteria, fungi, and plants, the cells of animals are not surrounded by a rigid cell wall that provides structure and support. However, animal cells secrete materials that form an extracellular matrix that serves a similar purpose. Certain animal cells are completely embedded within an extensive ECM, while other cells may only adhere to the ECM on one side. **Figure 10.1** illustrates the general features of the ECM and its relationship to cells. The major macromolecules of the ECM are proteins and polysaccharides. The most abundant proteins are those that form large fibers; the polysaccharides give the ECM a gel-like character.

As we will see, the ECM found in animals performs many important roles, including strength, structural support, organization, and cell signaling.

- **Strength:** The ECM is the "tough stuff" of animals' bodies. In the skin of mammals, the strength of the ECM prevents tearing. The ECM found in cartilage resists compression and provides protection to the joints. Similarly, the ECM protects the soft parts of the body, such as the internal organs.

- **Structural support:** The skeletons of many animals are composed primarily of ECM. Skeletons not only provide structural support but also facilitate movement via the functioning of the attached muscles.

- **Organization:** The attachment of cells to the ECM plays a key role in the proper arrangement of cells throughout the body. In addition, the ECM binds many body parts together, such as tendons to bones.

- **Cell signaling:** A newly discovered role of the ECM is cell signaling. One way that cells in multicellular organisms sense their environment is via changes in the ECM.

Let's now consider the synthesis and structure of ECM components found in animals.

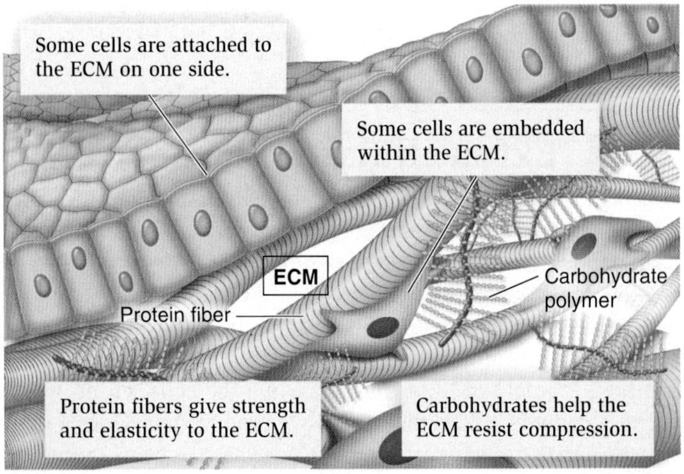

Adhesive and Structural Proteins Are Major Components of the ECM of Animals

The idea that fibers are important components of living organisms has a long history. In the Middle Ages, living organisms were thought to be composed of fibers rather than cells. This belief was eventually debunked by the cell theory, which is described in Chapter 4. In the 1850s, Rudolf Virchow, who proposed that all cells come from pre-existing cells, also suggested that all extracellular materials are made and secreted by cells. Around the same time, biologists realized that gelatin and glue, which are produced by the boiling of animal tissues, must contain a common fibrous substance. This substance was named **collagen**, from the Greek for "glue producing." Since that time, the advent of experimental techniques in the areas of chemistry, microscopy, and biophysics has enabled scientists to probe the structure of the ECM. We now understand that the ECM contains a mixture of several different components, including proteins such as collagen, that form fibers.

The proteins that are found in the ECM can be grouped into adhesive proteins such as fibronectin and laminin, and structural proteins, such as collagen and elastin (**Table 10.1**). Fibronectin and laminin have multiple binding sites that bind to other components in the ECM such as protein fibers and carbohydrates. These same proteins also have binding sites for receptors on the surfaces of cells. Therefore, adhesive proteins are so named because they adhere cells to the ECM.

Structural proteins such as collagen and elastin form large fibers that give the ECM its strength and elasticity. As described in Chapter 6, proteins that are secreted from eukaryotic cells

Some cells are attached to the ECM on one side.

Some cells are embedded within the ECM.

ECM

Protein fiber

Carbohydrate polymer

Protein fibers give strength and elasticity to the ECM.

Carbohydrates help the ECM resist compression.

Figure 10.1   Animal cells and the extracellular matrix (ECM).

Table 10.1	Proteins in the ECM of Animals	
General type	**Example**	**Function**
Adhesive	Fibronectin	Connects cells to the ECM and helps to organize components in the ECM.
	Laminin	Connects cells to the ECM and helps to organize components in the basal lamina, a specialized ECM found next to epithelial cells (described in Section 10.3).
Structural	Collagen	Forms large fibers and interconnected fibrous networks in the ECM. Provides tensile strength.
	Elastin	Forms elastic fibers in the ECM that can stretch and recoil.

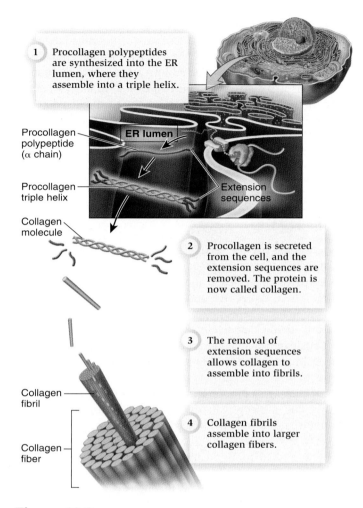

1. Procollagen polypeptides are synthesized into the ER lumen, where they assemble into a triple helix.

ER lumen

Procollagen polypeptide (α chain)

Procollagen triple helix

Extension sequences

Collagen molecule

2. Procollagen is secreted from the cell, and the extension sequences are removed. The protein is now called collagen.

3. The removal of extension sequences allows collagen to assemble into fibrils.

Collagen fibril

4. Collagen fibrils assemble into larger collagen fibers.

Collagen fiber

Figure 10.2   Formation of collagen fibers.

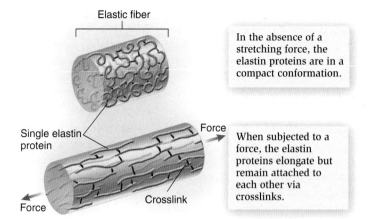

Elastic fiber

Single elastin protein

In the absence of a stretching force, the elastin proteins are in a compact conformation.

Force

When subjected to a force, the elastin proteins elongate but remain attached to each other via crosslinks.

Force

Crosslink

Figure 10.3   Structure and function of elastic fibers.

Biological inquiry: Let's suppose you started with an unstretched elastic fiber and treated it with a chemical that breaks the cross-links between adjacent elastin proteins. What would happen when the fiber is stretched?

are first directed to the endoplasmic reticulum (ER), then to the Golgi apparatus, and subsequently are secreted from the cell via vesicles that fuse with the plasma membrane. **Figure 10.2** depicts the synthesis and assembly of collagen. Individual procollagen polypeptides are first synthesized into the lumen (inside) of the ER. Three procollagen polypeptides then associate with each other to form a procollagen triple helix. The amino acid sequences at both ends of the polypeptides, termed extension sequences, serve to promote the formation of procollagen and prevent the formation of a much larger fiber. After procollagen is secreted out of the cell, extracellular enzymes remove the extension sequences. Once this occurs, the protein, now called collagen, can form larger structures. Collagen proteins assemble in a staggered way to form relatively thin collagen fibrils, which then align and create large collagen fibers.

A key function of collagen is to provide tensile strength, which is a measure of how much stretching force a material can bear without tearing apart. Collagen provides high tensile strength to many parts of the animal body. Collagen is the main protein found in bones, cartilage, tendons, and skin and is also found lining blood vessels and internal organs. In mammals, collagen consists of more than 25% of the total protein mass, much more than any other protein. Approximately 75% of the protein in

mammalian skin is composed of collagen. Leather is largely a pickled and tanned form of collagen.

In addition to tensile strength, elasticity is needed in regions of the body such as the lungs and blood vessels, which regularly expand and return to their original shape. In these places, the ECM contains an abundance of elastic fibers composed primarily of a protein called **elastin** (**Figure 10.3**). Elastin proteins form many covalent cross links to create a fiber. In the absence of a stretching force, each protein tends to adopt a compact conformation. When subjected to a stretching force, the compact proteins become more linear, with the covalent cross links holding the fiber together. When the stretching force has ended, the proteins naturally return to their compact conformation. In this way, elastic fibers behave much like a rubber band, stretching under tension and snapping back when the tension is released.

GENOMES & PROTEOMES

Collagens Are a Family of Proteins That Give Animal Cells a Variety of ECM Properties

As we have seen, proteins are important constituents of the ECM of animals. By analyzing genomes and the biochemical composition of cells, researchers have determined that many different types of collagen fibers are made. These are designated type I, type II, and so on. At least 27 different types of collagens have been identified in humans. Therefore, the human genome, as well as the genomes of other animals, has many different genes that encode collagen polypeptides.

Collagens have a common structure, in which three collagen polypeptides wind around each other to form a triple helix (see Figure 10.2). Each polypeptide is called an α chain. In some collagens, all three α chains are identical, while in others the α chains may be encoded by different collagen genes. Nevertheless, the triple helix structure remains common to all collagen proteins.

Each of the many different types of collagen polypeptides has a similar yet distinctive amino acid sequence that affects the structure and function of collagen fibers. For example, within the triple helix, the amino acid sequence of an α chain may cause collagen proteins to bind to each other very tightly, thereby creating a relatively stiff fiber. Such collagen fibers are found in bone and cartilage. In addition, amino acid side chains in the α chains influence the interactions between collagen proteins within a fiber. For example, the amino acid sequences of certain α chains may promote a looser interaction that produces a more bendable or thin fiber. More flexible collagen fibers support the lining of your lungs and intestines. In addition, domains within the collagen polypeptide may affect the spatial arrangement of collagen proteins. The collagen shown earlier in Figure 10.2 forms fibers in which the collagen proteins align themselves in parallel arrays. Not all collagen proteins form long fibers. For example, type IV collagen proteins interact with each other in a meshwork pattern. This meshwork acts as a filtration unit around capillaries.

Differential gene regulation controls which types of collagens are made throughout the body and in what amounts. Of the 27 types of collagens, Table 10.2 considers types I to IV, each of which varies with regard to where it is primarily synthesized and its structure and function. Collagen genes are regulated so that collagen is made in the correct sites of the body. In skin cells, for example, the genes that encode the collagen polypeptides that make up collagen types I, III, and IV are turned on, while the synthesis of type II collagen is minimal.

The regulation of collagen synthesis has received a great deal of attention due to the phenomenon of wrinkling. As we age, the amount of collagen that is synthesized in our skin significantly decreases. The underlying network of collagen fibers, which provides scaffolding for the surface of our skin, loosens and unravels. This is one of the factors that cause the skin of older people to sink, sag, and form wrinkles. Various therapeu-

tic and cosmetic agents have been developed to prevent or reverse the appearance of wrinkles, most with limited benefits. One approach is collagen injections, in which small amounts of collagen (from cows) are injected into areas where the body's collagen has weakened, filling the depressions to the level of the surrounding skin. Because collagen is naturally broken down in the skin, the injections are not permanent and last only about three to six months.

Animal Cells Also Secrete Polysaccharides into the ECM

Polysaccharides are the second major component of the extracellular matrix of animals. Among vertebrates, the most abundant types of polysaccharides in the ECM are **glycosaminoglycans (GAGs)**. These molecules are long, unbranched polysaccharides containing a repeating disaccharide unit (**Figure 10.4a**). GAGs are highly negatively charged molecules that tend to attract positively charged ions and water. The majority of GAGs in the ECM are linked to core proteins, forming **proteoglycans** (**Figure 10.4b**).

Providing resistance to compression is the primary function of GAGs and proteoglycans. Once secreted from cells, these macromolecules form a gel-like component in the ECM. Due to its high water content, the ECM is difficult to compress and thereby serves to protect cells. GAGs and proteoglycans are abundantly found in regions of the body that are subjected to harsh mechanical forces, such as the joints of the human body.

Table 10.2	Examples of Collagen Types	
Type	**Sites of synthesis***	**Structure and function**
I	Tendons, ligaments, bones, and skin	Forms a relatively rigid and thick fiber. Very abundant, provides most of the tensile strength to the ECM.
II	Cartilage, disks between vertebrae	Forms a fairly thick and rigid fiber but is more flexible than type I. Permits smooth movements of joints.
III	Arteries, skin, internal organs, and around muscles	Forms thin fibers, often arranged in a netlike pattern. Allows for greater elasticity in tissues.
IV	Skin, intestine, and kidneys; also found around capillaries	Does not form long fibers. Instead, the proteins are arranged in a meshwork pattern that provides organization and support to cell layers. Functions as a filter around capillaries.

*The sites of synthesis denote where a large amount of the collagen type is made.

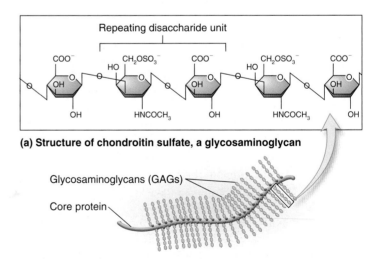

(a) Structure of chondroitin sulfate, a glycosaminoglycan

(b) General structure of a proteoglycan

Figure 10.4 Structures of glycosaminoglycans and proteoglycans. **(a)** Glycosaminoglycans (GAGs) are composed of repeating disaccharide units. The length can range from several dozen to 25,000 disaccharide units. The GAG shown here is chondroitin sulfate, which is commonly found in cartilage. **(b)** Proteoglycans are composed of a long, linear core protein with many GAGs attached. Each GAG is typically 80 disaccharide units long.

Two examples of GAGs are chondroitin sulfate, which is a major component of cartilage, and hyaluronic acid, which is found in the skin, eyes, and joint fluid.

Among many invertebrates, an important ECM component is **chitin**, a nitrogen-containing polysaccharide. Chitin forms the hard protective outer covering (called an exoskeleton) of insects, such as crickets and grasshoppers, and shellfish, such as lobsters and shrimp. In fact, the chitin exoskeleton is so rigid that as these animals grow, they must periodically shed this outer layer and secrete a new, larger one—a process called molting (look ahead to Figure 32.11).

The Cell Wall of Plants Provides Strength and Resistance to Compression

Let's now turn our attention to the extracellular matrix of plants. Plants cells are surrounded by an ECM called the **cell wall**, a protective layer that forms outside of the plasma membrane of the plant cell. Like animal cells, the cells of plants are surrounded by material that provides tensile strength and resistance to compression. The cell walls of plants, however, are usually thicker, stronger, and more rigid than the ECM found in animals. Plant cell walls provide rigidity for mechanical support, the maintenance of cell shape, and the direction of cell growth. As we learned in Chapter 5, the cell wall also prevents expansion when water enters the cell. During the evolution of plants, these structural features of the plant cell wall may have been key factors that caused plants to thrive in a sedentary lifestyle.

The cell walls of plants are composed of a primary cell wall and a secondary cell wall, so named according to the timing of their synthesis (**Figure 10.5**). The **primary cell wall** is made before the secondary cell wall. During cell division, the primary cell wall develops between two newly made daughter cells. It is usually very flexible and allows new cells to increase in size. The main macromolecule of the primary cell wall is **cellulose**, a polymer made of repeating molecules of glucose attached end to end. These glucose polymers associate with each other via hydrogen bonding to form microfibrils that provide great tensile strength (**Figure 10.6**).

Cellulose was discovered in 1838 by the French chemist Anselme Payen, who was the first scientist to try to separate wood into its component parts. After treating different types of wood with nitric acid, Payen obtained a fibrous substance that was also found in cotton and other plants. His chemical analysis revealed that the fibers were made of the carbohydrate glucose. Payen called this substance cellulose, which means consisting of cells. Cellulose is probably the single most abundant organic molecule on Earth. Wood consists mostly of cellulose, and cotton and paper (including the page you are reading now) are almost pure cellulose.

In addition to cellulose, other components found in the primary cell wall include hemicellulose, glycans, and pectins (see Figure 10.5). Hemicellulose is another linear carbohydrate with a structure similar to that of cellulose, but it contains sugars other than glucose in its structure and usually forms thinner microfibrils. Glycans, carbohydrates with branching structures, are also important in cell wall structure. The cross-linking glycans bind to cellulose and provide organization to the cellulose microfibrils. Pectins, which are highly negatively charged, attract water and have a gel-like character that provides the cell wall with the ability to resist compression. Besides carbohydrates, the primary cell wall also contains small amounts of protein. Some of these proteins may increase the strength of the cell wall, while others are enzymes involved in the synthesis and organization of the carbohydrate polymers. The mechanism of cellulose synthesis via extracellular enzymes is described in Chapter 30.

The **secondary cell wall** is synthesized and deposited between the plasma membrane and the primary cell wall after a plant cell matures and has stopped increasing in size. It is made in layers by the successive deposition of cellulose microfibrils and other components. While the primary wall structure is relatively similar in nearly all cell types and species, the structure of the secondary cell wall is more variable. The secondary cell wall

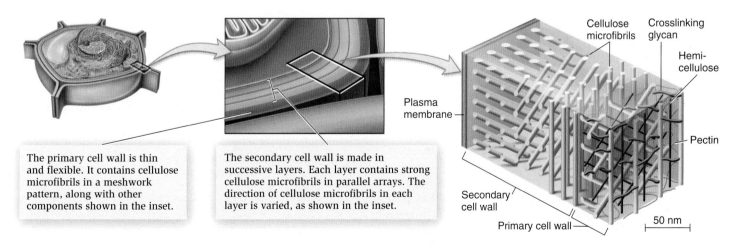

The primary cell wall is thin and flexible. It contains cellulose microfibrils in a meshwork pattern, along with other components shown in the inset.

The secondary cell wall is made in successive layers. Each layer contains strong cellulose microfibrils in parallel arrays. The direction of cellulose microfibrils in each layer is varied, as shown in the inset.

Cellulose microfibrils
Crosslinking glycan
Hemi-cellulose
Plasma membrane
Pectin
Secondary cell wall
Primary cell wall
50 nm

Figure 10.5 **Structure of the plant cell wall.** The primary cell wall is relatively thin and flexible. The secondary cell wall, which is produced only by certain plant cells, is made after the primary cell wall and is synthesized in successive layers.

Biological inquiry: With regard to cell growth, what would happen if the secondary cell wall was made too soon?

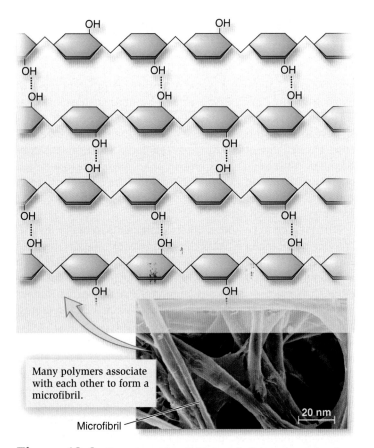

Many polymers associate with each other to form a microfibril.

Microfibril

20 nm

Figure 10.6 **Structure of cellulose.** Cellulose is made of repeating glucose units linked end to end that hydrogen-bond to each other to form microfibrils (SEM).

Table 10.3	Common Types of Cell Junctions
Type	**Description**
Animals	
Anchoring junctions	Cell junctions that hold adjacent cells together or bond cells to the ECM. Anchoring junctions are mechanically strong.
Tight junctions	Junctions between adjacent cells that prevent the leakage of material between cell layers.
Gap junctions	Channels that permit the direct exchange of ions and small molecules between the cytosol of adjacent cells.
Plants	
Middle lamella	A carbohydrate layer that cements together the cell walls of adjacent cells.
Plasmodesmata	Open passageways between the cell walls of adjacent cells that permit the direct diffusion of ions and molecules between cells.

often contains components in addition to those found in the primary cell wall. For example, phenolic compounds called lignins are very hard and impart considerable strength to the secondary wall structure. Lignin, a type of secondary metabolite described in Chapter 7, is found in the woody parts of plants.

10.2 Cell Junctions

Thus far, we have learned that the cells of animals and plants create an extracellular matrix that provides strength, support, and organization. For an organism to become a multicellular unit, cells within the body must be linked to each other and to the ECM. In animals and plants, this is accomplished by specialized structures called **cell junctions** (Table 10.3). In this section, we will examine different types of cell junctions in animal and plant cells.

Animal cells, which lack the structural support provided by the cell wall, have a more varied group of junctions than plant cells. In animal cells, junctions called anchoring junctions play a role in anchoring cells to each other or to the extracellular matrix. In other words, they hold cells in their proper place in the body. Other junctions, termed tight junctions, seal cells together to prevent small molecules from leaking through one

cell layer into another. Still another type of junction known as a gap junction allows cells to communicate directly with each other. In this section, we will examine all three of these types of junctions.

In plants, cellular organization is somewhat different because plants cells are surrounded by a rigid cell wall. As you will learn, plants cells are connected to each other by a component called the middle lamella, which cements their cell walls together. They also have junctions termed plasmodesmata that allow adjacent cells to communicate with each other.

Anchoring Junctions Link Animal Cells to Each Other and to the ECM

The advent of electron microscopy allowed researchers to explore the types of junctions that occur between cells and within the extracellular matrix. In the 1960s, Marilyn Farquhar, George Palade, and colleagues conducted several studies showing that various types of cellular junctions connect cells to each other. Over the past few decades, researchers have begun to unravel the functions and molecular structures of these junctions, called **anchoring junctions**, which attach cells to each other and to the extracellular matrix. Anchoring junctions rely on the functioning of membrane proteins called **cell adhesion molecules (CAMs)**. Two types of CAMs are cadherin and integrin.

Anchoring junctions are grouped into four main categories, according to their functional roles and their connections to cellular components (**Figure 10.7**)

1. **Adherens junctions** connect cells to each other via cadherins. In many cases, these junctions are organized into bands around cells. In the cytosol, adherens junctions bind to cytoskeletal filaments called actin filaments.

2. **Desmosomes** also connect cells to each other via cadherins. They are spotlike points of intercellular contact that rivet

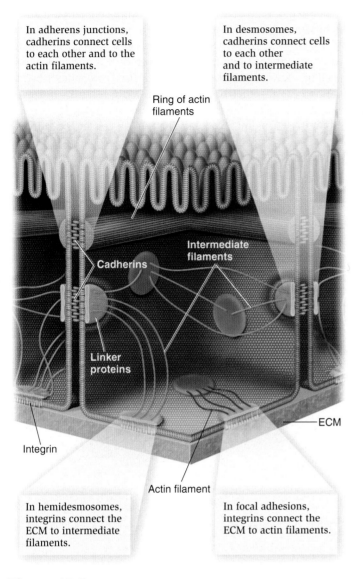

In adherens junctions, cadherins connect cells to each other and to the actin filaments.

In desmosomes, cadherins connect cells to each other and to intermediate filaments.

Ring of actin filaments

Intermediate filaments

Cadherins

Linker proteins

ECM

Integrin

Actin filament

In hemidesmosomes, integrins connect the ECM to intermediate filaments.

In focal adhesions, integrins connect the ECM to actin filaments.

Figure 10.7   Types of anchoring junctions.

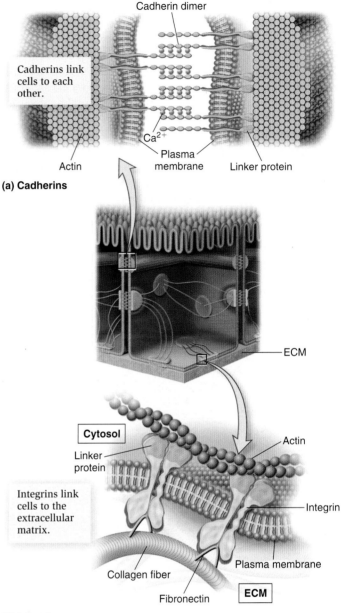

Cadherin dimer

Cadherins link cells to each other.

Ca^{2+}

Actin   Plasma membrane   Linker protein

(a) Cadherins

ECM

Cytosol   Actin

Linker protein

Integrins link cells to the extracellular matrix.

Integrin

Collagen fiber   Plasma membrane

Fibronectin   ECM

(b) Integrins

Figure 10.8   Types of cell adhesion molecules (CAMs). **(a)** A cadherin in one cell binds to a cadherin of an identical type in an adjacent cell. The binding requires Ca^{2+}. In the cytosol, cadherins bind to the actin or intermediate filaments of the cytoskeleton. **(b)** Integrins link cells to the extracellular matrix and form intracellular connections to actin or intermediate filaments.

cells together. Desmosomes are connected to cytoskeletal filaments called intermediate filaments.

3. **Focal adhesions** connect cells to the extracellular matrix via integrins. In the cytosol, focal adhesions bind to actin filaments.

4. **Hemidesmosomes** also connect cells to the extracellular matrix via integrins. Like desmosomes, they interact with intermediate filaments.

Let's now consider the molecular components of anchoring junctions. **Cadherins** are CAMs that create cell-to-cell junctions (**Figure 10.8a**). Each cadherin is a dimer of identical subunits. The extracellular domains of two cadherin dimers, each in adjacent cells, bind to each other to promote cell-to-cell adhesion. This binding requires the presence of calcium ions, which change the conformation of cadherin, so that cadherins in adjacent cells can bind to each other. On the inside of the cell, linker proteins connect cadherins to actin or intermediate filaments

of the cytoskeleton. This promotes a more stable interaction between two cells because their strong cytoskeletons are connected to each other.

Cadherins are the major CAMs in vertebrate species. The genomes of vertebrates contain several different cadherin genes, which encode slightly different cadherin proteins. Having different types of cadherins allows different types of cells to recognize each other. Dimer formation follows a homophilic, or like-to-like, binding mechanism. To understand the concept of homophilic binding, let's consider an example. One cadherin is called E-cadherin and another is N-cadherin. E-cadherin in one cell will bind to E-cadherin but not to N-cadherin in an adjacent cell. Similarly, N-cadherin will bind to N-cadherin but not to E-cadherin in an adjacent cell. By expressing only certain types of cadherins, each cell will bind only to other cells that express the same cadherin types. This phenomenon is important in the proper arrangement of cells throughout the body.

Integrins, a group of cell-surface receptor proteins, are a second type of CAM, one that creates connections between cells and the extracellular matrix. In the example shown in **Figure 10.8b**, an integrin is bound to fibronectin, an ECM protein that binds to other ECM components such as collagen fibers. Like cadherins, integrins also bind to actin or intermediate filaments in the cytosol of the cell, via linker proteins, to promote a strong association between the cytoskeleton of a cell and the extracellular matrix. Thus, integrins have an extracellular domain for the binding of ECM components and an intracellular domain for the binding of cytosolic proteins.

When these CAMs were first discovered, researchers imagined that cadherins and integrins played only a mechanical role in cell biology. In other words, their functions were described as holding cells together or to the ECM. More recently, however, experiments have shown that cadherins and integrins are important in cell communication. When cell-to-cell and cell-to-ECM junctions are formed or broken, this affects signaling pathways within the cell. Similarly, intracellular signaling pathways can affect cadherins and integrins in ways that alter intercellular junctions and the binding of cells to ECM components.

With regard to cell signaling, integrins are particularly interesting because they are capable of both outside-in and inside-out signaling (**Figure 10.9**). Integrins are so named because they integrate changes in the ECM to changes in the cytoskeleton. When the extracellular domain of an integrin binds to components of the ECM, this causes a conformational change

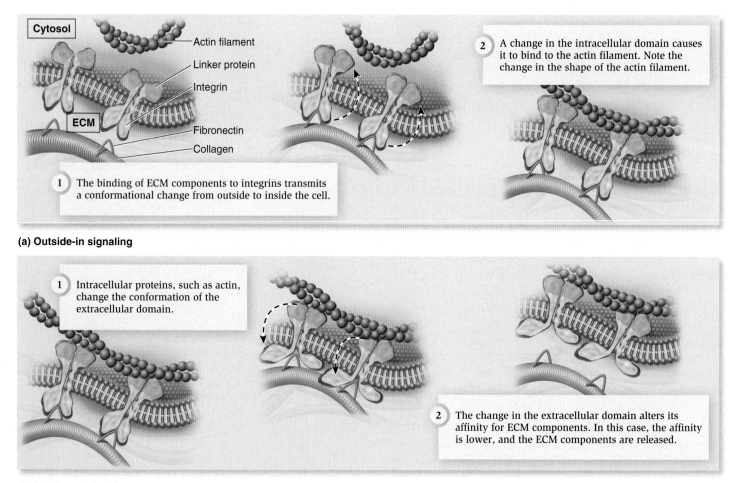

(a) Outside-in signaling

Cytosol

Actin filament
Linker protein
Integrin

ECM

Fibronectin
Collagen

1 The binding of ECM components to integrins transmits a conformational change from outside to inside the cell.

2 A change in the intracellular domain causes it to bind to the actin filament. Note the change in the shape of the actin filament.

1 Intracellular proteins, such as actin, change the conformation of the extracellular domain.

2 The change in the extracellular domain alters its affinity for ECM components. In this case, the affinity is lower, and the ECM components are released.

(b) Inside-out signaling

Figure 10.9 Cell signaling via integrins.   (a) Outside-in signaling occurs when an integrin binds to a component in the ECM, which transmits a signal to the cytosol, thereby affecting activities inside the cell. (b) Inside-out signaling occurs when the cytosol affects the structure of an integrin and thereby changes the integrin's ability to bind to components in the ECM. In the example shown here, the effect is to lower the affinity for an ECM component, causing it to be released from the integrin.

Biological inquiry: When an animal receives a wound, this causes outside-in signaling. Explain how.

that affects the structure of the intracellular domain, altering its interaction with cytoskeletal proteins. This outside-in signaling regulates cell adhesion (the ability of cells to adhere to the ECM), cell growth, and cell migration. In addition, signals generated inside the cell can alter integrins and affect their ability to bind to components in the ECM, either lowering or increasing their affinity. This phenomenon, termed inside-out signaling, is also important for cell adhesion and migration, and it provides a way for cells to contribute to the organization of the ECM.

Abnormalities in CAMs such as integrins are often associated with the ability of cancer cells to metastasize, that is, to move to other parts of the body. Cell adhesion molecules are critical for keeping cells in their correct locations. When their function becomes abnormal due to cancer-causing mutations, cells lose their proper connections with the ECM and adjacent cells and may spread to other parts of the body. This topic is considered in more detail in Chapter 14.

Tight Junctions Prevent the Leakage of Materials Across Animal Cell Layers

In animals, **tight junctions**, or occluding junctions, are a second type of junction, one that forms a tight seal between adjacent cells and thereby prevents extracellular material from leaking between cells. As an example, let's consider the intestine. The cells that line the intestine form a sheet that is one cell thick; one side faces the intestinal lumen, while the other faces the blood (**Figure 10.10**). Tight junctions between these cells ensure that nutrients pass through the plasma membranes of the intestinal cells before entering the blood and prevent the transport of materials from the blood into the intestine.

Tight junctions are made by membrane proteins, called occludin and claudin, that form interlaced strands in the plasma membrane (see inset to Figure 10.10). These strands of proteins,

each in adjacent cells, bind to each other and thereby form a tight seal between cells. Occluding junctions are not mechanically strong like anchoring junctions because they do not have strong connections with the cytoskeleton. Therefore, adjacent cells that have occluding junctions also have anchoring junctions to hold them in place.

The amazing ability of tight junctions to prevent the leakage of material across cell layers has been demonstrated by dye-injection studies. In 1972, Daniel Friend and Norton Gilula injected lanthanum, which is electron dense and can be visualized under the electron microscope, into the bloodstream of a rat. A few minutes later, a sample of a cell layer in the digestive tract was removed and visualized by electron microscopy. As seen in the micrograph in **Figure 10.11**, lanthanum diffused into the region between the cells that faces the blood, but it could not move past the tight junction to the side of the cell layer facing the lumen of the intestine.

Gap Junctions in Animal Cells Provide a Passageway for Intercellular Transport

A third type of junction found in animals is called a **gap junction** because a small gap occurs between the plasma membranes of cells connected by these junctions (**Figure 10.12**). In vertebrates, gap junctions are composed of a membrane protein called connexin. Invertebrates have a structurally similar protein called innexin. Six connexin proteins in one cell align with six connexin proteins in an adjacent cell to form a channel called a **connexon** (see inset to Figure 10.12).

The connexons allow the passage of ions and small molecules with a molecular mass that is less than 1,000 Daltons,

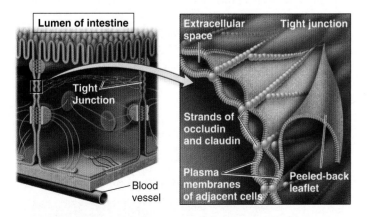

Figure 10.10 **Tight junctions between adjacent intestinal cells.** In this example, tight junctions form a seal that prevents the movement of material between cells, from the intestinal lumen into the blood, and vice versa. The inset shows the interconnected network of occludin and claudin that forms the tight junction.

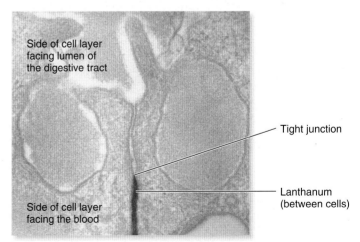

Figure 10.11 **An experiment demonstrating the function of a tight junction.** When lanthanum was injected into the bloodstream of a rat, it diffused between the cells in the region up to a tight junction, but could not diffuse past the junction to the other side of the cell layer.

Biological inquiry: What results would you expect if a rat was fed lanthanum and then a sample of intestinal cells was observed under the EM?

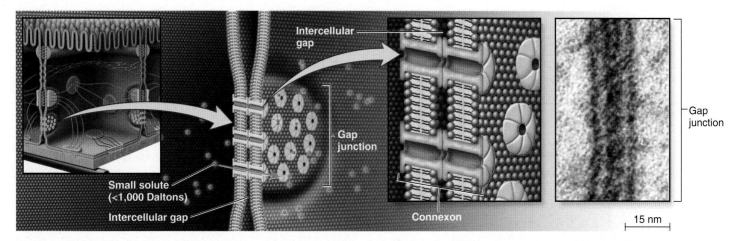

Figure 10.12   **Gap junctions between adjacent cells.**   Gap junctions form intercellular channels that allow the passage of small solutes with masses less than 1,000 Daltons. A transmembrane channel called a connexon consists of 12 proteins called connexins, 6 in each cell. The micrograph shows a gap junction between intestinal cells.

including amino acids, sugars, and signaling molecules like cAMP and IP_3. In this way, gap junctions allow adjacent cells to share metabolites and directly signal each other. At the same time, gap-junction channels are too small to allow the passage of RNA, proteins, or large carbohydrates. Therefore, cells that communicate via gap junctions still maintain their own distinctive set of macromolecules.

Because gap junctions allow the passage of ions, electrical changes in one cell are easily transmitted to an adjacent cell that is connected via gap junctions. In 1959, Edwin Furshpan and David Potter first postulated the existence of gap junctions. Their results, which showed that certain cells in the crayfish are electrically coupled, indicated that ions can directly move from the cytosol of one cell to the cytosol of an adjacent cell.

FEATURE INVESTIGATION

Loewenstein and Colleagues Followed the Transfer of Fluorescent Dyes to Determine the Size of Gap-Junction Channels

As we have seen, gap junctions allow the passage of small molecules, those with a mass up to about 1,000 Daltons. This property of gap junctions was determined in experiments involving the transfer of fluorescent dyes. During the 1960s, several research groups began using fluorescent dyes to study cell mor-

phology and function. As discussed in Chapter 4, the location of fluorescent dyes within cells can be seen via fluorescence microscopy. In 1964, Werner Loewenstein and colleagues observed that a fluorescent dye could move from one cell to an adjacent cell, which prompted them to investigate this phenomenon further.

In the experiment shown in **Figure 10.13**, Loewenstein and colleagues grew rat liver cells in the laboratory, where they formed a single layer (a monolayer). The adjacent cells formed

Figure 10.13   Use of fluorescent molecules to determine the size of gap-junction channels.

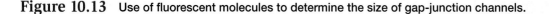

HYPOTHESIS Gap-junction channels allow the passage of ions and molecules, but there is a limit to how large the molecules can be.

STARTING MATERIALS Rat liver cells grown in laboratory.

Experimental level   Conceptual level

1   Grow rat liver cells in a laboratory on solid growth media until they become a monolayer. At this point, adjacent cells have formed gap junctions.

Tissue culture bottle

Gap junction

Rat liver cells

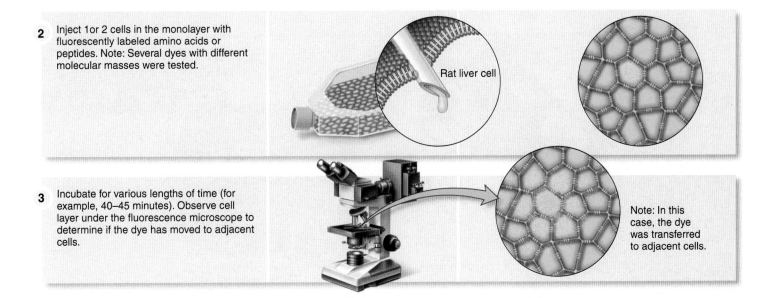

2 Inject 1or 2 cells in the monolayer with fluorescently labeled amino acids or peptides. Note: Several dyes with different molecular masses were tested.

Rat liver cell

3 Incubate for various lengths of time (for example, 40–45 minutes). Observe cell layer under the fluorescence microscope to determine if the dye has moved to adjacent cells.

Note: In this case, the dye was transferred to adjacent cells.

4 THE DATA

Mass of Dye	Transfer to adjacent cells *	Mass of Dye	Transfer to adjacent cells *
376	+ + + +	851**	−
464	+ + + +	901	+ + +
536	+ + +	946	−
559	+ + + +	1004	−
665	+	1158	−
688	+ + + +	1678	−
817	+ + +	1830	−

* The number of pluses indicates the relative speed of transfer. Four pluses denote fast transfer, while one plus is slow transfer. A minus indicates that transfer between cells did not occur. ** In some cases, molecules with less mass did not pass between cells compared to molecules with a higher mass. This may be due to differences in their structures (e.g., charges) that influence whether or not they can easily penetrate the channel.

gap junctions. The cells were injected with various dyes composed of fluorescently labeled amino acids or peptide molecules with different masses and then observed via fluorescence microscopy. As shown in the data, dyes with a molecular mass up to 901 Daltons were observed to pass from cell to cell. Larger dyes, however, did not move intercellularly. Loewenstein and other researchers subsequently investigated dye transfer in other cell types and species. Though some variation is found when comparing different cell types and species, the researchers generally observed that molecules with a mass greater than 1,000 Daltons do not pass through gap junctions.

The Middle Lamella Cements Adjacent Plant Cell Walls Together

In animal cells, we have seen that cell-to-cell contact, via anchoring junctions, tight junctions, and gap junctions, involves interactions between membrane proteins in adjacent cells. In plants, cell junctions are quite different. Rather than using membrane proteins to form cell-to-cell connections, an additional component in the ECM is made called the **middle lamella**. When plant cells are dividing, the middle lamella is the first layer that is formed. Next, the primary cell wall is made inside the middle lamella (**Figure 10.14**). The middle lamella is rich in pectins, negatively charged carbohydrate polymers that are also found in the primary cell wall (see Figure 10.5). These polymers attract water and make a hydrated gel. Ca^{2+} and Mg^{2+} interact with the negative charges in the carbohydrates and cement the cell walls of adjacent cells together.

The process of fruit ripening illustrates the importance of pectins in holding plant cells together. An unripened fruit, such as a green tomato, is very firm because the rigid cell walls of adjacent cells are firmly attached to each other. During ripening, the cells secrete a group of enzymes called pectinases, which digest pectins in the middle lamella as well as pectins in the primary cell wall. As this process continues, the attachments between cells are broken, and the cell walls become less rigid. For this reason, a red ripe tomato is much less firm than an unripe tomato.

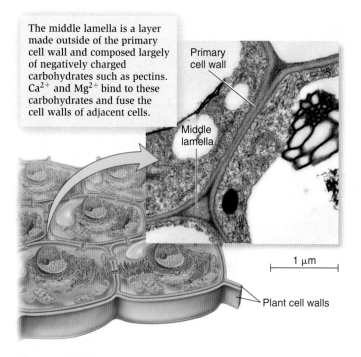

The middle lamella is a layer made outside of the primary cell wall and composed largely of negatively charged carbohydrates such as pectins. Ca^{2+} and Mg^{2+} bind to these carbohydrates and fuse the cell walls of adjacent cells.

Primary cell wall

Middle lamella

1 μm

Plant cell walls

Figure 10.14   Cell-to-cell junctions known as middle lamella.

Plasmodesmata Are Channels Connecting the Cytoplasm of Adjacent Plant Cells

In 1879, Eduard Tangl, a Russian botanist, observed intercellular cytoplasmic connections in the seeds of the strychnine tree and hypothesized that the cytoplasm of adjacent cells is connected by ducts in the cell walls. He first proposed the concept that direct cell-to-cell communication integrates the functioning of plant cells. The ducts or intercellular channels that Tangl observed are now known as **plasmodesmata** (singular, *plasmodesma*).

Plasmodesmata are functionally similar to gap junctions in animal cells, in that they allow the passage of ions, water, sugars, amino acids, and signaling molecules between cells. However, the structure of plasmodesmata is quite different from that of gap junctions. As shown in **Figure 10.15**, plasmodesmata are open channels in the cell walls of adjacent cells. At these sites, the plasma membrane of one cell is continuous with the plasma membrane of the other cell, which permits the diffusion of molecules from the cytosol of one cell to the cytosol of the other. In addition to a cytosolic connection, plasmodesmata also have a central tubule, called a desmotubule, connecting the ER membranes of adjacent cells.

10.3 Tissues

A **tissue** is a part of an animal or plant consisting of a group of cells having a similar structure and function. Because tissues are fundamental units within multicellular organisms, understanding the characteristics of tissues is essential to many areas of biology, particularly plant and animal development. For example,

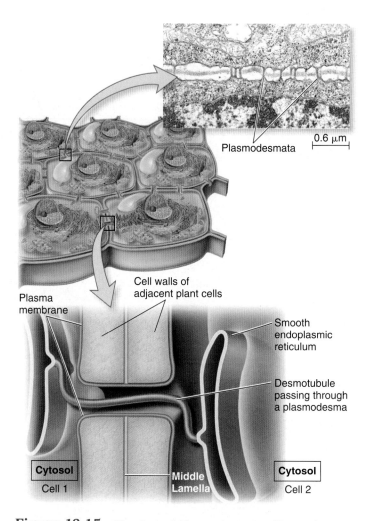

Plasmodesmata

0.6 μm

Plasma membrane

Cell walls of adjacent plant cells

Smooth endoplasmic reticulum

Desmotubule passing through a plasmodesma

Cytosol   Cell 1

Middle Lamella

Cytosol   Cell 2

Figure 10.15   Structure of plasmodesmata. Plasmodesmata are cell junctions connecting the cytosol of adjacent plant cells, allowing water and small molecules to pass from cell to cell. At these sites, the plasma membrane of one cell is continuous with the plasma membrane of an adjacent cell. In addition, the ER from one cell is connected to that of the adjacent cell via a desmotubule.

geneticists want to know how genes play a role in the formation and arrangement of tissues to create an entire organism. Likewise, plant and animal biologists are interested in understanding the genetic and morphological changes that occur as the tissues of young plants and animals develop from fertilized eggs into an adult organism. These topics will be examined in Chapters 19, 39, and 52.

In this section, we will view tissues from the perspective of cell biology. Animals and plants contain many different types of cells. Humans, for example, contain over 200 different cell types, each with a specific structure and function. Even so, these cells can be grouped into a few general categories. For example, muscle cells found in your heart (cardiac muscle cells), in your biceps (skeletal muscle cells), and around your arteries (smooth muscle cells) look somewhat different under the microscope and have unique roles in the body. Yet due to structural and functional similarities, all three types can be categorized as muscle tissue cells. In this section, we will begin by surveying the basic processes that cells undergo in order to create tissues. Next, we will examine the main categories of both animal and plant tissues. We will conclude by taking a more in-depth look at some similarities and differences between selected animal and plant tissues, focusing in particular on the functions of the ECM and cell junctions.

Six Basic Cell Processes Create Tissues and Organs

A multicellular organism such as a plant or animal contains many cells. For example, an adult human has somewhere between 10 trillion and 100 trillion cells in her or his body. Cells are organized into tissues, and tissues are organized into organs. An **organ** is a collection of two or more tissues that performs a specific function or set of functions. The heart is an organ found in the bodies of complex animals, while a leaf is an organ found in plants. We will examine the structures and functions of organs in Units VI and VII.

To create tissues and organs, cells undergo six basic processes that influence their morphology, arrangement, and number:

1. *Cell division*: As discussed in Chapters 9 and 15, eukaryotic cells progress through a cell cycle that leads to cell division.
2. *Cell growth*: Following cell division, cells take up nutrients and usually expand in volume.
3. *Differentiation*: Due to gene regulation, cells differentiate into specialized types of cells. Cell differentiation is described in Chapter 19.
4. *Migration*: During embryonic development in animals, cells migrate to their appropriate positions within the body. Cell migration does not occur during plant development.
5. *Apoptosis*: Cell death, also known as apoptosis, is necessary to produce certain morphological features of the body. For example, during development in mammals, the formation of individual fingers and toes requires the removal, by apoptosis, of the skin cells between them.
6. *Cell connections*: In the first two sections of this chapter, we learned that cells secrete an extracellular matrix that provides strength and support. In animals, the ECM serves to organize cells within tissues and organs. In plants, the cell wall forms the ECM that shapes plant tissues. Different types of cell junctions in both animal and plant cells enable cells to make physical contact and communicate with one another.

Animals Are Composed of Epithelial, Connective, Muscle, and Nervous Tissues

The body of an animal contains four general types of tissue—epithelial, connective, muscle, and nervous tissue—that serve very different purposes (**Figure 10.16**).

Epithelial Tissue Epithelial tissue is composed of cells that are joined together via tight junctions and form continuous

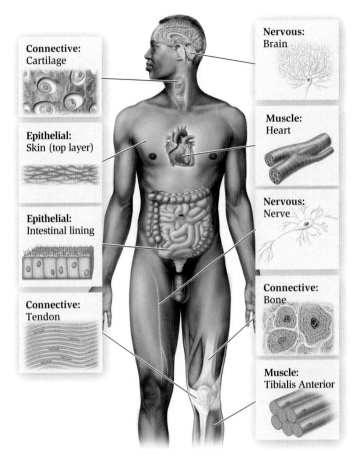

Figure 10.16 Examples of the four general types of tissues—epithelial, connective, muscle, and nervous—found in animals.

sheets. Epithelial tissue covers or forms the lining of all internal and external body surfaces. For example, epithelial tissue lines organs such as the lungs and digestive tract. In addition, epithelial tissue forms skin, a protective surface that shields the body from the outside environment.

Connective Tissue Most connective tissue provides support to the body and/or helps to connect different tissues to each other. Connective tissue is rich in extracellular matrix. In some cases, the tissue contains only a sparse population of cells that are embedded in the ECM. Examples of connective tissue include cartilage, tendons, bone, fat tissue, and the inner layers of the skin. Blood is also considered a form of connective tissue because it provides liquid connections to various regions of the body.

Muscle Tissue Muscle tissue can generate a force that facilitates movement. Muscle contraction is needed for bodily movements such as walking and running, and also plays a role in the movement of materials throughout the body. For example, contraction of heart muscle propels blood through your body. The properties of muscle tissue in animals are examined in Chapter 46.

Nervous Tissue Nervous tissue receives, generates, and conducts electrical signals throughout the body. In vertebrates, these electrical signals are integrated by nervous tissue in the brain and transmitted down the spinal cord to the rest of the body. Chapter 43 considers the cellular basis for nerve signals, known as action potentials, and Chapters 44 and 45 examine the organization of nervous systems in animals.

Plants Contain Dermal, Ground, and Vascular Tissues

The bodies of most plants contain three general types of tissue—dermal, ground, and vascular—each with a different structure suited to its functions, as shown in **Figure 10.17**.

Dermal Tissue The **dermal tissue** forms a covering on various parts of the plant. The **epidermis** refers to the newly made tissue on the surfaces of leaves, stems, and roots. Surfaces of leaves are usually coated with a waxy cuticle to prevent water loss and often have hairs, or trichomes, which are specialized types of epidermal cells that have diverse functions, including the secretion of oils and leaf protection. Epidermal cells called guard cells form pores in leaves known as stomata that permit gas exchange. The function of the root epidermis is the absorption of water and nutrients. The root epidermis does not have a waxy cuticle, because such a cuticle would inhibit water and nutrient absorption.

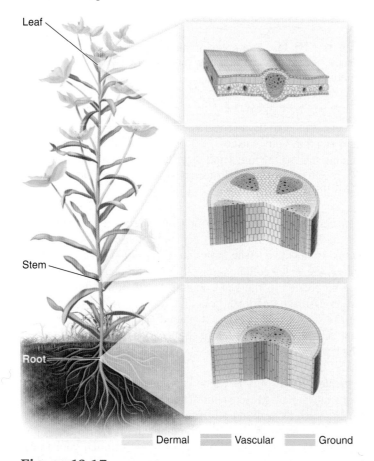

Figure 10.17 Locations of the three general types of tissues—dermal, ground, and vascular—found in plants.

Dermal   Vascular   Ground

Ground Tissue Most of a plant's body is made of **ground tissue**, which has a variety of functions, including photosynthesis, storage of carbohydrates, and support. Ground tissue can be subdivided into three types: parenchyma, collenchyma, and sclerenchyma. Let's look briefly at each of these types of ground tissue.

1. Parenchyma tissue is very active metabolically. The mesophyll, the central part of the leaf, which contains the cells that carry out the bulk of photosynthesis, is parenchyma tissue. Parenchyma tissue also functions in the storage of carbohydrates. The cells of parenchyma tissue usually lack a secondary cell wall or, if present, it is very thin.
2. Collenchyma tissue provides structural support to the plant body, particularly to growing regions such as the periphery of the stems and leaves. Collenchyma cells tend to have thick, secondary cell walls but do not contain much lignin. Therefore, they provide support but are also able to stretch.
3. Sclerenchyma tissue also provides structural support to the plant body, particularly to those parts that are no longer growing. The secondary cell walls of sclerenchyma cells tend to have large amounts of lignin and thereby provide rigid support. In many cases, sclerenchyma cells are dead at maturity but their cell walls continue to provide structural support during the life of the plant.

Vascular Tissue Most plants are vascular plants. In these species, which include ferns and flowering plants, the vascular tissue is composed of cells that are interconnected and form conducting vessels for water and nutrients. Two vascular tissues are found in flowering plants: xylem and phloem. The xylem transports water and mineral ions from the root to the rest of the plant, while the phloem distributes the products of photosynthesis and a variety of other nutrients throughout the plant. Some types of plants, such as mosses, are nonvascular plants that lack conducting vessels. These plants tend to be small and live in damp, shady places.

Plant and Animal Tissues Have Striking Differences and Similarities

Because plants and animals appear strikingly different, it is not too surprising that their cells and tissues show conspicuous differences. For example, the vascular tissue of plants (which transports water and nutrients) does not resemble any one tissue in animals. The blood vessels of animals (which transport blood carrying oxygen and nutrients throughout the body) are hollow tubes that contain both connective and muscle tissue. In addition, animals have two tissue types that are not found in plants: muscle and nervous tissue. Even so, plants are capable of movement and the transmission of signals via action potentials. For example, the Venus flytrap (*Dionaea muscipula*) has two modified leaves that resemble a clamshell. When an insect touches the surface of these leaves, an action potential is triggered across the leaf cells that causes the leaves to move closer to each other, thereby trapping the unsuspecting insect.

Although cellular differences are prominent between plants and animals, certain tissues show intriguing similarities. The epithelial tissue of animals and the dermal tissue of plants both form a protective covering over the organism. Also, the connective tissue of animals and the ground tissue of plants both play a role in structural support. Let's take a closer look at the similarities between these tissues in animals and plants.

A Comparison of Epithelial and Dermal Tissues Both the epithelial tissue in animals (also called an **epithelium**) and dermal tissue in plants form layers of cells. An epithelium can be classified according to its number of layers. Simple epithelium is one cell layer thick, while stratified epithelium has several layers (**Figure 10.18**). In both cases, the epithelium has a polarity, which is due to an asymmetry to its organization. The outer or apical side of an epithelium is exposed to air or to a watery fluid such as the lumen of the intestine. The inner or basal side rests on some type of support, such as another type of tissue or on a form of ECM called the basal lamina.

A hallmark of epithelial cells is that they form many connections with each other. For example, in the simple epithelium lining the intestine (**Figure 10.19**), adjacent cells form anchoring junctions with each other and with the basal lamina. These anchoring junctions hold the cells firmly in place. Tight junctions, found near the apical surface, prevent the leakage of materials from the lumen of the intestine into the blood. Instead, as described in Chapter 41, nutrients are selectively transported from the intestinal lumen into the cytosol of the epithelial cell and then are exported across the basal side of the cell into the blood. This phenomenon, called transepithelial transport, allows the body to take up the nutrients it needs, while preventing unwanted materials from getting into the bloodstream. Epithelial cells are also connected via gap junctions, which allow the exchange of nutrients and signaling molecules throughout the epithelium.

In flowering plants, the epidermis covers all of the newly made parts of a plant. For example, the upper and lower sides of leaves are covered by epidermis, which is usually a single layer of closely packed cells (**Figure 10.20a**). Epidermal cells have a thick primary cell wall and are tightly interlocked by their middle lamella.

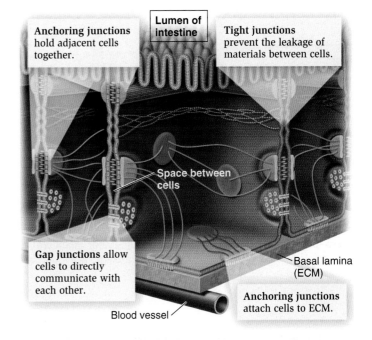

Figure 10.19 A closer look at a simple epithelium that lines the intestine. This figure emphasizes that the three major types of cell junctions are common in epithelial tissue.

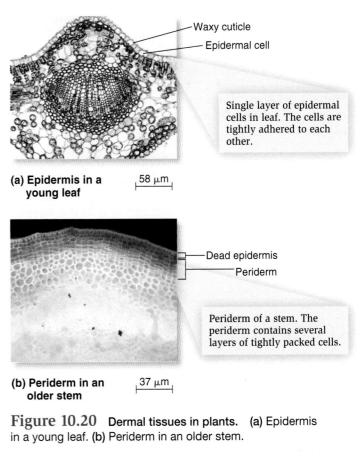

(a) Epidermis in a young leaf   58 μm

(b) Periderm in an older stem   37 μm

Figure 10.20 Dermal tissues in plants. (a) Epidermis in a young leaf. (b) Periderm in an older stem.

Biological inquiry: Do you notice any parallels between simple epithelium and epidermis, and between stratified epithelium and periderm?

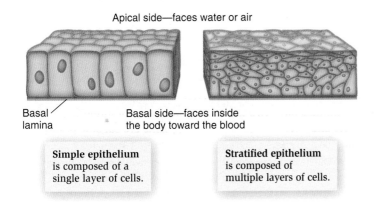

Apical side—faces water or air

Basal lamina

Basal side—faces inside the body toward the blood

Simple epithelium is composed of a single layer of cells.

Stratified epithelium is composed of multiple layers of cells.

Figure 10.18 Simple and stratified epithelia.

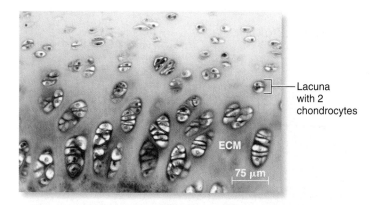

Figure 10.21 **An example of connective tissue in animals that is rich in extracellular matrix.** This micrograph of cartilage shows chondrocytes in the ECM. The chondrocytes, which are responsible for making the components of cartilage, are found in cavities called lacunae.

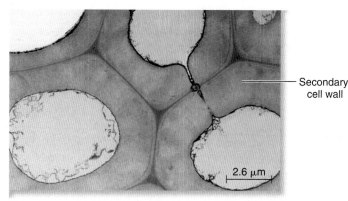

Figure 10.22 **An example of ground tissue in plants, which is rich in extracellular matrix.** This scanning electron micrograph shows cells from *Arabidopsis thaliana*. The cells themselves are dead; only their thick cell walls remain. These cell walls of the ECM provide support to the plant.

Thus, plant epidermal cells are tightly woven together, much like epithelial cell layers in animals. In leaves, the cells are often polygonal in shape, which means they are flattened cells that have several angles. The epidermal cells of new stems tend to have a more rectangular shape.

As a plant ages, the epidermis may be replaced by another type of dermal tissue called the periderm (**Figure 10.20b**). In woody plants, an example of periderm is the bark on trees. The tissue protects the plant from pathogens, prevents excessive water loss, and provides insulation. The periderm consists of interconnected cork cells, which may be several layers thick. The cork cells have extremely thick cell walls. When cork cells reach maturity, they die, but the cell walls continue to provide support.

A Comparison of Connective and Ground Tissue In contrast to epithelial tissue, which is almost entirely composed of cells, the connective tissue of animals is largely composed of extracellular matrix and has relatively few cells. In animal connective tissue, cell-to-cell contact is somewhat infrequent. Instead, cells are usually adhered to the ECM via integrins, as shown earlier in Figure 10.8b. In some cases, the primary function of cells within connective tissue is to synthesize the components of the ECM. For example, let's consider cartilage, a connective tissue found in joints such as your knees. The cells that synthesize cartilage, known as chondrocytes, actually represent a small proportion of the total volume of cartilage. As shown in **Figure 10.21**, the chondrocytes are found in small cavities within the cartilage called lacunae (singular, *lacuna*). In some types of cartilage, the chondrocytes represent only 1–2% of the total volume of the tissue! Chondrocytes are the only cells found in cartilage; they are solely responsible for the synthesis of protein fibers, such as collagen, as well as glycosaminoglycans and proteoglycans.

Similar to connective tissue in animals, ground tissue in plants provides structural support and is also rich in ECM. **Figure 10.22** shows a scanning electron micrograph of sclerenchyma cells found in *Arabidopsis thaliana*, a model plant studied by plant biologists. At maturity, the cells are dead, but the thick secondary cell walls continue to provide rigid support for the stem. However, not all ground tissue in plants is rich in ECM. For example, mesophyll cells are a type of parenchymal cell in the leaf that carry out photosynthesis. Because a thick cell wall would inhibit the transmission of light, mesophyll cells have thin cell walls with only a small amount of ECM.

Chapter Summary

10.1 Extracellular Matrix

- The extracellular matrix (ECM) is a network of material that is secreted from plant and animal cells and forms a complex meshwork outside of cells.
- In the ECM of animals, proteins and polysaccharides are the major constituents. These materials are involved in strength, structural support, organization, and cell signaling. (Figure 10.1)

- Adhesive proteins, such as fibronectin and laminin, help to adhere cells to the ECM. Structural proteins form fibers. Collagen fibers provide tensile strength, while elastic fibers allow regions of the body to stretch. (Table 10.1, Figures 10.2, 10.3)
- Differential gene regulation controls where in the body different types of collagen fibers are made. (Table 10.2)

- Glycosaminoglycans (GAGs) are polysaccharides of repeating disaccharide units that give a gel-like character to the ECM of animals. Proteoglycans consist of a core protein with attached GAGs. (Figure 10.4)

- Plant cells are surrounded by an ECM called the cell wall. The primary cell wall is made first. It is composed largely of cellulose. The secondary cell wall is made after the primary cell wall and is often quite thick and rigid. (Figures 10.5, 10.6)

10.2 Cell Junctions

- The three common types of cell junctions found in animals are anchoring, tight, and gap junctions. Plant junctions include middle lamella and plasmodesmata. (Table 10.3)

- Anchoring junctions involve cell adhesion molecules (CAMs), which bind cells to each other or to the ECM. The four types are adherens junctions, desmosomes, focal adhesions, and hemidesmosomes. (Figure 10.7)

- Two types of CAMs are cadherin and integrin. Cadherins link cells to each other, while integrins link cells to the ECM. In the cytosol, CAMs bind to actin or intermediate filaments. (Figure 10.8)

- Integrins can perform outside-in and inside-out signaling. This enables cells to communicate with the ECM. (Figure 10.9)

- Tight junctions, composed of occludin and claudin, prevent the leakage of materials between cells. (Figures 10.10, 10.11)

- Gap junctions form channels called connexons that permit the direct passage of materials between adjacent cells. (Figure 10.12)

- Experiments of Loewenstein and colleagues involving the transfer of fluorescent dyes showed that gap junctions permit the passage of substances with a molecular mass of less than 1,000 Daltons. (Figure 10.13)

- The cell walls of adjacent plant cells are cemented together via middle lamella. (Figure 10.14)

- Adjacent plant cells usually have direct connections called plasmodesmata, which are open channels in the cell walls. The ER of adjacent cells is also connected via plasmodesmata. (Figure 10.15)

10.3 Tissues

- A tissue is a group of cells that have a similar structure and function. An organ is composed of two or more tissues and carries out a particular function or functions.

- Six processes—cell division, cell growth, differentiation, migration, apoptosis, and cell connections—create tissues and organs.

- The four general kinds of tissues found in animals are epithelial, connective, muscle, and nervous tissues. (Figure 10.16)

- The three general kinds of tissues found in plants are dermal, ground, and vascular tissues. (Figure 10.17)

- Epithelial and dermal tissues form layers of cells that are highly interconnected. These layers can be one cell thick or several cells thick, and they serve as protective coverings for various parts of animal and plant bodies. (Figures 10.18, 10.19, 10.20)

- Connective and ground tissues are often rich in ECM and play a structural role in animals and plants. (Figures 10.21, 10.22)

TEST YOURSELF

1. The function of the extracellular matrix (ECM) in most multicellular organisms is
   a. to provide strength.
   b. to provide structural support.
   c. to organize cells and other body parts.
   d. cell signaling.
   e. all of the above.

2. The protein found in the ECM of animals that provides strength and resistance to tearing when stretched is
   a. elastin.
   b. cellulose.
   c. collagen.
   d. laminin.
   e. fibronectin.

3. The polysaccharide that forms the hard outer covering of many invertebrates is
   a. collagen.
   b. chitin.
   c. chondroitin sulfate.
   d. pectin.
   e. cellulose.

4. The single most abundant organic molecule of Earth is _____, and it is the main macromolecule of the _____.
   a. collagen, connective tissue of animals
   b. chitin, muscle tissue of animals
   c. cellulose, primary cell wall of plants
   d. integrins, cell junctions in plants
   e. pectin, secondary cell wall of plants

5. _____ are proteins that attach animal cells to the ECM.
   a. Cadherins
   b. Integrins
   c. Occludins
   d. Tight junctions
   e. Desmosomes

6. The gap junctions of animal cells differ from the plasmodesmata of plant cells in that
   a. gap junctions serve as communicating junctions and plasmodesmata serve as adhesion junctions.
   b. gap junctions prevent extracellular material from moving between adjacent cells, but the plasmodesmata do not.
   c. gap junctions allow for direct exchange of cellular material between cells, but plasmodesmata cannot allow the same type of exchange.
   d. gap junctions are formed by specialized proteins that form channels through the membranes of adjacent cells, but plasmodesmata are not formed by specialized proteins.
   e. all of the above.

7. Which of the following is involved in the process of tissue and organ formation in multicellular organisms?
   a. cell division
   b. cell differentiation
   c. cell connections
   d. cell growth
   e. all of the above

8. The tissue type common to animals that functions in the conduction of electrical signals is
   a. epithelial.
   b. dermal.
   c. muscle.
   d. nervous.
   e. ground.

9. Photosynthesis occurs mainly in the _____ tissue of plants.
   a. vascular
   b. dermal
   c. parenchyma
   d. collenchyma
   e. sclerenchyma

10. Which of the following is not a correct statement in the comparison of plant tissues to animal tissues?
   a. Nervous tissue of animals plays the same role as vascular tissue in plants.
   b. The dermal tissue of plants is similar to epithelial tissue of animals in that both provide a covering for the organism.
   c. The epithelial tissue of animals and the dermal tissue of plants have special characteristics that limit the movement of extracellular material between cell layers.
   d. The ground tissue of plants and the connective tissue of animals provide structural support for the organism.
   e. All of the above are correct comparisons between plant and animal tissues.

CONCEPTUAL QUESTIONS

1. List and explain the four characteristics of the extracellular matrix.
2. Distinguish between the primary cell wall and the secondary cell wall.
3. List and explain the four types of animal tissues.

EXPERIMENTAL QUESTIONS

1. What was the purpose of the study conducted by Loewenstein and colleagues?
2. Explain the experimental procedure used by Loewenstein to determine the size of gap-junction channels.
3. What did the results of Figure 10.13 indicate about the size of gap-junction channels?

COLLABORATIVE QUESTIONS

1. What role does the extracellular matrix play in animals?
2. What are the six basic cell processes required to create tissues and organs?

www.brookerbiology.com

This website includes answers to the Biological Inquiry questions found in the figure legends and all end-of-chapter questions.

11

NUCLEIC ACID STRUCTURE AND DNA REPLICATION

CHAPTER OUTLINE

A molecular model for the structure of a DNA double helix.

We begin our survey of genetics by examining DNA at the molecular level. Once we understand how DNA works at this level, it becomes easier to see how DNA function affects the properties of cells, and ultimately the characteristics or traits of both unicellular and multicellular organisms. The past several decades have seen exciting advances in techniques and approaches to investigate and even to alter the genetic material. Not only have these advances greatly expanded our understanding of molecular genetics, such technologies are also widely used in related disciplines such as biochemistry, cell biology, and microbiology.

To a large extent, our understanding of genetics comes from our knowledge of the molecular structure of DNA (deoxyribonucleic acid) and RNA (ribonucleic acid). In this chapter, we begin by considering some classic experiments that were consistent with the theory that DNA is the genetic material. We will then survey the molecular features of DNA, which will allow us to appreciate how DNA can store information and be accurately copied. Though this chapter is largely concerned with DNA, we will also consider the components of RNA, which bear some striking similarities to DNA.

11.1 Biochemical Identification of the Genetic Material

The genetic material functions as a blueprint for the construction of living organisms. In the case of multicellular organisms such as plants and animals, the information stored in the genetic material enables a fertilized egg to develop into an embryo, and eventually into an adult organism. In addition, the genetic material allows organisms to survive in their native environments. For example, an individual's DNA provides the blueprint to produce enzymes that are needed to metabolize food. To fulfill its role, the genetic material must meet several criteria:

1. **Information:** The genetic material must contain the information necessary to construct an entire organism.
2. **Transmission:** The genetic material must be passed from parent to offspring. It also must be passed from cell to cell during the process of cell division.
3. **Replication:** For transmission to occur, the genetic material must be accurately copied.
4. **Variation:** Differences in the genetic material must account for the known variation within each species and among different species.

The quest to identify the genetic material really began in the late 1800s, when a few scientists postulated that living organisms possess a blueprint that has a biochemical basis. In 1883, August Weismann and Karl Nägeli championed the idea that a chemical substance exists within living cells that is responsible for the transmission of traits from parents to offspring. During the next 30 years, experimentation along these lines centered on the behavior of **chromosomes**, which are structures found in cells that we now know contain the genetic material. The name *chromosome* literally means colored (*chromo*) body (*soma*), referring to the observation of early microscopists that the chromosomes are easily stained by colored dyes. By studying the transmission patterns of chromosomes from cell to cell and from parent to offspring, researchers were convinced that chromosomes carry the determinants that control the outcome of traits.

Ironically, the study of chromosomes initially misled researchers regarding the biochemical identity of the genetic material. Chromosomes contain two classes of macromolecules, namely, proteins and DNA. Scientists of this era viewed proteins as being more biochemically complex because they are made from 20 different amino acids. Furthermore, biochemists already knew that proteins perform an amazingly wide range of functions. Complexity seemed an important prerequisite for the blueprint of an organism. By comparison, DNA seemed less complex, because it contains only four types of repeating units called **nucleotides**, which will be described later in this chapter. In addition, the functional role of DNA in the nucleus had not been extensively investigated prior to the 1920s. Therefore, from the 1920s to the 1940s, most scientists were expecting that research studies would reveal that proteins are the genetic material. Contrary to this expectation, however, the experiments described in this section were pivotal in showing that DNA carries out this critical role.

Griffith's Bacterial Transformation Experiments Indicated the Existence of a Biochemical Genetic Material

Studies in microbiology were important in developing an experimental strategy to identify the genetic material. In the late 1920s, Frederick Griffith studied a type of bacterium known then as pneumococci and now classified as *Streptococcus pneumoniae*. Some strains of *S. pneumoniae* secrete a polysaccharide capsule, while other strains do not. When streaked on petri plates containing solid growth media, capsule-secreting strains have a smooth colony morphology, or form, and so look smooth to the naked eye, whereas those strains unable to secrete a capsule have a colony morphology that looks rough. In mammals, smooth strains of *S. pneumoniae* may cause pneumonia and other symptoms. In mice, such infections are often fatal.

As shown in **Figure 11.1**, Griffith injected live and/or heat-killed bacteria into mice and then observed whether or not the bacteria caused them to die. He investigated the effects of two strains of *S. pneumoniae*, type S for smooth and type R for rough. When injected into a live mouse, the smooth strain killed the mouse (Figure 11.1, step 1). A capsule is necessary for this to occur because it prevents the mouse's immune system from killing the bacterial cells. Following the death of the mouse, many type S bacteria were found in the mouse's blood. By comparison, when type R bacteria were injected into a mouse, they did not kill the mouse and living bacteria were not found in the live mouse's blood (Figure 11.1, step 2). In a follow-up to these results, Griffith also heat-killed the smooth bacteria and then injected them into the mouse. As expected, the mouse survived (Figure 11.1, step 3).

A surprising result occurred when Griffith mixed live type R bacteria with heat-killed type S bacteria, and then injected them into a mouse—the mouse died (Figure 11.1, step 4). The blood from the dead mouse contained living type S bacteria! These results indicated that a substance from dead type S bacteria was transforming the type R bacteria into type S. Griffith called this process **transformation**, and he termed the uniden-

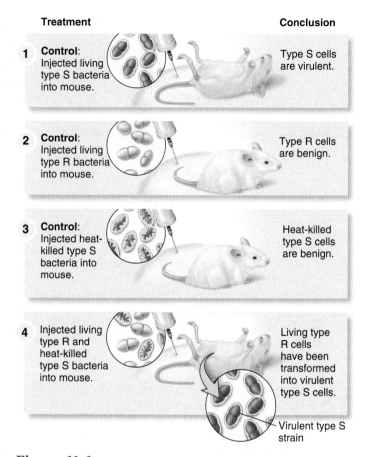

Figure 11.1   Griffith's experiments that showed the transformation of bacteria by a "transformation principle."

Biological inquiry: Let's suppose that the type R strain used by Griffith was resistant to killing by an antibiotic, while the type S strain lacked this trait. For the experiment described in step 4, would you expect the living type S bacteria found in the dead mouse's blood to be resistant to the antibiotic?

tified material that was responsible for this phenomenon the *transformation principle.*

Now that we have examined Griffith's experiments, it's helpful if we consider what these observations mean in genetic terms. Among different streptococcal strains, researchers knew that *variation* exists in the presence or absence of a capsule. According to Griffith's results, the transformed bacteria had acquired the *information* to make a capsule from the heat-killed cells. For the transformed bacteria to proliferate and thereby kill the mouse, the substance conferring the ability to make a capsule must be *replicated* so that it can be *transmitted* from mother to daughter cells during cell division. Taken together, these observations are consistent with the idea that the formation of a capsule is governed by genetic material, because it meets the four criteria described at the beginning of this section. Thus, the experiment of Figure 11.1, step 4 was consistent with the idea that some genetic material from the dead bacteria had been transferred to the living rough bacteria and provided those bacteria with a new trait. At the time of his studies, however, Griffith could not determine the biochemical composition of the transforming substance.

FEATURE INVESTIGATION

Avery, MacLeod, and McCarty Used Purification Methods to Reveal That DNA Is the Genetic Material

Exciting discoveries sometimes occur when researchers recognize that another scientist's experimental approach may be modified and then used to dig deeper into a scientific question. In the 1940s, Oswald Avery, Colin MacLeod, and Maclyn McCarty were also interested in the process of bacterial transformation. During the course of their studies, they realized that Griffith's observations could be used as part of an experimental strategy to biochemically identify the genetic material. They asked the question, What substance is being transferred from the dead type S bacteria to the live type R?

Avery, MacLeod, and McCarty used established biochemical procedures to purify classes of macromolecules, such as proteins, DNA, and RNA, from a type S streptococcal strain. Initially, they discovered that only the purified DNA could convert type R bacteria into type S. To further verify that DNA is

the genetic material, they performed the investigation outlined in **Figure 11.2**. They purified DNA from a type S strain and mixed it with type R bacteria. After allowing time for DNA uptake, they added an antibody that aggregated any nontransformed type R bacteria, which were removed by centrifugation. The remaining bacteria were incubated overnight on petri plates.

When they mixed their DNA extract with type R bacteria, some of the bacteria were converted to type S (see plate B in step 5 of Figure 11.2). As a control, if no DNA extract was added, no type S bacterial colonies were observed on the petri plates (see plate A in step 5). Though this result was consistent with the idea that DNA is the genetic material, a careful biochemist might argue that the DNA extract may not have been 100% pure. Realistically, any purified extract is likely to contain small traces of other substances. For this reason, the researchers realized that a small amount of contaminating material in the DNA extract could actually be the genetic material. The most likely contaminating substances in this case would be protein or RNA. To address this possibility, Avery, MacLeod,

Figure 11.2   The Avery, MacLeod, and McCarty experiments that identified DNA as Griffith's "transformation principle"—the genetic material.

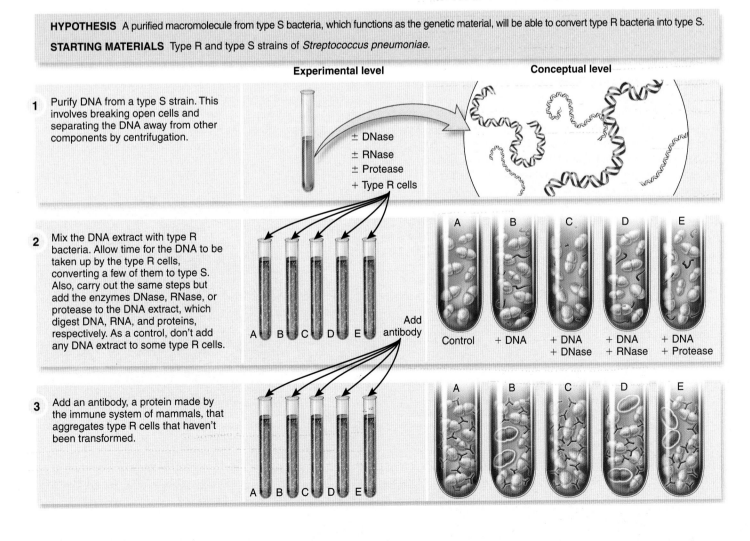

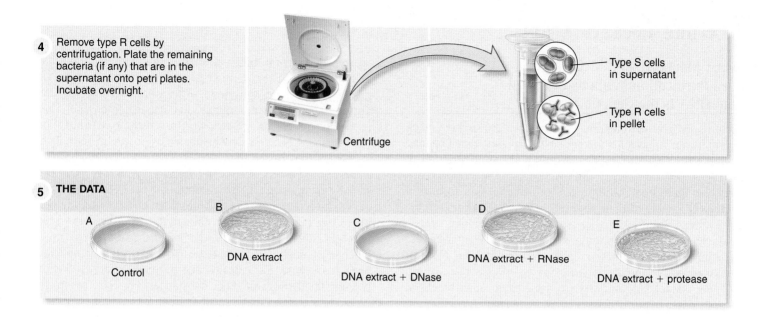

4 Remove type R cells by centrifugation. Plate the remaining bacteria (if any) that are in the supernatant onto petri plates. Incubate overnight.

Centrifuge

Type S cells in supernatant

Type R cells in pellet

5 THE DATA

A Control

B DNA extract

C DNA extract + DNase

D DNA extract + RNase

E DNA extract + protease

and McCarty treated the DNA extract with enzymes that digest DNA (called **DNase**), RNA (**RNase**), or protein (**protease**) (see step 2, Figure 11.2). When the DNA extracts were treated with RNase or protease, the type R bacteria were still converted into type S, suggesting that contaminating RNA or protein in the

extract was not acting as the genetic material (see plates D and E). Moreover, when the extract was treated with DNase, its ability to convert type R bacteria into type S was lost (see plate C). Taken together, these results indicated that DNA is the genetic material.

Hershey and Chase Determined That DNA Is the Genetic Material of T2 Bacteriophage

Although the work of Avery, MacLeod, and McCarty discussed in the Feature Investigation was consistent with the idea that DNA is the genetic material of streptococcal strains, further experimentation was needed to determine if the use of DNA as the genetic material is a widespread phenomenon in biology. In a second avenue of research conducted in 1952, the efforts of Alfred Hershey and Martha Chase centered on the study of a virus named T2. This virus infects bacterial cells, in this case *Escherichia coli*, and is therefore known as a **bacteriophage** or simply a **phage**. A T2 phage has an outer covering called the phage coat that contains a capsid, sheath, tail fibers, and baseplate (**Figure 11.3**). Biochemically, the phage coat is composed entirely of protein. DNA is found inside the T2 capsid. From a biochemical perspective, T2 is very simple because it is composed of only DNA and proteins.

The genetic material of T2 provides a blueprint to make new phages. However, as discussed in Chapter 18, all viruses must introduce their genetic material into the cytoplasm of a living host cell to replicate. In the case of T2, this first involves the attachment of its tail to the bacterial cell wall and then the injection of its genetic material into the cytoplasm (Figure 11.3).

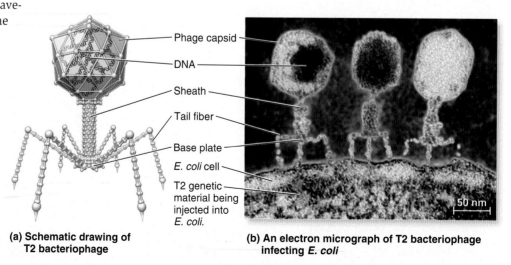

Phage capsid

DNA

Sheath

Tail fiber

Base plate

E. coli cell

T2 genetic material being injected into *E. coli*.

(a) Schematic drawing of T2 bacteriophage

50 nm

(b) An electron micrograph of T2 bacteriophage infecting *E. coli*

Figure 11.3 The structure of T2 bacteriophage. The colorized electron micrograph in part (b) shows T2 phages attached to an *E. coli* cell and injecting their genetic material into the cell.

To determine if DNA is the genetic material of T2, Hershey and Chase devised a method to separate the phage coat, which is attached to the outside of the bacterium, from the genetic material, which is injected into the cytoplasm. They reasoned that the attachment of T2 on the surface of the bacterium could be disrupted if the cells were subjected to high shear forces such as those produced by a blender. In addition, they needed a way to distinguish T2 DNA from T2 proteins. Hershey and Chase used radioisotopes, which are described in Chapter 2, as a way to label these molecules. Sulfur atoms are found in proteins but not in DNA, whereas phosphorus atoms are found in DNA but not in phage proteins. They exposed T2-infected bacterial cells to ^{35}S (a radioisotope of sulfur) or to ^{32}P (a radioisotope of phosphorus). These infected cells produced phages that had incorporated ^{35}S into their proteins or ^{32}P into their DNA, a process called labeling. The labeled phages were then used in the experiment shown in **Figure 11.4**.

Let's now consider the steps in this experiment. In separate tubes, they took samples of T2 phage, one labeled with ^{35}S and the other with ^{32}P, and mixed them with E. coli cells for a short period of time. This allowed the phages enough time to inject their genetic material into the bacterial cells. The samples were then subjected to a shearing force using a blender for up to 8 minutes. This treatment removed the phage coat from the surface of the bacterial cell without causing cell lysis. Each sample

was then subjected to centrifugation at a speed that caused the heavier bacterial cells to form a pellet at the bottom of the tube, while the lighter phage coats remained in the supernatant, which is the solution above the pellet. The amount of radioactivity in the supernatant (emitted from either ^{35}S or ^{32}P) was determined using an instrument called a scintillation counter.

As you can see in the data of Figure 11.4, most of the ^{35}S isotope (80%) was found in the supernatant. Because the shearing force removed only the phage coat, this result indicates that the empty phages contain primarily protein. In contrast, only about 35% of the ^{32}P was found in the supernatant following shearing. This indicates that most of the phage DNA was located within the bacterial cells in the pellet. Taken together, these results suggest that the phage DNA is injected into the bacterial cytoplasm during infection. This is the expected outcome if DNA is the genetic material of T2.

11.2 Nucleic Acid Structure

The two experiments we've examined by Avery, MacLeod, and McCarty and by Hershey and Chase provided compelling evidence to support the role of DNA as the genetic material. Since these pioneering studies, thousands of researchers have studied the role of the genetic material as the blueprint of life.

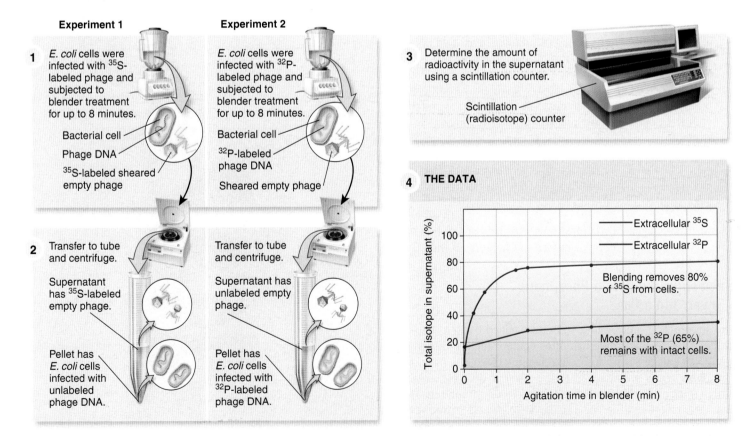

Figure 11.4 Hershey and Chase experiment showing that the genetic material of T2 phage is DNA.

Biological inquiry: In these experiments, which isotope do you expect to find in the pellet, ^{32}P or ^{35}S?

As discussed in other chapters, scientists have uncovered an important principle in biology: "structure determines function." When biologists want to understand the function of a material at the molecular and cellular level, they focus some of their efforts on the investigation of its biochemical structure. In this regard, an understanding of DNA structure has proven to be particularly informative because the structure makes it easy for us to understand how DNA can store information, how variation in structure can occur, and also how DNA can be replicated and transmitted to future generations.

DNA and its molecular cousin, RNA, are known as **nucleic acids**. This term is derived from the discovery of DNA by Friedrich Miescher in 1869. He isolated a novel phosphorus-containing substance from the nuclei of white blood cells found in waste surgical bandages. He named this substance nuclein. As the structure of DNA and RNA became better understood, it was found that they are acidic molecules, which means that they release hydrogen ions (H$^+$) in solution, and have a net negative charge at neutral pH. Thus, the name nucleic acid was coined.

DNA is a very large macromolecule composed of smaller building blocks. We can consider the structural features of DNA at different levels of complexity (**Figure 11.5**):

1. **Nucleotides** are the building blocks of DNA (and RNA).
2. A **strand** of DNA (or RNA) is formed by the covalent linkage of nucleotides in a linear manner.
3. Two strands of DNA can hydrogen-bond with each other to form a **double helix**. In a DNA double helix, two DNA strands are twisted together to form a structure that resembles a spiral staircase.
4. In living cells, DNA is associated with an array of different proteins to form **chromosomes**. The association of proteins with DNA organizes the long strands into a compact structure.
5. A **genome** is the complete complement of an organism's genetic material. For example, the genome of most bacteria is a single circular chromosome, while the genomes of eukaryotic species are sets of linear chromosomes.

The first three levels of complexity will be the focus of this section. Levels 4 and 5 will be discussed in Chapters 15 and 21.

Nucleotides Contain a Phosphate, a Sugar, and a Base

The structures of nucleotides were determined by P. A. Levene in the 1920s. A nucleotide has three components: a phosphate group, a pentose sugar, and a nitrogenous base. The base and phosphate group are attached to the sugar at different sites to form a nucleotide (**Figure 11.6**). The nucleotides found in DNA and RNA contain different sugars. Deoxyribose is found in DNA, while ribose is found in RNA. Compared to ribose, deoxyribose is missing a single oxygen atom; the prefix *deoxy-* (meaning without oxygen) refers to this missing atom.

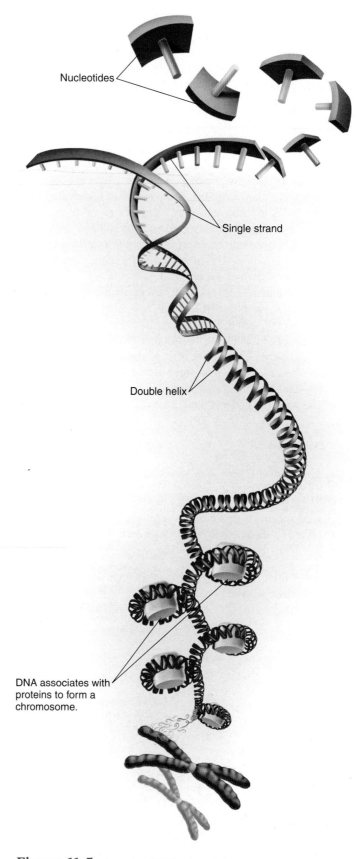

Figure 11.5   Levels of DNA structure to create a chromosome.

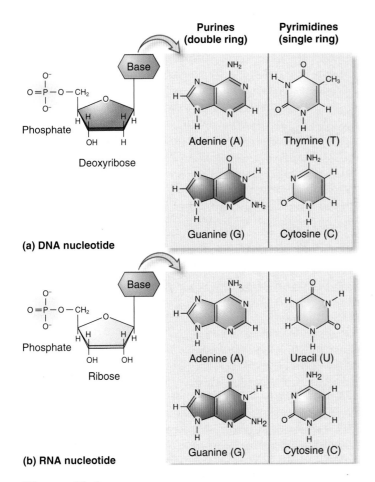

Purines (double ring) **Pyrimidines (single ring)**

Adenine (A) Thymine (T)

Guanine (G) Cytosine (C)

(a) DNA nucleotide

Adenine (A) Uracil (U)

Guanine (G) Cytosine (C)

(b) RNA nucleotide

Figure 11.6 Nucleotides and their components. For simplicity, the carbon atoms in the ring structures are not shown.

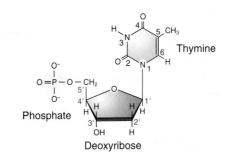

Thymine

Phosphate

Deoxyribose

Figure 11.7 Conventional numbering in a DNA nucleotide. The carbons in the sugar are given a prime designation, while those in the base are not.

A Strand Is a Linear Linkage of Nucleotides with Directionality

The next level of nucleotide structure is the formation of a strand of DNA or RNA in which nucleotides are covalently attached to each other in a linear fashion. **Figure 11.8** depicts a short strand of DNA with four nucleotides. The linkage is a phosphoester bond (a covalent bond between phosphorus and oxygen) involving a sugar molecule in one nucleotide and a phosphate group in the next nucleotide. Another way of viewing this linkage is to notice that a phosphate group connects two sugar molecules together. From this perspective, the linkage in DNA and RNA strands is called a **phosphodiester linkage**, which has two phosphoester bonds. The phosphates and sugar molecules form the **backbone** of a DNA or RNA strand, while the bases project from the backbone. The backbone is negatively charged due to the negative charges of the phosphate groups.

An important structural feature of a nucleic acid strand is the orientation of the nucleotides. Each phosphate in a phosphodiester linkage is covalently bound to the 5′ carbon in one nucleotide and to the 3′ carbon in the other. In a strand, all sugar molecules are oriented in the same direction. For example, in the strand shown in Figure 11.8, all of the 5′ carbons in every sugar molecule are above the 3′ carbons. A strand has a **directionality** based on the orientation of the sugar molecules within that strand. In Figure 11.8, the direction of the strand is said to be 5′ to 3′ when going from top to bottom. The 5′ end of a DNA strand has a phosphate group while the 3′ end has an —OH group.

From the perspective of function, a key feature of DNA and RNA structure is that a strand contains a specific sequence of bases. In Figure 11.8, the sequence of bases is thymine–adenine–cytosine–guanine. This sequence is abbreviated TACG. To indicate its directionality, the strand should be abbreviated 5′–TACG–3′. Because the nucleotides within a strand are attached to each other by stable covalent bonds, the sequence of bases cannot shuffle around and become rearranged. Therefore, the sequence of bases in a DNA strand will remain the same over time, except in rare cases when mutations occur. The base sequence of DNA and RNA is the critical feature that allows them to store and transmit information.

Five different bases are found in nucleotides, although any given nucleotide contains only one base. The five bases are subdivided into two categories, the **purines** and the **pyrimidines**, due to differences in their structures (Figure 11.6). The purine bases, **adenine** (**A**) and **guanine** (**G**), have a double-ring structure; the pyrimidine bases, **cytosine** (**C**), **thymine** (**T**), and **uracil** (**U**), have a single-ring structure. Thymine is found only in DNA, while uracil is found only in RNA. Adenine, guanine, and cytosine are found in both DNA and RNA.

A conventional numbering system has been created to describe the attachment sites of the base and phosphate to the sugar molecule (**Figure 11.7**). In the sugar ring, carbon atoms are numbered in a clockwise direction starting with the carbon atom to the right of the ring oxygen atom. The fifth carbon is outside the ring. The prime symbol (′) is used to distinguish the numbering of carbons in the sugar, while the atoms in the ring structures of the bases are not given the prime designation. The sugar carbons are designated 1′ (that is, "one prime"), 2′, 3′, 4′, and 5′. A base is attached to the 1′ carbon atom and a phosphate group is attached at the 5′ position.

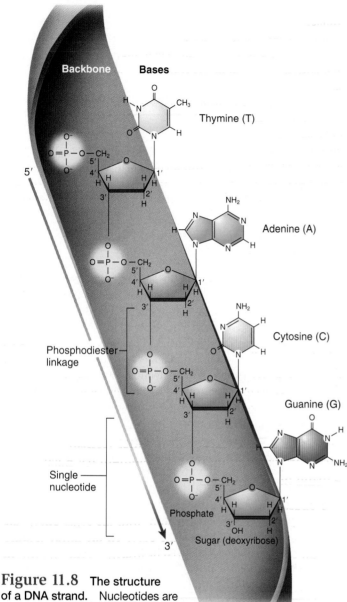

Figure 11.8 **The structure of a DNA strand.** Nucleotides are covalently bonded to each other in a linear manner. Notice the directionality of the strand, and that it carries a particular sequence of bases. An RNA strand has a very similar structure, except that the sugar is ribose rather than deoxyribose, and uracil is substituted for thymine.

A Few Key Experiments Paved the Way to Solving the Structure of DNA

In 1953, James Watson and Francis Crick, along with their collaborator Maurice Wilkins, proposed the structure of the DNA double helix. Prior to their work, DNA was already known to be composed of nucleotides. However, scientists did not know how the nucleotides are bonded together to form the structure of DNA. Watson and Crick wanted to determine the structure of

DNA because they thought this knowledge would provide insights regarding the function of genes. Before we examine the characteristics of the double helix, let's consider the events that led to the discovery of the double helix structure.

In the early 1950s, more information was known about the structure of proteins than that of nucleic acids. Linus Pauling correctly proposed that regions of proteins can fold into a structure known as an α helix (alpha helix). To determine the structure of the α helix, Pauling built large models by linking together simple ball-and-stick units. In this way, he could see if atoms fit together properly in a complicated three-dimensional structure. This approach is still widely used today, except that now researchers construct three-dimensional models on computers. Watson and Crick used a ball-and-stick approach to solve the structure of the DNA double helix.

X-ray diffraction was also a key experimental tool that led to the discovery of the DNA double helix. When a substance is exposed to radiation, such as X-rays, the atoms in the substance will cause the X-rays to be scattered (**Figure 11.9a**). If the substance has a repeating structure, the pattern of scattering, known as the diffraction pattern, is mathematically related to the structural arrangement of the atoms causing the scattering. The diffraction pattern is analyzed using mathematical theory to provide information regarding the three-dimensional structure of the molecule. Rosalind Franklin (**Figure 11.9b**), working in the same laboratory as Maurice Wilkins, was a gifted experimentalist who made marked advances in X-ray diffraction techniques involving DNA. The diffraction pattern of DNA fibers produced by Franklin is shown in **Figure 11.9c**. This pattern suggested a helical structure with a diameter that is relatively uniform and too wide to be a single-stranded helix. In addition, the pattern provided information regarding the number of nucleotides per turn, and was consistent with the 2-nm (nanometers) spacing between the strands in which a purine (A or G) bonds with a pyrimidine (T or C). These observations were instrumental in solving the structure of DNA.

Another piece of information that proved to be critical for the determination of the double helix structure came from the studies of Erwin Chargaff. He pioneered many of the biochemical techniques for the isolation, purification, and measurement of nucleic acids from living cells. Chargaff analyzed the base composition of DNA that was isolated from many different species. As shown in **Table 11.1**, he consistently observed from his experiments that the amount of adenine in each sample was similar to thymine and the amount of cytosine was similar to guanine. As we will see, this observation became crucial evidence that helped Watson and Crick develop the double helix model of DNA.

Watson and Crick Deduced the Double Helix Structure of DNA

Thus far, we have considered the experimental studies that led to the determination of the DNA double helix. These included the biochemical modeling approach of Pauling, the X-ray diffraction work of Franklin, and the base composition studies of Chargaff.

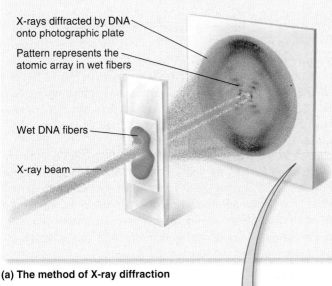

X-rays diffracted by DNA onto photographic plate

Pattern represents the atomic array in wet fibers

Wet DNA fibers

X-ray beam

(a) The method of X-ray diffraction

(b) Rosalind Franklin

(c) Franklin's X-ray diffraction pattern of wet DNA fibers

Figure 11.9 Rosalind Franklin and X-ray diffraction applied to DNA wet fibers.   (a) The exposure of X-rays to DNA causes the X-rays to be scattered. (b) Rosalind Franklin. (c) Because the DNA forms a repeating structure, the X-rays are scattered in a way that produces an X-ray diffraction pattern.

Table 11.1	Base Content in the DNA from a Variety of Organisms			
	% of bases			
Organism	**Adenine**	**Thymine**	**Guanine**	**Cytosine**
Escherichia coli (bacterium)	26.0	23.9	24.9	25.2
Streptococcus pneumoniae (bacterium)	29.8	31.6	20.5	18.0
Saccharomyces cerevisiae (yeast)	31.7	32.6	18.3	17.4
Turtle	28.7	27.9	22.0	21.3
Salmon	29.7	29.1	20.8	20.4
Chicken	28.0	28.4	22.0	21.6
Human	30.3	30.3	19.5	19.9

Watson and Crick assumed that nucleotides are linked together in a linear fashion and that the chemical linkage between two nucleotides is always the same. Then, they set out to build ball-and-stick models that incorporated all of the known experimental observations.

Modeling of chemical structures often involves trial and error. Watson and Crick initially considered several incorrect models. One model was a double helix in which the bases were on the outside of the helix. Another model showed the bases forming hydrogen bonds with the identical base in the opposite strand (that is, A to A, T to T, G to G, and C to C). However, the model-building revealed that the bases could not fit together this way. Finally, they realized that the hydrogen-bonding of adenine to thymine was structurally similar to that of cytosine to guanine. In both cases, a purine (A or G) bonds with a pyrimidine (T or C). With an interaction between A and T and between G and C, the ball-and-stick models showed that the two strands would form a double helix structure in which all atoms would fit together properly. This ball-and-stick model was consistent with the known data (**Figure 11.10**).

Watson, Crick, and Wilkins proposed the structure of DNA, which was published in the journal *Nature* in 1953. In 1962, they were awarded the Nobel Prize. Unfortunately, Rosalind Franklin had died before this time, and the Nobel Prize is not awarded posthumously.

DNA Has a Repeating, Antiparallel Helical Structure Formed by the Complementary Base Pairing of Nucleotides

The structure that Watson and Crick proposed is a double-stranded, repeating helical structure with the sugar-phosphate backbone on the outside and the bases on the inside (**Figure 11.11a**). This structure is stabilized by hydrogen bonding between

Figure 11.10 Watson and Crick and their model of the DNA double helix.

the bases in opposite strands to form **base pairs**. A distinguishing feature of base pairing is its specificity. An adenine base in one strand forms two hydrogen bonds with a thymine base in the opposite strand, or a guanine base forms three hydrogen bonds with a cytosine (**Figure 11.11b**). This **AT/GC rule** (also known as Chargaff's rule) explains the previous data of Chargaff in which there were equal amounts of A and T, and equal amounts of G and C, in DNA. According to the AT/GC rule, purines (A and G) always bond with pyrimidines (T and C) (recall that purines both have a double-ring structure, whereas pyrimidines have single rings). This keeps the width of the double helix relatively constant. One complete turn of the double helix is composed of 10 base pairs.

Due to the AT/GC rule, the base sequences of two DNA strands are **complementary** to each other. That is, you can predict the sequence in one DNA strand if you know the sequence in the opposite strand. For example, if one strand has the sequence of 5′–GCGGATTT–3′, the opposite strand must be 3′–CGCCTAAA–5′. With regard to their 5′ and 3′ directionality,

the two strands of a DNA double helix are **antiparallel**. If you look at Figure 11.11, one strand runs in the 5′ to 3′ direction from top to bottom, while the other strand is oriented 3′ to 5′ from top to bottom. Watson and Crick proposed an antiparallel structure in their original DNA model.

The DNA model we have been studying in Figure 11.11a is called a ribbon model, which clearly shows the components of the DNA molecule. However, other models are also used to visualize DNA. The model for the DNA double helix shown in **Figure 11.12** is a space-filling model in which the atoms are depicted as spheres. This type of structural model emphasizes the surface of DNA. As you can see in this model, the sugar-phosphate backbone is on the outermost surface of the double helix. In a living cell, the backbone has the most direct contact with water. The atoms of the bases are more internally located within the double-stranded structure. The indentations where the atoms of the bases make contact with the surrounding water are termed **grooves**. Two grooves, called the **major groove** and the **minor groove**, spiral around the double helix.

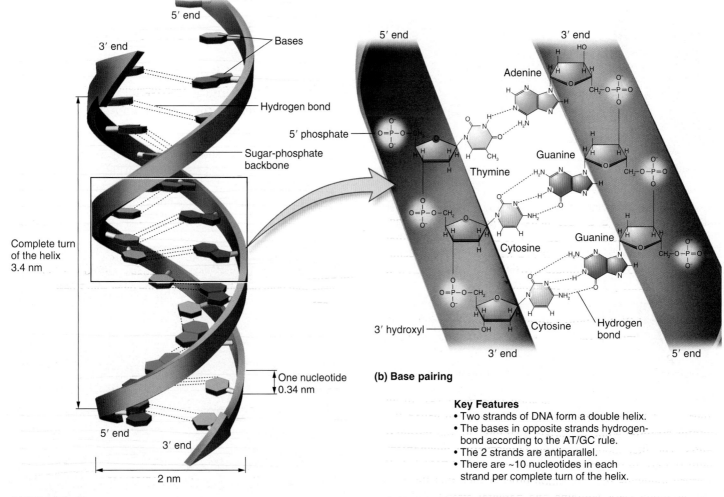

(b) **Base pairing**

Key Features
- Two strands of DNA form a double helix.
- The bases in opposite strands hydrogen-bond according to the AT/GC rule.
- The 2 strands are antiparallel.
- There are ~10 nucleotides in each strand per complete turn of the helix.

(a) **Double helix**

Figure 11.11 **Structure of the DNA double helix.** As seen in part **(a)**, DNA is a helix composed of two antiparallel strands. Part **(b)** shows the AT/GC base pairing that holds the strands together via hydrogen bonds.

Biological inquiry: If one DNA strand is 5′–GATTCGTTC–3′, what is the complementary strand?

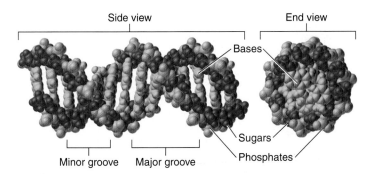

Figure 11.12   **A space-filling model of the DNA double helix.** In the sugar-phosphate backbone, sugar molecules are shown in blue and phosphate groups are shown in yellow. This backbone is on the outermost surface of the double helix. The atoms of the bases, shown in green, are more internally located within the double-stranded structure. Notice the major and minor grooves that are formed by this arrangement.

As discussed in later chapters, the major groove provides a location where a protein can bind to a particular sequence of bases and affect the expression of a gene.

11.3 DNA Replication

In the previous section, we considered the structure of the genetic material. DNA is a double helix that obeys the AT/GC rule. From this structure, Watson and Crick immediately were able to suggest a mechanism by which DNA can be copied. During this process, known as **DNA replication**, they proposed that the original DNA strands are used as templates for the synthesis of new DNA strands. In this section, we will begin by looking at an early experiment that helped to determine the method of DNA replication. Next, we will examine the structural characteristics that enable a double helix to be faithfully copied. Finally, we will consider the features of DNA replication that allow it to occur in living cells. As you will learn, several cellular proteins are needed to initiate DNA replication and allow it to proceed accurately and at a rapid pace.

Meselson and Stahl Used Density Measurements to Investigate Three Proposed Mechanisms of DNA Replication

Researchers in the late 1950s considered three different models for the mechanism of DNA replication (**Figure 11.13**). In all of these models, the two newly made strands are called the **daughter strands**, while the original strands are the **parental strands**. The first model is a **semiconservative mechanism** (Figure 11.13a). In this model, the double-stranded DNA is half conserved following the replication process such that the new double-stranded DNA contains one parental strand and one daughter strand. This mechanism is consistent with the ideas of Watson and Crick. Even so, other models were possible and had to be ruled out. According to a second model, called a **conservative mechanism**, both parental strands of DNA remain

together following DNA replication (Figure 11.13b). The original arrangement of parental strands is completely conserved, while the two newly made daughter strands are also together following replication. Finally, a third possibility, called a **dispersive mechanism**, proposed that segments of parental DNA and newly made DNA are interspersed in both strands following the replication process (Figure 11.13c).

In 1958, Matthew Meselson and Franklin Stahl devised an experimental approach to distinguish among these three mechanisms. An important feature of their research was the use of isotope labeling. Nitrogen, which is found in DNA, occurs in a common light (^{14}N) form and a rare heavy (^{15}N) form. Meselson and Stahl studied DNA replication in the bacterium *E. coli*. They grew *E. coli* cells for many generations in a medium that

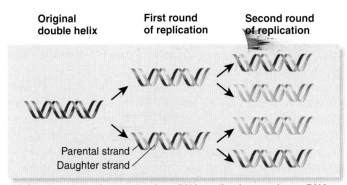

(a) Semiconservative mechanism. DNA replication produces DNA molecules with 1 parental strand and 1 newly made strand.

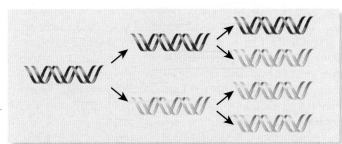

(b) Conservative mechanism. DNA replication produces 1 double helix with both parental strands, and the other with 2 new daughter strands.

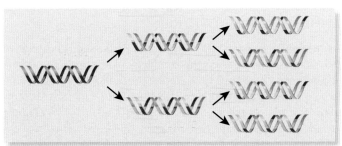

(c) Dispersive mechanism. DNA replication produces DNA strands in which segments of new DNA are interspersed with the parental DNA.

Figure 11.13   **Three mechanisms for DNA replication.** The strands of the original double helix are shown in red. New strands are produced for two rounds of replication; these new strands are shown in blue.

contained only the ^{15}N form of nitrogen (**Figure 11.14**). This produced a population of bacterial cells in which all of the DNA was heavy labeled. Then, they switched the bacteria to a medium that contained only ^{14}N as its nitrogen source. The cells were allowed to divide, and samples were collected after one generation (that is, one round of DNA replication), two generations, and so on. Because the bacteria were doubling in a medium that contained only ^{14}N, all of the newly made DNA strands would be labeled with light nitrogen, while the original strands would remain labeled with the heavy form.

Meselson and Stahl then used centrifugation to separate DNA molecules based on differences in density. Samples were placed on the top of a solution that contained a salt gradient, in this case cesium chloride. A double helix containing all heavy nitrogen has a higher density and will travel closer to the bottom of the gradient. By comparison, if both DNA strands contained ^{14}N, the DNA would have a light density and remain closer to the top of the gradient. If one strand contained ^{14}N and the other strand contained ^{15}N, the DNA would be half-heavy and have an intermediate density, ending up near the middle of the gradient.

After one cell doubling (that is, one round of DNA replication), all of the DNA exhibited a density that was half-heavy (Figure 11.14, step 5). These results are consistent with both the semiconservative and dispersive models. In contrast, the conservative mechanism predicts two different DNA types: a light type and a heavy type. Because the DNA was found in a single (half-heavy) band after one doubling, the conservative model was disproved. After two cell doublings, both light DNA and half-heavy DNA were observed. This result was also predicted by a semiconservative mechanism of DNA replication, because some DNA molecules should contain all light DNA while other molecules should be half-heavy (see Figure 11.13a). However, in a dispersive mechanism, all of the DNA strands after two generations would have been 1/4 heavy. This mechanism predicts that the heavy nitrogen would be evenly dispersed among four double helices, each strand containing 1/4 heavy nitrogen and 3/4 light nitrogen (see Figure 11.13c). This result was not obtained. Taken together, the results of the Meselson and Stahl experiment are consistent only with a semiconservative mechanism for DNA replication.

Semiconservative DNA Replication Proceeds According to the AT/GC Rule

As originally proposed by Watson and Crick, semiconservative DNA replication relies on the complementarity of DNA strands according to the AT/GC rule. During the replication process, the two complementary strands of DNA separate and serve as **template strands** for the synthesis of new strands of DNA (**Figure 11.15**). After the double helix has separated, individual nucleotides have access to the template strands. Hydrogen bonding between individual nucleotides and the template strands must obey the AT/GC rule. A covalent bond is formed between the phosphate of one nucleotide and the sugar of the previous nucleotide. The end result is that two double helices are made that have the same base sequence as the original DNA molecule. This is a critical feature of DNA replication, because it enables

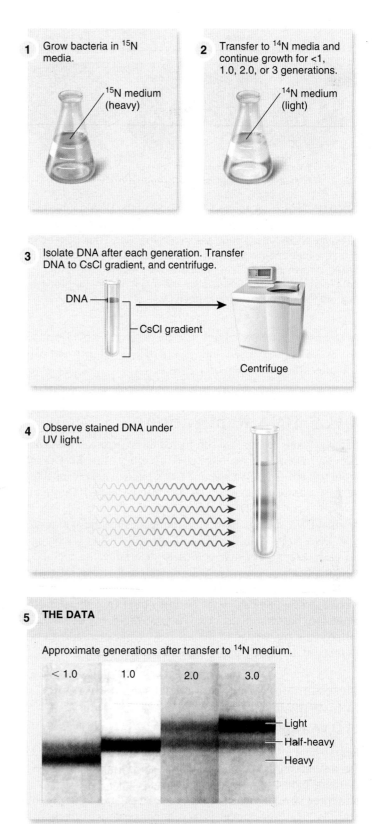

Figure 11.14   The Meselson and Stahl experiment showing that DNA replication is semiconservative.

Biological inquiry: If this experiment were conducted for four rounds of DNA replication (that is, four generations), what would be the expected fractions of light DNA and half-heavy DNA, according to the semiconservative model?

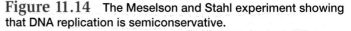

the replicated DNA molecules to retain the same information (that is, the same base sequence) as the original molecule. In this way, DNA has the striking ability to direct its own duplication.

DNA Replication Begins at an Origin of Replication, Where DNA Replication Forks Are Formed

An **origin of replication** is a site within a chromosome that serves as a starting point for DNA replication. At the origin, the two DNA strands unwind, and DNA replication proceeds outward from the origin in opposite directions, a process termed **bidirectional replication**. The number of origins of replication varies among different organisms. In bacteria, which have a small circular chromosome, a single origin of replication is found. Bidirectional replication starts at the origin of replication and proceeds until the new strands meet each other on the opposite side of the chromosome (**Figure 11.16a**). Eukaryotes have

larger chromosomes that are linear. They require multiple origins of replication so that the DNA can be replicated in a reasonable length of time. The newly made strands from each origin eventually make contact with each other to complete the replication process (**Figure 11.16b**).

An origin of replication provides an opening called a replication bubble that forms two DNA **replication forks**. DNA replication occurs near the opening that forms each replication fork (**Figure 11.17**, step 1). As discussed later, the synthesis of a strand always begins with a primer and the new DNA is made in the 5′ to 3′ direction. The manner in which the two daughter strands are synthesized is strikingly different (Figure 11.17, step 2). One strand, called the **leading strand**, is made in the

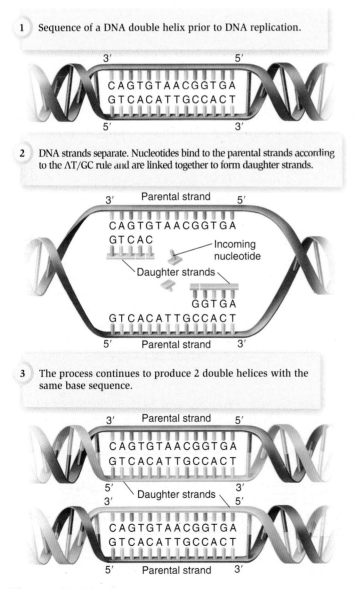

1 Sequence of a DNA double helix prior to DNA replication.

```
 3′              5′
   CAGTGTAACGGTGA
   GTCACATTGCCACT
 5′              3′
```

2 DNA strands separate. Nucleotides bind to the parental strands according to the AT/GC rule and are linked together to form daughter strands.

```
 3′   Parental strand   5′
   CAGTGTAACGGTGA
   GTCAC
        Incoming nucleotide
   Daughter strands
              GGTGA
   GTCACATTGCCACT
 5′   Parental strand   3′
```

3 The process continues to produce 2 double helices with the same base sequence.

```
 3′   Parental strand   5′
   CAGTGTAACGGTGA
   GTCACATTGCCACT
 5′                     3′
        Daughter strands
 3′                     5′
   CAGTGTAACGGTGA
   GTCACATTGCCACT
 5′   Parental strand   3′
```

Figure 11.15 DNA replication according to the AT/GC rule.

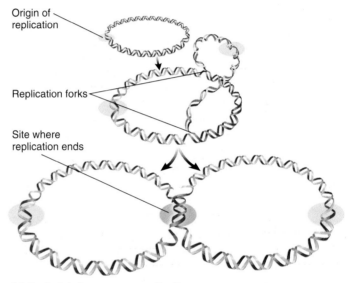

(a) Bacterial chromosome replication

Origin of replication

Replication forks

Site where replication ends

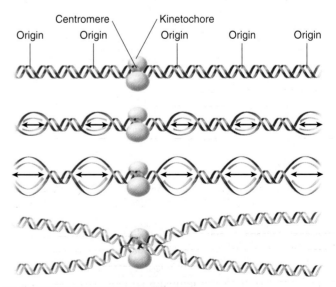

Centromere   Kinetochore
Origin   Origin   Origin   Origin   Origin

(b) Eukaryotic chromosome replication

Figure 11.16 Origins of replication in the chromosomes of different organisms.   (a) A single origin of replication is found in bacteria. (b) Multiple origins are found in a eukaryotic chromosome. A description of centromeres, kinetochores (proteins bound to centromeres) and sister chromatids can be found in Chapter 15.

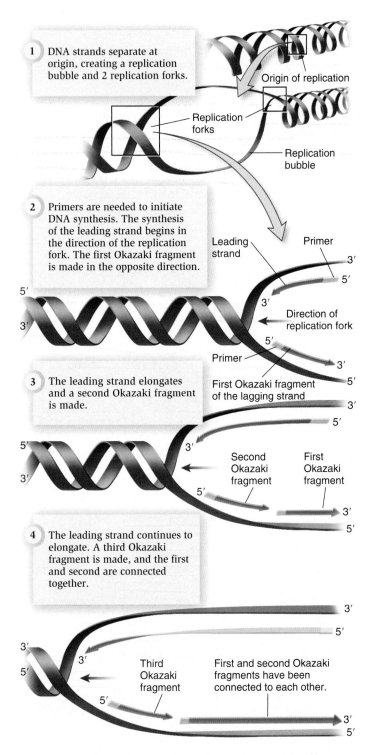

1. DNA strands separate at origin, creating a replication bubble and 2 replication forks.

Origin of replication

Replication forks

Replication bubble

2. Primers are needed to initiate DNA synthesis. The synthesis of the leading strand begins in the direction of the replication fork. The first Okazaki fragment is made in the opposite direction.

Leading strand

Primer

Direction of replication fork

Primer

First Okazaki fragment of the lagging strand

3. The leading strand elongates and a second Okazaki fragment is made.

Second Okazaki fragment

First Okazaki fragment

4. The leading strand continues to elongate. A third Okazaki fragment is made, and the first and second are connected together.

Third Okazaki fragment

First and second Okazaki fragments have been connected to each other.

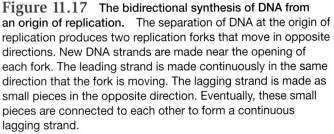

Figure 11.17 **The bidirectional synthesis of DNA from an origin of replication.** The separation of DNA at the origin of replication produces two replication forks that move in opposite directions. New DNA strands are made near the opening of each fork. The leading strand is made continuously in the same direction that the fork is moving. The lagging strand is made as small pieces in the opposite direction. Eventually, these small pieces are connected to each other to form a continuous lagging strand.

same direction that the fork is moving. The leading strand is synthesized as one long continuous molecule. By comparison, the other daughter strand, termed the **lagging strand**, is made as a series of small fragments that are eventually connected to each other to form a continuous strand. The synthesis of these fragments occurs in the direction away from the fork. For example, the single fragment seen in Figure 11.17, step 2, is synthesized from left to right. Notice that this fragment connects to a second fragment in step 4 of Figure 11.17. These DNA fragments are known as **Okazaki fragments**, after Reiji and Tuneko Okazaki, who initially discovered them in the late 1960s.

DNA Replication Requires the Action of Many Different Proteins

Thus far, we have considered how DNA replication occurs outward from an origin of replication in a region called a DNA replication fork. In all living species, a set of many different proteins is involved in this process. An understanding of the functions of these proteins is critical to explaining the replication process at the molecular level.

Helicase, Topoisomerase, and Single-Strand Binding Proteins: Formation and Movement of the Replication Fork

To act as a template for DNA replication, the strands of a double helix must separate and the resulting fork must move. As mentioned, an origin of replication serves as a site where this separation initially occurs. The strand separation at each fork then moves outward from the origin via the action of an enzyme called **DNA helicase**. At each fork, a helicase enzyme binds to one of the DNA strands and travels in the 5′ to 3′ direction toward the fork (**Figure 11.18**). It uses energy from ATP to separate the DNA strands and keeps the fork moving forward. The action of DNA helicase generates additional coiling just ahead of the replication fork that is alleviated by another enzyme called **DNA topoisomerase**.

After the two parental DNA strands have separated, they must remain that way until the complementary daughter strands have been made. The function of **single-strand binding proteins** is to bind to both of the single strands of parental DNA and prevent them from re-forming a double helix. In this way, the bases within the parental strands are kept exposed so that they can act as templates for the synthesis of complementary strands.

DNA Polymerase, Primase, and Ligase: Synthesis of the Leading and Lagging Strands

The enzyme **DNA polymerase** is responsible for covalently linking nucleotides together to form DNA strands. Arthur Kornberg originally identified this enzyme in the 1950s. The structure of DNA polymerase in action resembles a human hand with the DNA threaded through it (**Figure 11.19a**). As DNA polymerase slides along the DNA, free nucleotides with three phosphate groups, called **deoxynucleoside triphosphates**, hydrogen-bond to the exposed bases in the template strand according to the AT/GC rule. At the catalytic site, DNA polymerase breaks a bond at the middle phosphate and then attaches

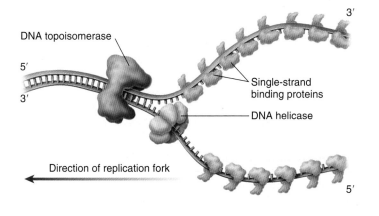

Figure 11.18 **Proteins that facilitate the formation and movement of a replication fork.**   DNA helicase travels along one DNA strand in the 5′ to 3′ direction and separates the DNA strands. After they are separated, single-strand binding proteins coat the DNA strands to prevent them from re-forming a double helix. Topoisomerase travels slightly ahead of the fork and alleviates coiling that is caused by the action of helicase.

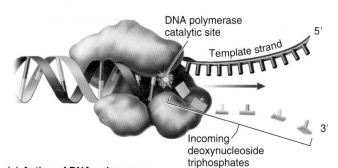

(a) Action of DNA polymerase

Figure 11.19 **Enzymatic synthesis of DNA.**   (a) DNA polymerase recognizes free nucleotides called deoxynucleoside triphosphates and attaches a deoxynucleoside monophosphate to the 3′ end of a growing strand. (b) DNA polymerase breaks the bond at the middle phosphate in a deoxynucleoside triphosphate, causing the release of pyrophosphate. This provides the energy to form a covalent phosphoester bond between the resulting deoxynucleoside monophosphate and the previous nucleotide in the growing strand.

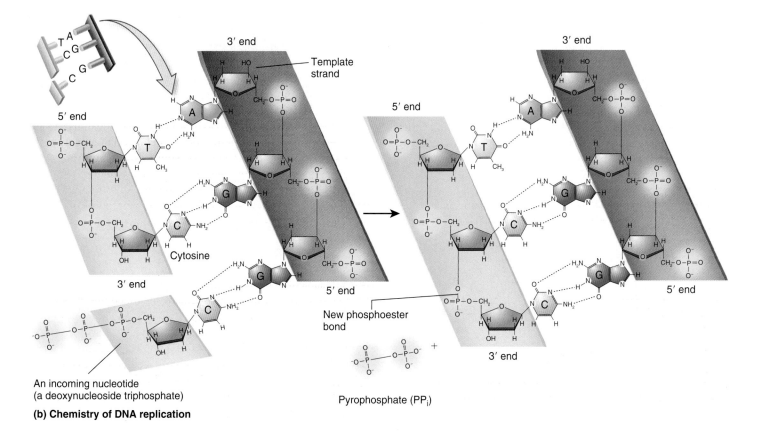

(b) Chemistry of DNA replication

the resulting nucleotide with one phosphate group (a deoxynucleoside monophosphate) to the 3′ end of a growing strand via a phosphoester bond. The breakage of the covalent bond that releases pyrophosphate (two phosphate groups) is an exergonic reaction that provides the energy to covalently connect adjacent nucleotides (**Figure 11.19b**). The rate of synthesis is truly remarkable. In bacteria, DNA polymerase can synthesize DNA at a rate of 500 nucleotides per second, while eukaryotic species can make DNA at a rate of about 50 nucleotides per second.

DNA polymerase has two specific enzymatic features that explain why the leading and lagging strands are made differently. First, DNA polymerase is unable to begin DNA synthesis on a bare template strand. However, if a DNA or RNA strand is already attached to a template strand, DNA polymerase can elongate such a pre-existing strand by making DNA. A different enzyme called **DNA primase** is required if the template strand is bare. DNA primase makes a complementary **primer** that is actually a short segment of RNA, typically 10 to 12 nucleotides in length. These short RNA strands start, or prime, the process of DNA replication (**Figure 11.20a**). At a later stage in DNA replication, the RNA primers are removed and replaced with DNA. A second feature of DNA polymerase is that once synthesis has begun, it can synthesize new DNA only in a 5′ to 3′ direction (**Figure 11.20b**).

In the leading strand, DNA primase makes one RNA primer at the origin and then DNA polymerase attaches nucleotides in a 5′ to 3′ direction as it slides toward the opening of the replication fork (**Figure 11.21**). In the lagging strand, DNA is also synthesized in a 5′ to 3′ direction, but this synthesis occurs in the direction away from the replication fork. In the lagging strand, short segments of DNA, the Okazaki fragments, are made. Each fragment contains a short RNA primer made by DNA primase at the 5′ end, and then the remainder of the fragment is a strand of DNA made by DNA polymerase.

To complete the synthesis of Okazaki fragments within the lagging strand, three additional events must occur: removal of the RNA primers, synthesis of DNA in the area where the primers have been removed, and the covalent joining of adjacent fragments of DNA (Figure 11.21, steps 3 and 4). The RNA primers are

removed by DNA polymerase, which can also digest the linkages between nucleotides in a 5′ to 3′ direction. After the RNA primer is removed, DNA polymerase then fills in the vacant region with DNA. However, once the DNA has been completely filled in, a covalent bond is missing between the last nucleotide added by DNA polymerase and the first nucleotide in the next DNA fragment. An enzyme known as **DNA ligase** catalyzes the formation of a covalent bond between these adjacent DNA fragments to complete the replication process in the lagging strand (Figure 11.21, step 4). Similarly, the single RNA primer of the leading strand is removed, DNA polymerase fills in the vacant region, and ligase completes the process. **Table 11.2** provides a summary of the actions of the proteins we've discussed in this section.

DNA Replication Is Very Accurate

Although errors can happen during DNA replication, permanent mistakes are extraordinarily rare. For example, during bacterial DNA replication, only one mistake per 100 million nucleotides is made. This high level of **fidelity** occurs for several reasons. First, hydrogen bonding between A and T or between G and C is more stable than between mismatched pairs. Second, the active site of DNA polymerase is unlikely to catalyze bond formation between adjacent nucleotides if a mismatched base pair is formed. A third way that DNA polymerase minimizes mistakes is by the enzymatic removal of mismatched nucleotides.

DNA polymerase can identify a mismatched nucleotide and remove it from the daughter strand. This event, called **proofreading**, occurs when DNA polymerase detects a mismatch, and then reverses its direction and digests the linkages between nucleotides at the end of a newly made strand in the 3′ to 5′ direction. Once it passes the mismatched base and removes it, DNA polymerase then changes direction again and continues to synthesize DNA in the 5′ to 3′ direction.

In addition to the proofreading ability of DNA polymerase, living cells also possess several other enzymes that function solely to repair abnormalities in DNA. If such lesions are not fixed, a permanent change in the genetic material, namely a mutation, can occur. These topics are considered in Chapter 14.

Table 11.2	Proteins Involved in DNA Replication
Common name	**Function**
DNA helicase	Separates double-stranded DNA into single strands
Single-strand binding protein	Binds to single-stranded DNA, and prevents it from re-forming a double helix
Topoisomerase	Removes tightened coils ahead of the replication fork
DNA primase	Synthesizes short RNA primers
DNA polymerase	Synthesizes DNA in the leading and lagging strands, removes RNA primers, and fills in gaps
DNA ligase	Covalently attaches adjacent Okazaki fragments

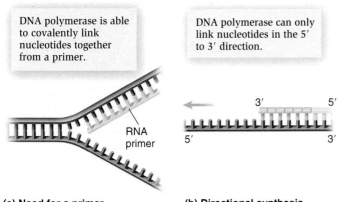

DNA polymerase is able to covalently link nucleotides together from a primer.

DNA polymerase can only link nucleotides in the 5′ to 3′ direction.

RNA primer

3′   5′

5′   3′

(a) Need for a primer

(b) Directional synthesis

Figure 11.20 Enzymatic feature of DNA polymerase. (a) DNA polymerase needs a primer to begin DNA synthesis and (b) once begun, it can synthesize DNA only in the 5′ to 3′ direction.

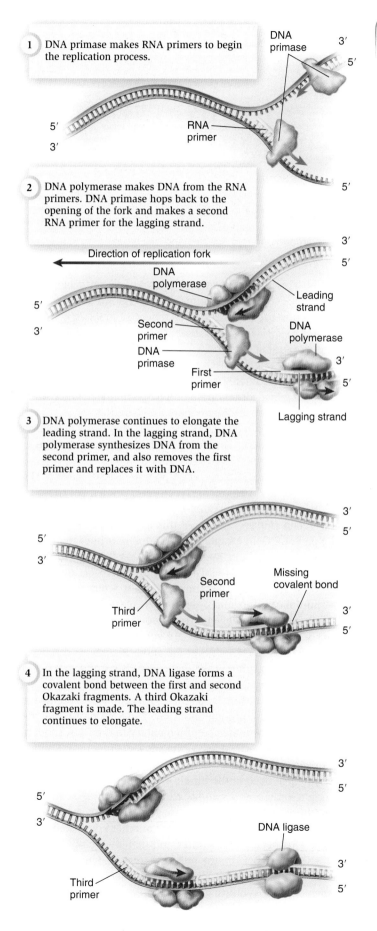

1 DNA primase makes RNA primers to begin the replication process.

DNA primase

RNA primer

2 DNA polymerase makes DNA from the RNA primers. DNA primase hops back to the opening of the fork and makes a second RNA primer for the lagging strand.

Direction of replication fork

DNA polymerase

Second primer

DNA primase

First primer

Leading strand

DNA polymerase

Lagging strand

3 DNA polymerase continues to elongate the leading strand. In the lagging strand, DNA polymerase synthesizes DNA from the second primer, and also removes the first primer and replaces it with DNA.

Second primer

Third primer

Missing covalent bond

4 In the lagging strand, DNA ligase forms a covalent bond between the first and second Okazaki fragments. A third Okazaki fragment is made. The leading strand continues to elongate.

DNA ligase

Third primer

GENOMES & PROTEOMES

DNA Polymerases Are a Family of Enzymes with Specialized Functions

Thus far, we have examined the general properties of DNA replication. Three important issues are speed, fidelity, and completeness. DNA replication must proceed quickly, with great accuracy, and gaps should not be left in the newly made strands. To ensure that these three requirements are met, nearly all living species produce more than one type of DNA polymerase. Each type may differ with regard to the rate of DNA synthesis, the accuracy of replication, and/or the ability to prevent the formation of DNA gaps. Let's first consider how evolution produced these different forms of DNA polymerase, and then examine how their functions are finely tuned to the process of DNA replication.

The genomes of most species have several DNA polymerase genes due to the phenomenon of gene duplication, which is described in Chapter 21. During evolution, independent genetic changes have altered the properties of each gene to produce a family of DNA polymerase enzymes with more specialized functions. These changes are suited to the organism in which they are found. For comparison, let's consider the bacterium *E. coli* and humans. *E. coli* has five different DNA polymerases, designated I, II, III, IV, and V. In humans, over a dozen different DNA polymerases have been identified (**Table 11.3**). Why does *E. coli* need 5 DNA polymerases while humans need 12 or more? The answer lies in specialization and the unique functional needs of each species.

In *E. coli*, DNA polymerase III is responsible for the majority of DNA replication. This enzyme is composed of multiple subunits, each with its own functional role. In addition to the catalytic subunit that actually synthesizes DNA, DNA polymerase III has other subunits that allow it to clamp onto the template DNA and synthesize new DNA very rapidly and with high fidelity. By comparison, DNA polymerase I is composed of a single subunit. Its role during DNA replication is to rapidly remove the RNA primers and fill in the short vacant regions with DNA. DNA polymerases II, IV, and V are involved in repairing DNA, and also in replicating DNA that has been damaged. DNA polymerases I and III become stalled when they encounter DNA damage and may be unable to make a complementary strand at such a site.

Figure 11.21 **Synthesis of the leading and lagging strands.** In the leading strand, DNA primase synthesizes an RNA primer at the origin, and then the leading strand is made in the same direction as the movement of the fork. Later, the RNA primer is removed, and the resulting gap is filled in by DNA polymerase, and connected by DNA ligase. In the lagging strand, primase periodically synthesizes an RNA primer near the opening of the replication fork, and then DNA polymerase synthesizes a short DNA fragment in a direction away from the fork. The RNA primers are removed by DNA polymerase, which also fills in the vacant region with DNA. Finally, DNA ligase connects adjacent Okazaki fragments.

Biological inquiry: Briefly describe the movement of primase in the lagging strand in this figure. When does it move from left to right, and when does it move from right to left?

Table 11.3	DNA Polymerases in *E. coli* and Humans
Polymerase types*	**Function**
E. coli:	
III	Replicates most of the DNA during cell division
I	Removes RNA primers and fills in the gaps
II, IV, and V	Repairs damaged DNA and replicates over DNA damage
Humans:	
α (alpha)	Primes DNA and synthesizes short DNA strands
δ (delta), ε (epsilon)	Displaces DNA polymerase α and then replicates DNA at a rapid rate
γ (gamma)	Replicates the mitochondrial DNA
η (eta), κ (kappa), ι (iota), ζ (zeta)	Replicates over damaged DNA
α, β (beta), δ, ε, σ (sigma), λ (lambda), μ (mu), φ (phi), θ (theta)	Repairs DNA or has other functions

*Certain DNA polymerases may have more than one function.

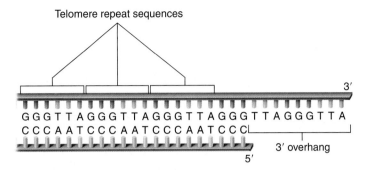

Figure 11.22 Telomere sequences at the end of a human chromosome. The telomere sequence shown here is found in humans and other mammals. The length of the 3′ overhang is variable among different species and cell types.

By comparison, DNA polymerases II, IV, and V don't stall. Therefore, though their rate of synthesis is not as rapid as DNA polymerases I and III, they ensure that DNA replication is complete.

In human cells, the DNA polymerases are designated with Greek letters (Table 11.3). DNA polymerase α has its own "built-in" primase subunit. It synthesizes short RNA regions followed by short DNA regions. Two other DNA polymerases, δ (delta) and ε (epsilon), then extend the DNA at a faster rate. DNA polymerase γ (gamma) functions in the mitochondria to replicate mitochondrial DNA.

Certain DNA polymerases function as lesion-replicating enzymes. Like DNA polymerases I and III in *E. coli*, when the general DNA polymerases (α, δ, or ε) encounter abnormalities in DNA structure (that is, a lesion), they may be unable to replicate over the damage. If this happens, lesion-replicating polymerases are attracted to the damaged DNA. These polymerases have special properties that enable them to synthesize a complementary strand over the abnormal region. Each type of lesion-replicating polymerase may be able to replicate over different kinds of DNA damage. Similarly, other human DNA polymerases play an important role in DNA repair. The need for multiple repair enzymes is rooted in the various ways that DNA can be damaged, as described in Chapter 14. Multicellular organisms must be particularly vigilant about repairing DNA or cancer may occur.

Telomerase Attaches DNA Sequences at the Ends of Eukaryotic Chromosomes

We will end our discussion of DNA replication by considering a specialized form of DNA replication that happens only in eukaryotic cells. This unusual form of DNA replication occurs at the ends of eukaryotic chromosomes. This region, called the **telomere**, is composed of a series of repeat sequences within the DNA and the special proteins that are bound to those sequences. Telomeric sequences consist of a short sequence that is repeated a few dozen to a few hundred times in a row (**Figure 11.22**). Also, a telomere has a region at the 3′ end that is termed a 3′ overhang, because it does not have a complementary strand. The repeat sequence shown here, 5′–GGGTTA–3′, is the sequence found in human telomeres. Other organisms have different repeat sequences. For example, the sequence found in the telomeres of maize is 5′–GGGTTTA–3′.

As discussed previously, DNA polymerase synthesizes DNA only in a 5′ to 3′ direction, and requires a primer. For these reasons, DNA polymerase cannot copy the tip of the DNA strand with a 3′ end (**Figure 11.23**). At this location, there is no place on the parental DNA strand for an upstream primer to be made. Conceivably, an RNA primer could be made that is complementary to the very tip of this DNA strand. However, after such a primer is removed by DNA polymerase, it would still leave a short, unreplicated region. Therefore, if this replication problem was not solved, a linear chromosome would become progressively shorter with each round of DNA replication.

In 1984, Carol Greider and Elizabeth Blackburn discovered an enzyme called **telomerase** that prevents chromosome shortening. It attaches many copies of a repeated DNA sequence to the ends of chromosomes (**Figure 11.24**). The telomerase enzyme contains both protein and RNA. The RNA part of telomerase has a sequence that is complementary to the DNA sequence found in the telomeric repeat. This allows telomerase to bind to the 3′ overhang region of the telomere. Following binding, the RNA sequence beyond the binding site functions as a template, allowing telomerase to synthesize a six-nucleotide sequence at the end of the DNA strand. The enzyme then moves to the new end of this DNA strand and attaches another six nucleotides to the end. This occurs many times and thereby greatly lengthens the 3′ end of the DNA in the telomeric region. This lengthening provides an upstream site for an RNA primer to be made. DNA polymerase then synthesizes the complementary DNA strand.

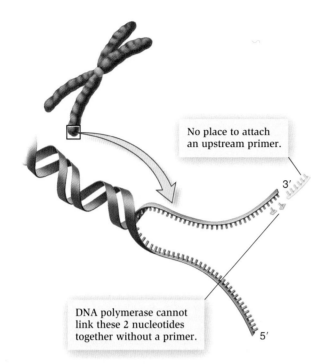

Figure 11.23 Enzymatic features of DNA polymerase that account for its inability to copy one of the DNA strands at the end of a linear chromosome.

In this way, the progressive shortening of eukaryotic chromosomes is prevented.

Researchers have discovered an interesting connection between telomeres and cellular aging. In humans and other mammals, the cells of the body have a predetermined life span. For example, if a small sample of skin is removed from a person's body and grown in the laboratory, the cells will double a finite number of times. Furthermore, the number of doublings depends on the age of the person from which the sample is taken. If a sample is from an infant, the cells will typically double about 80 times, while if a sample is from an older person, the cells will double only 10 to 20 times before division ceases. Cells that have doubled many times and have reached a point where they have lost the capacity to divide any further are termed **senescent**.

The progressive shortening of telomeres is correlated with cellular senescence, though the relationship between the two phenomena is not well understood. The telomerase enzyme is normally present in germ-line cells, which give rise to gametes, and also in many rapidly dividing somatic cells. However, telomerase function is typically reduced as an organism ages. In 1998, Andrea Bodnar and her colleagues inserted a gene that encodes a highly active telomerase into human cells grown in the laboratory, using techniques described in Chapter 20. The results were amazing. The expression of telomerase prevented telomere shortening and cellular senescence. The cells expressing telomerase continued to divide, just like younger, healthy cells!

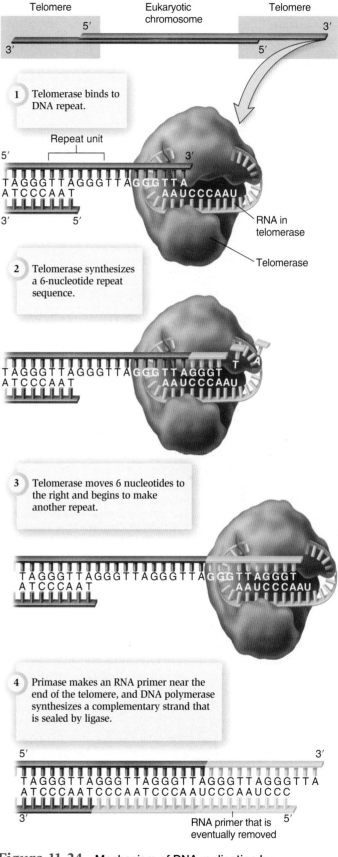

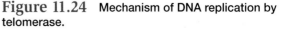

Figure 11.24 Mechanism of DNA replication by telomerase.

Telomerase function is also associated with cancer. When cells become cancerous, they continue to divide uncontrollably. In 90% of all types of human cancers, telomerase has been found to be present at high levels in the cancerous cells. This prevents telomere shortening and may play a role in the continued growth of cancer cells. The mechanism whereby cancer cells are able to increase the function of telomerase is not well understood and is a topic of active research.

CHAPTER SUMMARY

11.1 Biochemical Identification of the Genetic Material

- The genetic material carries information to produce the traits of organisms. It is replicated and transmitted from cell to cell and generation to generation, and it has differences that explain the variation among different organisms.

- Griffith's work with type R and type S bacteria was consistent with the transfer of genetic material, which he called the transformation principle. (Figure 11.1)

- Avery, MacLeod, and McCarty used biochemical methods to show that DNA is the transformation principle. (Figure 11.2)

- Hershey and Chase labeled T2 phage with ^{35}S and ^{32}P and determined that the ^{32}P-labeled DNA is the genetic material of this phage. (Figures 11.3, 11.4)

11.2 Nucleic Acid Structure

- DNA is composed of nucleotides, which covalently link together to form DNA strands. Two DNA strands are held together by hydrogen bonds between the bases. Chromosomes are made of DNA and proteins. (Figure 11.5)

- Nucleotides are composed of a phosphate, sugar, and nitrogenous base. The sugar can be deoxyribose (DNA) or ribose (RNA). The purine bases are adenine and guanine, while the pyrimidine bases are thymine (DNA only), cytosine, and uracil (RNA only). (Figure 11.6)

- The atoms in a nucleotide are numbered in a conventional way. (Figure 11.7)

- In a strand of DNA (or RNA), the sugars are connected by covalent bonds in a 5′ to 3′ direction. (Figure 11.8)

- Watson, Crick, and Wilkins used the X-ray diffraction data of Franklin and the biochemical data of Chargaff (that is, A = T, G = C), and constructed ball-and-stick models to reveal the double helix structure of DNA. (Figures 11.9, 11.10, Table 11.1)

- DNA is a double helix in which the DNA strands are antiparallel and obey the AT/GC rule. (Figures 11.11, 11.12)

11.3 DNA Replication

- Meselson and Stahl used ^{15}N- and ^{14}N-isotope labeling methods to show that DNA is replicated by a semiconservative mechanism in which the product of DNA replication is one original strand and one new strand. (Figures 11.13, 11.14)

- New DNA strands are made according to the AT/GC rule in which parental strands serve as templates for the synthesis of new daughter strands. (Figure 11.15)

- Within chromosomes, DNA replication begins at sites known as origins of replication, where the DNA unwinds to make replication forks. Bacterial chromosomes have a single origin, while eukaryotic chromosomes have many. (Figure 11.16)

- DNA synthesis occurs bidirectionally from an origin of replication. The synthesis of new DNA strands happens near each replication fork. The leading strand is made continuously, in the same direction that the fork is moving. The lagging strand is made in the opposite direction as short Okazaki fragments that are connected together. (Figure 11.17)

- DNA helicase separates DNA strands, single-strand binding proteins keep them separated, and DNA topoisomerase alleviates coiling ahead of the fork. (Figure 11.18, Table 11.2)

- Deoxynucleoside triphosphates bind to the template strands, and DNA polymerase catalyzes the formation of a phosphoester bond between the 3′ end of the strand and a deoxynucleoside monophosphate. Pyrophosphate is released. (Figure 11.19)

- DNA polymerase requires a primer and can make new DNA strands only in the 5′ to 3′ direction. (Figure 11.20)

- Living organisms have several different types of DNA polymerases with specialized functions. (Table 11.3)

- DNA primase makes one primer in the leading strand and multiple primers in the lagging strand. DNA polymerase extends these primers with DNA, and it removes the primers when they are no longer needed. DNA ligase connects adjacent Okazaki fragments. (Figure 11.21, Table 11.2)

- The ends of linear, eukaryotic chromosomes have telomeres composed of repeat sequences. DNA polymerase cannot synthesize the tip of one of these DNA strands (Figures 11.22, 11.23)

- Telomerase binds to the telomere repeat sequence and synthesizes a six-nucleotide repeat. This happens many times in a row to lengthen one DNA strand of the telomere. DNA primase, DNA polymerase, and DNA ligase are needed to synthesize the complementary DNA strand. (Figure 11.24)

TEST YOURSELF

1. Why did researchers initially believe the genetic material was protein?
    a. Proteins are more biochemically complex than DNA.
    b. Proteins are found only in the nucleus, but DNA is found in many areas of the cell.
    c. Proteins are much larger molecules and can store more information than DNA.
    d. All of the above.
    e. Both a and c.

2. Considering the components of a nucleotide, what component always determines whether the nucleotide would be incorporated into a DNA strand or an RNA strand?
    a. phosphate group
    b. pentose sugar
    c. nitrogenous base
    d. Both b and c

3. Which of the following equations would be appropriate when considering DNA base composition?
    a. %A + %T = %G + %C
    b. %A = %G
    c. %A = %G = %T = %C
    d. %A + %G = %T + %C

4. If the sequence of a section of DNA is 5′–CGCAACTAC–3′, what is the appropriate sequence for the opposite strand?
    a. 5′–GCGTTGATG–3′
    b. 3′–ATACCAGCA–5′
    c. 5′–ATACCAGCA–3′
    d. 3′–GCGTTGATG–5′

5. Of the following statements, which is correct when considering the process of DNA replication?
    a. New DNA molecules are composed of two completely new strands.
    b. New DNA molecules are composed of one strand from the old molecule and one new strand.
    c. New DNA molecules are composed of strands that are a mixture of sections from the old molecule and sections that are new.
    d. None of the above.

6. Meselson and Stahl were able to demonstrate semiconservative replication in *E. coli* by
    a. using radioactive isotopes of phosphorus to label the old strand and visually determining the relationship of old and new DNA strands.
    b. using different enzymes to eliminate old strands from DNA.
    c. using isotopes of nitrogen to label the DNA and determining the relationship of old and new DNA strands by density differences of the new molecules.
    d. labeling viral DNA before it was incorporated into a bacterial cell and visually determining the location of the DNA after centrifugation.

7. During replication of a DNA molecule, the daughter strands are not produced in exactly the same manner. One strand, the leading strand, is made toward the replication fork, while the lagging strand is made in fragments in the opposite direction. This difference in the synthesis of the two strands is the result of which of the following?
    a. DNA polymerase is not efficient enough to make two "good" strands of DNA.
    b. The two parental strands are antiparallel, and DNA polymerase makes DNA only in the 5′ to 3′ direction.
    c. The lagging strand is the result of DNA breakage due to UV light.
    d. The cell does not contain enough nucleotides to make two complete strands.

8. Considering the different proteins involved in DNA replication and their functions, which of the following *does not* represent an accurate relationship?
    a. DNA helicase unwinds the DNA molecule.
    b. DNA primase produces the Okazaki fragments of the lagging strand.
    c. DNA topoisomerase reduces the coiling ahead of the replication fork.
    d. Single-strand binding proteins prevent the double helix from re-forming.

9. Which of the following *does not* promote the fidelity of DNA replication?
    a. DNA polymerase will not form bonds between nucleotides if there is a mismatched base.
    b. DNA polymerase has the ability to recognize mismatched bases and remove them.
    c. Hydrogen bonds that hold the two strands together are more stable in correctly matched bases than between mismatched bases.
    d. All of the above are correct.

10. Most organisms have more than one form of DNA polymerase. Some forms play specific roles in replication, while others play roles in DNA repair. What is the proposed mechanism for the evolution of the multiple forms of DNA polymerase?
    a. Increased copies of polymerase genes due to chromosome number change over several generations.
    b. Species hybridization may introduce new forms of polymerases.
    c. Gene duplication followed by mutations alters the activity of the polymerase.
    d. Mutations change other genes into DNA polymerases.

CONCEPTUAL QUESTIONS

1. List and explain the four characteristics of the genetic material.

2. Explain or describe the essential features of the Watson and Crick model of the structure of DNA.

3. Explain the function of telomerase.

EXPERIMENTAL QUESTIONS

1. Avery, MacLeod, and McCarty worked with two strains of *Streptococcus pneumoniae* to determine the biochemical identity of the genetic material. Explain the characteristics of the *Streptococcus pneumoniae* strains that made them particularly well suited for such an experiment.

2. In the experiment of Avery, MacLeod, and McCarty, what was the purpose of using the protease, RNase, and DNase if only the DNA extract caused transformation?

3. The Hershey and Chase experiment used radioactive isotopes to track the DNA and protein of phages as they infected bacterial cells. Explain how this procedure allowed them to detect that DNA was the genetic material of this particular virus.

COLLABORATIVE QUESTIONS

1. Discuss the reasons why there are very few errors made during DNA replication.

2. What was Frederick Griffith's contribution to the study of DNA and why was it so important?

www.brookerbiology.com

This website includes answers to the Biological Inquiry questions found in the figure legends and all end-of-chapter questions.

12

GENE EXPRESSION AT THE MOLECULAR LEVEL

CHAPTER OUTLINE

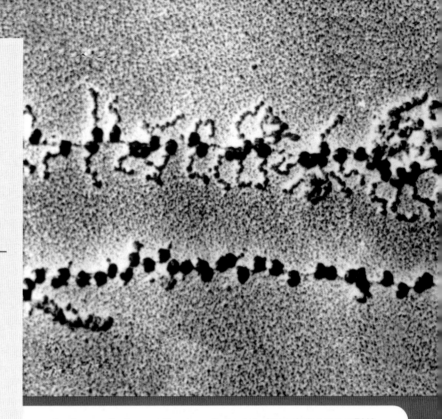

A micrograph of many ribosomes in the act of translating two mRNA molecules into polypeptides. The short polypeptides are seen emerging from the ribosomes.

A gene is defined as a unit of heredity. We can consider gene function at different biological levels. In Chapter 16, we will examine how genes affect the traits or characteristics of individuals. For example, we will consider how the transmission of genes from parents to offspring affects the color of the offspring's eyes and their likelihood of becoming bald. By comparison, in this chapter we will explore how genes work at the molecular level. You will learn how DNA sequences are organized to form genes, and how those genes are used as a template to make RNA copies, ultimately leading to the synthesis of a functional protein. The term **gene expression** can refer to gene function both at the level of traits and at the molecular level. In reality, the two phenomena are intricately woven together. The expression of genes at the molecular level affects the structure and function of cells, which, in turn, determine the traits that an organism expresses.

In this chapter, we will first consider how researchers came to realize that most genes store the information to make proteins. Then, we will explore the steps of gene expression as they occur at the molecular level. These steps include the use of a gene as a template to make an RNA molecule, the processing of the RNA into a functional molecule, and the use of RNA molecules to make proteins.

12.1 Overview of Gene Expression

Even before DNA was known to be the genetic material, scientists had asked the question, How do genes function to create the traits of living organisms? At the molecular level, a similar question can be asked, namely, How do genes affect the composition and/or function of molecules that are found within living

cells? The approach that was successful at answering this question involved the study of **mutations**, which are changes in the genetic material. Mutations can affect the genetic blueprint by altering gene function. For this reason, research that was focused on the effects of mutations proved instrumental in determining the molecular function of genes.

In this section, we will begin by considering two early experiments in which researchers studied the effects of mutations in humans and in a bread mold. Even though they involved different organisms, both studies led to the conclusion that the role of certain genes is to encode enzymes. Then we will examine the general features of gene expression at the molecular level.

The Study of Inborn Errors of Metabolism Suggested That Genes Encode Enzymes

In 1908, Archibald Garrod, a British physician, proposed a relationship between genes and the production of enzymes. Prior to his work, biochemists had studied many metabolic pathways that consist of a series of conversions of one molecule to another, each step catalyzed by an enzyme. **Figure 12.1** illustrates part of the metabolic pathway for the breakdown of phenylalanine, an amino acid commonly found in human diets. The enzyme phenylalanine hydroxylase catalyzes the conversion of phenylalanine to tyrosine, another amino acid. A different enzyme, tyrosine aminotransferase, converts tyrosine into the next molecule, called p-hydroxyphenylpyruvic acid. In each case, a specific enzyme catalyzes a single chemical reaction.

Garrod studied patients with defects in the ability to metabolize certain compounds. Much of his work centered on the inherited disease alkaptonuria, in which the patient's body accumulates abnormal levels of homogentisic acid (also called alkapton).

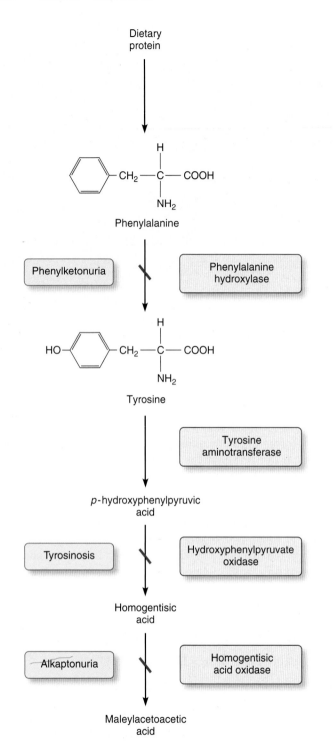

Dietary protein

Phenylalanine

Phenylketonuria — Phenylalanine hydroxylase

Tyrosine

Tyrosine aminotransferase

p-hydroxyphenylpyruvic acid

Tyrosinosis — Hydroxyphenylpyruvate oxidase

Homogentisic acid

Alkaptonuria — Homogentisic acid oxidase

Maleylacetoacetic acid

Figure 12.1 **The metabolic pathway that breaks down phenylalanine and its relationship to certain genetic diseases.** Each step in the pathway is catalyzed by a different enzyme, shown in the *right* boxes. If one of the enzymes is not functioning, the previous compound builds up, causing the conditions named in the *left* boxes. In this pathway, defects in phenylalanine hydroxylase, hydroxyphenylpyruvate oxidase, and homogentisic acid oxidase cause phenylketonuria, tyrosinosis, and alkaptonuria, respectively.

This compound, which is bluish black, results in discoloration of the skin and cartilage and causes the urine to appear black. Garrod hypothesized that the accumulation of homogentisic acid in these patients is due to a missing enzyme, namely homogentisic acid oxidase (Figure 12.1). Furthermore, he already knew that alkaptonuria is an inherited condition that follows what is called a recessive pattern of inheritance. As discussed in Chapter 16, if a disorder is inherited in a recessive manner, an individual with the disease has inherited the mutant (defective) gene that causes the disorder from each parent. Based on these observations, Garrod proposed a relationship between the inheritance of a mutant gene and a defect in metabolism. In the case of alkaptonuria, if an individual inherited the mutant gene from both parents, he or she would not produce any normal enzyme and would be unable to metabolize homogentisic acid. Garrod described alkaptonuria as an **inborn error of metabolism**. At the turn of the last century, this was a particularly insightful idea, because the structure and function of the genetic material were completely unknown.

Beadle and Tatum Proposed the One Gene–One Enzyme Hypothesis

Although Garrod published a book and several papers on inherited metabolic disorders, his work was largely ignored until the early 1940s, when it was rediscovered by George Beadle and Edward Tatum. Beadle and Tatum were also interested in the relationships among genes, enzymes, and traits. They decided to study the genetics of simple nutritional requirements in *Neurospora crassa*, a common bread mold. *Neurospora* is easily grown in the laboratory and has few nutritional requirements. The minimum requirements for growth are a carbon source (namely, sugar), inorganic salts, and one vitamin known as biotin. Otherwise, *Neurospora* has many different cellular enzymes that synthesize the small molecules, such as amino acids and many vitamins, that are essential for growth.

Hypothesizing that genes encode enzymes, Beadle and Tatum reasoned that a mutation or change in a gene might cause a defect in an enzyme that is needed for the synthesis of an essential molecule, such as a vitamin or amino acid. A mutant strain (one that carries such a mutation) would be unable to grow unless the vitamin or amino acid was supplemented in the growth medium. Strains without a mutation are called wild-type strains. In their original study of 1941, Beadle and Tatum isolated mutant strains that required vitamins for growth. In each case, a single mutation resulted in the requirement for a single type of vitamin in the growth medium. This early study by Beadle and Tatum led to additional research to study enzymes involved with the synthesis of other substances, such as the amino acid arginine. At that time, the pathway leading to arginine synthesis was thought to involve certain precursor molecules, including ornithine and citrulline. A simplified pathway for arginine synthesis is shown in **Figure 12.2a**. Each step is catalyzed by a different enzyme.

Researchers first isolated several different mutants that required arginine for growth. They hypothesized that each mutant strain might be blocked at only a single step in the consecutive series of reactions that lead to arginine synthesis. To test this hypothesis, the mutant strains were examined for their ability to grow in the presence of ornithine, citrulline, and/or arginine. A simplified depiction of the results is shown in **Figure 12.2b**. Based on their growth properties, the mutant strains that had been originally identified as requiring arginine for growth could be placed into three groups, designated 1, 2, and 3. Group 1 mutants were missing enzyme 1, needed for the conversion of a precursor molecule into ornithine. They could grow only if ornithine, citrulline, or arginine was added to the growth medium. Group 2 mutants were missing the second enzyme in this pathway that is needed for the conversion of ornithine into citrulline. The group 2 mutants would not grow if only ornithine was added but could grow if citrulline or arginine was added. Finally, the group 3 mutants were missing the enzyme needed for the conversion of citrulline into arginine. These mutants could grow only if arginine was added. Based on these results, the researchers were able to order the functions of the genes involved in arginine synthesis in the following way:

Group 1	Group 2	Group 3
⟶	⟶	⟶
Ornithine	Citrulline	Arginine

From these results and earlier studies, Beadle and Tatum concluded that a single gene controlled the synthesis of a single enzyme. This was referred to as the *one gene–one enzyme hypothesis*.

In later decades, this idea had to be modified in two ways. First, enzymes are only one category of cellular proteins. All proteins are encoded by genes, and many do not function as enzymes. Second, some proteins are composed of two or more different polypeptides. Therefore, a more accurate statement is that a gene encodes a polypeptide. The term **polypeptide** denotes structure; it is a linear sequence of amino acids. By comparison, the term **protein** denotes function. Some proteins are composed of one polypeptide. In such cases, a single gene does encode a single protein. In other cases, however, a functional protein is composed of two or more different polypeptides. An example is hemoglobin, the protein that carries oxygen in red blood cells, which is composed of two α-globin and two β-globin polypeptides. In this case, the expression of two genes (that is, the α-globin and β-globin genes) is needed to create a functional protein. Based on this more recent information, the original concept proposed by Beadle and Tatum has been modified to the **one gene–one polypeptide theory**.

Molecular Gene Expression Involves the Processes of Transcription and Translation

Thus far, we have considered two classic studies that led researchers to conclude that genes encode enzymes, or more broadly that genes encode polypeptides. Let's now consider the

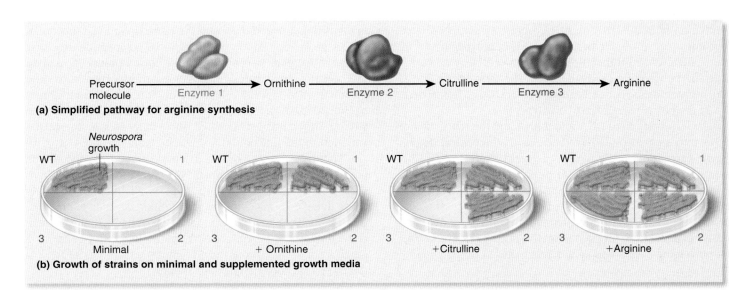

Figure 12.2 An example of an experiment that supported Beadle and Tatum's one gene–one enzyme hypothesis. (a) Simplified pathway for arginine synthesis. (b) Growth of wild-type (WT) and mutant (groups 1, 2, and 3) *Neurospora* strains in the presence of ornithine, citrulline, or arginine. As shown on the *left*, the mutant strains did not grow on minimal media lacking arginine. However, all mutants could grow if arginine was added, and certain mutants could grow if either ornithine or citrulline was added.

Biological inquiry: What type of enzyme function is missing in group 2 mutants?

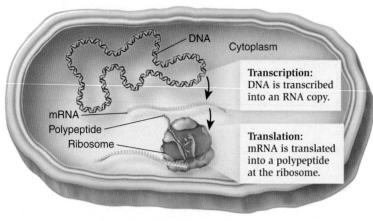

(a) Molecular gene expression in prokaryotes

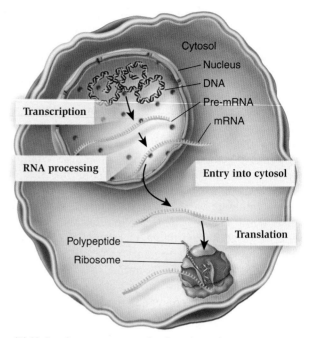

(b) Molecular gene expression in eukaryotes

Figure 12.3   The central dogma of gene expression at the molecular level.   (a) In bacteria that lack a cell nucleus, transcription and translation occur in the cytoplasm. (b) In eukaryotes, transcription and RNA processing occur in the nucleus, while translation takes place in the cytosol.

general steps of gene expression at the molecular level. The first step, known as **transcription**, produces an RNA copy of a gene, also called an RNA transcript (**Figure 12.3a**). The term transcription literally means the act of making a copy. Most genes, which are termed **structural genes**, produce an RNA molecule that contains the information to specify a polypeptide with a specific amino acid sequence. This type of RNA is termed **messenger RNA** (abbreviated **mRNA**), because its job is to carry information from the DNA to the ribosome. As discussed later, **ribosomes** play a key role in the synthesis of polypeptides. The second step, termed **translation**, is the process of synthesizing a specific polypeptide on a ribosome. The term translation is used because a nucleotide sequence in mRNA is "translated" into an amino acid sequence of a polypeptide.

The transcription of DNA into mRNA and the translation of mRNA into a polypeptide have been called the **central dogma** of gene expression at the molecular level (Figure 12.3a). This central dogma applies equally to prokaryotes and eukaryotes. However, in eukaryotes an additional step occurs between transcription and translation. This step, called **RNA processing**, occurs when the RNA transcript, termed **pre-mRNA**, is modified in ways that make it a functionally active mRNA (**Figure 12.3b**). The processing events will be described later in this chapter.

The Protein Products of Genes Determine an Organism's Characteristics

The genes that constitute the genetic material provide a blueprint for the characteristics of every organism. They contain the information necessary to create an organism and allow it to favorably interact with its environment. Each structural gene stores the information for the production of a polypeptide, which then

becomes a unit within a functional protein. The activities of proteins determine the structure and function of cells. Furthermore, the traits or characteristics of all organisms are rooted in the activities of cellular proteins. As an example, **Figure 12.4** shows two fruit flies, one with red eyes and one with white eyes. A normal gene necessary for red eye color encodes a protein that transports pigment precursor molecules into the cell. The white-eye trait is due to a mutant gene. Depending on where the mutation is found, it could prevent the transcription of a functional mRNA or the translation of a functional polypeptide. In either case, a functional transporter would not be produced, and so the precursor molecule would not be transported across the membrane and into the cell. Therefore, the red pigment cannot be synthesized and the eyes are white. In this example, the striking difference between red and white eyes is due to the variation in the function of a single protein!

The main purpose of the genetic material is to encode the production of cellular proteins in the correct cell, at the proper time, and in suitable amounts. This is an intricate task, because living cells make thousands of different kinds of proteins. Genetic analyses have shown that a typical bacterium can make a few thousand different proteins, and estimates for eukaryotes range from several thousand in simpler eukaryotes to tens of thousands in more complex eukaryotes.

12.2 Transcription

Thus far we have considered the central dogma of gene expression. According to this theory, DNA is used as a template to make mRNA, which contains the information to make a polypeptide. This section will focus on the first step of this process,

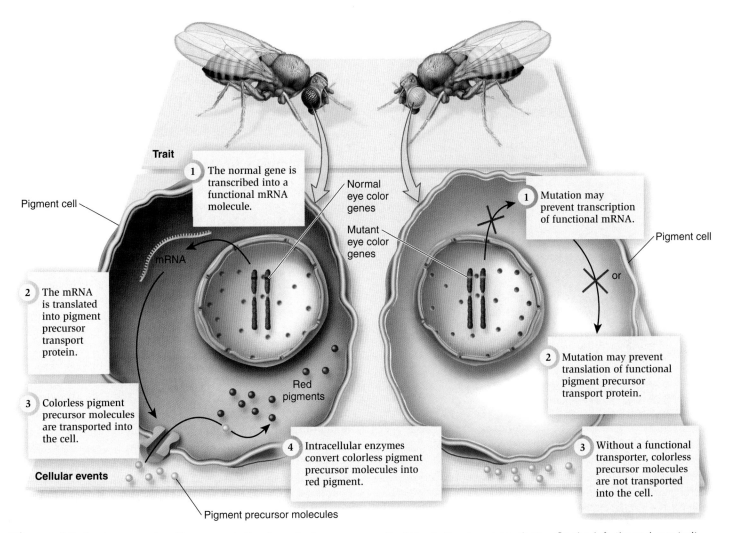

Trait

Pigment cell

1 The normal gene is transcribed into a functional mRNA molecule.

Normal eye color genes

Mutant eye color genes

mRNA

2 The mRNA is translated into pigment precursor transport protein.

3 Colorless pigment precursor molecules are transported into the cell.

Cellular events

Red pigments

4 Intracellular enzymes convert colorless pigment precursor molecules into red pigment.

Pigment precursor molecules

1 Mutation may prevent transcription of functional mRNA.

Pigment cell

or

2 Mutation may prevent translation of functional pigment precursor transport protein.

3 Without a functional transporter, colorless precursor molecules are not transported into the cell.

Figure 12.4   **An example of how a mutation in a single gene can affect the traits of an organism.**   On the *left*, the red-eye trait occurs when the normal eye color gene is expressed and produces a functional protein. However, as shown on the *right*, a mutation that renders the eye color gene inactive causes a fly to have white eyes.

namely transcription. Our understanding of transcription initially came from studies involving bacteria and bacteriophages (viruses that infect bacterial cells). In 1956, Eliot Volkin and Lazarus Astrachan first proposed that RNA is derived from the transcription of DNA. When they infected *Escherichia coli* cells with a bacteriophage (or phage for short), they discovered that the RNA made immediately after infection had a base composition that was very similar to the base composition of the phage DNA. They suggested that the bacteriophage DNA was used as a template to make bacteriophage RNA.

We now know that DNA is indeed an information storage unit. For genes to be expressed, the information in them must be accessed at the molecular level. Rather than accessing the information directly, however, a working copy of the DNA, composed of RNA, is made. This occurs by the process of transcription, in which a DNA sequence is copied into an RNA sequence. Importantly, transcription does not permanently alter the struc-

ture of DNA. Therefore, the same DNA can continue to store information even after an RNA copy has been made. In this section, we will examine the steps that are necessary for genes to act as transcriptional units. We will also consider some differences in these steps between prokaryotes and eukaryotes.

At the Molecular Level, a Gene Can Be Transcribed and Produces a Functional Product

A molecular definition of a **gene** is *an organized unit of DNA sequences that enables a segment of DNA to be transcribed into RNA and ultimately results in the formation of a functional product.* When a structural gene is transcribed, an mRNA is made that specifies the amino acid sequence of a polypeptide. In this case, the polypeptide is considered to be the functional product, while the mRNA is an intermediary in polypeptide synthesis. Among all species, over 90% of all genes are structural genes.

DNA

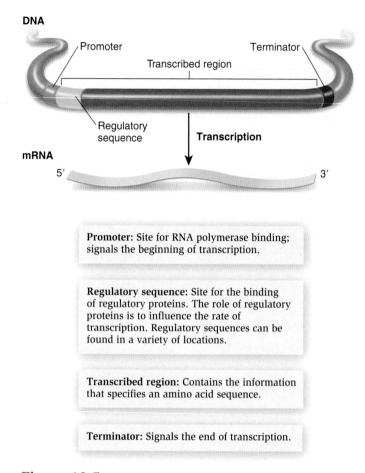

Promoter: Site for RNA polymerase binding; signals the beginning of transcription.

Regulatory sequence: Site for the binding of regulatory proteins. The role of regulatory proteins is to influence the rate of transcription. Regulatory sequences can be found in a variety of locations.

Transcribed region: Contains the information that specifies an amino acid sequence.

Terminator: Signals the end of transcription.

Figure 12.5 A structural gene as a transcriptional unit.

Biological inquiry: If a terminator was removed from a gene, how would this affect transcription? Where would transcription end?

However, for other genes, the functional product is the RNA itself. The RNA from a nonstructural gene is never translated. Two important products of nonstructural genes are transfer RNA and ribosomal RNA. **Transfer RNA (tRNA)** translates the language of mRNA into that of amino acids, while **ribosomal RNA (rRNA)** forms part of ribosomes, which provide the arena in which translation occurs. We'll learn more about these two types of RNA later in this chapter.

A gene is composed of specific base sequences organized in a way that allows the DNA to be transcribed into RNA. **Figure 12.5** shows the general organization of sequences that are needed to form a structural gene. Transcription begins near a site in the DNA called the **promoter**, while the **terminator** specifies the end of transcription. Therefore, these two sequences provide the boundaries for RNA to be synthesized within a defined location. Within this transcribed region is the information that will specify the amino acid sequence of the polypeptide when the mRNA is translated. As shown in Figure 12.5, the DNA is transcribed into mRNA from the end of the promoter through the coding sequence to the terminator.

Other DNA sequences are involved in the regulation of transcription. These **regulatory sequences** function as sites for

genetic regulatory proteins. When a regulatory protein binds to a regulatory sequence, this affects the rate of transcription. Some regulatory proteins enhance the rate of transcription, while others inhibit it. This topic will be considered in Chapter 13. Now let's look at the process of transcription in prokaryotes.

During Transcription, RNA Polymerase Uses a DNA Template to Make RNA

Transcription occurs in three stages called initiation, elongation, and termination, during which proteins interact with DNA sequences (**Figure 12.6a**). The **initiation stage** is a recognition step. In bacteria such as *E. coli*, a protein called **sigma factor** binds to **RNA polymerase**, which is the enzyme that synthesizes strands of RNA. Sigma factor also recognizes the base sequence of a promoter and binds there. In this way, sigma factor causes RNA polymerase to recognize a promoter sequence.

RNA polymerase synthesizes the RNA transcript during the **elongation stage**. For this to occur, sigma factor is released and RNA polymerase slides along the DNA in a way that separates the DNA strands, forming a small bubble-like structure known as the **open complex** (or transcription bubble) that is approximately 10 to 15 base pairs long. The DNA strand that is used as a template for RNA synthesis is called the **template** or **noncoding strand**. The opposite DNA strand is called the **coding strand**.

As shown in Figure 12.6, a new molecule of RNA is synthesized in the 5′ to 3′ direction. The complementarity rule used in this process is similar to the AT/GC rule of DNA replication, except that uracil substitutes for thymine in RNA. For example, an RNA with a sequence of nucleotides reading 5′–AUGUUACAUCGG–3′ will be transcribed from a DNA template with a sequence of 3′–TACAATGTAGCC–5′. In bacteria, the rate of RNA synthesis is about 40 nucleotides per second! Behind the open complex, the DNA rewinds back into a double helix. Eventually, RNA polymerase reaches a termination sequence, which causes it and the newly made RNA transcript to dissociate from the DNA. This event constitutes the **termination stage** of transcription.

The catalytic portion of RNA polymerase that is responsible for the synthesis of RNA has a similar structure in all species, including prokaryotes and eukaryotes. The structure of a bacterial RNA polymerase is shown in **Figure 12.6b**. RNA polymerase contains a cavity that allows it to slide along the DNA. The DNA strands enter at the side of the protein and are separated at a region where the cavity makes a right-angle turn. Nucleotides can enter this region through a small pore, and RNA is made in a 5′ to 3′ direction. Both the DNA and the newly made strand of RNA then exit from the top of the protein.

When considering the transcription of multiple genes within a chromosome, the direction of transcription and the DNA strand that is used as a template vary among different genes. **Figure 12.7** shows three genes that are adjacent to each other within a chromosome. Genes A and B are transcribed from left to right, using the bottom DNA strand as a template.

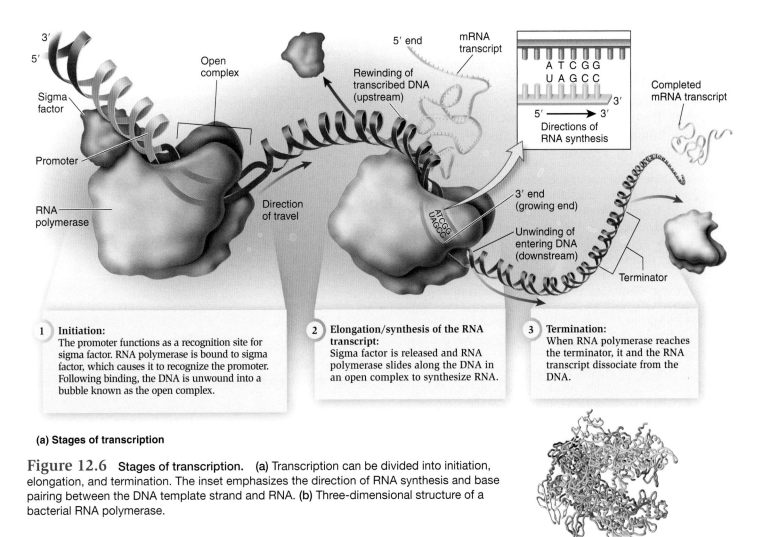

(a) Stages of transcription

1 **Initiation:**
The promoter functions as a recognition site for sigma factor. RNA polymerase is bound to sigma factor, which causes it to recognize the promoter. Following binding, the DNA is unwound into a bubble known as the open complex.

2 **Elongation/synthesis of the RNA transcript:**
Sigma factor is released and RNA polymerase slides along the DNA in an open complex to synthesize RNA.

3 **Termination:**
When RNA polymerase reaches the terminator, it and the RNA transcript dissociate from the DNA.

Figure 12.6 **Stages of transcription.** **(a)** Transcription can be divided into initiation, elongation, and termination. The inset emphasizes the direction of RNA synthesis and base pairing between the DNA template strand and RNA. **(b)** Three-dimensional structure of a bacterial RNA polymerase.

(b) Structure of a bacterial RNA polymerase

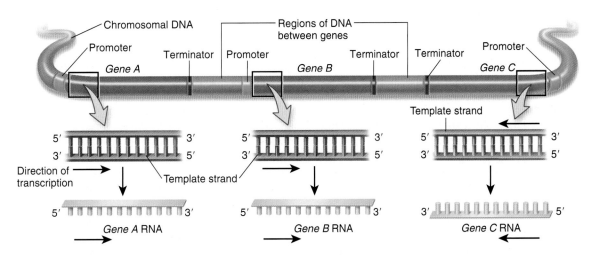

Figure 12.7 **The transcription of three different genes that are found in the same chromosome.** RNA polymerase synthesizes each RNA transcript in a 5′ to 3′ direction, sliding along a DNA template strand in a 3′ to 5′ direction. However, the use of the template strand can vary from gene to gene. For example, genes A and B use the bottom strand, while gene C uses the top strand.

By comparison, gene C is transcribed from right to left, using the top DNA strand as a template. Notice that in all three cases the synthesis of the RNA transcript occurs in a 5′ to 3′ direction, and the template strand is read in the 3′ to 5′ direction.

Eukaryotic Transcription Is Fundamentally Similar to Prokaryotic Except That More Proteins Are Involved in the Process

The basic features of transcription are identical between prokaryotic and eukaryotic organisms. Eukaryotic genes have promoters and the transcription process involves initiation, elongation, and termination. However, in eukaryotes, each step tends to involve a greater complexity of protein components. For example, three forms of RNA polymerase are found in eukaryotes, designated I, II, and III. RNA polymerase II is responsible for transcribing the mRNA from eukaryotic structural genes, while RNA polymerases I and III transcribe nonstructural genes such as the genes that encode tRNAs and rRNAs. By comparison, bacteria have a single type of RNA polymerase that transcribes all genes.

With regard to the initiation of transcription, the eukaryotic process is also more complex. Recall that bacteria such as *E. coli* use a single protein, sigma factor, to recognize the promoter of genes. By comparison, RNA polymerase II of eukaryotes always requires five general transcription factors to initiate transcription. **Transcription factors** are proteins that influence the ability of RNA polymerase to transcribe genes. In addition, the regulation of gene transcription in eukaryotes typically involves the function of several different proteins. This topic will be considered in Chapter 13.

12.3 RNA Processing in Eukaryotes

During the 1960s and 1970s, the physical structure of the gene became well established based largely on studies of bacterial genes. Most bacterial mRNAs can be translated into polypeptides as soon as they are made. By comparison, eukaryotic mRNA transcripts undergo RNA processing or modification that is needed for their proper translation. In eukaryotes, transcription produces a longer RNA, pre-mRNA, which undergoes certain processing events before it exits the nucleus. The final product is called a **mature mRNA**.

In the late 1970s, when the experimental tools became available to study eukaryotic genes at the molecular level, the scientific community was astonished by the discovery that the coding sequences within many eukaryotic structural genes are separated by DNA sequences that are transcribed but not translated into protein. These intervening sequences that are not translated are called **introns**, while coding sequences are found within **exons** contained in the mature mRNA. To create a functional mRNA, the RNA undergoes a process known as **splicing** in which the introns are removed and the remaining exons are

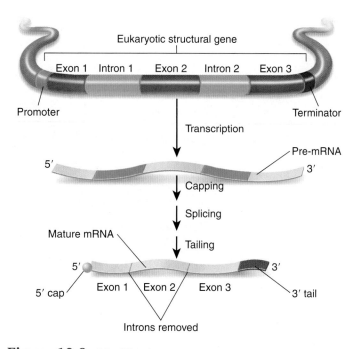

Figure 12.8   Modifications to eukaryotic pre-mRNA that are needed to create a functional (mature) mRNA molecule.

connected to each other (**Figure 12.8**). In addition to splicing, research has shown that pre-mRNA transcripts are modified in other ways, including the addition of caps and tails to the ends of the mRNA. After these modifications have been completed, the mRNA leaves the nucleus and enters the cytosol, where translation occurs. In this section, we will examine the molecular mechanisms that account for these RNA processing events and consider why they are functionally important.

Splicing Involves the Removal of Introns and the Linkage of Exons

In the late 1970s, several research groups, including those of Pierre Chambon, Bert O'Malley, and Phillip Leder, investigated the presence of introns in eukaryotic structural genes. Leder's experiments used electron microscopy to identify introns in the β-globin gene. β-globin is a polypeptide that is a subunit of hemoglobin. To identify introns, Leder used a strategy that involved the binding of mRNA to DNA. In this approach, the double-stranded DNA of the β-globin gene was first separated into single strands, and then mixed with mature mRNA from the β-globin gene that had been isolated from red blood cells. The mature mRNA, which already had its introns removed, was allowed to bind to the template strand of DNA, causing the intron in the DNA to loop out. Next, the coding strand was allowed to bind to the template strand, although it could not bind in places where the RNA was already bound. Leder and his colleagues realized that if mRNA binds to a region of a gene containing one intron, two single-stranded DNA loops

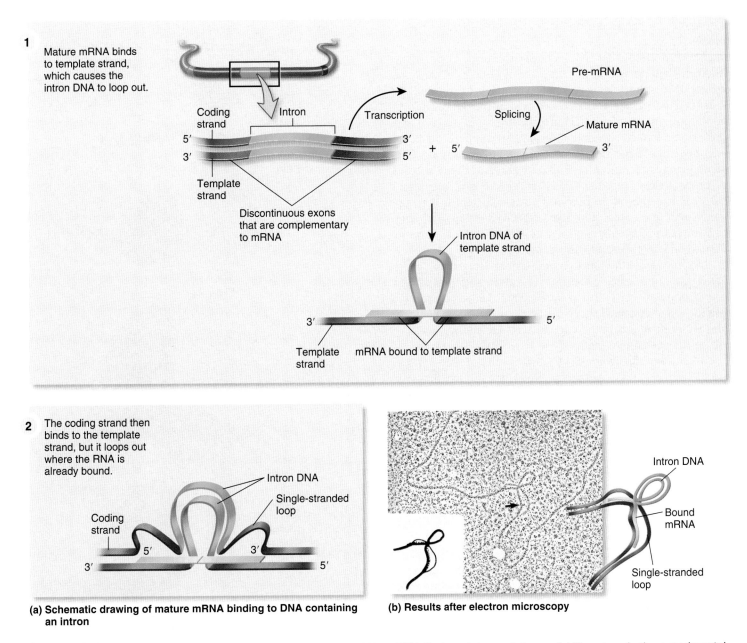

1 Mature mRNA binds to template strand, which causes the intron DNA to loop out.

Coding strand   Intron

Pre-mRNA

5′   3′
3′   5′

Transcription   Splicing   Mature mRNA

+   5′   3′

Template strand

Discontinuous exons that are complementary to mRNA

Intron DNA of template strand

3′   5′

Template strand   mRNA bound to template strand

2 The coding strand then binds to the template strand, but it loops out where the RNA is already bound.

Intron DNA

Single-stranded loop

Coding strand

5′   3′
3′   5′

Intron DNA

Bound mRNA

Single-stranded loop

(a) Schematic drawing of mature mRNA binding to DNA containing an intron

(b) Results after electron microscopy

Figure 12.9 Leder's experiment showing the binding of mRNA to DNA that contains an intron.   (a) The steps in the experimental procedure. (b) A micrograph of the results.

Biological inquiry: Explain why the double-stranded intron loops out of the structure shown in part (b).

will form that are separated by a double-stranded DNA region (**Figure 12.9a**, step 2). The intervening double-stranded region occurs because an intron has been spliced out of the mature mRNA, so that the mRNA cannot bind to this segment of the gene. In the actual experiment, samples were observed under the electron microscope. As seen in **Figure 12.9b**, the mRNA was bound to the DNA to form two single-stranded DNA loops separated by a double-stranded DNA region that is the intron. Chambon and O'Malley obtained similar results for other structural genes.

Since these initial discoveries, introns have been found in many eukaryotic genes. Splicing is less frequent among unicellular eukaryotic species, such as yeast, but is a widespread phenomenon among more complex eukaryotes. In mammals and flowering plants, most structural genes have one or more introns. As an extreme example, the human dystrophin gene, which harbors a mutation in people with Duchenne muscular dystrophy, has 79 exons punctuated by 78 introns. Introns are also found in a few bacterial genes, but they are very rare among all prokaryotic species.

Introns are removed from eukaryotic pre-mRNA by a large complex called a **spliceosome** that is composed of several subunits known as snRNPs (pronounced "snurps"). Each snRNP contains s̲mall n̲uclear R̲N̲A and a set of p̲roteins. This small nuclear RNA is the product of a nonstructural gene. Intron RNA is defined by particular sequences within the intron and at the intron-exon boundaries (**Figure 12.10**). These include a 5′ splice site, a branch site, and a 3′ splice site. Spliceosome subunits bind to specific sequences at these three locations. This binding causes the intron to loop outward, and the two exons are brought closer together. The 5′ splice site is then cut, and the 5′ end of the intron becomes covalently attached to the branch site. In the final step, the 3′ splice site is cut, and then the exons are covalently attached to each other. The intron is released and will be eventually degraded.

In some cases, the function of the spliceosome can be regulated so that the splicing of exons for a given mRNA can occur in two or more ways. This phenomenon, called **alternative splicing**, enables a single gene to encode two or more polypeptides with differences in their amino acid sequences. This topic will be considered in Chapter 13.

Although primarily in mRNAs, introns can be found in rRNA and tRNA molecules of certain species. These introns, however, are not removed by the action of a spliceosome. Instead, rRNA and tRNAs are **self-splicing**, which means that the RNA itself can catalyze the removal of its own intron. Portions of the RNA act like an enzyme to cleave the covalent bonds at the intron-exon boundaries and connect the exons together. An RNA molecule that catalyzes a chemical reaction is termed a **ribozyme**.

RNA Processing Also Involves Adding a 5′ Cap and a 3′ Poly A Tail to the Ends of Eukaryotic mRNAs

Mature mRNAs of eukaryotes have a 7-methylguanosine covalently attached at the 5′ end, an event known as **capping** (**Figure 12.11a**). Capping occurs while a pre-mRNA is being made by RNA polymerase, usually when the transcript is only 20 to 25 nucleotides in length. The 7-methylguanosine cap structure, called a **5′ cap**, is recognized by cap-binding proteins, which are needed for the proper exit of mRNA from the nucleus. Also, after an mRNA is in the cytosol, the cap structure is recognized by other cap-binding proteins that enable the mRNA to bind to a ribosome for translation.

At the 3′ end, most mature eukaryotic mRNAs have a string of adenine nucleotides, typically 100 to 200 nucleotides in length, referred to as a **poly A tail** (**Figure 12.11b**). A long poly A tail causes a eukaryotic mRNA to be more stable and thereby exist for a longer period of time in the cytosol. The poly A tail is not encoded in the gene sequence. Instead, the tail is added enzymatically after a pre-mRNA has been completely transcribed. Interestingly, new research has shown that some bacterial mRNAs also have poly A tails attached to them. However, the poly A tail has an opposite effect in bacteria; it causes the mRNA to be rapidly degraded. The importance of a poly A tail in bacterial mRNAs is not well understood.

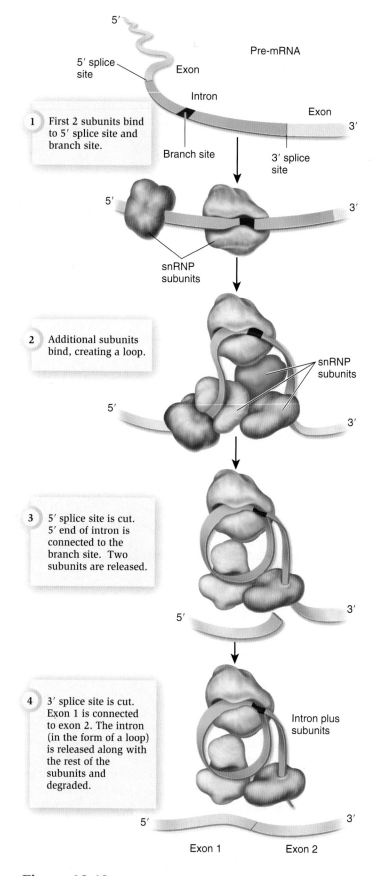

1. First 2 subunits bind to 5′ splice site and branch site.

2. Additional subunits bind, creating a loop.

3. 5′ splice site is cut. 5′ end of intron is connected to the branch site. Two subunits are released.

4. 3′ splice site is cut. Exon 1 is connected to exon 2. The intron (in the form of a loop) is released along with the rest of the subunits and degraded.

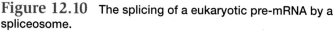

Figure 12.10   The splicing of a eukaryotic pre-mRNA by a spliceosome.

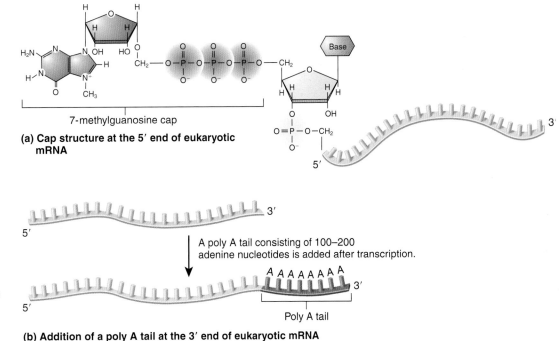

Figure 12.11

Modifications that occur at the ends of mRNA in eukaryotic cells. (a) A 7-methylguanosine cap is attached to the 5′ end. (b) A poly A tail is added to the 3′ end.

(a) Cap structure at the 5′ end of eukaryotic mRNA

A poly A tail consisting of 100–200 adenine nucleotides is added after transcription.

Poly A tail

(b) Addition of a poly A tail at the 3′ end of eukaryotic mRNA

12.4 Translation and the Genetic Code

In the two previous sections, we considered how an RNA transcript is made, and how eukaryotes process that transcript. Now we will begin to examine the next process, that of translation, at the molecular level. In 1960, Matthew Meselson and Francois Jacob found that proteins are synthesized on cellular structures known as ribosomes. One year later, Francois Jacob and Jacques Monod made an insightful hypothesis. They proposed that RNA, which is transcribed from DNA, provides the information for protein synthesis via ribosomes. This type of RNA, which they named messenger RNA (mRNA), carries information from the DNA to the ribosome, where proteins are made during the process of translation. Since these early studies, much has been learned about the details of translation.

Translation involves an interpretation of one language into another. In this case, the language of mRNA, which is a nucleotide sequence, is translated into the language of a polypeptide, which is an amino acid sequence. To understand the process of translation, we will first examine the **genetic code**, which specifies the relationship between the sequence of nucleotides in the mRNA and the sequence of amino acids in a polypeptide.

During Translation, the Genetic Code Is Used to Make a Polypeptide with a Specific Amino Acid Sequence

The ability of mRNA to be translated into a polypeptide relies on the genetic code. This code is contained in the sequence of bases in an mRNA molecule. The code is read in groups of three nucleotide bases known as **codons** (Table 12.1). The sequence of three bases in most codons specifies a particular amino acid. For example, the codon CCC specifies the amino acid proline, while the codon GGC encodes the amino acid glycine. From the analysis of many different species, including bacteria, protists, fungi, plants, and animals, researchers have found that the genetic code is nearly universal. Only a few rare exceptions to the genetic code have been noted (see note to Table 12.1).

Because there are 20 types of amino acids, at least 20 codons are needed to specify each type. With four types of bases in mRNA (A, U, G, and C), a genetic code containing two bases in a codon would not be sufficient, because only 4^2, or 16, different codons would be possible. By comparison, a three-base codon system can specify 4^3, or 64, different codons. Because the number of possible codons exceeds 20, the genetic code is said to be **degenerate**. This means that more than one codon can specify the same amino acid (Table 12.1). For example, the codons GGU, GGC, GGA, and GGG all specify the amino acid glycine. In most instances, the third base in the codon is the degenerate or variable base.

Let's look at the organization of a bacterial mRNA to see how translation occurs (**Figure 12.12**). The 5′ end of the mRNA contains a ribosomal-binding site. Beyond the 5′ end, a large portion of an mRNA functions as a coding sequence that specifies the amino acid sequence of a polypeptide. This coding sequence consists of a series of codons. The **start codon**, which is only a few nucleotides from the ribosomal-binding site, is usually AUG. This codon specifies methionine. The many codons that follow the start codon dictate the linear sequence of amino acids within a given polypeptide. A typical polypeptide is a few hundred amino acids in length. Finally, one of three **stop codons** signals the end of translation. These codons, also known as **termination** or **nonsense codons**, are UAA, UAG, and UGA.

Table 12.1   The Genetic Code*

					Second position							
		U		**C**		**A**		**G**				
First position	**U**	UUU, UUC	Phe	UCU, UCC, UCA, UCG	Ser	UAU, UAC	Tyr	UGU, UGC	Cys	U, C	**Third position**	
		UUA, UUG	Leu			UAA, UAG	Stop, Stop	UGA, UGG	Stop, Trp	A, G		
	C	CUU, CUC, CUA, CUG	Leu	CCU, CCC, CCA, CCG	Pro	CAU, CAC	His	CGU, CGC, CGA, CGG	Arg	U, C, A, G		
						CAA, CAG	Gln					
	A	AUU, AUC, AUA	Ile	ACU, ACC, ACA, ACG	Thr	AAU, AAC	Asn	AGU, AGC	Ser	U, C		
		AUG	Met/ start			AAA, AAG	Lys	AGA, AGG	Arg	A, G		
	G	GUU, GUC, GUA, GUG	Val	GCU, GCC, GCA, GCG	Ala	GAU, GAC	Asp	GGU, GGC, GGA, GGG	Gly	U, C		
						GAA, GAG	Glu			A, G		

*Exceptions to the genetic code are sporadically found among various species. A few examples are as follows: AUA encodes methionine in yeast and mammalian mitochondria; UGA encodes tryptophan in mammalian mitochondria; CUU, CUA, CUC, and CUG encode threonine in yeast mitochondria; AGA and AGG act as stop codons in ciliated protozoa and in yeast and mammalian mitochondria; and UAA and UAG encode glutamine in ciliated protozoa.

The start codon also defines the **reading frame** of an mRNA. Beginning at the start codon, each adjacent codon is read as a group of three bases, that is, a **triplet**, in the 5′ to 3′ direction. For example, look at the following two mRNA sequences and their corresponding amino acid sequences. The first sequence shows how the mRNA codons would be correctly translated into amino acids. In the second sequence, an additional U has been added to the same sequence after the start codon. This shifts the reading frame and thereby changes the codons as they occur in the 5′ to 3′ direction. This change results in the translation of the sequence into different amino acids.

	Ribosomal-binding site	Start codon

mRNA   5′–<u>AUAAGGAGG</u>UUACG (AUG) (CAG) (CAG) (GGC) (UUU) (ACC)–3′

Polypeptide   Met - Gln - Gln - Gly - Phe - Thr

	Ribosomal-binding site	Start codon

mRNA 5′–<u>AUAAGGAGG</u>UUACG (AUG) (UCA) (GCA) (GGG) (CUU) (UAC)C–3′

Polypeptide   Met - Ser - Ala - Gly - Leu - Tyr

From this comparison, we can also see that the genetic code is nonoverlapping, which means that each base functions within a single codon.

The relationships among the coding sequence of a gene, the codons in mRNA, and a polypeptide sequence are shown

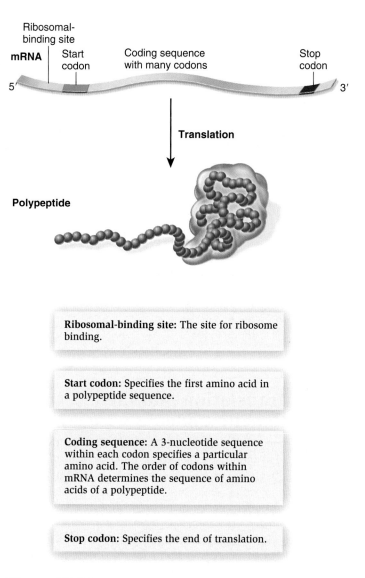

Ribosomal-binding site: The site for ribosome binding.

Start codon: Specifies the first amino acid in a polypeptide sequence.

Coding sequence: A 3-nucleotide sequence within each codon specifies a particular amino acid. The order of codons within mRNA determines the sequence of amino acids of a polypeptide.

Stop codon: Specifies the end of translation.

Figure 12.12   The organization of mRNA as a translational unit.

schematically in **Figure 12.13**. The coding strand of DNA corresponds to the mRNA strand, except that U in the mRNA is substituted for T in the DNA. The template strand is used to make mRNA. To translate a nucleotide sequence of mRNA into an amino acid sequence, recognition occurs between mRNA and transfer RNA (tRNA) molecules. Transfer RNA functions as the "translator" or intermediary between an mRNA codon and an amino acid. The **anticodon** allows a tRNA molecule to bind to a complementary codon in mRNA. Furthermore, the anticodon in a tRNA corresponds to the amino acid that it carries. For example, if the anticodon in a tRNA is 3′–AAG–5′, it is complementary to a 5′–UUC–3′ codon. According to the genetic code, a UUC codon specifies phenylalanine. Therefore, a tRNA with a 3′–AAG–5′ anticodon must carry phenylalanine. As another example, a tRNA with a 3′–GGG–5′ anticodon is complementary to a 5′– CCC–3′ codon, which specifies proline. This tRNA must carry proline.

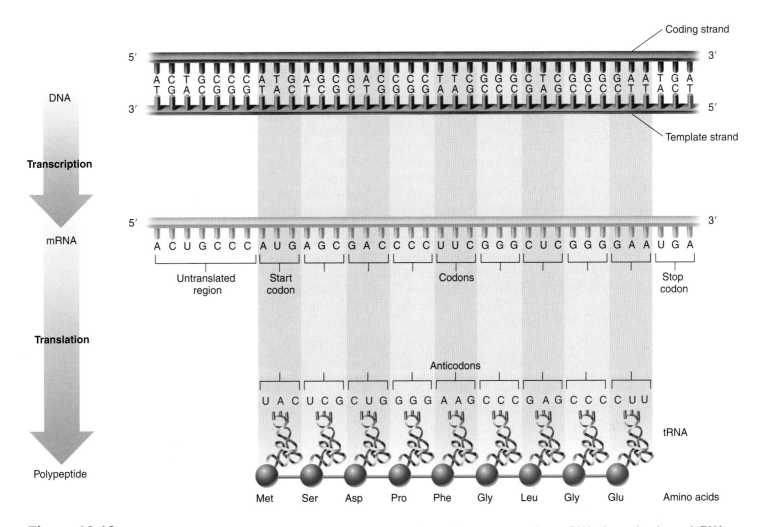

Figure 12.13 Relationships among the coding sequence of a gene, the codon sequence of an mRNA, the anticodons of tRNA, and the amino acid sequence of a polypeptide.

Biological inquiry: If an anticodon in a tRNA molecule has the sequence 3'–ACC–5', which amino acid does it carry?

Synthetic RNA Helped to Decipher the Genetic Code

Now let's look at some early experiments that allowed scientists to decipher the genetic code. During the early 1960s, the genetic code was determined by the collective efforts of several research groups headed by Marshall Nirenberg, Severo Ochoa, H. Gobind Khorana, and Phillip Leder. Prior to their studies, other scientists had discovered that bacterial cells can be broken open and components from the cytoplasm can synthesize polypeptides. This is termed an *in vitro* or cell-free translation system. Nirenberg and Ochoa made synthetic RNA molecules using an enzyme that covalently connects nucleotides together. Then, using this synthetic mRNA, they determined which amino acids were incorporated into polypeptides. For example, if an RNA molecule had only adenine-containing nucleotides (for example, 5'–AAAAAAAAAAAAAAAAAAA–3'), a polypeptide was produced that contained only lysine. This result indicated that the AAA codon specifies lysine.

Khorana and his collaborators developed a novel method to synthesize RNA. They first created short RNA molecules that were two to four nucleotides in length and had a defined sequence. For example, RNA molecules with the sequence 5'–AUC–3' were synthesized chemically. These short RNAs were then linked together enzymatically, in a 5' to 3' manner, to create a longer RNA molecule:

5'–AUCAUCAUCAUCAUCAUCAUCAUCAUC–3'

Depending on whether the reading frame begins with the first, second, or third base in this sequence, this RNA contains three different codons: AUC (isoleucine), UCA (serine), and CAU (histidine). Using a cell-free translation system, this RNA encoded a polypeptide containing only isoleucine, serine, and histidine.

Another method used to decipher the genetic code involved the chemical synthesis of short RNA molecules. This method is described in the Feature Investigation.

FEATURE INVESTIGATION

Nirenberg and Leder Found That RNA Triplets Can Promote the Binding of tRNA to Ribosomes

In 1964, Nirenberg and Leder discovered that RNA molecules containing any three nucleotides (that is, any triplet) can stimulate ribosomes to bind a tRNA molecule. In other words, an RNA triplet can act like a codon within an mRNA molecule. Ribosomes bind RNA triplets, and then a tRNA with the appropriate anticodon subsequently binds to the ribosome.

To establish the relationship between triplet sequences and specific amino acids, Nirenberg and Leder began with samples containing ribosomes and a particular triplet (**Figure 12.14**). For example, in one experiment they began with a sample of ribosomes that were mixed with 5′–CCC–3′ triplets. The triplets became bound to the ribosomes just like the binding of mRNA to a ribosome; we'll go through the whole process of translation

later in the chapter, but for now just keep in mind that tRNAs interact with mRNA on a ribosome during the synthesis of polypeptides. This sample was divided into 20 tubes. To each tube, they next added a mixture of cellular tRNAs that already had amino acids attached to them. However, each mixture of tRNAs had only one type of radiolabeled amino acid. For example, one mixture of tRNAs had only proline that was radiolabeled, a second mixture of tRNAs had only serine that was radiolabeled, and so on. The tRNA with the anticodon that was complementary to the added triplet would bind to the triplet, which was already bound to the ribosome. For example, if the triplet was 5′–CCC–3′, a tRNA with a 3′–GGG–5′ anticodon would be bound to the triplet/ribosome complex. This tRNA carries proline.

To determine which tRNA had bound, the samples from each tube were poured through a filter that trapped the large

Figure 12.14 Nirenberg and Leder's use of triplet binding assays to decipher the genetic code.

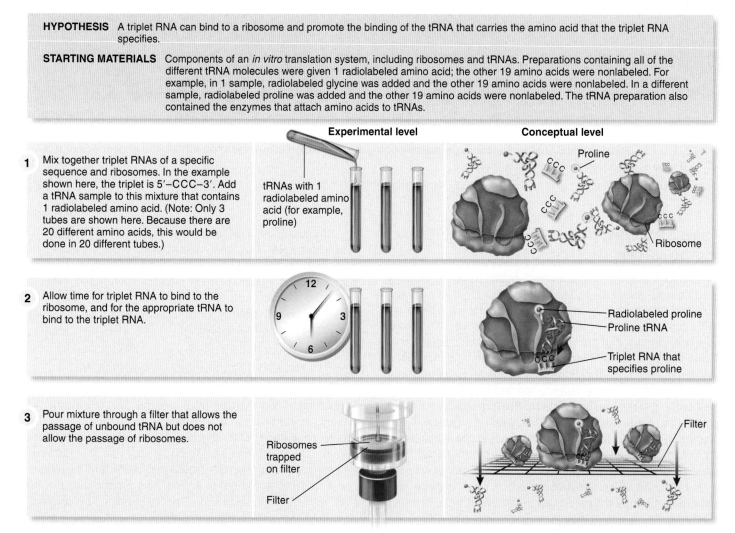

HYPOTHESIS A triplet RNA can bind to a ribosome and promote the binding of the tRNA that carries the amino acid that the triplet RNA specifies.

STARTING MATERIALS Components of an *in vitro* translation system, including ribosomes and tRNAs. Preparations containing all of the different tRNA molecules were given 1 radiolabeled amino acid; the other 19 amino acids were nonlabeled. For example, in 1 sample, radiolabeled glycine was added and the other 19 amino acids were nonlabeled. In a different sample, radiolabeled proline was added and the other 19 amino acids were nonlabeled. The tRNA preparation also contained the enzymes that attach amino acids to tRNAs.

Experimental level   **Conceptual level**

1. Mix together triplet RNAs of a specific sequence and ribosomes. In the example shown here, the triplet is 5′–CCC–3′. Add a tRNA sample to this mixture that contains 1 radiolabeled amino acid. (Note: Only 3 tubes are shown here. Because there are 20 different amino acids, this would be done in 20 different tubes.)

tRNAs with 1 radiolabeled amino acid (for example, proline)

Proline

Ribosome

2. Allow time for triplet RNA to bind to the ribosome, and for the appropriate tRNA to bind to the triplet RNA.

Radiolabeled proline
Proline tRNA

Triplet RNA that specifies proline

3. Pour mixture through a filter that allows the passage of unbound tRNA but does not allow the passage of ribosomes.

Ribosomes trapped on filter

Filter

Filter

4 Count radioactivity on the filter.

— Scintillation counter

---

5 THE DATA

Triplet	Radiolabeled amino acid trapped on the filter	Triplet	Radiolabeled amino acid trapped on the filter
5′ – AAA – 3′	Lysine	5′ – GAC – 3′	Aspartic acid
5′ – ACA – 3′, 5′ – ACC – 3′	Threonine	5′ – GCC – 3′	Alanine
5′ – AGA – 3′	Arginine	5′ – GGU – 3′, 5′ – GGC – 3′	Glycine
5′ – AUA – 3′, 5′ – AUU – 3′	Isoleucine	5′ – GUU – 3′	Valine
5′ – CCC – 3′	Proline	5′ – UAU – 3′	Tyrosine
5′ – CGC – 3′	Arginine	5′ – UGU – 3′	Cysteine
5′ – GAA – 3′	Glutamic acid	5′ – UUG – 3′	Leucine

ribosomes but did not trap tRNAs that were not bound to ribosomes. The researchers then determined the amount of radioactivity on each filter. If the tRNA carrying the radiolabeled amino acid was bound to the triplet/ribosome complex, radioactivity would be trapped on the filter. Because only one amino acid was radiolabeled in each tRNA sample, they could determine which triplet corresponded to which amino acid. In the example shown here, CCC corresponds to proline. Therefore, the tRNA sample containing radiolabeled proline showed a large amount of radioactivity on the filter. By studying triplets with different sequences, Nirenberg and Leder could identify many codons of the genetic code.

12.5 The Machinery of Translation

Let's now turn our attention to the components found in living cells that are needed to use the genetic code and translate mRNA into polypeptides. Earlier in this chapter, we considered the first step in gene expression, namely, transcription. To transcribe an RNA molecule, a pre-existing DNA template strand is used to make a complementary RNA strand. A single enzyme, RNA polymerase, can catalyze this reaction. By comparison, translation requires more components because the sequence of codons in an mRNA molecule must be translated into a sequence of amino acids according to the genetic code. A single protein cannot accomplish such a task. Instead, many different proteins and RNA molecules must interact in an intricate series of steps to achieve the synthesis of a polypeptide.

Due to its complexity, translation is a costly process from an energy point of view. A cell must make many different components, including mRNAs, tRNAs, ribosomes, and translation factors, so that polypeptides can be made (Table 12.2). The synthesis of these components uses a great deal of cellular energy. In addition, the mechanism of making a polypeptide, which we will examine later in this chapter, also uses a large amount of energy. Though the estimates vary from cell to cell

Table 12.2	Components of the Translation Machinery
Component	**Function**
mRNA	Contains the information for a polypeptide sequence according to the genetic code.
tRNA	A molecule with two functional sites. One site, termed the anticodon, recognizes a codon in mRNA. A second site has the appropriate amino acid attached to it.
Ribosomes	Composed of many proteins and rRNA molecules. The ribosome provides a location where mRNA and tRNA molecules can properly interact with each other. The ribosome also catalyzes the formation of covalent bonds between adjacent amino acids so that a polypeptide can be made.
Translation factors	Three categories of translation factors are needed for the three stages of translation that are described in the last section of this chapter. Initiation factors are required for the assembly of mRNA, the first tRNA, and ribosomal subunits. Elongation factors are needed to synthesize the polypeptide. And release factors are needed to recognize the stop codon and disassemble the translation machinery. Several translation factors use GTP as an energy source to carry out their functions.

and from species to species, most cells use a substantial amount of their energy to translate mRNA into polypeptides. In *E. coli*, for example, approximately 90% of the cellular energy is used for this process. This value underscores the complexity and importance of translation in living organisms. In this section, we will focus on the components of the translation machinery. The last section of the chapter will describe the steps of translation as they occur in living cells.

Transfer RNAs Share Common Structural Features

To understand how tRNAs act as carriers of the correct amino acids during translation, researchers have examined their structural characteristics. The cells of every organism make many different tRNA molecules, each encoded by a different gene. A tRNA is named according to the amino acid it carries. For example, tRNA^{ser} carries a serine. Because the genetic code contains six different serine codons, a cell produces more than one type of tRNA^{ser}.

The tRNAs of both prokaryotes and eukaryotes share common features. As originally proposed by Robert Holley in 1965, the two-dimensional structure of tRNAs exhibits a cloverleaf pattern. The structure has three stem-loops and a 3′ single-stranded region (**Figure 12.15a**). The stems are regions where the RNA is double stranded due to complementary base pairing, while the loops are regions without base pairing. The anticodon is located in the loop of the second stem-loop region. The 3′ single-stranded region is called the acceptor stem because it accepts the attachment of an amino acid. The three-dimensional structure of tRNA molecules involves additional folding of the secondary structure (**Figure 12.15b**).

Aminoacyl-tRNA Synthetases Charge tRNAs by Attaching an Appropriate Amino Acid

To perform its role during translation, a tRNA must have the appropriate amino acid attached to its 3′ end. The enzymes that catalyze the attachment of amino acids to tRNA molecules are known as **aminoacyl-tRNA synthetases**. Cells make 20 distinct aminoacyl-tRNA synthetase enzymes, one for each of the 20 different amino acids. Each aminoacyl-tRNA synthetase is named for the specific amino acid it attaches to tRNA. For example, alanyl-tRNA synthetase recognizes a tRNA with an alanine anticodon (that is, tRNA^{ala}) and attaches an alanine to it.

Aminoacyl-tRNA synthetases catalyze two chemical reactions involving an amino acid, a tRNA molecule, and an ATP (**Figure 12.16**). After an amino acid and ATP have bound to the enzyme, the amino acid is activated by the covalent attachment of an AMP molecule. Pyrophosphate is released. Next, the activated amino acid is covalently attached to the 3′ end of a tRNA molecule and AMP is released. Finally, the tRNA with its attached amino acid, called a **charged tRNA** or an **aminoacyl tRNA**, is released from the enzyme.

The ability of each aminoacyl-tRNA synthetase to recognize an appropriate tRNA has been called the second genetic code. A precise recognition process is necessary to maintain

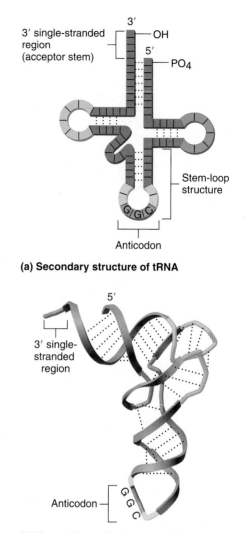

(a) Secondary structure of tRNA

(b) Three-dimensional structure of tRNA

Figure 12.15   Structure of tRNA.   (a) The two-dimensional or secondary structure is that of a cloverleaf. Notice the anticodon on the second of the three stem-loop structures. The single-stranded region (acceptor stem) is where an amino acid can attach. (b) The actual three-dimensional structure folds in on itself.

the fidelity of genetic information. If the wrong amino acid was attached to a tRNA, the amino acid sequence of the translated polypeptide would be incorrect. To prevent this from happening, aminoacyl-tRNA synthetases are amazingly accurate enzymes. The wrong amino acid is attached to a tRNA less than once in 100,000 times! The anticodon region of the tRNA is usually important for recognition by the correct aminoacyl-tRNA synthetase. In addition, the acceptor stem and base sequences in other stem-loop regions may facilitate binding to an aminoacyl-tRNA synthetase.

Ribosomes Are Assembled from rRNA and Proteins

As described earlier in Figure 12.13, the act of translation involves an interaction between mRNA and tRNA molecules. The ribosome can be thought of as a macromolecular arena where

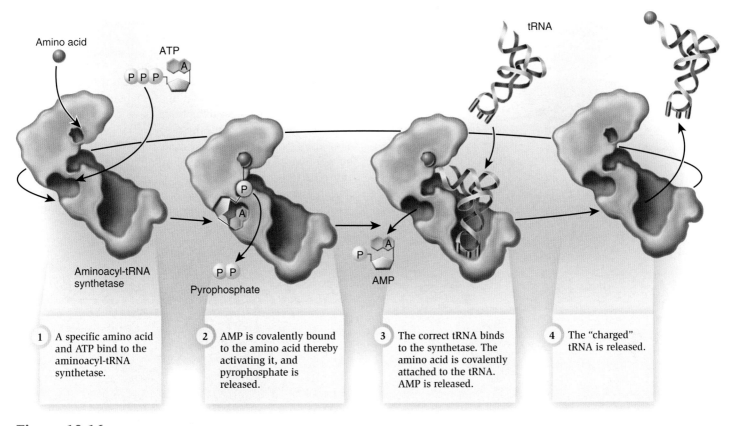

1. A specific amino acid and ATP bind to the aminoacyl-tRNA synthetase.

2. AMP is covalently bound to the amino acid thereby activating it, and pyrophosphate is released.

3. The correct tRNA binds to the synthetase. The amino acid is covalently attached to the tRNA. AMP is released.

4. The "charged" tRNA is released.

Figure 12.16 Aminoacyl-tRNA synthetase charging a tRNA.

translation takes place. Bacterial cells have one type of ribosome, which translates all mRNAs in the cytoplasm. However, because eukaryotic cells are compartmentalized into cellular organelles that are bounded by membranes, their translation machinery is more complex. Biochemically distinct ribosomes are found in different cellular compartments. The most abundant type of eukaryotic ribosome functions in the cytosol. In addition, all eukaryotic cells have ribosomes in their mitochondria. Plant and algal cells also have ribosomes in their chloroplasts. The compositions of mitochondrial and chloroplast ribosomes are more similar to bacterial ribosomes than they are to eukaryotic cytosolic ribosomes. Unless otherwise noted, the term eukaryotic ribosome refers to ribosomes in the cytosol, not to those found in organelles.

A ribosome is a large complex composed of structures called the large and small subunits. The term subunit is perhaps misleading, because each ribosomal subunit is itself formed from the assembly of many different proteins and RNA molecules. In ribosomes found in the bacterium *E. coli*, the small subunit is called 30S, and the large subunit is 50S (**Figure 12.17a**). The designations 30S and 50S refer to the rate at which these subunits sediment when subjected to a centrifugal force. This rate is described as a sedimentation coefficient in Svedberg units (S) in honor of Theodor Svedberg, who invented the ultracentrifuge. The 30S subunit is formed from the assembly of 21 different ribosomal proteins and one 16S rRNA molecule. The 50S subunit contains 34 different proteins and two different rRNA

molecules, called 5S and 23S (Figure 12.17a). Together, the 30S and 50S subunits form a 70S ribosome (Svedberg units don't add up linearly, because they are a function of both size and shape). In bacteria, the ribosomal proteins and rRNA molecules are synthesized in the cytoplasm, and the ribosomal subunits are assembled there as well.

Eukaryotic ribosomes consist of subunits that are slightly larger than their bacterial counterparts (**Figure 12.17b**). In eukaryotes, 40S and 60S subunits are combined to form an 80S ribosome. The 40S subunit is composed of 33 proteins and an 18S rRNA. The 60S subunit has 49 proteins and 5S, 5.8S, and 28S rRNAs. The synthesis of eukaryotic rRNA occurs in the nucleolus, a region of the nucleus that is specialized for that purpose. The ribosomal proteins are made in the cytosol and imported into the nucleus. The rRNAs and ribosomal proteins are then assembled to make the 40S and 60S subunits within the nucleolus. The 40S and 60S subunits are exported into the cytosol, where they associate to form an 80S ribosome during translation.

Due to structural differences between bacterial and eukaryotic ribosomes, certain chemicals may bind to bacterial ribosomes but not to eukaryotic ribosomes, and vice versa. Some **antibiotics**, which are chemicals that inhibit the growth of certain microorganisms, bind only to bacterial ribosomes and inhibit translation. Examples include erythromycin and chloramphenicol. Because these chemicals do not inhibit eukaryotic ribosomes, they have been effective drugs for the treatment of bacterial infections in humans and domesticated animals.

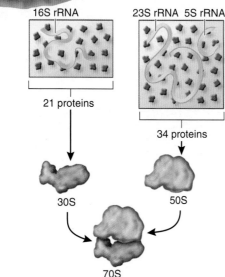

16S rRNA   23S rRNA   5S rRNA

21 proteins

34 proteins

30S   50S

70S

(a) Bacterial ribosomes

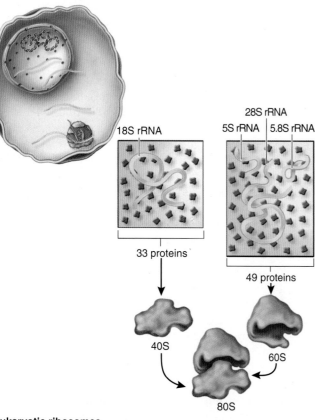

28S rRNA
5S rRNA   5.8S rRNA
18S rRNA

33 proteins

49 proteins

40S

60S

80S

(b) Eukaryotic ribosomes

Figure 12.17   **Composition of ribosomes.**   (a) Bacterial ribosomes. (b) Eukaryotic ribosomes found in the cytosol.

Components of Ribosomal Subunits Form Functional Sites for Translation

To understand the structure and function of the ribosome at the molecular level, researchers have determined the locations and functional roles of the individual ribosomal proteins and rRNAs. In recent years, a few research groups have succeeded in purifying ribosomes and causing them to crystallize in a test tube. Using the technique of X-ray diffraction, the crystallized ribosomes provide detailed information about ribosome structure. **Figure 12.18a** shows a model of a bacterial ribosome. The overall shape of each subunit is largely determined by the structure of the rRNAs, which constitute most of the mass of the ribosome.

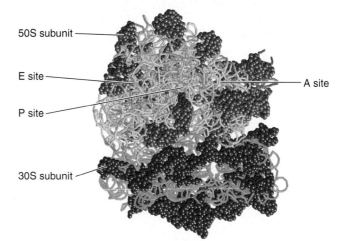

50S subunit

E site

P site

A site

30S subunit

(a) Bacterial ribosome model based on X-ray diffraction studies

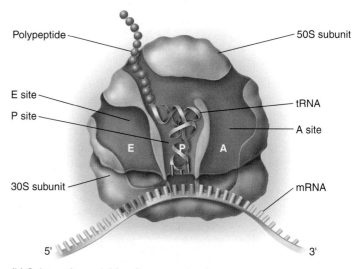

Polypeptide   50S subunit

E site

P site   tRNA

                                                     A site

E   P   A

30S subunit   mRNA

5'   3'

(b) Schematic model for ribosome structure

Figure 12.18   **Ribosome structure.**   (a) A model for the structure of the ribosome based on X-ray diffraction studies, showing the large and small subunits and the major binding sites. The rRNA is shown in gray (large subunit) and light blue (small subunit), while the ribosomal proteins are violet (large subunit) and dark blue (small subunit). (b) A schematic model emphasizing functional sites in the ribosome, also showing bound mRNA and tRNA with an attached polypeptide.

During bacterial translation, the mRNA lies on the surface of the 30S subunit, within a space between the 30S and 50S subunits (**Figure 12.18b**). As a polypeptide is synthesized, it exits through a hole within the 50S subunit. Ribosomes contain discrete sites where tRNAs bind and the polypeptide is synthesized. In 1964, James Watson proposed a two-site model for tRNA binding to the ribosome. These sites are known as the **peptidyl site** (**P site**) and **aminoacyl site** (**A site**). In 1984, Knud Nierhaus and Hans-Jorg Rheinberger expanded this to a three-site model (Figure 12.18b). The third site is known as the **exit site** (**E site**). Later, we will examine the roles of these sites in the synthesis of a polypeptide.

GENOMES & PROTEOMES

Comparisons of Small Subunit rRNAs Among Different Species Provide a Basis for Establishing Evolutionary Relationships

Translation is a fundamental process that is vital for the existence of all living species. The components that are needed for translation arose very early in the evolution of life on our planet. In fact, they arose in an ancestor that gave rise to all known living species. For this reason, all organisms have translational components that are evolutionarily related to each other. For example, the rRNA found in the small subunit of ribosomes is similar in all forms of life, though it is slightly larger in eukaryotic species (18S) than in bacterial species (16S). In other words, the gene for the small subunit rRNA (SSU rRNA) is found in the genomes of all organisms.

One way that geneticists explore evolutionary relationships is to compare the sequences of evolutionarily related genes. At the molecular level, gene evolution involves changes in DNA sequences. After two different species have diverged from each other during evolution, the genes of each species have an opportunity to accumulate changes, or mutations, that alter the sequences of those genes. After many generations, evolutionarily related species contain genes that are similar but not identical to each other, because each species will accumulate different mutations. In general, if a very long time has elapsed since two species diverged evolutionarily, their genes tend to be quite different. In contrast, if two species diverged relatively recently on an evolutionary time scale, their genes tend to be more similar.

Figure 12.19 compares a portion of the sequence of the small subunit rRNA gene from three mammalian and three bacterial species. The colors highlight different types of comparisons. The sequences shaded in yellow are identical in five or six species. Sequences that are identical in different species are said to be **evolutionarily conserved**. Presumably, these sequences were found in the primordial gene that gave rise to modern species and, because these sequences may have some critical function, have not been able to change over evolutionary time. Those sequences shaded in green are identical in all three mammals, but differ compared to one or more bacterial species. Actually, if you scan the mammalian species, you may notice that all three sequences are identical to each other in this region. The sequences shaded in red are identical in two or three bacterial species, but differ compared to the mammalian small subunit rRNA genes. The sequences from *E. coli* and *Serratia marcescens* are more similar to each other than the sequence from *Bacillus subtilis* is to either of them. This is consistent with the idea that *E. coli* and *S. marcescens* are more closely related evolutionarily than either of them is to *B. subtilis*.

12.6 The Stages of Translation

Like transcription, the process of translation occurs in three stages called initiation, elongation, and termination. **Figure 12.20** provides an overview of the process. During initiation, mRNA, the first tRNA, and ribosomal subunits assemble into a complex. Next, in the elongation stage, the ribosome moves from the start codon in the mRNA toward the stop codon, synthesizing a polypeptide according to the sequence of codons in the mRNA. Finally, the process is terminated when the ribosome reaches a stop codon and the complex disassembles, releasing the completed polypeptide. In this section, we will examine the steps in this process as they occur in living cells.

```
GATTAAGAGGGACGGCCGGGGGCATTCGTATTGCGCCGCTAGAGGTGAAATTC
```
Human
```
GATTAAGAGGGACGGCCGGGGGCATTCGTATTGCGCCGCTAGAGGTGAAATTC
```
Mouse
```
GATTAAGAGGGACGGCCGGGGGCATTCGTATTGCGCCGCTAGAGGTGAAATTC
```
Rat
```
CAAGCTTGAGTCTCGTAGAGGGGGGTAGAATTCCAGGTGTAGCGGTGAAATGC
```
E. coli
```
CAAGCTAGAGTCTCGTAGAGGGGGGTAGAATTCCAGGTGTAGCGGTGAAATGC
```
S. marcescens
```
GAGACTTGAGTACAGAAGAAGAGAGTGGAATTCCACGTGTAGCGGTGAAATGC
```
B. subtilis

Figure 12.19 Comparison of small subunit rRNA gene sequences from three eukaryotes and three bacterial species. Note the many similarities (yellow) and differences (green and red) among the sequences.

Translation Is Initiated with the Assembly of mRNA, tRNA, and the Ribosomal Subunits

During the **initiation stage**, mRNA, the first tRNA, and the ribosomal subunits assemble into a functional complex. In all species, this assembly requires the help of proteins called **ribosomal initiation factors** that facilitate the interactions between these components (see Table 12.2). The assembly also requires an input of energy. Guanosine triphosphate (GTP) is hydrolyzed by the initiation factors to provide the necessary energy.

In the absence of translation, the small and large ribosomal subunits exist separately. To begin assembly in bacteria, mRNA binds to the small ribosomal subunit (**Figure 12.21**). The binding of mRNA to this subunit is facilitated by a short ribosomal-binding sequence near the 5′ end of the mRNA. This sequence is complementary to a portion of the 16S rRNA within the small ribosomal subunit. For this reason, the mRNA and rRNA hydrogen-bond to each other by base pairing. The start codon is usually just a few nucleotides downstream (that is, toward the 3′ end) from the ribosomal-binding sequence. A specific tRNA, which functions as the **initiator tRNA**, recognizes the start codon AUG in mRNA and binds to it. To complete the initiation stage, the large ribosomal subunit associates with the small subunit.

At the end of this stage, the initiator tRNA is located in the P site of the ribosome.

In eukaryotic species, the initiation phase has two differences compared to their bacterial counterparts. First, instead of a ribosomal-binding sequence, eukaryotic mRNAs have a 7-methylguanosine cap at their 5′ end. This cap is recognized by cap-binding proteins that promote the binding of the mRNA to the small ribosomal subunit. Also, unlike bacteria, in which the start codon is very close to a ribosomal-binding sequence, the location of start codons in eukaryotes is more variable. To identify a start codon, Marilyn Kozak proposed in 1978 that the small ribosomal subunit begins at the 5′ end and then scans along the mRNA in the 3′ direction in search of an AUG sequence. In many, but not all, cases the first AUG codon is used as a start codon. By analyzing the sequences of many eukaryotic mRNAs, Kozak and her colleagues discovered that the sequence around an AUG codon is important for it to be used as a start codon. The sequence for optimal start codon recognition is shown here:

Upstream of start codon	Start codon	Downstream coding region

. . . G C C (A or G) C C (**A U G**) G . . . .

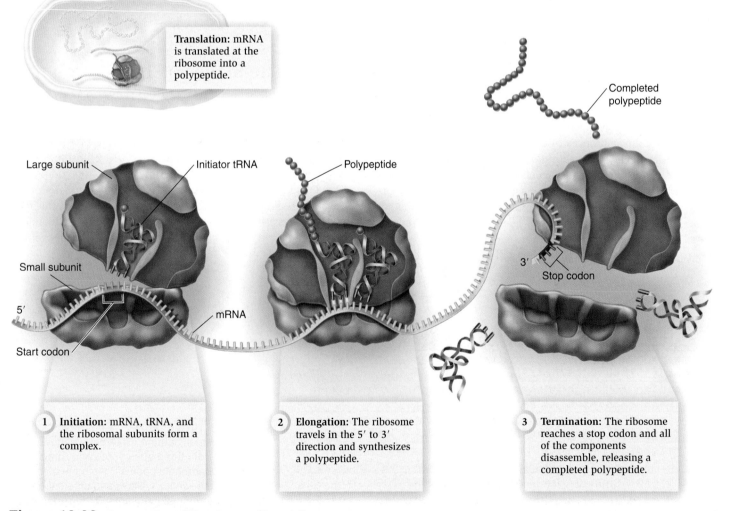

Translation: mRNA is translated at the ribosome into a polypeptide.

Large subunit   Initiator tRNA   Polypeptide

Completed polypeptide

Small subunit

5′

Start codon   mRNA

3′   Stop codon

1. **Initiation:** mRNA, tRNA, and the ribosomal subunits form a complex.

2. **Elongation:** The ribosome travels in the 5′ to 3′ direction and synthesizes a polypeptide.

3. **Termination:** The ribosome reaches a stop codon and all of the components disassemble, releasing a completed polypeptide.

Figure 12.20   An overview of the stages of translation.

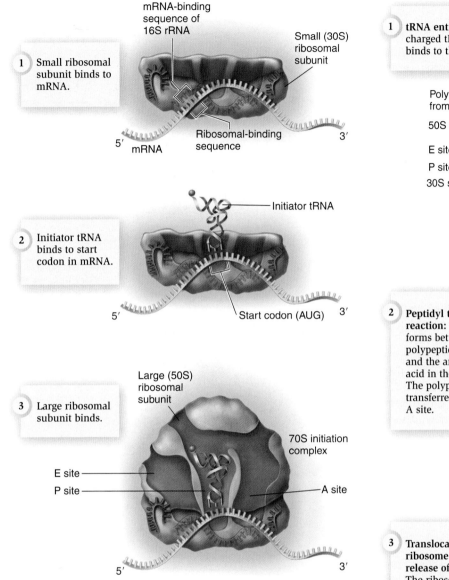

Figure 12.21 Initiation stage of translation in bacteria.

Aside from an AUG codon itself, a guanosine just past the start codon and the sequence of six bases directly upstream from the start codon are important for start codon selection. If an AUG codon is within a site that deviates markedly from this optimal sequence, the small subunit may skip this codon and instead use another AUG codon farther downstream. Once the small subunit selects a start codon, an initiator tRNA binds to the start codon, and then the large ribosomal subunit associates with the small subunit to complete the assembly process.

Polypeptide Synthesis Occurs During the Elongation Stage

As its name suggests, the **elongation stage** involves the covalent bonding of amino acids to each other, one at a time, to create a polypeptide. Even though this process involves several different

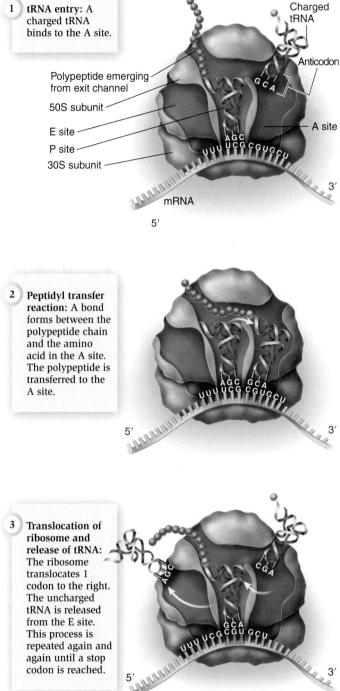

Figure 12.22 Elongation stage of translation in bacteria.

components, translation occurs at a remarkable rate. Under normal cellular conditions, the translation machinery can elongate a polypeptide chain at a rate of 15 to 18 amino acids per second in bacteria and 6 amino acids per second in eukaryotes!

To elongate a polypeptide by one amino acid, a tRNA brings a new amino acid to the ribosome, where it is attached to the end of a growing polypeptide chain. In step 1 of **Figure 12.22**, translation has already proceeded to a point where a short polypeptide is attached to the tRNA located in the P site of the ribosome. This is called peptidyl tRNA. In the first step of elongation, an aminoacyl tRNA carrying a single amino acid binds to the A site. This binding occurs because the anticodon in the tRNA is complementary to the codon in the mRNA. The hydrolysis of GTP by proteins that function as **elongation factors** provides the energy for the binding of the tRNA to the A site (see Table 12.2). At this stage of translation, a peptidyl tRNA is at the P site and an aminoacyl tRNA is at the A site. This is how the P and A sites came to be named.

In the second step, a **peptide bond** is formed between the amino acid at the A site and the growing polypeptide chain, thereby lengthening the chain by one amino acid. As this occurs, the polypeptide is removed from the tRNA in the P site and transferred to the amino acid at the A site, an event termed a **peptidyl transfer reaction**. This reaction is catalyzed by a region of the 50S subunit known as the peptidyltransferase complex, which is composed of several proteins and rRNA. Thomas Steitz, Peter Moore, and their colleagues proposed that the rRNA is responsible for catalyzing bond formation between adjacent amino acids. In other words, the ribosome is a ribozyme!

After the peptidyl transfer reaction is complete, the third step involves the movement or translocation of the ribosome toward the 3′ end of the mRNA by exactly one codon. This shifts the tRNAs at the P and A sites to the E and P sites, respectively. Notice that the next codon in the mRNA is now exposed in the unoccupied A site. The uncharged tRNA exits the E site. At this point, the next charged tRNA can enter the empty A site, and the same series of steps will add the next amino acid to the polypeptide chain.

Termination Occurs When a Stop Codon Is Reached in the mRNA

When a stop codon is found in the A site of a ribosome, translation is ended. The three stop codons, UAA, UAG, and UGA, are not recognized by a tRNA with a complementary sequence. Instead, they are recognized by proteins known as **release factors** (see Table 12.2). Interestingly, the three-dimensional structure of a release factor protein mimics the structure of tRNAs.

Figure 12.23 illustrates the termination stage of translation. In step 1 of this figure, the completed polypeptide chain is attached to a tRNA in the P site. A stop codon is located at the A site. In the first step, a release factor binds to the stop codon at the A site. In the second step, the bond between the polypeptide and the tRNA is hydrolyzed, causing the polypeptide and tRNA to be released from the ribosome. Finally, in the third step the ribosomal subunits, mRNA and release factor dissociate.

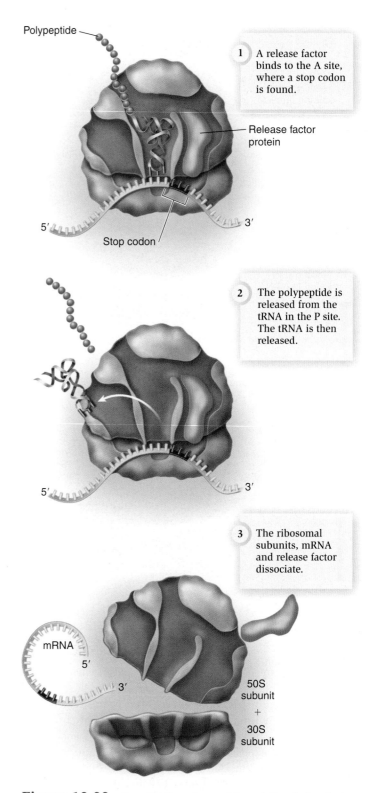

Figure 12.23 Termination stage of translation in bacteria.

A Polypeptide Chain Has Directionality from Its Amino End to Its Carboxyl End

Polypeptide synthesis has a directionality that parallels the 5′ to 3′ orientation of mRNA. **Figure 12.24** compares the sequence of a very short polypeptide with the mRNA that encodes it. The first amino acid is said to be at the **N-terminus** or **amino terminus** of the polypeptide. The term N-terminus refers to the presence of a nitrogen atom (N) at this end, while amino terminus indicates the presence of a free amino group. Peptide bonds connect the amino acids together. These bonds form between the carboxyl group of the previous amino acid and the amino group of the next amino acid. The last amino acid in a completed polypeptide does not have another amino acid attached to its carboxyl group. This last amino acid is said to be located at the **C-terminus** or **carboxyl terminus**. A carboxyl group (COOH) is always found at this end of the polypeptide chain. Note: At neutral pH, the amino group is positively charged (NH_3^+), while the carboxyl group is negatively charged (COO^-).

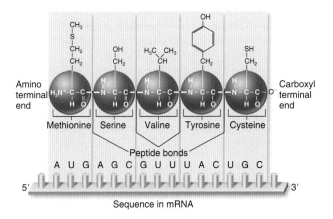

Figure 12.24 The directionality of a polypeptide compared to an mRNA.

Biological inquiry: Is the codon for the last amino acid in a polypeptide closer to the 5′ or 3′ end of the mRNA?

CHAPTER SUMMARY

12.1 Overview of Gene Expression

- Based on his studies of inborn errors of metabolism, Garrod hypothesized that genes encode enzymes. (Figure 12.1)
- By studying the nutritional requirements of bread mold, Beadle and Tatum proposed the one gene–one enzyme hypothesis. Later, this idea was expanded to the one gene–one polypeptide theory. (Figure 12.2)
- A polypeptide is a unit of structure. A protein, composed of one or more polypeptides, is a unit of function.
- At the molecular level, the central dogma of genetics states that most genes are transcribed into mRNA, and then the mRNA is translated into polypeptides. Eukaryotes modify their RNA transcripts to make them functional. (Figure 12.3)
- The molecular expression of genes is fundamental to the characteristics of an organism's traits. (Figure 12.4)

12.2 Transcription

- The promoter of a gene signals the beginning of transcription while the terminator specifies where transcription will end for a given gene. Regulatory sequences control whether a gene is turned on or off. (Figure 12.5)
- In bacteria, sigma factor binds to RNA polymerase and to a promoter, thereby promoting the initiation of transcription. The RNA transcript is made during the elongation stage due to base pairing of nucleotides to the template strand of DNA. RNA polymerase is released from the DNA at the termination site. (Figure 12.6)
- The genes along a chromosome are transcribed in different directions. (Figure 12.7)

12.3 RNA Processing in Eukaryotes

- Eukaryotic mRNA is first made as a pre-mRNA that is capped, spliced, and given a poly A tail. (Figures 12.8, 12.11)
- Eukaryotic genes often have introns. These are intervening sequences that are found in between the coding sequences, which are called exons. Microscopy studies with mature mRNA revealed the presence of introns in eukaryotic genes. (Figure 12.9)
- Introns are removed from eukaryotic pre-mRNA by a spliceosome. The components of a spliceosome first recognize the intron boundaries and the branch site, and then remove the intron and connect the adjacent exons. (Figure 12.10)

12.4 Translation and the Genetic Code

- The genetic code is found in mRNA molecules and determines the amino acid sequences of polypeptides. Each of the 64 codons specifies a start codon (methionine), other amino acids, or a stop codon. (Table 12.1, Figure 12.12)
- The template strand of DNA is used to make mRNA with a series of codons. Recognition between mRNA and many tRNA molecules determines the amino acid sequence of a polypeptide. (Figure 12.13)
- Nirenberg and Leder used the ability of RNA triplets to promote the binding of tRNA to ribosomes as a way to determine many of the codons of the genetic code. (Figure 12.14)

12.5 The Machinery of Translation

- Translation requires mRNA, tRNAs, ribosomes, and many translation factors. (Table 12.2)
- tRNA molecules have a cloverleaf structure. Two important sites are the 3′ end, which covalently binds an amino acid, and the anticodon, which base-pairs with a codon in mRNA. (Figure 12.15)

- The enzyme aminoacyl-tRNA synthetase attaches the correct amino acid to a tRNA molecule. (Figure 12.16)
- Ribosomes are composed of rRNA molecules and many proteins to produce a small and large subunit. Bacterial and eukaryotic ribosomes differ in their composition. (Figure 12.17)
- Ribosomes have three sites called the A, P, and E sites, which are locations for the binding and release of tRNA molecules. (Figure 12.18)
- The gene that encodes the small subunit rRNA has been extensively used in the evolutionary comparisons of different species. (Figure 12.19)

12.6 The Stages of Translation

- Translation occurs in three stages called initiation, elongation, and termination. (Figure 12.20)
- During initiation of translation, the mRNA assembles with the ribosomal subunits and the first tRNA molecule. (Figure 12.21)
- Polypeptide synthesis occurs during the elongation stage, one amino acid at a time. (Figure 12.22)
- During the termination of translation, the binding of a release factor to the stop codon causes the release of the completed polypeptide and the disassembly of the mRNA and ribosomal subunits. (Figure 12.23)
- A polypeptide has a directionality in which the first amino acid is at the N- or amino-terminus, while the last amino acid is at the C- or carboxyl-terminus. (Figure 12.24)

TEST YOURSELF

1. Which of the following best represents the central dogma of gene expression?
   a. During transcription, DNA codes for polypeptides.
   b. During transcription, DNA codes for RNA, which codes for polypeptides during translation.
   c. During translation, DNA codes for RNA, which codes for polypeptides during transcription.
   d. None of the above.

2. Transcription of a gene begins at a site on DNA called _____ and ends at a site on DNA known as _____.
   a. an initiation codon, the termination codon
   b. a promoter, the termination codon
   c. an initiation codon, the terminator
   d. a promoter, the terminator
   e. an initiator, the terminator

3. The product of transcription of a structural gene is
   a. tRNA.   d. polypeptide.
   b. mRNA.   e. a, b, and c.
   c. rRNA.

4. During eukaryotic RNA processing, the nontranslated sequences that are removed are called
   a. exons.   d. codons.
   b. introns.   e. ribozymes.
   c. promoters.

5. Ribozymes are
   a. the organelles where translation takes place.
   b. the RNA molecules that are components of ribosomes.
   c. the proteins that are components of ribosomes.
   d. the portions of the pre-mRNA that are removed.
   e. RNA molecules that catalyze chemical reactions.

6. The _____ is the organelle where the translation process takes place.
   a. mitochondria   d. lysosome
   b. nucleus   e. ribozyme
   c. ribosome

7. The region of the tRNA that is complementary to the triplet on the mRNA is
   a. the acceptor stem.   d. the anticodon.
   b. the codon.   e. the adaptor loop.
   c. the peptidyl site.

8. During the initiation step of translation, the first codon, ____, will enter the _____ and associate with the initiator tRNA.
   a. UAG, A site   d. AUG, P site
   b. AUG, A site   e. AUG, E site
   c. UAG, P site

9. The movement of the polypeptide from the tRNA in the P site to the tRNA in the A site is referred to as
   a. peptide bonding.   d. peptidyl transfer reaction.
   b. aminoacyl binding.   e. elongation.
   c. translation.

10. The synthesis of a polypeptide occurs during which stage of translation?
    a. initiation   c. termination
    b. elongation   d. splicing

CONCEPTUAL QUESTIONS

1. Define mutation.
2. Explain the one gene–one enzyme hypothesis and the more modern modifications of this hypothesis.
3. What is the function of an aminoacyl-tRNA synthetase?

EXPERIMENTAL QUESTIONS

1. Briefly explain how studying the pathway that leads to arginine synthesis allowed Beadle and Tatum to conclude that one gene encodes one enzyme.
2. What was the benefit of using radiolabeled amino acids in the Nirenberg and Leder experiment?
3. Predict the results that Nirenberg and Leder would have found for the following triplets: AUG, UAA, UAG, UGA.

COLLABORATIVE QUESTIONS

1. Discuss RNA processing in eukaryotes.
2. How can we analyze rRNA to establish evolutionary relationships between different species of organisms?

www.brookerbiology.com

This website includes answers to the Biological Inquiry questions found in the figure legends and all end-of-chapter questions.

13

GENE REGULATION

CHAPTER OUTLINE

13.1 Overview of Gene Regulation

13.2 Regulation of Transcription in Bacteria

13.3 Regulation of Transcription in Eukaryotes

13.4 Regulation of RNA Processing and Translation in Eukaryotes

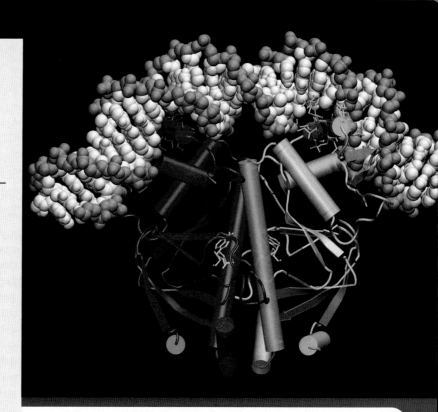

A model for a protein that binds to DNA and regulates genes.
The protein shown in red and pink is binding to the DNA double helix, shown in orange and white. This protein, described later in Figure 13.10, activates gene transcription.

The term **gene regulation** refers to the ability of cells to control their level of gene expression. The majority of structural genes in all species are regulated so that the proteins they encode are produced at certain times and in specific amounts. By comparison, unregulated genes have essentially constant levels of expression in all conditions over time. These are called **constitutive genes**. Frequently, constitutive genes, also referred to as housekeeping genes, encode proteins that are always necessary for the survival of an organism.

The importance of gene regulation is underscored by the number of genes devoted to this process in an organism. A sizeable portion of the genomes of complex organisms such as plants and animals is devoted to the process of gene regulation. For example, in *Arabidopsis thaliana*, a plant that has been studied as a model organism by plant geneticists, over 5% of its genome is involved with regulating gene transcription. This species has more than 1,500 different genes that encode proteins that regulate the transcription of other genes.

In this chapter, we will begin with an overview that emphasizes the benefits of gene regulation and the general mechanisms that achieve such regulation. Later sections will describe the ways that prokaryotes and eukaryotes regulate their genes at the molecular level.

13.1 Overview of Gene Regulation

Living organisms derive many benefits from regulating genes. This process ensures that proteins will be produced only when they are required. For example, certain proteins help an organism to survive environmental stress such as drastic changes in temperature or exposure to ultraviolet (UV) light. Because these proteins are required only when the organism is confronted with the stress, a cell conserves energy and resources if such proteins are made just when the stressor is present. In addition, some proteins function in the metabolism of small molecules that may or may not be present in the environment. For example, certain proteins are needed for a bacterium to take up and metabolize particular sugars. These proteins are required only when the bacterium is exposed to such sugars in its environment.

Like bacteria, eukaryotic organisms such as protists, fungi, plants, and animals need to adapt to changes in their environment. For example, all eukaryotic species can respond to environmental stresses such as UV radiation by turning on genes that provide protection against this harmful environmental agent. In humans, exposure to UV light turns on genes that result in a tanning response.

In addition to responding to environmental conditions, eukaryotes also regulate genes when cells are dividing and specializing, and when they are going through developmental stages. For example, some proteins are needed just for cell division. These are necessary only when a cell is preparing to divide and during the division process. In multicellular organisms, certain proteins are made only in particular cell types, or their amounts may vary from cell to cell. In humans, for example, some proteins are needed only in muscle cells but not in nerve cells, and vice versa. Similarly, in multicellular organisms that progress through different developmental stages (fertilized egg, embryo, and adult), certain proteins are needed only at particular stages of development. In this section, we will examine a few examples that illustrate the important consequences of gene regulation.

We will also survey the points in the gene expression process at which genes are regulated in prokaryotic and eukaryotic cells.

Bacteria Regulate Genes to Respond to Nutrients in Their Environment

As mentioned, gene regulation in prokaryotes is often used to respond to changes in the environment. To fully appreciate how genetic regulation helps bacteria survive, let's look at an example. The bacterium *Escherichia coli* can use many types of sugars as food sources, thereby increasing its chances of survival. Let's consider the process of how it uses lactose, which is the sugar found in milk. *E. coli* can use lactose because it carries genes that code for proteins that enable it to take up lactose and metabolize it.

Figure 13.1 illustrates the effects of lactose on the regulation of those genes. Lactose use requires a transporter, called the lactose permease, that facilitates the uptake of lactose into the cell, and an enzyme, called β-galactosidase, that catalyzes the breakdown of lactose. When lactose is not present in the environment, an *E. coli* cell makes very little of these proteins. However, when lactose becomes available, gene regulation causes the synthesis of many more of these proteins, enabling the bacterium to efficiently use lactose from its environment. Eventually, all of the lactose in the environment will be used up. At this point, the genes encoding these proteins will be shut off, and most of the proteins will be degraded. In the case of lactose use,

the main benefit of gene regulation is that the proteins that are needed for this process are made only when lactose is present in the environment. Therefore, *E. coli* does not waste valuable energy making these proteins when they are not needed.

Eukaryotic Gene Regulation Produces Different Cell Types in a Single Organism

One of the most amazing examples of gene regulation is the phenomenon of **cell differentiation**. In multicellular organisms such as plants and animals, cells become specialized, or differentiated, into particular types. In humans, for example, cells may become specialized into muscle cells, nerve cells, skin cells, or other types. **Figure 13.2** shows micrographs of three types of cells found in humans. As seen here, their morphologies are strikingly different. Likewise, their functions within the body are also quite different. Muscle cells are important in body movements, nerve cells function in cell signaling, and skin cells form a protective outer surface to the body.

Gene regulation is responsible for creating different types of cells within a multicellular organism. The three cell types shown in Figure 13.2 contain the same genome, meaning they carry the same set of genes. However, their proteomes are quite different due to gene regulation. Certain proteins are found in particular cell types, but not in others. Alternatively, a protein may be present in all three cell types, but the relative amounts of the protein may be different. The amount of a given protein depends

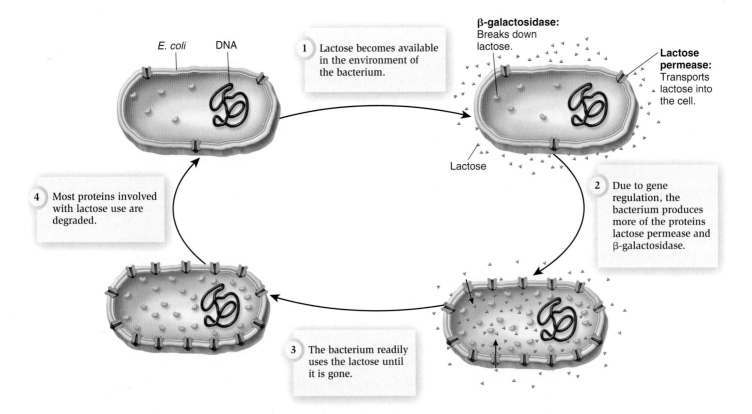

Figure 13.1   Gene regulation of lactose utilization in *E. coli.*

Biological inquiry: What is the advantage to E. coli *for regulating the genes involved with lactose utilization?*

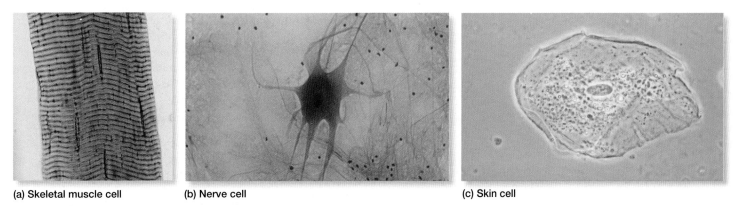

(a) Skeletal muscle cell   (b) Nerve cell   (c) Skin cell

Figure 13.2   Examples of different cell types in humans.

on many factors, including how strongly the corresponding gene is turned on and how much protein is synthesized from mRNA. Gene regulation at the level of transcription and translation plays a major role in determining the proteome of each cell type. These forms of gene regulation will be discussed later in this chapter. In addition, factors described in Unit II such as protein degradation, post-translational modifications, and feedback inhibition are important in controlling the functions and relative amounts of proteins in a cell.

Eukaryotic Gene Regulation Enables Multicellular Organisms to Progress Through Developmental Stages

In multicellular organisms that progress through developmental stages, certain genes are expressed at one stage of development but not another. This topic is discussed in Chapter 19. Let's consider an example of such developmental gene regulation in mammals. Early stages of mammalian development occur in the uterus of female mammals. Following fertilization, an embryo develops inside the uterus. In humans, the embryonic stage is from fertilization to 8 weeks. During this stage, major developmental changes produce the various body parts. The fetal stage occurs from 8 weeks to birth (41 weeks). This stage is characterized by continued refinement of body parts and a large increase in size.

Because of this internal development, a system has evolved to provide both the embryonic and fetal cells with the oxygen they require for cellular respiration. The oxygen demands of a rapidly growing embryo and fetus are quite different from the needs of the mother. Gene regulation plays a vital role in ensuring that an embryo and fetus get the proper amount of oxygen. As described in Chapter 48, hemoglobin is the main protein that delivers oxygen to the cells of a mammal's body. The genomes of mammals carry several genes (designated with Greek letters) that encode slightly different globin polypeptides. A hemoglobin protein is composed of four globin polypeptides, two encoded by one globin gene and two encoded by another globin gene (**Figure 13.3**). During the embryonic stage of development, the epsilon (ϵ)-globin and zeta (ζ)-globin genes are turned on. At the fetal stage, these genes are turned off, and the alpha (α)-globin and gamma (γ)-globin genes are turned on. Finally, at birth, the γ-globin gene is turned off, and the beta (β)-globin gene is turned

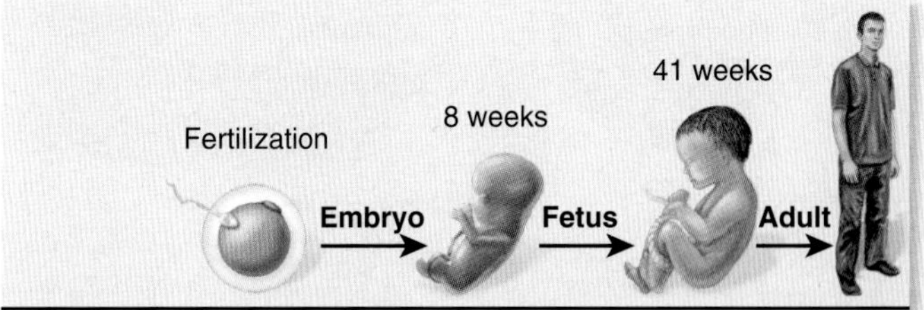

	Embryo	Fetus	Adult
Hemoglobin protein	2 ζ-globins 2 ϵ-globins	2 α-globins 2 γ-globins	2 α-globins 2 β-globins
Oxygen affinity	highest	high	moderate
Gene expression α-globin gene β-globin gene γ-globin gene ζ-globin gene ϵ-globin gene	off off off on on	on off on off off	on on off off off

Figure 13.3   Developmental regulation of human globin genes.

on. The hemoglobin that is produced during the embryonic and fetal stages has a much higher binding affinity for oxygen than does the hemoglobin that is produced after birth. Therefore, the embryo and fetus can remove oxygen from the mother's bloodstream and use that oxygen for their own needs. This occurs across the placenta, where the mother's bloodstream is adjacent to the bloodstream of the embryo or fetus. In this way, gene regulation enables mammals to develop internally, even though the embryo and fetus are not breathing on their own.

Gene Regulation Can Occur at Different Points in the Process from DNA to Protein

Thus far we have learned that gene regulation can have a dramatic impact on the ability of organisms to respond to environmental changes, differentiate cells, and progress through developmental stages. Because structural genes encode proteins, the regulation of gene expression can occur at any of the steps that are needed to produce a functional protein.

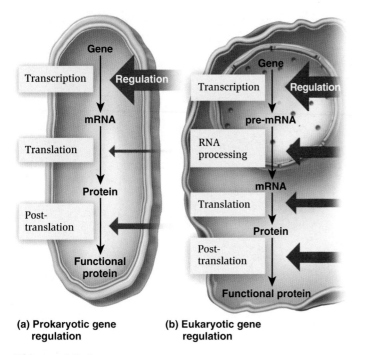

(a) Prokaryotic gene regulation

(b) Eukaryotic gene regulation

Figure 13.4 Overview of points of control for gene regulation in (a) prokaryotes and (b) eukaryotes. The relative sizes of the red arrows indicate the prominence with which regulation at that point is used to control the production of functional proteins.

In bacteria, gene regulation most commonly occurs at the level of transcription, which means that bacteria regulate how much mRNA is made from most genes (**Figure 13.4a**). When geneticists say a gene is "turned off," they mean that no mRNA is made from that gene, while a gene that is "turned on" is transcribed into mRNA. A second way for bacteria to regulate gene expression is to control the rate at which mRNA is translated into protein. This form of gene regulation is relatively uncommon in bacteria. Finally, gene expression can be regulated at the protein or post-translational level. Various types of protein regulation are best understood within the context of cell biology, so they are discussed in the unit on cells. For example, in Chapter 7, we learned that enzymes are regulated by feedback inhibition.

In eukaryotes, gene regulation occurs at many levels, including transcription, RNA processing, translation, and post-translation (**Figure 13.4b**). As in their bacterial counterparts, transcriptional regulation is a prominent form of eukaryotic gene regulation. As discussed later in this chapter, eukaryotic genes are transcriptionally regulated in several different ways, some of which are not found in bacteria. Also, as discussed in Chapter 12, eukaryotes process their mRNA transcripts in ways that do not commonly occur in bacteria. For example, RNA splicing is a widespread phenomenon in eukaryotes. Later in this chapter, we will examine how this process is regulated to create two or more different types of mRNA from a single gene. Eukaryotes also regulate mRNA after its modification. The amount of mRNA may be regulated by controlling its degra-

dation. In addition, the translation of an mRNA may be regulated by small, inhibitory RNA molecules or by RNA-binding proteins that prevent translation from occurring. As in prokaryotes, eukaryotic proteins can be regulated in a variety of ways including feedback inhibition, post-translational modification, and degradation via the proteasome. These mechanisms are discussed in Unit II. Now let's take a look at transcriptional regulation in bacteria.

13.2 Regulation of Transcription in Bacteria

Due to gene regulation, bacteria can respond specifically to changes in their cellular and environmental conditions. As we have seen, when a bacterium is exposed to a particular nutrient in its environment such as a sugar, the genes are expressed that encode proteins that are needed for the uptake and metabolism of the sugar. Alternatively, bacteria have enzymes that synthesize cellular molecules such as particular amino acids. It is to their advantage for bacteria to turn off the genes that encode those enzymes when a sufficient amount of the amino acid is present in the cytoplasm. In this section, we will examine the underlying molecular mechanisms that bring about transcriptional regulation in bacteria.

Transcriptional Regulation Usually Involves Regulatory Transcription Factors and Small Effector Molecules

In most cases, transcriptional regulation involves the actions of **regulatory transcription factors**, which are proteins that bind to DNA in the vicinity of a promoter and affect the rate of transcription of one or more nearby genes. These transcription factors can either decrease or increase the rate of transcription of a gene. **Repressors** are transcription factors that bind to DNA and inhibit transcription, whereas **activators** increase the rate of transcription. The term **negative control** refers to transcriptional regulation by repressor proteins; **positive control** refers to regulation by activator proteins (**Figure 13.5a**).

In conjunction with regulatory transcription factors, molecules called **small effector molecules** often play a critical role in transcriptional regulation. A small effector molecule exerts its effects by binding to a regulatory transcription factor and causing a conformational change in the protein. In some cases, the effect of the conformational change determines whether or not the protein can bind to the DNA. **Figure 13.5b** illustrates an example. When the small effector molecule is not present in the cytoplasm, the repressor binds to the DNA and inhibits transcription. However, at a later time when the small effector molecule is found in the cytoplasm, it will bind to the repressor and cause a conformational change that inhibits the ability of the protein to bind the DNA. The gene is turned on because the repressor is released. Regulatory transcription factors that respond

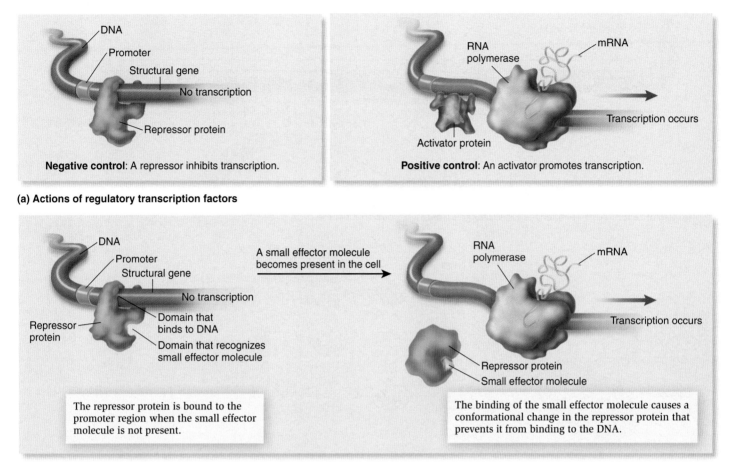

(a) Actions of regulatory transcription factors

Negative control: A repressor inhibits transcription.

Positive control: An activator promotes transcription.

The repressor protein is bound to the promoter region when the small effector molecule is not present.

The binding of the small effector molecule causes a conformational change in the repressor protein that prevents it from binding to the DNA.

(b) Action of a small effector molecule on a repressor protein

Figure 13.5 Actions of regulatory transcription factors and small effector molecules.   (a) Regulatory transcription factors may exert positive or negative control. (b) One way that a small effector molecule may exert its effects is by preventing a repressor protein from binding to the DNA.

to small effector molecules have two functional regions called **domains**. One domain is a site where the protein binds to the DNA. The other domain is the binding site for the small effector molecule.

The *lac* Operon Contains Genes That Encode Proteins Involved in Lactose Metabolism

In bacteria, genes are sometimes clustered together in units that are under the transcriptional control of a single promoter. This arrangement, known as an **operon**, is transcribed into an mRNA, called a **polycistronic mRNA**, that contains the coding sequences for two or more structural genes. The transcription of the structural genes occurs as a single unit. An operon organization allows a bacterium to coordinately regulate a group of genes that encode proteins with a common function.

The genome of *E. coli* carries an operon called the ***lac* operon** that contains the genes for the enzymes we discussed earlier that allow it to metabolize lactose (see Figure 13.1). **Figure 13.6a** shows the organization of this operon as it is found in the *E. coli* chromosome, as well as the polycistronic mRNA that is

transcribed from it. The *lac* operon contains a promoter that is used to transcribe three structural genes, *lacZ*, *lacY*, and *lacA*. *LacZ* encodes β-galactosidase, which as you may remember is an enzyme that breaks down lactose (**Figure 13.6b**). As a side reaction, β-galactosidase also converts a small percentage of lactose into allolactose, a structurally similar sugar or lactose analogue. As described later, allolactose is important in the regulation of the *lac* operon. The *lacY* gene encodes lactose permease, which is a membrane protein required for the transport of lactose into the cytoplasm of the bacterium. The *lacA* gene encodes galactoside transacetylase, which covalently modifies lactose and lactose analogues by attaching an acetyl group (—COCH$_3$). Although the functional necessity of this enzyme remains unclear, the attachment of acetyl groups to nonmetabolizable lactose analogues may prevent their toxic buildup in the cytoplasm.

Near the *lac* promoter are two regulatory sites designated the operator and the CAP site (see Figure 13.6a). The **operator**, or *lacO* site, is a sequence of nucleotides that provides a binding site for a repressor protein. The **CAP site** is a DNA sequence recognized by an activator protein.

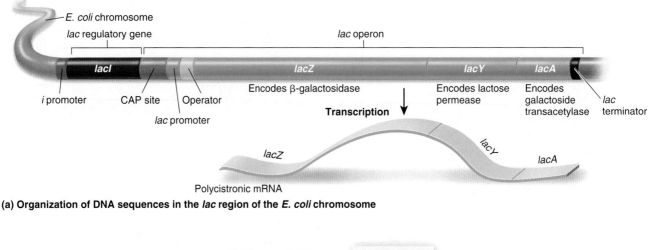

(a) Organization of DNA sequences in the *lac* region of the *E. coli* chromosome

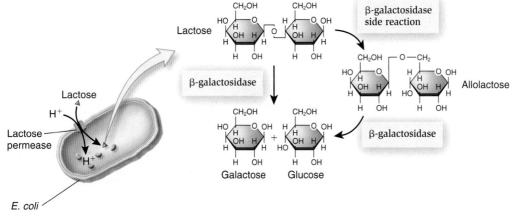

(b) Functions of lactose permease and β-galactosidase

Figure 13.6 **The *lac* operon.** (a) This diagram depicts a region of the *E. coli* chromosome that contains the *lacI* regulatory gene and the adjacent *lac* operon, as well as the polycistronic mRNA transcribed from the operon. (b) Function of lactose permease and β-galactosidase. Lactose permease cotransports H⁺ with lactose. Bacteria maintain an H⁺ gradient across their cytoplasmic membrane that drives the active transport of lactose into the cytoplasm. β-galactosidase cleaves lactose into galactose and glucose. As a side reaction, it can also convert lactose into allolactose.

Adjacent to the *lac* operon is the *lacI* gene, which encodes the **lac repressor**. This repressor protein is important for the regulation of the *lac* operon. The *lacI* gene, which is constitutively expressed at fairly low levels, has its own promoter called the *i* promoter. It is considered to be a **regulatory gene** because the sole function of the encoded protein is to regulate the expression of other genes. The *lacI* gene has its own promoter and is not considered a part of the *lac* operon. Let's now take a look at how the *lac* operon is regulated by the *lac* repressor.

The *lac* Operon Is Under Negative Control by a Repressor Protein

In the late 1950s, the first researchers to investigate genetic regulation were Francois Jacob and Jacques Monod at the Pasteur Institute in Paris, France. Their focus on gene regulation stemmed from an interest in the phenomenon known as enzyme adaptation, which had been identified early in the 20th century. Enzyme adaptation refers to the observation that a particular enzyme appears within a living cell only after the cell has been exposed to the substrate for that enzyme. Jacob and Monod studied lactose metabolism in *E. coli* to investigate this phenomenon. When they exposed bacteria to lactose, the levels of lactose-using enzymes in the cells increased by 1,000- to 10,000-fold. After lactose was removed, the synthesis of the enzymes abruptly stopped.

The first mechanism of regulation that Jacob and Monod discovered involved the *lac* repressor protein, which binds to the sequence of nucleotides found at the *lac* operator site. Once bound, the *lac* repressor prevents RNA polymerase from transcribing the *lacZ*, *lacY*, and *lacA* genes (**Figure 13.7a**). RNA polymerase can bind to the promoter when the *lac* repressor is bound to the operator site, but RNA polymerase cannot move past the operator to transcribe the *lacZ*, *lacY*, and *lacA* genes.

Whether or not the *lac* repressor binds to the operator site depends on allolactose, which is the previously mentioned side product of the β-galactosidase enzyme (see Figure 13.6b).

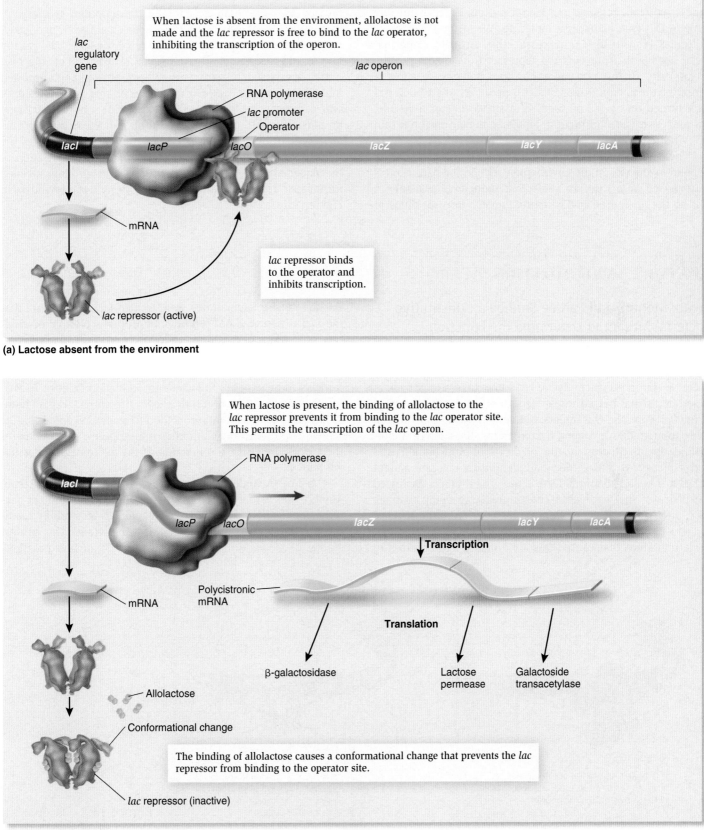

(a) Lactose absent from the environment

When lactose is absent from the environment, allolactose is not made and the *lac* repressor is free to bind to the *lac* operator, inhibiting the transcription of the operon.

lac regulatory gene

lac operon

RNA polymerase

lac promoter

Operator

lacI   lacP   lacO   lacZ   lacY   lacA

mRNA

lac repressor binds to the operator and inhibits transcription.

lac repressor (active)

(b) Lactose present

When lactose is present, the binding of allolactose to the *lac* repressor prevents it from binding to the *lac* operator site. This permits the transcription of the *lac* operon.

RNA polymerase

lacI   lacP   lacO   lacZ   lacY   lacA

Transcription

mRNA

Polycistronic mRNA

Translation

β-galactosidase   Lactose permease   Galactoside transacetylase

Allolactose

Conformational change

The binding of allolactose causes a conformational change that prevents the *lac* repressor from binding to the operator site.

lac repressor (inactive)

Figure 13.7   Negative control of an inducible set of genes: function of the *lac* repressor in regulating the *lac* operon.

Allolactose is an example of a small effector molecule. The *lac* repressor protein contains four identical subunits, each one recognizing a single allolactose molecule. When four allolactose molecules bind to the *lac* repressor, this causes a conformational change that prevents the repressor from binding to the operator site. Under these conditions, RNA polymerase is free to transcribe the operon (**Figure 13.7b**). Because transcription has been turned on by the presence of a small effector molecule, this process is called **induction**. Similarly, the *lac* operon is said to be **inducible**.

The regulation of the *lac* operon enables *E. coli* to efficiently respond to changes in the environment. When the bacterium is not exposed to lactose, no allolactose **inducer** is available to bind to the *lac* repressor. Therefore, the *lac* repressor binds to the operator site and inhibits transcription. In reality, the repressor does not completely inhibit transcription, so that very small amounts of β-galactosidase, lactose permease, and galactoside transacetylase are made. Even so, the levels are far too low for the bacterium to readily use lactose. When the bacterium is exposed to lactose, a small amount can be transported into the cytoplasm via lactose permease, and β-galactosidase will convert some of it to allolactose. The cytoplasmic level of allolactose will gradually rise until allolactose binds to the *lac* repressor, which induces the *lac* operon and promotes the transcription of the *lacZ*, *lacY*, and *lacA* genes. Translation of the encoded polypeptides will produce the proteins needed for lactose uptake and metabolism as described previously in Figure 13.1.

FEATURE INVESTIGATION

Jacob, Monod, and Pardee Studied a Constitutive Bacterial Mutant to Determine the Function of the *lac* Repressor

Thus far, we have learned that the *lac* repressor binds to the *lac* operator site to exert its effects. Let's now take a look back at experiments that helped researchers to determine the function of the *lac* repressor. Our understanding of *lac* operon regulation came from studies involving *E. coli* strains that showed abnormalities in the process. In the 1950s, Jacob and Monod, and their colleague Arthur Pardee, had identified a few rare mutant bacteria that had abnormal lactose use. The mutants expressed the genes of the *lac* operon constitutively, meaning that the *lacZ*, *lacY*, and *lacA* genes were expressed even in the absence of lactose in the environment. The researchers discovered that some mutations that caused this abnormality had occurred in the *lacI* region. Such strains were termed *lacI*⁻ to indicate that the *lacI* region was not functioning properly. Normal or wild-type *lacI* strains of *E. coli* are called *lacI*⁺ in comparison.

The researchers hypothesized that the *lacI*⁻ mutation resulted in the synthesis of an internal inducer, making it unnecessary for cells to be exposed to lactose for induction (**Figure 13.8a**). By comparison, **Figure 13.8b** shows the correct explanation. A loss-of-function mutation in the *lacI* gene prevented the *lac* repressor protein from inhibiting transcription. At the time of their work, however, the function of the *lac* repressor was not yet known.

To further understand the nature of this mutation, Jacob, Monod, and Pardee applied a genetic approach. Although bacterial conjugation is described in Chapter 18, let's briefly examine this process to understand this experiment. The earliest studies of Jacob, Monod, and Pardee in 1959 involved matings

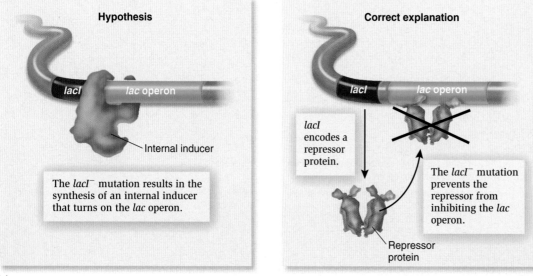

Hypothesis

lacI   *lac* operon

Internal inducer

The *lacI*⁻ mutation results in the synthesis of an internal inducer that turns on the *lac* operon.

(a)

Correct explanation

lacI   *lac* operon

lacI encodes a repressor protein.

The *lacI*⁻ mutation prevents the repressor from inhibiting the *lac* operon.

Repressor protein

(b)

Figure 13.8 Jacob, Monod, and Pardee's hypothesis for the function of the *lacI* region.

between recipient cells, termed F⁻, and donor cells, which were Hfr strains that transferred a portion of the bacterial chromosome. Later experiments in 1961 involved the transfer of circular segments of DNA known as F factors. We will consider this later type of experiment here. Sometimes an F factor also carries genes that were originally found within the bacterial chromosome. These types of F factors are called F′ factors (F prime factors). A strain of bacteria containing F′ factor genes is called a **merozygote**, or partial diploid. The production of merozygotes was instrumental in allowing Jacob, Monod, and Pardee to elucidate the function of the *lacI* gene.

As shown in **Figure 13.9**, these researchers studied the *lac* operon in a bacterial strain carrying a *lacI⁻* mutation that caused constitutive expression of the *lac* operon. In addition, the mutant strain was subjected to mating to create a merozygote that also carried a normal *lac* operon and normal *lacI⁺* gene on an F′ factor. The constitutive mutant and corresponding merozygote were allowed to grow separately in liquid media and then divided into two tubes each. In half of the tubes, the cells were incubated with lactose to determine if lactose was needed to induce the expression of the operon. In the other tubes, lactose was omitted.

To monitor the expression of the *lac* operon, the cells were broken open and then tested for the amount of β-galactosidase they released by measuring its ability to convert a colorless compound into a yellow product.

The data table of Figure 13.9 summarizes the effects of this constitutive mutation and its analysis in a merozygote. As Jacob, Monod, and Pardee already knew, the *lacI⁻* mutant strain expressed the *lac* operon constitutively, in the presence and absence of lactose. However, when a normal *lac* operon and *lacI⁺* gene were introduced on an F′ factor into a cell harboring the mutant *lacI⁻* gene on the chromosome, the normal *lacI⁺* gene could regulate both operons. In the absence of lactose, both operons were shut off. This occurred because a single *lacI⁺* gene on the F′ factor can produce enough repressor protein to bind to both operator sites. Furthermore, this protein is diffusible (can spread through the cytoplasm) and can bind to *lac* operons that are on the F′ factor and on the bacterial chromosome. The hypothesis that the *lacI⁻* mutation resulted in the synthesis of an internal inducer was rejected. If this hypothesis had been correct, the inducer would have been made in the absence of lactose, and the *lac* operon would have been expressed.

Figure 13.9   The experiment performed by Jacob, Monod, and Pardee to study a *lacI* constitutive mutant.

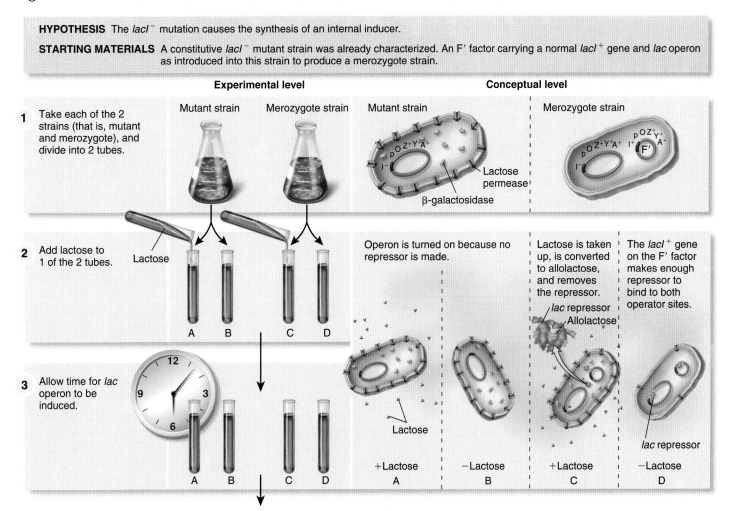

HYPOTHESIS The *lacI⁻* mutation causes the synthesis of an internal inducer.

STARTING MATERIALS A constitutive *lacI⁻* mutant strain was already characterized. An F′ factor carrying a normal *lacI⁺* gene and *lac* operon as introduced into this strain to produce a merozygote strain.

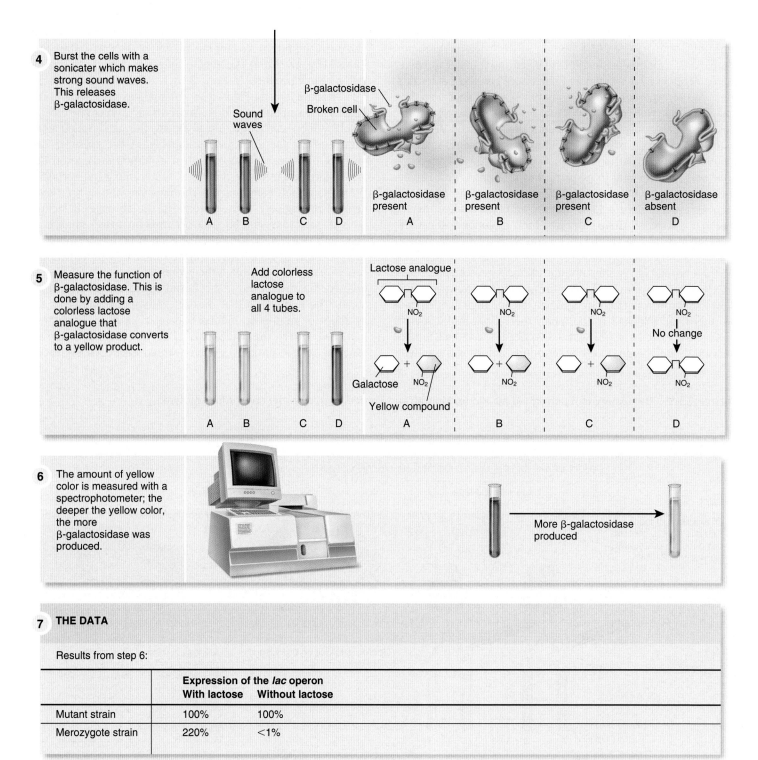

4 Burst the cells with a sonicater which makes strong sound waves. This releases β-galactosidase.

Sound waves

β-galactosidase

Broken cell

β-galactosidase present

A

β-galactosidase present

B

β-galactosidase present

C

β-galactosidase absent

D

A B C D

5 Measure the function of β-galactosidase. This is done by adding a colorless lactose analogue that β-galactosidase converts to a yellow product.

Add colorless lactose analogue to all 4 tubes.

Lactose analogue

NO₂

Galactose   NO₂

Yellow compound

No change

A B C D

A B C D

6 The amount of yellow color is measured with a spectrophotometer; the deeper the yellow color, the more β-galactosidase was produced.

More β-galactosidase produced

7 THE DATA

Results from step 6:

	Expression of the *lac* operon	
	With lactose	Without lactose
Mutant strain	100%	100%
Merozygote strain	220%	<1%

This result was not obtained. Taken together, the data indicated that the normal *lacI* gene encodes a diffusible protein that represses the *lac* operon.

The interactions between regulatory proteins and DNA sequences illustrated in this experiment have led to the definition of two genetic terms. In both prokaryotes and eukaryotes, a **trans-effect** is a form of genetic regulation that can occur even though two DNA segments are not physically adjacent. The action of the *lac* repressor on the *lac* operon is a *trans*-effect. In contrast, a **cis-effect** or a **cis-acting element** is a DNA segment that must be adjacent to the gene(s) that it regulates. The *lac* operator site is an example of a *cis*-acting element. A *trans*-effect is mediated by genes that encode diffusible regulatory proteins, whereas a *cis*-effect is mediated by DNA sequences that bind regulatory proteins.

The *lac* Operon Is Also Under Positive Control by an Activator Protein

In addition to negative control by a repressor protein, the *lac* operon is also positively regulated by an activator protein called the **catabolite activator protein (CAP)**, also known as the cAMP receptor protein (CRP). In a process known as **catabolite repression**, this form of transcriptional regulation is influenced by the presence of glucose, which is a catabolite (it is broken down—**catabolized**—inside the cell). The ability of glucose to repress the *lac* operon depends on a small effector molecule, **cyclic AMP (cAMP)**, that is produced from ATP via an enzyme known as adenylyl cyclase. When cAMP is made, it binds to CAP, which causes CAP to bind to the DNA. Because CAP is needed for activation of the *lac* operon, the operon is turned off when CAP is not bound to the DNA. Glucose inhibits the production of cAMP and thereby prevents the binding of CAP to the DNA.

The genetic regulation involving CAP and cAMP is an example of positive control (**Figure 13.10**). When cAMP binds to CAP, the cAMP-CAP complex binds to the CAP site near the *lac* promoter. This causes a bend in the DNA that enhances the ability of RNA polymerase to bind to the promoter. In this way, the rate of transcription is increased. Though it may seem puzzling, the term catabolite repression was coined before the action of the

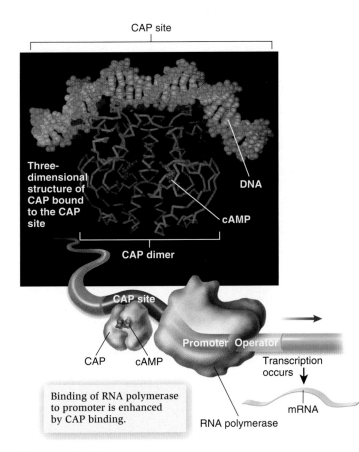

Figure 13.10 Positive regulation of the *lac* operon by the catabolite activator protein (CAP). When cAMP is bound to CAP, CAP binds to the DNA and causes it to bend. This bend facilitates the binding of RNA polymerase.

cAMP-CAP complex was understood at the molecular level. Historically, the primary observation of researchers was that glucose (a catabolite) inhibited (repressed) lactose metabolism. Further experimentation revealed that CAP is an activator protein.

Let's now consider how the concerted actions of the *lac* repressor and CAP allow bacteria to use the sugars in their environment most efficiently. When a bacterium is exposed to glucose, the uptake of glucose into the cell inhibits adenylyl cyclase, which causes cAMP levels to drop. Without cAMP, CAP cannot bind to the CAP site. When a bacterium is exposed to lactose, allolactose levels in the cytoplasm rise. The binding of allolactose prevents the lac repressor from binding to the operator site. With these ideas in mind, **Figure 13.11** considers the four possible environmental conditions that an *E. coli* bacterium might experience with regard to these sugars. When both lactose and glucose levels are high, the *lac* operon is shut off, because CAP does not activate transcription. Under these conditions, the bacterium uses glucose rather than lactose. Greater efficiency is achieved if the bacterium uses one type of sugar at a time. At another time, if lactose levels are high and glucose is low, the *lac* operon is turned on because CAP is bound to the CAP site and the *lac* repressor is not bound to the operator site. Under these conditions, the bacterium metabolizes lactose. When lactose levels are low, the *lac* repressor prevents transcription of the *lac* operon, whether glucose levels are high or low.

The *trp* Operon Is Also Under Negative Control by a Repressor Protein

So far in this section, we have examined the regulation of the *lac* operon. Let's now consider an example of an operon that encodes enzymes involved in biosynthesis rather than degradation. Our example is the ***trp* operon** of *E. coli*, which encodes enzymes that are required to make the amino acid tryptophan, a building block of cellular proteins. More specifically, the *trpE*, *trpD*, *trpC*, *trpB*, and *trpA* genes encode enzymes that are involved in a pathway that leads to tryptophan synthesis.

The *trp* operon is regulated by a repressor protein that is encoded by the *trpR* gene. The binding of the repressor to the *trp* operator site inhibits transcription. The ability of the *trp* repressor to bind to the *trp* operator is controlled by tryptophan, which is the product of the enzymes that are encoded by the operon. When tryptophan levels within the cell are very low, the *trp* repressor cannot bind to the operator site. Under these conditions, RNA polymerase readily transcribes the operon (**Figure 13.12a**). In this way, the cell expresses the genes that encode enzymes that result in the synthesis of tryptophan, which is in short supply. Alternatively, when the tryptophan levels within the cell are high, tryptophan turns off the *trp* operon. Tryptophan acts as a small effector molecule or **corepressor** by binding to the *trp* repressor protein. This causes a conformational change in the repressor that allows it to bind to the *trp* operator site, inhibiting the ability of RNA polymerase to transcribe the operon (**Figure 13.12b**). Therefore, the bacterium does not waste energy making tryptophan when it is abundant.

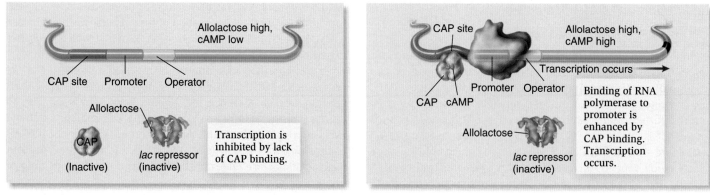

(a) Lactose high, glucose high

(b) Lactose high, glucose low

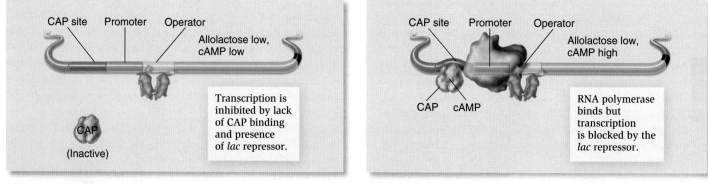

(c) Lactose low, glucose high

(d) Lactose low, glucose low

Figure 13.11   Effects of lactose and glucose on the expression of the *lac* operon.

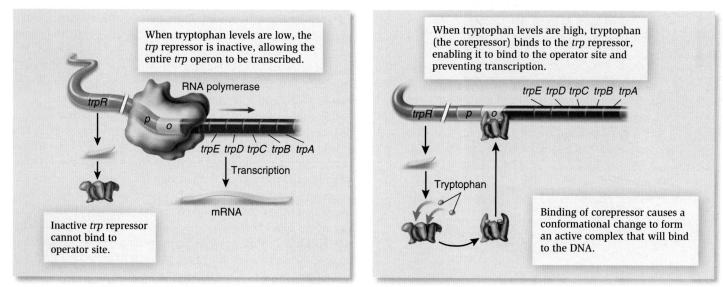

(a) Low tryptophan

(b) High tryptophan

Figure 13.12   Negative control of a repressible set of genes: function of the *trp* repressor and corepressor in regulating the *trp* operon.

Biological inquiry: How are the functions of the lac repressor and trp repressor similar to each other, and how are they different?

When comparing the *lac* and *trp* operons, the actions of their small effector molecules are quite different. The *lac* repressor binds to its operator in the absence of its small effector molecule, while the *trp* repressor binds to its operator only in the presence of its small effector molecule. The *lac* operon is categorized as inducible because its small effector molecule, namely allolactose, induces transcription. By comparison, the *trp* operon is considered to be **repressible** because its small effector molecule, namely tryptophan, represses transcription.

Repressible Operons Usually Encode Anabolic Enzymes, and Inducible Operons Encode Catabolic Enzymes

Thus far, we have seen that bacterial operons can be transcriptionally regulated in a positive or negative way—or sometimes both. By comparing the mechanisms of regulation among many bacterial operons, geneticists have noticed a general trend. When the genes in an operon encode proteins that function in the breakdown, or **catabolism**, of a substance, the substance to be broken down (or a related compound) often acts as the inducer. This keeps the genes turned off unless the appropriate substance is available. For example, allolactose, which is a product of lactose metabolism, acts as an inducer of the *lac* operon. An inducible form of regulation allows the bacterium to express the appropriate genes only when they are needed to metabolize lactose.

Other cellular enzymes are important for synthesizing organic molecules, a process termed **anabolism**. Because these molecules are generally needed for the functioning of the cell, the genes that encode these anabolic enzymes tend to be regulated by a repressible mechanism, allowing the genes to be transcribed unless they are turned off. The small effector molecule is commonly a product of the enzymes' biosynthetic activities. For example, as we learned, tryptophan is produced by the enzymes that are encoded by the *trp* operon. When enough of this amino acid has been made, tryptophan itself acts as a corepressor, turning off the genes required for tryptophan biosynthesis. Therefore, a repressible form of regulation provides the bacterium with a way to prevent the overproduction of the product of a biosynthetic pathway. Now let's take a look at how transcription is regulated in eukaryotes.

13.3 Regulation of Transcription in Eukaryotes

Transcriptional regulation in eukaryotes follows some of the same principles as those that are found in prokaryotes. For example, activator and repressor proteins are involved in regulating genes by influencing the ability of RNA polymerase to initiate transcription. In addition, many eukaryotic genes are regulated by small effector molecules. However, some important differences

also occur. In eukaryotic species, genes are almost always organized individually, not in operons. In addition, eukaryotic gene regulation tends to be more intricate, because eukaryotes are faced with complexities that differ from their prokaryotic counterparts. For example, eukaryotes have more complicated cell structures that include many more proteins and a variety of cell organelles. Many eukaryotes such as animals and plants are multicellular and contain different cell types. As discussed earlier in this chapter, animal cells may differentiate into nerve cells, muscle cells, and skin cells, among others. Furthermore, animals and plants progress through developmental stages that require changes in gene expression. For these reasons, gene regulation in eukaryotes requires much more coordination and integration.

By studying transcriptional regulation, researchers have discovered that most eukaryotic genes, particularly those found in multicellular species, are regulated by many factors. This phenomenon is called **combinatorial control** because the combination of many factors determines the expression of any given gene. At the level of transcription, common factors that contribute to combinatorial control include the following:

1. One or more activator proteins may stimulate the ability of RNA polymerase to initiate transcription.
2. One or more repressor proteins may inhibit the ability of RNA polymerase to initiate transcription.
3. The function of activators and repressors may be modulated in a variety of ways. These include the binding of small effector molecules, protein–protein interactions, and covalent modifications.
4. Activator proteins may promote the loosening up of the region in the chromosome where a gene is located, thereby making it easier for the gene to be recognized and transcribed by RNA polymerase.
5. DNA methylation (usually) inhibits transcription, either by preventing the binding of an activator protein or by recruiting proteins that cause the DNA to become more compact.

All five of these factors may contribute to the regulation of a single gene, or possibly only three or four will play a role. In most cases, transcriptional regulation is aimed at controlling the initiation of transcription at the promoter. In this section, we will survey these basic kinds of gene regulation in eukaryotic species.

Eukaryotic Structural Genes Have a Core Promoter and Response Elements

To understand gene regulation in eukaryotes, we first need to consider the DNA sequences that are needed to initiate transcription. For eukaryotic structural genes that encode proteins, three features are found in most promoters: a **transcriptional start site**, a **TATA box**, and **response elements** (**Figure 13.13**).

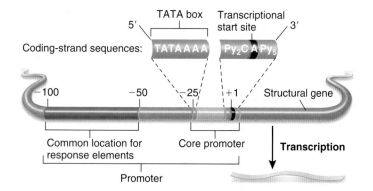

Figure 13.13 A common organization of sequences for the promoter of a eukaryotic structural gene. The core promoter is made up of a TATA box and a transcriptional start site. The TATA box sequence is 5′–TATAAAA–3′. The sequence of the transcriptional start site in the 5′ to 3′ direction is two pyrimidines (C or T), then CA, then five pyrimidines. The A marks the site of the first A in the RNA transcript. Note: The sequences shown for the TATA box and transcriptional start site are those found in the coding strand. *Py* refers to either pyrimidine, cytosine or thymine. Upstream from the core promoter, response elements such as enhancers and silencers are usually found.

The TATA box and transcriptional start site form the **core promoter**. The transcriptional start site is the place in the DNA where transcription actually begins. The TATA box, which is a 5′–TATAAAA–3′ sequence, is usually about 25 base pairs upstream from a transcriptional start site. The TATA box is important in determining the precise starting point for transcription. If it is missing from the core promoter, the transcriptional start site does not function properly, and transcription may start at a variety of different locations, producing unusable transcripts. The core promoter, by itself, results in a low level of transcription that is termed **basal transcription**.

Response elements (or regulatory elements) are DNA segments that regulate eukaryotic genes. As described later, response elements are recognized by regulatory proteins that control the ability of RNA polymerase to initiate transcription at the core promoter. Some response elements, known as **enhancers**, play a role in the ability of RNA polymerase to begin transcription and thereby enhance the rate of transcription. When enhancers are not functioning, most eukaryotic genes have very low levels of basal transcription. Other response elements, known as **silencers**, prevent transcription of a given gene when its expression is not needed. When these sequences function, the rate of transcription is decreased.

A common location for response elements is the region that is 50 to 100 base pairs upstream from the transcriptional start site (Figure 13.13). However, the locations of response elements are quite variable among different eukaryotic genes. Response elements can be quite distant from the promoter, even 100,000 base pairs away, yet exert strong effects on the ability of RNA polymerase to initiate transcription at the core promoter! Response elements were first discovered by Susumu Tonegawa and coworkers in the 1980s. While studying genes that play a

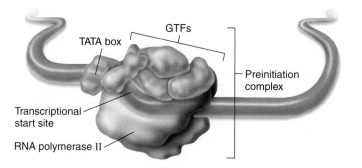

Figure 13.14 The preinitiation complex. General transcription factors (GTFs) and RNA polymerase II assemble into the preinitiation complex at the core promoter in eukaryotic structural genes.

role in immunity, they identified a region that was far away from the core promoter but was needed for high levels of transcription to take place.

RNA Polymerase II, General Transcription Factors, and Mediator Are Needed to Transcribe Eukaryotic Structural Genes

By studying transcription in a variety of eukaryotic species, researchers have identified three types of proteins that play a role in initiating transcription at the core promoter of structural genes. These are RNA polymerase II, five different proteins called **general transcription factors (GTFs)**, and a large protein complex called **mediator**. GTFs are needed for DNA binding at the core promoter and for initiation of transcription. Mediator is also needed for initiation.

The GTFs and RNA polymerase II must come together at the core promoter before transcription can be initiated. A series of interactions must occur between these proteins so that RNA polymerase II can bind to the DNA. **Figure 13.14** shows the structure of the completed assembly of GTFs and RNA polymerase II at the TATA box, which is known as the **preinitiation complex**. *In vitro*, when researchers mix together GTFs, RNA polymerase II, and a DNA sequence containing a TATA box and a transcriptional start site, the DNA is transcribed into RNA. Therefore, these components are referred to as the **basal transcription apparatus**. In a living cell, however, additional components such as regulatory transcription factors control the assembly of GTFs and RNA polymerase II at the core promoter, and are responsible for causing transcription to begin at a fast or slow rate.

A third component needed for transcription in eukaryotes is the mediator protein complex. Mediator is composed of several proteins that bind to each other to form an elliptical-shaped complex that partially wraps around RNA polymerase II and the GTFs. Mediator derives its name from the observation that it mediates interactions between RNA polymerase II and regulatory transcription factors such as activators or repressors that bind to enhancers or silencers. The function of mediator is to control the rate at which RNA polymerase can begin to transcribe RNA at the transcriptional start site.

Activators and Repressors May Influence the Function of GTFs or Mediator

In eukaryotes, regulatory transcription factor proteins called activators and repressors bind to enhancers or silencers, respectively, and regulate the rate of transcription of a nearby gene. Most activators and repressors do not bind directly to RNA polymerase II. Instead, some of them regulate the function of RNA polymerase II by interacting with other proteins that bind to RNA polymerase II. Certain activators and repressors exert their effects by binding to GTFs or mediator.

As shown in **Figure 13.15a**, some activators bind to an enhancer and then influence the function of GTFs. For example, an activator may improve the ability of a GTF called TFIID to initiate transcription. The function of TFIID is to recognize the TATA box and begin the assembly process. An activator may recruit TFIID to the TATA box, thereby promoting the assembly of GTFs and RNA polymerase II into the preinitiation complex. In contrast, repressors may inhibit the function of TFIID. Certain repressors exert their effects by preventing the binding of TFIID to the TATA box, or by inhibiting the ability of TFIID to assemble other GTFs and RNA polymerase II at the core promoter.

A second way that regulatory transcription factors control RNA polymerase II is via mediator. Activators stimulate the function of mediator and thereby cause RNA polymerase II to proceed to the elongation phase of transcription more quickly (**Figure 13.15b**). Alternatively, repressors have the opposite effect to those seen in this part of the figure. When a repressor inhibits mediator, RNA polymerase II cannot progress to the elongation stage.

A third way that regulatory transcription factors influence transcription is by recruiting proteins that affect the compaction or packing of DNA to the promoter region. As described in Chapter 15, the DNA found in eukaryotic chromosomes is highly compacted. To be accessible to RNA polymerase II and GTFs, the DNA must be loosened up. This topic is described next.

Gene Accessibility Is Controlled by Changes in Chromatin Structure

The regulation of gene transcription is not simply dependent on the activity of activators and repressors that interact with GTFs and mediator. In eukaryotes, DNA is associated with proteins to form a compact structure called **chromatin**. The

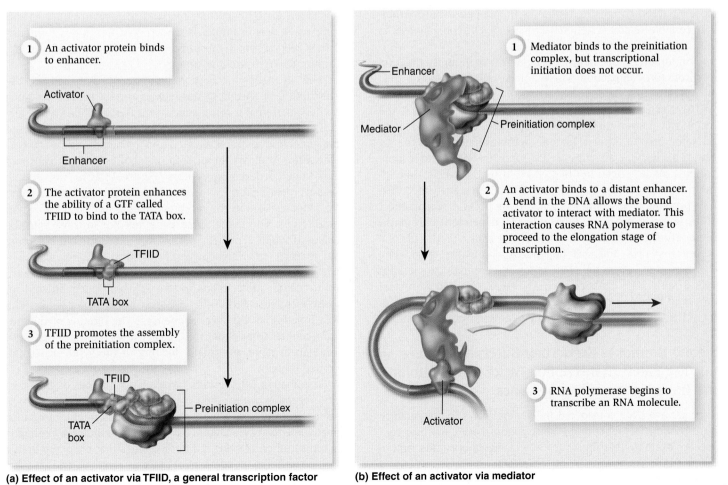

(a) Effect of an activator via TFIID, a general transcription factor

1. An activator protein binds to enhancer.

Activator

Enhancer

2. The activator protein enhances the ability of a GTF called TFIID to bind to the TATA box.

TFIID

TATA box

3. TFIID promotes the assembly of the preinitiation complex.

TFIID

TATA box

Preinitiation complex

(b) Effect of an activator via mediator

Enhancer

Mediator

Preinitiation complex

1. Mediator binds to the preinitiation complex, but transcriptional initiation does not occur.

2. An activator binds to a distant enhancer. A bend in the DNA allows the bound activator to interact with mediator. This interaction causes RNA polymerase to proceed to the elongation stage of transcription.

3. RNA polymerase begins to transcribe an RNA molecule.

Activator

Figure 13.15 Two mechanisms of eukaryotic activators.

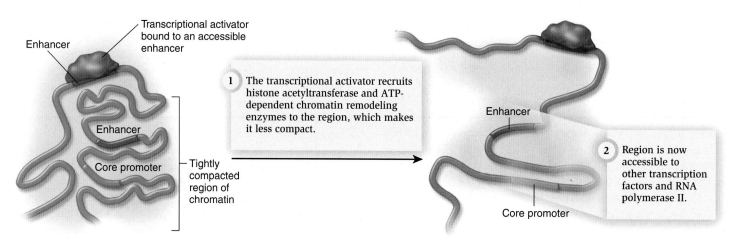

Figure 13.16   Effects of an activator on chromatin compaction.

three-dimensional packing of chromatin is an important parameter affecting gene expression. If the chromatin is very tightly packed, in what is called a **closed conformation**, transcription may be difficult or impossible. Transcription requires changes in chromatin structure that allow transcription factors to gain access to and bind to the DNA in the promoter region. Such loosely packed chromatin, said to be in an **open conformation**, is accessible to transcription factors and RNA polymerase so that transcription can take place.

An important role of some activators is to diminish the level of chromatin compaction where a gene is located. To do this, an activator first binds to an accessible enhancer site (**Figure 13.16**). Next, the binding of the activator recruits proteins to the region that loosen the level of compaction. In some cases, an activator protein attracts **histone acetyltransferase** to the region. This enzyme attaches acetyl groups (—COCH$_3$) to histone proteins. As described in Chapter 15, histone proteins are critical in the compaction of eukaryotic DNA. When acetylated, histone proteins do not bind as tightly to the DNA. A second effect of an activator protein is to recruit **ATP-dependent chromatin remodeling enzymes** to the site. The overall effect of histone acetyltransferase and ATP-dependent chromatin remodeling enzymes is to loosen the compaction of chromatin, sometimes over a fairly long distance such as several hundred or several thousand base pairs of DNA. This loosening facilitates the ability of RNA polymerase II to recognize and transcribe a gene.

Steroid Hormones Exert Their Effects by Binding to a Regulatory Transcription Factor and Controlling the Transcription of Nearby Genes

Thus far, we have considered the general ways that regulatory transcription factors control transcription. Let's now turn to a specific example that illustrates how a regulatory transcription factor functions within living cells. Our example involves a transcriptional activator that responds to steroid hormones. This factor is known as a **steroid receptor**, because it binds

directly to the steroid hormone. The hormone is an example of a small effector molecule.

As discussed in Chapter 50, steroid hormones are a category of hormones that are synthesized by specialized cells of many organisms, including the endocrine glands of mammals, and then secreted into the bloodstream. The hormones are then taken up by cells that respond to the hormones in different ways. For example, glucocorticoid hormones influence nutrient metabolism in most body cells by promoting the metabolism of glucose, proteins, and fats.

The effect of glucocorticoid hormones is to activate the transcription of specific genes. Under certain conditions, such as when an animal is fasting and needs to regulate its blood levels of glucose, fats, and amino acids, glucocorticoids are released from endocrine cells and secreted into the bloodstream. The hormone diffuses across the plasma membrane of target cells and binds to glucocorticoid receptors (**Figure 13.17**). This binding releases proteins called chaperones and thereby exposes an amino acid sequence within the receptor called a nuclear localization signal (NLS). This signal directs the receptor to travel into the nucleus through a nuclear pore. Two glucocorticoid receptors bind to each other noncovalently to form a dimer and then travel through the nuclear pore into the nucleus. The glucocorticoid receptor dimer binds to two adjacent glucocorticoid response elements (GREs) that are next to particular genes. The GREs function as enhancer sequences. The binding of the glucocorticoid receptor dimer to GREs activates the transcription of the adjacent gene, eventually leading to the synthesis of the encoded protein.

Mammalian cells usually have a large number of glucocorticoid receptors within their cytosol. Because GREs are located near several different genes, the uptake of hormone molecules can activate many glucocorticoid receptors and thereby enhance the transcription of several different genes that encode proteins involved with the metabolism of glucose, proteins, and fats. For this reason, glucocorticoid hormones facilitate the coordinated expression of genes that play a role in nutrient metabolism.

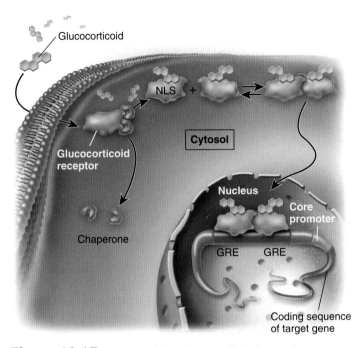

Figure 13.17 Action of the glucocorticoid receptor as a transcriptional activator.

Biological inquiry: If a GRE next to a gene was deleted, how would that affect the regulation of the gene?

Structural Features of Regulatory Transcription Factors Allow Them to Bind to DNA

Thus far, we have considered various ways that transcription factors bind to DNA to influence transcription. Let's now take a closer look at the structural features that are common to many types of transcription factor proteins. General and regulatory transcription factors have been identified from many organisms, including prokaryotes, yeast, plants, and animals. Transcription factor proteins contain domains that have specific functions. When a domain or portion of a domain has a very similar structure in many different proteins, such a structure is called a **motif**.

Figure 13.18 depicts several different motif structures commonly found in transcription factor proteins. An α helix is frequently important in the recognition of the DNA double helix. In helix-turn-helix and helix-loop-helix motifs, an α helix called the recognition helix makes contact with and recognizes a base sequence along the major groove of the DNA (Figure 13.18a,b). As discussed in Chapter 11, the major groove is a region of the DNA double helix where the bases contact the surrounding water in the cell. Hydrogen bonding between an α helix and nucleotide bases enables the protein to recognize a specific DNA sequence. Similarly, a zinc finger motif is composed of one α helix and a β pleated sheet structure that are held together by a zinc metal ion (Zn^{2+}) (Figure 13.18c). The zinc finger also can recognize DNA sequences within the major groove. The glucocorticoid receptor described in Figure 13.17 binds to GREs via zinc fingers.

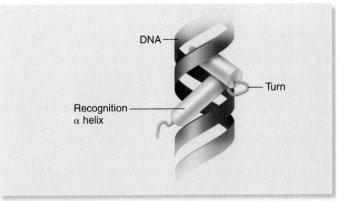

(a) Helix-turn-helix motif

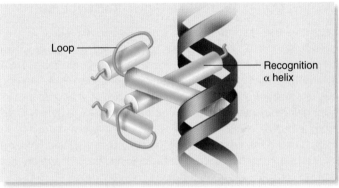

(b) Helix-loop-helix motif

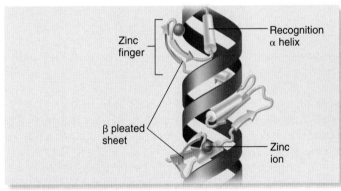

(c) Zinc finger motif

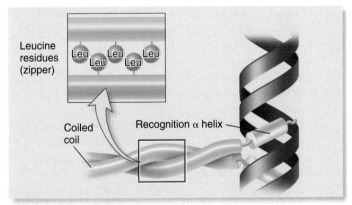

(d) Leucine zipper motif

Figure 13.18 Motif structures of transcription factors. In this figure, α helices are represented as cylinders, while β pleated sheets are represented as flattened arrows.

An interesting feature of certain motifs is that they promote protein dimerization, two proteins binding to each other as did the glucocorticoid receptors we discussed earlier. The leucine zipper and helix-loop-helix motifs mediate protein dimerization. For example, Figure 13.18d depicts the dimerization and DNA binding of two proteins that have leucine zippers. Alternating leucine residues in both proteins interact ("zip up"), resulting in protein dimerization. In some cases, two identical transcription factor proteins will come together to form a **homodimer**, or two different transcription factors will form a **heterodimer**. Dimerization of transcription factors can be an important way to modulate their function.

DNA Methylation Inhibits Gene Transcription

Let's now turn our attention to a mechanism that silences gene expression. DNA structure can be modified by the covalent attachment of methyl groups ($-CH_3$) by an enzyme called **DNA methylase**. Such **DNA methylation** is common in some eukaryotic species but not all. For example, yeast and *Drosophila* have little or no detectable methylation of their DNA, whereas DNA methylation in vertebrates and plants is relatively abundant. In mammals, approximately 5% of the DNA is methylated. Eukaryotic DNA methylation occurs on the cytosine base. The sequence that is methylated is shown here:

$$
\begin{array}{c}
CH_3 \\
| \\
5' - CG - 3' \\
3' - GC - 5' \\
| \\
CH_3
\end{array}
$$

Note that this sequence contains cytosines in both strands. When only one strand is methylated, this is called hemimethylation; the methylation of the cytosines in both strands is termed full methylation.

DNA methylation usually inhibits the transcription of eukaryotic genes, particularly when it occurs in the vicinity of the promoter. In vertebrates and plants, many genes contain sequences called **CpG islands** near their promoters. CpG refers to the nucleotides of <u>C</u> and <u>G</u> in DNA that are connected by a phosphodiester linkage. A CpG island is a cluster of CpG sites. Unmethylated CpG islands are correlated with active genes, while repressed genes contain methylated CpG islands. In this way, DNA methylation may play an important role in the silencing of particular genes.

Methylation can inhibit transcription in two general ways. First, methylation of CpG islands may prevent an activator from binding to an enhancer element. The inability of an activator protein to bind to the DNA would inhibit the initiation of transcription. A second way that methylation inhibits transcription is by converting chromatin from an open to a closed conformation. Proteins known as **methyl-CpG-binding proteins** bind methylated sequences. Once bound to the DNA, the methyl-CpG-binding protein recruits other proteins to the region that cause the chromatin to become very compact.

13.4 Regulation of RNA Processing and Translation in Eukaryotes

In the first three sections of this chapter, we have mostly focused on the regulation of gene transcription in prokaryotes and eukaryotes. Unlike bacteria, eukaryotic gene expression is commonly regulated at the levels of RNA processing and translation. These added levels of regulation provide benefits that are important to eukaryotic species. First, by regulating RNA processing, eukaryotes can produce more than one mRNA transcript from a single gene. This allows a gene to encode two or more polypeptides, and thereby increases the complexity of eukaryotic proteomes. A second issue is timing. Transcriptional regulation in eukaryotes takes a fair amount of time before its effects are observed at the cellular level. During transcriptional regulation, (1) the chromatin must be converted to an open conformation, (2) the gene must be transcribed, (3) the RNA must be processed and exported from the nucleus, and (4) the protein must be made via translation. All of these steps take time, on the order of several minutes. One way to achieve faster regulation is to control steps that occur after an RNA transcript is made. In eukaryotes, translational regulation provides a faster way to regulate the levels of gene products, namely proteins. Translation can be regulated by controlling the stability of an mRNA transcript, causing it to remain in the cytosol for a long time or causing it to be rapidly degraded. Alternatively, small RNA molecules or RNA-binding proteins can bind to mRNAs and control whether or not a ribosome can translate the mRNA into a polypeptide.

During the past few decades, many critical advances have been made regarding our knowledge of the regulation of RNA processing and translation. Even so, molecular geneticists are still finding new forms of regulation, making this an exciting area of modern research. In this section, we will survey a few of the known mechanisms of RNA processing and translational regulation.

Alternative Splicing of pre-mRNAs Creates Protein Diversity

In eukaryotes, a pre-mRNA transcript is processed before it becomes a mature mRNA. When a pre-mRNA has multiple introns and exons, splicing may occur in more than one way. Such **alternative splicing** causes mRNAs to contain different patterns of exons. Alternative splicing is a form of gene regulation that allows an organism to use the same gene to make different proteins at different stages of development, in different cells types, and/or in response to a change in the environmental conditions. Alternative splicing is an important form of gene regulation in complex eukaryotes such as animals and plants.

As an example of how alternative splicing occurs, let's suppose a human pre-mRNA contains seven exons (**Figure 13.19**). In nerve cells, it is spliced to contain the following pattern of exons: 1-2-3-5-6-7. In muscle cells, it is alternatively spliced to have a different pattern: 1-2-4-5-6-7. The linear sequence of the exons is maintained in both patterns. In this example, the mRNA

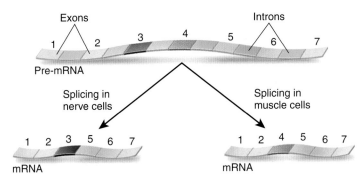

Figure 13.19 **Alternative splicing as a regulatory method for RNA processing.** In this example, the pre-mRNA transcript can be spliced to contain exon 3 (in nerve cells) or exon 4 (in muscle cells), but not both.

from nerve cells contains exon 3, while the mRNA from muscle cells contains exon 4. When alternative splicing occurs, proteins with significant differences in their amino acid sequences are produced.

In most cases, the alternative versions of a protein will have similar functions, because much of their amino acid sequences will be identical to each other. Nevertheless, alternative splicing produces differences in amino acid sequences that will provide each protein with its own unique characteristics. The alternatively spliced versions tend to be expressed in different cell types (for example, nerve versus muscle cells) or at different stages of development (for example, embryonic versus adult). This provides a way for multicellular organisms to "fine-tune" a given protein to function optimally in a given cell type or stage of development. The advantage of alternative splicing is that two (or more) different polypeptides can be derived from a single gene, thereby increasing the size of the proteome while minimizing the size of the genome. A small genome size is beneficial because less energy is spent replicating the DNA and the DNA more easily fits within the nucleus of the cell.

Table 13.1	Genome Size and Biological Complexity			
Species	**Level of complexity**	**Genome size (million bp)**	**Approximate number of genes**	**Percentage of genes alternatively spliced**
Escherichia coli	A unicellular prokaryote	4.2	4,000	0
Saccharomyces cerevisiae	A unicellular eukaryote	12	6,000	<1
Caenorhabditis elegans	A tiny worm (about 1,000 cells)	97	19,000	2
Drosophila melanogaster	An insect	137	14,000	7
Arabidopsis thaliana	A flowering plant	142	26,000	11
Homo sapiens	A complex mammal	3,000	25,000	70

fruit fly (*Drosophila melanogaster*), a small flowering plant (*Arabidopsis thaliana*), and humans (*Homo sapiens*). A general trend is that less complex organisms tend to have fewer genes. For example, unicellular organisms have only a few thousand genes, whereas multicellular species have tens of thousands. Even so, the trend is by no means a linear one. If we compare *C. elegans* and *D. melanogaster*, the fly actually has fewer genes even though it is biologically more complex. A second trend you can see in Table 13.1 concerns alternative splicing. This phenomenon does not occur in bacteria and is rare in *S. cerevisiae*. The frequency of alternative splicing increases from worms to flies to humans. For example, the level of alternative splicing is ten-fold higher in humans compared to *Drosophila*. This trend can partially explain the increase in biological complexity among these species.

GENOMES & PROTEOMES

Increases in Biological Complexity Are Correlated with Greater Sizes of Genomes and Proteomes

As we have just seen, alternative splicing can increase the proteome size without increasing the total number of genes. For organisms to become more complex, as in higher plants and animals, evolution has produced more complex proteomes. In the past few decades, many technical advances have improved our ability to analyze the genomes and proteomes of many different species. Researchers have been able to determine the amount of DNA from several species and estimate the total number of genes. In addition, scientists can also estimate the number of polypeptides if information is available concerning the degree of alternative splicing in a given species.

Table 13.1 compares six species, a bacterium (*Escherichia coli*), a eukaryotic single-celled organism (yeast—*Saccharomyces cerevisiae*), a small nematode worm (*Caenorhabditis elegans*), a

MicroRNAs Inhibit mRNA by Translational Repression or mRNA Degradation

Let's now turn our attention to regulatory mechanisms that affect translation. One method involves the silencing of pre-existing mRNA. **MicroRNAs** (abbreviated **miRNAs**) are small RNA molecules, typically 22 nucleotides in length, that silence the expression of specific mRNAs. In 1993, Victor Ambros and his colleagues, who were interested in the developmental stages that occur in the worm *C. elegans*, determined that the transcription of a particular gene produced a small RNA, now called a microRNA, that does not encode a protein. Instead, this miRNA was found to be complementary to an mRNA and inhibit its translation.

Insight into the mechanism of miRNA inhibition came from the surprising results of Andrew Fire and Craig Mello who showed that double-stranded RNA is very potent at silencing the expression of specific genes. In 2006, Fire and Mello were awarded the Nobel Prize for their studies aimed at understanding the effects

Pre-miRNA

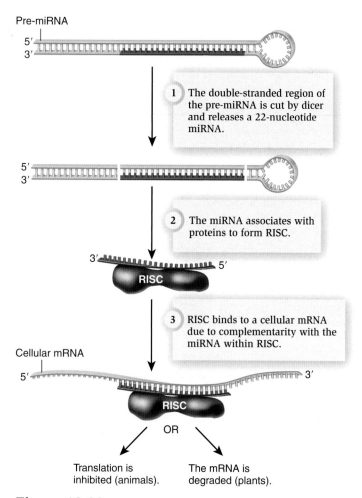

1 The double-stranded region of the pre-miRNA is cut by dicer and releases a 22-nucleotide miRNA.

2 The miRNA associates with proteins to form RISC.

RISC

3 RISC binds to a cellular mRNA due to complementarity with the miRNA within RISC.

Cellular mRNA

RISC

OR

Translation is inhibited (animals).

The mRNA is degraded (plants).

Figure 13.20   Mechanism of action of microRNA (miRNA).

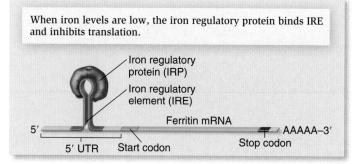

When iron levels are low, the iron regulatory protein binds IRE and inhibits translation.

Iron regulatory protein (IRP)
Iron regulatory element (IRE)
Ferritin mRNA
5′ UTR   Start codon   Stop codon   AAAAA–3′

(a) Low iron levels

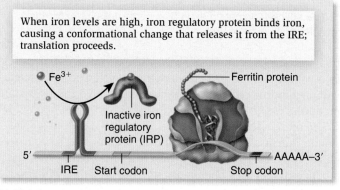

When iron levels are high, iron regulatory protein binds iron, causing a conformational change that releases it from the IRE; translation proceeds.

Fe^{3+}
Ferritin protein
Inactive iron regulatory protein (IRP)
5′   IRE   Start codon   Stop codon   AAAAA–3′

(b) High iron levels

Figure 13.21   Translational regulation of ferritin mRNA by the iron regulatory protein (IRP).

Biological inquiry: Poisoning may occur when a young child finds a bottle of vitamins, such as those that taste like candy, and eats a large number of them. One of the toxic effects involves the ingestion of too much iron. How does the IRP protect people from the toxic effects of too much iron?

of miRNA (**Figure 13.20**). MiRNAs are first synthesized as a pre-miRNA, which forms a hairpin structure. The double-stranded region is cut by an enzyme called dicer. This releases a miRNA that associates with cellular proteins to become part of a complex called the **RNA-induced silencing complex (RISC)**. The miRNA, which is complementary to specific cellular mRNAs, will direct the RISC to those mRNAs. Upon binding, two different things may happen. In some cases, the binding inhibits translation. This is the common effect of miRNAs in animals. In plants, the RISC directs the degradation of the mRNA. In either case, the expression of the mRNA is silenced. Fire and Mello called this **RNA interference (RNAi)**, because the miRNA interferes with the proper expression of an mRNA.

Since this study, researchers have discovered that genes encoding miRNAs are widely found in animals and plants. In humans, for example, approximately 200 different genes encode miRNAs. MiRNAs represent an important mechanism of mRNA silencing.

The Prevention of Iron Toxicity in Mammals Involves the Regulation of Translation

Another way to regulate translation involves RNA-binding proteins that directly affect translational initiation. The regulation of iron absorption provides a well-studied example. Iron is a

vital cofactor for many cellular enzymes. However, it is toxic at high levels. To prevent toxicity, mammalian cells synthesize a protein called ferritin, which forms a hollow, spherical complex that can store excess iron.

The mRNA that encodes ferritin is controlled by an RNA-binding protein known as the **iron regulatory protein (IRP)**. When iron levels in the cytosol are low and more ferritin is not needed, IRP binds to a response element within the ferritin mRNA known as the **iron regulatory element (IRE)**. The IRE is located upstream from the start codon in a region called the 5′-untranslated region (5′-UTR). Due to base pairing, it forms a stem-loop structure. The binding of IRP to the IRE inhibits translation of the ferritin mRNA (**Figure 13.21a**). However, when iron is abundant in the cytosol, the iron binds directly to IRP and prevents it from binding to the IRE. Under these conditions, the ferritin mRNA is translated to make more ferritin protein (**Figure 13.21b**). This mechanism of translational control allows cells to rapidly respond to changes in their environment. When cells are confronted with high levels of iron, they can quickly make more ferritin protein to prevent the toxic buildup of iron. This mechanism is faster than transcriptional regulation, which would require the activation of the ferritin gene and the transcription of ferritin mRNA prior to the synthesis of more ferritin protein.

CHAPTER SUMMARY

13.1 Overview of Gene Regulation

- Constitutive genes are expressed at constant levels while most genes are regulated so that the levels of gene expression can vary under different conditions.

- Organisms regulate genes so that gene products are made only when they are needed. An example is the synthesis of the gene products needed for lactose utilization in bacteria. (Figure 13.1)

- Multicellular eukaryotes regulate genes to produce different cell types, such as muscle, nerve, and skin cells. (Figure 13.2)

- Eukaryotes also regulate genes so the gene products are produced at different developmental stages. An example is the group of globin genes in mammals. (Figure 13.3)

- All organisms regulate gene expression at a variety of levels, including transcription, translation, and post-translation. Eukaryotes also regulate RNA processing. (Figure 13.4)

13.2 Regulation of Transcription in Bacteria

- Activators and repressors are regulatory proteins that bind to the DNA and regulate the expression of genes. Small effector molecules control the ability of regulatory proteins to bind to DNA. (Figure 13.5)

- The *lac* operon found in *E. coli* is an arrangement of three structural genes controlled by a single promoter. The operon is transcribed into a polycistronic mRNA. The operator and CAP site are involved with gene regulation via the *lac* repressor and CAP, respectively. (Figure 13.6)

- The *lac* repressor binds to the operator site and prevents RNA polymerase from transcribing the operon. When allolactose binds to the repressor, this causes a conformational change that prevents the repressor from binding to the operator site. (Figure 13.7)

- Mutations in *lacI* can result in the constitutive expression of the *lac* operon. Jacob, Monod, and Pardee proposed an alternate hypothesis to explain this observation. (Figure 13.8)

- By constructing a merozygote, Jacob, Monod, and Pardee determined that *lacI* encodes a diffusible protein that represses the *lac* operon. (Figure 13.9)

- The catabolite activator protein (CAP) binds to the CAP site in the presence of cAMP. This causes a bend in the DNA, which allows RNA polymerase to bind to the promoter. (Figure 13.10)

- Glucose inhibits cAMP production. This inhibits the expression of the *lac* operon because CAP cannot bind to the CAP site. This form of regulation provides bacteria with a more efficient utilization of their resources because the bacteria use one sugar at a time. (Figure 13.11)

- The *trp* operon is repressible. The presence of tryptophan causes the *trp* repressor to bind to the *trp* operator and stop transcription. This prevents the excessive buildup of tryptophan in the cell, which would be a waste of energy. (Figure 13.12)

13.3 Regulation of Transcription in Eukaryotes

- Eukaryotic genes exhibit combinatorial control, meaning that many factors control the expression of a single gene. (See list on p. 267.)

- Eukaryotic promoters consist of a core promoter and response elements, such as enhancers or silencers, that regulate the rate of transcription. (Figure 13.13)

- Eukaryotes have multiple forms of RNA polymerase. The form named RNA polymerase II transcribes structural genes.

- General transcription factors (GTFs) are needed for RNA polymerase II to bind to the core promoter. (Figure 13.14)

- Activators and repressors may regulate RNA polymerase II by interacting with GTFs, such as TFIID, or via mediator, a protein complex that wraps around RNA polymerase II. (Figure 13.15)

- A loosening of chromatin is needed for eukaryotic genes to be transcribed. (Figure 13.16)

- Steroid hormones bind to receptors that function as activator proteins. (Figure 13.17)

- General and regulatory transcription factors have motifs, which are common structural features that enable them to bind to the DNA and to each other. (Figure 13.18)

- DNA methylation, which occurs at CpG islands near promoters, usually inhibits transcription by preventing the binding of activator proteins or by promoting the compaction of chromatin.

13.4 Regulation of RNA Processing and Translation in Eukaryotes

- Alternative splicing occurs when a single type of pre-mRNA can be spliced in more than one way, producing polypeptides with somewhat different sequences. This is a common way for complex eukaryotes to increase the size of their proteomes. (Figure 13.19, Table 13.1)

- MicroRNAs (miRNAs) inhibit mRNAs, either by inhibiting translation or by promoting the degradation of mRNAs. (Figure 13.20)

- RNA-binding proteins can regulate the translation of specific mRNAs. An example is iron regulatory protein (IRP), which regulates the translation of ferritin mRNA. (Figure 13.21)

TEST YOURSELF

1. Genes that are expressed at all times at relatively constant levels are known as ____ genes.
   a. induced
   b. repressed
   c. positive
   d. constitutive
   e. structural

2. Which of the following is not considered a level of gene regulation in prokaryotes?
   a. transcriptional
   b. RNA processing
   c. translational
   d. post-translational
   e. All of the above are levels at which prokaryotes are able to regulate gene expression.

3. Transcription factors that bind to DNA and stimulate transcription are
   a. repressors.
   b. small effector molecules.
   c. activators.
   d. promoters.
   e. operators.

4. In prokaryotes the unit of DNA that contains multiple structural genes under the control of a single promoter is called _____. The mRNA produced from this unit is referred to as _____ mRNA.
   a. an operator, a polycistronic
   b. a template, a structural
   c. an operon, a polycistronic
   d. an operon, a monocistronic
   e. a template, a monocistronic

5. In the *lac* operon, the repressor protein binds to _____ to inhibit transcription.
   a. the promoter
   b. the operator
   c. the CAP site
   d. the enhancer
   e. lactose

6. The presence of _____ in the medium prevents the CAP from binding to the DNA, resulting in _____ in transcription of the *lac* operon.
   a. lactose, an increase
   b. glucose, an increase
   c. cAMP, a decrease
   d. glucose, a decrease
   e. lactose, a decrease

7. The *trp* operon is considered _____ operon because the structural genes necessary for tryptophan synthesis are not expressed when the levels of tryptophan in the cell are high.
   a. an inducible
   b. a positive
   c. a repressible
   d. a negative
   e. both c and d

8. Response elements that function to increase transcription levels in eukaryotes are called
   a. promoters.
   b. silencers.
   c. enhancers.
   d. transcriptional start sites.
   e. activators.

9. DNA methylation in many eukaryotic organisms seems to relate to
   a. increased translation levels.
   b. introns that will be removed.
   c. regions of DNA that do not contain structural genes.
   d. decreased transcription levels.
   e. response elements that are not necessary for transcription.

10. _____ refers to the phenomenon where a single type of pre-mRNA may give rise to multiple mRNAs due to different patterns of intron and exon removal.
    a. Spliceosomes
    b. Variable expression
    c. Alternative splicing
    d. Polycistronic mRNA
    e. Induced silencing

CONCEPTUAL QUESTIONS

1. List the components of the *lac* operon and explain the function of each.

2. Explain the difference between inducible and repressible operons.

3. Explain how DNA methylation inhibits transcription.

EXPERIMENTAL QUESTIONS

1. What were the key observations made by Jacob, Monod, and Pardee that led to the development of their hypothesis regarding the *lacI* gene and the regulation of the *lac* operon?

2. What was the hypothesis proposed by the researchers to explain the function of the *lacI* and the regulation of the *lac* operon?

3. How did Jacob, Monod, and Pardee test the hypothesis? What were the results of the experiment? How do these results support the idea that the *lacI* gene produces a repressor protein and does not function as an operator?

COLLABORATIVE QUESTIONS

1. Discuss the different points at which gene regulation can take place to produce a functional protein.

2. Discuss several factors that control the regulation of transcription in eukaryotes.

www.brookerbiology.com

This website includes answers to the Biological Inquiry questions found in the figure legends and all end-of-chapter questions.

14

MUTATION, DNA REPAIR, AND CANCER

Tanning and cigarette smoking are two activities that expose people to agents that cause mutations. Such mutations may lead to cancer.

As we have learned in Chapters 11 through 13, DNA stores the information for the synthesis of proteins. When a gene is expressed, the DNA itself does not normally change. However, on rare occasions, a mutation may occur. A **mutation** is a heritable change in the genetic material. The structure of DNA, its base sequence, has been changed permanently and this alteration can be passed from mother to daughter cells during cell division. Likewise, if a mutation occurs in a reproductive cell, it may also be passed from parent to offspring.

On the positive side, mutations are essential to the continuity of life. They supply the variation that enables species to evolve and become better adapted to their environments. Mutations provide the foundation for evolutionary change. On the negative side, however, new mutations are more likely to be harmful than beneficial to the individual. The genes within modern species are the products of billions of years of evolution and have evolved to work properly. Random mutations are more likely to disrupt genes than to enhance their function. Many forms of inherited diseases, such as cystic fibrosis and sickle-cell anemia, are caused by gene mutations. For this and many other reasons, understanding the molecular nature of mutations is a compelling area of research.

Because mutations can be quite harmful, all species have evolved several ways to repair damaged DNA. Such **DNA repair systems** reverse DNA damage before a permanent mutation can occur. DNA repair systems are vital to the survival of all organisms. If these systems did not exist, mutations would be so prevalent that few species, if any, would survive. In this chapter, we will examine a few ways that these systems operate.

The final topic we discuss in this chapter is **cancer**, a disease that is caused by gene mutations that lead to uncontrolled cell growth. DNA repair systems may prevent the occurrence of cancer by minimizing the number of mutations in any given cell. As we will learn later, multicellular organisms also possess other safeguards to help prevent cancer. Unfortunately, however, many individuals develop cancer due to naturally occurring mutations that are unavoidable or due to exposure to environmental agents such as cigarette smoke. At the end of this chapter, we will explore the various types of genetic changes that lead to the progression of cancer.

14.1 Mutation

To appreciate why mutations are beneficial or detrimental, we must understand how changes in DNA structure can ultimately affect DNA function. Most of our understanding of mutation has come from the study of experimental organisms, such as bacteria and *Drosophila*. Researchers can expose these organisms to environmental agents that cause mutations and then study the consequences of the mutations that arise. In addition, because these organisms have a short generation time, researchers can investigate the effects of mutations when they are passed from parent to offspring.

The structure and amount of genetic material can be altered in a variety of ways. For example, chromosome structure and number can change. These types of genetic changes will be considered in Chapter 15. In this section, we will focus our attention on **gene mutations**, which are relatively small changes in DNA structure that alter a particular gene. We will be primarily concerned with the ways that mutations may affect the expression of single genes. We will also consider how the timing of mutations during an organism's development has important consequences.

In addition, we will explore how environmental agents may bring about mutations, and examine a testing method that can determine if an agent causes mutations.

Gene Mutations Alter the DNA Sequence of a Gene

Mutations can cause two basic types of changes to a gene: (1) the base sequence within a gene can be changed; and (2) one or more nucleotides can be added to or removed from the gene. A **point mutation** affects only a single base pair within the DNA. For example, the DNA sequence shown here has been altered by a **base substitution**:

$$5'—CCCGCTAGATA—3' \longrightarrow 5'—CCCGC\mathbf{G}AGATA—3'$$
$$3'—GGGCGATCTAT—5' \qquad\qquad 3'—GGGCG\mathbf{C}TCTAT—5'$$

A point mutation could also involve the addition or deletion of a single base pair to a DNA sequence. For example, in the following sequence, a single base pair has been added to the DNA:

$$5'—GGCGCTAGATC—3' \longrightarrow 5'—GGC\mathbf{A}GCTAGATC—3'$$
$$3'—CCGCGATCTAG—5' \qquad\qquad 3'—CCG\mathbf{T}CGATCTAG—5'$$

Though point mutations may seem like small changes to a DNA sequence, they can have important consequences when genes are expressed. This topic is discussed next.

Gene Mutations May Affect the Amino Acid Sequence of a Polypeptide

If a mutation occurs within the region of a structural gene that specifies the amino acid sequence, such a mutation may alter that sequence in a variety of ways. **Table 14.1** considers the potential effects of point mutations. **Silent mutations** do not alter the amino acid sequence of the polypeptide, even though the nucleotide sequence has changed. Because the genetic code is degenerate, silent mutations can occur in the third base within most codons without changing the type of amino acid it encodes. Silent mutations are also considered **neutral mutations**, because they do not affect the function of the encoded protein.

A **missense mutation** is a base substitution that changes a single amino acid in a polypeptide sequence. Missense mutations may not alter protein function because they change only a single amino acid within polypeptides that typically are hundreds of amino acids in length. When a missense mutation has no detectable effect on protein function, it is also referred to as a neutral mutation. A missense mutation that substitutes an amino acid with a chemistry similar to the original amino acid is likely to be neutral. For example, a missense mutation that substitutes a glutamic acid for an aspartic acid may not alter protein function because both amino acids are negatively charged and have similar side chain structures.

Alternatively, some missense mutations have a dramatic effect on protein function. A striking example of such a missense mutation occurs in the human disease known as **sickle-cell**

Table 14.1	Consequences of Point Mutations Within the Coding Sequence of a Structural Gene

Mutation in the DNA	Effect on polypeptide	Example*
None	None	ATGGCCGGCCCGAAAGAGACC — Met–Ala–Gly–Pro–Lys–Glu–Thr
Base substitution	Silent— causes no change	ATGGCCGGCCC**C**AAAGAGACC — Met–Ala–Gly–Pro–Lys–Glu–Thr
Base substitution	Missense— changes one amino acid	ATG**C**CCGGCCCGAAAGAGACC — Met–Pro–Gly–Pro–Lys–Glu–Thr
Base substitution	Nonsense— changes to a stop codon	ATGGCCGGCCCG**T**AAGAGACC — Met–Ala–Gly–Pro–STOP
Addition (or deletion) of single base	Frameshift— produces a different amino acid sequence	ATGGCCGGC**A**CCGAAAGAGACC — Met–Ala–Gly–Thr–Glu–Arg–Ala

*DNA sequence in the coding strand. This sequence is the same as the mRNA sequence, except that RNA contains uracil (U) instead of thymine (T).

anemia. This disease involves a mutation in the β-globin gene, which encodes one of the polypeptide subunits that make up hemoglobin, the oxygen-carrying protein in red blood cells. In the most common form of this disease, a missense mutation alters the polypeptide sequence so that the sixth amino acid is changed from a glutamic acid to a valine. Because glutamic acid is hydrophilic but valine is hydrophobic, this single amino acid substitution alters the structure and function of the hemoglobin protein. The mutant hemoglobin subunits tend to stick to one another when the oxygen concentration is low. The aggregated proteins form fiber-like structures within red blood cells, which causes the cells to lose their normal morphology and become sickle-shaped (**Figure 14.1**). It's rather amazing that a single amino acid substitution could have such a profound effect on the structure of cells.

Two other types of point mutations cause more dramatic changes to a polypeptide sequence. A **nonsense mutation** involves a change from a normal codon to a stop or termination codon. This causes translation to be terminated earlier than expected, producing a truncated polypeptide (see Table 14.1). A shorter polypeptide is much less likely than a normal polypeptide to function properly. Finally, a **frameshift mutation** involves the addition or deletion of nucleotides that are not in multiples of three nucleotides. For example, a frameshift mutation could involve the addition or deletion of one, two, four, or five nucleotides. Because the codons are read in multiples of three, this shifts the reading frame so that a completely different

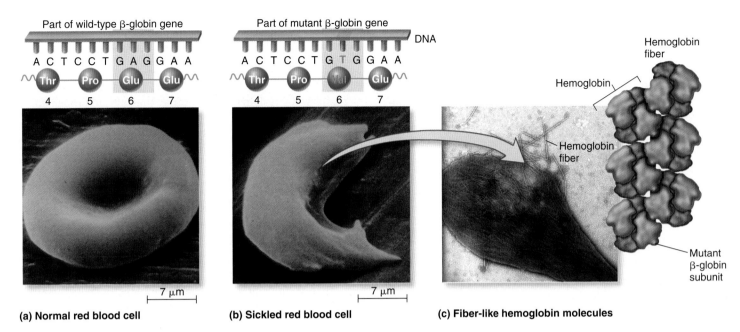

Figure 14.1 **A missense mutation that causes red blood cells to sickle in sickle-cell anemia.** Scanning electron micrographs of (a) normal red blood cells and (b) sickled red blood cells. As shown above the micrographs, a missense mutation in the β-globin gene (which codes for a subunit of hemoglobin) changes the sixth amino acid in the β-globin polypeptide from a glutamic acid to a valine. (c) This micrograph shows how this alteration to the structure of β-globin causes the formation of abnormal fiber-like structures.

Biological inquiry: Based on the fiber-like structures seen in part (c), what aspect of hemoglobin structure does a glutamic acid at the sixth position in β-globin prevent? Speculate on how the charge of this amino acid may play a role.

amino acid sequence occurs downstream from the mutation (see Table 14.1). Such a large change in polypeptide structure is also likely to inhibit protein function.

Except for silent mutations, new mutations are more likely to produce polypeptides that have reduced rather than enhanced function. However, mutations can occasionally produce a polypeptide that has a better ability to function. While these favorable mutations are relatively rare, they may result in an organism with a greater likelihood to survive and to reproduce. The favorable effect of a mutation may cause it to increase in frequency in a population over the course of many generations. This topic is discussed in Chapter 24.

Gene Mutations Can Occur Outside of Coding Sequences and Influence Gene Expression

Thus far, we have focused our attention on mutations in the coding regions of structural genes. In Chapters 12 and 13, we learned how other DNA sequences play important roles during gene expression. A mutation can occur within noncoding sequences and affect gene expression (**Table 14.2**). For example, a mutation may alter the sequence within the promoter of a gene. A mutation that causes an increase in the rate of transcription is called an up promoter mutation. Such a mutation may improve the ability of transcription factors and RNA polymerase to recognize the promoter. In contrast, a down promoter mutation causes a decrease in transcription.

Table 14.2	Effects of Mutations Outside of the Coding Sequence of a Gene
Sequence	**Effect of mutation**
Promoter	May increase or decrease the rate of transcription
Transcriptional response element/operator site	May alter the regulation of transcription
Splice junctions	May alter the ability of pre-mRNA to be properly spliced
Translational response element	May alter the ability of mRNA to be translationally regulated

Mutations in transcriptional response elements or operator sites can alter the regulation of gene transcription. For example, in Chapter 13, we considered the roles of regulatory sequences such as the *lac* operator site in *E. coli*, which is recognized by the *lac* repressor protein (refer back to Figure 13.7). Mutations in the *lac* operator site can disrupt the proper regulation of the *lac* operon. An operator mutation may change the DNA sequence so that the *lac* repressor protein does not bind to it. This mutation would cause the operon to be constitutively expressed even in the absence of lactose.

Mutations can also occur in other noncoding regions of a gene and alter gene expression in a way that may affect the encoded polypeptide sequence or its expression. For example, in Chapter 12 we discussed the splicing of RNA during RNA processing in eukaryotes (refer back to Figure 12.10). Mutations can occur at the boundaries between introns and exons and prevent proper splicing. When this occurs, the mRNA may not contain the correct order and number of exons. In Chapter 13, we also discussed response elements that are recognized by proteins that control translation. For example, the iron regulatory protein recognizes response elements within the mRNA that encodes ferritin (refer back to Figure 13.21). A mutation that alters this response element may prevent the proper translational regulation of ferritin.

FEATURE INVESTIGATION

The Lederbergs Used Replica Plating to Show That Mutations Are Random Events

Thus far, we have considered how mutations can affect the expression of genes. Let's now ask the question, How do mutations occur? For many centuries, biologists have wondered whether mutations that affect the traits of an individual occur as a result of pre-existing circumstances or whether they are unplanned events that happen randomly in any gene of any individual. This question has an interesting history. In the 19th century, French naturalist Jean Baptiste Lamarck proposed that physiological events (for example, use and disuse) determine whether traits are passed along to offspring. For example, his hypothesis suggested that an individual who practiced and became adept at a physical activity, such as the long jump, would pass that quality on to his or her offspring. Alternatively, geneticists in the early 1900s suggested that genetic variation occurs as a matter of chance and that natural selection results in the differential reproductive success of organisms that are better adapted to their environments. According to this view, those individuals whose genes happen to contain beneficial mutations are more likely to survive and to pass those genes to their offspring.

These opposing views were tested in bacterial studies in the 1940s and 1950s. One such study, by Joshua and Ester Lederberg, focused on the occurrence of mutations in bacteria (**Figure 14.2**). First, they placed a large number of E. coli bacteria onto a master plate that was incubated overnight, so that each bacterial cell divided many times to form a bacterial colony composed of millions of cells. Using a technique known as **replica plating**, a sterile piece of velvet cloth was lightly touched to this plate to pick up a few bacterial cells from each colony. They then transferred this replica to two secondary plates that contained an agent that selected for the growth of bacterial cells with a particular mutation.

Figure 14.2   The experiment performed by the Lederbergs that used replica plating to show that mutations are random events.

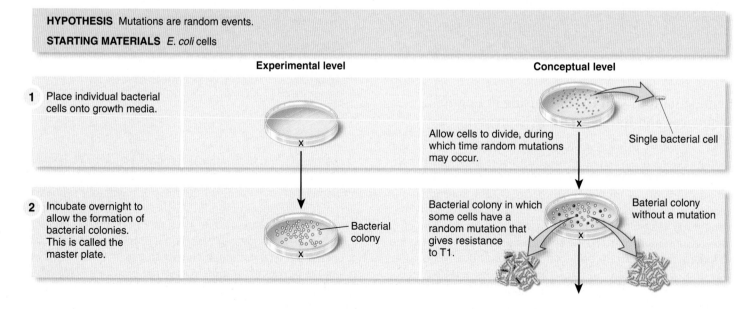

HYPOTHESIS Mutations are random events.
STARTING MATERIALS E. coli cells

Experimental level   Conceptual level

1 Place individual bacterial cells onto growth media.

Allow cells to divide, during which time random mutations may occur.   Single bacterial cell

2 Incubate overnight to allow the formation of bacterial colonies. This is called the master plate.

Bacterial colony

Bacterial colony in which some cells have a random mutation that gives resistance to T1.   Bacterial colony without a mutation

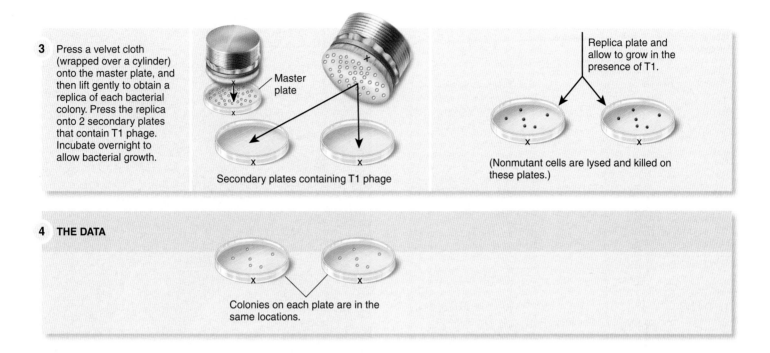

3 Press a velvet cloth (wrapped over a cylinder) onto the master plate, and then lift gently to obtain a replica of each bacterial colony. Press the replica onto 2 secondary plates that contain T1 phage. Incubate overnight to allow bacterial growth.

Master plate

Secondary plates containing T1 phage

Replica plate and allow to grow in the presence of T1.

(Nonmutant cells are lysed and killed on these plates.)

4 THE DATA

Colonies on each plate are in the same locations.

In the example shown in Figure 14.2, the secondary plates contained T1 bacteriophages, or phages, which are viruses that infect bacteria and cause them to burst or lyse. On these plates, only those rare cells that had acquired a mutation conferring resistance to T1, termed *ton^r*, could grow. All other cells were lysed by the proliferation of phages in the bacterial cytoplasm. Therefore, only a few colonies were observed on the secondary

plates. Strikingly, these colonies occupied the same locations on each plate. These results indicated that the *ton^r* mutations occurred randomly while the bacterial cells were forming colonies on the nonselective master plate. The presence of T1 bacteriophages in the secondary plates simply selected for the growth of *ton^r* mutants that had occurred previously. These results supported the idea that mutations are random events.

Mutations Can Occur in Germ-Line or Somatic Cells

Let's now consider how the timing of a mutation may have an important impact on its potential effects. Multicellular organisms begin their lives as a single fertilized egg cell and by cell division produce the many cells of the adult organism. For example, humans begin as a single fertilized egg and become an adult with approximately 10 trillion cells. These cells carry copies of the genome. When a mutation occurs in a single gene, however, the mutation may or may not be found in all the cells in a multicellular organism. A mutation can occur in any cell of the body, either very early in life, such as in a gamete (eggs or sperm) or a fertilized egg, or later in life, such as in the embryonic or adult stages. The exact time and location of a mutation is critical both to the severity of the genetic effect and to whether the mutation can be passed on to offspring.

Geneticists classify the cells of animals into two types: germ-line and somatic cells. The term **germ line** refers to cells that give rise to gametes such as egg and sperm cells. A germ-line mutation can occur directly in a sperm or egg cell, or it can

occur in a precursor cell that produces the gametes. If a mutant gamete participates in fertilization, all the cells of the resulting offspring will contain the mutation (**Figure 14.3a**). Likewise, when such an individual produces gametes, the mutation may be transmitted to future generations of offspring. Because an individual carries two copies of most genes, a new mutation in a single gene has a 50% chance of being transmitted from parent to offspring.

The **somatic cells** constitute all cells of the body excluding the germ-line cells. Examples include skin cells, muscle cells, and so forth. Mutations can also occur within somatic cells at early or late stages of development. **Figure 14.3b** illustrates the consequences of a mutation that happened during the embryonic stage. In this example, a mutation occurred within a single embryonic cell. This single cell was the precursor for many cells of the adult organism. Therefore, in the adult, a patch of tissue was made up of cells that carried the mutation. The size of any patch would depend on the timing of a mutation. In general, the earlier a mutation occurs during development, the larger the patch. An individual with somatic regions that are genetically different from each other is called a **genetic mosaic**.

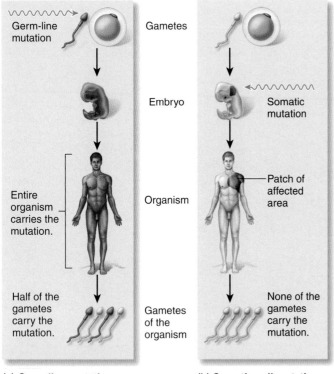

(a) Germ-line mutation   **(b) Somatic cell mutation**

Figure 14.3 **The effects of germ-line versus somatic cell mutations.** Because humans have two copies of each gene, a germ-line mutation in one of those two copies will be transmitted to only half of the gametes.

Figure 14.4 **Example of a somatic mutation.** This person has a patch of gray hair because a somatic mutation occurred in a single cell during embryonic development that prevented pigmentation of the hair. This cell continued to divide to produce a patch of gray hair.

Figure 14.4 illustrates an individual who had a somatic mutation during an early stage of development. In this case, the person has a patch of gray hair while the rest of the hair is pigmented. Presumably, this individual initially had a single mutation happen in an embryonic cell that ultimately gave rise to a patch of scalp that produced the gray hair. Although a patch of gray hair is not a harmful consequence, mutations during early stages of life can be quite harmful, especially if they disrupt essential developmental processes. For example, a somatic mutation during embryonic development is the cause of a human disease called McCune-Albright syndrome, which is characterized by bone defects, hormonal imbalances, and skin abnormalities. Therefore, even though it is sensible to avoid environmental agents that cause mutations during all stages of life, the possibility of somatic mutations is a rather compelling reason to avoid such agents during the early stages of life such as embryonic and fetal development, infancy, and early childhood. Let's now consider mutations caused by agents in our environment as well as those that occur naturally.

Mutations May Be Spontaneous or Induced

Biologists categorize the causes of mutation as being either spontaneous or induced. **Spontaneous mutations** result from abnormalities in biological processes (**Table 14.3**). Spontaneous mutations reflect the fact that biology isn't perfect. Enzymes, for example, can function abnormally. In Chapter 11, we learned that DNA polymerase can make a mistake during DNA replication by putting the wrong base in a newly synthesized daughter strand. Though such errors are rare, they do occur at a low rate. In addition, normal metabolic processes within the cell may produce toxic chemicals such as free radicals that can react directly with the DNA and alter its structure. Finally, the structure of nucleotides is not absolutely stable. On occasion, the structure may spontaneously isomerize, and such a change may cause a mutation if it occurs immediately prior to DNA replication. (Structural isomers are discussed in Chapter 3.)

The rates of mutations vary from species to species, and from gene to gene. Larger genes are usually more likely to incur a mutation than are smaller genes. Each time the DNA replicates, a common rate of spontaneous mutation is approximately 1 mutation for every 1 million genes, which equals 1 in 10^6, or simply 10^{-6}. This is the expected rate of "background" mutation, which creates the variation that is the raw material of evolution.

Induced mutations are brought about by environmental agents that enter the cell and then alter the structure of DNA. They cause the mutation rate to be higher than the spontaneous mutation rate. Such agents that are known to cause mutation are called **mutagens**. These can be chemical substances or physical agents that ultimately lead to changes in DNA structure. We will consider their effects next.

Table 14.3	Some Common Causes of Gene Mutations

Common causes of mutations	Description
Spontaneous:	
Errors in DNA replication	A mistake by DNA polymerase may cause a point mutation.
Toxic metabolic products	The products of normal metabolic processes may be reactive chemicals such as free radicals that can alter the structure of DNA.
Spontaneous changes in nucleotide structure	On rare occasions, the linkage between purines and deoxyribose can spontaneously break. Also, changes in base structure (isomerization) may cause mispairing during DNA replication.
Transposons	As discussed in Chapter 21, transposons are small segments of DNA that can insert at various sites in the genome. If they insert into a gene, they may inactivate the gene.
Induced:	
Chemical agents	Chemical substances, such as benzo(a)-pyrene, a chemical found in cigarette smoke, may cause changes in the structure of DNA.
Physical agents	Physical agents such as UV (ultraviolet) light and X-rays can damage the DNA.

Table 14.4	Examples of Mutagens

Mutagen	Effect(s) on DNA structure
Chemical:	
Nitrous acid	Deaminates bases
5-Bromouracil	Acts as a base analogue
2-Aminopurine	Acts as a base analogue
Nitrogen mustard	Alkylates bases
Ethyl methanesulfonate	Alkylates bases
Proflavin	Inserts next to bases in the DNA double helix and causes additions or deletions
Physical:	
X-rays	Causes base deletions, single nicks in DNA strands, cross-linking, and chromosomal breaks
UV light	Promotes pyrimidine dimer formation, which involves covalent bonds between adjacent pyrimidines (C or T)

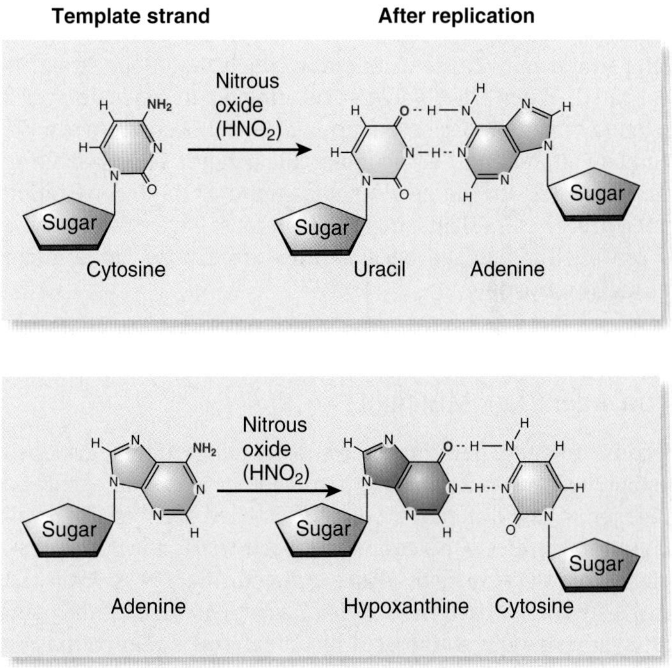

Mutagens Alter DNA Structure in Different Ways

Researchers have discovered that an enormous array of agents can act as mutagens to permanently alter the structure of DNA. We often hear in the news media that we should avoid these agents in our foods and living environments. We even use products such as sunscreens that help us avoid the mutagenic effects of ultraviolet (UV) light from the sun. The public is concerned about mutagens for two important reasons. First, mutagenic agents are often involved in the development of human cancers. Second, because new mutations may be deleterious, people want to avoid mutagens to prevent gene mutations that may have harmful effects in their future offspring. Mutagenic agents are categorized as **chemical** or **physical mutagens** (**Table 14.4**).

Mutagens can alter the structure of DNA in various ways. Some chemical mutagens act by covalently modifying the structure of nucleotides. For example, nitrous acid (HNO_2) deaminates bases by replacing amino groups with keto groups ($-NH_2$ to $=O$). This can change cytosine to uracil, and adenine to a base called hypoxanthine. When this altered DNA replicates, the modified bases do not pair with the appropriate nucleotides in the newly made strand. Instead, uracil pairs with adenine, and hypoxanthine pairs with cytosine (**Figure 14.5**). Similarly, 5-bromouracil and 2-aminopurine, which are called base analogues, have structures that are similar to particular bases. When incorporated into DNA, they also cause errors in DNA replication. Other chemical mutagens can also disrupt the appropriate

Figure 14.5 Deamination and mispairing of modified bases by a chemical mutagen. Nitrous acid changes cytosine to uracil and adenine to hypoxanthine by replacing NH_2 with an oxygen. During DNA replication, uracil will pair with adenine, and hypoxanthine will pair with cytosine. These incorrect bases will create mutations in the newly replicated strand.

pairing between nucleotides by alkylating bases within the DNA. During alkylation, methyl or ethyl groups are covalently attached to the bases. Examples of alkylating agents include nitrogen mustards (used as a chemical weapon during World War I) and ethyl methanesulfonate (EMS).

Some chemical mutagens exert their effects by interfering with DNA replication. For example, acridine dyes, such as proflavin, contain flat planar structures that insert in between the bases of the double helix, thereby distorting the helical structure. When DNA containing these mutagens is replicated, single nucleotide additions and/or deletions can be incorporated into the newly made strands.

DNA molecules are also sensitive to physical agents such as radiation. In particular, radiation of short wavelength and high energy, known as ionizing radiation, is known to alter DNA structure. Ionizing radiation includes X-rays and gamma rays. This type of radiation can penetrate deeply into biological materials, where it creates free radicals. These molecules can alter the structure of DNA in a variety of ways. Exposure to high doses of ionizing radiation can cause base deletions, breaks in one DNA strand, or even a break in both DNA strands.

Nonionizing radiation, such as UV light, contains less energy, and so it only penetrates the surface of biological materials such as the skin. Nevertheless, UV light is known to cause DNA mutations. For example, UV light can cause the formation of a **thymine dimer**, one type of pyrimidine dimer, which is a site where two adjacent thymine bases become covalently cross-linked to each other (**Figure 14.6**). A thymine dimer within a DNA strand may cause a mutation when that DNA strand is replicated. When DNA polymerase attempts to replicate over a thymine dimer, proper base pairing does not occur between the template strand and the incoming nucleotides. This mispairing can cause gaps in the newly made strand or the incorporation of incorrect bases. Plants, in particular, must have effective ways to prevent UV damage because they are exposed to sunlight throughout the day.

Testing Methods Can Determine If an Agent Is a Mutagen

Because of their harmful effects, researchers have developed testing methods that can ascertain whether or not an agent is a mutagen. Many different kinds of tests have been used to evaluate mutagenicity. One commonly used test is the **Ames test**, which was developed by Bruce Ames in the 1970s. This test uses a strain of a bacterium, *Salmonella typhimurium*, that cannot synthesize the amino acid histidine. This strain contains a point mutation within a gene that encodes an enzyme required for histidine biosynthesis. The mutation renders the enzyme inactive. The bacteria cannot grow unless histidine has been added to the growth medium. However, a second mutation in a bacterial cell may correct the first mutation and thereby restore the ability to synthesize histidine. The Ames test monitors the rate at which this second mutation occurs and thereby indicates whether an agent increases the mutation rate above the spontaneous rate.

Figure 14.7 outlines the steps in the Ames test. The suspected mutagen is mixed with a rat liver extract and the bacterial strain of *S. typhimurium*, which cannot synthesize histidine. Because some potential mutagens may require activation by

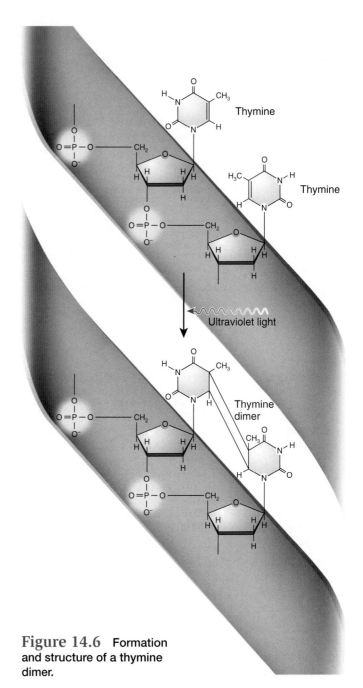

Figure 14.6 **Formation and structure of a thymine dimer.**

cellular enzymes, the rat liver extract provides a mixture of enzymes that may cause such activation. This step improves the ability to identify agents that may cause mutation in mammals. As a control, bacteria that have not been exposed to the mutagen are also tested. After an incubation period in which mutations may occur, a large number of bacteria are plated on a growth medium that does not contain histidine. The *Salmonella* strain is not expected to grow on these plates. However, if a mutation has occurred that allows a cell to synthesize histidine, the bacterium harboring this second mutation will proliferate during an overnight incubation period to form a visible bacterial colony.

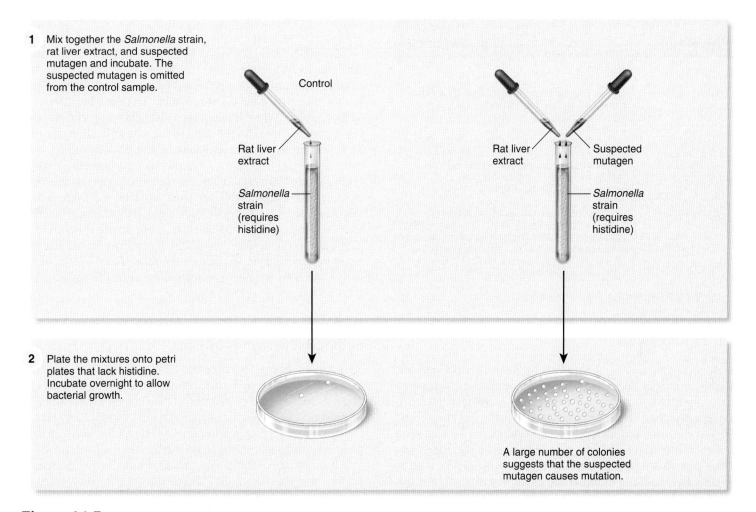

1 Mix together the *Salmonella* strain, rat liver extract, and suspected mutagen and incubate. The suspected mutagen is omitted from the control sample.

Control

Rat liver extract

Salmonella strain (requires histidine)

Rat liver extract

Suspected mutagen

Salmonella strain (requires histidine)

2 Plate the mixtures onto petri plates that lack histidine. Incubate overnight to allow bacterial growth.

A large number of colonies suggests that the suspected mutagen causes mutation.

Figure 14.7   **The Ames test for mutagenicity.**   In this example, two million bacterial cells were placed on each plate. Two colonies were observed from the control sample while 44 were observed from the sample exposed to a suspected mutagen.

Biological Inquiry: Based on the results seen in this figure, what is the rate of mutation that is caused by the suspected mutagen?

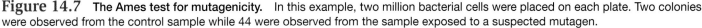

To estimate the mutation rate, the colonies that grow in the absence of histidine are counted and compared with the total number of bacterial cells that were originally placed on the plate for both the test and the control. The control condition is a measure of the spontaneous mutation rate, while the test condition monitors the rate of mutation in the presence of the suspected mutagen. As an example, let's suppose that 2 million bacteria were placed in both the test and control tubes. In the control experiment, 2 bacterial colonies were observed. The spontaneous mutation rate is calculated by dividing 2 (the number of mutants) by 2 million (the number of original cells). This equals 1 in 1 million, or 1×10^{-6}. By comparison, 44 colonies were observed in the test condition (Figure 14.7). In this case, the mutation rate would be 44 divided by 2 million, which equals 2.2×10^{-5}. The mutation rate in the presence of the mutagen is over 20 times higher than the spontaneous mutation rate. These results suggest that the suspected mutagen is causing mutations to occur. Now let's look at how organisms can repair changes in DNA structure.

14.2 DNA Repair

In the previous section, we considered the causes and consequences of mutation. As we have seen, mutations are random events that often have negative consequences. For this reason, all living organisms must have the ability to repair changes that occur in the structure of DNA in order to minimize mutation. Such DNA repair systems have been studied extensively in many organisms, particularly *E. coli*, yeast, and mammals. The diverse ways of repairing DNA underscore the extreme necessity for the structure of DNA to be maintained properly. The importance of these systems becomes evident when they are missing. For example, bacteria contain several different DNA repair systems. When even a single system is absent, the bacteria have a much higher rate of mutation. In fact, the rate of mutation can be so high that these bacterial strains are sometimes called "mutator strains"!

Living cells contain several DNA repair systems that can fix different types of DNA alterations (**Table 14.5**). Each repair

Table 14.5	Common Types of DNA Repair Systems*
System	**Description**
Direct repair	A repair enzyme recognizes an incorrect structure in the DNA and directly converts it back to a correct structure.
Base excision and nucleotide excision repair	An abnormal base or nucleotide is recognized and a portion of the strand containing the abnormality is removed. The complementary DNA strand is then used as a template to synthesize a normal DNA strand.
Methyl-directed mismatch repair	Similar to excision repair except that the DNA defect is a base pair mismatch in the DNA, not an abnormal nucleotide. The mismatch is recognized, and a strand of DNA in this region is removed. The complementary strand is used to synthesize a normal strand of DNA.

*Other types of repair systems exist; these are common examples.

system is composed of one or more proteins that play specific roles in the repair mechanism. DNA repair requires two coordinated events. In the first step, one or more proteins in the repair system detect an irregularity in DNA structure. In the second step, the abnormality is repaired. In some cases, the change in DNA structure can be directly repaired. For example, sometimes DNA is modified by the attachment of an alkyl group, such as $—CH_2CH_3$, to a base. In **direct repair**, an enzyme removes this alkyl group, thereby restoring the structure of the original base. More commonly, however, the altered DNA is removed and a new segment of DNA is synthesized. In this section, we will examine two of these repair systems, namely nucleotide excision repair and methyl-directed mismatch repair, as examples of how such systems operate. These two systems, which are found in all species, represent important mechanisms of DNA repair.

Nucleotide Excision Repair Systems Remove Segments of Damaged DNA

One of the most common types of DNA repair systems is **nucleotide excision repair (NER)**, which can fix many different types of DNA damage, including UV-induced damage, chemically modified bases, missing bases, and various types of cross-links (such as pyrimidine dimers). In NER, a region encompassing several nucleotides in the damaged strand is removed from the DNA, and the intact undamaged strand is used as a template for resynthesis of a normal complementary strand. NER is found in all eukaryotes and prokaryotes, although its molecular mechanism is better understood in prokaryotic species.

In *E. coli*, the NER system is composed of four key proteins: UvrA, UvrB, UvrC, and UvrD. They are named Uvr because they are involved in ultraviolet light repair of pyrimidine dimers, although these proteins are also important in repairing chemically

damaged DNA. In addition, DNA polymerase and DNA ligase are required to complete the repair process.

During DNA repair by the NER system, two UvrA proteins and one UvrB protein form a complex that tracks along the DNA in search of a damaged site (**Figure 14.8**). Such DNA will have a distorted double helix, which is sensed by the UvrA-UvrB complex. When it identifies a damaged segment, the two UvrA proteins are released and UvrC binds to UvrB at the site. The UvrC protein makes incisions in one DNA strand on both sides of the damaged site. After this incision process, UvrC is released. UvrD, which is a helicase, recognizes the region, binds to UvrB, and UvrB is released. UvrD unravels the DNA, which removes a short DNA strand that contains the damaged region. UvrD is released. After the damaged DNA strand is removed, a gap is left in the double helix. DNA polymerase fills in the gap using the undamaged strand as a template. Finally, DNA ligase makes the final covalent connection between the newly made DNA and the original DNA strand.

Human Genetic Diseases Occur When a Component of the NER System Is Missing

Thus far, we have considered the NER system in *E. coli*. In humans, NER systems were discovered by the analysis of genetic diseases that affect DNA repair. These include xeroderma pigmentosum (XP), Cockayne's syndrome (CS), and PIBIDS. (PIBIDS is an acronym for a syndrome with symptoms that include photosensitivity [increased sensitivity to sunlight], ichthyosis [a skin abnormality], brittle hair, impaired intelligence, decreased fertility, and short stature.) Photosensitivity is a common characteristic in all three syndromes because of an inability to repair UV-induced lesions. Therefore, people with these syndromes must avoid prolonged exposure to sunlight. **Figure 14.9** shows a photograph of a person with XP. Such individuals may have pigmentation abnormalities, many precancerous lesions, and a high predisposition to developing skin cancer.

Researchers have found that XP, CS, and PIBIDS result from defects in different genes that encode NER proteins. For example, XP can be caused by defects in seven different NER genes. In all cases, individuals have a defective NER system. In recent years, several human NER genes have been successfully identified. Although more research is needed to completely understand the mechanisms of DNA repair, the identification of NER genes has helped to unravel the complexities of NER systems in human cells.

Mismatch Repair Systems Recognize and Correct a Base Pair Mismatch

Another type of abnormality that should not occur in DNA is a base mismatch, in which the structure of the DNA double helix does not obey the AT/GC rule of base pairing. During the normal course of DNA replication, DNA polymerase may add an incorrect nucleotide to the growing strand by mistake. This creates a mismatch between a nucleotide in the parental

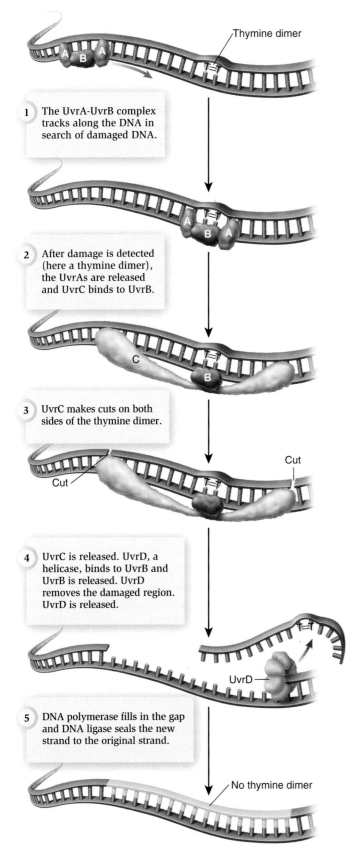

1. The UvrA-UvrB complex tracks along the DNA in search of damaged DNA.

2. After damage is detected (here a thymine dimer), the UvrAs are released and UvrC binds to UvrB.

3. UvrC makes cuts on both sides of the thymine dimer.

4. UvrC is released. UvrD, a helicase, binds to UvrB and UvrB is released. UvrD removes the damaged region. UvrD is released.

5. DNA polymerase fills in the gap and DNA ligase seals the new strand to the original strand.

Figure 14.8   Nucleotide excision repair in *E. coli*.

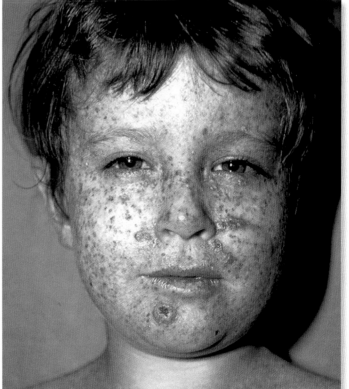

Figure 14.9   An individual affected by xeroderma pigmentosum.

Biological inquiry: Why is this person so sensitive to the sun?

strand and one in the newly made strand. Various DNA repair mechanisms may recognize and remove this mismatch. For example, as we learned in Chapter 11, DNA polymerase itself has a proofreading ability that can detect mismatches and remove them. However, if this proofreading ability fails, cells contain additional DNA repair systems that can detect base mismatches and fix them.

One of these DNA repair systems is the **methyl-directed mismatch repair** system, which exists in all species. The molecular mechanism of mismatch repair has been studied extensively in *E. coli*. This system involves the participation of several proteins that detect the mismatch and specifically remove a segment from the newly made strand. Keep in mind that if DNA polymerase makes a mistake, the new strand contains the incorrect base, while the parental strand is normal. Therefore, a critical aspect of methyl-directed mismatch repair is that it specifically repairs the new strand rather than the parental template strand.

Three proteins in *E. coli*, designated MutH, MutL, and MutS, detect the mismatch and direct the removal of a portion of the new strand carrying the mismatched base (**Figure 14.10**). These proteins are named Mut because their absence leads to a much higher <u>mut</u>ation rate than that of normal strains of *E. coli*.

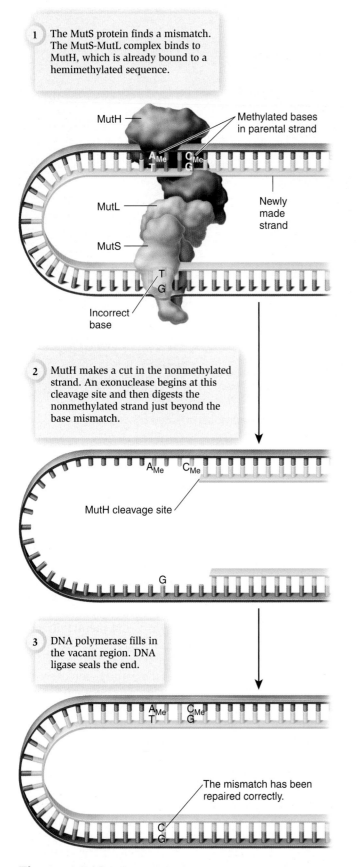

1 The MutS protein finds a mismatch. The MutS-MutL complex binds to MutH, which is already bound to a hemimethylated sequence.

MutH
Methylated bases in parental strand
MutL
MutS
Newly made strand
Incorrect base

2 MutH makes a cut in the nonmethylated strand. An exonuclease begins at this cleavage site and then digests the nonmethylated strand just beyond the base mismatch.

A_{Me} C_{Me}
MutH cleavage site
G

3 DNA polymerase fills in the vacant region. DNA ligase seals the end.

A_{Me} C_{Me}
T   G
The mismatch has been repaired correctly.
C
G

Figure 14.10   Methyl-directed mismatch repair in *E. coli.*

The role of the MutH protein is to distinguish between the parental and new DNA strands. In *E. coli*, adenine and cytosine bases are commonly methylated. Therefore, prior to DNA replication, the parental DNA has already been methylated. Immediately after DNA replication, the new strand is not yet methylated. Therefore, the MutH protein can use the presence or absence of methylation to distinguish between the parental strand and the new strand. MutH binds to a hemimethylated sequence in which the parental strand is methylated and the new strand is not methylated.

The role of MutS is to find mismatches. When MutS has located a DNA mismatch, it forms a complex with MutL. This MutS-MutL complex interacts with the previously bound MutH; the DNA must loop for this to occur. This interaction stimulates MutH to make a cut in the nonmethylated DNA strand. After the strand is cut, an exonuclease is attracted to the site where MutH is bound. The exonuclease then digests the nonmethylated DNA strand in the direction of the mismatch and proceeds just beyond the mismatch site where the MutS-MutL complex is bound. The MutS-MutL complex is released. This leaves a gap in the new strand that is repaired by DNA polymerase and DNA ligase. The net result is that the mismatch has been corrected by removing the incorrect region in the new strand and re-synthesizing the correct sequence using the parental DNA as a template.

Similar to defects in NER systems, mutations in the human mismatch repair systems are associated with particular types of cancer. Mutations in two human mismatch repair genes, *hMSH2* and *hMLH1*, play a role in the development of a type of colon cancer. These mutations cause the mismatch repair system to function improperly, thereby increasing the rate of mutation.

14.3 Cancer

Cancer is a disease of multicellular organisms that is characterized by uncontrolled cell division. Over 1 million Americans are diagnosed with cancer each year, and about half that number will die from the disease. In about 10% of cancers, a higher predisposition to develop the disease is an inherited trait. Most cancers, though, perhaps 90%, do not involve genetic changes that are passed from parent to offspring. Rather, cancer is usually an acquired condition that typically occurs later in life. At least 80% of all human cancers are related to exposure to **carcinogens**, which are agents that increase the likelihood of developing cancer. Most carcinogens, such as UV light and certain chemicals in cigarette smoke, are mutagens that promote genetic changes in somatic cells. These DNA alterations can lead to effects on gene expression that ultimately affect cell division, and thereby lead to cancer. In this section, we will explore such genetic abnormalities.

In most cases, the development of cancer is a multistep process (**Figure 14.11**). Cancers originate from a single cell. This single cell and its line of daughter cells undergo a series of mutations that causes the cells to grow abnormally. At an early

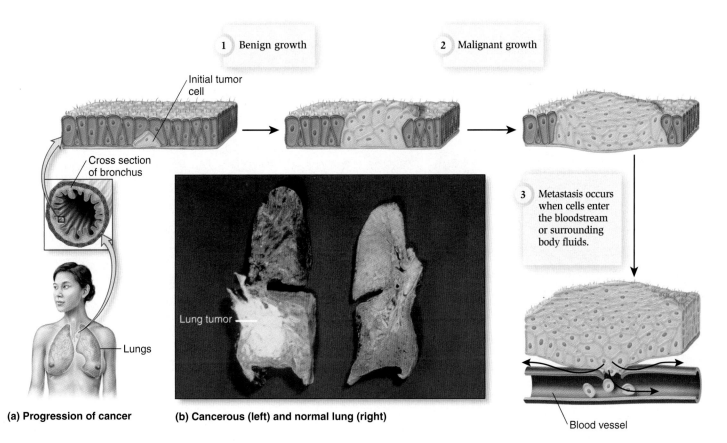

Figure 14.11 **Cancer: Its progression and effects.** (a) In a healthy individual, an initial gene mutation converts a normal cell into a tumor cell. This tumor cell divides to produce a benign tumor. Additional genetic changes in the tumor cells may occur, leading to a malignant tumor. At a later stage in malignancy, the tumor cells will invade surrounding tissues, and some malignant cells may metastasize by traveling through the bloodstream to other parts of the body. (b) On the *right* is a photo of a human lung that was obtained from a healthy nonsmoker. The lung shown on the *left* has been ravaged by lung cancer. This lung was taken from a person who was a heavy smoker.

stage, the cells form a **tumor**, which is an overgrowth of cells that serves no useful purpose. For most types of cancer, a tumor begins as a precancerous or **benign** growth. This may be followed by additional mutations that cause cells in the tumor to progress to the cancerous stage. The tumor is now **malignant** because the cancer cells have lost their normal growth regulation and are **invasive** (that is, they can invade healthy tissues) and **metastatic** (that is, they can migrate to other parts of the body). The final stage of cancer involves these cancerous cells spreading through the bloodstream or surrounding body fluids and establishing more tumors elsewhere. If left untreated, malignant cells will cause the death of the organism.

Over the past few decades, researchers have identified many genes that promote cancer when they are mutant. By comparing the function of each mutant, cancer-causing gene with the corresponding nonmutant gene found in healthy cells, these genes have been placed into two categories. In some cases, a mutation causes a gene to be overactive. This overactivity contributes to the uncontrolled cell growth that is observed in cancer cells. This type of mutant gene is called an **oncogene**.

Alternatively, when a **tumor-suppressor gene** is normal (that is, not mutant), it encodes a protein that prevents cancer. However, when a mutation eliminates its function, cancer may occur. Thus, the two categories of cancer-causing genes are based on the effects of mutations. Oncogenes are the result of mutations that cause overactivity, while loss-of-function mutations in tumor-suppressor genes promote cancer. In this section, we will begin with a discussion of oncogenes and then consider tumor-suppressor genes.

Oncogenes Cause the Overactivity of Proteins That Promote Cell Division

Over the past three decades, researchers have identified many oncogenes. In parallel with our increased understanding of cell division, it has become apparent that most oncogenes encode proteins that function in cell growth signaling pathways. Cell division is regulated, in part, by hormones known as growth factors that bind to cell surface receptors and initiate a cascade of cellular events that lead eventually to cell division (**Figure 14.12**).

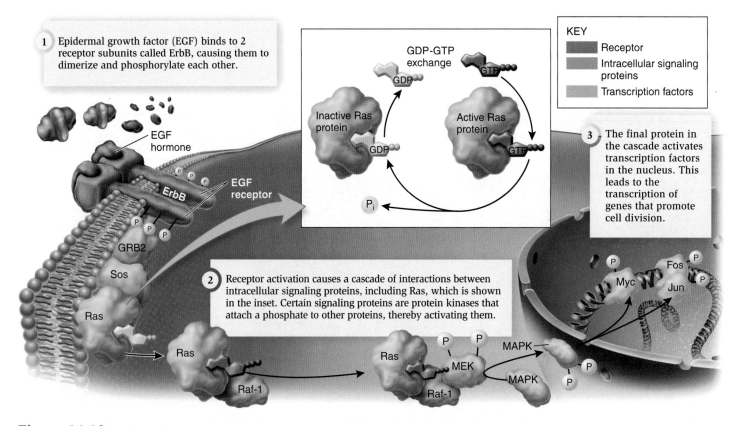

Figure 14.12   **The epidermal growth factor signaling pathway.**   (*Inset*) When GTP is bound, the activated Ras protein promotes cell division. When GTP is hydrolyzed to GDP and P$_i$, Ras is inactivated and cell division is inhibited.

In this example, a growth hormone known as epidermal growth factor (EGF) binds to its receptor, ErbB, thereby leading to the activation of an intracellular signaling pathway. This involves the sequential activation of six proteins called GRB2, Sos, Ras, Raf-1, MEK, and MAPK. This pathway, also known as a signal cascade, ultimately transmits a change in gene transcription by activating the transcription factors Myc, Fos, and Jun. The transcription of specific genes is activated in response to the growth hormone. The activated gene products promote the cell's progression through the cell cycle. Figure 14.12 is just one example of a pathway between a growth factor and gene activation. Eukaryotic species produce many different growth factors, and the signaling pathways are often more complex than the one shown here.

Mutations in the genes that produce these cell growth signaling proteins can change them into oncogenes. For example, **Table 14.6** lists oncogenes that can affect the EGF signaling pathway, which was shown in Figure 14.12. Oncogenes result in an abnormally high level of activity in these proteins, which can include growth factor receptors, intracellular signaling proteins, and transcription factors.

An oncogene may promote cancer by keeping the cell division signaling pathway in a permanent "on" position. This can occur in two ways. First, in some cancers the amount of gene product is abnormally high; in other words, the affected cell makes too much of the protein encoded by the oncogene.

Table 14.6	Examples of Genes Encoding Proteins of the Epidermal Growth Factor Signaling Pathway That Can Mutate to Become Oncogenes
Gene*	**Cellular function**
erbB	Growth factor receptor for EGF (epidermal growth factor)
ras	GTP/GDP-binding protein
raf	Serine kinase
myc	Transcription factor
fos	Transcription factor
jun	Transcription factor

* The genes described in this table are found in humans as well as other vertebrate species. Most of the genes have been given three-letter names that are abbreviations for the type of cancer the oncogene causes or the type of virus in which the gene was first identified.

In 1982, research groups headed by Robert Gallo and Mark Groudine showed that a *myc* gene, which codes for a transcription factor in the EGF signaling pathway, was present in 10 times as many copies as normal in a form of leukemia (a cancer of white blood cells) called promyelocytic leukemia (hence the name *myc*). Since that time, researchers have found that *myc* genes are overexpressed in many forms of cancer, including those of the breast, lung, and colon. The overexpression of this transcription factor leads to the transcriptional activation of genes that promote cell division.

A second way oncogenes can keep cell division turned on is by producing a functionally hyperactive protein. For example, mutations that alter the amino acid sequence of the Ras protein (an intracellular signaling protein in the EGF signaling pathway) have been shown to cause functional abnormalities. The Ras protein is a GTPase that hydrolyzes GTP to GDP + P_i (see Figure 14.12, inset). When GTP is bound, the activated Ras protein promotes cell division. After it has been activated, the Ras protein returns to its inactive state by hydrolyzing its bound GTP, and cell division is inhibited. Mutations that convert the normal *ras* gene into an oncogenic *ras* either decrease the GTPase activity of the Ras protein or increase the rate of exchange of bound GDP for GTP. Both of these functional changes result in a greater amount of the active GTP-bound form of the Ras protein. In this way, these mutations keep the signaling pathway turned on.

Mutations in Proto-Oncogenes Convert Them to Oncogenes

Thus far, we have examined the functions of proteins that cause cancer when they become overactive. In some cases the amount of the protein increases, while in other cases the activity of the protein increases. Such overactivity promotes uncontrolled cell division. Let's now consider the common types of genetic changes that create such oncogenes. A **proto-oncogene** is a normal gene that, if mutated, can become an oncogene. Four common genetic changes may convert a proto-oncogene into an oncogene (**Figure 14.13**). These changes are categorized as missense mutations, gene amplifications, chromosomal translocations, and retroviral insertions.

A missense mutation (Figure 14.13a), which we considered earlier in this chapter, is responsible for the conversion of the *ras* gene into an oncogene. For example, a change of a glycine to a valine in the Ras protein converts the *ras* gene into an oncogene (**Figure 14.14**). This mutation decreases the ability of the Ras protein to hydrolyze its GTP. Experimentally, chemical mutagens have been shown to cause this missense mutation and thereby lead to cancer.

Another genetic event that occurs in cancer cells is **gene amplification**, in this case an increase in the copy number of a proto-oncogene (see Figure 14.13b). An abnormal increase in copy number results in too much of the encoded protein. As mentioned previously, Gallo and Groudine discovered that *myc* was amplified in a human leukemia. Many human cancers are associated with the amplification of particular proto-oncogenes. In some cases, the extent of gene amplification is correlated with the progression of tumors to increasing malignancy. A malignancy may become more difficult to treat as the copy number of a proto-oncogene increases. In other types of malignancies, gene amplification is more random and may be a secondary event that increases the expression of oncogenes previously activated by other genetic changes.

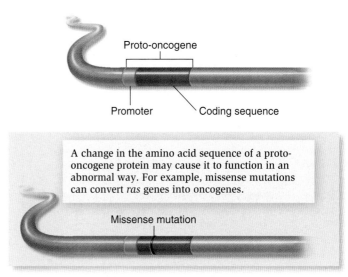

A change in the amino acid sequence of a proto-oncogene protein may cause it to function in an abnormal way. For example, missense mutations can convert *ras* genes into oncogenes.

Missense mutation

(a) Missense mutation

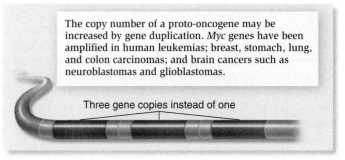

The copy number of a proto-oncogene may be increased by gene duplication. *Myc* genes have been amplified in human leukemias; breast, stomach, lung, and colon carcinomas; and brain cancers such as neuroblastomas and glioblastomas.

Three gene copies instead of one

(b) Gene amplification

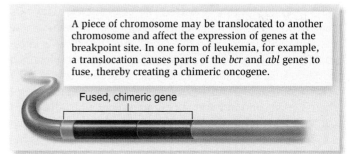

A piece of chromosome may be translocated to another chromosome and affect the expression of genes at the breakpoint site. In one form of leukemia, for example, a translocation causes parts of the *bcr* and *abl* genes to fuse, thereby creating a chimeric oncogene.

Fused, chimeric gene

(c) Chromosomal translocation

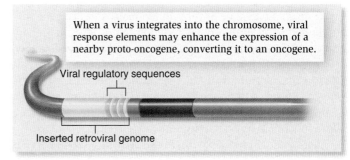

When a virus integrates into the chromosome, viral response elements may enhance the expression of a nearby proto-oncogene, converting it to an oncogene.

Viral regulatory sequences

Inserted retroviral genome

(d) Retroviral insertion

Figure 14.13 Genetic changes that convert proto-oncogenes to oncogenes.

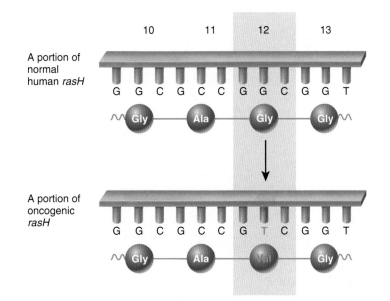

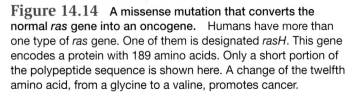

Figure 14.14 A missense mutation that converts the normal *ras* gene into an oncogene. Humans have more than one type of *ras* gene. One of them is designated *rasH*. This gene encodes a protein with 189 amino acids. Only a short portion of the polypeptide sequence is shown here. A change of the twelfth amino acid, from a glycine to a valine, promotes cancer.

Biological inquiry: How does this mutation affect Ras protein function? How does this alteration in function contribute to cancer?

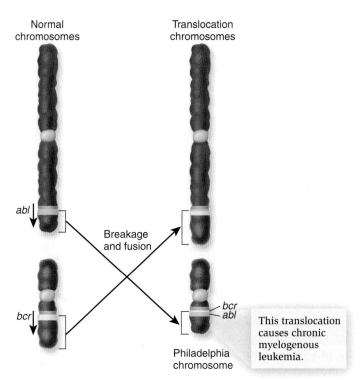

Figure 14.15 The formation of a chimeric gene that is found in people with certain forms of leukemia. The fusion of the *abl* and *bcr* genes creates a chimeric gene that encodes an abnormal fusion protein, leading to leukemia. The blue regions are the promoters for the *abl* and *bcr* genes.

A third type of genetic alteration that can lead to cancer is a chromosomal translocation (see Figure 14.13c). This occurs when two different chromosomes break, and the ends of the broken chromosomes fuse with each other incorrectly. Very specific types of chromosomal translocations have been identified in certain types of tumors. In 1960, Peter Nowell discovered that a certain form of leukemia was correlated with the presence of a shortened version of a human chromosome, which he called the Philadelphia chromosome after the city where it was discovered. Rather than a deletion, this shortened chromosome is the result of a chromosome translocation in which pieces of two different chromosomes fused with each other. Later studies revealed that this activates a proto-oncogene, *abl*, in an unusual way (**Figure 14.15**). In healthy individuals, the *bcr* gene and the *abl* gene are located on different chromosomes. In certain forms of leukemia, these chromosomes break and fuse together, an abnormal event that causes the promoter and the first part of *bcr* to fuse with part of *abl*. This fusion creates a **chimeric gene** composed of two gene fragments. This chimeric gene acts as an oncogene that encodes an abnormal fusion protein, whose functional hyperactivity leads to leukemia.

Finally, certain types of viruses can convert proto-oncogenes into oncogenes during the viral replication cycle (see Figure 14.13d). Retroviruses insert their DNA into the chromosomal DNA of their host cell. The viral genome contains promoter and response elements that cause a high level of expression of viral genes. On occasion, the viral DNA may insert into a host chromosome in such a way that a viral promoter and response elements are next to a proto-oncogene. This may result in the overexpression of the proto-oncogene, thereby promoting cancer. This is one way for a virus to cause cancer. Alternatively, a virus may cause cancer because it carries an oncogene in the viral genome. This phenomenon is described next.

Some Types of Cancer Are Caused by Viruses

The great majority of cancers are caused by mutagens that alter the structure and expression of genes that are found in cells. A few viruses, however, are known to cause cancer in plants, animals, and humans (**Table 14.7**).

In 1911, the first cancer-causing virus to be discovered was isolated from chicken sarcomas by Peyton Rous. A **sarcoma** is a tumor of connective tissue such as bone or cartilage. During the 1970s, Rous sarcoma virus (RSV) research led to the identification of a viral gene that acts as an oncogene. Researchers investigated RSV by using it to infect chicken cells grown in the laboratory. This causes the chicken cells to grow like cancer cells, continuously and in an uncontrolled manner. Researchers identified mutant RSV strains that infected and proliferated within chicken cells without transforming them into malignant cells. These RSV strains were missing a gene that is found in the form

Table 14.7	Examples of Viruses That Cause Cancer
Virus	**Description**
Rous sarcoma virus	Causes sarcomas in chickens
Simian sarcoma virus	Causes sarcomas in monkeys
Abelson leukemia virus	Causes leukemia in mice
Hardy-Zuckerman-4 feline sarcoma virus	Causes sarcomas in cats
Hepatitis B	Causes liver cancer in several species, including humans
Papillomavirus	Causes benign tumors and malignant carcinomas in several species including humans; causes cervical cancer in humans
Epstein-Barr virus	Causes Burkitt's lymphoma, which primarily occurs in immunosuppressed individuals such as AIDS patients

Table 14.8	Functions of Selected Tumor-Suppressor Genes
Gene	**Function of encoded protein**
	Maintain genome integrity:
p53	p53 is a transcription factor that positively regulates a few specific target genes and negatively regulates others in a general manner. It acts as a sensor of DNA damage. It can prevent the progression through the cell cycle and also can promote apoptosis.
BRCA-1 BRCA-2	Brca-1 and Brca-2 proteins are both involved in the cellular defense against DNA damage. They may play a role in sensing DNA damage, or they may act to facilitate DNA repair. These genes are sometimes mutant in persons with inherited forms of breast cancer.
XPD	This represents several different genes whose products function in DNA repair. These genes are defective in patients with xeroderma pigmentosum.
	Inhibit cell division (negative regulators):
Rb	The Rb protein is a negative regulator that represses the transcription of genes required for DNA replication and cell division.
NF1	The NF1 protein stimulates Ras to hydrolyze its GTP to GDP. Loss of NF1 function causes the Ras protein to be overactive, which promotes cell division.
p16	A negative regulator of cyclin-dependent protein kinase.

of the virus that does cause cancer. This gene was called the *src* gene because it causes sarcoma.

Harold Varmus and Michael Bishop, in collaboration with Peter Vogt, later discovered that normal (non-viral-infected) cells contain a copy of the *src* gene in their chromosomes. It is a proto-oncogene. Researchers have speculated that RSV may have acquired the *src* gene during the viral replication cycle. The viral DNA integrates into the host cell DNA and later excises itself. An RSV may have integrated next to the *src* gene in a host cell, and later brought the *src* gene along with the viral DNA during the excision process. Once incorporated into a viral genome, the *src* gene is overexpressed because it is transcribed from a very active viral promoter. This ultimately produces too much of the Src protein and promotes uncontrolled cell division.

Tumor-Suppressor Genes Prevent Mutation or Cell Proliferation

Thus far, we have examined the first category of genes that can promote cancer, namely oncogenes. We now turn our attention to the second category of genes, those called tumor-suppressor genes. The role of a normal (nonmutant) tumor-suppressor gene is to prevent cancerous growth. The proteins encoded by tumor-suppressor genes usually have one of two functions (Table 14.8). Some tumor-suppressor genes encode proteins that maintain the integrity of the genome. Such proteins monitor and/or repair alterations in the genome. The proteins encoded by these genes are vital for the prevention of abnormalities such as DNA breaks and improperly segregated chromosomes. Therefore, when these proteins are functioning properly, they minimize the chance that a cancer-causing mutation will occur. In some cases, the proteins will prevent a cell from progressing through the cell cycle. These are termed **checkpoint proteins** because their role is to <u>check</u> the integrity of the genome and prevent a

cell from progressing past a certain <u>point</u> in the cell cycle if a genetic abnormality is detected. Checkpoint proteins are not usually required to regulate normal, healthy cell division, but they can stop cell division if an abnormality is detected. A second type of tumor-suppressor gene encodes proteins that are negative regulators or inhibitors of cell division. Their function is necessary to properly halt cell division. If their function is lost, cell division is abnormally accelerated. Let's now consider one example of a checkpoint protein and one example of a negative regulator. The function of checkpoint proteins is also discussed in Chapter 9 and shown in Figure 9.23.

Proteins called cyclins and cyclin-dependent protein kinases (cdks) are responsible for advancing a cell through the four phases of the cell cycle (**Figure 14.16**). The formation of activated cyclin/cdk complexes can be stopped by checkpoint proteins. A specific example of a tumor-suppressor gene that encodes a checkpoint protein is *p53*. Its name refers to the molecular mass of the p53 <u>p</u>rotein, which is <u>53</u> kDa (kilodaltons). About 50% of all human cancers are associated with defects in this gene, including malignant tumors of the lung, breast, esophagus, liver, bladder, and brain, as well as leukemias and lymphomas (cancer of the lymphatic system).

As shown in Figure 14.16, p53 is a G_1 checkpoint protein. The expression of the *p53* gene is induced when DNA is damaged. The p53 protein is a regulatory transcription factor that activates several different genes, leading to the synthesis of proteins that stop the cell cycle and other proteins that repair the DNA.

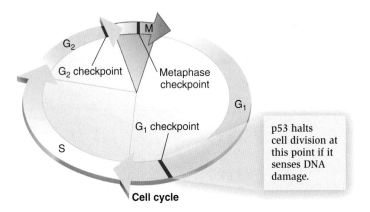

Figure 14.16 **The cell cycle and checkpoints.** As discussed in Chapter 9, eukaryotic cells progress through a cell cycle composed of G_1, S, G_2, and M phases. The red bars indicate common checkpoints that will stop the cell cycle if genetic abnormalities are detected. The p53 protein will stop a cell at the G_1 checkpoint if it senses DNA damage.

When p53 is activated, a cell cannot progress from G_1 to the S, or synthesis, phase of the cell cycle. If the DNA is eventually repaired, a cell may later proceed through the cell cycle.

Alternatively, if the DNA damage is too severe, the p53 protein will also activate other genes that promote programmed cell death. This process, called **apoptosis**, involves cell shrinkage and DNA degradation. Enzymes known as **caspases** are activated during apoptosis. They function as proteases that are sometimes called the "executioners" of the cell. Caspases digest selected cellular proteins such as microfilaments, which are components of the cytoskeleton. This causes the cell to break down into small vesicles that are eventually phagocytized by cells of the immune system. It is beneficial for a multicellular organism to kill an occasional cell with cancer-causing potential.

When checkpoint genes are rendered inactive by mutation, the division of normal healthy cells may not be adversely affected. For example, mice that are missing the *p53* gene are born healthy. This indicates that cell division leading to normal growth is regulated properly. Therefore, checkpoint proteins such as p53 are not necessary for normal cell growth and division. However, these mice are very sensitive to mutagens such as UV light and easily develop cancer. This result suggests that checkpoint proteins are needed to stop cell division only when there is a strong likelihood that a mutation may have occurred. The loss of checkpoint protein function makes it more likely that undesirable genetic changes will occur that could cause cancerous growth.

Our second example of a tumor-suppressor gene is the *Rb* gene, which is a negative regulator of cell division. It was the first tumor-suppressor gene to be identified in humans by studying patients with a disease called retinoblastoma, a cancerous tumor that occurs in the retina of the eye. Some people have an inherited predisposition to develop this disease within the first few years of life. By comparison, the noninherited form of retinoblastoma, which is caused by environmental agents, is more likely to occur later in life. Based on these differences, in 1971 Alfred Knudson proposed a "two-hit" model for retinoblastoma. According to this idea, retinoblastoma requires two mutations to occur. People have two copies of the *Rb* gene, one from each parent. Individuals with the inherited form of the disease already have received one mutant gene from one of their parents. They need only one additional mutation to develop the disease. Because the retina has more than 1 million cells, it is somewhat likely that a mutation may occur in one of these cells at an early age, leading to the disease. However, people with the noninherited form of the disease must have two mutations in the same retinal cell to cause the disease. Because two mutations are less likely than a single mutation, the noninherited form of this disease is expected to occur much later in life, and only rarely. Since Knudsen's original hypothesis, molecular studies have confirmed the two-hit hypothesis.

The Rb protein negatively controls a regulatory transcription factor called E2F that activates genes required for cell cycle progression from G_1 to S phase. The binding of the Rb protein to E2F inhibits its activity and prevents cell division (**Figure 14.17**). When a normal cell is supposed to divide, cyclins bind to cyclin-dependent protein kinases. This binding activates the kinases, which then leads to the phosphorylation of the Rb protein. The phosphorylated form of the Rb protein is released from E2F, thereby allowing E2F to bind to enhancers and activate genes needed to progress through the cell cycle. By comparison, we can imagine how the cell cycle becomes unregulated without functional Rb protein. When both copies of Rb are defective, the E2F protein is always active. This explains why uncontrolled cell division occurs in retinoblastoma.

Gene Mutations, Chromosome Loss, and DNA Methylation Can Inhibit the Expression of Tumor-Suppressor Genes

Cancer biologists would also like to understand how tumor-suppressor genes are inactivated, because this knowledge may ultimately help them to prevent cancer. The function of tumor-suppressor genes is lost in three common ways. First, a mutation can occur specifically within a tumor-suppressor gene to inactivate its function. For example, a mutation could abolish the function of the promoter for a tumor-suppressor gene or introduce an early stop codon in its coding sequence. Either of these would prevent the expression of a functional protein. Chromosome loss is a second way that the function of a tumor-suppressor gene is lost. Chromosome loss may contribute to the progression of cancer if the missing chromosome carries one or more tumor-suppressor genes.

Recently, researchers have discovered a third way that these genes may be inactivated. Tumor-suppressor genes found in cancer cells are sometimes abnormally methylated. As discussed in Chapter 13, transcription is inhibited when CpG islands near

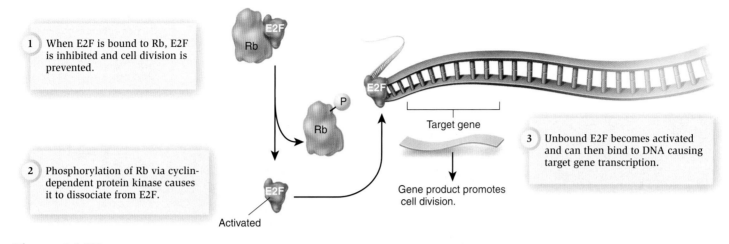

1 When E2F is bound to Rb, E2F is inhibited and cell division is prevented.

2 Phosphorylation of Rb via cyclin-dependent protein kinase causes it to dissociate from E2F.

Activated

Target gene

Gene product promotes cell division.

3 Unbound E2F becomes activated and can then bind to DNA causing target gene transcription.

Figure 14.17 Function of the Rb protein.   The Rb protein inhibits the function of a regulatory transcription factor called E2F that turns on genes that cause a cell to divide. When cells are supposed to divide, Rb is phosphorylated by cyclin-dependent protein kinases, which allows E2F to function. If Rb protein is not made or is not functional due to a mutation, E2F will always be active, and the cell will be stimulated to divide uncontrollably.

a promoter region are methylated. Such methylation near the promoters of tumor-suppressor genes has been found in many types of tumors, suggesting that this form of gene inactivation plays an important role in the formation and/or progression of malignancy.

Most Forms of Cancer Are Caused by a Series of Genetic Changes That Progressively Alter the Growth Properties of Cells

The discovery of oncogenes and tumor-suppressor genes has allowed researchers to study the progression of certain forms of cancer at the molecular level. Cancer usually requires multiple genetic changes to the same cell, perhaps in the range of 10 or more. Many cancers begin with a benign genetic alteration that, over time and with additional mutations, leads to malignancy. Furthermore, a malignancy can continue to accumulate genetic changes that make it even more difficult to treat because the cells divide faster or invade surrounding tissues more readily.

Lung cancer is a type of cancer that progresses through different stages of abnormal cell proliferation. Lung cancer is diagnosed in approximately 170,000 men and women each year in the U.S. Worldwide, more than 1.2 million cases are diagnosed. Nearly 90% of these cases are caused by smoking and are thus preventable. Unlike other cancers for which early diagnosis is possible, lung cancer is usually detected only after it has become advanced and is difficult if not impossible to cure. The five-year survival rate for lung cancer patients is less than 15%.

Most cancers in the lung are **carcinomas**—cancers of epithelial cells (**Figure 14.18a**). Epithelial cells are described in Chapter 10. The top images in this figure show the normal epithelium found in a healthy lung. The rest of the figure shows the progression of cancer that is due to mutations in basal cells—

a type of epithelial cell. As mutations accumulate in basal cells, their numbers increase dramatically. This causes a thickening of the epithelium, a condition called hyperplasia. The proliferation of basal cells causes the loss of the ciliated, columnar epithelial cells that normally line the airways. As additional mutations accumulate, the basal cells develop more abnormal morphologies, a condition known as dysplasia. In the early stages of dysplasia, the abnormal basal cells are precancerous. If the source of chronic irritation (usually cigarette smoke) is eliminated, the abnormal cells are likely to disappear. Alternatively, if smoking continues, these abnormal cells may accumulate additional genetic changes and lose the ability to stop dividing. Such cells have become cancerous—the person has basal cell carcinoma.

The basement membrane is a sheetlike layer of extracellular matrix components that provides a barrier between the lung cells and the bloodstream. If the cancer cells have not yet metastasized by penetrating the basement membrane, they will not have spread into the blood and to other parts of the body. If the tumor is removed at this stage, the patient should be cured. The lower images in Figure 14.18a show a tumor that has broken through the basement membrane. The metastasis of these cells to other parts of the body will likely kill the patient, usually within a year of being diagnosed.

The cellular changes that lead to lung cancer are correlated with genetic changes (**Figure 14.18b**). These include the occurrence of mutations that create oncogenes and inhibit tumor-suppressor genes. The order of mutations is not absolute. Rather, it is the total number of genetic changes, not their exact order, that is important. It takes time for such multiple changes to accumulate, so cancer is usually a disease of older people. Reducing your exposure to mutagens throughout your lifetime will minimize the risk of mutations to your genes that could promote cancer.

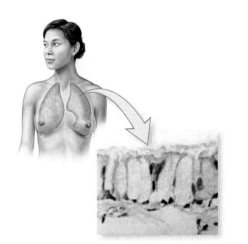

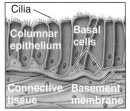

Normal lung epithelium

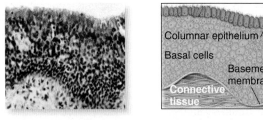

Hyperplasia

Loss of ciliated cells

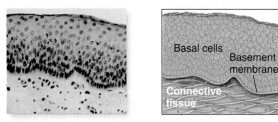

Dysplasia (initially precancerous, then cancerous)

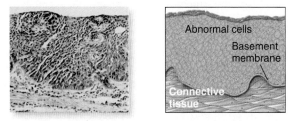

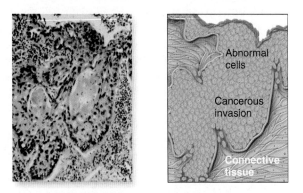

Invasive cancerous cells that can metastasize

(a) Cellular changes

Genes that are commonly mutated in lung cancer include:

Oncogenes	Tumor-suppressor genes
erbB – epidermal growth factor receptor	*p53* – checkpoint
ras – cell signaling	*Rb* – negative regulator
myc – transcription factor	*p16* – negative regulator
Cyclin D1 – promotes the cell cycle	*XPD* – DNA repair

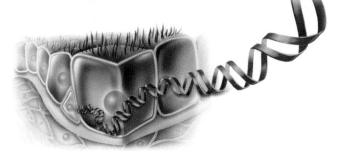

(b) Genetic changes

Figure 14.18 Progression of changes leading to lung cancer. In addition to connective tissue, lung tissue is largely composed of different types of epithelial cells, including columnar and basal cells. The general properties of connective and epithelial tissues are described in Chapter 10. **(a)** A progression of cellular changes in basal cells, caused by the accumulation of mutations, leads to basal cell carcinoma, a common type of lung cancer. **(b)** Mutations in several different genes contribute to lung cancer.

GENOMES & PROTEOMES

Mutations in Approximately 300 Human Genes May Promote Cancer

Researchers have identified a large number of genes that are mutated in cancer cells. Though not all of these mutant genes have been directly shown to affect the growth rate of cells, such mutations are likely to be found in tumors because they provide some type of growth advantage for the cell population from which the cancer developed. For example, certain mutations may enable cells to metastasize to neighboring locations. These mutations may not affect growth rate, but they provide the growth advantage that cancer cells are not limited to growing in a particular location. They can migrate to new locations.

Researchers have estimated that about 300 different genes may play a role in the development of human cancer. With an approximate genome size of 20,000 to 25,000 genes, this observation indicates that over 1% of our genes have the potential to promote cancer if their function is altered by a mutation.

In addition to mutations within specific genes, another common genetic change associated with cancer is abnormalities in chromosome structure and number. **Figure 14.19** compares the chromosome composition of a normal male cell and a tumor cell taken from the same person. The normal composition for this person is 22 pairs of chromosomes plus two sex chromosomes (X and Y). By comparison, the chromosome composition of the tumor cell is quite bizarre, including the fact that the

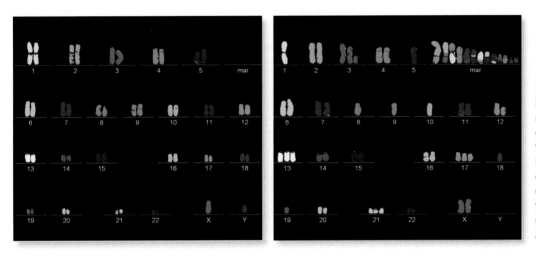

Figure 14.19 A comparison between chromosomes found in a normal human cell and a cancer cell from the same person. The set found in the cancer cell on the right is highly abnormal, with extra copies of some chromosomes and lost copies of others. Chromosomes made of fused pieces of chromosomes (designated *mar* in this figure) are also common in cancer cells.

tumor cell has two X chromosomes, which is characteristic of females. The tumor cells are missing several chromosomes. If tumor-suppressor genes were located on these missing chromosomes, their function is lost as well. Figure 14.19 also shows a few cases of extra chromosomes. If these chromosomes contain proto-oncogenes, the expression of those genes may be over-active. Finally, tumor cells often contain chromosomes that have translocations. This is the most common class of mutation that creates cancer-causing genes. Such translocations may create chimeric genes (as in the case of the Philadelphia chromosome discussed earlier in this chapter) or they may place a gene next to the regulatory sequences of another gene.

CHAPTER SUMMARY

14.1 Mutation

- A mutation is a heritable change in the genetic material.
- Point mutations, which affect a single nucleotide, can alter the coding sequence of genes in several ways. These include silent, missense, nonsense, and frameshift mutations. (Table 14.1)
- Sickle-cell anemia is caused by a missense mutation that changes a single amino acid in hemoglobin. (Figure 14.1)
- Gene mutations can also alter gene function by changing DNA sequences that are not within the coding region. (Table 14.2)
- The Lederbergs used replica plating and showed that mutations conferring resistance to T1 bacteriophage occurred randomly while the bacteria were not being exposed to the phage. (Figure 14.2)
- Germ-line mutations affect gametes while somatic mutations affect only a part of the body. (Figures 14.3. 14.4)
- Spontaneous mutations are the result of errors in natural biological processes, while induced mutations are due to agents in the environment that cause changes in DNA structure. (Table 14.3)
- Mutagens are agents that lead to mutations in the DNA. These can be chemical or physical agents. (Table 14.4, Figures 14.5, 14.6)
- Testing methods, such as the Ames test, can determine whether an agent is a mutagen. (Figure 14.7)

14.2 DNA Repair

- DNA repair systems involve proteins that sense DNA damage and repair it before a mutation occurs. (Table 14.5)

- Nucleotide excision repair systems recognize various types of DNA damage, such as thymine dimers. This type of system excises the damaged strand, and then a new strand is made. (Figure 14.8)
- Certain inherited diseases in humans are due to defects in nucleotide excision repair. (Figure 14.9)
- Mismatch repair systems recognize a base mismatch that is due to an error in DNA replication. A portion of the daughter strand is removed, and then a new region without the error is made. (Figure 14.10)

14.3 Cancer

- Cancer is due to the accumulation of mutations in a line of cells that leads to uncontrolled cell growth. (Figure 14.11)
- Mutations in proto-oncogenes that result in overactivity produce cancer-causing genes called oncogenes.
- Oncogenes often encode proteins involved in cell-signaling pathways that promote cell division. (Figure 14.12, Table 14.6)
- Four common types of genetic changes, namely missense mutations, gene amplifications, chromosomal translocations, and retroviral insertions, can change proto-oncogenes into oncogenes. (Figure 14.13)
- A mutation that inhibits the ability of the Ras protein to hydrolyze its GTP converts the *ras* gene into an oncogene. (Figure 14.14)
- A chromosome translocation that fuses parts of the *bcr* gene and the *abl* gene also creates an oncogene that causes leukemia. (Figure 14.15)
- Some types of cancer are caused by viruses. (Table 14.7)

- The normal function of tumor-suppressor genes is to prevent cancer. Loss-of-function mutations in such genes can promote cancer. Tumor-suppressor genes often encode proteins that are checkpoint proteins or negative regulators of cell division. (Table 14.8)

- Checkpoint proteins monitor the integrity of the genome and prevent the cell from progressing through the cell cycle if abnormalities are detected. (Figure 14.16)

- The Rb protein is a negative regulator of cell division because it inhibits E2F, a transcription factor that promotes cell division. (Figure 14.17)

- Gene mutations, chromosome loss, and DNA methylation are common ways that tumor-suppressor genes are inactivated.

- Most forms of cancer, such as lung cancer, involve multiple genetic changes that lead to malignancy. (Figure 14.18)

- Over 300 human genes are known to be associated with cancer when they become mutant. In addition, changes in chromosome number and structure are commonly found in cancer cells. (Figure 14.19)

TEST YOURSELF

1. Point mutations that do not alter the amino acid sequence of the resulting gene product are called _____ mutations.
   a. frameshift
   b. natural
   c. silent
   d. nonsense
   e. missense

2. Some point mutations will lead to an mRNA that produces a much shorter polypeptide. This type of mutation is known as a _____ mutation.
   a. neutral
   b. silent
   c. missense
   d. nonsense
   e. chromosomal

3. The type of mutation that alters the entire amino acid sequence from the site of the mutation is known as a _____ mutation.
   a. neutral
   b. silent
   c. missense
   d. nonsense
   e. frameshift

4. Mutagens can cause mutations by
   a. chemically altering DNA nucleotides.
   b. disrupting DNA replication.
   c. altering the genetic code of an organism.
   d. all of the above.
   e. a and b only.

5. The mutagenic effect of UV light is
   a. the alteration of cytosine bases to adenine bases.
   b. the formation of purine dimers that interfere with genetic expression.
   c. the breaking of the sugar-phosphate backbone of the DNA molecule.
   d. the formation of pyrimidine dimers that disrupt DNA replication.
   e. the deletion of thymine bases along the DNA molecule.

6. The Ames test
   a. provides a way to determine if any type of cell has experienced a mutation.
   b. provides a way to determine the mutagenic effect of certain types of agents.
   c. allows researchers to experimentally disrupt gene activity by causing a mutation in a specific gene.
   d. provides a way to repair mutations in bacterial cells.
   e. all of the above.

7. Xeroderma pigmentosum
   a. is a genetic disorder that results in uncontrolled cell growth.
   b. is a genetic disorder where normal NER systems are not fully functional.
   c. is a genetic disorder that results in the loss of pigment in certain patches of skin.
   d. results from the lack of DNA polymerase proofreading.
   e. both b and d.

8. During mismatch repair, the parental strand is distinguishable from the new strand by
   a. the lack of mutations in the parental strand.
   b. the presence of methyl groups on the new strand.
   c. the presence of methyl groups on the parental strand.
   d. the 3′ to 5′ orientation of the strand.
   e. the AUG codon on the new strand.

9. Cancer cells are said to be metastatic when they
   a. begin to divide uncontrollably.
   b. invade healthy tissue.
   c. migrate to other parts of the body.
   d. cause mutations in other healthy cells.
   e. all of the above.

10. Oncogenes are
    a. mutations in genes that normally inhibit the progression of a cell through the cell cycle.
    b. mutations that cause the overexpression of genes that normally stimulate cell division.
    c. viruses that cause cancer.
    d. mutations in genes that cause metastasis.
    e. all of the above.

CONCEPTUAL QUESTIONS

1. Explain the difference between a missense mutation and a nonsense mutation.

2. Explain how a frameshift mutation can lead to premature termination of a polypeptide.

3. Define oncogene, tumor-suppressor gene, and proto-oncogene.

EXPERIMENTAL QUESTIONS

1. Explain the difference between the opposing views of mutation prior to the Lederbergs' study.

2. What hypothesis was being tested by the Lederbergs? What were the results of the experiment?

3. How did the results of the Lederbergs support the idea that mutations are random events?

COLLABORATIVE QUESTIONS

1. Discuss the pros and cons of mutation.

2. Discuss three ways that alterations in DNA structure can be repaired.

www.brookerbiology.com

This website includes answers to the Biological Inquiry questions found in the figure legends and all end-of-chapter questions.

15

EUKARYOTIC CHROMOSOMES, MITOSIS, AND MEIOSIS

CHAPTER OUTLINE

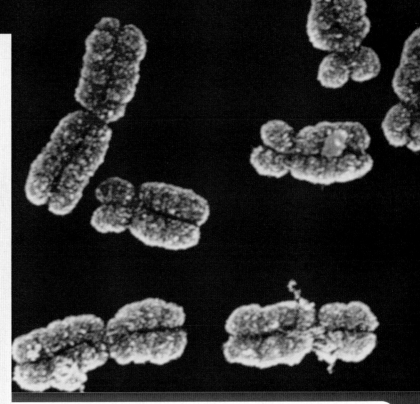

A scanning electron micrograph of highly compacted chromosomes found in a dividing cell.

The **chromosomes** are structures in living cells that contain the genetic material. Genes are physically located within the chromosomes. Biochemically, chromosomes are composed of a very long molecule of DNA, which is the genetic material, and proteins, which are bound to the DNA and provide it with an organized structure.

The primary function of the genetic material in the chromosomes is to store the information needed to produce the characteristics of an organism. To fulfill their role at the molecular level, DNA sequences facilitate four important processes. These are (1) the synthesis of RNA and cellular proteins, (2) the replication of chromosomes, (3) the compaction of chromosomes so they can fit within living cells, and (4) the proper segregation of chromosomes between dividing cells. In this chapter, we will examine the last two of these topics.

This chapter begins with a discussion of the structure of eukaryotic chromosomes at the molecular level. Then we will turn to the process of reproduction in eukaryotic species at the cellular level. In these discussions we will be concerned with two phenomena. First, we will consider how cells divide to produce new daughter cells. Second, we will examine sexual reproduction from a cellular and genetic perspective. We will pay close attention to the sorting of chromosomes during cell division. Lastly, we will examine variation in the structure and number of chromosomes. As you will learn, a variety of mechanisms that alter chromosome structure and number can have important consequences for the organisms that carry them.

15.1 Molecular Structure of Eukaryotic Chromosomes

We now turn our attention to the ways that eukaryotic chromosomes are folded to fit in a living cell. A typical eukaryotic chromosome contains a single, linear, double-stranded DNA molecule that may be hundreds of millions of base pairs in length. If the DNA from a single set of human chromosomes were stretched from end to end, the length would be over 1 meter! By comparison, most eukaryotic cells are only 10–100 μm (micrometers) in diameter, and the cell nucleus is only about 2–4 μm in diameter. Therefore, to fit inside the nucleus, the DNA in a eukaryotic cell must be folded and packaged by a staggering amount.

Before biologists understood chromosome structure, they described the genetic material according to its appearance under the microscope. When a cell is dividing, the nuclear membrane is no longer present, and the chromosomes become very compact. Such chromosomes are readily stained with colored dyes. The term chromosome literally means colored body. This is the form of the genetic material that we are accustomed to seeing in photomicrographs. The term chromatin was first used to describe the genetic material that is found in the nucleus of nondividing cells. While in the nucleus, the genetic material is much less compact and appears to be in a twisted, spaghetti-like configuration.

Over the past couple of decades, as researchers have gained a more complete understanding of genetic material, the meaning of these two terms, namely chromosome and chromatin, have changed. The term chromosome is now used to describe a discrete unit of genetic material. For example, a human somatic cell contains 46 chromosomes. It would also be correct to say there are 46 chromosomes in the nucleus of a nondividing cell.

By comparison, the term chromatin has taken on a biochemical meaning. **Chromatin** is now used to describe the DNA-protein complex that makes up eukaryotic chromosomes. The chromosomes found in the nucleus are composed of chromatin, as are the highly condensed chromosomes found in dividing cells. Chromosomes are very dynamic structures that alternate between tight and loose compaction states in response to changes in protein composition. In this section, we will focus our attention on two issues of chromosome structure. First, we will consider how chromosomes are compacted and organized within the cell nucleus. Then, we will examine the additional compaction that is necessary to produce the highly condensed chromosomes that occur during cell division.

DNA Wraps Around Histone Proteins to Form Nucleosomes

The first way that DNA is compacted is by wrapping itself around a group of proteins called **histones**. As shown in **Figure 15.1**, a repeating structural unit of eukaryotic chromatin is the **nucleosome**, which is 11 nm (nanometers) in diameter and composed of double-stranded DNA wrapped around an octamer of histone proteins. Each octamer contains eight histone subunits. Different kinds of histone proteins form the histone octamer, two each of four kinds called H2A, H2B, H3, and H4. H2A and H2B are so named because they are similar in structure. Histone proteins are very basic proteins because they contain a large number of positively charged lysine and arginine amino acids. The negative charges that are found in the phosphate of DNA are attracted to the positive charges on histone proteins. The DNA lies on the surface of the histone octamer and makes 1.65 turns around it. The amount of DNA that is required to wrap around the histone octamer is 146 or 147 bp (base pairs). The amino terminal tail of each histone protein protrudes from the histone octamer. As discussed later, covalent modifications of these tails is one way to control the degree of chromatin compaction.

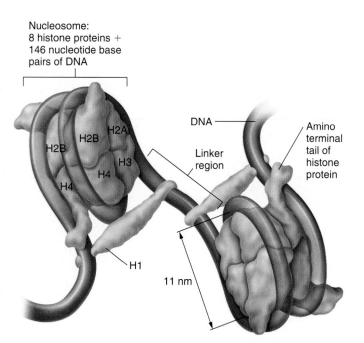

Figure 15.1   **Structure of a nucleosome.**   A nucleosome is composed of double-stranded DNA wrapped around an octamer of histone proteins. A linker region connects two adjacent nucleosomes. Histone H1 is bound to the linker region, as are other proteins not shown in this figure.

The nucleosomes are connected by linker regions of DNA that vary in length from 20 to 100 bp, depending on the species and cell type. A particular histone named histone H1 is bound to the linker region, as are other types of proteins. The overall structure of connected nucleosomes resembles beads on a string. This structure shortens the length of the DNA molecule about sevenfold. Evidence for the beads-on-a-string structure is described next.

FEATURE INVESTIGATION

Noll Confirmed Kornberg's Beads-on-a-String Model by Digestion of the Linker Region

The beads-on-a-string model of nucleosome structure was originally proposed by Roger Kornberg in 1974. Markus Noll decided to test Kornberg's model by digesting chromatin with DNase-I, an enzyme that cuts the DNA backbone. He reasoned that if the model was correct, the linker region of DNA would be more accessible to DNase-I than would the 146-bp region that is tightly bound to the histones. Therefore, he expected incubation with DNase-I to make cuts in the linker region and produce DNA pieces that would be approximately 200 bp in length, given that the DNA used had linker regions of about 50 bp.

(Note: The size of the DNA fragments was expected to vary somewhat, since the linker region is not of constant length and the cut within the linker region may occur at different sites.)

Figure 15.2 describes Noll's experimental protocol. He began with nuclei from rat liver cells and incubated them with low, medium, or high concentrations of DNase-I. At high concentrations, the enzyme should make a cut in every linker region, while at medium and low concentrations, DNase-I may occasionally miss cutting a linker region, which would produce larger DNA fragments in multiples of 200 bp. Following digestion, the DNA was extracted from the cell nuclei, and then analyzed by gel electrophoresis to determine the sizes of the DNA fragments. (Gel electrophoresis is described in Chapter 20.)

Figure 15.2 Noll's DNase-I digestion experiment, which verified the beads-on-a-string model of DNA compaction.

HYPOTHESIS DNA wraps around histone proteins in a regular, repeating pattern.

STARTING MATERIAL Nuclei from rat liver cells.

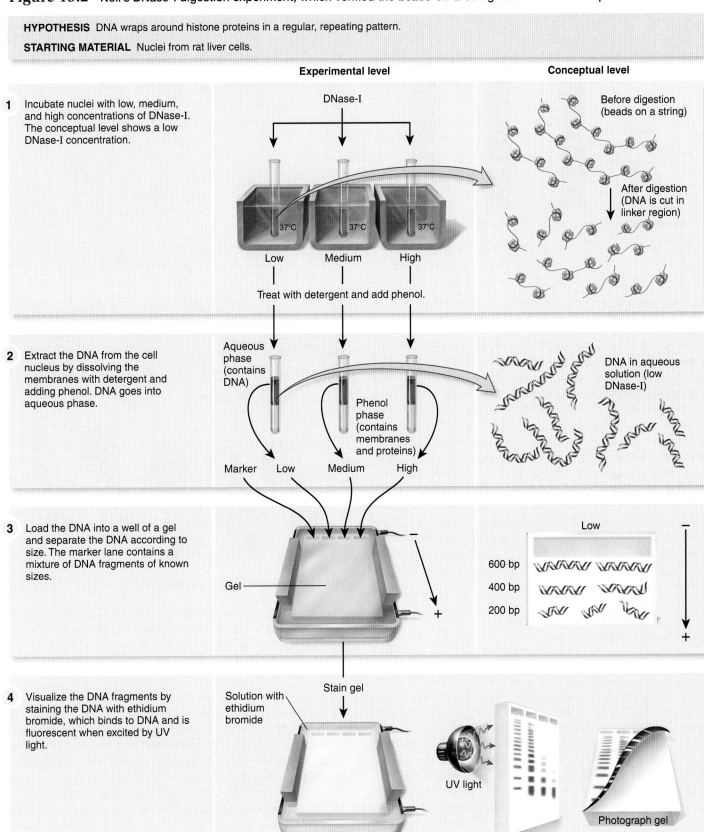

Experimental level

Conceptual level

1 Incubate nuclei with low, medium, and high concentrations of DNase-I. The conceptual level shows a low DNase-I concentration.

DNase-I

37°C   37°C   37°C

Low   Medium   High

Treat with detergent and add phenol.

Before digestion (beads on a string)

After digestion (DNA is cut in linker region)

2 Extract the DNA from the cell nucleus by dissolving the membranes with detergent and adding phenol. DNA goes into aqueous phase.

Aqueous phase (contains DNA)

Phenol phase (contains membranes and proteins)

Marker   Low   Medium   High

DNA in aqueous solution (low DNase-I)

3 Load the DNA into a well of a gel and separate the DNA according to size. The marker lane contains a mixture of DNA fragments of known sizes.

Gel

–

+

Low

600 bp

400 bp

200 bp

–

+

4 Visualize the DNA fragments by staining the DNA with ethidium bromide, which binds to DNA and is fluorescent when excited by UV light.

Solution with ethidium bromide

Stain gel

UV light

Photograph gel

5   THE DATA

Results from the experiment:

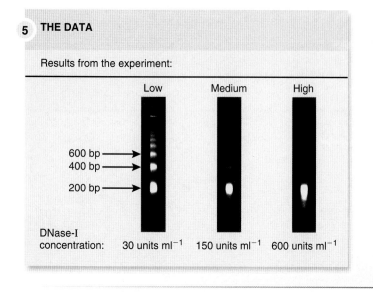

As shown in the data, at high DNase-I concentrations, the chromosomal DNA was digested into fragments of approximately 200 bp in length. This result is predicted by the beads-on-a-string model. At lower DNase-I concentrations, longer pieces were observed that were in multiples of 200 bp (400, 600, and so on). These longer pieces are explained by occasional uncut linker regions. For example, a DNA piece might contain two nucleosomes and be 400 bp in length. Taken together, these results strongly supported the nucleosome model for chromatin structure.

Nucleosomes Compact to Form a 30-nm Fiber

Nucleosome units are organized into a more compact structure that is 30 nm in diameter, known as the **30-nm fiber** (**Figure 15.3a**). Histone H1 and other proteins are important in the formation of the 30-nm fiber, which shortens the nucleosome structure another sevenfold. The structure of the 30-nm fiber has proven difficult to determine because the conformation of the DNA may be substantially altered when extracted from living cells. A current model for the 30-nm fiber was proposed by Rachel Horowitz and Christopher Woodcock in the 1990s (**Figure 15.3b**). According to their model, linker regions in the 30-nm structure are variably bent and twisted, and little direct contact is observed between nucleosomes. The 30-nm fiber forms an asymmetric, three-dimensional zigzag of nucleosomes. At this level of compaction, the overall picture of chromatin that emerges is an irregular, fluctuating structure with stable nucleosome units connected by bendable linker regions.

Chromatin Loops Are Anchored to the Nuclear Matrix

Thus far, we have examined two mechanisms that compact eukaryotic DNA, the formation of nucleosomes and their arrangement into a 30-nm fiber. Taken together, these two events shorten the folded DNA about 49-fold. A third level of compaction involves interactions between the 30-nm fibers and a filamentous network of proteins in the nucleus called the **nuclear matrix**. This matrix consists of the **nuclear lamina**, which is a collection of protein fibers that line the inner nuclear membrane, and an internal nuclear matrix that is connected to the lamina (**Figure 15.4a**). The internal matrix is an intricate network of irregular protein fibers plus many other proteins that bind to these fibers.

The nuclear matrix is involved in the compaction of the 30-nm fiber by participating in the formation of **radial loop domains**. These loops, often 25,000 to 200,000 base pairs in size,

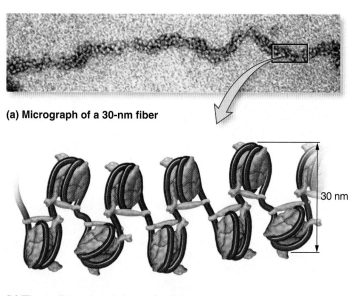

(a) Micrograph of a 30-nm fiber

(b) Three-dimensional zigzag model

Figure 15.3 The 30-nm fiber. (a) A photomicrograph of the 30-nm fiber. (b) In this three-dimensional zigzag model, the linker DNA forms a bendable structure with little contact between adjacent nucleosomes.

are anchored to the nuclear matrix (**Figure 15.4b**). In this way, the nuclear matrix organizes the chromosomes within the nucleus.

Each chromosome in the cell nucleus is located in a discrete and nonoverlapping **chromosome territory**, which can be experimentally viewed in nondividing cells (refer back to Chapter 4, Figure 4.17). **Figure 15.5** shows a model comparing human chromosomes in their fully compacted state, when a cell is preparing to divide, with chromosomes in nondividing cells. Each chromosome in nondividing cells occupies its own discrete region in the cell nucleus that usually does not overlap with the territory of adjacent chromosomes. In other words, different chromosomes are not substantially intertwined with each other, even when they are in a noncompacted condition.

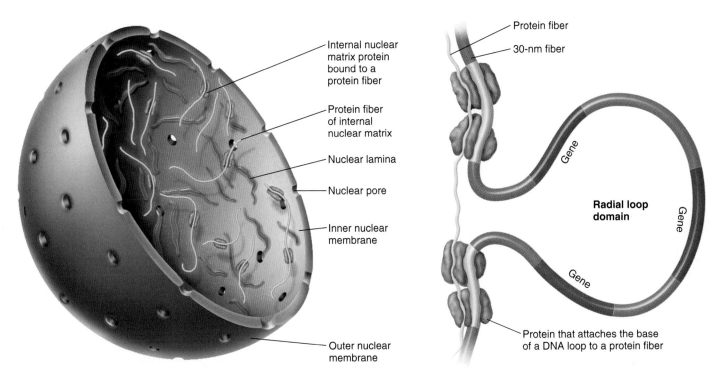

(a) **Proteins that form the nuclear matrix**

(b) **Radial loop domain bound to a protein fiber**

Figure 15.4 Structure of the nuclear matrix and its attachment to the 30-nm fiber. (a) This schematic drawing shows the arrangement of the matrix within a cell nucleus. The nuclear lamina is a collection of fibrous proteins that line the inner nuclear membrane. The internal nuclear matrix is also composed of protein fibers and many other proteins associated with them. (b) The radial loops are attached to the protein fibers of the nuclear matrix.

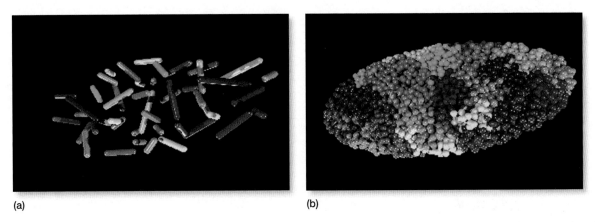

(a)

(b)

Figure 15.5 A molecular model showing chromosome territories in the cell nucleus of humans. Each of the 23 pairs of human chromosomes is labeled with a different color. (a) Compacted chromosomes in a cell that is preparing to divide. (b) Chromosomes in the cell nucleus of a nondividing cell. Each of these chromosomes occupies its own distinct, nonoverlapping territory within the nucleus.

The compaction level of chromosomes in the cell nucleus is not completely uniform. This variability can be seen with a light microscope and was first observed by the German cytologist E. Heitz in 1928. He used the term **heterochromatin** to describe the highly compacted regions of chromosomes. In general, these regions are transcriptionally inactive because of their tight conformation, which prevents transcription factors and RNA polymerase from gaining access to genes. By compar-

ison, the less condensed regions, known as **euchromatin**, reflect areas that are capable of gene transcription. Euchromatin is the form of chromatin in which the 30-nm fiber forms radial loop domains. In heterochromatin, these radial loop domains are compacted even further. In nondividing cells, most chromosomal regions are euchromatic and some localized regions are heterochromatic.

During Cell Division, Chromosomes Undergo Maximum Compaction

When cells prepare to divide, the chromosomes become even more compacted or condensed. This aids in their proper sorting and movement during metaphase, which is a stage of cell division described in the next section. **Figure 15.6** illustrates the levels of compaction that contribute to the formation of a metaphase chromosome. DNA in the nucleus is always compacted by forming nucleosomes and condensing into a 30-nm fiber

(Figure 15.6a,b,c). In euchromatin, the 30-nm fibers are arranged in radial loop domains that are relatively loose, meaning that a fair amount of space is between the 30-nm fibers (Figure 15.6d). The average width of such loops is about 300 nm.

By comparison, heterochromatin involves a much tighter packing of the loops, so little space is between the 30-nm fibers (Figure 15.6e). Heterochromatic regions tend to be wider, in the range of 700 nm. When cells prepare to divide, all of the euchromatin is converted to heterochromatin. Because the 30-nm fibers are much closer together in heterochromatin, the conversion of

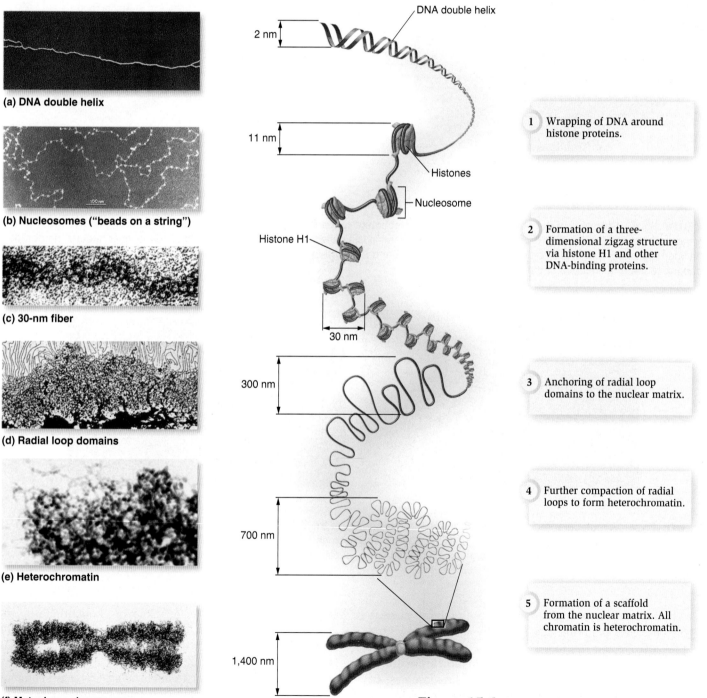

(a) DNA double helix

(b) Nucleosomes ("beads on a string")

(c) 30-nm fiber

(d) Radial loop domains

(e) Heterochromatin

(f) Metaphase chromosome

2 nm — DNA double helix

11 nm — Histones — Nucleosome

Histone H1

30 nm

300 nm

700 nm

1,400 nm

1 Wrapping of DNA around histone proteins.

2 Formation of a three-dimensional zigzag structure via histone H1 and other DNA-binding proteins.

3 Anchoring of radial loop domains to the nuclear matrix.

4 Further compaction of radial loops to form heterochromatin.

5 Formation of a scaffold from the nuclear matrix. All chromatin is heterochromatin.

Figure 15.6   The steps in eukaryotic chromosomal compaction leading to the metaphase chromosome.

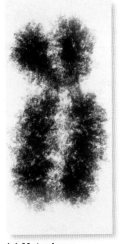

(a) Metaphase chromosome

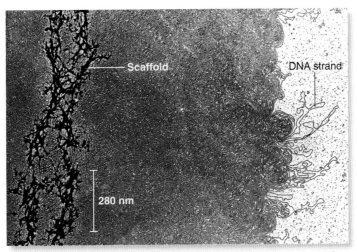

(b) Metaphase chromosome treated with high salt to remove proteins

Figure 15.7 The importance of histones and scaffolding proteins in the compaction of eukaryotic chromosomes. (a) Transmission electron micrograph of a metaphase chromosome. (b) This photomicrograph shows a metaphase chromosome following treatment with a high salt concentration to remove the histone proteins. The label on the *left* points to the scaffold that anchors the bases of the radial loops. The *right* label points to an elongated strand of DNA.

Biological inquiry: After they have replicated and become compacted in preparation for cell division, chromosomes are often shaped like an X, as in part (a) of this figure. Which proteins are primarily responsible for this X shape?

euchromatin to heterochromatin greatly shortens the chromosomes. In a metaphase chromosome, which contains two copies of the DNA (Figure 15.6f), the width averages about 1,400 nm, but the length of a metaphase chromosome is much shorter than the same chromosome in the cell nucleus during interphase. These highly condensed chromosomes undergo little gene transcription because it is difficult for transcription proteins to gain access to the compacted DNA. Therefore, most transcriptional activity ceases during cell division, which usually lasts for a relatively short time.

In metaphase chromosomes, the highly compacted radial loops remain anchored to a **scaffold**, which is formed from proteins in the nuclear matrix (**Figure 15.7**). Experimentally, the scaffold proteins that hold the loops in place can be separated into an observable form. If a metaphase chromosome is treated with a high concentration of salt to remove histone proteins, the highly compact configuration is lost, but the bottoms of the elongated DNA loops remain attached to the scaffold. In the photomicrograph shown in Figure 15.7b, a label points to an elongated DNA strand emanating from the darkly stained scaffold. The scaffold retains the shape of the original metaphase chromosome even though the DNA strands have become greatly elongated. These remarkable results illustrate the importance of both nuclear matrix proteins (which form the scaffold) and histones (which are needed to compact the DNA) for the structure of chromosomes.

GENOMES & PROTEOMES

The Histone Code Controls Chromatin Compaction

In this section, we have learned that the genomes of eukaryotic species are greatly compacted to fit inside the cell nucleus. Even euchromatin, which is looser than heterochromatin, is still compacted to such a degree that it is difficult for transcription fac-

tors and RNA polymerase to access and transcribe genes. As described in Chapters 12 and 13, chromatin must be loosened up so that genes can be transcribed into RNA.

As discussed earlier, each of the histone proteins consists of a globular domain and a flexible, charged amino terminus called an amino terminal tail. The DNA wraps around the globular domains, as depicted in Figure 15.1, and the amino terminal tails protrude from the chromatin. In recent years, researchers have discovered that particular amino acids in the amino terminal tails are subject to several types of covalent modifications, including acetylation, methylation, and phosphorylation. Over 50 different enzymes have been identified in mammals that selectively modify amino terminal tails. **Figure 15.8** shows examples of sites in the tails of H2A, H2B, H3, and H4 that can be modified.

These tail modifications can have two effects. First, they may directly influence interactions between nucleosomes. Second, histone modifications provide binding sites that are recognized by proteins. According to the **histone code hypothesis**, proposed by Brian Strahl and David Allis in 2000, the pattern of histone modification is recognized by particular proteins, much like a language or code. For example, one pattern might involve phosphorylation of the serine at the first amino acid in H2A and acetylation of the fifth and eighth amino acids in H4, which are lysines. A different pattern could involve acetylation of the fifth amino acid, a lysine, in H2B and methylation of the third amino acid in H4, which is an arginine.

The pattern of covalent modifications of amino terminal tails provides binding sites for proteins that subsequently affect the degree of chromatin compaction. One pattern of histone modification may attract proteins that cause the chromatin to become even more compact. This would silence the transcription of genes in the region. Alternatively, a different combination of histone modifications may attract proteins, such as chromatin remodeling enzymes discussed in Chapter 13, that serve to loosen the chromatin and thereby promote gene transcription.

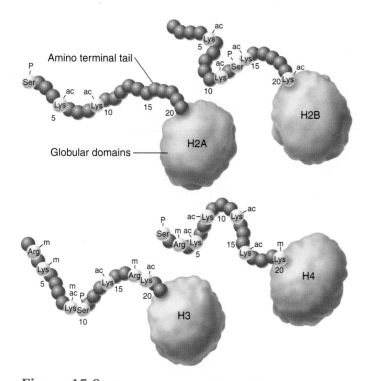

Figure 15.8 **Examples of covalent modifications that occur to the amino terminal tails of histone proteins.** The amino acids are numbered from the amino terminus. The modifications shown here are *m* for methylation, *p* for phosphorylation, and *ac* for acetylation. Many more modifications can occur to the amino terminal tails; the ones shown here represent common examples.

In this way, the histone code plays a key role in accessing the information within the genomes of eukaryotic species. Researchers are trying to unravel which patterns of histone modifications promote compaction, and which promote a loosening of chromatin structure. In other words, they are trying to decipher the histone code.

15.2 Mitotic Cell Division

We now turn our attention to the mechanism of cell division and its relationship to chromosome structure and replication. During the process of **mitotic cell division**, a cell divides to produce two new cells that are genetically identical to the original cell. By convention, the original cell is usually called the mother cell, and the new cells are the two daughter cells. Mitotic cell division involves **mitosis**, which is the division of the nucleus into two nuclei, and **cytokinesis**, which is the division of one mother cell into two daughter cells. One purpose of mitotic cell division is **asexual reproduction**. Certain unicellular eukaryotic organisms, such as baker's yeast (*Saccharomyces cerevisiae*) and the amoeba, increase their numbers in this manner.

A second important reason for mitotic cell division is the production and maintenance of multicellularity. Organisms such as plants, animals, and most fungi are derived from a single cell that subsequently undergoes repeated cellular divisions to become a multicellular organism. Humans, for example, begin as a single fertilized egg and repeated cellular divisions produce an adult with several trillion cells. As you might imagine, the precise transmission of chromosomes is critical during every cell division so that all cells of the body receive the correct amount of genetic material.

In this section, we will explore how the process of mitotic cell division requires the duplication, organization, and sorting of chromosomes. We will also examine how a single cell is separated into two distinct cells by cytokinesis. But first, we need to consider some general features of chromosomes in eukaryotic species.

Eukaryotic Chromosomes Are Inherited in Sets

To understand the chromosomal composition of cells and the behavior of chromosomes during cell division, scientists observe chromosomes with the use of microscopes. **Cytogenetics** is the field of genetics that involves the microscopic examination of chromosomes. As discussed earlier in this chapter, when a cell prepares to divide the chromosomes become more tightly compacted, which shortens them and thereby increases their diameter. A consequence of this shortening is that distinctive shapes and numbers of chromosomes become visible with a light microscope.

Figure 15.9 shows the general procedure for preparing and viewing chromosomes from a eukaryotic cell. In this example, the cells are obtained from a sample of human blood. In particular, the chromosomes within lymphocytes (a type of white blood cell) are examined. A sample of the blood cells is obtained and treated with drugs that stimulate the cells to divide. The actively dividing cells are centrifuged to concentrate them and then mixed with a hypotonic solution that makes the cells swell. The expansion in cell structure causes the chromosomes to spread out from each other, making it easier to see each individual chromosome. Next, the cells are concentrated by a second centrifugation and treated with a fixative, which chemically freezes them so that the chromosomes will no longer move around. The cells are then exposed to a chemical dye that binds to the chromosomes and stains them. As we will learn later, this gives chromosomes a distinctive banding pattern that greatly enhances their contrast and ability to be uniquely identified. The cells are then placed on a slide and viewed with a light microscope. In a cytogenetics laboratory, the microscopes are equipped with an electronic camera to photograph the chromosomes. On a computer screen, the chromosomes can be organized in a standard way, usually from largest to smallest. A photographic representation of the chromosomes, as in the photo in step 5 of Figure 15.9, is called a **karyotype**. A karyotype reveals how many chromosomes are found within an actively dividing cell.

By studying the karyotypes of many species, scientists have discovered that eukaryotic chromosomes occur in sets; each set is composed of several different types of chromosomes. For example, one set of human chromosomes contains 23 different types of chromosomes. By convention, the chromosomes are

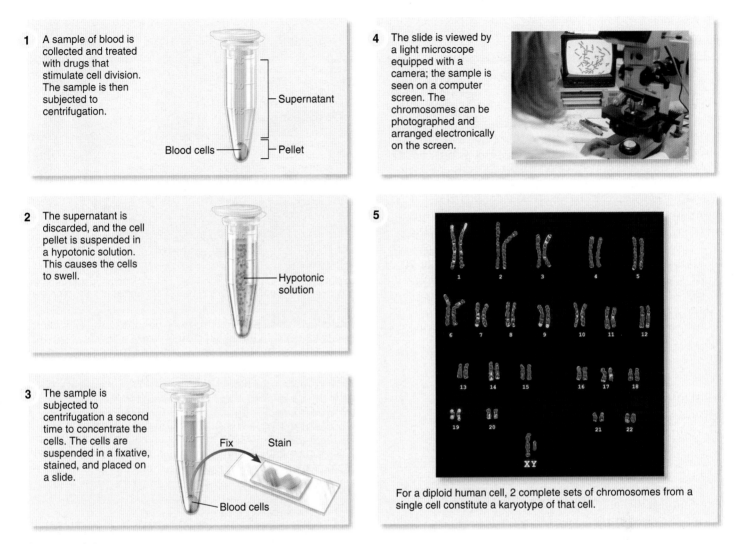

1. A sample of blood is collected and treated with drugs that stimulate cell division. The sample is then subjected to centrifugation.

Supernatant

Blood cells — Pellet

2. The supernatant is discarded, and the cell pellet is suspended in a hypotonic solution. This causes the cells to swell.

Hypotonic solution

3. The sample is subjected to centrifugation a second time to concentrate the cells. The cells are suspended in a fixative, stained, and placed on a slide.

Fix   Stain

Blood cells

4. The slide is viewed by a light microscope equipped with a camera; the sample is seen on a computer screen. The chromosomes can be photographed and arranged electronically on the screen.

5.

For a diploid human cell, 2 complete sets of chromosomes from a single cell constitute a karyotype of that cell.

Figure 15.9   The procedure for making a karyotype.

Biological inquiry: Researchers usually treat cells with drugs that stimulate them to divide prior to the procedure for making a karyotype. Why would this be useful?

numbered according to size, with the largest chromosomes having the smallest numbers. For example, human chromosomes 1, 2, and 3 are relatively large, whereas 21 and 22 are the two smallest (Figure 15.9). This numbering system does not apply to the sex chromosomes, which determine the sex of the individual. Sex chromosomes are designated with the letters X and Y in humans.

A second feature of many eukaryotic species is that an individual has two sets of chromosomes. Again, if we consider humans as an example, the karyotype shown in Figure 15.9 contains two sets of chromosomes, with 23 different chromosomes in each set. Therefore, this human cell contains a total of 46 chromosomes. A person's cells have 46 chromosomes each because the individual inherited one set from the father and one set from the mother. When the cells of an organism carry two sets of chromosomes, that organism is said to be **diploid**. Geneticists use the letter n to represent a set of chromosomes, so diploid organisms are referred to as $2n$. For example, humans are $2n$,

where $n = 23$. Most human cells are diploid. An exception involves **gametes**, namely sperm and egg cells. Gametes are **haploid** or $1n$, which means they contain one set of chromosomes.

When a species is diploid, the members of a pair of chromosomes are called **homologues**. As you can see in Figure 15.9, a cell has two copies of chromosome 1, two copies of chromosome 2, and so forth. Within each pair, the chromosome on the left is a homologue to the one on the right and vice versa. In the case of animals, one of each of these pairs comes from an organism's mother, and one comes from the father; these are referred to as maternal and paternal chromosomes, respectively.

Homologous chromosomes are very similar to each other. Each of the two chromosomes in a homologous pair is nearly identical in size and contains a similar composition of genetic material. A particular gene found on one copy of a chromosome is also found on the homologue. Because one homologue is received from each parent, the two homologues may vary with regard to the way that a gene affects an organism's traits.

As an example, let's consider an eye color gene in humans. One chromosome might carry the form of an eye color gene that confers brown eyes, while the gene on the homologue could confer blue eyes. This topic will be considered in Chapter 16.

The DNA sequences on homologous chromosomes are very similar. In most cases, the sequence of bases on one homologue would differ by less than 1% from the sequence on the other homologue. For example, the DNA sequence of chromosome 1 that you inherited from your mother would be greater than 99% identical to the DNA sequence of chromosome 1 that you inherited from your father. Nevertheless, keep in mind that the sequences are not identical. The slight differences in DNA sequence provide important variation in gene function. Again, if we use an eye color gene as an example, a minor difference in DNA sequence distinguishes two forms of the gene, brown versus blue.

The striking similarity between homologous chromosomes does not apply to pairs of sex chromosomes (for example, X and Y). These chromosomes differ in size and genetic composition. Certain genes that are found on the X chromosome are not found on the Y chromosome, and vice versa. The X and Y chromosomes are not considered homologous chromosomes, although they do have short regions of homology.

In Preparation for Cell Division, Eukaryotic Chromosomes Are Replicated and Compacted to Produce Pairs Called Sister Chromatids

Now that we understand that chromosomes are found in sets, and that many eukaryotic species are diploid, we will now turn our attention to how those chromosomes are replicated and sorted during cell division. Let's begin with the process of chromosome replication. In Chapter 11, we examined the molecular process of DNA replication. **Figure 15.10** describes the process at the chromosomal level. Prior to DNA replication, the DNA of each eukaryotic chromosome consists of a linear DNA double helix that is found in the nucleus and is not highly compacted. When the DNA is replicated, two identical copies of the original double helix are created. These copies, along with associated proteins, lie side by side and are termed **sister chromatids**. When a cell prepares to divide, the sister chromatids become highly compacted and readily visible under the microscope. As shown in the inset to Figure 15.10, the two sister chromatids are tightly associated at a region called the **centromere**. The centromere serves as an attachment site for a group of proteins that form the **kinetochore**, which is necessary for sorting each chromosome.

With regard to the cell cycle, which is described in Chapter 9, **Figure 15.11** provides an overview that relates chromosome replication and cell division. In the G₁ phase, the original cell had three pairs of chromosomes, for a total of six individual chromosomes. Such a cell is diploid $(2n)$ and contains three chromosomes per set $(n = 3)$. The paternal set is shown in blue, and the homologous maternal set is shown in red. In G₁, the chromosomes are not highly compacted. During the S, or synthesis, phase, these chromosomes replicate to yield 12 chromatids (that is, 6 pairs of sister chromatids). At the start of mito-

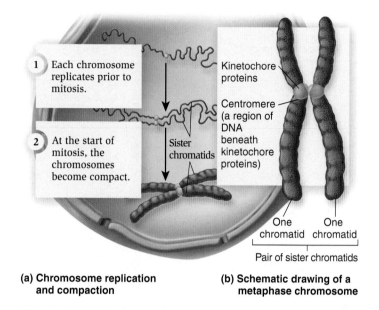

(a) Chromosome replication and compaction

(b) Schematic drawing of a metaphase chromosome

Figure 15.10   Replication and compaction of chromosomes into pairs of sister chromatids.   (a) Chromosomal replication producing a pair of sister chromatids. While the chromosomes are elongated, they are replicated to produce two copies that are connected and lie parallel to each other. This is a pair of sister chromatids. Later, when the cell is preparing to divide, the sister chromatids condense into more compact structures that are easily seen with a light microscope. (b) A schematic drawing of a metaphase chromosome. This structure has two chromatids that lie side by side. The two chromatids are held together by kinetochore proteins that bind to each other and to the centromeres on each chromatid.

sis, the chromatids become highly compacted, and during the process of cell division they are divided equally into two daughter cells. The term **M phase** refers to the sequential events of mitosis and cytokinesis. After cell division is completed, these two daughter cells each contain six chromosomes (three pairs of homologues).

The Transmission of Chromosomes Requires a Sorting Process Known as Mitosis

Mitosis is the sorting process that ensures that each daughter cell will obtain the correct number and types of chromosomes. Mitosis was first observed microscopically in the 1870s by a German biologist named Walter Flemming, who coined the term mitosis (from the Greek *mitos*, meaning thread). He studied the large, transparent skin cells of salamander larvae, as they were dividing, and noticed that chromosomes are constructed of "threads" that are doubled in appearance along their length. These double threads divided and moved apart, one going to each of the two daughter nuclei. By this mechanism, Flemming pointed out, the two daughter cells receive an identical group of threads, the same as the number of threads in the mother cell.

We now know that the **mitotic spindle apparatus** (also known simply as the **mitotic spindle**) is responsible for orga-

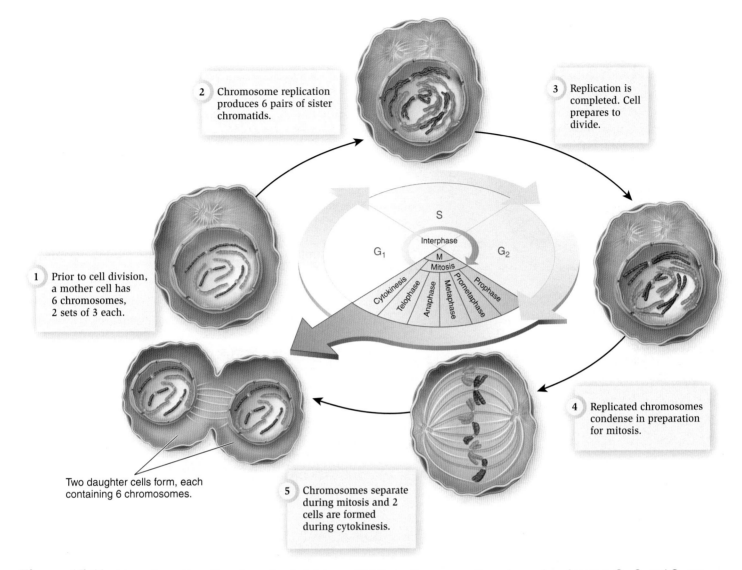

2 Chromosome replication produces 6 pairs of sister chromatids.

3 Replication is completed. Cell prepares to divide.

1 Prior to cell division, a mother cell has 6 chromosomes, 2 sets of 3 each.

4 Replicated chromosomes condense in preparation for mitosis.

Two daughter cells form, each containing 6 chromosomes.

5 Chromosomes separate during mitosis and 2 cells are formed during cytokinesis.

S

Interphase

G_1 G_2

M

Mitosis

Cytokinesis   Telophase   Anaphase   Metaphase   Prometaphase   Prophase

Figure 15.11   **The eukaryotic cell cycle and cell division.**   Dividing cells progress through a series of stages. G_1, S, and G_2 are collectively known as interphase, and M phase includes mitosis and cytokinesis. This diagram shows the progression of a cell through the cell cycle to produce two daughter cells. Note: The width of the phases shown in this figure is not meant to reflect their actual length. As discussed in Chapter 9, G_1 is typically the longest phase of the cell cycle, while M phase is relatively short.

nizing and sorting the chromosomes during mitosis. The structure of the mitotic spindle in animal cells is shown in **Figure 15.12**. The mitotic spindle is formed from two structures called the **centrosomes**. Because they are a site for organizing microtubules, centrosomes are also referred to as microtubule organizing centers (MTOCs). A single centrosome duplicates during interphase. After they separate from each other during mitosis, each centrosome defines a **pole** of the spindle apparatus, one within each of the future daughter cells. In animal cells, a conspicuous structure found in the centrosome is a pair of centrioles. However, centrioles are not found in many other eukaryotic species, such as plants, and are not required for spindle formation.

The spindle is formed from protein fibers called microtubules that are rooted in the centrosomes. Each centrosome organizes the construction of the microtubules by rapidly poly-

merizing tubulin proteins. The three types of spindle microtubules are termed astral, polar, and kinetochore microtubules. The astral microtubules, which extend away from the chromosomes, are important for positioning the spindle apparatus within the cell. The polar microtubules project into the region between the two poles. Polar microtubules that overlap with each other play a role in the separation of the two poles. Finally, the kinetochore microtubules are attached to kinetochores, which are bound to the centromere of each chromosome.

Now that we understand the structure of the mitotic spindle, we can examine the sequence of events that occurs during mitosis. **Figure 15.13** depicts the process of mitosis in an animal cell, though the process is quite similar in a plant cell. In the simplified diagrams shown along the bottom of this figure, the original mother cell contains six chromosomes, as in Figure 15.11.

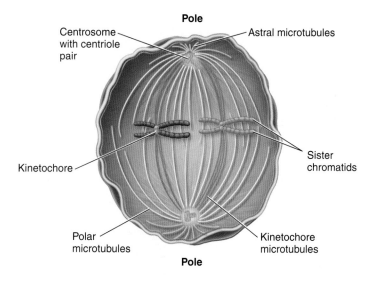

Figure 15.12   **The structure of the mitotic spindle.**   The mitotic spindle is formed by the centrosomes from three types of microtubules. The astral microtubules emanate away from the region between the poles. The polar microtubules project into the region between the two poles. The kinetochore microtubules are attached to the kinetochores of sister chromatids.

One set of chromosomes is again depicted in red, while the homologous set is blue; remember that these represent maternal and paternal chromosomes. Mitosis is subdivided into phases known as prophase, prometaphase, metaphase, anaphase, and telophase. Prior to mitosis, the cells are in **interphase**, a phase of the cell cycle during which the chromosomes are decondensed and found in the nucleus (Figure 15.13a). At the start of mitosis, in **prophase**, the chromosomes have already replicated to

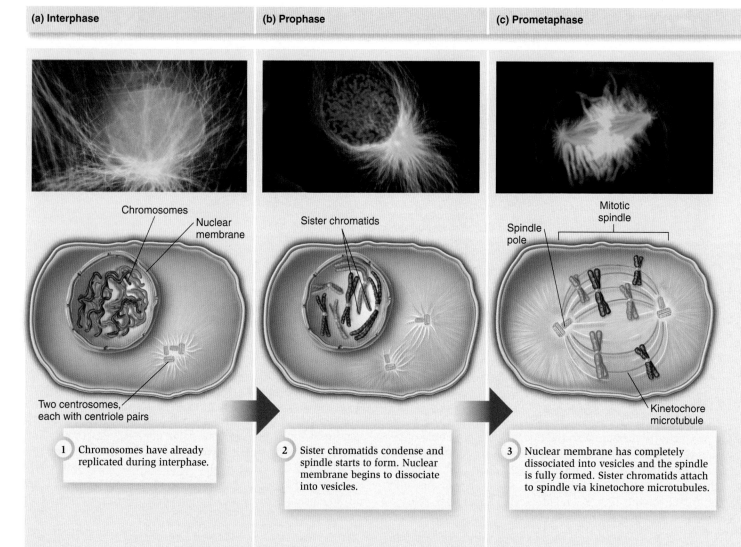

Figure 15.13   **The process of mitosis in an animal cell.**   The *top* panels illustrate the cells of a newt progressing through mitosis. The *bottom* panels are schematic drawings that emphasize the sorting and separation of the chromosomes in which the original diploid cell had 6 chromosomes (3 in each set). At the start of mitosis, these have already replicated into 12 chromatids. The final result is 2 daughter cells each containing 6 chromosomes.

produce 12 chromatids, joined as six pairs of sister chromatids (Figure 15.13b). As prophase proceeds, the nuclear membrane begins to dissociate into small vesicles. At the same time, the chromatids condense into highly compacted structures that are readily visible by light microscopy.

The mitotic spindle is completely formed during **prometaphase** (Figure 15.13c). As mitosis progresses, the centrosomes move apart and demarcate the two poles. Once the nuclear membrane has dissociated, the spindle fibers can interact with the sister chromatids. Initially, kinetochore microtubules are rapidly formed and can be seen under a microscope growing out from the two poles. As it grows, if a kinetochore microtubule happens to make contact with a kinetochore, it is said to be "captured" and remains firmly attached to the kinetochore. This seemingly random process is how sister chromatids become attached to kinetochore microtubules. Alternatively, if a kinetochore microtubule does not collide with a kinetochore, the microtubule will eventually depolymerize and retract to the centrosome. As the end of prometaphase nears, the two kinetochores on each pair of sister chromatids are attached to kinetochore microtubules from opposite poles. As these events are occurring, the sister chromatids are seen under the microscope to undergo jerky movements as they are tugged, back and forth, between the two poles by the kinetochore microtubules.

Eventually, the pairs of sister chromatids are aligned along a plane halfway between the poles called the **metaphase plate**. The pairs of sister chromatids have become organized into a single row within this plane. When this alignment is complete, the cell is in **metaphase** of mitosis (Figure 15.13d). The chromatids can then be equally distributed into two daughter cells.

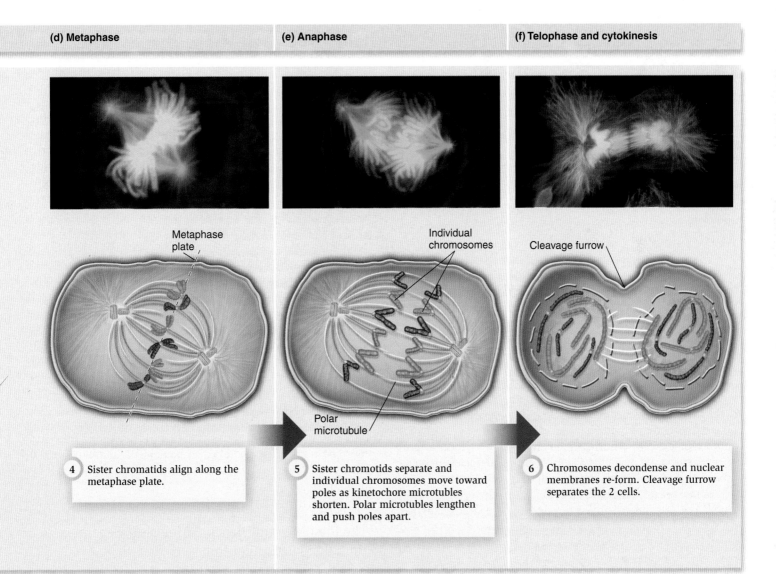

(d) Metaphase

Metaphase plate

4 Sister chromatids align along the metaphase plate.

(e) Anaphase

Individual chromosomes

Polar microtubule

5 Sister chromotids separate and individual chromosomes move toward poles as kinetochore microtubles shorten. Polar microtubules lengthen and push poles apart.

(f) Telophase and cytokinesis

Cleavage furrow

6 Chromosomes decondense and nuclear membranes re-form. Cleavage furrow separates the 2 cells.

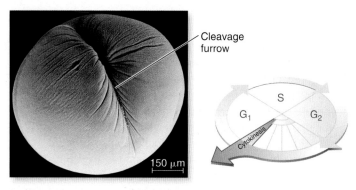

(a) Cleavage of an animal cell

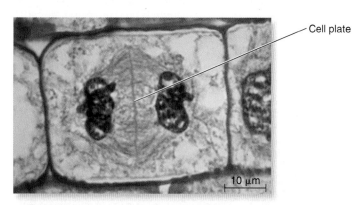

(b) Formation of a cell plate in a plant cell

Figure 15.14   Micrographs showing cytokinesis in animal and plant cells.

The next step in the sorting process occurs during **anaphase** (Figure 15.13e). At this stage, the connections between the pairs of sister chromatids are broken. Each chromatid, now an individual chromosome, is linked to only one of the two poles by one or more kinetochore microtubules. As anaphase proceeds, the kinetochore microtubules shorten, pulling the chromosomes toward the pole to which they are attached. In addition, the two poles move farther away from each other. This occurs because the overlapping polar microtubules lengthen and push against each other, thereby pushing the poles farther apart. During **telophase**, the chromosomes have reached their respective poles and decondense. The nuclear membranes now re-form to produce two separate nuclei. In Figure 15.13f, two nuclei are being produced that contain six chromosomes each.

In most cases, mitosis is quickly followed by cytokinesis, in which the two nuclei are segregated into separate daughter cells. While the stages of mitosis are similar between plant and animal cells, the process of cytokinesis is quite different. In animal cells, cytokinesis involves the formation of a **cleavage furrow**, which constricts like a drawstring to separate the cells (**Figure 15.14a**). In plants, the two daughter cells are separated by the formation of a **cell plate** (**Figure 15.14b**), which forms a cell wall between the two daughter cells.

Mitosis and cytokinesis ultimately produce two daughter cells having the same number of chromosomes as the mother cell. Barring rare mutations, the two daughter cells are genetically identical to each other and to the mother cell from which they were derived. Thus, the critical consequence of this sorting process is to ensure genetic consistency from one cell to the next. The development of multicellularity relies on the repeated process of mitosis and cytokinesis. For diploid organisms that are multicellular, most of the somatic cells are diploid and genetically identical to each other. Next, we will consider how diploid cells can divide to produce haploid cells.

15.3 Meiosis and Sexual Reproduction

We now turn our attention to sexual reproduction. As discussed earlier, a diploid cell contains two homologous sets of chromosomes, while a haploid cell contains a single set. For example, a diploid human cell contains 46 chromosomes, but a human gamete (sperm or egg cell) is a haploid cell that contains only 23 chromosomes, one from each of the 23 pairs. **Sexual reproduction** requires a **fertilization** event in which two haploid gametes unite to create a diploid cell called a **zygote**. In the case of many multicellular species, the zygote then grows and divides by mitosis into a multicellular organism with many diploid cells.

Meiosis is the process by which haploid cells are produced from a cell that was originally diploid. The term *meiosis*, which means "to make smaller," refers to the fewer chromosomes found in cells following this process. For this to occur, the chromosomes must be correctly sorted and distributed in a way that reduces the chromosome number to half its original diploid value. In the case of human gametes, for example, each gamete must receive half the total number of chromosomes, but not just any 23 chromosomes will do. A gamete must receive one chromosome from each of the 23 pairs. For this to happen, two rounds of divisions are necessary, termed meiosis I and meiosis II (**Figure 15.15**). When a cell begins meiosis, it contains chromosomes that are found in homologous pairs. When meiosis is completed, a single diploid cell with homologous pairs of chromosomes has produced four haploid cells. In this section, we will examine the cellular events of meiosis that reduce the chromosome number from diploid to haploid. In addition, we will briefly consider how this process plays a role in the life cycles of fungi, plants, and animals.

The First Meiotic Division, Meiosis I, Separates Homologous Chromosomes

Like mitosis, meiosis begins after a cell has progressed through the G_1, S, and G_2 phases of the cell cycle. However, two key events occur at the beginning of meiosis that do not occur in mitosis. First, homologous pairs of sister chromatids associate

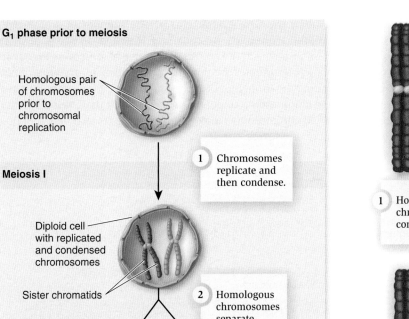

G₁ phase prior to meiosis

Homologous pair of chromosomes prior to chromosomal replication

Meiosis I

1 Chromosomes replicate and then condense.

Diploid cell with replicated and condensed chromosomes

Sister chromatids

2 Homologous chromosomes separate.

Haploid cells with pairs of sister chromatids

Meiosis II

3 Sister chromatids separate.

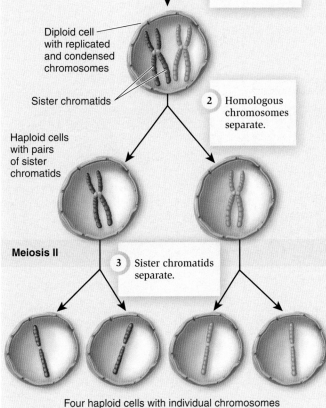

Four haploid cells with individual chromosomes

Figure 15.15 **How the process of meiosis reduces chromosome number.** This simplified diagram emphasizes the reduction in chromosome number as a diploid cell divides by meiosis to produce 4 haploid cells.

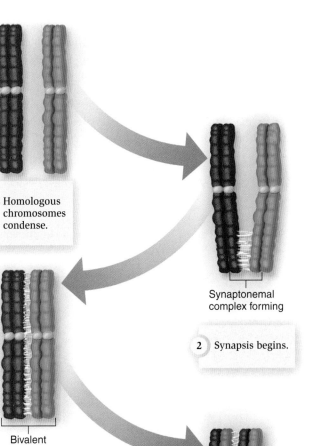

1 Homologous chromosomes condense.

Synaptonemal complex forming

2 Synapsis begins.

Bivalent

3 Bivalents form.

Chiasma

4 Crossing over occurs.

5 Chiasma becomes visible as chromosome arms separate during late prophase.

Figure 15.16 **Formation of a bivalent and crossing over during meiosis I.** At the beginning of meiosis, homologous chromosomes pair with each other to form a bivalent, usually with a synaptonemal complex between them. Crossing over then occurs between homologous chromatids within the bivalent. During this process, homologues exchange segments of chromosomes.

with each other, lying side by side to form a **bivalent**, also called a **tetrad** (**Figure 15.16**). The process of forming a bivalent is termed **synapsis**. In most eukaryotic species, a synaptonemal complex is formed between homologous chromosomes. However, the synaptonemal complex is not required for the pairing of homologous chromosomes because some species of fungi completely lack such a complex, yet their chromosomes associate with each other correctly. At present, the precise role of the synaptonemal complex is not clearly understood.

The second event that occurs at the beginning of meiosis, but not usually during mitosis, is **crossing over**, which involves a physical exchange between chromosome pieces of the bivalent (Figure 15.16). As discussed in Chapter 17, crossing over may increase the genetic variation of a species. After crossing over occurs, the arms of the chromosomes tend to separate but remain adhered at a crossover site. This connection is called a **chiasma** (plural, chiasmata), because it physically resembles the Greek letter chi, χ. The number of crossovers is carefully controlled by cells and depends on the size of the chromosome and the species. The range of crossovers for eukaryotic chromosomes is typically one or two to a couple dozen. During the formation of sperm in humans, for example, an average chromosome undergoes slightly more than two crossovers, while chromosomes in certain plant species may undergo 20 or more crossovers.

Figure 15.17   The stages of meiosis in an animal cell.

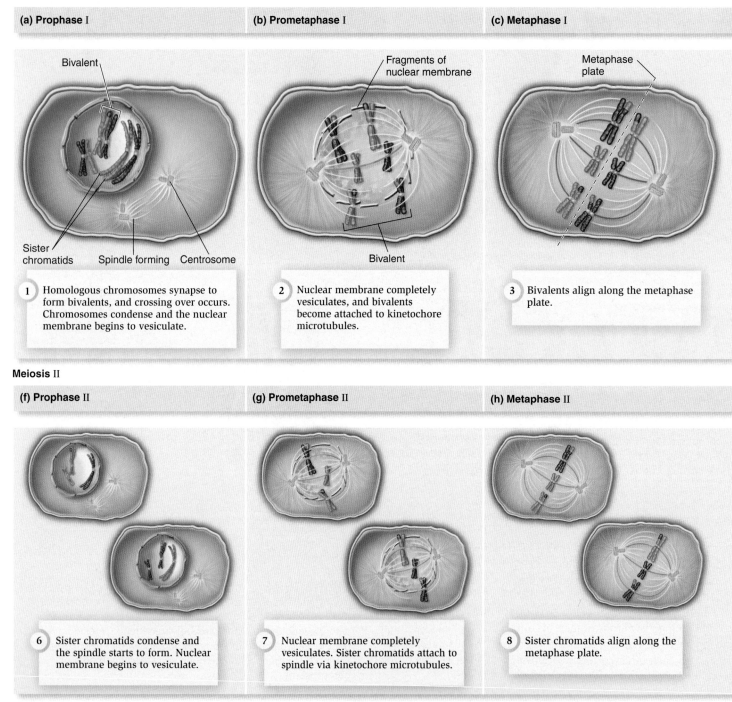

Meiosis I

(a) Prophase I

Bivalent

Sister chromatids   Spindle forming   Centrosome

1   Homologous chromosomes synapse to form bivalents, and crossing over occurs. Chromosomes condense and the nuclear membrane begins to vesiculate.

(b) Prometaphase I

Fragments of nuclear membrane

Bivalent

2   Nuclear membrane completely vesiculates, and bivalents become attached to kinetochore microtubules.

(c) Metaphase I

Metaphase plate

3   Bivalents align along the metaphase plate.

Meiosis II

(f) Prophase II

6   Sister chromatids condense and the spindle starts to form. Nuclear membrane begins to vesiculate.

(g) Prometaphase II

7   Nuclear membrane completely vesiculates. Sister chromatids attach to spindle via kinetochore microtubules.

(h) Metaphase II

8   Sister chromatids align along the metaphase plate.

Now that we have an understanding of bivalent formation and crossing over, we are ready to consider the phases of meiosis (**Figure 15.17**). These simplified diagrams depict a diploid cell ($2n$) that contains a total of six chromosomes (as in our look at mitosis in Figure 15.13). Prior to meiosis, the chromosomes are replicated in S phase to produce pairs of sister chromatids. This single replication event is then followed by the two sequential cell divisions called meiosis I and II. Like mitosis, each of these is subdivided into prophase, prometaphase, metaphase, anaphase, and telophase.

The sorting that occurs during **meiosis I** separates homologues from each other (Figure 15.17). In prophase I, the replicated chromosomes condense and the homologous chromosomes form bivalents as the nuclear membrane starts to vesiculate (fragment into small vesicles). In prometaphase I, the spindle apparatus is complete, and the chromatids are attached to kinetochore microtubules. At metaphase I, the bivalents are organized along the metaphase plate. However, notice how this pattern of alignment is strikingly different from that observed during mitosis (see Figure 15.13d). In particular, the sister chromatids are aligned in a double row rather than a single row (as in mitosis). Furthermore, the arrangement of sister chromatids within this double row is random with regard to the (blue and red) homologues. In Figure 15.17c, one of the red homologues is to the left of the metaphase plate and the other two are to the right, while two of the blue homologues are to the left of the metaphase plate and other one is to the right.

In other cells, homologues could be arranged differently along the metaphase plate (for example, three blues to the left and none to the right, or none to the left and three to the right). Because eukaryotic species typically have many chromosomes per set, homologues can be randomly aligned along the metaphase plate in a variety of ways. For example, consider that humans have 23 chromosomes per set. The possible number of different, random alignments equals 2^n, where n equals the number of chromosomes per set. Thus, in humans, this equals 2^{23}, or over 8 million possibilities. Because the homologues are genetically similar but not identical, we see from this calculation that the random alignment of homologous chromosomes provides a mechanism to promote a vast amount of genetic diversity among the resulting haploid cells. When meiosis is complete, it is very unlikely that any two human gametes will have the same combination of homologous chromosomes.

The segregation of homologues occurs during anaphase I (Figure 15.17d). The connections between bivalents break, but not the connections that hold sister chromatids together. Each joined pair of chromatids migrates to one pole, and the homologous pair of chromatids moves to the opposite pole, both pulled by kinetochore microtubules. Finally, at telophase I, the sister chromatids have reached their

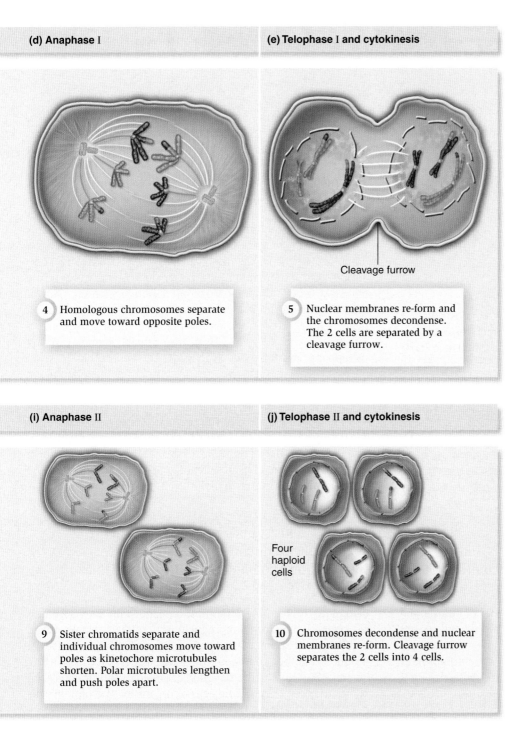

(d) Anaphase I

4 Homologous chromosomes separate and move toward opposite poles.

(e) Telophase I and cytokinesis

Cleavage furrow

5 Nuclear membranes re-form and the chromosomes decondense. The 2 cells are separated by a cleavage furrow.

(i) Anaphase II

9 Sister chromatids separate and individual chromosomes move toward poles as kinetochore microtubules shorten. Polar microtubules lengthen and push poles apart.

(j) Telophase II and cytokinesis

Four haploid cells

10 Chromosomes decondense and nuclear membranes re-form. Cleavage furrow separates the 2 cells into 4 cells.

Table 15.1	A Comparison of Mitosis, Meiosis I, and Meiosis II		
Event	**Mitosis**	**Meiosis I**	**Meiosis II**
Synapsis during prophase:	No	Yes, bivalents are formed	No
Crossing over during prophase:	Rarely	Commonly	Rarely
Attachment to poles at prometaphase:	A pair of sister chromatids is attached to both poles.	A pair of sister chromatids is attached to just one pole.	A pair of sister chromatids is attached to both poles.
Alignment along the metaphase plate:	Sister chromatids	Bivalents	Sister chromatids
Type of separation at anaphase:	Sister chromatids separate. A single chromatid, now called a chromosome, moves to each pole.	Bivalents separate. A pair of sister chromatids moves to each pole.	Sister chromatids separate. A single chromatid, now called a chromosome, moves to each pole.

respective poles, and they then decondense. The nuclear membranes now re-form to produce two separate nuclei. If we consider the end result of meiosis I, we see that two cells are produced, each with three pairs of sister chromatids; it is a reduction division. The original diploid cell had its chromosomes in homologous pairs, while the two cells produced at the end of meiosis I are considered haploid—they do not have pairs of homologous chromosomes.

The Second Meiotic Division, Meiosis II, Separates Sister Chromatids

Meiosis I is followed by cytokinesis and then **meiosis II** (see Figure 15.17). An S phase does not occur between meiosis I and meiosis II. The sorting events of meiosis II are similar to those of mitosis, but the starting point is different. For a diploid cell with six chromosomes, mitosis begins with 12 chromatids that are joined as six pairs of sister chromatids (see Figure 15.13). By comparison, the two cells that begin meiosis II each have six chromatids that are joined as three pairs of sister chromatids. Otherwise, the steps that occur during prophase, prometaphase, metaphase, anaphase, and telophase of meiosis II are analogous to a mitotic division. Sister chromatids are separated during anaphase II, unlike anaphase I in which bivalents are separated.

Changes in a Few Key Steps in Meiosis and Mitosis Account for the Different Outcomes of These Two Processes

If we compare the outcome of meiosis with that of mitosis, the results are quite different. Mitosis produces two diploid daughter cells that are genetically identical. In our previous example shown in Figure 15.13, the starting cell had six chromosomes (three homologous pairs of chromosomes) and both daughter cells had copies of the same six chromosomes. By comparison, meiosis reduces the number of sets of chromosomes. In the

example shown in Figure 15.17, the starting cell also had six chromosomes while the four daughter cells had only three chromosomes. But the daughter cells did not contain a random mix of three chromosomes. Each haploid daughter cell contained one complete set of chromosomes, while the original diploid mother cell had two complete sets.

Table 15.1 emphasizes the differences between certain key steps in meiosis and mitosis that account for the different outcomes of these two processes. During prophase of meiosis I, the homologues synapse to form bivalents. This explains why crossing over occurs commonly during meiosis, but rarely during mitosis. During prometaphase of mitosis and meiosis II, pairs of sister chromatids are attached to both poles. In contrast, during meiosis I, each pair of sister chromatids (within a bivalent) is attached to a single pole. This affects their alignment during metaphase. Bivalents align along the metaphase plate during metaphase of meiosis I, whereas sister chromatids align along the metaphase plate during metaphase of mitosis and meiosis II. At anaphase of meiosis I, the homologues separate, while the sister chromatids remain together. In contrast, sister chromatid separation occurs during anaphase of mitosis and meiosis II. Taken together, the steps of meiosis result in a process in which two sequential cell divisions create four haploid cells, while the steps of mitosis create two diploid cells.

Mitosis, Meiosis, and Fertilization Allow Sexually Reproducing Species to Produce Haploid and Diploid Cells at Different Times in Their Life Cycles

Let's now turn our attention to the relationship between mitosis, meiosis, and sexual reproduction in animals, plants, fungi, and protists. For any given species, the sequence of events that produces another generation of organisms is known as a **life cycle**. For sexually reproducing organisms, this involves an alternation between haploid cells or organisms and diploid cells or organisms (**Figure 15.18**).

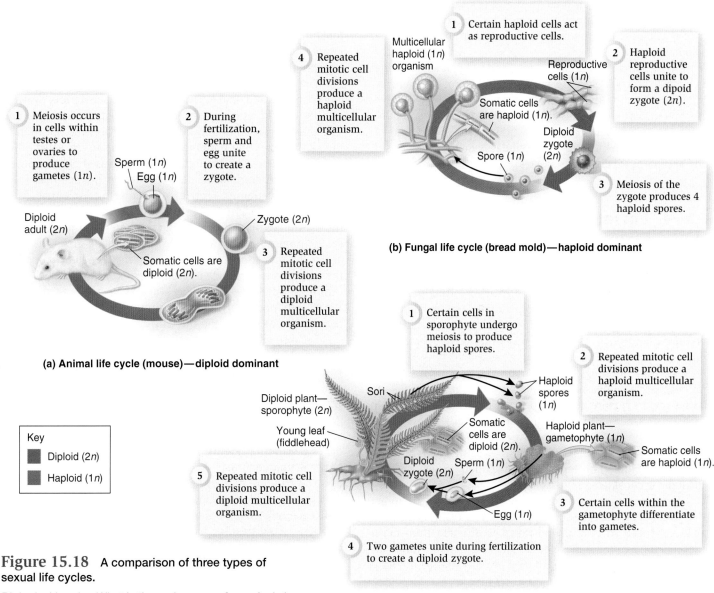

1 Meiosis occurs in cells within testes or ovaries to produce gametes (1*n*).

2 During fertilization, sperm and egg unite to create a zygote.

Sperm (1*n*)
Egg (1*n*)

Diploid adult (2*n*)

Zygote (2*n*)

Somatic cells are diploid (2*n*).

3 Repeated mitotic cell divisions produce a diploid multicellular organism.

(a) Animal life cycle (mouse)—diploid dominant

1 Certain haploid cells act as reproductive cells.

Multicellular haploid (1*n*) organism

Reproductive cells (1*n*)

2 Haploid reproductive cells unite to form a diploid zygote (2*n*).

Somatic cells are haploid (1*n*).

Diploid zygote (2*n*)

Spore (1*n*)

4 Repeated mitotic cell divisions produce a haploid multicellular organism.

3 Meiosis of the zygote produces 4 haploid spores.

(b) Fungal life cycle (bread mold)—haploid dominant

Key

■ Diploid (2*n*)

■ Haploid (1*n*)

1 Certain cells in sporophyte undergo meiosis to produce haploid spores.

Diploid plant— sporophyte (2*n*)

Sori

Young leaf (fiddlehead)

Haploid spores (1*n*)

2 Repeated mitotic cell divisions produce a haploid multicellular organism.

Somatic cells are diploid (2*n*).

Haploid plant— gametophyte (1*n*)

Somatic cells are haploid (1*n*).

Diploid zygote (2*n*)

Sperm (1*n*)

3 Certain cells within the gametophyte differentiate into gametes.

5 Repeated mitotic cell divisions produce a diploid multicellular organism.

Egg (1*n*)

4 Two gametes unite during fertilization to create a diploid zygote.

(c) Plant life cycle (fern)—alternation of generations

Figure 15.18 A comparison of three types of sexual life cycles.

Biological inquiry: What is the main reason for meiosis in animals? What is the reason for mitosis in animals?

Most species of animals are diploid, and their haploid gametes are considered to be a specialized type of cell. For this reason, animals are viewed as **diploid-dominant species** (Figure 15.18a). By comparison, many fungi and some protists are just the opposite, they are **haploid-dominant species** (Figure 15.18b). In fungi, such as bread mold, the multicellular organism is haploid (1*n*). Haploid fungal cells are most commonly produced by mitosis. During sexual reproduction, haploid cells unite to create a diploid zygote, which then immediately proceeds through meiosis to create four haploid cells called spores. Each spore goes through mitotic cellular divisions to create a multicellular haploid organism.

Plants and some algae have life cycles that are intermediate between the extreme cases of diploid or haploid dominance. Such species exhibit an **alternation of generations** (Figure 15.18c). The species alternate between multicellular diploid organisms, called **sporophytes**, and multicellular haploid organisms, called **gametophytes**. Meiosis in certain cells within the sporophyte produces haploid spores, which divide by mitosis to produce the gametophyte. Particular cells within the gametophyte differentiate into haploid gametes. Fertilization occurs between two gametes, producing a diploid zygote that then undergoes repeated mitotic cell divisions to create a sporophyte.

Among different plant species, the relative sizes of the haploid and diploid organisms vary greatly. In bryophytes, such as mosses, the haploid gametophyte is a visible multicellular organism, while the diploid sporophyte is tiny and survives within the haploid organism. In other plants, such as ferns, as shown in Figure 15.18c, both the diploid sporophyte and haploid gametophyte can grow independently. The sporophyte is considerably larger and is the organism that we commonly think of as a fern.

In seed-bearing plants, such as roses and oak trees, the diploid sporophyte is the large multicellular plant, while the gametophyte is composed of only a few cells and is formed within the sporophyte.

When comparing animals, plants, and fungi, it's interesting to consider how gametes are made. Animals produce gametes by meiosis. In contrast, plants and fungi produce reproductive cells by mitosis. The gametophyte of plants is a haploid multicellular organism that is created by mitotic cellular divisions of a haploid spore. Within the multicellular gametophyte, certain cells become specialized as gametes.

15.4 Variation in Chromosome Structure and Number

In the three previous sections of this chapter, we have examined two important features of chromosomes. First, we considered their structure, and second, we explored two sorting processes that determine the chromosome number following cell division. In this section, we will examine how the structures and numbers of chromosomes can vary between different species and within the same species.

The study of chromosomal variation is important for several reasons. Geneticists have discovered that variations in chromosome structure and number can have major effects on the characteristics of an organism. For example, we now know that several human genetic diseases are caused by such changes. In addition, changes in chromosome structure and number have been an important force in the evolution of new species. This topic is considered in Chapter 25.

Chromosome variation can be viewed in two ways. On relatively rare occasions, the structure or number of chromosomes can be changed so that an individual is different from most other members of the same species. This is generally viewed as an abnormality. Alternatively, some types of variation are normal. For example, the structure and number of chromosomes among different species tend to show wide variation. In this section, we will examine both abnormal and normal types of variation. Let's begin with natural (normal) variation.

Natural Variation Exists in Chromosome Structure and Number

Before we begin to examine chromosome variation, we need to have a reference point for a normal set of chromosomes. To determine what the normal chromosomes of a species look like, a cytogeneticist microscopically examines the chromosomes from several members of the species. Chromosome composition within a given species tends to remain relatively constant. In most cases, normal individuals of the same species will have the same number and types of chromosomes.

For example, as mentioned previously, the normal chromosome composition of human cells is two sets of 23 chromosomes for a total of 46. Other diploid species may have different numbers of chromosomes. The dog has 78 chromosomes (39 per set), the fruit fly has 8 chromosomes (4 per set), and the tomato has 24 chromosomes (12 per set). When comparing distantly related species, such as humans and fruit flies, major differences in chromosomal composition are observed.

The chromosomes of a given species can also vary considerably in size and shape. Cytogeneticists have various ways to classify and identify chromosomes in their metaphase form. The three most commonly used features are size, location of the centromere, and banding patterns that are revealed when the chromosomes are treated with stains. Based on centromere location, each chromosome is classified as **metacentric** (near the middle), **submetacentric** (off center), **acrocentric** (near one end), or **telocentric** (at the end) (**Figure 15.19a**). Because the centromere is not exactly in the center of a chromosome, each chromosome has a short arm and a long arm. The short arm is designated with the letter *p* (for the French *petite*), while the long arm is designated with the letter *q*. In the case of telocentric chromosomes, the short arm may be nearly nonexistent. When preparing a karyotype, the chromosomes are aligned with the short arms on top and the long arms on the bottom.

Because different chromosomes often have similar sizes and centromeric locations, cytogeneticists must use additional methods to accurately identify each type of chromosome within a karyotype. For detailed identification, chromosomes are treated with stains to produce characteristic banding patterns. Cytogeneticists use several different staining procedures to identify specific chromosomes. An example is Giemsa stain, which produces **G banding** (**Figure 15.19b**). The alternating pattern of G bands is unique for each type of chromosome.

The banding pattern of eukaryotic chromosomes is useful in two ways. First, individual chromosomes can be distinguished from each other, even if they have similar sizes and centromeric locations. Also, banding patterns are used to detect changes in chromosome structure that occur as a result of mutation.

Mutations Can Alter Chromosome Structure

Let's now consider how the structures of normal chromosomes can be modified by mutations. These chromosomal mutations are categorized as deficiencies, duplications, inversions, and translocations (**Figure 15.20**).

Deficiencies and duplications are changes in the total amount of genetic material in a single chromosome. When a **deficiency** occurs, a segment of chromosomal material is missing. In other words, the affected chromosome is deficient in a significant amount of genetic material. The term **deletion** is also used to describe a missing region of a chromosome. In contrast, a **duplication** occurs when a section of a chromosome occurs two or more times in a row.

The consequences of a deficiency or duplication depend on their size and whether they include genes or portions of genes that are vital to the development of the organism. When deficiencies or duplications have an effect, they are usually detrimental. Larger changes in the amount of genetic material

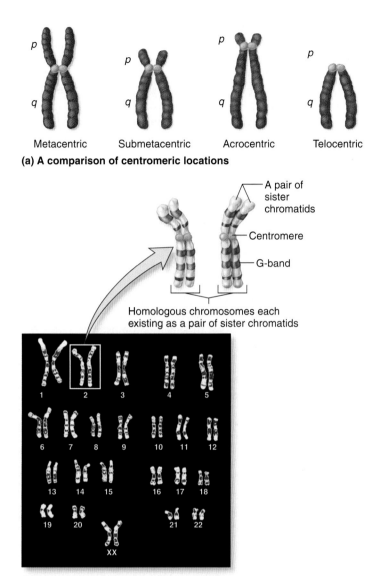

(a) A comparison of centromeric locations

Metacentric   Submetacentric   Acrocentric   Telocentric

A pair of sister chromatids

Centromere

G-band

Homologous chromosomes each existing as a pair of sister chromatids

(b) Giemsa staining of human chromosomes

Figure 15.19   Features of normal metaphase chromosomes. **(a)** A comparison of centromeric locations. **(b)** Illustrations of human chromosomes that have been stained with Giemsa to show G bands.

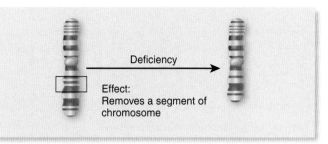

(a) Deficiency (deletion)

Deficiency

Effect: Removes a segment of chromosome

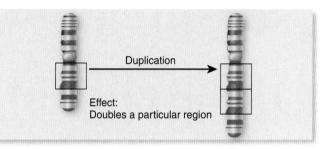

(b) Duplication

Duplication

Effect: Doubles a particular region

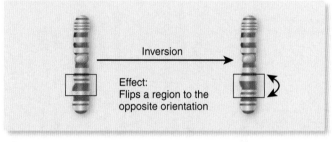

(c) Inversion

Inversion

Effect: Flips a region to the opposite orientation

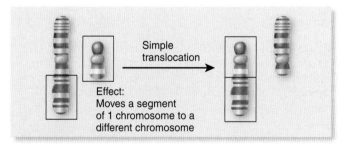

(d) Simple translocation

Simple translocation

Effect: Moves a segment of 1 chromosome to a different chromosome

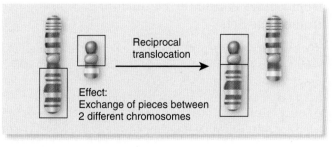

(e) Reciprocal translocation

Reciprocal translocation

Effect: Exchange of pieces between 2 different chromosomes

Figure 15.20   Types of changes in chromosome structure.

tend to be more harmful because more genes are missing or duplicated.

Inversions and translocations are chromosomal rearrangements. An **inversion** is a change in the direction of the genetic material along a single chromosome. When a segment of one chromosome has been inverted, the order of G bands is opposite to that of a normal chromosome (see Figure 15.20). A **translocation** occurs when one segment of a chromosome becomes attached to a different chromosome. In a **simple translocation**, a single piece of chromosome is attached to another chromosome. In a **reciprocal translocation**, two different types of chromosomes exchange pieces, thereby producing two abnormal chromosomes carrying translocations.

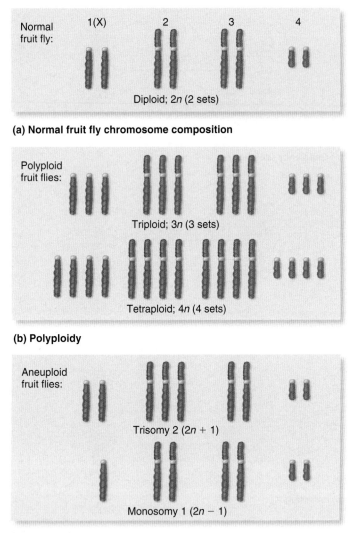

Normal
fruit fly:

1(X)   2   3   4

Diploid; 2*n* (2 sets)

(a) Normal fruit fly chromosome composition

Polyploid
fruit flies:

Triploid; 3*n* (3 sets)

Tetraploid; 4*n* (4 sets)

(b) Polyploidy

Aneuploid
fruit flies:

Trisomy 2 (2*n* + 1)

Monosomy 1 (2*n* − 1)

(c) Aneuploidy

Figure 15.21 Types of variation in chromosome number.
(a) The normal diploid number of chromosomes in *Drosophila*.
The X chromosome is also called chromosome 1. Examples of
chromosomes of **(b)** polyploid flies and **(c)** aneuploid flies.

Variation Occurs in the Number of Individual Chromosomes and in the Number of Chromosome Sets

Let's now turn our attention to changes in chromosome number. Variations in chromosome number can be categorized in two ways: variation in the number of sets of chromosomes, and variation in the number of particular chromosomes within a set. The suffix "ploid" or "ploidy" refers to a complete set of chromosomes. Organisms that are **euploid** (the prefix "eu-" means true) have a chromosome number that is viewed as the normal number. For example, in a species that is diploid, a euploid organism would have two sets of chromosomes in its somatic cells. In *Drosophila melanogaster*, for example, a normal indi-

vidual has eight chromosomes. The species is diploid, having two sets of four chromosomes each (**Figure 15.21a**). Organisms can vary with regard to the number of sets of chromosomes they have. For example, on rare occasions, an abnormal fruit fly can be produced with 12 chromosomes, containing three sets of 4 chromosomes each (**Figure 15.21b**). Organisms with three or more sets of chromosomes are called **polyploid**. A diploid organism is referred to as 2*n*, a **triploid** organism as 3*n*, a **tetraploid** organism as 4*n*, and so forth.

The second way that chromosome number can vary is a phenomenon called **aneuploidy**. This refers to an alteration in the number of particular chromosomes, so that the total number of chromosomes is not an exact multiple of a set. For example, an abnormal fruit fly could contain nine chromosomes instead of eight because it had three copies of chromosome 2 instead of the normal two copies (**Figure 15.21c**). Such an animal is said to have trisomy 2 or to be **trisomic**. Instead of being perfectly diploid, a trisomic animal is 2*n* + 1. By comparison, a fruit fly could be lacking a single chromosome, such as chromosome 1, and contain a total of seven chromosomes (2*n* − 1). This animal is **monosomic** and would be described as having monosomy 1.

Nondisjunction and Interspecies Breeding Are Common Mechanisms That Change Chromosome Number

Variations in chromosome number are fairly widespread and have a significant impact on the characteristics of plants and animals. For these reasons, researchers have wanted to understand the mechanisms that cause these variations. In some cases, a change in chromosome number is the result of the abnormal sorting of chromosomes during cell division. The term **nondisjunction** refers to an event in which the chromosomes do not sort properly during cell division. **Meiotic nondisjunction** can occur during meiosis I or meiosis II and produce haploid cells that have too many or too few chromosomes (**Figure 15.22**). If such a gamete fuses with a normal gamete during fertilization, the resulting organism will be aneuploid and will have an abnormal chromosomal composition in all of its cells.

Another mechanism that leads to variation in chromosome number is interspecies breeding. An **alloploid** organism contains a least one set of chromosomes from two or more different species (**Figure 15.23a**). This term refers to the occurrence of chromosome sets (*ploidy*) from the genomes of different (*allo-*) species. An alloploid that has only one set of chromosomes from two different species is an **allodiploid**. Species that are close evolutionary relatives are most likely to breed and produce allodiploid offspring. For example, closely related species of grasses may interbreed to produce allodiploids. When an organism contains two or more complete sets of chromosomes from two or more different species, such an organism is called an **allopolyploid**. An **allotetraploid** is a type of allopolyploid that contains two complete sets of chromosomes from two species for a total of four sets (**Figure 15.23b**).

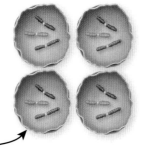

Diploid cell in metaphase

Normal meiosis

Nondisjunction in meiosis I (Homologues fail to separate)

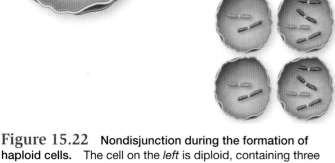

Figure 15.22 Nondisjunction during the formation of haploid cells. The cell on the *left* is diploid, containing three pairs of chromosomes. During the formation of haploid cells, one copy of each chromosome should be sorted into each cell. Nondisjunction, however, can produce a cell with too many or too few chromosomes.

Biological inquiry: Could nondisjunction produce a polyploid individual?

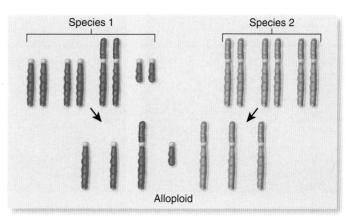

(a) Alloploidy (allodiploid)

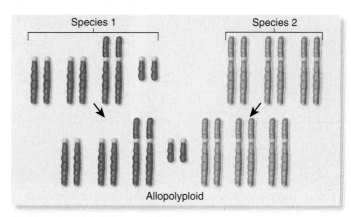

(b) Allopolyploidy (allotetraploid)

Figure 15.23 A comparison of alloploidy and allopolyploidy.

Changes in Chromosome Number Have Important Consequences

Let's now consider how changes in chromosome number may impact the characteristics of animals and plants. In many cases, animals do not tolerate deviations from diploidy well. For example, polyploidy in mammals is generally a lethal condition. However, a few cases of naturally occurring variations from diploidy do occur in animals. Male bees, which are called drones, contain a single set of chromosomes. They are produced from unfertilized eggs. By comparison, female bees are diploid. A few examples of vertebrate polyploid animals have been discovered. Interestingly, on rare occasions, animals that are morphologically very similar to each other can be found as a diploid species as well as a separate polyploid species. This situation occurs among certain amphibians and reptiles. **Figure 15.24** shows photographs of a diploid and a tetraploid frog. As you can see, they look very similar to each other. Their differences in chromosome number can be revealed only by a microscopic examination of the chromosomes in the somatic cells of the animals.

In contrast to animals, plants commonly exhibit polyploidy. Among ferns and flowering plants, about 30–35% of species are

polyploid. Polyploidy is also important in agriculture. In many instances, polyploid strains of plants display characteristics that are helpful to humans. They are often larger in size and more robust. These traits are clearly advantageous in the production of food. Many of the fruits and grains we eat are produced from polyploid plants. For example, the species of wheat that we use to make bread, *Triticum aestivum*, is an allohexaploid (containing six sets of chromosomes) that arose from the union of diploid genomes from three closely related species (**Figure 15.25a**). During the course of its cultivation, two diploid species must have interbred to create an allotetraploid, and then a third species interbred with the allotetraploid to create an allohexaploid. Plant polyploids tend to exhibit a greater adaptability, which allows them to withstand harsher environmental conditions. Polyploid ornamental plants commonly produce larger flowers than their diploid counterparts (**Figure 15.25b**).

Although polyploidy is often beneficial in plants, aneuploidy in all eukaryotic species usually has detrimental consequences on the characteristics of an organism. To understand why, consider the relationship between gene expression and chromosome number. For many, but not all, genes, the level of gene expression is correlated with the number of genes per cell.

(a) *Hyla chrysorelis*   **(b)** *Hyla versicolor*

Figure 15.24 **Differences in chromosome number in two closely related frog species.** The frog in **(a)** is diploid, whereas the frog in **(b)** is tetraploid. These frogs are in the act of performing their mating calls, which is why the skin under their mouths is protruding as a large bubble.

(a) Wheat, *Triticum aestivum*   **(b)** Diploid daylily (left) and tetraploid (right)

Figure 15.25 **Examples of polyploid plants.** **(a)** Cultivated wheat, *Triticum aestivum*, is an allohexaploid. It was derived from three different diploid species of grasses that originally were found in the Middle East and were cultivated by ancient farmers in this region. Modern varieties of wheat have been produced from this allohexaploid species. **(b)** Differences in euploidy may exist in two closely related daylily species. The flower stems on the *left* are diploid, whereas the stems with the larger flowers on the *right* are tetraploid.

Compared to a diploid cell, if a gene is carried on a chromosome that is present in three copies instead of two, approximately 150% of the normal amount of gene product will be made. Alternatively, if only one copy of that gene is present due to a missing chromosome, only 50% of the gene product is usually made. Therefore, in trisomic and monosomic individuals, an imbalance in the level of gene expression occurs. This imbalance interferes with the proper functioning of cells.

One important reason that geneticists are so interested in aneuploidy is its relationship to certain inherited disorders in humans. Even though most people are born with a normal number of chromosomes, alterations in chromosome number occur at a surprising frequency during gamete formation. About 5–10% of all fertilized human eggs result in an embryo with an abnormality in chromosome number. In most cases, these abnormal embryos do not develop properly and result in a spontaneous abortion very early in pregnancy. Approximately 50% of all spontaneous abortions are due to alterations in chromosome number.

In some cases, an abnormality in chromosome number produces an offspring that can survive. Several human disorders are the result of abnormalities in chromosome number. The most common are trisomies of chromosomes 21, 18, or 13, or abnormalities in the number of the sex chromosomes (**Table 15.2**). Most of the known trisomies involve chromosomes that are relatively small, so they carry fewer genes. Trisomies of the other human chromosomes and most monosomies are presumed to be lethal and have been found in spontaneously aborted embryos and fetuses.

Table 15.2	Aneuploid Conditions in Humans		
Condition	**Frequency (# of live births)**	**Syndrome**	**Characteristics**
Autosomal:			
Trisomy 21	1/800	Down	Mental retardation, abnormal pattern of palm creases, slanted eyes, flattened face, short stature
Trisomy 18	1/6,000	Edward	Mental and physical retardation, facial abnormalities, extreme muscle tone, early death
Trisomy 13	1/15,000	Patau	Mental and physical retardation, wide variety of defects in organs, large triangular nose, early death
Sex chromosomal:			
XXY	1/1,000 (males)	Klinefelter	Sexual immaturity (no sperm), breast swelling (males)
XYY	1/1,000 (males)	Jacobs	Tall
XXX	1/1,500 (females)	Triple X	Tall and thin, menstrual irregularity
XO	1/5,000 (females)	Turner	Short stature, webbed neck, sexually undeveloped

Human abnormalities in chromosome number are influenced by the age of the parents. Older parents are more likely to produce children with abnormalities in chromosome number, because meiotic nondisjunction is more likely to occur in older cells. **Down syndrome**, which was first described by the English physician John Langdon Down in 1866, provides an example. This disorder is caused by the inheritance of three copies of chromosome 21 (Table 15.2). The incidence of Down syndrome rises with the age of either parent. In males, however, the rise occurs relatively late in life, usually past the age when most men have children. By comparison, the likelihood of having a child with Down syndrome rises dramatically during the later reproductive ages of women.

CHAPTER SUMMARY

15.1 Molecular Structure of Eukaryotic Chromosomes

- Chromosomes are structures in living cells that carry the genetic material.
- Chromatin is the name given to the DNA-protein complex that makes up chromosomes.
- In eukaryotic chromosomes, the DNA is wrapped around histone proteins to form nucleosomes. (Figure 15.1)
- Noll tested the nucleosome model for chromatin structure by digesting chromosomes with DNase-I and showing that the DNA was obtained in multiples of 200 bp. (Figure 15.2)
- Nucleosomes are further compacted into 30-nm fibers because the linker regions are variably twisted and bent into a zigzag pattern. (Figure 15.3)
- A third level of compaction of eukaryotic chromosomes involves the formation of radial loop domains in which the bases of 30-nm fibers are anchored to a network of proteins called the nuclear matrix. This level of compaction is called euchromatin. In heterochromatin, the loops are even more closely packed together. (Figure 15.4)
- Within the nucleus of a nondividing cell, each chromosome occupies its own nonoverlapping chromosome territory. (Figure 15.5)
- Chromosome compaction to produce a metaphase chromosome also involves the conversion of all euchromatin into heterochromatin. (Figure 15.6)
- In a metaphase chromosome, the DNA loops are anchored to a protein scaffold that is formed from the nuclear matrix. (Figure 15.7)
- Gene expression is controlled by the level of compaction of the chromatin. The pattern of covalent modification of the amino terminal tails of histone proteins, also called the histone code, is one way to control the level of compaction. (Figure 15.8).

15.2 Mitotic Cell Division

- The process of mitosis involves the sorting of chromosomes to produce two nuclei with the same number and types of chromosomes.
- Cytogeneticists examine cells microscopically to determine their chromosome composition. A micrograph that shows the alignment of chromosomes from a given cell is called a karyotype. (Figure 15.9)
- Diploid cells have two sets of chromosomes while haploid cells have only one set.
- During S phase, eukaryotic chromosomes are replicated to produce a pair of identical sister chromatids that remain attached to each other. (Figure 15.10)

- The cell cycle is a series of stages needed for cell division. G_1, S, and G_2 are known as interphase. Mitosis and cytokinesis occur during M phase. (Figure 15.11)
- The mitotic spindle is composed of astral, kinetochore, and polar microtubules. The spindle organizes the process of cell division and plays a central role in chromosome sorting. (Figure 15.12)
- Mitosis occurs in five phases called prophase, prometaphase, metaphase, anaphase, and telophase. During prophase, the chromosomes condense and the nuclear membrane fragments. The spindle apparatus is completely formed by the end of prometaphase. During metaphase, the chromosomes are aligned in a single row along the metaphase plate. At anaphase, the sister chromatids separate from each other and move to opposite poles; the poles themselves also move farther apart. During telophase, the chromosomes decondense and the nuclear membranes re-form. (Figure 15.13)
- Cytokinesis in animal cells involves the formation of a cleavage furrow. In plant cells, two separated cells are produced by the formation of a cell plate. (Figure 15.14)

15.3 Meiosis and Sexual Reproduction

- The process of meiosis begins with a diploid cell and produces four haploid cells with one set of chromosomes each. (Figure 15.15)
- During prophase of meiosis, homologous pairs of sister chromosomes synapse, and crossing over occurs. After crossing over, chiasmata are observed. (Figure 15.16)
- Meiosis is divided into meiosis I and II, each composed of prophase, prometaphase, metaphase, anaphase, and telophase. During meiosis I, the homologues are separated to different cells, and during meiosis II the sister chromatids are separated to different cells. (Figure 15.17, Table 15.1)
- The life cycle of animals is diploid dominant, while many fungi and protists show a haploid dominant life cycle. Plants alternate between diploid and haploid forms. (Figure 15.18)

15.4 Variation in Chromosome Structure and Number

- Chromosomes are named metacentric, submetacentric, acrocentric, and telocentric according to their centromere location. Each type of chromosome can be uniquely identified by its banding pattern after staining. (Figure 15.19)
- Deficiencies, duplications, inversions, and translocations are different ways to alter chromosome structure. (Figure 15.20)
- A euploid organism has the correct number of chromosomes. A polyploid organism has three or more sets of chromosomes. An aneuploid organism has one too many (trisomy) or one too few (monosomy) chromosomes. (Figure 15.21)
- Nondisjunction and interspecies matings can alter chromosome number. An alloploid has chromosome sets from two or more different species. (Figures 15.22, 15.23)

- Polyploid animals are relatively rare, but polyploid plants are common and tend to be more robust than their diploid counterparts. (Figures 15.24, 15.25)
- Aneuploidy in humans is responsible for several types of human genetic diseases including Down syndrome. (Table 15.2)

TEST YOURSELF

1. In eukaryotic cells, chromosomes consist of
   a. DNA and RNA.   d. DNA and proteins.
   b. DNA only.   e. RNA only.
   c. RNA and proteins.

2. A nucleosome is
   a. a dark-staining body composed of RNA and proteins found in the nucleus.
   b. a protein that helps organize the structure of chromosomes.
   c. another word for a chromosome.
   d. a structure composed of eight histones wrapped with DNA.
   e. the short arm of a chromosome.

3. Compaction of chromosomes is not uniform. Some areas are more compact than others. Which of the following statements is true about the differing levels of organization in the eukaryotic chromosome?
   a. More compact areas of the chromosome are called euchromatin and contain genes that are not expressed.
   b. More compact areas of the chromosome are called heterochromatin and contain genes that are not expressed.
   c. More compact areas of the chromosome are called heterochromatin and contain genes that are expressed at high levels.
   d. More compact areas of the chromosome are called euchromatin and contain genes that are expressed at high levels.
   e. None of the above is correct.

4. Which of the following is a reason for mitotic cell division?
   a. reproduction
   b. gamete formation in animals
   c. multicellularity
   d. all of the above
   e. both a and c

5. A replicated chromosome is composed of
   a. two homologous chromosomes held together at the centromere.
   b. four sister chromatids held together at the centromere.
   c. two sister chromatids held together at the centromere.
   d. four homologous chromosomes held together at the centromere.
   e. one chromosome with a centromere.

6. Which of the following is not an event of anaphase of mitosis?
   a. The nuclear envelope breaks down.
   b. Sister chromatids separate.
   c. Kinetochore microtubules shorten, pulling the chromosomes to the pole.
   d. Polar microtubules push against each other, moving the poles farther apart.
   e. All of the above occur during anaphase.

7. A student is looking at cells under the microscope. The cells are from an organism that has a diploid number of 14. For one particular slide, the cell has seven replicated chromosomes aligned at the metaphase plate of the cell. Which of the following statements accurately describes this particular cell?
   a. The cell is in metaphase of mitosis.
   b. The cell is in metaphase I of meiosis.
   c. The cell is in metaphase II of meiosis.
   d. All of the above are correct.
   e. Both b and c are correct.

8. Which of the following statements accurately describes a difference between meiosis and mitosis?
   a. Mitosis produces diploid cells and meiosis produces haploid cells.
   b. Homologous chromosomes synapse during meiosis but do not synapse during mitosis.
   c. Crossing over commonly occurs during meiosis, but it does not commonly occur during mitosis.
   d. All of the above are correct.
   e. Both a and c are correct.

9. During crossing over in meiosis I,
   a. homologous chromosomes are not altered.
   b. homologous chromosomes exchange genetic material.
   c. chromosomal damage occurs.
   d. genetic information is lost.
   e. cytokinesis occurs.

10. Aneuploidy is the result of
    a. duplication of a region of a chromosome.
    b. inversion of a region of a chromosome.
    c. nondisjunction during meiosis.
    d. interspecies breeding.
    e. all of the above.

CONCEPTUAL QUESTIONS

1. Distinguish between heterochromatin and euchromatin.
2. Explain the histone code hypothesis.
3. Distinguish between homologous chromosomes and sister chromatids.

EXPERIMENTAL QUESTIONS

1. For the Feature Investigation by Markus Noll, what is DNase-I and why was it used in this experiment?
2. Noll hypothesized that DNA fragments of 200 bp would support the beads-on-a-string model of nucleosome structure for this particular species. Explain the rationale for this particular fragment length. The experiment consisted of three assays with differing concentrations of DNase-I. How did the different concentrations of DNase-I affect the results?
3. Explain the results of the experiment shown in Figure 15.7b.

COLLABORATIVE QUESTIONS

1. Discuss several variations in chromosome structure and number.
2. What is crossing over, when does it occur, and what is its significance?

---

www.brookerbiology.com

This website includes answers to the Biological Inquiry questions found in the figure legends and all end-of-chapter questions.

16

SIMPLE PATTERNS OF INHERITANCE

CHAPTER OUTLINE

The pea plant, *Pisum sativum*, studied by Mendel.

Long before people knew anything about cells or chromosomes, they observed patterns of heredity and speculated about them. The ancient Greek physician Hippocrates, famous for his authorship of the physician's oath, provided the first known explanation for the transmission of hereditary traits (ca. 400 BCE). He suggested that "seeds" produced by all parts of the body are collected and transmitted to offspring at the time of conception, and that these seeds cause offspring to resemble their parents. This idea, known as **pangenesis**, influenced the thinking of scientists for many centuries.

The first systematic studies of genetic crosses were carried out by the plant breeder Joseph Kolreuter between 1761 and 1766. In crosses between different strains of tobacco plants, Kolreuter found that the offspring were usually intermediate in appearance between the two parents. He concluded that parents make equal genetic contributions to their offspring and that their genetic material blends together as it is passed to the next generation. This interpretation was consistent with the concept known as **blending inheritance**, which was widely accepted at that time. In the late 1700s, Jean-Baptiste Lamarck, a French naturalist, hypothesized that species change over the course of many generations by adapting to new environments. According to Lamarck, behavioral changes modify traits, and such modified traits were inherited by offspring. For example, an individual who became adept at archery would pass that skill to his or her offspring. Overall, the prevailing view prior to the 1800s was that hereditary traits were rather malleable and could change and blend over the course of one or two generations.

In the last chapter, we considered the process of cell division and how chromosomes are transmitted during mitosis and meiosis. Observations of these processes in the second half of the 19th century provided compelling evidence for **particulate inheritance**—the idea that the determinants of hereditary traits are transmitted intact from one generation to the next. Remarkably, this idea was first proposed in the 1860s by a researcher who knew nothing about chromosomes (**Figure 16.1**). Gregor Mendel, remembered today as the "father of genetics," used statistical analysis of carefully designed breeding experiments to arrive at the concept of a gene. Forty years later, through the convergence of Mendel's work and that of cell biologists, this concept became the foundation of the modern science of genetics.

In this chapter we will consider inheritance patterns, and how the transmission of genes is related to the transmission of chromosomes. In the first section, we consider the fundamental genetic patterns known as Mendelian inheritance and the relationship of these patterns to the behavior of chromosomes during meiosis. In the second section, we examine the distinctive inheritance patterns of genes located on the X chromosome, paying special attention to the work of Thomas Hunt Morgan, whose investigation of these patterns confirmed that genes are on chromosomes. Finally, drawing on what you have already learned about genes and their expression, we will discuss the molecular basis of Mendelian inheritance and its variations.

16.1 Mendel's Laws and the Chromosome Theory of Inheritance

Gregor Johann Mendel (1822–1884) grew up on a small farm in northern Moravia, then a part of the Austrian Empire and now in the Czech Republic. At the age of 21 he entered the Augustinian monastery of St. Thomas in Brno, and he was ordained as a priest in 1847. Mendel then worked for a short time as a substitute teacher, but to continue teaching he needed a license.

Figure 16.1   Gregor Johann Mendel, the father of genetics.

Surprisingly, he failed the licensing exam due to poor answers in physics and natural history, so he enrolled at the University of Vienna to expand his knowledge in these two areas. Mendel's training in physics and mathematics taught him to perceive the world as an orderly place, governed by natural laws that could be stated as simple mathematical relationships.

In 1856, Mendel began his historic studies on pea plants. For eight years, he analyzed thousands of pea plants that he grew on a small plot in his monastery garden. He kept meticulously accurate records that included quantitative data concerning the outcome of his studies. He published his work, entitled "Experiments on Plant Hybrids," in 1866. This paper was largely ignored by scientists at that time, partly because of its title and because it was published in a rather obscure journal (*The Proceedings of the Brünn Society of Natural History*). Also, Mendel was clearly ahead of his time. During this period, biology had not yet become a quantitative, experimental science. In addition, the behavior of chromosomes during mitosis and meiosis, which provides a framework for understanding inheritance patterns, had yet to be studied. Prior to his death in 1884, Mendel reflected, "My scientific work has brought me a great deal of satisfaction and I am convinced that it will be appreciated before long by the whole world." Sixteen years later, in 1900, Mendel's work was independently rediscovered by three biologists with an interest in plant genetics: Hugo de Vries of Holland, Carl Correns of Germany, and Erich von Tschermak of Austria. Within a few years, the impact of Mendel's studies was felt around the world.

In this section, we will examine Mendel's experiments and how they led to the formulation of the basic genetic principles we call Mendel's laws. We will see that these principles apply not only to the pea plants Mendel studied but also to a wide variety of sexually reproducing organisms, including humans. Next, we will consider how the study of chromosomes in the late 19th and early 20th centuries provided a physical explanation for Mendel's laws. At the end of the section, we will briefly examine the laws of probability on which Mendel based his analysis and how they are used to predict simple patterns of inheritance.

Analysis of Inheritance Patterns in Pea Plants Led Mendel to Formulate Two Basic Laws of Genetics

When two individuals with different characteristics are mated or crossed to each other, this is called a **hybridization** experiment, and the offspring are referred to as hybrids. For example, a hybridization experiment could involve a cross between a purple-flowered plant and a white-flowered plant. Mendel was particularly intrigued by the consistency with which offspring of such crosses showed characteristics of one or the other parent in successive generations. His intellectual foundation in physics and the natural sciences led him to consider that this regularity might be rooted in natural laws that could be expressed mathematically. To uncover these laws, he carried out quantitative hybridization experiments in which he carefully analyzed the numbers of offspring carrying specific traits. This analysis led him to formulate two fundamental genetic principles, known today as the law of segregation and the law of independent assortment.

Mendel chose the garden pea, *Pisum sativum*, to investigate the natural laws that govern plant hybrids. Several properties of this species were particularly advantageous for studying plant hybridization. First, it had many readily available varieties that differed in visible characteristics such as the appearance of seeds, pods, flowers, and stems. Such features of an organism are called **characters** or **traits**. **Figure 16.2** illustrates the seven characteristics that Mendel eventually chose to follow in his breeding experiments. Each of these traits was found in two variants. For example, one trait he followed was height, which had the variants known as tall and dwarf. Another was seed color, which had the variants yellow and green.

A second important feature of garden peas is that they are normally self-fertilizing. In **self-fertilization**, a female gamete is fertilized by a male gamete from the same plant. Like many flowering plants, peas have male and female sex organs in the same flower (**Figure 16.3**). Male gametes (sperm cells) are produced within pollen grains, which are formed in the male structures called stamens. Female gametes (egg cells) are produced in structures called ovules that form within an organ called an ovary. For fertilization to occur, a pollen grain must land on the receptacle called a stigma, enabling a sperm to migrate to an ovule and fuse with an egg cell. In peas, the stamens and the ovaries are enclosed by a modified petal, an arrangement that greatly favors self-fertilization. Self-fertilization makes it easy to produce plants that breed true for a given trait, meaning that the trait does not vary from generation to generation.

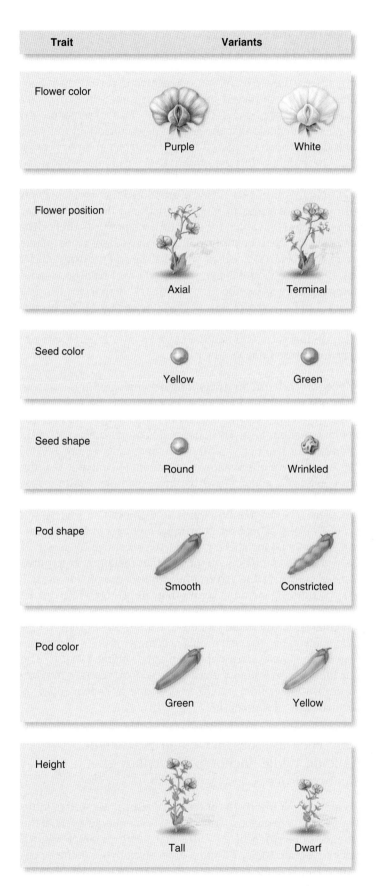

Trait	Variants	
Flower color	Purple	White
Flower position	Axial	Terminal
Seed color	Yellow	Green
Seed shape	Round	Wrinkled
Pod shape	Smooth	Constricted
Pod color	Green	Yellow
Height	Tall	Dwarf

Figure 16.2   The seven traits that Mendel studied.

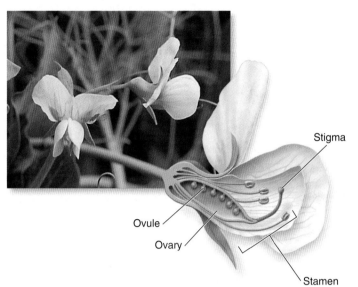

Figure 16.3   Flower structure in pea plants. The pea flower produces both male and female gametes. Sperm form in the pollen produced within the stamens; egg cells form in ovules within the ovary. A modified petal encloses the stamens and stigma, encouraging self-fertilization.

For example, if a pea plant with yellow seeds breeds true for seed color, all the plants that grow from these seeds will also produce yellow seeds. A variety that continues to exhibit the same trait after several generations of self-fertilization is called a **true-breeding line**. Prior to conducting the studies described in this chapter, Mendel had already established that the seven traits that he chose to study were true-breeding in the strains of pea plants he had obtained.

A third reason for using garden peas in hybridization experiments is the ease of making crosses: the flowers are quite large and easy to manipulate. In some cases Mendel wanted his pea plants to self-fertilize, but in others he wanted to cross plants that differed with respect to some trait, a process called **cross-fertilization** or hybridization. In garden peas, cross-fertilization requires placing pollen from one plant on the stigma of another plant's flower. Mendel's cross-fertilization procedure is shown in **Figure 16.4**. He would pry open an immature flower and remove the stamens before they produced pollen, so that the flower could not self-fertilize. He then used a paintbrush to transfer pollen from another plant to the stigma of the flower that had its stamens removed. In this way, Mendel was able to cross-fertilize any two of his true-breeding pea plants and obtain any type of hybrid he wanted.

By Following the Inheritance Pattern of Single Traits, Mendel's Work Revealed the Law of Segregation

Mendel began his investigations by studying the inheritance patterns of pea plants that differed with regard to a single trait.

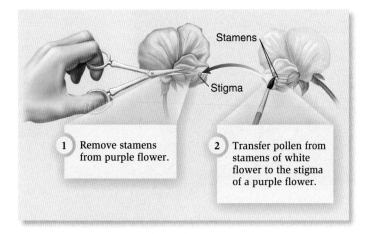

Figure 16.4   A procedure for cross-fertilizing pea plants.

A cross in which an experimenter follows the variants of only one trait is called a **single-factor cross**. As an example, we will consider a single-factor cross in which Mendel followed the tall and dwarf variants for height (**Figure 16.5**). The left-hand side of Figure 16.5a shows his experimental approach. The true-breeding parents are termed the **P generation** (parental generation), and a cross of these plants is called a P cross. The first-generation offspring of a P cross constitute the **F₁ generation** (first filial generation, from the Latin *filius*, son). When the true-breeding parents differ with regard to a single trait, their F₁ offspring are called single-trait hybrids, or **monohybrids**. When Mendel crossed true-breeding tall and dwarf plants, he observed that all plants of the F₁ generation were tall.

Next, Mendel followed the transmission of this trait for a second generation. To do so, he allowed the F₁ monohybrids to self-fertilize, producing a generation called the **F₂ generation** (second filial generation). The dwarf trait reappeared in the F₂ offspring: three-fourths of the plants were tall and one-fourth were dwarf. Mendel obtained similar results for each of the traits that he studied, as shown in the data of Figure 16.5b. A quantitative analysis of his data allowed Mendel to postulate three important ideas regarding the properties and transmission of these traits from parents to offspring:

Dominant and Recessive Traits   Perhaps the most surprising outcome of Mendel's work was that the data argued strongly against the prevailing notion of a blending mechanism of heredity. In all seven cases, the F₁ generation displayed traits distinctly like one of the two parents rather than intermediate traits. Using genetic terms that Mendel originated, we describe the alternative traits as dominant and recessive. The term **dominant** describes the displayed trait, while the term **recessive** describes a trait that is masked by the presence of a dominant trait. Tall stems and green pods are examples of dominant traits; dwarf stems and yellow pods are examples of recessive traits. We say that tall is dominant over dwarf and green is dominant over yellow.

Genes and Alleles   Mendel's results were consistent with a particulate mechanism of inheritance, in which the determinants of traits are inherited as unchanging, discrete units. In all seven cases, the recessive trait reappeared in the F₂ generation: some F₂ plants displayed the dominant trait, while a smaller proportion showed the recessive trait. This observation led Mendel to conclude that the genetic determinants of traits are "unit factors" that are passed intact from generation to generation. These unit factors are what we now call **genes** (from the Greek *genos*, birth), a term coined by the Danish botanist Wilhelm Johannsen in 1911. Mendel postulated that every individual carries two genes for a given trait, and that the gene for each trait has two variant forms, which we now call **alleles**. For example, the gene controlling height in Mendel's pea plants occurs in two variants, called the tall allele and the dwarf allele. The right-hand side of Figure 16.5a shows Mendel's conclusions, using genetic symbols (letters) that were adopted later. The letters *T* and *t* represent the alleles of the gene for plant height. By convention, the uppercase letter represents the dominant allele (in this case, tall) and the same letter in lowercase represents the recessive allele (dwarf).

Segregation of Alleles   When Mendel compared the numbers of F₂ offspring exhibiting dominant and recessive traits, he noticed a recurring pattern. Although there was some experimental variation, he always observed approximately a 3:1 ratio between the dominant and the recessive trait (Figure 16.5b). This quantitative observation allowed him to conclude that the two copies of a gene carried by an F₁ plant **segregate** (separate) from each other, so that each sperm or egg carries only one allele. The diagram in **Figure 16.6** shows that segregation of the F₁ alleles should result in equal numbers of gametes carrying the dominant allele (*T*) and the recessive allele (*t*). If these gametes combine with one another randomly at fertilization, as shown in the figure, this would account for the 3:1 ratio of the F₂ generation. Note that the genotype *Tt* can be produced by two different combinations of alleles—the *T* allele can come from the male gamete and the *t* allele from the female gamete, or vice versa. This accounts for the fact that the *Tt* genotype is produced twice as often as either *TT* or *tt*. The idea that *the two copies of a gene segregate from each other during transmission from parent to offspring* is known today as Mendel's **law of segregation**.

Genotype Describes an Organism's Genetic Makeup, While Phenotype Describes Its Characteristics

To continue our discussion of Mendel's results, we need to introduce a few more genetic terms. The term **genotype** refers to the genetic composition of an individual. In the example shown earlier in Figure 16.5a, *TT* and *tt* are the genotypes of the P generation and *Tt* is the genotype of the F₁ generation. In a P cross, both parents are true-breeding plants, which means that each has identical copies of the gene for height. An individual with

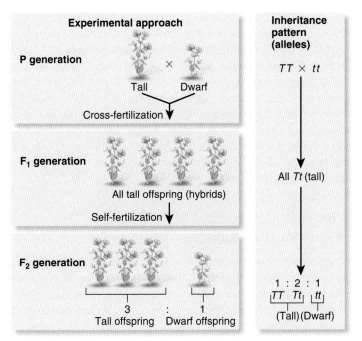

(a) Mendel's protocol for making single-factor crosses

THE DATA

P cross	F₁ generation	F₂ generation	Ratio
Purple × white flowers	All purple	705 purple, 224 white	3.15:1
Axial × terminal flowers	All axial	651 axial, 207 terminal	3.14:1
Yellow × green seeds	All yellow	6,022 yellow, 2,001 green	3.01:1
Round × wrinkled seeds	All round	5,474 round, 1,850 wrinkled	2.96:1
Smooth × constricted pods	All smooth	882 smooth, 299 constricted	2.95:1
Green × yellow pods	All green	428 green, 152 yellow	2.82:1
Tall × dwarf stem	All tall	787 tall, 277 dwarf	2.84:1
Total	**All dominant**	**14,949 dominant, 5,010 recessive**	**2.98:1**

(b) Mendel's observed data for all 7 traits

Figure 16.5 Mendel's analyses of single-factor crosses.

two identical copies of a gene is said to be **homozygous** with respect to that gene. In the specific P cross we are considering, the tall plant is homozygous for *T* and the dwarf plant is homozygous for *t*. In contrast, a **heterozygous** individual carries two different alleles of the same gene. Plants of the F₁ generation are heterozygous, with the genotype *Tt*, because every individual carries one copy of the tall allele and one copy of the dwarf allele.

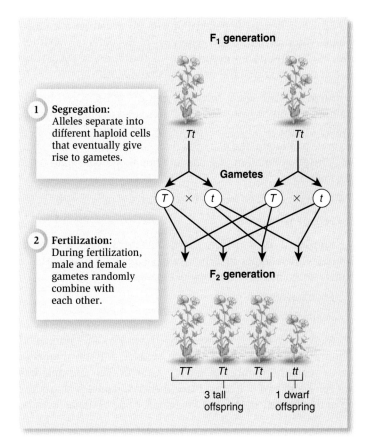

Figure 16.6 How the law of segregation explains Mendel's observed ratios. The segregation of alleles in the F₁ generation gives rise to gametes that carry just one of the two alleles. These gametes combine randomly during fertilization, producing the allele combinations *TT*, *Tt*, and *tt* in the F₂ offspring. The combination *Tt* occurs twice as often as either of the other two combinations because it can be produced in two different ways. The *TT* and *Tt* offspring are tall, while the *tt* offspring are dwarf.

The F₂ generation includes both homozygous individuals (homozygotes) and heterozygous individuals (heterozygotes).

The term **phenotype** refers to the characteristics of an organism that are the result of the expression of its genes. In the example in Figure 16.5a, one of the parent plants is phenotypically tall and the other is phenotypically dwarf. Although the F₁ offspring are heterozygous *(Tt)*, their phenotypes are tall because each of them has a copy of the dominant tall allele. In contrast, the F₂ plants display both phenotypes in a ratio of 3:1. Later in the chapter we will examine the underlying molecular mechanisms that produce phenotypes, but in our discussion of Mendel's results the term simply refers to a visible trait such as flower color or height.

A Punnett Square Can Be Used to Predict the Outcome of Crosses

A common way to predict the outcome of simple genetic crosses is to make a **Punnett square**, a method originally proposed by the British geneticist Reginald Punnett. To construct a Punnett square, you must know the genotypes of the parents. What follows is a step-by-step description of the Punnett square approach using a cross of heterozygous tall plants.

Step 1. *Write down the genotypes of both parents.* In this example, a heterozygous tall plant is crossed to another heterozygous tall plant. The plant providing the pollen is considered the male parent and the plant providing the eggs, the female parent. (In self-pollination, a single individual produces both types of gametes.)

Male parent: *Tt*

Female parent: *Tt*

Step 2. *Write down the possible gametes that each parent can make.* Remember that the law of segregation tells us that a gamete contains only one copy of each gene.

Male gametes: *T* or *t*

Female gametes: *T* or *t*

Step 3. *Create an empty Punnett square.* The number of columns equals the number of male gametes, and the number of rows equals the number of female gametes. Our example has two rows and two columns. Place the male gametes across the top of the Punnett square and the female gametes along the side.

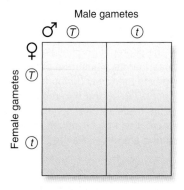

Step 4. *Fill in the possible genotypes of the offspring by combining the alleles of the gametes in the empty boxes.*

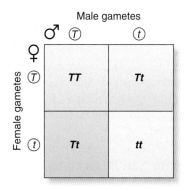

Step 5. *Determine the relative proportions of genotypes and phenotypes of the offspring.* The genotypes are obtained directly from the Punnett square. In this example, the geno-

types are *TT*, *Tt*, and *tt* in a 1:2:1 ratio. To determine the phenotypes, you must know which allele is dominant. For plant height, *T* (tall) is dominant to *t* (dwarf). The genotypes *TT* and *Tt* are tall, whereas the genotype *tt* is dwarf. Therefore, our Punnett square shows us that the ratio of phenotypes is expected to be 3:1, or 3 tall plants to 1 dwarf plant. Keep in mind, however, that these are predicted ratios for large numbers of offspring. If only a few offspring are produced, the observed ratios could deviate significantly from the predicted ratios. We will examine the question of sample size and genetic prediction later in this chapter.

A Testcross Can Be Used to Determine an Individual's Genotype

When a trait has two variants, one of which is dominant over the other, we know that an individual with a recessive phenotype is homozygous for the recessive allele. A dwarf pea plant, for example, must have the genotype *tt*. But an individual with a dominant phenotype may be either homozygous or heterozygous—a tall pea plant may have the genotype *TT* or *Tt*. To distinguish between these two possibilities, Mendel devised a method called a **testcross** that is still used today. In a testcross, the researcher crosses the individual of interest to a homozygous recessive individual and observes the phenotypes of the offspring. **Figure 16.7** shows how this procedure can be used to determine the genotype of a tall pea plant. If the testcross produces some dwarf offspring, as shown on the right side of the figure, these offspring must have two copies of the recessive allele, one inherited from each parent. Therefore, the tall parent must be a heterozygote, with the genotype *Tt*. Alternatively, if all of the offspring are tall, as shown on the left, the tall parent is likely to be a homozygote, with the genotype *TT*.

Analyzing the Inheritance Pattern of Two Traits Simultaneously Demonstrated the Law of Independent Assortment

Mendel's analysis of single-factor crosses suggested that traits are inherited as discrete units and that the alleles for a given gene segregate during the formation of haploid cells. To obtain additional insights into how genes are transmitted from parents to offspring, Mendel conducted crosses in which he simultaneously followed the inheritance of two different traits. A cross of this type is called a **two-factor cross**. We will examine a two-factor cross in which Mendel simultaneously followed the inheritance of seed color and seed shape (**Figure 16.8**). He began by crossing pea plants from strains that bred true for both traits. The plants of one strain had yellow, round seeds and plants of the other strain had green, wrinkled seeds. He then allowed the F_1 offspring to self-fertilize and observed the phenotypes of the F_2 generation.

Before we discuss Mendel's results, let's consider possible patterns of inheritance for two traits. One possibility is that the two genes are linked in some way, so that variants that occur together in the parents are always inherited as a unit. In our

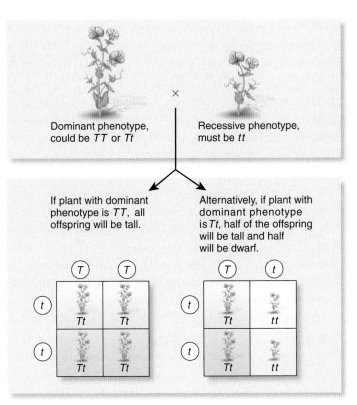

Figure 16.7 A testcross. The purpose of this experiment is to determine if the organism with the dominant phenotype, in this case a tall pea plant, is a homozygote (*TT*) or a heterozygote (*Tt*).

Biological inquiry: Let's suppose we had a plant with purple flowers and unknown genotype and conducted a testcross to determine its genotype. We obtained 41 plants, 20 of which had white flowers and 21 with purple flowers. What was the genotype of the original purple-flowered plant?

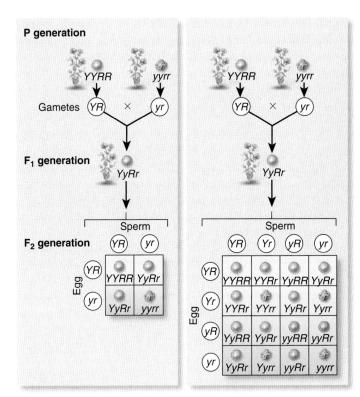

(a) Hypothesis: Linked assortment

(b) Hypothesis: Independent assortment

P Cross	F_1 generation	F_2 generation
Yellow, round seeds × Green, wrinkled seeds	Yellow, round seeds	315 yellow, round seeds 101 yellow, wrinkled seeds 108 green, round seeds 32 green, wrinkled seeds

(c) The data observed by Mendel

Figure 16.8 Two hypotheses for the assortment of two different genes. This figure shows a cross between two true-breeding pea plants, one with yellow, round seeds and one with green, wrinkled seeds. All of the F_1 offspring have yellow, round seeds. When the F_1 offspring self-fertilize, the two hypotheses predict different phenotypes in the F_2 generation. **(a)** The linkage hypothesis proposes that the two parental alleles always stay associated with each other. In this case, all of the F_2 offspring will have either yellow, round seeds or green, wrinkled seeds. **(b)** The independent assortment hypothesis proposes that each allele assorts independently. In this case, the F_2 generation will display four different phenotypes. **(c)** Mendel's observations supported the independent assortment hypothesis.

example, the allele for yellow seeds (*Y*) would always be inherited with the allele for round seeds (*R*), as shown in Figure 16.8a. The recessive alleles are green seeds (*y*) and wrinkled seeds (*r*). A second possibility is that the two genes are independent of one another, so that their alleles are randomly distributed into gametes (Figure 16.8b). By following the transmission pattern of two traits simultaneously, Mendel could determine whether the genes that determine seed shape and seed color **assort** (are distributed) together as a unit or independently of each other.

What experimental results could Mendel predict for each of these two models? The homozygous P generation can produce only two kinds of gametes, *YR* and *yr*, so in either case the F_1 offspring would be heterozygous for both traits; that is, they would have the genotypes *YyRr*. Because Mendel knew from his earlier experiments that yellow was dominant over green and round over wrinkled, he could predict that all the F_1 plants would have yellow, round seeds. In contrast, as shown in Figure 16.8, the ratios he obtained in the F_2 generation would depend on whether the traits assort together or independently.

If the parental traits are linked, as in Figure 16.8a, the F_1 plants could only produce gametes that are *YR* or *yr*. These gametes would combine to create offspring with the genotypes

YYRR (yellow, round), *YyRr* (yellow, round), or *yyrr* (green, wrinkled). The ratio of phenotypes would be 3 yellow, round to 1 green, wrinkled. Every F_2 plant would be phenotypically like one P-generation parent or the other; none would display a new combination of the parental traits. However, if the alleles assort independently, the F_2 generation would show a wider range of genotypes and phenotypes, as shown by the large

Punnett square in Figure 16.8b. In this case, each F_1 parent produces four kinds of gametes—*YR, Yr, yR,* and *yr*—instead of two, so the square is constructed with four rows on each side and shows 16 possible genotypes. The F_2 generation includes plants with yellow, round seeds; yellow, wrinkled seeds; green, round seeds; and green, wrinkled seeds, in a ratio of 9:3:3:1.

The actual results of this two-factor cross are shown in Figure 16.8c. Crossing the true-breeding parents produced **dihybrid** offspring—offspring that are hybrids with respect to both traits. These F_1 dihybrids all had yellow, round seeds, confirming that yellow and round are dominant traits. This result was consistent with either hypothesis. However, the data for the F_2 generation were consistent only with the independent assortment hypothesis. Mendel observed four phenotypically different types of F_2 offspring, in a ratio that was reasonably close to 9:3:3:1.

Mendel's results were similar for every pair of traits he studied. His data supported the idea, now called the **law of independent assortment**, that *the alleles of different genes assort independently of each other during gamete formation.* Independent assortment means that a specific allele for one gene may be found in a gamete regardless of which allele for a different gene is found in the same gamete. In our example, the yellow and green alleles assort independently of the round and wrinkled alleles. The union of gametes from F_1 plants carrying these alleles produces the F_2 genotypic and phenotypic ratios shown in Figure 16.8b.

The Chromosome Theory of Inheritance Relates Mendel's Observations to the Behavior of Chromosomes

Mendel's studies with pea plants led to the concept of a gene, which is the foundation for our understanding of inheritance. However, at the time of Mendel's work, the physical nature and location of genes was a complete mystery. In fact, the idea that inheritance has a physical basis was not even addressed until 1883, when the German biologist August Weismann and the Swiss botanist Carl Nägeli championed the idea that a substance in living cells is responsible for the transmission of hereditary traits. Nägeli also suggested that parents contribute equal amounts of this substance to their offspring. This idea challenged other researchers to identify the genetic material. Several scientists, including the German biologists Eduard Strasburger and Walter Flemming, observed dividing cells under the microscope and suggested that the chromosomes are the carriers of the genetic material. As we now know, the genetic material is the DNA within chromosomes.

In the early 1900s, the idea that chromosomes carry the genetic material dramatically unfolded as researchers continued to study the processes of fertilization, mitosis, and meiosis. It became increasingly clear that the characteristics of organisms are rooted in the continuity of cells during the life of an organism and from one generation to the next. Several scientists noted striking parallels between the segregation and assortment

of traits noted by Mendel and the behavior of chromosomes during meiosis. Among these scientists were the German biologist Theodor Boveri and the American biologist Walter Sutton, who independently proposed the chromosome theory of inheritance. According to this theory, the inheritance patterns of traits can be explained by the transmission of chromosomes during meiosis and fertilization. Thomas Hunt Morgan's studies of fruit flies, which we will examine later in this chapter, were instrumental in supporting this theory.

The **chromosome theory of inheritance** consists of a few fundamental principles:

1. Chromosomes contain the genetic material, which is transmitted from parent to offspring and from cell to cell. Genes are found in the chromosomes.
2. Chromosomes are replicated and passed from parent to offspring. They are also passed from cell to cell during the multicellular development of an organism. Each type of chromosome retains its individuality during cell division and gamete formation.
3. The nucleus of a diploid cell contains two sets of chromosomes, which are found in homologous pairs. One member of each pair is inherited from the mother and the other from the father. The maternal and paternal sets of homologous chromosomes are functionally equivalent; each set carries a full complement of genes.
4. At meiosis, one member of each chromosome pair segregates into one daughter nucleus and its homologue segregates into the other daughter nucleus. Each of the resulting haploid cells contains only one set of chromosomes. During the formation of haploid cells, the members of different chromosome pairs segregate independently of each other.
5. Gametes are haploid cells that combine to form a diploid cell during fertilization, with each gamete transmitting one set of chromosomes to the offspring. In animals, one set comes from the mother and the other set comes from the father.

Now that you have an understanding of the basic tenets of the chromosome theory, let's relate these ideas to Mendel's laws of inheritance.

Chromosomes and Segregation Mendel's law of segregation can be explained by the pairing and segregation of homologous chromosomes during meiosis. Before we examine this idea, it will be helpful to introduce another genetic term. The physical location of a gene on a chromosome is called the gene's **locus** (plural, loci). As shown in **Figure 16.9**, each member of a homologous chromosome pair carries an allele of the same gene at the same locus. The individual in this example is heterozygous (*Tt*), so each homologue has a different allele.

Figure 16.10 follows a homologous chromosome pair through the events of meiosis. This example involves a pea plant, heterozygous for height, *Tt*. The top of Figure 16.10 shows the two homologues prior to DNA replication. When a cell prepares

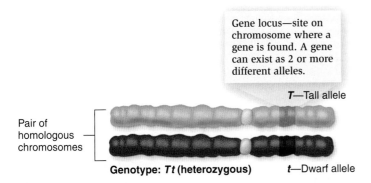

Figure 16.9 **A gene locus.** The locus (location) of a gene is the same for each member of a homologous pair, whether the individual is homozygous or heterozygous for that gene. This individual is heterozygous (*Tt*) for a gene for plant height in peas.

to divide, the homologues replicate to produce pairs of sister chromatids. Each chromatid carries a copy of the allele found on the original homologue, either *T* or *t*. The homologues, each consisting of two sister chromatids, pair during metaphase I and then segregate into two daughter cells during later phases of meiosis I. One of these cells has two copies of the *T* allele, and the other has two copies of the *t* allele. The sister chromatids separate during meiosis II, which produces four haploid cells. The end result of meiosis is that each haploid cell has a copy of just one of the two original homologues. Two of the cells have a chromosome carrying the *T* allele, while the other two have a chromosome carrying the *t* allele at the same locus. If the haploid cells shown at the bottom of Figure 16.10 combine randomly during fertilization, they produce diploid offspring with the genotypic and phenotypic ratios shown earlier in Figure 16.6.

Chromosomes and Independent Assortment The law of independent assortment can also be explained by the behavior of chromosomes during meiosis. **Figure 16.11** shows the segregation of two pairs of homologous chromosomes in a pea plant. One pair carries the gene for seed color: the yellow allele (*Y*) is on one chromosome, and the green allele (*y*) is on its homologue. The other pair of chromosomes carries the gene for seed shape: one member of the pair has the round allele (*R*), while its homologue carries the wrinkled allele (*r*). Thus, this individual is heterozygous for both genes, with the genotype *YyRr*.

When meiosis begins, each of the chromosomes has already replicated and consists of two sister chromatids. At metaphase I of meiosis, the two pairs of chromosomes randomly align themselves along the metaphase plate. This alignment can occur in two equally probable ways, shown on the two sides of the figure. On the left, the chromosome carrying the *y* allele is aligned on the same side of the metaphase plate as the chromosome carrying the *R* allele; *Y* is aligned with *r*. On the right, the opposite has occurred: *Y* is aligned with *R* and *y* is with *r*. In each case, the chromosomes that aligned on the same side of the metaphase plate segregate into the same daughter cell. In this way, the random alignment of chromosome pairs during meiosis I leads to the independent assortment of alleles found

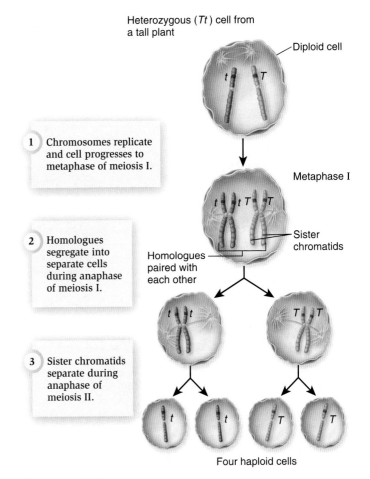

Figure 16.10 **The chromosomal basis of allele segregation.** This example shows a pair of homologous chromosomes in a cell of a pea plant. The blue chromosome was inherited from the male parent and the red chromosome was inherited from the female parent. This individual is heterozygous for a height gene (*Tt*). The two homologues segregate from each other during meiosis, leading to segregation of the tall allele (*T*) and the dwarf allele (*t*) into different haploid cells.

Biological inquiry: When we say that alleles segregate, what does the word segregate *mean? How is this related to meiosis, described in Chapter 15?*

on different chromosomes. For two loci found on different chromosomes, each with two variant alleles, meiosis produces four allele combinations in equal numbers, as seen at the bottom of the figure.

If a *YyRr* (dihybrid) plant undergoes self-fertilization, any two gametes can combine randomly during fertilization. Because four kinds of gametes are made, this allows for 16 possible allele combinations in the offspring. These genotypes, in turn, produce four phenotypes in a 9:3:3:1 ratio, as seen earlier in Figure 16.8. This ratio is the expected outcome when a heterozygote for two genes on different chromosomes undergoes self-fertilization.

But what if two genes are located on the same chromosome? In this case, the transmission pattern may not conform to the law of independent assortment. We will discuss this phenomenon, known as linkage, in Chapter 17.

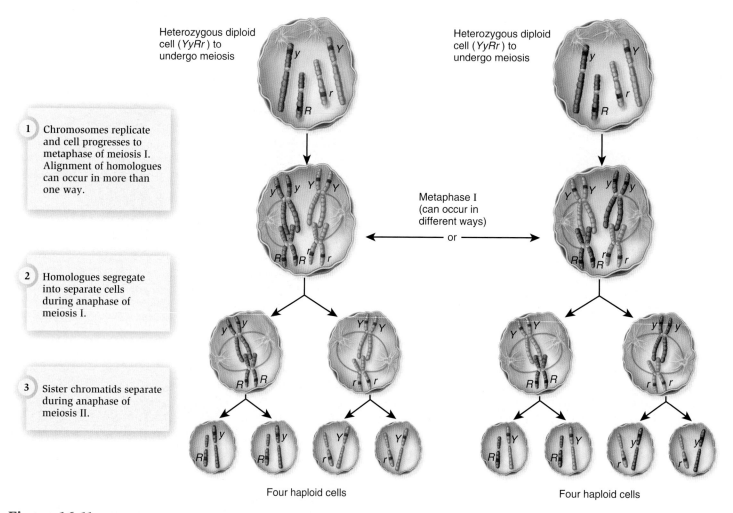

Heterozygous diploid cell (*YyRr*) to undergo meiosis

Heterozygous diploid cell (*YyRr*) to undergo meiosis

1 Chromosomes replicate and cell progresses to metaphase of meiosis I. Alignment of homologues can occur in more than one way.

Metaphase I (can occur in different ways)

— or —

2 Homologues segregate into separate cells during anaphase of meiosis I.

3 Sister chromatids separate during anaphase of meiosis II.

Four haploid cells

Four haploid cells

Figure 16.11 **The chromosomal basis of independent assortment.** The genes for seed color (*Y* or *y*) and seed shape (*R* or *r*) in peas are on different chromosomes. During metaphase of meiosis I, different arrangements of the two chromosome pairs can lead to different combinations of the alleles in the resulting haploid cells. On the *left,* the chromosome carrying the dominant *R* allele has segregated with the chromosome carrying the recessive *y* allele; on the *right*, the two chromosomes carrying the dominant alleles (*R* and *Y*) have segregated together.

Genetic Predictions Are Based on Probability

As you have seen, Mendel's laws of inheritance can be used to predict the outcome of genetic crosses. This is useful in many ways. In agriculture, for example, plant and animal breeders use predictions about the types of offspring their crosses will produce to develop commercially important crops and livestock. In addition, people are often interested in the potential characteristics of their future children. This information is particularly important to individuals who may carry alleles that cause inherited diseases. Of course, no one can see into the future and definitively predict what will happen. Nevertheless, genetic counselors can often help couples predict the likelihood of having an affected child. This probability is one factor that may influence a couple's decision about whether to have children.

Genetic predictions are based on the mathematical rules of probability. The chance that an event will have a particular outcome is called the **probability** of that outcome. The probability of a given outcome depends on the number of possible outcomes. For example, if you draw a card at random from a 52-card deck, the probability that you will get the jack of diamonds is 1 in 52, because there are 52 possible outcomes for the draw. In contrast, only two outcomes are possible when you flip a coin, so the probability is one in two (1/2, or 0.5, or 50%) that the heads side will be showing when the coin lands. The general formula for the probability (*P*) that a random event will have a specific outcome is

$$P = \frac{\text{Number of times an event occurs}}{\text{Total number of possible outcomes}}$$

Thus, for a single coin toss, the chance of getting heads is

$$P_{heads} = \frac{1 \text{ heads}}{(1 \text{ heads} + 1 \text{ tails})} = \frac{1}{2}$$

Earlier in this chapter, we considered the use of Punnett squares to predict the fractions of offspring with a given genotype or phenotype. Our example was self-fertilization of a pea plant that was heterozygous for the height gene (*Tt*), and our Punnett square predicted that one-fourth of the offspring would be dwarf. We can make the same prediction by using a probability calculation.

$$P_{dwarf} = \frac{1 \ tt}{(1 \ TT + 2 \ Tt + 1 \ tt)} = \frac{1}{4}$$

Probability and Sample Size A probability calculation allows us to predict the likelihood that a future event will have a specific outcome. However, the accuracy of this prediction depends to a great extent on the number of events we observe, or in other words, on the size of our sample. For example, if we toss a coin six times, the calculation we just presented for P_{heads} suggests that we should get heads three times and tails three times. However, each coin toss is an independent event, meaning that every time we toss the coin there is a random chance that it will come up heads or tails, regardless of the outcome of the previous toss. With only six tosses, we would not be too surprised if we got four heads and two tails. The deviation between the observed and expected outcomes is called the **random sampling error**. With a small sample, the random sampling error may cause the observed data to be quite different from the expected outcome. By comparison, if we flipped a coin 1,000 times, the percentage of heads would be fairly close to the predicted 50%. With a larger sample, we expect the sampling error to be smaller.

Earlier in this chapter, we examined Mendel's data for pea plants and learned that his observations were very close to the outcome that we would predict from a Punnett square. Mendel counted a large number of pea plants, which made his sampling error quite small. However, when we apply probability to humans, the small size of human families may cause the observed data to be quite different from the expected outcome. Consider, for example, a couple who are both heterozygous for an allele that affects eye color (*Bb*). The dominant allele is brown, so both of these parents have brown eyes. We can use a Punnett square to predict that one-fourth of their offspring will have blue eyes, which is the recessive phenotype for a *bb* homozygote. However, the genotype or phenotype of each child is independent of each other. Each child has one chance in four of having the recessive phenotype; the birth of a blue-eyed child does not make it more or less likely that the next child will have blue eyes. Thus, in a family with four children, we would not be too surprised if two of them had blue eyes—twice the predicted number. In this case, a large deviation occurs between the observed outcome (50% blue-eyed children) and the expected outcome (25% blue-eyed children). This large random sampling error can be attributed to the small sample size.

The Product Rule and the Sum Rule Punnett squares allow us to predict the likelihood that a genetic cross will produce an offspring with a particular genotype or phenotype. To predict the likelihood of producing multiple offspring with particular genotypes or phenotypes, we can use the **product rule**, which states that *the probability that two or more independent events will occur is equal to the product of their individual probabilities.* As we have already discussed, events are independent if the outcome of one event does not affect the outcome of another. In our previous coin-toss example, each toss is an independent event— if one toss comes up heads, another toss still has an equal chance of coming up either heads or tails. If we toss a coin twice, what is the probability that we will get heads both times? The product rule says that it is equal to the probability of getting heads on the first toss (1/2) times the probability of getting heads on the second toss (1/2), or one in four (1/2 × 1/2 = 1/4).

To see how the product rule can be applied to a genetics problem, let's consider a rare, recessive human trait known as congenital analgesia. (Congenital refers to a condition present at birth; analgesia means insensitivity to pain.) People with this trait can distinguish between sensations such as sharp and dull, or hot and cold, but they do not perceive extremes of sensation as painful. The first known case of congenital analgesia, described in 1932, was a man who made his living entertaining the public as a "human pincushion." For a phenotypically normal couple, each heterozygous for the recessive allele causing congenital analgesia, we can ask, What is the probability that their first three offspring will have the disorder? To answer this question, we must first determine the probability of a single offspring having the abnormal phenotype. By using a Punnett square, we would find that the probability of an individual offspring being homozygous recessive is 1/4. Thus, each of this couple's children has one chance in four of having the disorder.

We can now use the product rule to calculate the probability of this couple having three affected offspring in a row. The phenotypes of the first, second, and third offspring are independent events; that is, the phenotype of the first offspring does not affect the phenotype of the second or third offspring. The product rule tells us that the probability of all three children having the abnormal phenotype is

$$\frac{1}{4} \times \frac{1}{4} \times \frac{1}{4} = \frac{1}{64} = 0.016$$

The probability of the first three offspring having the disorder is 0.016, or 1.6%. In other words, we can say that this couple's chance of having three children in a row with congenital analgesia is very small—only 1.6 out of 100. The phenotypes of the first, second, and third child are independent of each other.

Let's now consider a second way to predict the outcome of particular crosses. In a cross between two heterozygous (*Tt*) pea plants, we may want to know the probability of a particular offspring being a homozygote. In this case we are asking, What is the chance that this individual will be either homozygous *TT* or homozygous *tt*? To answer an "either/or" question we use the sum rule, which applies to events with mutually exclusive outcomes. When we say that outcomes are mutually exclusive, we mean that they cannot occur at the same time. A pea plant can be tall or dwarf, but not both at the same time: the tall and dwarf phenotypes are mutually exclusive. Similarly, a plant with the genotype *TT* cannot be *Tt* or *tt*. Each of these genotypes is mutually exclusive with the other two. According to the **sum rule**, *the probability that one of two or more mutually exclusive outcomes will occur is the sum of the probabilities of the possible outcomes*. This means that to find the probability that an offspring will be either homozygous *TT* or homozygous *tt*, we add together the probability that it will be *TT* and the probability that it will be *tt*. Using a Punnett square, we find that the probability for each of these genotypes is one in four. We can now use the sum rule to determine the probability of an individual having one of these genotypes.

$$\frac{1}{4} + \frac{1}{4} = \frac{1}{2}$$

(probability of *TT*)   (probability of *tt*)   (probability of either *TT* or *tt*)

This calculation predicts that in crosses of two *Tt* parents, half of the offspring will be homozygotes—either *TT* or *tt*.

Pedigree Analysis Examines the Inheritance of Human Traits

As we have seen, Mendel conducted experiments by making selective crosses of pea plants and analyzing large numbers of offspring. Later geneticists also relied on crosses of experimental organisms, especially fruit flies. Obviously, geneticists studying human traits cannot use this approach, for ethical and practical reasons. Instead, human geneticists must rely on information from family trees, or pedigrees. In this approach, called **pedigree analysis**, an inherited trait is analyzed over the course of a few generations in one family. The results of this method may be less definitive than the results of breeding experiments because the small size of human families may lead to large sampling errors. Nevertheless, a pedigree analysis can often provide important clues concerning human inheritance.

Pedigree analysis has been used to understand the inheritance of human genetic diseases that follow simple Mendelian patterns. Many genes that play a role in disease exist in two forms—the normal allele and an abnormal allele that has arisen by mutation. The disease symptoms are associated with the mutant allele. Pedigree analysis allows us to determine whether the mutant allele is dominant or recessive and to predict the likelihood of an individual being affected.

We have already considered one human abnormality caused by a recessive allele, congenital analgesia. We will use another recessive condition to illustrate pedigree analysis. The pedigree in **Figure 16.12** concerns a human genetic disease known as cystic fibrosis (CF). Approximately 3% of Americans of European descent are heterozygous carriers of the recessive CF allele. These carriers are phenotypically normal. Individuals who are homozygous for the CF allele exhibit the disease symptoms, which include abnormalities of the pancreas, intestine, sweat glands, and lungs. (You will see how a single gene can have such far-reaching effects when we discuss the molecular basis

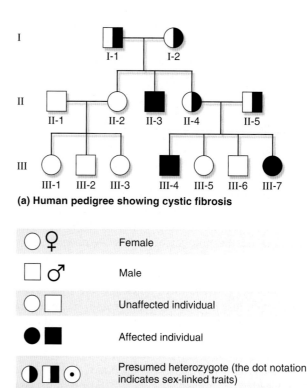

(a) Human pedigree showing cystic fibrosis

○♀	Female
□♂	Male
○ □	Unaffected individual
● ■	Affected individual
◐ ◪ ⊙	Presumed heterozygote (the dot notation indicates sex-linked traits)

(b) Symbols used in a human pedigree

Figure 16.12   **A family pedigree of a recessive trait.** Some members of the family in this pedigree are affected with cystic fibrosis. Phenotypically normal individuals I-1, I-2, II-4, and II-5 are presumed to be heterozygotes because they have produced affected offspring.

Biological inquiry: Let's suppose a genetic disease is caused by a mutant allele. If two affected parents produce an unaffected offspring, can the mutant allele be recessive?

of Mendelian inheritance later in the chapter.) A human pedigree like the one in Figure 16.12 shows the oldest generation (designated by the Roman numeral I) at the top, with later generations (II and III) below it. A man (represented by a square) and a woman (represented by a circle) who produce offspring are connected by a horizontal line; a vertical line connects parents with their offspring. Siblings (brothers and sisters) are denoted by downward projections from a single horizontal line, from left to right in the order of their birth. For example, individuals I-1 and I-2 are the parents of individuals II-2, II-3, and II-4, who are all siblings. Individuals affected by the disease, such as individual II-3, are depicted by filled symbols.

The pattern of affected and unaffected individuals in this pedigree is consistent with a recessive mode of inheritance for CF: two unaffected individuals can produce an affected offspring. Such individuals are presumed to be heterozygotes (designated by a half-filled symbol). However, the same unaffected parents can also produce unaffected offspring, because an individual must inherit two copies of the mutant allele to exhibit the disease. A recessive mode of inheritance is also characterized

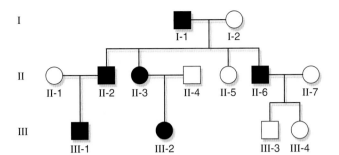

Figure 16.13 A family pedigree of a dominant trait. Huntington disease is caused by a dominant allele. Note that each affected offspring in this pedigree has an affected parent.

by the observation that all of the offspring of two affected individuals will be affected. However, for genetic diseases like CF that limit survival or fertility, there may rarely or never be cases where two affected individuals produce offspring.

Although many of the alleles causing human genetic diseases are recessive, some are known to be dominant. **Figure 16.13** shows a family pedigree involving Huntington disease, a condition that causes the degeneration of brain cells involved in emotions, intellect, and movement. If you examine this pedigree, you will see that every affected individual has one affected parent. This pattern is characteristic of most dominant disorders. The symptoms of Huntington disease, which usually begin to appear when people are 30 to 50 years old, include uncontrollable jerking movements of the limbs, trunk, and face; progressive loss of mental abilities; and the development of psychiatric problems. In 1993, researchers identified the gene involved in this disorder. The normal allele encodes a protein called huntingtin, which functions in nerve cells. The mutant allele encodes an abnormal form of the protein, which aggregates within nerve cells in the brain. Further research is needed to determine how this aggregation contributes to the disease.

Most human genes are found on the paired chromosomes known as **autosomes**, which are the same in both sexes. Mendelian inheritance patterns involving these autosomal genes are described as autosomal inheritance patterns. Huntington disease is an example of a trait with an autosomal dominant inheritance pattern, while cystic fibrosis illustrates the pattern called autosomal recessive. However, some human genes are located on sex chromosomes, which are different in males and females. These genes have their own characteristic inheritance patterns, which we will consider in the next section.

16.2 Sex Chromosomes and X-Linked Inheritance Patterns

In the first part of this chapter we discussed Mendel's experiments that established the basis for understanding how traits are transmitted from parents to offspring. We also examined the chromosome theory of inheritance, which provided a framework for explaining Mendel's observations. Mendelian patterns of gene transmission are observed for many genes located on autosomes in a wide variety of eukaryotic species.

We will now turn our attention to genes located on **sex chromosomes**. As you learned in Chapter 15, this term refers to a distinctive pair of chromosomes that are different in males and females. Sex chromosomes are found in many but not all species with two sexes. The study of sex chromosomes proved pivotal in confirming the chromosome theory. The distinctive transmission patterns of genes on sex chromosomes helped early geneticists show that particular genes are located on particular chromosomes. Later, other researchers became interested in these genes because some of them were found to cause inherited diseases.

In this section, we will consider several mechanisms by which the presence or composition of sex chromosomes determines an individual's sex in different species. We will then examine some of the early research involving sex chromosomes that provided convincing evidence for the chromosome theory of inheritance. Finally, we will consider the inheritance patterns of genes on sex chromosomes and why recessive alleles are expressed more frequently in males than in females.

In Many Species, Sex Differences Are Due to the Presence of Sex Chromosomes

According to the chromosome theory of inheritance, chromosomes carry the genes that determine an organism's traits. Some early evidence supporting this theory involved a consideration of sex determination. In 1901, the American biologist C. E. McClung, who studied fruit flies, suggested that male and female sexes in insects are due to the inheritance of particular chromosomes. Following McClung's initial observations, several mechanisms of sex determination were found in different species of animals. Some examples are described in **Figure 16.14**. All of these mechanisms involve chromosomal differences between the sexes, and most involve a difference in a single pair of sex chromosomes.

In the X-Y system of sex determination, which operates in mammals, the somatic cells of males have one X and one Y chromosome, while female somatic cells contain two X chromosomes (Figure 16.14a). For example, the 46 chromosomes carried by human cells consist of one pair of sex chromosomes (either XY or XX) and 22 pairs of autosomes. The presence of the Y chromosome causes maleness in mammals. This is known from the analysis of rare individuals who carry chromosomal abnormalities. For example, mistakes that occasionally occur during meiosis may produce an individual who carries two X chromosomes and one Y chromosome. Such an individual develops into a male. A gene called the *SRY* gene located on the Y chromosome of mammals plays a key role in the developmental pathway that leads to maleness.

The X-O system operates in many insects (Figure 16.14b). Females in this system have a pair of sex chromosomes and are designated XX. In some insect species that follow the X-O system, the male has only one sex chromosome, the X, and is designated XO. In other X-O insect species, such as *Drosophila melanogaster*, the male has both an X chromosome and a Y chromosome and is designated XY. Unlike the Y chromosome of mammals, the Y chromosome in the X-O system does not determine maleness. In both types of insect species (XO or XY males), the individual's sex is determined by the ratio between its X chromosomes and its sets of autosomes. If a fly has one X chromosome and is diploid for the autosomes (2n), this ratio is 1/2, or 0.5. This fly will become a male whether or not it receives a Y chromosome. On the other hand, if a diploid fly receives two X chromosomes, the ratio is 2/2, or 1.0, and the fly becomes a female.

Thus far, we have considered examples where females have two similar copies of a sex chromosome, the X. However, in some animal species, such as birds and some fish, the male carries two similar chromosomes (Figure 16.14c). This is called the Z-W system to distinguish it from the X-Y system found in mammals. The male is ZZ and the female is ZW.

Not all chromosomal mechanisms of sex determination involve a special pair of sex chromosomes. An interesting mechanism known as the haplo-diploid system is found in bees (Figure 16.14d). The male bee, or drone, is produced from an unfertilized haploid egg. Thus, male bees are haploid individuals. Females, both worker bees and queen bees, are produced from fertilized eggs and therefore are diploid.

Although sex in many species of animals is determined by chromosomes, other mechanisms are also known. In certain reptiles and fish, sex is controlled by environmental factors such as temperature. For example, in the American alligator (*Alligator mississippiensis*), temperature controls sex development. When eggs of this alligator are incubated at 33°C, 100% of them produce male individuals. When the eggs are incubated at a temperature below 33°C, they produce 100% females, while at a temperature above 33°C, they produce 95% females.

Most species of flowering plants, including pea plants, have a single type of diploid plant, or sporophyte, that makes both male and female gametophytes. However, the sporophytes of some species have two sexually distinct types of individuals, one with flowers that produce male gametophytes, and the other with flowers that produce female gametophytes. Examples include hollies, willows, poplars, and date palms. Sex chromosomes, designated X and Y, are responsible for sex determination in many such species. The male plant is XY, while the female plant is XX. However, in some species with separate sexes, microscopic examination of the chromosomes does not reveal distinct types of sex chromosomes.

(a) The X-Y system in mammals

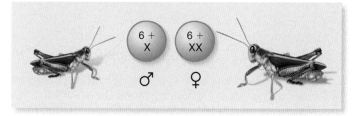

(b) The X-O system in certain insects

(c) The Z-W system in birds

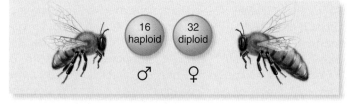

(d) The haplo-diploid system in bees

Figure 16.14   Different mechanisms of sex determination in animals.   The numbers shown in the circles indicate the numbers of autosomes.

Biological inquiry: If a person is born with only one X chromosome and no Y chromosome, would you expect that person to be a male or a female? Explain your answer.

In Humans, Recessive X-Linked Traits Are More Likely to Occur in Males

In humans, the X chromosome is rather large and carries many genes, while the Y chromosome is quite small and has relatively few genes. Therefore, many genes are found on the X chromosome but not on the Y; these are known as **X-linked genes**. By comparison, only a few genes are known to be Y linked, meaning that they are found on the Y chromosome but not on the X. The term **sex linked** refers to genes that are found on one sex chromosome but not on the other. Because few genes are found on the Y chromosome, the term usually refers to X-linked genes.

In mammals, a male cannot be described as being homozygous or heterozygous for an X-linked gene, because these terms describe genes that are present in two copies. Instead, the term **hemizygous** is used to describe the single copy of an X-linked gene in a male. Many recessive X-linked alleles cause diseases in humans, and these diseases occur more frequently in males than in females.

As an example, let's consider the X-linked recessive disorder called classical hemophilia (hemophilia A). In individuals with hemophilia, blood does not clot normally and a minor cut may bleed for a long time. Small bumps can lead to large bruises because broken capillaries may leak blood profusely into surrounding tissues before the capillaries are repaired. Common accidental injuries pose a threat of severe internal or external bleeding for hemophiliacs. Hemophilia A is caused by a recessive X-linked gene that encodes a defective form of a clotting protein. If a mother is a heterozygous carrier of hemophilia A, each of her children has a 50% chance of inheriting the recessive allele. A Punnett square shows a cross between a normal father and a heterozygous mother. X^{h-A} is the chromosome that carries the recessive allele for hemophilia A.

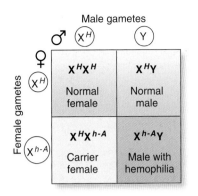

Although each child has a 50% chance of inheriting the hemophilia allele from the mother, only sons will exhibit the disorder. Because sons do not inherit an X chromosome from their fathers, a son who inherits the abnormal allele from his mother will have hemophilia. However, a daughter who inherits the hemophilia allele from her mother will also inherit a normal allele from her father. This daughter will have a normal phenotype, but if she passes the abnormal allele to her sons they will have hemophilia.

FEATURE INVESTIGATION

Morgan's Experiments Showed a Correlation Between a Genetic Trait and the Inheritance of a Sex Chromosome in *Drosophila*

The distinctive inheritance pattern of X-linked recessive alleles provides a way of demonstrating that a specific gene is on an X chromosome. In fact, an X-linked gene was the first gene to be located on a specific chromosome. In 1910, the American geneticist Thomas Hunt Morgan began work on a project in which he reared large populations of fruit flies, *Drosophila melanogaster*, in the dark to determine if their eyes would atrophy from disuse and disappear in future generations. Even after many consecutive generations, the flies showed no noticeable changes. After two years, however, Morgan finally obtained an

interesting result: a true-breeding line of *Drosophila* produced a male fly with white eyes rather than the normal red eyes. The white-eye trait must have arisen from a new mutation that converted a red-eye allele into a white-eye allele.

Using an approach similar to Mendel's, Morgan studied the inheritance of the white-eye trait by making crosses and quantitatively analyzing their outcome. In the experiment described in **Figure 16.15**, Morgan crossed his white-eyed male to a red-eyed female. All of the F_1 offspring had red eyes, indicating that red is dominant to white. The F_1 offspring were then mated to each other to obtain an F_2 generation. As seen in the data table, this cross produced 2,459 red-eyed females, 1,011 red-eyed males, and 782 white-eyed males. Surprisingly, no white-eyed females were observed in the F_2 generation.

Figure 16.15 Morgan's crosses of red-eyed and white-eyed *Drosophila*.

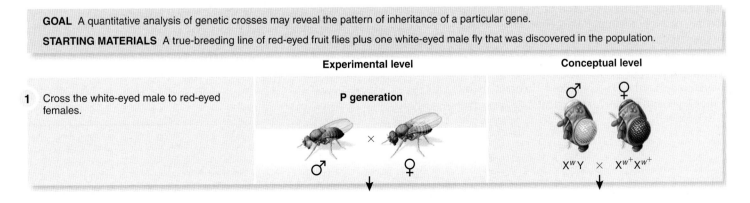

GOAL A quantitative analysis of genetic crosses may reveal the pattern of inheritance of a particular gene.

STARTING MATERIALS A true-breeding line of red-eyed fruit flies plus one white-eyed male fly that was discovered in the population.

Experimental level | Conceptual level

1 Cross the white-eyed male to red-eyed females.

P generation

$X^w Y$ × $X^{w^+} X^{w^+}$

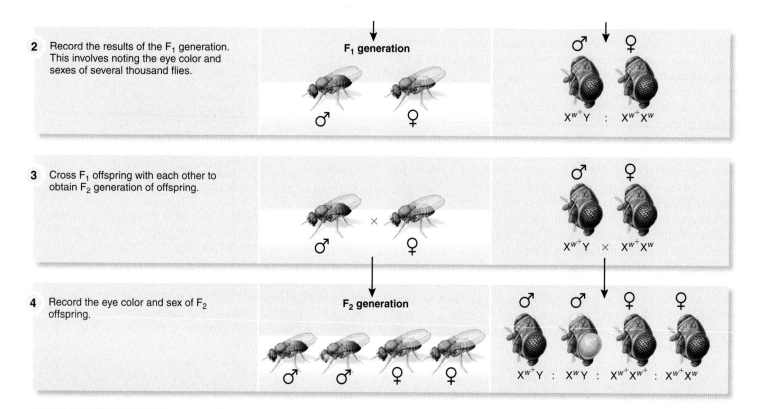

2 Record the results of the F₁ generation. This involves noting the eye color and sexes of several thousand flies.

F₁ generation

$X^{w^+}Y$: $X^{w^+}X^w$

3 Cross F₁ offspring with each other to obtain F₂ generation of offspring.

$X^{w^+}Y$ × $X^{w^+}X^w$

4 Record the eye color and sex of F₂ offspring.

F₂ generation

$X^{w^+}Y$: X^wY : $X^{w^+}X^{w^+}$: $X^{w^+}X^w$

5 THE DATA

Cross	Results
Original white-eyed male to red-eyed females	F₁ generation All red-eyed flies
F₁ males to F₁ females	F₂ generation 2,459 red-eyed females 1,011 red-eyed males 0 white-eyed females 782 white-eyed males

Morgan's results suggested a connection between the alleles for eye color and the sex of the offspring. As shown in the conceptual column of Figure 16.15 and in the Punnett square, the data are consistent with the idea that the eye-color alleles in *Drosophila* are located on the X chromosome. X^{w^+} is the chromosome carrying the normal allele for red eyes and X^w is the chromosome carrying the mutant allele for white eyes.

The Punnett square predicts that the F₂ generation will not have any white-eyed females, a prediction that was confirmed by Morgan's experimental data. However, it should also be pointed out that the experimental ratio of red eyes to white eyes in the F₂ generation is (2,459 + 1,011):782, which equals 4.4:1.

This ratio deviates significantly from the ratio of 3:1 predicted in the Punnett square. The lower than expected number of white-eyed flies is explained by a decreased survival of white-eyed flies.

F₁ male is $X^{w^+}Y$
F₁ female is $X^{w^+}X^w$

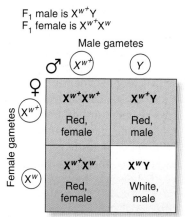

Following this initial discovery, Morgan carried out many experimental crosses that located specific genes on the *Drosophila* X chromosome. This research provided some of the most persuasive evidence for the Mendelian gene concept and the chromosome theory of inheritance, which are the foundations of modern genetics. In 1933, Morgan became the first geneticist to receive a Nobel Prize.

16.3 Variations in Inheritance Patterns and Their Molecular Basis

The term **Mendelian inheritance** describes the inheritance patterns of genes that segregate and assort independently. In the first section of this chapter, we considered the inheritance pattern of traits affected by a single gene that is found in two variants, one of which is completely dominant over the other. This pattern is called **simple Mendelian inheritance** because the phenotypic ratios in the offspring clearly demonstrate Mendel's laws. In the second section, we examined **X-linked inheritance**, the pattern displayed by pairs of dominant and recessive alleles located on X chromosomes. Early geneticists observed these Mendelian inheritance patterns without knowing why one trait was dominant over another.

In this section we will discuss the molecular basis of dominant and recessive traits, and we will see how the molecular expression of a gene can have widespread effects on an organism's phenotype. In addition, we will examine the inheritance patterns of genes that segregate and assort independently but do not display a simple dominant/recessive relationship. The transmission of these genes from parents to offspring does not usually produce the ratios of phenotypes we would expect on the basis of Mendel's observations. This does not mean that Mendel was wrong. Rather, the inheritance patterns of many traits are more intricate and interesting than the simple patterns he chose to study. As described in Table 16.1, our understanding of gene function at the molecular level explains both simple Mendelian inheritance and other, more complex inheritance patterns that conform to Mendel's laws. This modern knowledge also sheds light on the role of the environment in producing an organism's phenotype, which we will discuss at the end of the section.

Protein Function Explains the Phenomenon of Dominance

As we learned at the beginning of this chapter, Mendel studied seven traits that were found in two variants each. The dominant variants are the common alleles for these traits in pea plants. For any given gene, geneticists refer to a prevalent allele in a population as a **wild-type allele** (see Figure 16.2). In most cases, a wild-type allele encodes a protein that is made in the proper amount and functions normally. By comparison, alleles that have been altered by mutation are called **mutant alleles**; these tend to be rare in natural populations. In the case of Mendel's seven traits, the recessive alleles are due to rare mutations.

By studying genes and their gene products at the molecular level, researchers have discovered that a mutant allele is often defective in its ability to express a functional protein. In other words, mutations that produce mutant alleles are likely to decrease or eliminate the synthesis or functional activity of a protein. Such mutations are often inherited in a recessive fashion. To understand why many defective alleles are recessive, we need to take a quantitative look at protein function.

Table 16.1	Different Types of Mendelian Inheritance Patterns and Their Molecular Basis
Type	**Description**
Simple Mendelian inheritance	**Inheritance pattern:** Pattern of traits determined by a pair of alleles that display a dominant/recessive relationship and are located on an autosome. The presence of the dominant allele masks the presence of the recessive allele.
	Molecular basis: In many cases, the amount of protein produced by a heterozygote, which may be 50% of that produced by a dominant homozygote, is sufficient to produce the dominant trait.
X-linked inheritance	**Inheritance pattern:** Pattern of traits determined by genes that display a dominant/recessive relationship and are located on the X chromosome. In mammals and fruit flies, males are hemizygous for X-linked genes. In these species, X-linked recessive traits occur more frequently in males than in females.
	Molecular basis: In a female with one recessive X-linked allele (a heterozygote), the protein encoded by the dominant allele is sufficient to produce the dominant trait. A male with a recessive X-linked allele (a hemizygote) does not have a dominant allele and does not make any of the functional protein.
Incomplete dominance	**Inheritance pattern:** Pattern that occurs when the heterozygote has a phenotype intermediate to the phenotypes of the dominant and recessive homozygotes, as when a cross between red-flowered and white-flowered plants produces pink-flowered offspring.
	Molecular basis: 50% of the protein encoded by the normal (wild-type) allele is not sufficient to produce the normal trait.
Codominance	**Inheritance pattern:** Pattern that occurs when the heterozygote expresses both alleles simultaneously. For example, a human carrying the A and B alleles for the ABO antigens of red blood cells produces both the A and the B antigens (has an AB blood type).
	Molecular basis: The codominant alleles encode proteins that function slightly differently from each other. In a heterozygote, the function of each protein affects the phenotype uniquely.
Sex-influenced inheritance	**Inheritance pattern:** Pattern that occurs when an allele is recessive in one sex and dominant in the other. An example is pattern baldness in humans.
	Molecular basis: Sex hormones affect the molecular expression of genes, which can have an impact on the phenotype.

In a simple dominant/recessive relationship, the recessive allele does not affect the phenotype of the heterozygote. In this type of relationship, a single copy of the dominant (wild-type) allele is sufficient to mask the effects of the recessive allele. But if the recessive allele cannot produce a functional protein, how do we explain the dominant phenotype of the heterozygote? **Figure 16.16** considers the example of flower color in a pea plant. As shown at the top of this figure, the gene encodes an enzyme that is needed to make a purple pigment. The *P* allele is dominant because one *P* allele encodes enough of the functional protein—50% of the amount found in a normal homozygote—to provide a normal phenotype. Thus, the *PP* homozygote and the *Pp* heterozygote both make enough of the purple pigment to yield purple flowers. The *pp* heterozygote cannot make any of the functional enzyme required for pigment synthesis, so its flowers are white.

The explanation "50% of the normal protein is enough" is true for many dominant alleles. In such cases, the normal homozygote is making much more of the wild-type protein than necessary, so if the amount is reduced to 50%, as it is in the heterozygote, the individual still has plenty of this protein to accomplish whatever cellular function it performs. In other cases, however, an allele may be dominant because the heterozygote actually produces more than 50% of the normal amount of functional protein. This increased production is due to the phenomenon of gene regulation, which is discussed in Chapter 13. The normal gene is "up-regulated" in the heterozygote to compensate for the lack of function of the defective allele.

GENOMES & PROTEOMES

Single-Gene Mutations Cause Many Inherited Diseases and Have Pleiotropic Effects

The idea that recessive alleles usually cause a substantial decrease in the expression of a functional protein is supported by analyses of many human genetic diseases. Keep in mind that a genetic disease is caused by a rare mutant allele. **Table 16.2** lists several examples of human genetic diseases in which a recessive allele fails to produce a specific cellular protein in its active form.

Over 7,000 human disorders are caused by mutations in single genes. With a human genome size of 20,000 to 25,000 genes, this means that roughly one-third of our genes are known to cause some kind of abnormality when mutations alter their expression. Any particular single-gene disorder is relatively rare.

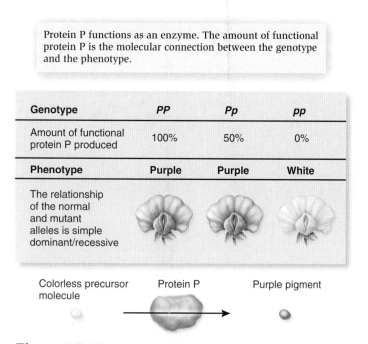

Protein P functions as an enzyme. The amount of functional protein P is the molecular connection between the genotype and the phenotype.

Genotype	*PP*	*Pp*	*pp*
Amount of functional protein P produced	100%	50%	0%
Phenotype	Purple	Purple	White
The relationship of the normal and mutant alleles is simple dominant/recessive			

Colorless precursor molecule   Protein P   Purple pigment

Figure 16.16   **How genes give rise to traits in a simple dominant/recessive relationship.**   In many cases, the amount of protein encoded by a single dominant allele is sufficient to produce the normal phenotype. In this example, the normal phenotype is purple flower color in a pea plant. The normal allele (*P*) encodes protein P, an enzyme needed for the synthesis of purple pigment. A plant with one or two copies of the normal allele produces enough pigment to produce purple flowers. In a *pp* homozygote, the complete lack of the normal protein results in white flowers.

Table 16.2	Examples of Recessive Human Genetic Diseases	
Disease	**Protein produced by the normal gene***	**Description**
Phenylketonuria	Phenylalanine hydroxylase	Inability to metabolize phenylalanine. Can lead to severe mental retardation and physical degeneration. The disease can be prevented by following a phenylalanine-free diet beginning early in life.
Cystic fibrosis	A chloride-ion transporter	Inability to regulate ion balance in epithelial cells. Leads to a variety of abnormalities, including production of thick lung mucus and chronic lung infections.
Tay-Sachs disease	Hexosaminidase A	Defect in lipid metabolism. Leads to paralysis, blindness, and early death.
Alpha-1 antitrypsin deficiency	Alpha-1 antitrypsin	Inability to prevent the activity of protease enzymes. Causes liver damage and emphysema.
Hemophilia A	Coagulation Factor VIII	A defect in blood clotting due to a missing clotting factor. An accident may cause excessive bleeding or internal hemorrhaging.

* Individuals who exhibit the disease are homozygous (or hemizygous) for a recessive allele that results in a defect in the amount or function of the normal protein.

But taken together, about one individual in 100 has a disorder that is due to a single-gene mutation. Such diseases generally have simple inheritance patterns in family pedigrees. Although the majority of these diseases follow a recessive inheritance pattern, some are known to be dominant. We have already discussed Huntington disease as an example of a dominant human disorder (see Figure 16.13). Other examples of diseases caused by dominant alleles include achondroplasia (a form of dwarfism) and osteogenesis imperfecta (brittle bone disease).

Single-gene disorders illustrate the phenomenon of **pleiotropy**, which means that a mutation in a single gene can have multiple effects on an individual's phenotype. Pleiotropy occurs for several reasons, including:

1. The expression of a single gene can affect cell function in more than one way. For example, a defect in a microtubule protein may affect cell division and cell movement.
2. A gene may be expressed in different cell types in a multicellular organism.
3. A gene may be expressed at different stages of development.

In this genetics unit, we tend to discuss genes as they affect a single trait. This educational approach allows us to appreciate how genes function, and how they are transmitted from parents to offspring. However, this focus may also obscure how amazing genes really are. In all or nearly all cases, the expression of a gene is pleiotropic with regard to the characteristics of an organism. The expression of any given gene influences the expression of many other genes in the genome, and vice versa. Pleiotropy is revealed when researchers study the effects of gene mutations.

As an example of a pleiotropic mutation, let's consider cystic fibrosis (CF), which we considered earlier as an example of a recessive human disorder (see Figure 16.12). In the late 1980s, the gene for CF was identified. The normal allele encodes a protein called the cystic fibrosis transmembrane conductance regulator (CFTR) that regulates ionic balance by allowing the transport of chloride ions (Cl^-) across epithelial-cell membranes. The mutation that causes CF diminishes the function of this Cl^- transporter, affecting several parts of the body in different ways. Because the movement of Cl^- affects water transport across membranes, the most severe symptom of CF is thick mucus in the lungs, which occurs because of a water imbalance. In sweat glands, the normal Cl^- transporter has the function of recycling salt out of the glands and back into the skin before it can be lost to the outside world. Persons with CF have excessively salty sweat due to their inability to recycle salt back into their skin cells. A common test for CF is measurement of salt on the skin. Another effect is seen in the reproductive systems of males who are homozygous for the CF allele. Most males with CF are infertile because the vas deferens, the tubules that transport sperm from the testes, are absent or undeveloped. Presumably, a normally functioning Cl^- transporter is needed for the proper development of the vas deferens in the embryo. Taken together, we can see that a defect in CFTR has multiple effects throughout the body.

Incomplete Dominance Results in an Intermediate Phenotype

For certain traits, a heterozygote that carries two different alleles exhibits a phenotype that is intermediate between the corresponding homozygous individuals. This phenomenon is known as **incomplete dominance**. In 1905, Carl Correns discovered this pattern of inheritance for alleles affecting flower color in the four-o'clock plant (*Mirabilis jalapa*). **Figure 16.17** shows a cross between two four-o'clock plants, a red-flowered homozygote and a white-flowered homozygote. The allele for red flower color is designated C^R and the white allele is C^W. These alleles are designated with superscripts rather than upper- and lowercase letters because neither allele is dominant.

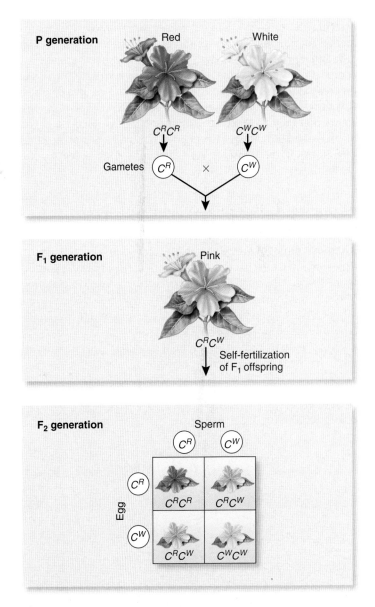

Figure 16.17 Incomplete dominance in the four-o'clock plant. When red-flowered and white-flowered homozygotes (C^RC^R and C^WC^W) are crossed, the resulting heterozygote (C^RC^W) has an intermediate phenotype of pink flowers.

The offspring of this cross have pink flowers—that is, they are $C^R C^W$ heterozygotes with an intermediate phenotype. If these F_1 offspring are allowed to self-fertilize, the F_2 generation shows a ratio of 1/4 red-flowered plants, 1/2 pink-flowered plants, and 1/4 white-flowered plants. As noted in the Punnett square, this is a 1:2:1 phenotypic ratio rather than the 3:1 ratio observed for simple Mendelian inheritance. In this case, 50% of the protein encoded by the C^R gene is not sufficient to produce the red-flower phenotype.

The degree to which we judge an allele to exhibit incomplete dominance may depend on how closely we examine an individual's phenotype. An example is an inherited human disease called phenylketonuria (PKU). This disorder is caused by a rare mutation in a gene that encodes an enzyme called phenylalanine hydroxylase. This enzyme is needed to metabolize the amino acid phenylalanine. If left untreated, homozygotes carrying the mutant allele suffer severe symptoms, including mental retardation, seizures, microcephaly (small head), poor development of tooth enamel, and decreased body growth. By comparison, heterozygotes appear phenotypically normal. For this reason, geneticists consider PKU to be a recessive disorder. However, biochemical analysis of the blood of heterozygotes shows that they typically have a phenylalanine blood level double that of an individual carrying two normal copies of the gene. Therefore, at this closer level of examination, heterozygotes exhibit incomplete dominance.

ABO Blood Type Is an Example of Multiple Alleles and Codominance

Although diploid individuals have only two copies of most genes, many genes have three or more variants in one population. We describe such a gene as occurring in **multiple alleles**. The phenotype depends on which two alleles an individual inherits. ABO blood types in humans are an example of phenotypes produced by multiple alleles.

As shown in **Table 16.3**, human red blood cells have structures on their plasma membrane known as surface antigens,

which are constructed from several sugar molecules that are connected to form a carbohydrate tree. Antigens are substances (in this case, carbohydrates) that are recognized as foreign by antibodies produced by the immune system. Three types of surface antigens, known as A, B, and O, are found on red blood cells. The synthesis of these antigens is determined by enzymes that are encoded by a gene that exists in three alleles designated I^A, I^B, and i, respectively. The i allele is recessive to both I^A and I^B. A person who is ii homozygous will have red blood cells with the surface antigen O (blood type O). The red blood cells of an $I^A I^A$ homozygous or $I^A i$ heterozygous individual will have surface antigen A (blood type A). Similarly, a homozygous $I^B I^B$ or heterozygous $I^B i$ individual will produce surface antigen B (blood type B). A person who is $I^A I^B$ heterozygous makes both antigens, A and B, on every red blood cell (blood type AB). The phenomenon in which a single individual expresses both alleles is called **codominance**.

What is the molecular explanation for codominance? Biochemists have analyzed the carbohydrate tree produced in people of differing blood types. The differences are shown schematically in Table 16.3. In type O, the carbohydrate tree is smaller than in type A or type B because a sugar has not been attached to a specific site on the tree. People with blood type O have a loss-of-function mutation in the gene that encodes the enzyme that attaches a sugar at this site. This enzyme, called a glycosyl transferase, is inactive in type O individuals. In contrast, the type A and type B antigens have sugars attached to this site, but each of them has a different sugar. This difference occurs because the enzymes encoded by the I^A allele and the I^B allele have slightly different active sites. As a result, the enzyme encoded by the I^A allele attaches a sugar called N-acetylgalactosamine to the carbohydrate tree, while the enzyme encoded by the I^B allele attaches galactose. N-acetylgalactosamine is represented by an orange hexagon in Table 16.3, and galactose by a green triangle.

The attachment of two different sugars gives surface antigens A and B significantly different molecular structures. Such differences in shape allow antibodies to recognize and bind very

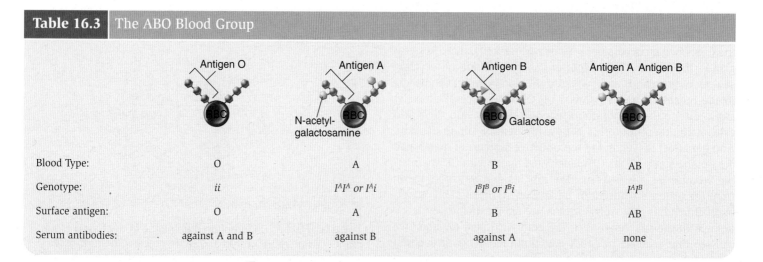

Table 16.3	The ABO Blood Group			
	Antigen O	Antigen A	Antigen B	Antigen A Antigen B
Blood Type:	O	A	B	AB
Genotype:	ii	$I^A I^A$ or $I^A i$	$I^B I^B$ or $I^B i$	$I^A I^B$
Surface antigen:	O	A	B	AB
Serum antibodies:	against A and B	against B	against A	none

specifically to certain antigens. The blood of type A individuals has antibodies, called serum antibodies, that bind to the B antigen. Similarly, type B individuals produce antibodies against the A antigen. Type O individuals produce both kinds of antibodies and type AB individuals produce neither. (No antibodies are produced against antigen O.) When a person receives a blood transfusion, the donor's blood must be an appropriate match with the recipient's blood to avoid a dangerous antigen-antibody reaction. For example, if a person with type O blood is given type A blood, the recipient's anti-A antibodies will react with the donated blood cells and cause them to agglutinate (clump together). This situation is life-threatening because it will cause the blood vessels to clog. Identification of the donor and recipient blood types, called blood typing, is essential for safe transfusions.

The Expression of Certain Traits Is Influenced by the Sex of the Individual

Certain autosomal genes are expressed differently in heterozygous males and females. The term **sex-influenced inheritance** refers to the phenomenon in which an allele is dominant in one sex but recessive in the other. Pattern baldness is an example of a sex-influenced trait in humans. This trait is characterized by a balding pattern in which hair loss occurs on the front and top but not on the sides (**Figure16.18**). A male who is heterozygous for the pattern-baldness allele (designated *B*) will become bald, but a heterozygous female will not. In other words, the baldness allele is dominant in males but recessive in females:

Genotype	Phenotype	
	Females	Males
BB	bald	bald
Bb	nonbald	bald
bb	nonbald	nonbald

A woman who is homozygous for the baldness allele will develop the trait (although in women it is usually characterized by a significant thinning of the hair that occurs relatively late in life).

As you can see from the pedigree in Figure 16.18, a bald male may have inherited the baldness allele from either parent. Thus, a striking observation is that fathers with pattern baldness can pass this trait to their sons. This could not occur if the trait was X linked, because fathers transmit only Y chromosomes to their sons.

The sex-influenced nature of pattern baldness is related to the production of the male sex hormone testosterone. The gene that affects pattern baldness encodes an enzyme called 5 α-reductase, which converts testosterone to 5-α-dihydrotestosterone (DHT). DHT binds to cellular receptors and affects the expression of many genes, including those in the cells of the scalp. The allele that causes pattern baldness results in an overexpres-

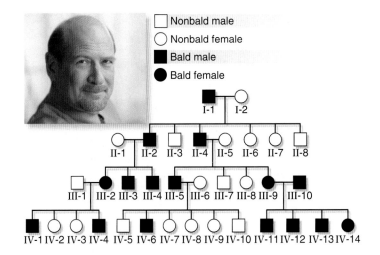

Figure 16.18 **Pattern baldness.** Pattern baldness, shown in an adult male in the photograph, is an example of the sex-influenced expression of an autosomal gene. Bald individuals are represented by filled symbols in the pedigree.

Biological inquiry: With regard to baldness, which phenotypes have a single genotype?

sion of this enzyme. Because mature males normally make more testosterone than females, this allele has a greater phenotypic impact in males. However, a rare tumor of the adrenal gland can cause the secretion of abnormally large amounts of testosterone in females. If this occurs in a woman who is heterozygous *Bb*, she will become bald. If the tumor is removed surgically, her hair will return to its normal condition.

The Environment Plays a Vital Role in the Making of a Phenotype

In this chapter, we have been concerned mainly with the effects of genes on phenotypes. However, phenotypes are shaped by an organism's environment as well as by its genes. An organism cannot exist without its genes or without an environment in which to live. Both are indispensable for life. An organism's genotype provides the information for environmental conditions to create a phenotype.

The term **norm of reaction** refers to the effects of environmental variation on a phenotype. Specifically, it is the phenotypic range seen in individuals with a particular genotype. To evaluate the norm of reaction, researchers study members of true-breeding strains that have the same genotypes, and subject them to different environmental conditions. For example, **Figure 16.19** shows the norm of reaction for genetically identical plants raised at different temperatures. As shown in the figure, these plants attain a maximal height when raised at 75°F. At 55°F and 85°F, the plants are substantially shorter. Growth cannot occur below 40°F or above 95°F.

The norm of reaction can be quite dramatic when we consider environmental influences on certain inherited diseases.

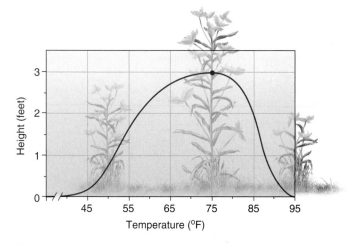

Figure 16.19 **The norm of reaction.** The norm of reaction is the range of phenotypes that an organism with a particular genotype exhibits under different environmental conditions. In this example, genetically identical plants were grown at different temperatures in a greenhouse and then measured for height.

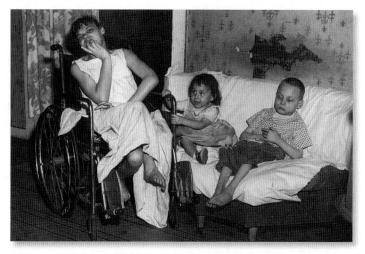

Figure 16.20 **Environmental influences on the expression of PKU within a single family.** All three children in this photo have inherited the alleles that cause PKU. The child in the middle was raised on a phenylalanine-free diet and developed normally. The other two children, born before the benefits of such a diet were known, were raised on diets containing phenylalanine. These two children have symptoms of PKU, including mental impairment.

Photo from the March of Dimes Birth Defects Foundation.

A striking example is the human genetic disease phenylketonuria (PKU). As we discussed earlier in the chapter, this disorder is caused by a rare mutation in the gene for phenylalanine hydroxylase, an enzyme needed to metabolize the amino acid phenylalanine. People with one or two functional copies of the gene can eat foods containing the amino acid phenylalanine and metabolize it correctly. However, homozygotes for the defective allele cannot metabolize phenylalanine. When these individuals eat a standard diet containing phenylalanine, this amino acid accumulates within their bodies and becomes highly toxic. Under these conditions, PKU homozygotes manifest a variety of detrimental symptoms, including mental impairment, underdeveloped teeth, and foul-smelling urine.

In contrast, when these individuals are identified at birth and given a restricted diet that is free of phenylalanine, they develop normally (**Figure 16.20**). This is a dramatic example of how genes and the environment can interact to determine an individual's phenotype. In the U.S., most newborns are tested for PKU, which occurs in about 1 in 10,000 babies. A newborn who is found to have this disorder can be raised on a phenylalanine-free diet and develop normally.

Chapter Summary

16.1 Mendel's Laws and the Chromosome Theory of Inheritance

- Mendel focused his attention on seven traits found in garden peas that existed in two variants each. (Figures 16.1, 16.2)

- Mendel could allow his peas to self-fertilize or he could carry out cross-fertilization, also known as hybridization. (Figures 16.3, 16.4)

- By following the inheritance pattern of a single trait (a single-factor cross) for two generations, Mendel determined the law of segregation. This law tells us that two alleles of a gene segregate from each other when passed from parents to offspring. (Figures 16.5, 16.6)

- The genotype is the genetic makeup of an organism. Alleles are alternative versions of the same gene. Phenotype is a description of the traits that an organism displays.

- A Punnett square can be constructed to predict the outcome of crosses.

- A testcross can be conducted to determine if an individual displaying a dominant trait is a homozygote or heterozygote. (Figure 16.7)

- By conducting a dihybrid cross, Mendel determined the law of independent assortment, which says that the alleles for two different genes assort independently of each other. In a dihybrid cross, this yields a 9:3:3:1 ratio in the F_2 generation. (Figure 16.8)

- The chromosome theory of inheritance explains how the steps of meiosis account for the inheritance patterns observed by Mendel. Each gene is located at a particular locus on a chromosome. (Figures 16.9, 16.10, 16.11)

- The product rule and sum rule allow us to predict the outcome of crosses based on probability. Random sampling error is the deviation between observed and predicted values.

- Instead of conducting crosses, the inheritance patterns in humans are determined from a pedigree analysis. (Figures 16.12, 16.13)

16.2 Sex Chromosomes and X-Linked Inheritance Patterns

- Many species of animals and a few species of plants have separate male and female sexes. In many cases, sex is determined by differences in sex chromosomes. (Figure 16.14)

- In mammals, recessive X-linked traits are more likely to occur in males. An example is hemophilia.

- Morgan carried out crosses that showed that an eye-color gene in *Drosophila* is located on the X chromosome. (Figure 16.15)

16.3 Variations in Inheritance Patterns and Their Molecular Basis

- Several inheritance patterns have been discovered that obey Mendel's laws but yield differing ratios of offspring compared to Mendel's crosses. (Table 16.1)

- Recessive inheritance is often due to a loss-of-function allele. The heterozygote has a dominant phenotype because 50% of the normal protein is sufficient to produce that phenotype. (Figure 16.16)

- Mutant genes are responsible for many inherited diseases in humans. In many cases, the effects of a mutant gene are pleiotropic, meaning that the gene affects several different aspects of bodily structure and function. (Table 16.2)

- Incomplete dominance occurs when a heterozygote has a phenotype that is intermediate between either homozygote. This occurs because 50% of the normal protein is not enough to give the same phenotype as a normal homozygote. (Figure 16.17)

- ABO blood type is an example of multiple alleles in which a gene exists in three alleles in a population. The A and B alleles are codominant, which means that both are expressed in the same individual. These alleles encode enzymes with different specificities for attaching sugar molecules to make antigens. (Table 16.3)

- Pattern baldness in people is a sex-influenced trait that is dominant in males and recessive in females. This pattern occurs because sex hormones influence the expression of certain genes. (Figure 16.18)

- All traits are influenced by the environment. The norm of reaction is a description of how a trait may change depending on the environmental conditions. (Figures 16.19, 16.20)

TEST YOURSELF

1. Based on Mendel's experimental crosses, what is the expected F_2 phenotypic ratio of a monohybrid cross?
   a. 1:2:1   d. 9:3:3:1
   b. 2:1   e. 4:1
   c. 3:1

2. During which phase of cellular division does Mendel's law of segregation physically occur?
   a. mitosis   d. all of the above
   b. meiosis I   e. b and c only
   c. meiosis II

3. An individual that has two different alleles of a particular gene is said to be
   a. dihybrid.   d. heterozygous.
   b. recessive.   e. hemizygous.
   c. homozygous.

4. Which of Mendel's laws cannot be observed in a monohybrid cross?
   a. segregation
   b. dominance/recessiveness
   c. independent assortment
   d. codominance
   e. All of the above can be observed in a monohybrid cross.

5. During a ____ cross, an individual with the dominant phenotype and unknown genotype is crossed with a _____ individual to determine the unknown genotype.
   a. monohybrid, homozygous recessive
   b. dihybrid, heterozygous
   c. test, homozygous dominant
   d. monohybrid, homozygous dominant
   e. test, homozygous recessive

6. In humans, males are said to be _____ at X-linked loci.
   a. dominant   d. heterozygous
   b. homozygous   e. hemizygous
   c. recessive

7. A gene that affects more than one phenotypic trait is said to be
   a. dominant.   d. pleiotropic.
   b. wild type.   e. heterozygous.
   c. dihybrid.

8. A hypothetical flowering plant species produces red, pink, and white flowers. To determine the inheritance pattern, the following crosses were conducted with the results indicated:
   red × red → all red
   white × white → all white
   red × white → all pink

   What type of inheritance pattern does this represent?
   a. dominance/recessiveness   d. incomplete dominance
   b. X-linked   e. pleiotropy
   c. codominance

9. Genes located on a sex chromosome are said to be
   a. X-linked.   d. sex-linked.
   b. dominant.   e. sex-influenced.
   c. hemizygous.

10. Genes that are expressed differently depending on whether the individual is male or female are
   a. sex-linked.   d. incomplete dominant.
   b. pleiotropic.   e. hemizygous.
   c. sex-influenced.

CONCEPTUAL QUESTIONS

1. Define genotype and phenotype.

2. Define autosome.

3. Explain why recessive X-linked traits in humans are more likely to occur in males.

EXPERIMENTAL QUESTIONS

1. Prior to the Feature Investigation, what was the original purpose of Morgan's experiments with *Drosophila*?

2. How was Morgan able to demonstrate that red-eye color is dominant to white-eye color?

3. What results led Morgan to conclude that eye color was associated with the sex of the individual?

COLLABORATIVE QUESTIONS

1. Discuss Mendel's two laws and why they are important.

2. What are the fundamental principles of the chromosome theory of inheritance?

---

www.brookerbiology.com

This website includes answers to the Biological Inquiry questions found in the figure legends and all end-of-chapter questions.

17

COMPLEX PATTERNS
OF INHERITANCE

CHAPTER OUTLINE

Snail shells (*Lymnaea peregra*) that coil to the right or left. The direction of coiling of a snail's shell is an example of a complex inheritance pattern.

In Chapter 16, we examined inheritance patterns in which the outcome of a single trait was governed by a single gene. In the cases we considered, the alleles segregated and assorted independently, allowing us to predict the phenotypes of offspring from the genotypes of their parents. These phenotypes occurred in definite ratios and they did not overlap—a pea plant was either tall or dwarf; a blood type was either A, B, or O. The inheritance patterns of most traits are more complex, however, and in this chapter we will examine some of the factors that complicate the prediction of phenotypes.

In the first section of the chapter, we will consider how two or more different genes may affect the outcome of a single trait. For example, we examine continuously varying traits like human skin color, and you will see how the interaction of multiple genes and environmental influences can produce such a continuum. In the rest of the chapter we will consider inheritance patterns that defy Mendel's laws of inheritance. First we discuss genes that are linked on the same chromosome and therefore do not assort independently. Next we consider the genes found in chloroplasts and mitochondria, which defy the law of segregation. Don't worry, Mendel's laws do describe most inheritance patterns, and they accurately reflect the behavior of chromosomes during meiosis. However, as you will learn in this chapter, they simply don't apply to all of the genes that eukaryotic organisms possess. We will end the chapter with a discussion of three inheritance patterns, X inactivation, genomic imprinting, and maternal effect, that were not easily explained until researchers began to unravel genetic events that occur at the cellular and molecular levels. As you will learn, males and females don't always regulate their genes in the same way, and this can lead to seemingly bizarre patterns of inheritance that are distinct from X-linked and sex-influenced inheritance patterns,

which we considered in the previous chapter. An exciting advance over the past few decades has been a better understanding of such unusual patterns of inheritance.

Studies of complex inheritance patterns such as those described in this chapter have helped us appreciate more fully how genes influence phenotypes. These studies have revealed an astounding variety in the ways that inheritance occurs. The picture that emerges is of a wonderful web of diverse mechanisms by which genes give rise to phenotypes. **Table 17.1** provides a summary of the most common patterns of inheritance.

17.1 Gene Interactions

The study of single genes was pivotal in establishing the science of genetics. This focus allowed Mendel to formulate the basic laws of inheritance for traits with a simple dominant/recessive inheritance pattern. Likewise, this approach helped later researchers understand inheritance patterns involving incomplete dominance and codominance, as well as traits that are influenced by an individual's sex. In reality, however, all or nearly all traits are influenced by many genes. For example, in both plants and animals, height is affected by genes that encode proteins involved in the production of growth hormones, cell division, the uptake of nutrients, metabolism, and many other functions. A defect in any of these genes is likely to have a negative impact on an individual's height.

If height is controlled by many genes, you may be wondering how Mendel was able to study the effects of a single gene that produced tall or dwarf pea plants. The answer lies in the genotypes of his strains. Although many genes affect the height of pea plants, Mendel chose true-breeding strains that differed

Table 17.1	Different Types of Inheritance Patterns
Type	**Description**
Mendelian	Inheritance patterns in which a single gene affects a single trait, and the alleles segregate and assort independently. These patterns include simple dominant/recessive traits, X-linked traits controlled by a single gene, incomplete dominance, codominance, and sex-influenced traits (refer back to Table 16.1).
Epistasis	A type of gene interaction in which the alleles of one gene mask the effects of a dominant allele of another gene.
Continuous variation	Inheritance pattern in which the offspring display a continuous range of phenotypes. This pattern is produced by the additive interactions of several genes, together with environmental influences.
Linkage	Inheritance patterns involving two or more genes that are close together on the same chromosome. These genes do not assort independently.
Extranuclear inheritance	Transmission pattern of genes found in the DNA of mitochondria or chloroplasts, which are inherited independently of genes in the nucleus and do not segregate during meiosis. Usually these genes are inherited from the mother.
X inactivation	Phenomenon of female mammals in which one X chromosome is inactivated in every somatic cell, producing a mosaic phenotype. Most genes on the inactivated X chromosome are not expressed.
Genomic imprinting	Inheritance pattern in which an allele from one parent is inactivated in the somatic cells of the offspring, while the allele from the other parent is expressed.
Maternal effect	Inheritance pattern in which the genotype of the mother determines the phenotype of the offspring. This occurs because maternal effect genes of the mother provide gene products to developing egg cells.

with regard to only one of these genes. As a hypothetical example, let's suppose that pea plants have 10 genes affecting height, which we will call *K, L, M, N, O, P, Q, R, S,* and *T*. The genotypes of two hypothetical strains of pea plants may be:

Tall strain: *KK LL MM NN OO PP QQ RR SS TT*

Dwarf strain: *KK LL MM NN OO PP QQ RR SS tt*

In this example, the tall and dwarf strains differ at only a single locus. One strain is *TT* and the other is *tt*, and this accounts for the difference in their height. If we make crosses of tall and dwarf plants, the genotypes of the F$_2$ offspring will differ with regard to only one gene; the other nine genes will be identical in all of them. This approach allows a researcher to study the effects of a single gene even though many genes may affect a single trait.

In this section, we will examine situations in which a single trait is controlled by two or more genes, each of which has two

or more alleles. This phenomenon is called a **gene interaction**. As you will see, allelic variation at two or more loci may affect the outcome of traits in different ways. First we will look at interactions in which an allele of one gene prevents the expression of an allele of a different gene. Then we will discuss interactions in which multiple genes have additive effects on a single trait. These additive effects, together with environmental influences, account for the continuous phenotypic variation that we see for most traits.

An Epistatic Gene Interaction Occurs When the Allele of One Gene Masks the Phenotypic Effects of a Different Gene

In some gene interactions, the alleles of one gene mask the expression of the alleles of another gene. This phenomenon is called **epistasis** (Greek *ephistanai*, stopping). An example is the unexpected gene interaction discovered by William Bateson and Reginald Punnett in the early 1900s, when they were studying crosses involving the sweet pea, *Lathyrus odoratus*. A cross between a true-breeding purple-flowered plant and a true-breeding white-flowered plant produced an F$_1$ generation with all purple-flowered plants and an F$_2$ generation with a 3:1 ratio of purple- to white-flowered plants. Of course, Mendel's laws predicted this result. The surprise came when the researchers crossed two different varieties of white-flowered sweet peas (**Figure 17.1**). All of the F$_1$ generation plants had purple flowers! When these plants were allowed to self-fertilize, the F$_2$ generation had purple-flowered and white-flowered plants in a 9:7 ratio. From these results, Bateson and Punnett deduced that two different genes were involved. To have purple flowers, a plant must have one or two dominant alleles for each of these genes. The relationships among the alleles are as follows:

C (one allele for purple) is dominant to *c* (white)

P (an allele for purple of a different gene) is dominant to *p* (white)

cc masks *P*, or *pp* masks *C*, in either case producing white flowers

A plant that was homozygous for either *c* or *p* would have white flowers even if it had a purple-producing allele at the other locus.

Epistatic interactions often arise because two or more different proteins are involved in a single cellular function. For example, two or more proteins may be part of an enzymatic pathway leading to the formation of a single product. This is the case for the formation of a purple pigment in the sweet pea strains we have been discussing:

	Enzyme C		Enzyme P	
Colorless precursor	$\longrightarrow$	Colorless intermediate	$\longrightarrow$	Purple pigment

In this example, a colorless precursor molecule must be acted on by two different enzymes to produce the purple pig-

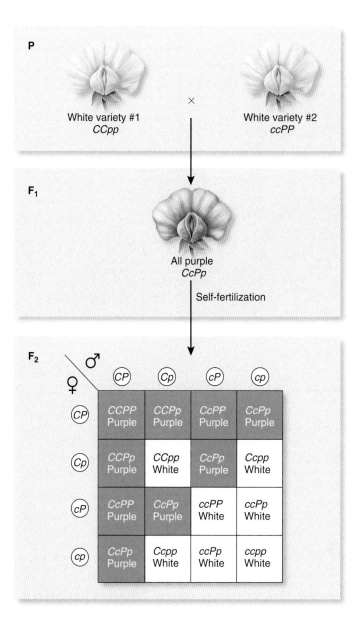

Figure 17.1 Epistasis in the sweet pea. The color of the sweet pea flower is controlled by two genes, each with a dominant and a recessive allele. Each of the dominant alleles (*C* and *P*) encodes an enzyme required for the synthesis of purple pigment. A plant that is homozygous recessive for either gene (*cc* or *pp*) cannot synthesize the pigment and will have white flowers.

Biological inquiry: In a Ccpp individual, which functional enzyme is missing? Is it the enzyme encoded by the C or P gene?

ment. Gene *C* encodes a functional protein, enzyme C, that converts the colorless precursor into a colorless intermediate. The recessive *c* allele results in a lack of production of enzyme C in the homozygote. Gene *P* encodes the functional enzyme P, which converts the colorless intermediate into the purple pigment. Like the *c* allele, the *p* allele results in an inability to produce a functional protein. A plant that is homozygous for either of the recessive alleles will not make any functional enzyme C

or enzyme P. When either of these enzymes is missing, the plant cannot make the purple pigment and has white flowers.

Polygenic Inheritance and Environmental Influences Produce Continuous Phenotypic Variation

As we have just seen, an epistatic interaction causes the alleles of one gene to mask the effects of a different gene. Let's now turn to another way that the alleles of different genes may affect the phenotype of a single trait. In many cases, the effects of alleles may be additive. This has been observed for many traits, particularly those that are quantitative in nature.

Until now we have discussed the inheritance of traits with clearly defined phenotypic variants, such as red or white eyes in fruit flies. These are known as discrete traits, or **discontinuous traits**, because the phenotypes do not overlap. For most traits, however, the phenotypes cannot be sorted into discrete categories. The majority of traits in all organisms are **continuous traits**, also called **quantitative traits**, which show continuous variation over a range of phenotypes. In humans, quantitative traits include height, weight, skin color, metabolic rate, and heart size, to mention a few. In the case of domestic animals and plant crops, many of the traits that people consider desirable are quantitative in nature, such as the number of eggs a chicken lays, the amount of milk a cow produces, and the number of apples on an apple tree. Consequently, much of our modern understanding of quantitative traits comes from agricultural research.

Quantitative traits are **polygenic**, which means that several or many genes contribute to the outcome of the trait. For many polygenic traits, genes contribute to the phenotype in an additive way. Another important factor is the environment. As we saw in Chapter 16, the environment plays a vital role in the phenotypic expression of genes. Environmental factors often have a major impact on quantitative traits. For example, an animal's diet affects its weight, and the amount of rain and sunlight that fall on an apple tree affect how many apples it produces.

Because quantitative traits are polygenic and greatly influenced by environmental conditions, the phenotypes among different individuals may vary substantially in any given population. As an example, let's consider skin pigmentation in people. This trait is influenced by several genes that tend to interact in an additive way. As a simplified example, let's consider a population in which this trait is controlled by three genes, which we will designate *A*, *B*, and *C*. Each gene has a dark allele, designated A^D, B^D, or C^D, and a light allele, designated A^L, B^L, or C^L, respectively. All of the alleles encode enzymes that cause the synthesis of skin pigment, but the enzymes encoded by dark alleles cause more pigment synthesis than the enzymes encoded by light alleles. **Figure 17.2** considers a hypothetical case in which people who were heterozygous for all three genes produced a large population of offspring. The bar graph shows the genotypes of the offspring, grouped according to the total number of dark alleles. As shown by the shading of the figure, skin pigmentation increases as the number of dark alleles increases.

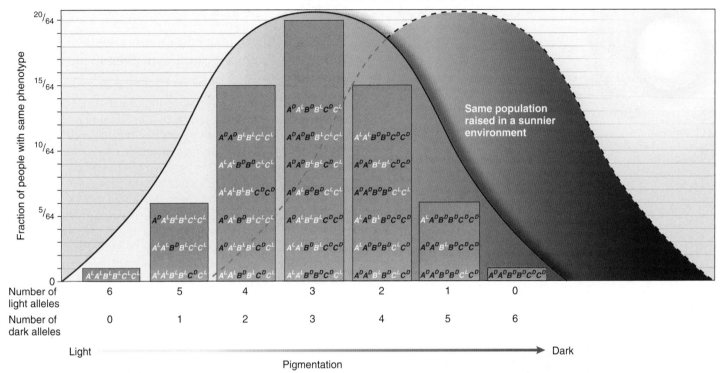

Figure 17.2   Continuous variation in a polygenic trait.   Skin color is a polygenic trait that can display a continuum of phenotypes. The bell curve on the *left* (solid line) shows the range of skin pigmentation in a hypothetical human population. The bar graphs below the curve show the additive effects of three genes that affect pigment production in this population; each bar shows the fraction of people with a particular number of dark alleles (A^D, B^D, and C^D) and light alleles (A^L, B^L, and C^L). The bell curve on the *right* (dashed line) represents the expected range of phenotypes if the same population was raised in a sunnier environment.

Offspring who have no dark alleles or no light alleles—that is, who are homozygous for all three genes—are fewer in number than those with some combination of dark and light alleles. As seen in the bell-shaped curve above the bar graph, the phenotypes of the offspring fall along a continuum. This continuous phenotypic variation, which is typical of quantitative traits, is produced by genotypic differences together with environmental effects. A second bell-shaped curve (the dashed line) depicts the expected phenotypic range if the same population of offspring had been raised in a sunnier environment, which increases pigment production. These two curves illustrate how the environment can also have a significant influence on the range of phenotypes.

In our discussion of genetics, we tend to focus on discrete traits because this makes it easier to relate a specific genotype with a phenotype. This is usually not possible for continuous traits. For example, as depicted in the middle bar of Figure 17.2, seven different genotypes can produce individuals with a medium amount of pigmentation. Nevertheless, it is important to emphasize that the majority of traits in all organisms are continuous, not discrete. Most traits are influenced by multiple genes, and the environment has an important impact on the phenotypic outcome.

17.2 Genes on the Same Chromosome: Linkage, Recombination, and Mapping

In all of the inheritance patterns we have studied so far, the alleles segregate and assort independently as predicted by Mendel's laws. As we have seen, phenotypes can be influenced by a variety of factors, including gene interactions and environmental effects, that make it difficult to relate genotype to phenotype. Even so, if we understand all of these factors and take them into account, we can see that each of the genes is transmitted according to Mendel's laws.

In the rest of this chapter, we will consider inheritance patterns in which the outcome of a cross violates one of Mendel's laws. In this section, we focus on transmission patterns that do not conform to the law of independent assortment. We will begin by examining the first experimental cross that demonstrated this pattern. You will learn that this pattern was explained by Thomas Hunt Morgan, who proposed that genes located close to each other on the same chromosome tend to be inherited as a group. Finally, we will see how crossing over between such genes provided the first method of mapping genes on chromosomes.

FEATURE INVESTIGATION

Bateson and Punnett's Crosses of Sweet Peas Showed That Genes Do Not Always Assort Independently

In Chapter 16, we learned that the independent assortment of alleles is due to the random alignment of homologous chromosomes during meiosis (refer back to Figure 16.11). But what happens when the alleles of different genes are on the same chromosome? A typical chromosome contains many hundreds or even a few thousand different genes. When two genes are close together on the same chromosome, they tend to be transmitted as a unit, a phenomenon known as **linkage**. A group of genes that usu-

ally stay together during meiosis is called a **linkage group**, and the genes in the group are said to be linked. In a two-factor cross, linked genes do not follow the law of independent assortment.

The first study showing linkage between two different genes was a cross of sweet peas carried out by William Bateson and Reginald Punnett in 1905. A surprising result occurred when they conducted a cross involving two different traits, flower color and pollen shape (**Figure 17.3**). One of the parent plants had purple flowers (*PP*) and long pollen (*LL*); the other had red flowers (*pp*) and round pollen (*ll*). As Bateson and Punnett expected, the F_1 plants all had purple flowers and long pollen (*PpLl*). The unexpected result came in the F_2 generation.

Figure 17.3 A cross of sweet peas showing that independent assortment does not always occur.

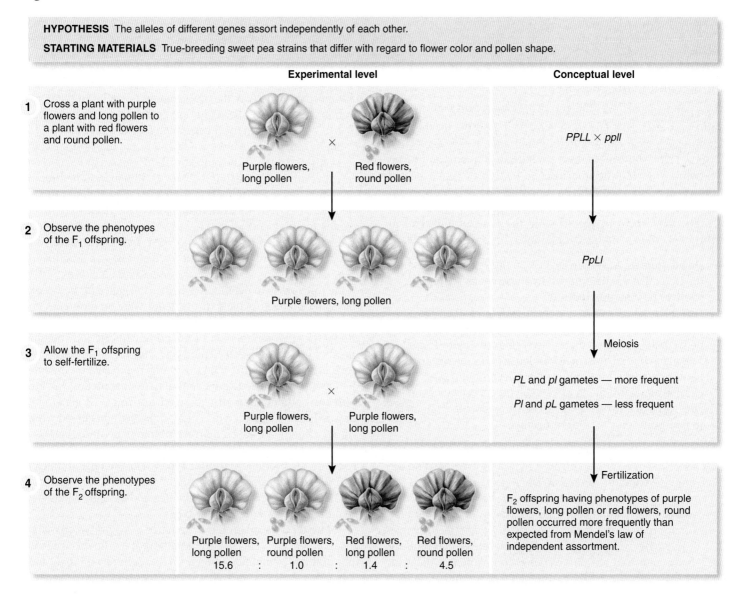

HYPOTHESIS The alleles of different genes assort independently of each other.

STARTING MATERIALS True-breeding sweet pea strains that differ with regard to flower color and pollen shape.

	Experimental level	Conceptual level
1 Cross a plant with purple flowers and long pollen to a plant with red flowers and round pollen.	Purple flowers, long pollen × Red flowers, round pollen	$PPLL \times ppll$
2 Observe the phenotypes of the F_1 offspring.	Purple flowers, long pollen	$PpLl$
3 Allow the F_1 offspring to self-fertilize.	Purple flowers, long pollen × Purple flowers, long pollen	Meiosis *PL* and *pl* gametes — more frequent *Pl* and *pL* gametes — less frequent
4 Observe the phenotypes of the F_2 offspring.	Purple flowers, long pollen Purple flowers, round pollen Red flowers, long pollen Red flowers, round pollen 15.6 : 1.0 : 1.4 : 4.5	Fertilization F_2 offspring having phenotypes of purple flowers, long pollen or red flowers, round pollen occurred more frequently than expected from Mendel's law of independent assortment.

5   THE DATA

Phenotypes of F₂ offspring	Observed number	Observed ratio	Expected number	Expected ratio
Purple flowers, long pollen	296	15.6	240	9
Purple flowers, round pollen	19	1.0	80	3
Red flowers, long pollen	27	1.4	80	3
Red flowers, round pollen	85	4.5	27	1

Although the offspring displayed the four phenotypes predicted by Mendel's laws, the observed numbers of offspring did not conform to the predicted 9:3:3:1 ratio. Rather, as seen in the data in Figure 17.3, the F₂ generation had a much higher proportion of the two phenotypes found in the parental generation: purple flowers with long pollen, and red flowers with round pollen.

These results did not support the law of independent assortment. Bateson and Punnett suggested that the transmission of flower color and pollen shape was somehow coupled, so that these traits did not always assort independently. Although the law of independent assortment applies to many other genes, in this example, the hypothesis of independent assortment was rejected.

Linkage and Crossing Over Produce Parental and Recombinant Phenotypes

Although Bateson and Punnett realized their results did not conform to Mendel's law of independent assortment, they did not provide a clear explanation for their data. A few years later, Thomas Hunt Morgan obtained similar ratios in crosses of fruit flies while studying the transmission pattern of genes located on the X chromosome. Like Bateson and Punnett, Morgan observed many more F₂ offspring with the parental combination of traits than would be predicted on the basis of independent assortment. To explain his data, Morgan proposed these ideas:

1. When different genes are located on the same chromosome, the traits that are determined by those genes are most likely to be inherited together.
2. Due to crossing over during meiosis, homologous chromosomes can exchange pieces of chromosomes and create new combinations of alleles (refer back to Figure 15.17).
3. The likelihood of crossing over depends on the distance between two genes. Crossovers between homologous chromosomes are much more likely to occur between two genes that are farther apart in the chromosome compared to two genes that are closer together.

To illustrate the first two of these ideas, **Figure 17.4** considers a series of crosses involving two genes that are linked on the same chromosome in *Drosophila*. The P generation cross is between flies that are homozygous for alleles that affect body color and wing shape. The female is homozygous for the wild-type alleles that produce gray body color (b^+b^+) and straight wings (c^+c^+); the male is homozygous for mutant alleles that produce black body color (bb) and curved wings (cc). Note that the symbols for the genes are based on the name of the mutant allele; the wild-type allele is indicated by a superscript plus sign ($+$).

The chromosomes next to the flies in Figure 17.4 show the arrangement of these alleles. If the two genes are on the same chromosome, we know the arrangement of alleles in the P generation flies because these flies are homozygous for both genes ($b^+b^+c^+c^+$ or $bbcc$). In the P generation female on the left, b^+ and c^+ are linked, while b and c are linked in the male on the right.

Let's now look at the outcome of the crosses in Figure 17.4. As expected, the F₁ offspring (b^+bc^+c) all had gray bodies and straight wings, confirming that these are the dominant traits. In the next cross, F₁ females were mated to males that were homozygous for both recessive alleles ($bbcc$). A cross in which an individual with a dominant phenotype is mated with a homozygous recessive individual is called a **testcross**, as described in Chapter 16. In the crosses we are discussing here, the purpose of the testcross is to determine whether the genes for body color and wing shape are linked. If the genes were on different chromosomes and assorted independently, this testcross should have produced equal numbers of F₂ offspring with the four possible phenotypes. The observed numbers, shown above the F₂ phenotypes, clearly conflict with this prediction based on independent assortment. The two most abundant phenotypes are those with the combinations of characteristics in the P generation: gray bodies and straight wings or black bodies and curved wings. These offspring are called **nonrecombinants** because their combination of traits has not changed from the parental generation. They are also termed **parental types**. The smaller number of offspring that have a different combination of traits—gray bodies and curved wings or black bodies and straight wings—are **recombinants** or **nonparental types**.

How do we explain the occurrence of recombinants when genes are linked on the same chromosome? As shown beside the flies of the F₂ generation in Figure 17.4, each recombinant individual has a chromosome that is the product of a crossover. The crossover occurred while the F₁ female fly was making egg cells.

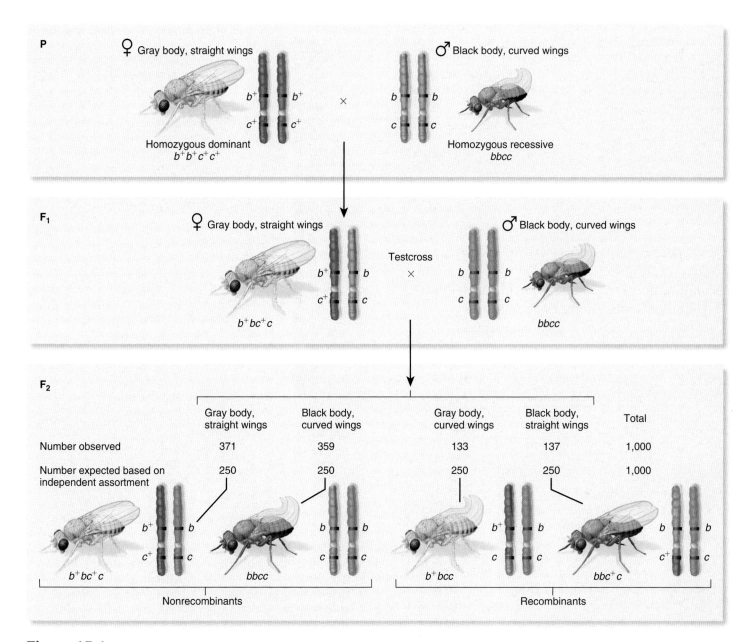

Figure 17.4 Linkage and recombination of alleles. An experimenter crossed $b^+b^+c^+c^+$ and $bbcc$ flies to produce F_1 heterozygotes. F_1 females were then testcrossed to $bbcc$ males. The large number of parental phenotypes in the F_2 generation suggests that the two genes are linked on the same chromosome. F_2 recombinant phenotypes occur because the alleles can be rearranged by crossing over.

As shown below, four different egg cells are possible:

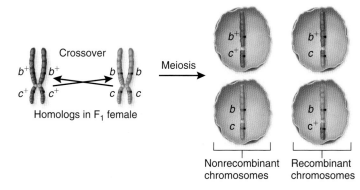

Due to crossing over, two of the four egg cells produced by meiosis have recombinant chromosomes. What happens when eggs containing such chromosomes are fertilized in the test-cross? Each of the male fly's sperm cells carries a chromosome with the two recessive alleles. If the egg contains the recombinant chromosome carrying the b^+ and c alleles, the testcross will produce an F_2 offspring with a gray body and curved wings. If the egg contains the recombinant chromosome carrying the b and c^+ alleles, F_2 offspring will have a black body and straight wings. Therefore, crossing over in the F_1 female can explain the occurrence of both types of F_2 recombinant offspring.

Morgan's ideas about linkage and crossing over were based on similar data, derived from his studies of genes on the X chromosome. The idea that linked genes tend to be inherited together explained the high frequency of parental combinations of traits in certain crosses. The suggestion that crossing over produces chromosomes with new allele combinations accounted for the occurrence of recombinant phenotypes. Morgan's third idea regarding linkage was that the frequency of crossing over between linked genes depends on the distance between them. This suggested a method for determining the relative positions of genes on a chromosome, as we will see next.

GENOMES & PROTEOMES

Recombination Frequencies Provide a Method for Mapping Genes Along Chromosomes

The oldest approach to studying the arrangement of genes in a species' genome is called **genetic linkage mapping** (also known as gene mapping or chromosome mapping). This experimental method is used to determine the linear order of genes that are linked to each other along the same chromosome. As depicted in **Figure 17.5**, this linear arrangement is shown in a chart known as a **genetic linkage map**. Each gene has its own unique locus at a particular site within a chromosome. For example, the gene for black body color (*b*) that we discussed earlier is located near the middle of the chromosome, while the gene for curved wings (*c*) is closer to one end. The first genetic linkage map, showing five genes on the *Drosophila* X chromosome, was constructed in 1911 by Alfred Sturtevant, an undergraduate who spent time in Morgan's laboratory.

Genetic linkage mapping allows us to estimate the relative distances between linked genes based on the likelihood that a crossover will occur between them. This likelihood is proportional to the distance between the genes, as Morgan first proposed. If the genes are very close together, a crossover is unlikely to begin in the region between them. However, if the genes are very far apart, a crossover is more likely to be initiated between them and thereby recombine their alleles. Therefore, in a cross involving two genes on the same chromosome, the percentage of recombinant offspring is correlated with the distance between the genes. This correlation provides the experimental basis for gene mapping. If a two-factor testcross produces many recombinant offspring, the experimenter concludes that the genes are far apart. If very few recombinant offspring are observed, the genes must be close together.

To find the distance between two genes, the experimenter must determine the frequency of crossing over between them, called their **recombination frequency**. This is accomplished by conducting a testcross. As an example, let's refer back to the *Drosophila* testcross described in Figure 17.4. As we discussed, the genes for body color and wing shape are on the same chromosome; the recombinant offspring are the result of crossing

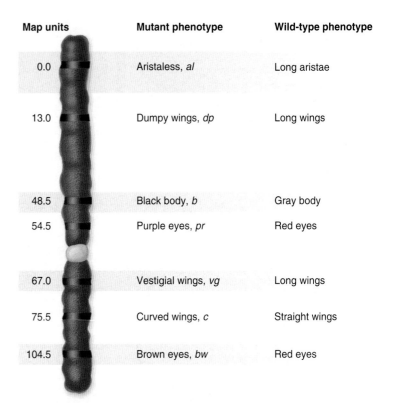

Figure 17.5 A simplified genetic linkage map. This map shows the relative locations of a few genes along a chromosome in *Drosophila melanogaster*. The name of each gene is based on the mutant phenotype. The numbers on the left are map units (mu). The distance between two genes, in map units, corresponds to their recombination frequency in testcrosses.

Map units	Mutant phenotype	Wild-type phenotype
0.0	Aristaless, *al*	Long aristae
13.0	Dumpy wings, *dp*	Long wings
48.5	Black body, *b*	Gray body
54.5	Purple eyes, *pr*	Red eyes
67.0	Vestigial wings, *vg*	Long wings
75.5	Curved wings, *c*	Straight wings
104.5	Brown eyes, *bw*	Red eyes

over during egg formation in the F$_1$ female. We can use the data from the testcross shown in Figure 17.4 to estimate the distance between these two genes. The **map distance** between two linked genes is defined as the number of recombinant offspring divided by the total number of offspring times 100.

$$\text{Map distance} = \frac{\text{Number of recombinant offspring}}{\text{Total number of offspring}} \times 100$$

$$= \frac{133 + 137}{371 + 359 + 133 + 137} \times 100$$

$$= 27.0 \text{ map units}$$

The units of distance are called **map units (mu)**, or sometimes **centiMorgans (cM)** in honor of Thomas Hunt Morgan. One map unit is equivalent to a 1% recombination frequency. In this example, 270 out of 1,000 offspring are recombinants, so the recombination frequency is 27% and the two genes are 27.0 mu apart.

Genetic linkage mapping has been useful for analyzing the genes of organisms that are easily crossed and produce many offspring in a short time. It has been used to map the genes of several plant species and of certain species of animals, such as

Drosophila. However, for most organisms, including humans, linkage mapping is impractical due to long generation times or the inability to carry out experimental crosses. Fortunately, many alternative methods of gene mapping have been developed in the past few decades that are faster and do not depend on crosses. These newer cytological and molecular approaches, which we will discuss in Chapter 20, are now used to map genes in a wide variety of organisms.

17.3 Extranuclear Inheritance: Organelle Genomes

In the previous section, we examined the inheritance patterns of linked genes that violate the law of independent assortment. In this section, we will explore inheritance patterns that violate the law of segregation. Gene transmission may defy this law because some genes are not found on the chromosomes in the cell nucleus. The segregation of genes is explained by the pairing and segregation of homologous chromosomes during meiosis; genes found elsewhere in the cell do not segregate in the same way. The transmission of genes that are located outside the cell nucleus is called **extranuclear inheritance**.

Two important types of extranuclear inheritance patterns involve genes that are found in mitochondria and chloroplasts (**Figure 17.6**). Extranuclear inheritance is also called **cytoplasmic inheritance** because these organelles are in the cytoplasm of the cell. As we discussed in Chapter 6, mitochondria and chloroplasts are found in eukaryotic cells because of an ancient endosymbiotic relationship. They contain their own genetic material, or genomes. Although these **organelle genomes** are much smaller than nuclear genomes, researchers have discovered that they are critically important in the phenotypes of organisms. In plants, for example, the chloroplast genome carries many genes that are vital for photosynthesis. Mitochondrial genes are critical for respiration. In humans, mutations in the mitochondrial genome may cause inherited diseases. In this section, we will examine the transmission patterns observed for genes found in the chloroplast and mitochondrial genomes and consider how mutations in these genes may affect an individual's traits.

Chloroplast Genomes Are Often Maternally Inherited

One of the first experiments showing an extranuclear inheritance pattern was carried out by Carl Correns in 1909. Correns discovered that leaf pigmentation in the four-o'clock plant (*Mirabilis jalapa*) follows a pattern of inheritance that does not obey Mendel's law of segregation. Four-o'clock leaves may be green, white, or variegated, as shown in **Figure 17.7**. Correns observed that the pigmentation of the offspring depended solely on the pigmentation of the maternal parent, a phenomenon called **maternal inheritance**. If the female parent had white leaves,

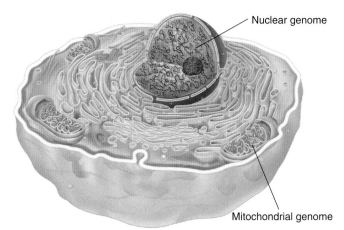

(a) An animal cell

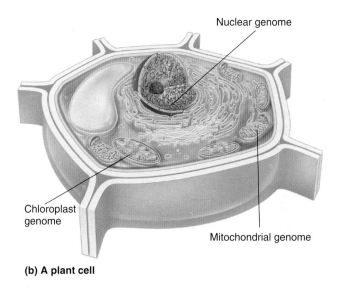

(b) A plant cell

Figure 17.6   The locations of genetic material in animal and plant cells.   The chromosomes in the cell nucleus are collectively known as the nuclear genome. Mitochondria and chloroplasts have small circular chromosomes, which are called the mitochondrial and chloroplast genomes.

all of the offspring had white leaves. Similarly, if the female was green, so were all of the offspring. The offspring of a variegated female parent could be green, white, or variegated.

At the time, Correns did not understand that chloroplasts contain some genes. We now know that the pigmentation of four-o'clock leaves can be explained by the occurrence of genetically different types of chloroplasts in the leaf cells. As discussed in Chapter 8, chloroplasts are the site of photosynthesis, and their green color is due to the presence of the pigment called chlorophyll. Certain genes required for chlorophyll synthesis are found within the chloroplast DNA. The green phenotype is due to the presence of chloroplasts that have normal genes and synthesize the usual quantity of chlorophyll. The white phenotype is caused by a mutation in a gene within the chloroplast DNA that prevents the synthesis of most of the chlorophyll.

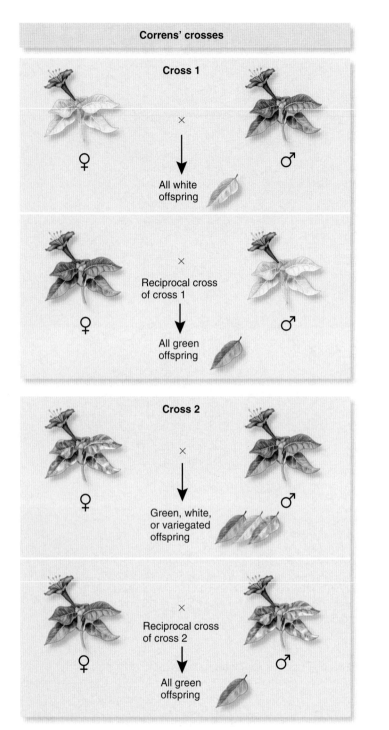

Figure 17.7 **Maternal inheritance in the four-o'clock plant.** The genes for green pigment synthesis in plants are part of the chloroplast genome. The white phenotype in four o'clocks is due to chloroplasts with a mutant allele that greatly reduces green pigment production. The variegated phenotype is due to a mixture of normal and mutant chloroplasts. In four o'clocks, the egg contains all of the plastids that are inherited by the offspring, so the phenotype of the offspring is determined by the female parent.

Biological inquiry: In this example, where is the gene located that causes the green color of four-o'clock leaves? How is this gene transmitted from parent to offspring?

(Enough chlorophyll is made for the plant to survive.) The variegated phenotype occurs in leaves that have a mixture of the two types of chloroplasts.

Leaf pigmentation follows a maternal inheritance pattern because the chloroplasts in four o'clocks are inherited only through the cytoplasm of the egg (**Figure 17.8**). During plant fertilization, a sperm cell from a pollen grain fertilizes an egg cell to create a zygote, which eventually develops into a plant. In four o'clocks, the egg cell contains several proplastids that are inherited by the offspring, while the sperm cell does not contribute any proplastids. As discussed in Chapter 4, proplastids develop into various types of plastids, including chloroplasts. Thus, the phenotype of a four-o'clock plant depends on the types of proplastids it inherits from the maternal parent. If the maternal parent transmits only normal proplastids, all offspring will have green leaves (Figure 17.8a). Alternatively, if the maternal parent transmits only mutant proplastids, all offspring will have white leaves (Figure 17.8b). The genetic composition of the paternal parent does not affect the outcome. Because an egg cell contains several proplastids, an offspring from a variegated maternal parent may inherit only normal proplastids, only

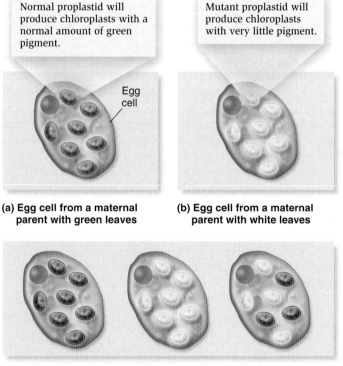

(c) Possible egg cells from a maternal parent with variegated leaves

Figure 17.8 **Plastid composition of egg cells from green, white, and variegated four-o'clock plants.** In this drawing of four-o'clock egg cells, normal proplastids are represented as green and mutant proplastids as white. Proplastids do not differentiate into chloroplasts in egg cells, and they are not actually green. **(a)** A green plant produces eggs carrying normal proplastids. **(b)** A white plant produces eggs carrying mutant proplastids. **(c)** A variegated plant produces eggs that may contain either or both types of proplastids.

mutant proplastids, or a mixture of normal and mutant proplastids. Consequently, the offspring of a variegated maternal parent can be green, white, or variegated individuals (Figure 17.8c).

The variegated phenotype is due to segregation events that occur after fertilization. As a zygote containing both types of chloroplasts divides to produce a multicellular plant, some cells may receive mostly normal chloroplasts. Further division of these cells gives rise to a patch of green tissue. Alternatively, as a matter of chance, other cells may receive mostly mutant chloroplasts that are defective in chlorophyll synthesis. This results in a patch of tissue that is white.

In most species of plants, the egg cell provides most of the zygote's cytoplasm, while the much smaller male gamete often provides little more than a nucleus. Therefore, chloroplasts are most often inherited via the egg. In seed-bearing plants, maternal inheritance of chloroplasts is the most common transmission pattern. However, certain species exhibit a pattern called **biparental inheritance**, in which both the pollen and the egg contribute chloroplasts to the offspring. Others exhibit **paternal inheritance**, in which only the pollen contributes these organelles. For example, most types of pine trees show paternal inheritance of chloroplasts.

Mitochondrial Genomes Are Maternally Inherited in Humans and Most Other Species

Mitochondria are found in nearly all eukaryotic species. Similar to the transmission of chloroplasts in plants, maternal inheritance is the most common pattern of mitochondrial transmission in eukaryotic species, although some species do exhibit biparental or paternal inheritance. The mitochondrial genome of many mammalian species has been analyzed and usually contains a total of 37 genes. Twenty-four genes encode tRNAs and rRNAs, which are needed for translation inside the mitochondrion. Thirteen genes encode proteins that are involved in oxidative phosphorylation. As discussed in Chapter 7, the primary function of the mitochondrion is the synthesis of ATP via oxidative phosphorylation.

In humans, as in most species, mitochondria are maternally inherited. Researchers have discovered that mutations in human mitochondrial genes can cause a variety of rare diseases (**Table 17.2**). These are usually chronic degenerative disorders that affect the brain, eyes, heart, muscle, kidney, and endocrine glands. For example, Leber's hereditary optic neuropathy (LHON) affects the optic nerve. It may lead to the progressive loss of vision in one or both eyes. LHON can be caused by a mutation in one of several different mitochondrial genes.

17.4 X Inactivation, Genomic Imprinting, and Maternal Effect

We will end our discussion of complex inheritance patterns by considering examples in which the timing and control of gene expression create inheritance patterns that are determined by

Table 17.2	Examples of Human Mitochondrial Diseases
Disease	**Description**
Leber's hereditary optic neuropathy	Caused by a mutation in one of several mitochondrial genes that encode electron-transport proteins. The main symptom is loss of vision.
Neurogenic muscle weakness	Caused by a mutation in a mitochondrial gene that encodes a subunit of mitochondrial ATP synthase, which is required for ATP synthesis. Symptoms involve abnormalities in the nervous system that affect the muscles and eyes.
Mitochondrial encephalomyopathy, lactic acidosis, and strokelike episodes	Mutations in mitochondrial genes that encode tRNAs for leucine and lysine. Symptoms include strokelike episodes, secretion of lactic acid into the bloodstream, seizures, migraine headaches, and lack of coordination.
Maternal myopathy and cardiomyopathy	A mutation in a mitochondrial gene that encodes a tRNA for leucine. The primary symptoms involve muscle abnormalities, most notably in the heart.
Myoclonic epilepsy and ragged-red muscle fibers	A mutation in a mitochondrial gene that encodes a tRNA for lysine. Symptoms include epilepsy, dementia, blindness, deafness, and heart and kidney malfunctions.

the sex of the individual or by the sex of the parents. The first two patterns, called X inactivation and genomic imprinting, are types of **epigenetic inheritance**. In epigenetic inheritance, modification of a gene or chromosome during egg formation, sperm formation, or early stages of embryo growth alters gene expression in a way that is fixed during an individual's lifetime. Epigenetic changes permanently affect the phenotype of the individual, but they are not permanent over the course of many generations and they do not change the actual DNA sequence. For example, a gene may undergo an epigenetic change that inactivates it for an individual's entire life, so it is never expressed in that individual. However, when the same individual makes gametes, the gene may become activated and remain active during the lifetime of an offspring that inherits the gene.

At the end of this section, we will also consider genes that exhibit a bizarre inheritance pattern called the **maternal effect**, in which the genotype of the mother directly determines the phenotype of her offspring. Surprisingly, for maternal effect genes, the genotypes of the father and of the offspring themselves do not affect the offspring's phenotype. As you will learn, this phenomenon is explained by the accumulation of gene products that the mother provides to her developing eggs.

In Female Mammals, One X Chromosome Is Inactivated in Each Somatic Cell

In 1961, the British geneticist Mary Lyon proposed the phenomenon of **X inactivation**, in which one X chromosome in the somatic cells of female mammals is inactivated, meaning that its genes

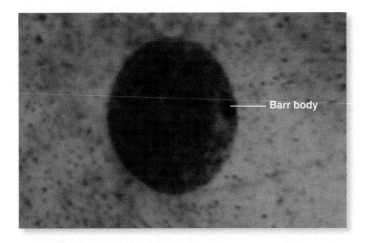

Figure 17.9 X-chromosome inactivation in female mammals. This light micrograph shows the nucleus of a human female cell. The label shows the Barr body, a condensed, inactivated X chromosome found just inside the nuclear envelope in the somatic cells of female mammals.

are not expressed. The **Lyon hypothesis**, as X inactivation also came to be known, was based on two lines of evidence. The first evidence came from microscopic studies of mammalian cells. In 1949, Murray Barr and Ewart Bertram identified a highly condensed structure in the cells of female cats that was not found in the cells of male cats. This structure was named a **Barr body** after one of its discoverers (**Figure 17.9**). In 1960, Susumu Ohno correctly proposed that a Barr body is a highly condensed X chromosome. Lyon's second line of evidence was the inheritance pattern of variegated coat colors in certain mammals. A classic case is the calico cat, which has randomly distributed patches of black and orange fur (**Figure 17.10a**).

According to the Lyon hypothesis, the calico pattern is explained by the permanent inactivation of one X chromosome in each cell that forms a patch of the cat's skin, as shown in **Figure 17.10b**. The gene involved is an X-linked gene that occurs as an orange allele, X^O, and a black allele, X^B. A female cat that is heterozygous for this gene will be calico. (The white underside is due to a dominant allele of a different autosomal gene.) At an early stage of embryonic development, one of the two X chromosomes is randomly inactivated in each of the cat's somatic cells, including those that will give rise to the hair-producing skin cells. As the embryo grows and matures, the pattern of X inactivation is maintained during subsequent cell divisions. For example, skin cells derived from a single embryonic cell in which the X^B-carrying chromosome has been inactivated will produce a patch of orange fur, because they express only the X^O allele that is carried on the active chromosome. Alternatively, a group of skin cells in which the chromosome carrying X^O has been inactivated will express only the X^B allele, producing a patch of black fur. Because the primary event of X inactivation is a random process that occurs at an early stage of development, the result is an animal with randomly distributed patches of black and orange fur.

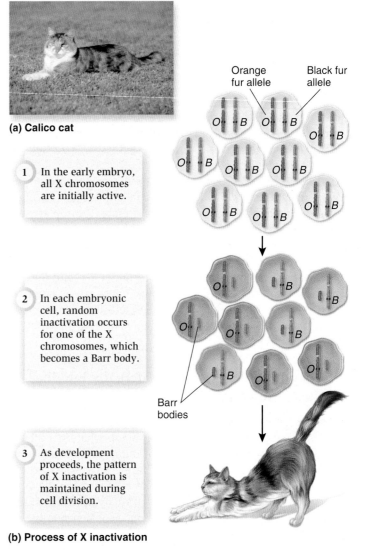

(a) Calico cat

1 In the early embryo, all X chromosomes are initially active.

Orange fur allele   Black fur allele

2 In each embryonic cell, random inactivation occurs for one of the X chromosomes, which becomes a Barr body.

Barr bodies

3 As development proceeds, the pattern of X inactivation is maintained during cell division.

(b) Process of X inactivation

Figure 17.10 Random X-chromosome inactivation in a calico cat. (a) A calico cat. (b) X inactivation during embryonic development. The calico pattern is due to random X-chromosome inactivation in a female that is heterozygous for the X-linked gene with black and orange alleles. The cells at the *top* of this figure represent a small mass of cells making up the very early embryo. In these cells, both X chromosomes are active. At an early stage of embryonic development, one X chromosome is randomly inactivated in each cell. The initial inactivation pattern is maintained in the descendents of each cell as the embryo matures into an adult. The pattern of orange and black fur in the adult cat reflects the pattern of X inactivation in the embryo.

Biological inquiry: If a female cat is homozygous for the orange allele, would it show a calico phenotype?

In female mammals that are heterozygous for X-linked genes, approximately half of their somatic cells will express one allele, while the rest of their somatic cells will express the other allele. These heterozygotes are called **mosaics** because they are composed of two types of cells, analogous to the different-colored pieces in the pictures called mosaics. The phenomenon of mosaicism is readily apparent in calico cats, in which the alleles affect fur color. Likewise, human females who are heterozygous for X-linked genes are mosaics, with one allele expressed in some cells and the alternative allele in other cells. Women who are heterozygous for recessive X-linked alleles usually show the dominant trait because the expression of the dominant allele in 50% of their cells is sufficient to produce the dominant phenotype.

On rare occasions, a female who is heterozygous for a recessive X-linked disease-causing allele may show mild or even severe disease symptoms. Because the pattern of X-chromosome inactivation is random, there will be a small percentage of heterozygous women who happen to inactivate the X chromosome carrying the normal allele in a large percentage of their cells, as a matter of bad luck. As an example, let's consider the recessive X-linked form of hemophilia that we discussed in Chapter 16. This type of hemophilia is caused by a defect in a gene that encodes a blood-clotting factor, called factor VIII, that is made by cells in the liver and secreted into the bloodstream. X inactivation in humans occurs when an embryo is 10 days old. At this stage, the liver contains only about a dozen cells. In most females who are heterozygous for the normal and hemophilia alleles, roughly half of their liver cells will express the normal allele. However, on rare occasions, all or most of the dozen embryonic liver cells might happen to inactivate the X chromosome carrying the dominant normal allele. Following growth and development, such a female will have a very low level of factor VIII and as a result will show symptoms of hemophilia.

At this point, you may be wondering why X inactivation occurs. Researchers have proposed that X inactivation achieves **dosage compensation** between male and female mammals. The X chromosome carries many genes, while the Y chromosome has only a few. The inactivation of one X chromosome in the female reduces the number of expressed copies (doses) of X-linked genes from two to one. As a result, the expression of X-linked genes in females and males is roughly equal.

The X Chromosome Has an X Inactivation Center That Controls Compaction into a Barr Body

After the Lyon hypothesis was confirmed, researchers became interested in the genetic control of X inactivation. The cells of humans and other mammals have the ability to count their X chromosomes and allow only one of them to remain active. Additional X chromosomes are converted to Barr bodies. In normal females, two X chromosomes are counted and one is inactivated. In normal males, one X chromosome is counted and none inactivated. On occasion, however, people are born with abnormalities in the number of their sex chromosomes. In these disorders, known as Turner syndrome, Triple X syndrome, and Klinefelter syndrome, the cells inactivate the number of X chromosomes necessary to leave a single active chromosome.

Phenotype	Chromosome Composition	Number of Barr Bodies
Normal female	XX	1
Normal male	XY	0
Turner syndrome (female)	XO	0
Triple X syndrome (female)	XXX	2
Klinefelter syndrome (male)	XXY	1

Although the genetic control of inactivation is not entirely understood at the molecular level, a short region on the X chromosome called the **X inactivation center (Xic)** is known to play a critical role. Eeva Therman and Klaus Patau identified Xic from its key role in X inactivation. The counting of human X chromosomes is accomplished by counting the number of Xics. The Xic on each X chromosome is necessary for inactivation to occur. Therman and Patau found that in cells with two X chromosomes, if one of them is missing its Xic due to a chromosome mutation, neither X chromosome will be inactivated. This is a lethal condition for a human female embryo.

The expression of a specific gene within the X inactivation center is required for compaction of the X chromosome into a Barr body. This gene, discovered in 1991, is named *Xist* (for X inactive specific transcript). The *Xist* gene product is a long RNA molecule that does not encode a protein. Instead, the role of *Xist* RNA is to coat one of the two X chromosomes during the process of X inactivation. The *Xist* gene on the inactivated X chromosome continues to be expressed after other genes on this chromosome have been silenced.

The process of X inactivation can be divided into three phases: initiation, spreading, and maintenance (**Figure 17.11**). During initiation, one of the X chromosomes is targeted for inactivation. This chromosome is inactivated during the spreading phase, so called because inactivation begins near the X inactivation center and spreads in both directions along the chromosome. Spreading requires the transcription of the *Xist* gene and coating of the X chromosome with *Xist* RNA. After coating, proteins associate with the *Xist* RNA and promote compaction of the chromosome into a Barr body. Maintenance refers to replication of the compacted chromosome during subsequent cell divisions. While initiation and spreading occur only during embryonic development, maintenance occurs throughout the individual's life. Continued activity of the *Xist* gene on an inactivated X chromosome maintains this chromosome as a Barr body during cell division. Whenever a somatic cell divides in a female mammal, the Barr body is replicated to produce two Barr bodies.

The Transcription of an Imprinted Gene Depends on the Sex of the Parent

As we have seen, X inactivation is a type of epigenetic inheritance in which a chromosome is modified in the early embryo, permanently altering gene expression in that individual. Other types of epigenetic inheritance occur in which genes or chromosomes

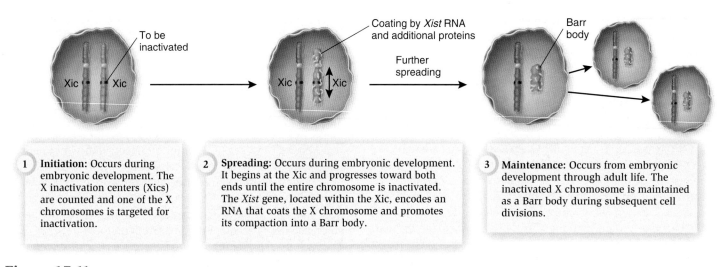

1. **Initiation:** Occurs during embryonic development. The X inactivation centers (Xics) are counted and one of the X chromosomes is targeted for inactivation.

2. **Spreading:** Occurs during embryonic development. It begins at the Xic and progresses toward both ends until the entire chromosome is inactivated. The *Xist* gene, located within the Xic, encodes an RNA that coats the X chromosome and promotes its compaction into a Barr body.

3. **Maintenance:** Occurs from embryonic development through adult life. The inactivated X chromosome is maintained as a Barr body during subsequent cell divisions.

Figure 17.11 The process of X inactivation.

are modified in the gametes of a parent, permanently altering gene expression in the offspring. **Genomic imprinting** refers to a phenomenon in which a segment of DNA is **imprinted**, or marked, in a way that affects gene expression throughout the life of the individual who inherits that DNA.

Genomic imprinting occurs in numerous species, including insects, plants, and mammals. Imprinting may involve a single gene, a part of a chromosome, an entire chromosome, or even all of the chromosomes inherited from one parent. It is permanent in the somatic cells of a given individual, but the marking of the DNA is altered from generation to generation. Imprinted genes do not follow a Mendelian pattern of inheritance because imprinting causes the offspring to distinguish between maternally and paternally inherited alleles. Depending on how a par-

ticular gene is marked by each parent, the offspring will express either the maternal or the paternal allele, but not both.

Let's consider a specific example of imprinting that involves a gene called *Igf-2* that is found in mice and other mammals. This gene encodes a growth hormone called insulin-like growth factor 2 that is needed for proper growth. If a normal copy of this gene is not expressed, a mouse will be dwarf. The *Igf-2* gene is known to be located on an autosome, not on a sex chromosome. Because mice are diploid, they have two copies of this gene, one from each parent.

Researchers have discovered that mutations can occur in the *Igf-2* gene that block the function of the Igf-2 hormone. When mice carrying normal or mutant alleles are crossed to each other, a bizarre result is obtained (**Figure 17.12**). If the male parent

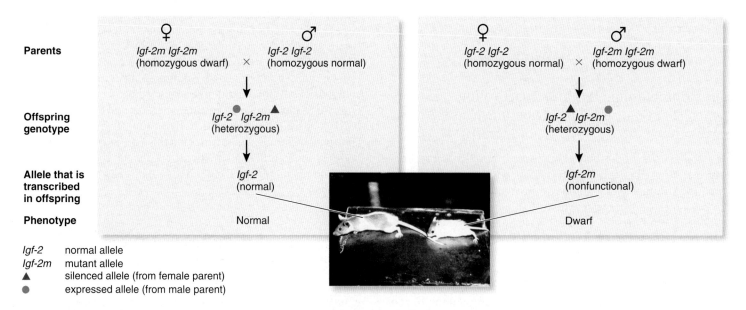

Igf-2	normal allele
Igf-2m	mutant allele
▲	silenced allele (from female parent)
●	expressed allele (from male parent)

Figure 17.12 An example of genomic imprinting in the mouse. In the cross on the *left*, a homozygous male with the normal *Igf-2* allele is crossed to a homozygous female carrying a defective allele, *Igf-2m*. An offspring is phenotypically normal because the paternal allele is expressed. In the cross on the *right*, a homozygous male carrying the defective allele is crossed to a homozygous normal female. In this case, an offspring is dwarf because the paternal allele is defective and the maternal allele is not expressed.

is homozygous for the normal allele and the female is homozygous for the mutant allele, all the offspring grow to a normal size. In contrast, if the male is homozygous for the mutant allele and the female is homozygous for the normal allele, all the offspring are dwarf. The reason this result is so surprising is that the normal and dwarf offspring have the same genotype but different phenotypes! These phenotypes are not the result of any external influence on the offspring's development. Rather, the allele that is expressed in their somatic cells depends on which parent contributed which allele. In mice, the *Igf-2* gene inherited from the mother is imprinted in such a way that it cannot be transcribed into mRNA. Therefore, only the paternal gene is expressed. The mouse on the left side of Figure 17.12 is normal because it expresses a functional paternal gene. In contrast, the mouse on the right is dwarf because the paternal gene is a mutant allele that results in a nonfunctional hormone. In both cases, the maternal gene is inactive due to imprinting.

Why is the maternal gene not transcribed into mRNA? To answer this question we need to consider the molecular function of genes. As discussed in Chapter 13, the attachment of methyl (—CH$_3$) groups to the bases of DNA can alter gene transcription. For most genes, methylation silences gene expression by causing the DNA to become more compact. For a few genes, methylation may enhance gene expression by attracting activator proteins to the promoter. Researchers have discovered that DNA methylation is the marking process that occurs during the imprinting of certain genes, including the *Igf-2* gene.

Figure 17.13 shows the imprinting process in which a maternal gene is methylated. The left side of the figure follows the marking process during the life of a female individual; the right side follows the same process in a male. Both individuals received a methylated gene from their mother and a nonmethylated copy of the same gene from their father. Via cell division, the zygote develops into a multicellular organism. Each time a somatic cell divides, enzymes in the cell maintain the methylation of the maternal gene, while the paternal gene remains unmethylated. If the methylation inhibits transcription of this gene, only the paternal copy will be expressed in the somatic cells of both the male and female offspring.

The methylation state of an imprinted gene may be altered when individuals make gametes. First, as shown in Figure 17.13, the methylation is erased. Next, the gene may be methylated again, but that depends on whether the individual is a female or male. In females making eggs, both copies of the gene are methylated; in males making sperm, neither copy is methylated. When we consider the effects of methylation over the course of two or more generations, we can see how this phenomenon creates an epigenetic transmission pattern. The male in Figure 17.13 has inherited a methylated gene from his mother that is transcriptionally silenced in his somatic cells. Although he does not express this gene during his lifetime, he can pass on an active, nonmethylated copy of this exact same gene to his offspring.

Genomic imprinting is a recently discovered phenomenon that has been shown to occur for a few genes in mammals. For some genes, such as *Igf-2*, the maternal allele is silenced, while for other genes the paternal allele is silenced. Biologists are still trying to understand the reason for this curious marking process.

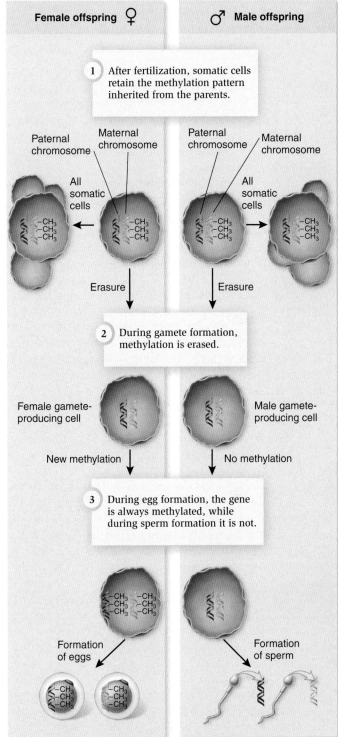

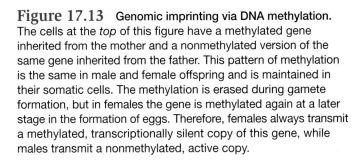

Figure 17.13 Genomic imprinting via DNA methylation. The cells at the *top* of this figure have a methylated gene inherited from the mother and a nonmethylated version of the same gene inherited from the father. This pattern of methylation is the same in male and female offspring and is maintained in their somatic cells. The methylation is erased during gamete formation, but in females the gene is methylated again at a later stage in the formation of eggs. Therefore, females always transmit a methylated, transcriptionally silent copy of this gene, while males transmit a nonmethylated, active copy.

For Maternal Effect Genes, the Genotype of the Mother Determines the Phenotype of the Offspring

In epigenetic inheritance, genes are altered in ways that affect their expression in an individual or the individual's offspring. As we have seen, some of these alterations produce strange inheritance patterns, in which organisms with the same genotype have different phenotypes. Another strange inheritance pattern, with a very different explanation, involves a category of genes called **maternal effect genes**.

Inheritance patterns due to maternal effect genes were first identified in the 1920s by A. E. Boycott, in his studies of the freshwater snail *Lymnaea peregra*. In this species, the shell and internal organs can be arranged in either a right-handed (dextral) or a left-handed (sinistral) direction. The dextral orientation is more common and is dominant to the sinistral orientation. Whether a snail's body curves in a dextral or a sinistral direction depends on the pattern of cell division immediately following fertilization. **Figure 17.14** shows the results of Boycott's crosses of true-breeding strains of snails with either a dextral or a sinistral orientation. When a dextral female (*DD*) was crossed to a sinistral male (*dd*), all of the offspring were dextral. However, crossing a sinistral female (*dd*) to a dextral male (*DD*) produced the opposite result: all of the offspring were sinistral. These seemingly contradictory outcomes could not be explained in terms of Mendelian inheritance.

Alfred Sturtevant later suggested that snail coiling is due to a maternal effect gene that exists as a dextral (*D*) and a sinistral (*d*) allele. In the cross shown on the left, the P generation female is dextral (*DD*) and the male is sinistral (*dd*). In the cross on the right, the female is sinistral (*dd*) and the male is dextral (*DD*). In either case, the F₁ offspring are *Dd*. When the F₁ individuals from these two crosses are mated to each other, a genotypic ratio of 1 *DD* : 2 *Dd* : 1 *dd* is predicted for the F₂ generation. Because the *D* allele is dominant to the *d* allele, a Mendelian inheritance pattern would produce a 3:1 phenotypic ratio of dextral to sinistral snails. Instead, the snails of the F₂ generation were all dextral. To explain this observed result, Sturtevant proposed that the phenotype of the F₂ offspring depended solely on the genotype of the F₁ mother. Because the F₁ mothers were *Dd*, and the *D* allele is dominant, the F₂ offspring were dextral even if their genotype was *dd*!

Sturtevant's hypothesis is supported by the ratio of phenotypes seen in the F₃ generation. When members of the F₂ generation were crossed, the F₃ generation exhibited a 3:1 ratio of dextral to sinistral snails. These F₃ phenotypes reflect the genotypes of the F₂ mothers. The ratio of genotypes for the F₂ females was 1 *DD* : 2 *Dd* : 1 *dd*. The *DD* and *Dd* females produced dextral offspring, while the *dd* females produced sinistral offspring. This is consistent with the 3:1 phenotypic ratio in the F₃ generation.

The peculiar inheritance pattern of maternal effect genes can be explained by the process of egg maturation in female animals (**Figure 17.15**). Maternal cells called nurse cells surround a de-

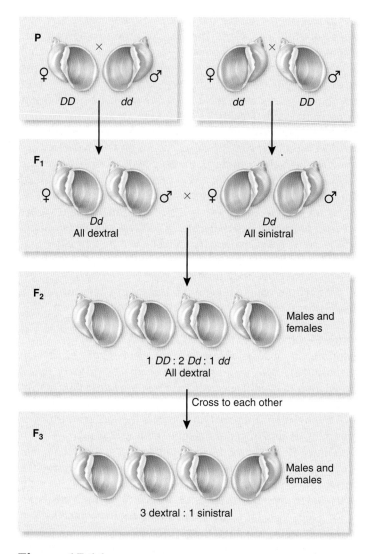

Figure 17.14   **The inheritance of snail coiling direction as an example of a maternal effect gene.**   In the snails shown in this experiment, the direction of body coiling is controlled by a single pair of genes. *D* (dextral, or right-handed) is dominant to *d* (sinistral, or left-handed). The genotype of the mother determines the phenotype of the offspring. A *DD* or *Dd* mother will produce dextral offspring and a *dd* mother will produce sinistral offspring, regardless of the genotypes of the father and of the offspring themselves.

Biological inquiry: An offspring has a genotype of Dd and coils to the left. What is the genotype of its mother?

veloping egg cell and provide it with nutrients. Within these diploid nurse cells, both copies of a maternal effect gene are activated to produce their gene products. The gene products are transported into the egg, where they persist for a significant time during embryonic development. The *D* and *d* gene products influence the pattern of cell division during the early stages of the snail's embryonic development. If an egg receives only the *D* gene product, the snail will develop a dextral orientation, while an egg that receives only the *d* gene product will produce a

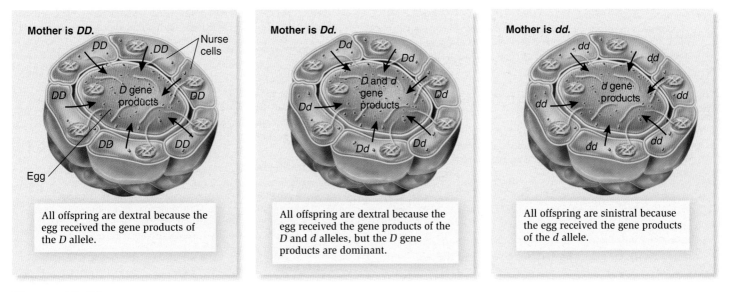

Figure 17.15 **The mechanism of maternal effect in snail coiling.** In this simplified diagram, the mother's diploid nurse cells transfer gene products to the egg as it matures. These gene products persist after fertilization, affecting development of the early embryo. If the nurse cells are *DD* or *Dd*, they will transfer the dominant *D* gene product to the egg, causing the offspring to be dextral. If the nurse cells are *dd*, only the *d* gene product will be transferred to the egg and the offspring will be sinistral.

snail with a sinistral orientation. If an egg receives both *D* and *d* gene products, the snail will be dextral because the *D* gene product is dominant over *d*. In this way, the gene products of nurse cells, which are determined by the mother's genotype, influence the development of the offspring.

Several dozen maternal effect genes have been identified in experimental organisms, such as *Drosophila*. Recently, they have also been found in mice and humans. As we will discuss in Chapter 19, the products of maternal effect genes are critically important in the early stages of animal development.

CHAPTER SUMMARY

- A variety of inheritance patterns are more complex than Mendel had realized. Many of these do not obey one or both of his laws of inheritance. (Table 17.1)

17.1 Gene Interactions

- When the alleles of one gene mask the effects of the alleles of a different gene, this type of gene interaction is called epistasis. (Figure 17.1)
- Quantitative traits such as height and weight are polygenic, which means that several genes govern the trait. Often, the alleles of such genes contribute in an additive way to the phenotype. This produces continuous variation in the trait, which is graphed as a bell curve. (Figure 17.2)

17.2 Genes on the Same Chromosome: Linkage, Recombination, and Mapping

- When two different genes are on the same chromosome, they are said to be linked. Linked genes tend to be inherited as a unit, unless crossing over separates them. (Figures 17.3, 17.4)
- The percentage of offspring produced in a two-factor testcross can be used to map the relative locations of genes along a chromosome. (Figure 17.5)

17.3 Extranuclear Inheritance: Organelle Genomes

- Mitochondria and chloroplasts carry a small number of genes. The inheritance of such genes is called extranuclear inheritance. (Figure 17.6)
- Chloroplasts in the four-o'clock plant are transmitted via the egg, a pattern called maternal inheritance. (Figures 17.7, 17.8)
- Several human diseases are known to be caused by mutations in mitochondrial genes, which follow a maternal inheritance pattern. (Table 17.2)

17.4 X Inactivation, Genomic Imprinting, and Maternal Effect

- Epigenetic inheritance refers to patterns in which a gene is inactivated during the life of an organism, but not over the course of many generations.
- X inactivation in mammals occurs when one X chromosome is randomly inactivated in females. If the female is heterozygous for an X-linked gene, this can lead to a variegated phenotype. (Figures 17.9, 17.10)
- X inactivation occurs in three phases: initiation, spreading, and maintenance. (Figure 17.11)
- Imprinted genes are inactivated by one parent but not both. The offspring expresses only one of the two alleles. (Figure 17.12)
- During gamete formation, methylation of a gene from one parent is a mechanism to achieve imprinting. (Figure 17.13)

- For maternal effect genes, the genotype of the mother determines the phenotype of the offspring. This is explained by the phenomenon that the mother's nurse cells contribute gene products to egg cells that are needed for early stages of development. (Figures 17.14, 17.15)

TEST YOURSELF

1. Quantitative traits such as height and weight are governed by several genes that usually contribute in an additive way to the trait. This is called
   a. independent assortment.
   b. discontinuous inheritance.
   c. maternal inheritance.
   d. linkage.
   e. polygenic inheritance.

2. When two genes are located on the same chromosome they are said to be
   a. homologous.   d. linked.
   b. allelic.   e. polygenic.
   c. epistatic.

3. Based on the ideas proposed by Morgan, which of the following statements concerning linkage is *not* true?
   a. Traits determined by genes located on the same chromosome are likely to be inherited together.
   b. Crossing over between homologous chromosomes can create new gene combinations.
   c. Crossing over is more likely to occur between genes that are closer together.
   d. The probability of crossing over depends on the distance between the genes.
   e. All but one of the above statements are correct.

4. In genetic linkage mapping, 1 map unit is equivalent to
   a. 100 base pairs.
   b. 1 base pair.
   c. 10% recombination frequency.
   d. 1% recombination frequency.
   e. 1% the length of the chromosome.

5. Organelle heredity is possible because
   a. gene products may be stored in organelles.
   b. mRNA may be stored in organelles.
   c. some organelles contain genetic information.
   d. conjugation of nuclei occurs before cellular division.
   e. both a and c.

6. In many organisms, organelles such as the mitochondria are contributed by only the egg. This phenomenon is known as
   a. biparental inheritance.
   b. paternal inheritance.
   c. maternal effect.
   d. maternal inheritance.
   e. both c and d.

7. Modification of a gene during gamete formation or early development that alters the way the gene is expressed during the individual's lifetime is called
   a. maternal inheritance.
   b. epigenetic inheritance.
   c. epistasis.
   d. multiple allelism.
   e. alternative splicing.

8. When a gene is inactivated during gamete formation and that gene is maintained in an inactivated state in the somatic cells of offspring, such an inheritance pattern is called
   a. linkage.
   b. X inactivation.
   c. maternal effect.
   d. genomic imprinting.
   e. polygenic inheritance.

9. Calico coat pattern in cats is the result of
   a. X inactivation.   d. genomic imprinting.
   b. epistasis.   e. maternal inheritance.
   c. organelle heredity.

10. Maternal effect inheritance can be explained by
    a. gene products that are given to an egg by the nurse cells.
    b. the methylation of genes during gamete formation.
    c. the spreading of X inactivation from the Xic locus.
    d. the inheritance of alleles that contribute additively to a trait.
    e. none of the above.

CONCEPTUAL QUESTIONS

1. Define linkage and linkage group.
2. Explain extranuclear inheritance and give two examples.
3. Define genomic imprinting.

EXPERIMENTAL QUESTIONS

1. What hypothesis were Bateson and Punnett testing when conducting the crosses in the sweet pea?
2. What were the expected results of Bateson and Punnett's cross?
3. How did the observed results differ from the predicted results? How did Bateson and Punnett explain the results of this particular cross?

COLLABORATIVE QUESTIONS

1. Discuss two types of gene interactions.
2. Discuss the concept of linkage.

---

www.brookerbiology.com

This website includes answers to the Biological Inquiry questions found in the figure legends and all end-of-chapter questions.

18

GENETICS OF BACTERIA AND VIRUSES

CHAPTER OUTLINE

An electron micrograph showing the budding of human immunodeficiency viruses from the surface of a human cell.

I n Chapters 15 through 17, we focused on the genetics of eukaryotic species. We considered the structure of chromosomes and their transmission during meiosis and mitosis, and examined the inheritance patterns that occur when genes are passed from parents to offspring. Plants and animals are amenable to inheritance studies for two reasons. First, their allelic differences give rise to readily discernable phenotypic differences between individuals, such as white versus red eyes in *Drosophila* or tall versus dwarf pea plants. Second, because eukaryotic species reproduce sexually, researchers can make crosses and analyze the transmission patterns of traits that are passed from parents to offspring. The ability to follow allelic differences in a genetic cross is a basic tool in the genetic examination of eukaryotic species.

In this chapter, we turn our attention to the genetic analyses of bacteria and viruses. From a human perspective, this is a compelling topic due to its impact on health. Infectious diseases caused by bacteria and viruses are a leading cause of human death, accounting for a quarter to a third of deaths worldwide. The spread of infectious diseases results from human behavior, and in recent times it has been accelerated by changes in land use patterns, increased trade and travel, and the inappropriate use of antibiotic drugs. Although the incidence of fatal infectious diseases in the U.S. is low compared to the worldwide average, an alarming increase in more deadly strains of bacteria and viruses has occurred over the past few decades. Since 1980, the number of deaths in the U.S. due to infectious diseases has approximately doubled.

In the first part of this chapter, we will examine the bacterial genome and the methods used in its investigation. Like their eukaryotic counterparts, bacteria have allelic differences that affect their cellular traits, and the techniques of modern microbiology make many of these differences easy to detect. Readily discernable differences among bacteria include sensitivity to antibiotics and differences in nutritional requirements for growth. Although bacteria reproduce asexually by cell division, their genetic variety is enhanced by the phenomenon called genetic transfer, in which genes are passed from one bacterial cell to another. Like sexual reproduction in eukaryotes, genetic transfer enhances the genetic diversity observed among different species and within the same species. A transferred gene may be beneficial to the recipient. For example, a bacterial cell carrying a gene that provides antibiotic resistance may transfer this gene to another bacterium, allowing the recipient cell to survive exposure to the antibiotic. In this chapter, we will explore three interesting ways that bacteria can transfer genetic material.

In the second part of this chapter, we examine viruses and other nonliving particles that infect living cells. All organisms are susceptible to infection by one or more types of viruses, which use cellular machinery to replicate their own genomes. Once a cell is infected, the genetic material of a virus orchestrates a series of events that ultimately leads to the production of new virus particles. In this chapter, we will consider the biological complexity of viruses and explore viral reproductive cycles.

18.1 Bacterial Genetics

Prokaryotes, which include bacteria and archaea, are usually unicellular organisms that lack cell nuclei. Individual prokaryotic cells may exist as single units or remain associated with each other after cell division, forming pairs, chains, or clumps.

Bacteria are widespread on Earth, and several species are known to cause many types of infectious diseases. Many species of archaea are also known, though they are less common than bacteria. In this chapter, we focus on bacteria. We will compare the genomes of bacteria, archaea, and eukaryotes in Chapter 21.

Much of our knowledge of bacterial genetics comes from studies of *Escherichia coli*, which is a natural inhabitant of the colon of humans and other species. As you learned in previous chapters, this bacterium has been an important model organism for studying molecular genetics. Indeed, we know more about the molecular biology of this species than any other species on our planet.

We begin this section by exploring the structure and replication of the bacterial genome. We will examine the organization of DNA sequences along a bacterial chromosome, how the chromosome is compacted to fit inside a bacterium, and how it is transmitted by asexual reproduction. We will then look at three types of genetic transfer called conjugation, transformation, and transduction in which DNA from one bacterium is transferred to another. Finally, we will consider gene transfer between different species, a process called horizontal gene transfer.

Bacteria Typically Have One or Several Copies of a Single Circular Chromosome

The genes of bacteria are found within structures known as bacterial chromosomes. Although a bacterial cell usually has a single type of chromosome, it may have more than one copy of that chromosome. The number of copies depends on the bacterial species and on growth conditions, but a bacterium typically has one to four identical chromosomes. Each bacterial chromosome is tightly packed within a distinct **nucleoid** region of the cell (**Figure 18.1**). Unlike the eukaryotic nucleus, the bacterial nucleoid is not a separate cellular compartment bounded by a membrane. The DNA in the nucleoid is in direct contact with the cytoplasm of the cell.

Like eukaryotic chromosomes, bacterial chromosomes are molecules of double-stranded DNA. Unlike eukaryotic chromosomes, however, bacterial chromosomes are usually circular and tend to be much shorter, typically only a few million nucleotides long. For example, the chromosome of *E. coli* has approximately 4.6 million base pairs, and the *Hemophilus influenzae* chromosome has roughly 1.8 million. By comparison, an average eukaryotic chromosome may be 100 million bp in length.

A typical bacterial chromosome contains a few thousand unique genes that are found throughout the chromosome (**Figure 18.2**). Structural gene sequences (nucleotide sequences that encode proteins) account for the largest part of bacterial DNA. Other nucleotide sequences in the chromosome influence DNA replication, gene expression, and chromosome structure. One of these sequences is the origin of replication, which is a few hundred nucleotides long. Bacterial chromosomes have a single origin of replication that functions as an initiation site for the assembly of several proteins that are required for DNA replication (refer back to Figure 11.16a).

A variety of short, repetitive sequences (less than 10 nucleotides long) have been identified in many bacterial species. These sequences are found in multiple copies and are usually interspersed within the intergenic regions throughout the bacterial chromosome, as shown in Figure 18.2. Some of these short, repetitive sequences appear to play no useful role, and they are not transcribed into RNA. Others may play a role in a variety of genetic processes, including DNA folding, DNA replication, and gene expression.

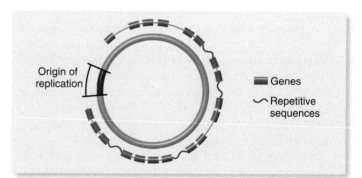

Key features

- Most, but not all, bacterial species contain circular chromosomal DNA.

- A typical chromosome is a few million base pairs in length.

- Most bacterial species contain a single type of chromosome, but it may be present in multiple copies.

- Several thousand different genes are interspersed throughout the chromosome.

- One origin of replication is required to initiate DNA replication.

- Short, repetitive sequences may be interspersed throughout the chromosome.

Figure 18.2   The organization of nucleotide sequences in bacterial chromosomal DNA.

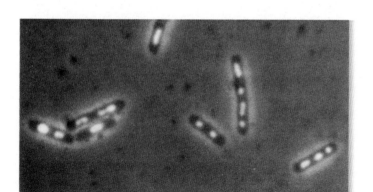

Figure 18.1   Nucleoids within the bacterium *Bacillus subtilis*.   In the light micrograph shown here, the nucleoids are fluorescently labeled and seen as bright, oval-shaped regions within the bacterial cytoplasm. Two or more nucleoids are usually found within each cell.

The Formation of Chromosomal Loops and DNA Supercoiling Makes the Bacterial Chromosome Compact

Bacterial cells are much smaller than most eukaryotic cells (refer back to Figure 4.1). *E. coli* cells, for example, are approximately 1 μm wide and 2 μm long. To fit within a bacterial cell, the DNA of a typical bacterial chromosome must be compacted about 1,000-fold. The compaction process, shown in **Figure 18.3**, involves the formation of loops and the supercoiling of the looped DNA.

Unlike eukaryotic DNA, bacterial DNA is not wound around histone proteins to form nucleosomes. However, the binding of proteins to bacterial DNA is important in the formation of **loop domains**, which are chromosomal segments that are folded into loops. As seen in Figure 18.3, DNA-binding proteins anchor the bases of the loops in place. The number of loops varies according to the size of a bacterial chromosome and the species. The *E. coli* chromosome has 50 to 100 loop domains, each with about 40,000 to 80,000 bp. This looping compacts the circular chromosome about 10-fold. A similar process of loop-domain formation occurs in eukaryotic chromatin compaction, which is described in Chapter 15.

DNA **supercoiling** is a second important way to compact the bacterial chromosome. Because DNA is a long, thin molecule, twisting can dramatically change its conformation. This compaction is similar to what happens to a rubber band if you twist it in one direction. Because the two strands of DNA already coil around each other, the formation of additional coils due to twisting is referred to as supercoiling. Bacterial enzymes called topoisomerases twist the DNA and control the degree of DNA supercoiling.

Plasmids Are Small Pieces of Extrachromosomal DNA

In addition to chromosomal DNA, bacterial cells commonly contain **plasmids**, which are small, circular pieces of DNA that exist independently of the bacterial chromosome (**Figure 18.4**). Plasmids occur naturally in many strains of bacteria and in a few types of eukaryotic cells, such as yeast. The smallest plasmids consist of just a few thousand base pairs and carry only a gene or two; the largest are in the range of 100,000 to 500,000 bp and carry several dozen or even hundreds of genes. A plasmid has its own origin of replication that allows it to be replicated independently of the bacterial chromosome. The DNA sequence of the origin of replication influences how many copies of the plasmid are found within a cell. Some origins are said to be very strong because they result in many copies of the plasmid, perhaps as many as 100 per cell. Other origins of replication have sequences that are much weaker, so that the number of copies is relatively low, such as one or two per cell.

Why do bacteria have plasmids? Plasmids are not usually necessary for survival. However, in many cases, certain genes within a plasmid provide some type of growth advantage to the cell. By studying plasmids in many different species, researchers have discovered that most plasmids fall into five different categories:

1. Resistance plasmids, also known as R factors, contain genes that confer resistance against antibiotics and other types of toxins.
2. Degradative plasmids carry genes that enable the bacterium to digest and utilize an unusual substance. For example, a degradative plasmid may carry genes that allow a bacterium to digest an organic solvent such as toluene.
3. Col-plasmids contain genes that encode colicines, which are proteins that kill other bacteria.
4. Virulence plasmids carry genes that turn a bacterium into a pathogenic strain.
5. Fertility plasmids, also known as F factors, allow bacteria to mate with each other, a topic described later in this chapter.

On occasion, a plasmid may integrate into the bacterial chromosome. Plasmids that can integrate are termed **episomes**.

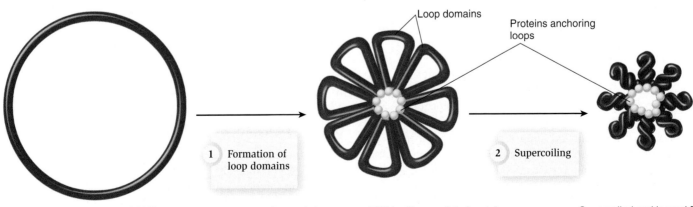

Figure 18.3 **The compaction of a bacterial chromosome.** As a way to compact the large, circular chromosome, segments are organized into smaller loop domains by binding to proteins at the bases of the loops. These loops are made more compact by DNA supercoiling.

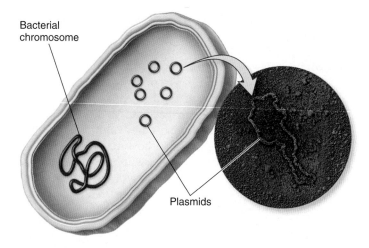

Bacterial
chromosome

Plasmids

Figure 18.4   **Plasmids in a bacterial cell.**   Plasmids are small, circular DNA molecules that exist independently of the bacterial chromosome.

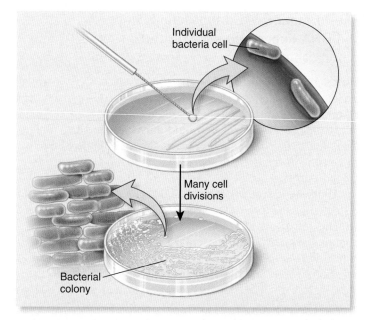

Individual
bacteria cell

Many cell
divisions

Bacterial
colony

Bacteria Reproduce Asexually by Binary Fission

Thus far, we have considered the genetic material of bacteria, and how the bacterial chromosome is compacted to fit inside the cell. Let's now turn our attention to the process of cell division. The capacity of bacteria to divide is really quite astounding. The cells of some species, such as *E. coli*, can divide every 20–30 minutes. When placed on a solid growth medium in a petri dish, an *E. coli* cell and its daughter cells will undergo repeated cellular divisions and form a clone of genetically identical cells called a **bacterial colony** (**Figure 18.5**). Starting with a single cell that is invisible to the naked eye, a visible bacterial colony containing 10 to 100 million cells will form in less than a day!

As described in Chapter 15, the division of eukaryotic cells requires a sorting process called mitosis, because eukaryotic chromosomes occur in sets, and each daughter cell must receive the correct number and types of chromosomes. By comparison, a bacterial cell has only a single type of chromosome and a much simpler division process called **binary fission**. **Figure 18.6** shows this process for a cell with a single chromosome. Before it divides, the cell replicates its DNA. This produces two identical copies of the chromosome. Next, the cell's plasma membrane is drawn inward and deposits new cell-wall material, separating the two daughter cells. Each daughter cell receives one of the copies of the original chromosome. Therefore, except when a mutation occurs, each daughter cell contains an identical copy of the mother cell's genetic material. Like other types of asexual reproduction, binary fission does not involve genetic contributions from two different parents.

If a bacterial cell contains plasmids, these will replicate independently of the bacterial chromosome. When a plasmid is found in multiple copies in a cell, binary fission usually results in each daughter cell containing one or more copies of the plasmid.

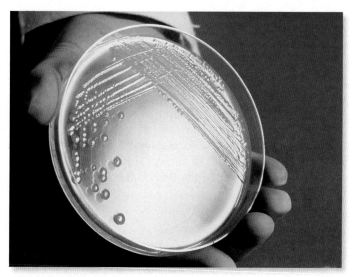

Figure 18.5   **Growth of a bacterial colony.**   Through successive cell divisions, a single bacterial cell of *E. coli* forms a genetically identical group of cells called a bacterial colony.

Biological inquiry: Let's suppose a bacterial strain divides every 30 minutes. If a single cell is placed on a plate, how many cells will be in the colony after 16 hours?

Three Modes of Genetic Transfer Enhance Genetic Diversity in Bacteria

Even though bacteria reproduce asexually, they exhibit a great deal of genetic diversity. Within a given bacterial species, the term **strain** refers to a lineage that has genetic differences compared to another strain. For example, one strain of *E. coli* may be resistant to an antibiotic while another strain may be sensitive to the same antibiotic. Genetic diversity in bacteria comes

Mother cell

Plasma membrane — Cell wall

Chromosome

Bacterial chromosome

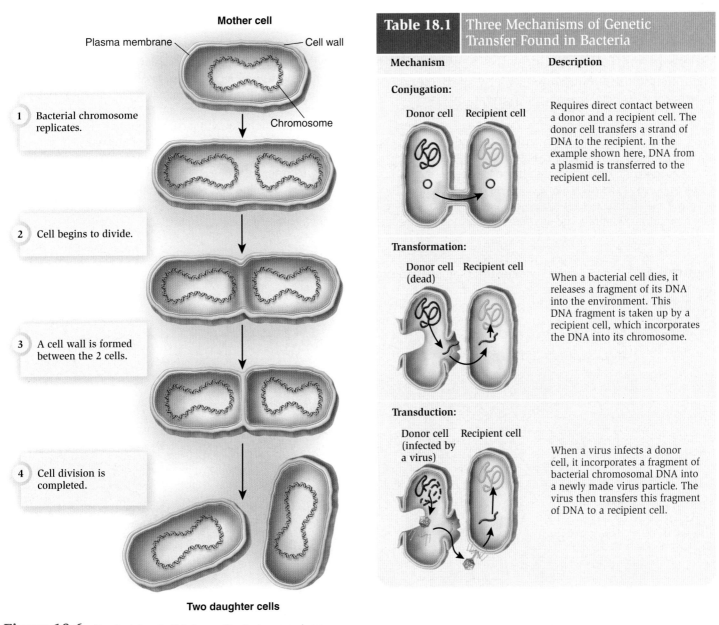

1   Bacterial chromosome replicates.

2   Cell begins to divide.

3   A cell wall is formed between the 2 cells.

4   Cell division is completed.

Two daughter cells

Figure 18.6 Bacterial cell division.   Bacteria reproduce by a type of cell division called binary fission. Before a bacterium divides, the bacterial chromosome is replicated to produce two identical copies. These two copies segregate from each other during cell division, with one copy going to each daughter cell.

Table 18.1	Three Mechanisms of Genetic Transfer Found in Bacteria	
Mechanism		**Description**
Conjugation: Donor cell Recipient cell		Requires direct contact between a donor and a recipient cell. The donor cell transfers a strand of DNA to the recipient. In the example shown here, DNA from a plasmid is transferred to the recipient cell.
Transformation: Donor cell Recipient cell (dead)		When a bacterial cell dies, it releases a fragment of its DNA into the environment. This DNA fragment is taken up by a recipient cell, which incorporates the DNA into its chromosome.
Transduction: Donor cell Recipient cell (infected by a virus)		When a virus infects a donor cell, it incorporates a fragment of bacterial chromosomal DNA into a newly made virus particle. The virus then transfers this fragment of DNA to a recipient cell.

primarily from two sources. First, mutations can occur that alter the bacterial genome and affect the traits of bacterial cells. For example, a mutation may give rise to a bacterial strain that requires the amino acid histidine from an outside source for growth, although other strains of the same species can make this amino acid.

The second way that diversity can arise is by **genetic transfer**, in which genetic material is transferred from one bacterial cell to another. For example, a plasmid carrying an antibiotic-resistance gene may be transferred from a resistant strain to a sensitive strain. Genetic transfer occurs in three very different ways (**Table 18.1**). The process known as **conjugation** involves a direct physical interaction between two bacterial cells. During conjugation, one bacterium acts as a donor and transfers DNA to a recipient cell. In the process of **transformation**, DNA that is released into the environment when a bacterium dies is taken up by another bacterial cell. **Transduction** occurs when a virus infects a bacterial cell and then transfers some of that cell's DNA to another bacterium. These three types of genetic transfer have been extensively investigated in research laboratories, and their molecular pathways continue to be studied with great interest. Let's now examine these mechanisms in greater detail and consider the experiments that led to their discovery.

FEATURE INVESTIGATION

Lederberg and Tatum's Work with *E. coli* Demonstrated Genetic Transfer Between Bacteria and Led to the Discovery of Conjugation

In 1946, Joshua Lederberg and Edward Tatum carried out the first experiment that clearly demonstrated genetic transfer from one bacterial strain to another (**Figure 18.7**). The researchers had been studying strains of *E. coli* that had different nutritional requirements for growth. They designated one strain *met⁻ bio⁻ thr⁺ leu⁺ thi⁺* because its growth required that the amino acid methionine (met) and the vitamin biotin (bio) be added to the growth medium. This strain did not require the amino acids threonine (thr) and leucine (leu) or the vitamin thiamine (thi) for growth. Another strain, designated *met⁺ bio⁺ thr⁻ leu⁻ thi⁻*, had just the opposite requirement. It needed threonine, leucine, and thiamine in its growth medium, but not methionine or biotin. These differences in nutritional requirements correspond to allelic differences between the two strains. The *met⁻ bio⁻ thr⁺ leu⁺ thi⁺* strain had defective genes encoding enzymes necessary for methionine and biotin synthesis, while *met⁺ bio⁺ thr⁻ leu⁻ thi⁻* had defective genes for the enzymes required to make threonine, leucine, and thiamine.

Figure 18.7   Experiment of Lederberg and Tatum demonstrating genetic transfer in *E. coli*.

HYPOTHESIS Genetic material can be transferred from one bacterial strain to another.
STARTING MATERIALS Two bacterial strains, one that was *met⁻ bio⁻ thr⁺ leu⁺ thi⁺* and the other that was *met⁺ bio⁺ thr⁻ leu⁻ thi⁻*.

Experimental level / Conceptual level

1 In 3 separate tubes, add either the *met⁻ bio⁻ thr⁺ leu⁺ thi⁺* strain, the *met⁺ bio⁺ thr⁻ leu⁻ thi⁻* strain, or a mixture of both strains.

Incubate several hours.

met⁻ bio⁻ thr⁺ leu⁺ thi⁺   *met⁺ bio⁺ thr⁻ leu⁻ thi⁻*

2 Remove 10^8 cells from each tube and spread onto plates that lack methionine, biotin, threonine, leucine, and thiamine.

10^8 10^8 10^8

Nutrient agar plates lacking amino acids, biotin, and thiamine.

3 Incubate overnight to allow growth of bacterial colonies.

No colonies   Bacterial colonies (*met⁺ bio⁺ thr⁺ leu⁺ thi⁺*)   No colonies

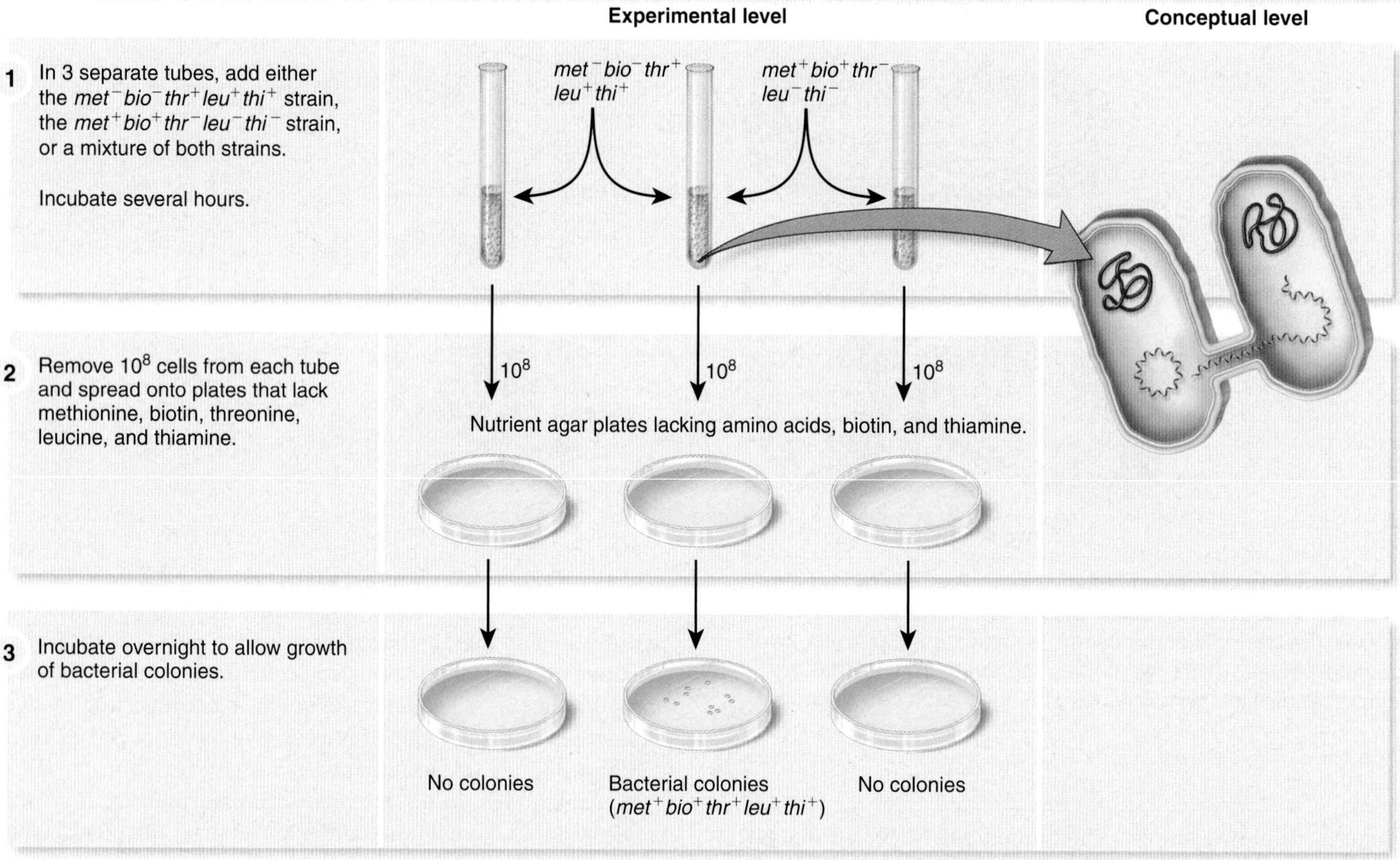

4 **THE DATA**

Strain	Number of colonies after overnight growth
met⁻ bio⁻ thr⁺ leu⁺ thi⁺	0
met⁺ bio⁺ thr⁻ leu⁻ thi⁻	0
Both strains together	~10

Figure 18.7 compares the results of mixing the two *E. coli* strains with the results when they were not mixed. The tube shown on the left contains only *met⁻ bio⁻ thr⁺ leu⁺ thi⁺* cells, and the tube on the right has only *met⁺ bio⁺ thr⁻ leu⁻ thi⁻* cells. The middle tube contains a mixture of the two kinds of cells. In each case, the researchers applied about 100 million (10^8) cells to plates containing a growth medium lacking amino acids, biotin, and thiamine. When the unmixed strains were applied to

these plates, no colonies were observed to grow. This result is expected because the plates did not contain the methionine and biotin that the *met⁻ bio⁻ thr⁺ leu⁺ thi⁺* cells need for growth, or the threonine, leucine, and thiamine that the *met⁺ bio⁺ thr⁻ leu⁻ thi⁻* cells require. The striking result occurred when the researchers plated 10^8 cells from the tube containing the mixture of the two strains. In this case, approximately 10 cells multiplied and formed visible bacterial colonies on the plates. Because these cells were able to reproduce without supplemental amino acids or vitamins, their genotype must have been *met⁺ bio⁺ thr⁺ leu⁺ thi⁺*. Mutation cannot account for the occurrence of this new genotype because colonies were not observed on the other two plates, which had the same number of cells and also could have incurred mutations.

To explain the results of their experiment, Lederberg and Tatum hypothesized that some genetic material was transferred between the two strains when they were mixed. Note that this transfer could have occurred in two ways. One possibility is that the genes providing the ability to synthesize threonine, leucine, and thiamine (*thr⁺ leu⁺ thi⁺*) were transferred to the *met⁺ bio⁺ thr⁻ leu⁻ thi⁻* strain. Alternatively, the genes providing the ability to synthesize methionine and biotin (*met⁺ bio⁺*) may have been transferred to the *met⁻ bio⁻ thr⁺ leu⁺ thi⁺* cells. The experimental results cannot distinguish between these two possibilities, but they provide compelling evidence that at least one of them occurred.

How did the bacteria in Lederberg and Tatum's experiment transfer genes between strains? Two mechanisms seemed plausible. Either genetic material was released from one strain and taken up by the other, or cells of the two different strains made contact with each other and directly transferred genetic material. To distinguish these two scenarios, Bernard Davis conducted experiments using the same two strains of *E. coli*. The apparatus he used, known as a U-tube, is shown in **Figure 18.8**. The tube had a filter with pores big enough for pieces of DNA to pass through, but too small to permit the passage of bacteria. After filling the tube with a liquid medium, Davis added *met⁻ bio⁻ thr⁺ leu⁺ thi⁺* bacteria on one side of the filter and *met⁺ bio⁺ thr⁻ leu⁻ thi⁻* bacteria on the other. The application of pressure or suction promoted the movement of liquid through the

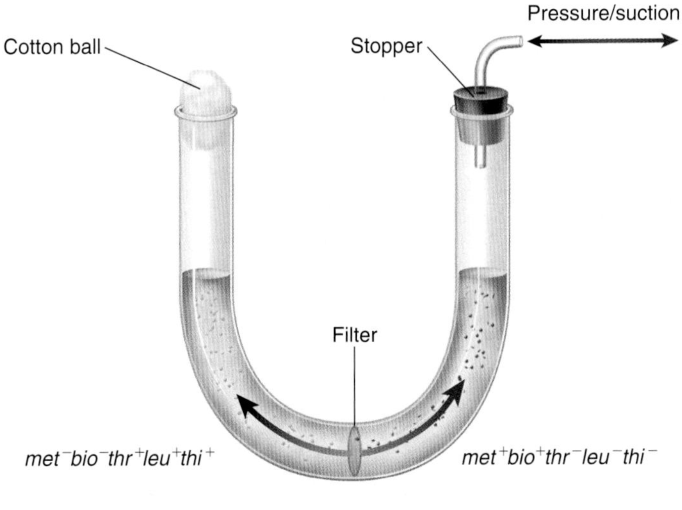

Figure 18.8   A U-tube apparatus like the one used by Bernard Davis.   Bacteria of two different strains are suspended in the liquid in the tube and separated by a filter. The liquid is forced through the filter by alternating suction and pressure. The pores in the filter are too small for the passage of bacteria, but they allow the passage of DNA.

Biological inquiry: Would the results have been different if the pore size was larger and allowed the passage of bacterial cells?

pores. Although the two kinds of bacteria could not mix, any genetic material released by one of them would be available to the other.

After allowing the bacteria to incubate in the U-tube, Davis placed cells from each side of the tube on growth plates lacking methionine, biotin, threonine, leucine, and thiamine. No bacterial colonies grew on these plates. This result showed that without physical contact, the two *E. coli* strains do not transfer genetic material from one to the other. The conceptual level of Figure 18.7 shows the physical connection that explains Lederberg and Tatum's results. Genetic transfer that requires direct cell-to-cell contact is now called conjugation, and it was subsequently observed in other species of bacteria. Many, but not all, species of bacteria can conjugate.

During Conjugation, DNA Is Transferred from a Donor Cell to a Recipient Cell

In the early 1950s, Joshua and Esther Lederberg, William Hayes, and Luca Cavalli-Sforza independently discovered that only certain bacterial strains can donate genetic material during conjugation. For example, only about 5% of *E. coli* strains found in nature can act as donor strains. Further research showed that a strain that is incapable of acting as a donor can acquire this ability after being mixed with a donor strain. Hayes correctly proposed that donor strains contain a fertility factor, or **F factor**, that can be transferred to recipient strains. Also, other donor *E. coli* strains were later identified that can transfer portions of

the bacterial chromosome at high frequencies. After a segment of the chromosome is transferred, it then inserts, or recombines, into the chromosome of the recipient cell. Such donor strains were named Hfr (for <u>H</u>igh <u>f</u>requency of <u>r</u>ecombination).

The micrograph in **Figure 18.9a** shows two conjugating *E. coli* cells. The cell on the left is designated F^+, meaning that it has an F factor. This donor cell is transferring genetic material to the recipient cell on the right, which lacks an F factor and is designated F^-. As discussed earlier, the F factor, or fertility factor, is a type of plasmid found in many bacterial species. F factors carry several genes that are required for conjugation and also may carry genes that confer a growth advantage for the bacterium.

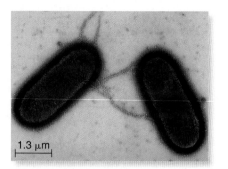

(a) Micrograph of conjugating cells

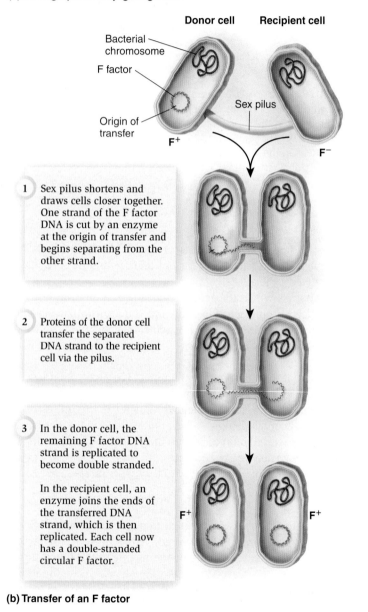

(b) Transfer of an F factor

Donor cell   Recipient cell

Bacterial chromosome

F factor

Sex pilus

Origin of transfer

F⁺   F⁻

1   Sex pilus shortens and draws cells closer together. One strand of the F factor DNA is cut by an enzyme at the origin of transfer and begins separating from the other strand.

2   Proteins of the donor cell transfer the separated DNA strand to the recipient cell via the pilus.

3   In the donor cell, the remaining F factor DNA strand is replicated to become double stranded.

In the recipient cell, an enzyme joins the ends of the transferred DNA strand, which is then replicated. Each cell now has a double-stranded circular F factor.

F⁺   F⁺

Figure 18.9   Bacterial conjugation.   (a) Two *E. coli* cells conjugating. The cell on the *left*, designated *F⁺*, is the donor; the cell on the *right*, designated *F⁻*, is the recipient. The two cells make contact via sex pili made by the *F⁺* cell. **(b)** The transfer of an F factor during conjugation. At the end of conjugation, both the donor cell and the recipient cell are *F⁺*.

Figure 18.9b describes the events that occur during conjugation in *E. coli*. The process is similar in other bacteria that are capable of conjugating, although the details vary somewhat from one species to another. Contact between donor and recipient cells is a key step that initiates the conjugation process. Some bacteria have hairlike appendages called pili (singular, pilus), which are hollow tubes that promote the attachment of bacteria to surfaces and to each other.

Sex pili are pili made by F^+ cells that bind specifically to F^- cells. They are so named because conjugation has sometimes been called bacterial mating or bacterial sex. However, these terms are a bit misleading because the process does not involve equal genetic contributions from two gametes and it does not produce offspring. Instead, bacterial mating is a form of genetic transfer that alters the genetic composition of the recipient cell. The genes responsible for the formation of sex pili are located on the F factor. In *E. coli* and some other species, F^+ cells make very long pili that attempt to make contact with nearby F^- cells. Once contact is made, the pili shorten, drawing the donor and recipient cells closer together.

Successful contact between a donor and a recipient cell stimulates the donor cell to begin the transfer process. Genes within the F factor encode proteins that promote the transfer of one strand of F factor DNA. This DNA strand is cut at the origin of transfer, and then the strand travels through the pilus into the recipient cell. The other strand remains in the donor cell and is replicated, restoring the F factor DNA to its original double-stranded condition. In the recipient cell, the two ends of the newly acquired F factor DNA strand are joined to form a circular molecule, which is then replicated to become double stranded. The end result of conjugation is that the recipient cell has acquired an F factor, converting it from an F^- to an F^+ cell. The genetic composition of the donor strain has not been changed.

In Transformation, Bacteria Take Up DNA from the Environment

In contrast to conjugation, the process of genetic transfer known as bacterial transformation does not require direct contact between bacterial cells. Frederick Griffith first discovered this process in 1928 while working with strains of *Streptococcus pneumoniae*. He called the unknown substance that transferred properties of one strain to another the transforming principle. Experimental work in the 1940s and 1950s showed that the transforming principle is DNA and that transformation involves the transfer of genes. We discussed some of these important experiments in Chapter 11.

During transformation, a living bacterial cell imports a strand of DNA that another bacterium released into the environment when it died. This DNA strand may then insert or recombine into the bacterial chromosome, so that the live bacterium is now carrying genes from the dead bacterium. Not all bacterial strains have the ability to take up DNA. Those that do have this ability are described as **naturally competent**, and they have genes that encode proteins called competence factors.

Competence factors facilitate the binding of DNA fragments to the bacterial cell surface, the uptake of DNA into the cytoplasm, and the incorporation of the imported DNA into the bacterial chromosome. Temperature, ionic conditions, and the availability of nutrients also affect whether or not a bacterium will be competent to take up genetic material. Thus, environmental conditions influence the expression of competence genes.

In recent years, biologists have unraveled some of the steps that occur when competent bacterial cells are transformed by taking up genetic material released from a dead bacterium. In the example shown in **Figure 18.10**, the foreign DNA carries a gene, tet^R, that confers resistance to the antibiotic tetracycline. First, a large fragment of the foreign DNA binds to a surface receptor on the outside of a bacterial cell. Enzymes secreted by the bacterium cut this large fragment into fragments small enough to enter the cell. The next step is for a small DNA fragment to begin its entry into the bacterial cytoplasm. One of the two DNA strands of this fragment is degraded. The other strand enters the bacterial cytoplasm via a DNA uptake system that transports the DNA across the plasma membrane. Finally, the imported DNA strand is incorporated into the bacterial chromosome and the complementary strand is synthesized. Following transformation, the recipient cell has been transformed from a tetracycline-sensitive to a tetracycline-resistant cell.

In Transduction, Viruses Transfer Genetic Material from One Bacterium to Another

Perhaps the most curious method of genetic transfer is transduction, in which viruses that infect bacteria transfer bacterial genes from one bacterium to another. As you will learn in the second part of this chapter, the viruses that we will consider are DNA-containing particles that use the bacterial cellular machinery for their own replication. The new viral particles made in this way usually contain only viral genes. In some cases, however, a virus may pick up a piece of DNA from the bacterial chromosome. When a virus carrying a segment of bacterial DNA infects another bacterium, it transfers this segment into the chromosome of its new bacterial host. This is the genetic transfer process called transduction. We will consider transduction in more detail when we look at viral reproductive cycles in the second part of this chapter.

GENOMES & PROTEOMES

Horizontal Gene Transfer Is the Transfer of Genes Between Different Species

So far we have considered gene transfer from one bacterial strain to another strain of the same species. In addition, conjugation, transformation, and transduction occasionally occur between cells of different bacterial species. Gene transfer between different species is termed **horizontal gene transfer**, in contrast to the

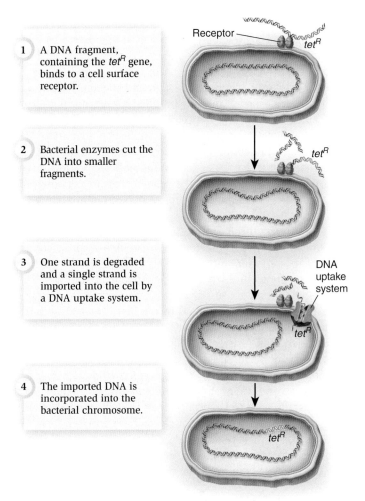

1 A DNA fragment, containing the tet^R gene, binds to a cell surface receptor.

2 Bacterial enzymes cut the DNA into smaller fragments.

3 One strand is degraded and a single strand is imported into the cell by a DNA uptake system.

4 The imported DNA is incorporated into the bacterial chromosome.

Transformed cell that is resistant to the antibiotic tetracycline

Figure 18.10 **The steps of bacterial transformation.** This process has transformed a bacterium that was sensitive to the antibiotic tetracycline into one that can grow in the presence of this antibiotic.

vertical gene transfer that occurs when genes are passed from one generation to the next among individuals of the same species.

In recent years, analyses of bacterial genomes have shown that a sizeable fraction of their genes are derived from horizontal gene transfer. For example, roughly 17% of the genes of *E. coli* and of *Salmonella typhimurium* have been acquired by horizontal transfer during the past 100 million years. Many of these acquired genes are for traits that give cells a selective advantage, including genes that confer antibiotic resistance, the ability to degrade toxic compounds, and the ability to withstand extreme environments. Some horizontally transferred genes confer pathogenicity, turning a harmless bacterial strain into one that can cause disease. Geneticists have suggested that horizontal gene transfer has played a major role in the evolution of different bacterial species. In many cases, the acquisition of new genes allows a bacterium to survive in a new type of environment and can eventually lead to the formation of a new species.

The medical relevance of horizontal gene transfer is profound. Antibiotics are widely prescribed to treat bacterial infections in humans and domestic animals. They are also used in agriculture to increase the growth of livestock and to control bacterial diseases of high-value fruits and vegetables. Unfortunately, the widespread and uncontrolled use of antibiotics has greatly increased the prevalence of antibiotic-resistant strains of bacteria because these strains have a selective advantage over those that are susceptible to antibiotics. Resistant strains carry genes that counteract the action of antibiotics in various ways. A resistance gene may encode a protein that breaks down the drug, pumps it out of the cell, or prevents it from inhibiting cellular processes.

The term **acquired antibiotic resistance** refers to the common phenomenon of a previously susceptible strain becoming resistant to a specific antibiotic. This change may result from genetic alterations in the bacteria's own genome, but it is often due to the horizontal transfer of resistance genes from a resistant strain. As often mentioned in the news media, antibiotic resistance has increased dramatically worldwide over the past few decades, with resistant strains reported in almost all pathogenic strains of bacteria. For example, the most common cause of pneumonia is infection by *Streptococcus pneumoniae*. In many countries, nearly 50% of all *S. pneumoniae* strains are now penicillin resistant, with resistance to other drugs increasing as well. Some of the most severe antibiotic resistance problems occur in hospitals. Resistant strains of *Klebsiella pneumoniae* and *Enterococcus faecium* are significant causes of infection and death among critically ill patients in intensive care units.

Figure 18.11   A plant infected with tobacco mosaic virus.

18.2 Viral Genetics, Viroids, and Prions

In earlier chapters and the first section of this chapter, we examined the replication and expression of eukaryotic and prokaryotic genes. Because all living organisms are either eukaryotes or prokaryotes, you may be thinking that we have considered every type of genome. Strangely, however, nonliving things also have genomes. Viruses and viroids are nonliving particles with nucleic acid genomes that require the assistance of living cells to reproduce.

The first virus to be discovered was tobacco mosaic virus (TMV). This virus infects several species of plants and causes mosaic-like patterns in which normal-colored patches are interspersed with light green or yellowish patches on the leaves (**Figure 18.11**). TMV damages leaves, flowers, and fruit but almost never kills the plant. In 1883, the German scientist Adolf Mayer determined that this disease could be spread by spraying the sap from one plant onto another. By subjecting this sap to filtration, the Russian scientist Dmitri Ivanovski demonstrated that the disease-causing agent was not a bacterium. Sap that had been passed through filters with pores small enough to prevent the passage of bacterial cells was still able to spread the disease. At first, some researchers suggested that the agent was a chemical

toxin. However, the Dutch botanist Martinus Beijerinck ruled out this possibility by showing that sap could continue to transmit the disease after many plant generations. A toxin would have been diluted after many generations, but Beijerinck's results indicated that the disease agent was multiplying in the plant. Around the same time, animal viruses were discovered in connection with a disease of cattle called foot-and-mouth disease. In 1900, the first human virus, the virus that causes yellow fever, was identified. Since these early studies, microbiologists, geneticists, and molecular biologists have taken a great interest in the study of viruses.

In this section, we will consider the structure, genetic composition, and replication of viruses. We will examine viral reproductive cycles, paying particular attention to HIV (human immunodeficiency virus), the virus that causes AIDS (acquired immunodeficiency syndrome) in humans. We will also look briefly at simpler nucleic acid particles called viroids that cause diseases in plants. Finally, we will discuss the infectious proteins known as prions, which cause devastating neurological diseases in humans and other mammals. Unlike other agents of infection, prions have no genes and cannot be copied by the replication machinery of a cell. Instead, they increase their numbers by inducing changes in other protein molecules within living cells.

Viruses Are Remarkably Varied, Despite Their Simple Structure

A **virus** is a small infectious particle that consists of nucleic acid enclosed in a protein coat. Researchers have identified and studied over 4,000 different types of viruses. Although all viruses share some similarities, such as small size and the reliance on a living cell for replication, they vary greatly in their characteristics, including their host range, structure, and genome composition. Some of the major differences are described next and in Table 18.2.

Differences in Host Range A cell that is infected by a virus is called a **host cell**, and a species that can be infected by a specific

Table 18.2 | Hosts and Characteristics of Selected Viruses

Virus or group of viruses	Host	Effect on host	Nucleic acid*	Genome size (kb)†	Number of genes†
Phage fd	E. coli	Slows growth	ssDNA	6.4	10
Phage λ	E. coli	Causes lysis	dsDNA	48.5	36
Phage T4	E. coli	Causes lysis	dsDNA	169	>190
Phage Qβ	E. coli	Slows growth	ssRNA	4.2	4
Tobacco mosaic virus	Many plants	Causes mottling and necrosis of leaves and other plant parts	ssRNA	6.4	6
Cauliflower mosaic virus	A few plants	Similar to tobacco mosaic virus	dsDNA	8.0	1
Baculoviruses	Insects	Most baculoviruses are species specific; they usually kill the insect	dsDNA	133.9	154
Parvovirus	Mammals	Causes respiratory, flulike symptoms	ssDNA	5.0	5
Influenza virus	Mammals	Causes classical "flu," with fever, cough, sore throat, and headache	ssRNA	13.5	12
Epstein-Barr virus	Humans	Causes mononucleosis, with fever, sore throat, and fatigue	dsDNA	172	80
Adenoviruses	Humans	Cause respiratory symptoms and diarrhea	dsDNA	34	35
Herpes simplex type II	Humans	Causes blistering sores around the genital region	dsDNA	155	74
HIV (type I)	Humans	Causes AIDS, an immunodeficiency syndrome eventually leading to death	ssRNA	9.2	9

* The abbreviations *ss* and *ds* refer to single stranded and double stranded, respectively.

† Several of the viruses listed in this table are found in different strains that show some variation with regard to genome size and number of genes. The numbers reported in this table are typical values. The abbreviation *kb* refers to kilobase, which equals 1,000 bases.

virus is called a host species for that virus. Viruses differ greatly in their **host range**, meaning the number of species and cell types they can infect. Table 18.2 lists a few examples of viruses with widely different ranges of host species. Tobacco mosaic virus, which we discussed earlier, has a broad host range. TMV is known to infect over 150 different species of plants. By comparison, other viruses have a limited host range. Some viruses infect only a single species, such as humans. Furthermore, a virus may infect only a specific cell type in that species. **Figure 18.12** shows some viruses that infect particular human cells and cause disease.

Structural Differences Although the existence of viruses was postulated in the 1890s, viruses were not observed until the 1930s, when the electron microscope was invented. Viruses cannot be resolved by even the best light microscope. In fact, most of them are smaller than the wavelength of visible light. Viruses range in size from about 20–400 nm in diameter (1 nanometer $= 10^{-9}$ meters). For comparison, a typical bacterium is 1,000 nm in diameter, and the diameter of most eukaryotic cells is 10 to 1,000 times that of a bacterium. Over 50 million adenoviruses could fit into an average-sized human cell.

Figure 18.13 shows basic structural variations of viruses. All viruses have a protein coat called a **capsid** enclosing a genome that consists of one or more molecules of nucleic acid. Capsids are composed of one or several different proteins and may be a variety of shapes, including helical and polyhedral.

Figure 18.13a shows the structure of tobacco mosaic virus, which has a helical capsid made of identical protein subunits. Figure 18.13b shows an adenovirus, which has a polyhedral capsid. Many viruses that infect animal cells, such as the influenza virus shown in Figure 18.13c, have a **viral envelope** enclosing the capsid. The envelope consists of a membrane that is derived from the plasma membrane of the host cell and is embedded with virally encoded spike glycoproteins. You will see how such viruses acquire their envelopes when we examine viral reproductive cycles later in the chapter.

In addition to encasing and protecting the genetic material, the capsid enables viruses to infect their hosts. The capsids of many viruses have specialized proteins that help them bind to the surface of a host cell. The adenovirus in Figure 18.13b has glycoproteins called spike glycoproteins that serve this purpose. Similarly, the envelope and spike glycoproteins of the influenza virus are important in the process of viral attachment to the host cell. Viruses that infect bacteria, called **bacteriophages**, or **phages**, may have much more complex capsids, with accessory structures used for anchoring the virus to a host cell and injecting the viral nucleic acid (Figure 18.13d). As we will discuss later, the tail fibers of such bacteriophages are needed to attach the virus to the bacterial cell wall.

Genome Differences The genetic material in a virus is called a **viral genome**. The composition of viral genomes varies markedly among different types of viruses, as suggested by the examples

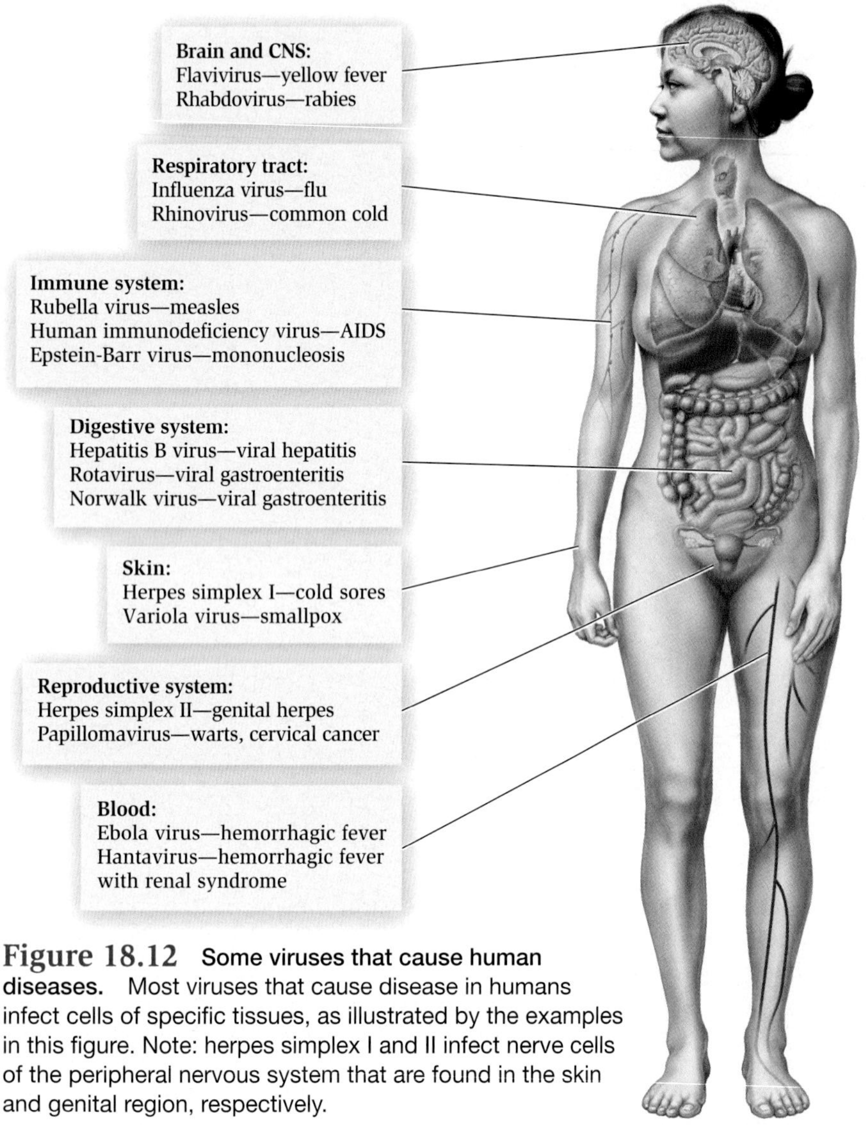

Brain and CNS:
Flavivirus—yellow fever
Rhabdovirus—rabies

Respiratory tract:
Influenza virus—flu
Rhinovirus—common cold

Immune system:
Rubella virus—measles
Human immunodeficiency virus—AIDS
Epstein-Barr virus—mononucleosis

Digestive system:
Hepatitis B virus—viral hepatitis
Rotavirus—viral gastroenteritis
Norwalk virus—viral gastroenteritis

Skin:
Herpes simplex I—cold sores
Variola virus—smallpox

Reproductive system:
Herpes simplex II—genital herpes
Papillomavirus—warts, cervical cancer

Blood:
Ebola virus—hemorrhagic fever
Hantavirus—hemorrhagic fever
with renal syndrome

Figure 18.12   **Some viruses that cause human diseases.**   Most viruses that cause disease in humans infect cells of specific tissues, as illustrated by the examples in this figure. Note: herpes simplex I and II infect nerve cells of the peripheral nervous system that are found in the skin and genital region, respectively.

in Table 18.2. The nucleic acid of some viruses is DNA, while in others it is RNA. These are referred to as DNA viruses and RNA viruses, respectively. It is striking that some viruses use RNA for their genome, whereas all living organisms use DNA. In some viruses the nucleic acid is single stranded, whereas in others it is double stranded. The genome can be linear or circular, depending on the type of virus. Some kinds of viruses have more than one copy of the genome.

Viral genomes also vary considerably in size, ranging from a few thousand to more than a hundred thousand nucleotides in length (see Table 18.2). For example, the genomes of some simple viruses, such as phage Qβ, are only a few thousand nucleotides in length and contain only a few genes. Other viruses, particularly those with a complex structure, contain many more genes. Phage T4 is an example of a complex virus having many genes. These extra genes are largely involved in the formation of the elaborate protein coat shown in Figure 18.13d.

Viruses Reproduce by Mobilizing Their Host Cells to Produce New Viruses

Viruses do not exhibit the properties that are found in living organisms, as described in Chapter 1. They are not cells or composed of cells, and they cannot carry out metabolism on their own. When a virus infects a cell, the expression of viral genes leads to a series of steps, called a **viral reproductive cycle**, that result in the production of new viruses. The details of the steps may be quite different among various types of viruses, and even the same virus may have the capacity to follow alternative cycles. However, by studying the reproductive cycles of hundreds of different viruses, researchers have determined that most viruses follow five or six basic steps.

To illustrate general features of viral reproductive cycles, **Figure 18.14** considers these steps for two types of viruses. Figure 18.14a shows the cycle of phage λ (lambda), a bacteriophage with double-stranded DNA as its genome. Figure 18.14b shows the cycle of HIV, an enveloped animal virus containing single-stranded RNA. The descriptions that follow compare the reproductive cycles of these two very different viruses. Later in the chapter, we will further examine HIV, which is the causative agent of AIDS.

Step 1: Attachment In the first step of a viral reproductive cycle, the virus must attach to the surface of a host cell. This attachment is usually specific for one kind of cell because proteins in the virus recognize and bind to specific molecules on the cell surface. In the case of phage λ, the phage tail fibers bind to proteins in the outer bacterial cell membrane of *E. coli*. In the case of HIV, spike glycoproteins in the viral envelope bind to protein receptors in the plasma membrane of human blood cells called T cells.

Step 2: Entry After attachment, the viral genome must enter the host cell. Attachment of phage λ stimulates a conformational change in the phage coat proteins, so that the shaft contracts and the phage injects its DNA into the bacterial cytoplasm. In contrast, the envelope of HIV fuses with the plasma membrane of the host cell, so that both the capsid and its contents are released into the cytosol. Some of the HIV capsid proteins are then removed by host cell enzymes, allowing the viral RNA and reverse transcriptase to escape from the capsid.

Once a viral genome has entered the cell, one or several viral genes are expressed immediately due to the action of host cell enzymes and ribosomes. Expression of these key genes leads quickly to either step 3 or step 4 of the reproductive cycle, depending on the specific virus. The genome of some viruses, including both phage λ and HIV, can integrate into a chromosome of the host cell.

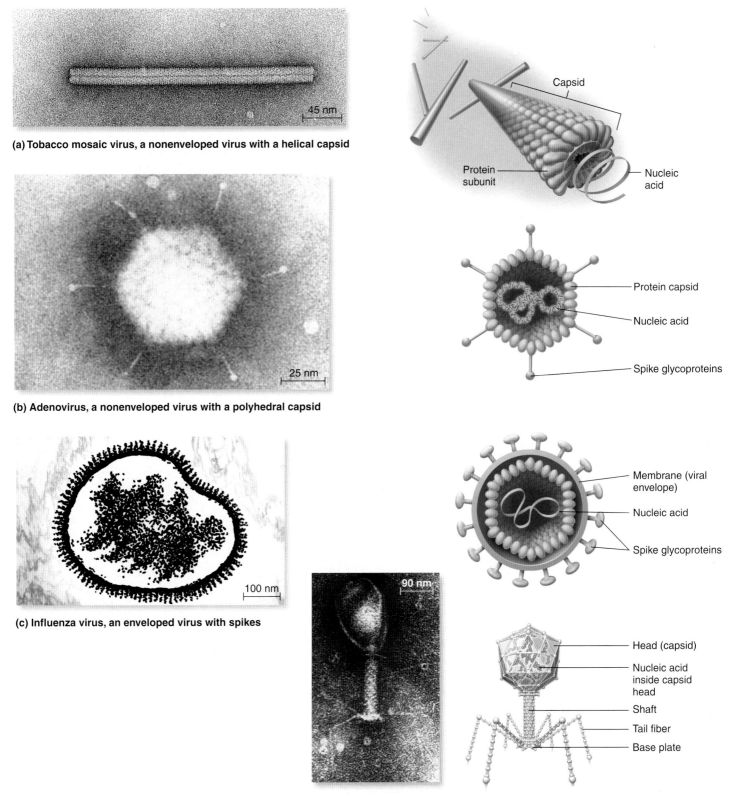

(a) Tobacco mosaic virus, a nonenveloped virus with a helical capsid

45 nm

Capsid

Protein subunit

Nucleic acid

(b) Adenovirus, a nonenveloped virus with a polyhedral capsid

25 nm

Protein capsid

Nucleic acid

Spike glycoproteins

(c) Influenza virus, an enveloped virus with spikes

100 nm

Membrane (viral envelope)

Nucleic acid

Spike glycoproteins

(d) T4, a bacteriophage

90 nm

Head (capsid)

Nucleic acid inside capsid head

Shaft

Tail fiber

Base plate

Figure 18.13 Variations in the structure of viruses. All viruses contain nucleic acid surrounded by a protein capsid. They may or may not have an outer envelope surrounding the capsid. **(a)** Tobacco mosaic virus (TMV) has a capsid made of 2,130 identical protein subunits, helically arranged around a strand of nucleic acid (RNA). **(b)** Adenoviruses have polyhedral capsids with glycoprotein spikes. **(c)** Many animal viruses, including the influenza virus, have an envelope composed of a lipid bilayer and spike proteins. The lipid bilayer is obtained from the host cell when the virus buds through the plasma membrane. **(d)** Some bacteriophages, such as T4, have capsids with accessory structures that facilitate invasion of a bacterial cell.

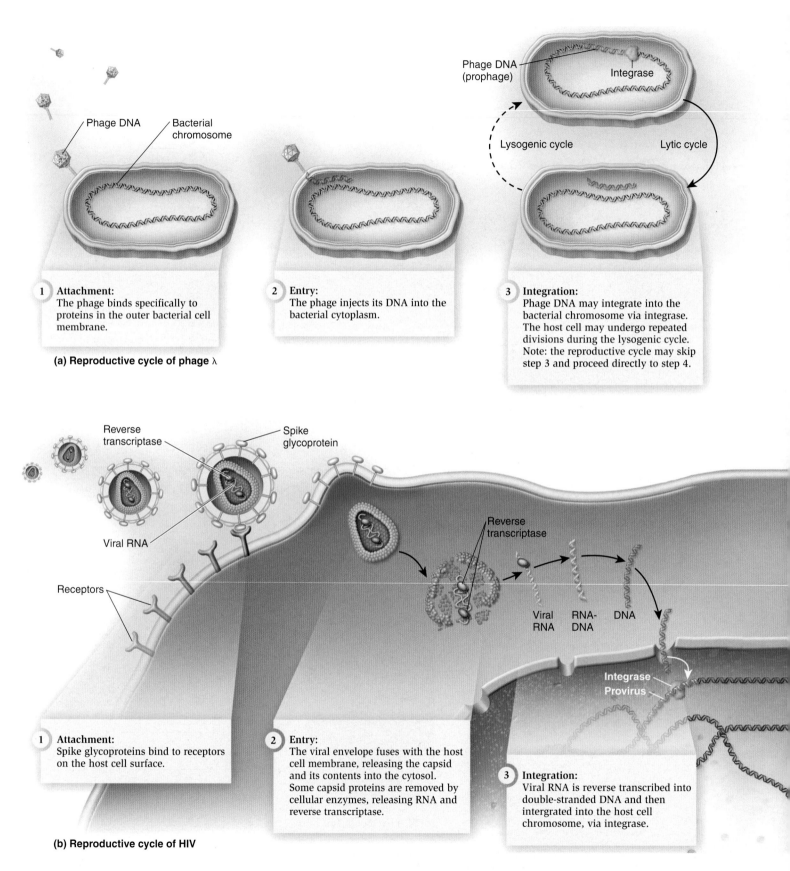

(a) Reproductive cycle of phage λ

1 Attachment:
The phage binds specifically to proteins in the outer bacterial cell membrane.

2 Entry:
The phage injects its DNA into the bacterial cytoplasm.

3 Integration:
Phage DNA may integrate into the bacterial chromosome via integrase. The host cell may undergo repeated divisions during the lysogenic cycle. Note: the reproductive cycle may skip step 3 and proceed directly to step 4.

(b) Reproductive cycle of HIV

1 Attachment:
Spike glycoproteins bind to receptors on the host cell surface.

2 Entry:
The viral envelope fuses with the host cell membrane, releasing the capsid and its contents into the cytosol. Some capsid proteins are removed by cellular enzymes, releasing RNA and reverse transcriptase.

3 Integration:
Viral RNA is reverse transcribed into double-stranded DNA and then intergrated into the host cell chromosome, via integrase.

Figure 18.14 Comparison of the steps of two viral reproductive cycles. (a) The reproductive cycle of phage λ, a bacteriophage with a double-stranded DNA genome. (b) The reproductive cycle of HIV, an enveloped animal virus with a single-stranded RNA genome.

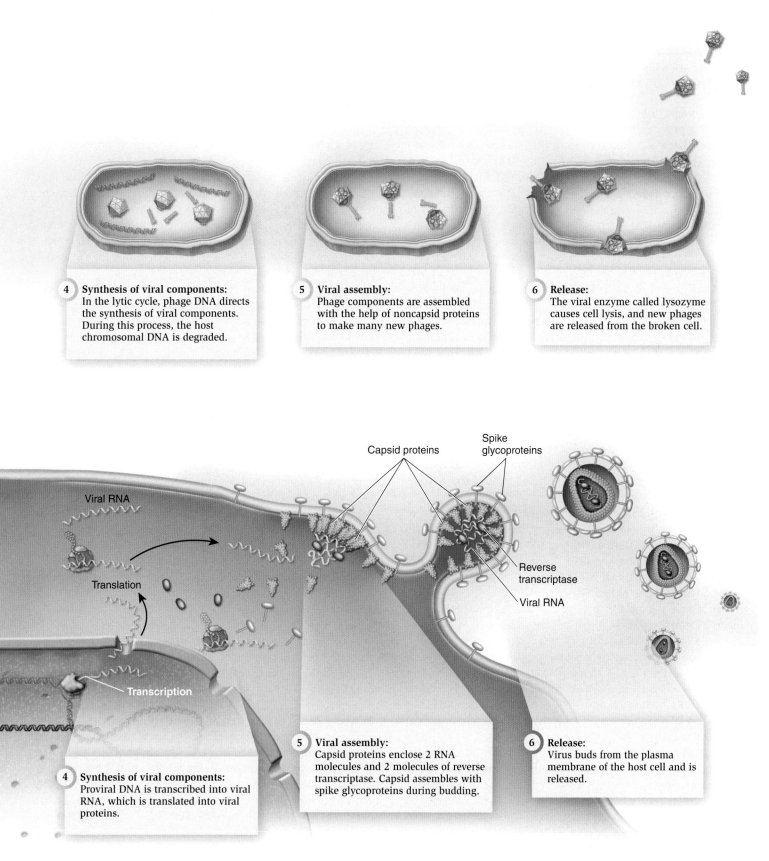

4 **Synthesis of viral components:**
In the lytic cycle, phage DNA directs the synthesis of viral components. During this process, the host chromosomal DNA is degraded.

5 **Viral assembly:**
Phage components are assembled with the help of noncapsid proteins to make many new phages.

6 **Release:**
The viral enzyme called lysozyme causes cell lysis, and new phages are released from the broken cell.

Capsid proteins

Spike glycoproteins

Viral RNA

Translation

Reverse transcriptase

Viral RNA

Transcription

4 **Synthesis of viral components:**
Proviral DNA is transcribed into viral RNA, which is translated into viral proteins.

5 **Viral assembly:**
Capsid proteins enclose 2 RNA molecules and 2 molecules of reverse transcriptase. Capsid assembles with spike glycoproteins during budding.

6 **Release:**
Virus buds from the plasma membrane of the host cell and is released.

For such viruses, the cycle may proceed from step 2 to step 3 as described next, delaying the production of new viruses. Alternatively, the cycle may proceed directly from step 2 to step 4 and quickly lead to the production of new viruses.

Step 3: Integration Viruses that are capable of integration carry a gene that encodes an enzyme called **integrase**. For integration to occur, this gene is expressed soon after entry so that integrase protein is made. Integrase cuts the host's chromosomal DNA and inserts the viral genome into the chromosome. In the case of the bacteriophage, the double-stranded DNA that entered the cell can be directly integrated into the double-stranded DNA of the chromosome. Once integrated, the phage DNA in a bacterium is called a **prophage**. At a later time, the prophage may be excised from the bacterial chromosome and the cycle then proceeds to step 4.

In the case of an RNA virus, the viral genome must be copied into DNA for integration to occur. HIV accomplishes this by means of a viral enzyme called **reverse transcriptase**, which is carried within the capsid and released into the host cell along with the viral RNA. Reverse transcriptase uses the viral RNA strand to make a complementary copy of DNA, and it then uses the DNA strand as a template to make double-stranded viral DNA. This process is called reverse transcription because it is the reverse of the usual transcription process, in which a DNA strand is used to make a complementary strand of RNA. The viral double-stranded DNA enters the host cell nucleus and is inserted into a host chromosome via integrase. Once integrated, the viral DNA in a eukaryotic cell is called a **provirus**.

Step 4: Synthesis of Viral Components The production of new viruses by a host cell always involves replication of the viral genome and synthesis of the capsid proteins that make up the viral coat. For some viruses, including HIV, the host cell must synthesize additional viral proteins, such as integrase, that facilitate the viral reproductive cycle.

In the case of a bacteriophage that has been integrated into the host chromosome, the prophage must be excised as described in step 3 before synthesis of new viral components can occur. Following excision, host cell enzymes such as DNA polymerase make many copies of the phage DNA and transcribe the genes within these copies into mRNA. Host cell ribosomes translate this viral mRNA into viral proteins. For many bacteriophages, the expression of phage genes leads to the degradation of the host chromosomal DNA.

In the case of HIV, the DNA provirus is not excised from the host chromosome. Instead, it is transcribed in the nucleus to produce many copies of viral RNA. These RNA molecules enter the cytosol, where they serve two purposes. First, they are translated to make viral proteins using host cell ribosomes. These proteins include capsid proteins, reverse transcriptase, and HIV spike proteins that are inserted into the cell's plasma membrane. The spike glycoproteins are destined to be part of the HIV viral envelope. Second, the RNA molecules serve as the genome for new viral particles.

Step 5: Viral Assembly After all of the necessary components have been synthesized, they must be assembled into new viruses. Some viruses with a simple structure self-assemble, meaning that viral components made by the host cell spontaneously bind to each other to form a complete virus particle. An example of a self-assembling virus is the tobacco mosaic virus we examined earlier (see Figure 18.13a). TMV capsid proteins assemble around a TMV RNA molecule, which becomes trapped inside the hollow capsid.

Other viruses, including the two shown in Figure 18.14, have more complicated structures that do not self-assemble. The correct assembly of phage λ requires the help of noncapsid proteins that are not found in the completed phage particle. Some of these noncapsid proteins function as enzymes that modify capsid proteins, while others serve as scaffolding for the assembly of the capsid.

The assembly of an HIV virus occurs in two stages. First, capsid proteins assemble around two molecules of HIV RNA and two molecules of reverse transcriptase. Next, the newly formed capsid acquires its outer envelope in a budding process. This second phase of assembly occurs during step 6, as the virus is released from the cell.

Step 6: Release The final stage of a viral reproductive cycle is the release of new viruses from the host cell. The release of bacteriophages is a dramatic event. Because bacteria are surrounded by a rigid cell wall, the phages must lyse their host cell in order to escape. After the phages have been assembled, a phage-encoded enzyme called lysozyme digests the bacterial cell wall, causing the cell to burst. Lysis releases many new phages into the environment, where they can infect other bacteria and begin the cycle again.

The release of enveloped viruses from an animal cell is far less dramatic. This type of virus escapes by a mechanism called budding that does not lyse the cell. In the case of HIV, a newly assembled virus particle associates with a portion of the plasma membrane containing HIV spike glycoproteins. The membrane enfolds the viral capsid and eventually buds from the surface of the cell. This is how the virus acquires its envelope, which is a piece of host cell membrane studded with viral glycoproteins. The photograph at the beginning of this chapter shows the budding of new HIV particles from a T cell.

Latency As we saw in step 3, some viruses can integrate their genomes into a host chromosome. In some cases the prophage or provirus may remain inactive, or **latent**, for a long time. Most of the viral genes are silent during latency, and the viral reproductive cycle does not progress to step 4. Examples of latent viruses include phage λ as well as human viruses such as HIV and different types of herpes viruses, which cause genital herpes and cold sores.

Latency in bacteriophages is usually called **lysogeny**, and both the prophage and its host cell are said to be lysogenic. When a lysogenic bacterium prepares to divide, it copies the prophage DNA along with its own, so that each daughter cell inherits a copy of the prophage. A prophage can be replicated repeatedly in this way without killing the host cell or producing new phage particles. However, a lysogenic prophage may eventually become activated, and it is then excised from the host chromosome. This sequence of events—integration, prophage replication, and excision—is called a bacteriophage **lysogenic cycle** (see Figure 18.14a, step 3). The excised phage proceeds to steps 4, 5, and 6—synthesis, assembly, and release—of the reproductive cycle. For bacteriophages, these steps are called a **lytic cycle** because the production and release of new viruses lyses the host cell.

Many bacteriophages can alternate between lysogenic and lytic cycles (**Figure 18.15**). A bacteriophage that may spend some of its time in the lysogenic cycle is called a **temperate phage**. Phage λ is an example of a temperate phage. Upon infection, it can either enter the lysogenic cycle or proceed directly to the lytic cycle. Other phages, called **virulent phages**, have only lytic cycles. The genome of a virulent phage is not capable of integration into a host chromosome. Phage T2, which we examined in Chapter 11, is a virulent phage that infects *E. coli*. Unlike phage λ, which may coexist harmlessly with *E. coli*, T2 always lyses the infected cell.

For phages such as λ that can follow either cycle, environmental conditions influence whether or not viral DNA is integrated into a host chromosome and how long the virus remains latent or lysogenic. If nutrients are readily available, phage λ usually proceeds directly to the lytic cycle after its DNA enters the cell. Alternatively, if nutrients are in short supply, latency is often favored because there may not be sufficient material to make new viruses. If more nutrients become available later, this may cause the prophage to become activated. At this point, the viral reproductive cycle will switch to the lytic cycle, and new viruses will be made, assembled, and released.

Bacterial Transduction Results from an Error in Phage Replication

Earlier in this chapter we considered transduction, in which gene transfer occurs between two bacterial cells with the aid of a bacteriophage. Now that we have examined the steps in the viral reproductive cycle, let's take a closer look at this type of gene transfer.

The mechanism of transduction is actually an error in a phage lytic cycle, as shown in **Figure 18.16**. During the synthesis of phage DNA and proteins, the bacterial chromosome is degraded into small pieces. When the new viruses are assembled, coat proteins occasionally surround a piece of bacterial DNA instead of phage genetic material. This mistake creates an abnormal phage carrying bacterial chromosomal DNA. In the example shown here, a phage called P1 infects an *E. coli* cell that has a gene (*his*+) for histidine synthesis. Phage P1 causes the host cell chromosome to break up, and one of the new phages encloses a piece of host DNA that carries this gene.

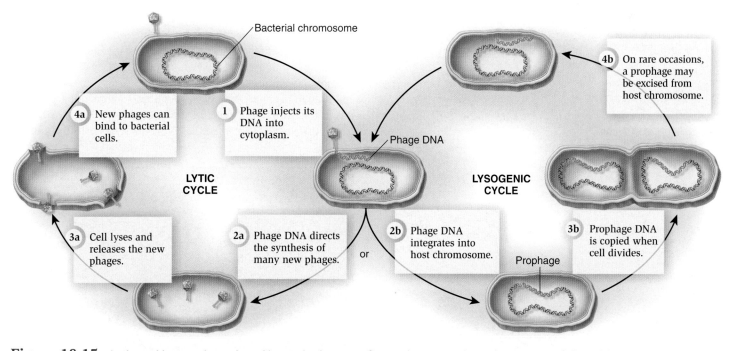

Figure 18.15 Lytic and lysogenic cycles of bacteriophages. Some phages, such as phage λ, may follow either a lytic or a lysogenic reproductive cycle. During the lysogenic cycle, the integrated phage DNA, or prophage, is replicated along with the DNA of the host cell. Other phages, such as T2, follow only lytic cycles.

Biological inquiry: From the perspective of the virus, what are the primary advantages of the lytic and lysogenic cycles?

When this abnormal phage is released, it binds to an *E. coli* cell that lacks the *his+* gene. It inserts the bacterial DNA fragment into the recipient cell, which then incorporates this fragment into its own chromosome by crossing over. In this example, genetic transfer by transduction converts a *his−* strain of *E. coli* to a *his+* strain.

AIDS Is a Fatal Immunodeficiency Syndrome Caused by HIV

A primary reason researchers have been interested in viral reproductive cycles is the ability of many viruses to cause diseases in

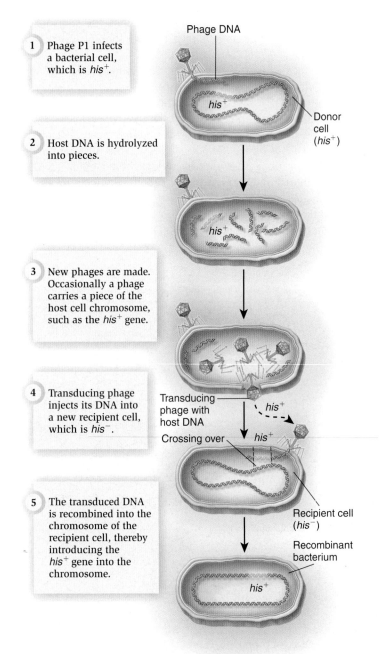

1 Phage P1 infects a bacterial cell, which is *his+*.

2 Host DNA is hydrolyzed into pieces.

3 New phages are made. Occasionally a phage carries a piece of the host cell chromosome, such as the *his+* gene.

4 Transducing phage injects its DNA into a new recipient cell, which is *his−*.

5 The transduced DNA is recombined into the chromosome of the recipient cell, thereby introducing the *his+* gene into the chromosome.

Phage DNA

his+

Donor cell (*his+*)

his+

Transducing phage with host DNA

Crossing over   *his+*

his+

Recipient cell (*his−*)

Recombinant bacterium

his+

The recombinant bacterium has a genotype (*his+*) that is different from the original recipient bacterial cell (*his−*).

Figure 18.16 Bacterial transduction by P1 phage.

humans and other mammalian hosts. Some examples of human disease-causing viruses were presented in Figure 18.12. Newly arising viruses, called **emerging viruses**, often cause public alarm and may lead to a significant loss of human life. New strains of influenza virus arise fairly regularly due to new mutations. In the U.S., despite attempts to minimize influenza deaths by vaccination, over 30,000 people die annually from this disease. Another emerging virus causes a severe acute respiratory syndrome (SARS). This enveloped RNA virus was first identified by researchers in Hong Kong, the U.S., and Germany in 2003. The origin of SARS is not understood, but the virus is likely to have arisen from viruses that infect humans and livestock.

During the past few decades, the most devastating example of an emerging virus has been **human immunodeficiency virus (HIV)**, the causative agent of **acquired immune deficiency syndrome (AIDS)**. AIDS is primarily spread by sexual contact between infected and uninfected individuals, but it can also be spread by the transfusion of HIV-infected blood, by the sharing of needles among drug users, and from mother to unborn child. The total number of AIDS deaths between 1981 and the end of 2003 was 20 million; approximately 0.5 million of these deaths occurred in the U.S. During 2004, around 5 million adults and children became infected with HIV. Worldwide, nearly 1 in every 100 adults between 15 and 49 is infected. In the U.S. about 40,000 new HIV infections occur each year. Seventy percent of these new infections are in men and 30% in women.

The devastating effects of AIDS result from viral destruction of a type of white blood cell termed a helper T cell, which plays an essential role in the immune system of mammals. **Figure 18.17** shows HIV virus particles invading a T cell. As described in Chapter 53, T cells interact with other cells of the immune system to facilitate the production of antibodies and other molecules that target and kill foreign invaders of the body. When large numbers of T cells are destroyed by HIV, the function of the immune system is seriously compromised and the individual becomes highly susceptible to infectious diseases called opportunistic infections that would not normally occur in a healthy person. For example, *Pneumocystis carinii*, a fungus

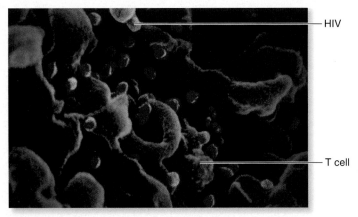

HIV

T cell

Figure 18.17 Micrograph of HIV invading a human T cell. This is a colorized scanning electron micrograph. The surface of the T cell is purple and HIV particles are red.

that causes pneumonia, is easily destroyed by a normal immune system. However, in people with AIDS, infection by this fungus can be fatal.

The mechanism by which HIV kills T cells is not well understood. However, a great deal is known about how the virus invades a T cell and uses it to make more HIV particles. **Figure 18.18** takes a closer look at the HIV reproductive cycle we considered earlier. The structure of HIV consists of two copies of single-stranded RNA and two molecules of reverse transcriptase, which are packaged into a capsid surrounded by a viral envelope. This structure is characteristic of the RNA viruses called **retroviruses**, which utilize reverse transcription to produce viral DNA that can be integrated into the host cell genome.

Let's take a detailed look at the events of the HIV reproductive cycle, as shown in Figure 18.18. The cycle begins when a viral particle binds to the surface of a T cell. Spike glycoproteins

in the viral envelope specifically recognize receptors that are abundant on human T cells, making these cells targets for HIV infection. The viral envelope fuses with the plasma membrane of the T cell, and the capsid is released into the cytosol. The capsid proteins are partially removed by host cell enzymes, releasing the viral RNA and reverse transcriptase. Reverse transcription then occurs, producing HIV DNA that enters the nucleus and is integrated into the host chromosome. Later, the proviral DNA is transcribed into viral RNA in the nucleus. The HIV RNA molecules enter the cytosol, where they serve as the genetic material for new virus particles and are also translated into HIV proteins. When the viral components have been synthesized, capsid proteins surround viral RNA and reverse transcriptase molecules. The newly assembled particles bud from the T cell, wrapped in pieces of T-cell membrane studded with viral spike glycoproteins.

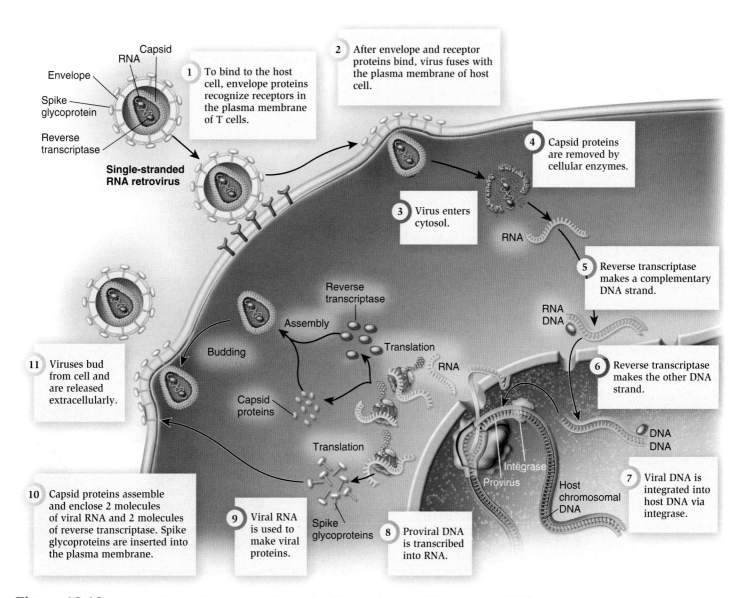

Figure 18.18 A closer look at the reproductive cycle of human immunodeficiency virus (HIV).

An insidious feature of reverse transcriptase is that it lacks a proofreading function. In Chapter 11, we learned that DNA polymerase has a proofreading function that causes it to replicate DNA with few errors. Because reverse transcriptase lacks this function, it makes more errors and thereby tends to create mutant strains of HIV. This undermines the ability of the body to combat HIV because mutant strains may be resistant to the body's defenses, and also makes it difficult to create a vaccine that would provide immunity to all strains of the virus. In addition, mutant strains of HIV may be resistant to antiviral drugs.

Drugs Have Been Developed to Combat the Proliferation of HIV

A compelling reason to understand the reproductive cycle of HIV and other disease-causing viruses is that such knowledge may be used to invent drugs that stop viral proliferation. In the U.S., the estimated annual number of AIDS-related deaths fell approximately 14% from 1998 to 2002, owing in part to the development of new antiviral drugs. These drugs inhibit viral proliferation, though they don't completely eliminate the virus from the body.

One approach to the design of antiviral treatments is to create drugs that specifically bind to proteins encoded by the viral genome. For example, azidothymidine (AZT) mimics the structure of a normal nucleotide and can bind to reverse transcriptase. In this way, AZT inhibits reverse transcription, thereby retarding viral replication. Other drugs used to combat HIV inhibit other proteins that are critical in the HIV reproductive cycle. Proteases, which are enzymes that cut other proteins, are needed during the assembly of the HIV capsid. Certain proteases cleave capsid proteins, making them smaller, active HIV proteins. If the larger capsid proteins are not broken apart, they cannot assemble themselves into new functional HIV particles. Several drugs known as protease inhibitors have been developed that bind to the HIV proteases and inhibit their function.

A major challenge in AIDS research is to discover drugs that inhibit viral proteins without also binding to host cell proteins and inhibiting normal cellular functions. Researchers are optimistic, however, that as we learn more about the molecular structures and functions of viral proteins we will be able to design antiviral drugs with a minimum of harmful side effects. A second challenge is to develop drugs to which mutant strains will not become resistant. As we discussed earlier, HIV readily accumulates mutations during viral replication. A current strategy is to treat HIV patients with a cocktail of three or more drugs, making it less likely that any mutant strain will overcome all of the inhibitory effects.

Another approach to fighting HIV and other infectious diseases is vaccination. A vaccine is a substance or group of substances that causes the immune system to respond to and destroy infectious agents such as bacteria or viruses. The ideal HIV vaccine should be both inexpensive and easy to store and administer, and it must confer long-lasting protection against HIV infection by sexual contact or by exposure to infected blood. Importantly, due to the high mutation rate of HIV, the vaccine must protect against exposure to many different strains of the virus. Despite great advances in our understanding of HIV and the human immune system, such a vaccine has yet to be developed. However, many clinical trials are underway, and researchers hope eventually to create a vaccine that will prevent the spread of AIDS.

Viroids Are RNA Particles That Infect Plant Cells

Thus far in this section, we have examined the reproductive cycles of viruses. Some nonliving infectious agents are even simpler than viruses. In 1971, Theodor Diener discovered that the agent of potato spindle tuber disease is a small RNA molecule devoid of any protein. He coined the term **viroid** for this newly discovered infectious particle.

Viroids are composed solely of a single-stranded circular RNA molecule a few hundred nucleotides in length. They infect plant cells, where they depend entirely on host enzymes for their replication. Some viroids are replicated in the host cell nucleus, others in the chloroplast. In contrast to viral genomes, the RNA genomes of viroids do not code for any proteins. The RNA of some viroids is known to possess ribozyme activity, and some researchers think that this activity may damage plants by interfering with the function of host cell molecules. However, the mechanism by which viroids induce disease is not well understood.

Since Diener's initial discovery, many more viroids have been characterized as the agents of diseases affecting many economically important plants, including potato, tomato, cucumber, orange, coconut, grape, avocado, peach, apple, pear, and plum (**Figure 18.19**). Some viroids have devastating effects, as

Figure 18.19 Effects of the tomato apical stunt viroid. The plant on the left is infected, while the one on the right is healthy.

illustrated by the case of the coconut cadang-cadang viroid, which has killed more than 20 million coconut trees in Southeast Asia. Other viroids cause less severe damage, causing necrosis on leaves, shortening of stems, bark cracking, and delays in foliation, flowering, and fruit ripening. A few viroids induce mild symptoms or no symptoms at all.

Several Hypotheses Have Been Proposed to Explain the Origin of Viruses

Because viruses and viroids are such small particles, there is no fossil record of their evolution, and researchers must rely on analyses of modern viruses to propose hypotheses about their origin. Viral genomes follow the same rules of gene expression as the genomes of their host cells. Viral genes have promoter sequences that are similar to those of their host cells, and the translation of viral proteins relies on the genetic code. Viruses depend entirely on host cells for their proliferation. No known virus makes its own ribosomes or generates the energy it requires to make new viruses. Therefore, many biologists have argued that cells must have evolved before viruses.

A common hypothesis for the origin of viruses is that they evolved from macromolecules inside living cells. The precursors of the first viruses may have been plasmids or similar small nucleic acid fragments. Biologists have hypothesized that such fragments may have acquired genes that code for proteins that facilitate their own replication. Though many biologists favor the idea that viruses originated from primitive plasmids or other chromosomal elements, some have suggested that they are an example of regressive evolution. This hypothesis proposes that viruses are degenerate cells that have retained the minimal genetic information essential for reproduction.

A new and interesting hypothesis is that viruses did not evolve from living cells but instead evolved in parallel with cellular organisms. As discussed in Chapter 22, the precursors of cellular DNA genomes may have been RNA molecules that could replicate independently of cells. This stage of evolution, termed the RNA world, could have involved the parallel evolution of both viruses and cellular organisms.

Prions Are Infectious Proteins That Cause Neurodegenerative Diseases

Before we end our discussion of infectious particles, let's consider an unusual mechanism in which agents known as **prions** cause a group of rare, fatal brain diseases affecting humans and other mammals. Until the 1980s, biologists thought that any infectious agent, whether living or nonliving, must have genetic material. It would seem that genetic material is needed to store the information to create new infectious particles.

In the 1960s, British researchers Tikva Alper and J. S. Griffith discovered that preparations from animals with certain neurodegenerative diseases remained infectious even after exposure to radiation that would destroy any DNA or RNA. They suggested that the infectious agent was a protein. Furthermore, Alper and Griffith speculated that the protein usually preferred one folding pattern, but it could sometimes misfold and then catalyze other proteins to misfold, creating a chain reaction of misfolding. In the early 1970s, Stanley Prusiner, moved by the death of a patient from a neurodegenerative disease, began to search for the causative agent. In 1982, Prusiner isolated a disease-causing particle composed entirely of protein, which he called a prion. The term was based on his characterization of the particle as a proteinaceous infectious agent. Further experiments carried out by Prusiner and others conclusively demonstrated the particle's infectious character. In 1997, Prusiner was awarded the Nobel Prize in Physiology or Medicine for his work on prions.

Prion diseases arise from the ability of the prion protein to induce abnormal folding in normal protein molecules (**Figure 18.20**). The prion protein exists in a disease-causing conformation designated PrP^{Sc}. The superscript Sc refers to scrapie, an example of a prion disease. A normal conformation of this same protein, which does not cause disease, is termed PrP^C. The superscript C stands for cellular. The normal protein is encoded by an individual's genome, and the protein is expressed at low levels in certain types of nerve cells. A healthy person may become "infected" with the abnormal protein by eating meat of an animal with the disease. Unlike most other proteins in the diet, the prion protein escapes digestion in the stomach and small intestine and is absorbed into the bloodstream. The prion gradually converts the cell's normal proteins to the abnormal conformation. As a prion disease progresses, the PrP^{Sc} protein is deposited as dense aggregates in the cells of the brain and peripheral nervous tissues, causing the disease symptoms. Some of the abnormal prion protein is excreted from infected cells, where it travels through the bloodstream. In this way, a prion disease can spread through the body like many viral diseases.

Prions are now known to cause several types of neurodegenerative diseases affecting humans and livestock (**Table 18.3**).

Table 18.3	Examples of Neurodegenerative Diseases Caused by Infectious Prions
Disease	**Description**
Kuru	A human disease that was once common in certain native tribes of New Guinea. It begins with a loss of coordination, usually followed by dementia. Infection was spread by ritual cannibalism, a practice that ended in 1958.
Scrapie	A disease of sheep and pigs characterized by intense itching in which the animals tend to scrape themselves against trees or other objects, followed by neurodegeneration.
Mad cow disease	Begins with changes in posture and temperament, followed by loss of coordination and neurodegeneration.

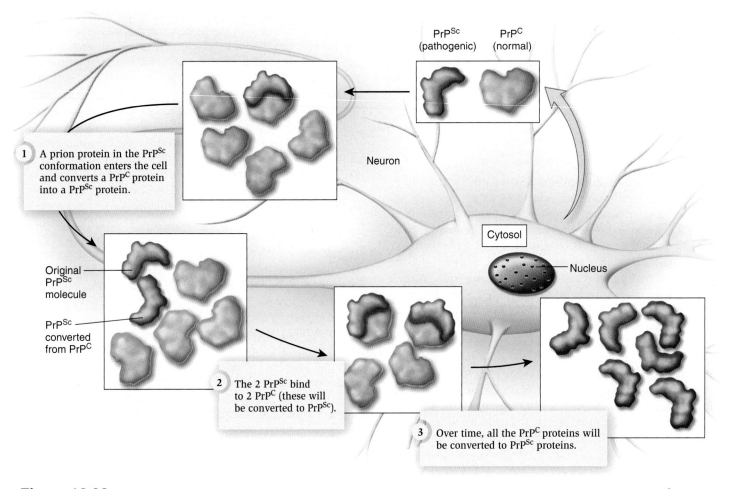

PrP^Sc (pathogenic)   PrP^C (normal)

Neuron

1 A prion protein in the PrP^Sc conformation enters the cell and converts a PrP^C protein into a PrP^Sc protein.

Original PrP^Sc molecule

PrP^Sc converted from PrP^C

2 The 2 PrP^Sc bind to 2 PrP^C (these will be converted to PrP^Sc).

3 Over time, all the PrP^C proteins will be converted to PrP^Sc proteins.

Cytosol

Nucleus

Figure 18.20 **A proposed molecular mechanism of prion diseases.**   A healthy neuron normally contains only the PrP^C conformation of the prion protein. The abnormal PrP^Sc conformation catalyzes the conversion of PrP^C into PrP^Sc. Over time, the PrP^Sc protein accumulates at high levels, leading to symptoms of prion disease.

Biological inquiry: Researchers are trying to discover drugs that prevent prion diseases. What are possible effects of a drug that would prevent the spread of the disease?

As a group, prion diseases are termed transmissible spongiform encephalopathies (TSE). The postmortem examination of the brains of affected individuals reveals a substantial destruction of brain tissue; the brain has a spongy appearance. Most prion diseases progress fairly slowly. Over the course of several months to a few years, symptoms proceed from a loss of motor control to dementia, paralysis, wasting, and eventually death.

These symptoms are correlated with an increase in the level of prion protein in the nerve cells of infected individuals. No current treatment can halt progression of any of the TSEs. For this reason, great public alarm occurs when an outbreak of a TSE is reported. For example, in the U.S., a single report of mad cow disease in 2003 prompted several countries to restrict the import of American beef.

CHAPTER SUMMARY

18.1 Bacterial Genetics

- Bacteria typically have a single type of circular chromosome found in the nucleoid region of the cell. The chromosome contains many genes, an origin of replication, and repetitive sequences. (Figures 18.1, 18.2)

- The bacterial chromosome is made more compact by the formation of loops and DNA supercoiling. (Figure 18.3)

- Plasmids are small DNA molecules that exist independently of the bacterial chromosome. Examples are resistant, degradative, col-, virulence, and fertility plasmids. (Figure 18.4)

- When placed on solid growth media, a single bacterial cell will divide many times to produce a colony composed of many cells. (Figure 18.5)

- Bacterial cells reproduce by a process called binary fission. (Figure 18.6)

- Three common modes of genetic transfer among bacteria are conjugation, transformation, and transduction. (Table 18.1)
- Beadle and Tatum's work demonstrated the transfer of bacterial genes between different strains of *E. coli*. (Figure 18.7)
- Using a U-tube, Davis showed that direct contact was needed for the type of genetic transfer observed by Beadle and Tatum. (Figure 18.8)
- During the mechanism of conjugation, a strand of DNA from an F factor is transferred from a donor to a recipient cell through a sex pilus. (Figure 18.9)
- Transformation is the process in which a segment of DNA from the environment is taken up by a competent cell and incorporated into the bacterial chromosome. (Figure 18.10)
- Horizontal gene transfer is the transfer of genes between different species.

18.2 Viral Genetics, Viroids, and Prions

- Tobacco mosaic virus was the first virus to be discovered. It infects many species of plants. (Figure 18.11)
- Viruses vary with regard to their host range, structure, and genome composition. (Table 18.2, Figures 18.12, 18.13)
- Viruses follow a series of steps called a reproductive cycle to make new viruses. These steps typically include attachment, entry, integration, synthesis, assembly, and release. (Figure 18.14)
- Some bacteriophages can alternate between the lytic and lysogenic cycles. (Figure 18.15)
- Bacterial transduction is a form of genetic transfer in which a bacteriophage transfers a segment of bacterial chromosomal DNA to another cell. (Figure 18.16)
- The disease AIDS is caused by a virus called HIV. The virus is a retrovirus whose reproductive cycle involves the integration of the viral genome into a chromosome in the host cell. (Figures 18.17, 18.18)
- Drugs to combat viral diseases are often directed against viral proteins.
- Viroids are RNA molecules that infect plant cells. (Figure 18.19)
- Prions are proteins that exist in an abnormal conformation that can cause disease. (Figure 18.20, Table 18.3)

TEST YOURSELF

1. Genetic diversity is maintained in bacterial populations by all of the following *except*
   a. binary fission.
   b. mutation.
   c. transformation.
   d. transduction.
   e. conjugation.

2. Bacterial cells divide by a process known as
   a. mitosis.
   b. cytokinesis.
   c. meiosis.
   d. binary fission.
   e. glycolysis.

3. Genetic transfer, whereby a bacterial cell takes up bacterial DNA from the environment, is
   a. conjugation.
   b. binary fission.

   c. recombination.
   d. transformation.
   e. transduction.

4. A bacterial cell can donate DNA during conjugation when it
   a. produces competence factors.
   b. contains an F factor.
   c. is virulent.
   d. has been infected by a bacteriophage.
   e. all of the above.

5. A bacterial species that becomes resistant to certain antibiotics may have acquired the resistance genes from another bacterial species. This phenomenon of acquiring genes from another species is known as
   a. hybridization.
   b. conjugation.
   c. horizontal gene transfer.
   d. vertical gene transfer.
   e. competence.

6. The _____ is the protein coat of a virus.
   a. host
   b. prion
   c. capsid
   d. viroid
   e. capsule

7. Among the viruses identified, the characteristics of their genomes show many variations. Which of the following does not describe a typical characteristic of viral genomes?
   a. The genetic material may be DNA or RNA.
   b. The nucleic acid may be single stranded or double stranded.
   c. The genome may carry just a few genes or several dozen.
   d. The number of copies of the genome may vary.
   e. All of the above describe typical variation in viral genomes.

8. During viral infection, attachment is usually specific to a particular cell type because
   a. the virus is attracted to the appropriate host cells by proteins secreted into the extracellular fluid.
   b. the virus recognizes and binds to specific molecules in the cytoplasm of the host cell.
   c. the virus recognizes and binds to specific molecules on the surface of the host cell.
   d. the host cell produces channel proteins that provide passageways for viruses to enter the cytoplasm.
   e. the virus releases specific proteins that make holes in the membrane large enough for the virus to enter.

9. HIV, a retrovirus, has a high mutation rate because
   a. the DNA of the viral genome is less stable than other viral genomes.
   b. the viral enzyme reverse transcriptase has a high likelihood of making replication errors.
   c. the viral genome is altered every time it is incorporated into the host genome.
   d. antibodies produced by the host cell mutate the viral genome when infection occurs.
   e. all of the above.

10. A _____ is an infectious agent solely composed of RNA, whereas a _____ is an infectious agent solely composed of protein.
    a. retrovirus, bacteriophage
    b. viroid, virus
    c. prion, virus
    d. retrovirus, prion
    e. viroid, prion

CONCEPTUAL QUESTIONS

1. Define plasmid.
2. Explain acquired antibiotic resistance and its significance for disease.
3. Explain the steps in a viral reproductive cycle.

COLLABORATIVE QUESTIONS

1. Discuss the three mechanisms of genetic transfer in bacteria. What is the significance of horizontal genetic transfer?
2. Discuss the steps of viral reproduction.

EXPERIMENTAL QUESTIONS

1. What was the hypothesis tested by Lederberg and Tatum?

2. During the Lederberg and Tatum experiment, the researchers compared the growth of mutant strains under two scenarios: mixed strains or unmixed strains. When the unmixed strains were plated on the experimental growth medium, why were no colonies observed to grow? When the mixed strains were plated on the experimental growth medium, a number of colonies were seen to grow. What was the significance of the growth of these colonies?

3. The genetic transfer seen in the Lederberg and Tatum experiment could have occurred in one of two ways: taking up DNA released into the environment or contact between two bacterial cells allowing for direct transfer. Bernard Davis conducted an experiment to determine the correct process. Explain how his results indicated the correct gene transfer process.

www.brookerbiology.com

This website includes answers to the Biological Inquiry questions found in the figure legends and all end-of-chapter questions.

19

DEVELOPMENTAL GENETICS

CHAPTER OUTLINE

Experimental organisms for studying development. *Drosophila melanogaster*, the fruit fly, has served as a model organism to study animal development. This is a scanning electron micrograph.

In biology, the term **development** refers to a series of changes in the state of a cell, tissue, organ, or organism. Development is the underlying process that gives rise to the structure and function of living organisms. The structure or form of an organism is called its morphology. Researchers interested in morphology are fascinated by the changes that give rise to the shapes of cells, tissues, organs, and entire organisms. As we have learned throughout this textbook, an important paradigm in biology is that structure (morphology) determines function. Scientists interested in biological functions are called physiologists. Development is also of great importance to physiologists because the functions of a specialized cell, like a nerve cell, and a complex organ, such as a heart, are the result of development.

Biologists came to realize that plants and animals undergo amazing changes in development to create the structure and function that is found in adult organisms. Such observations motivated researchers to uncover the underlying processes that promote these developmental changes. Since the 1940s, genetics has emerged as the fundamental force behind development. It is currently one of the most intensely researched fields of biology.

Developmental genetics is aimed at understanding how gene expression controls the process of development. In this chapter, we will learn how the sequential actions of genes provide a program for the development of an organism from a fertilized egg to an adult. The last couple of decades have seen exciting advances in our understanding of developmental genetics at the molecular level. Scientists have chosen a few experimental organisms, such as the fruit fly, a nematode worm, the mouse, and the plant *Arabidopsis*, and worked toward the identification and characterization of the genes required for running their developmental programs. In certain organisms, notably the fruit fly, most of the genes that play a critical role in the embryonic stages of development have been identified. Researchers are now exploring how the proteins encoded by these genes control the course of development. In this chapter, we will begin with an overview that emphasizes the general principles of development. We will then examine the details of development in plants and animals, with an emphasis on genetics. Chapters 39 and 51 also consider plant and animal development, with an emphasis on structure and function.

19.1 General Themes in Development

Plants and animals usually begin to develop when sperm and egg unite to produce a **zygote**, a diploid cell that divides and develops into an **embryo**, and eventually into an adult organism. During the early stages of development, cells divide and begin to arrange themselves into ordered units. As this occurs, each cell also becomes **determined**, which means it is destined to become a particular cell type, such as a muscle or intestinal cell. This commitment occurs long before a cell becomes **differentiated**, meaning that the cell's morphology and function have changed, usually permanently, into a highly specialized cell type. In an adult, each cell type plays its own particular role for the good of the entire individual. In animals, for example, muscle cells allow an organism to move, while intestinal cells facilitate the absorption of nutrients. This division of labor among various cells of an organism works collectively to promote its survival.

The genomes of living organisms contain a set of genes that constitute a program of development. In unicellular species, the program controls the structure and function of the cell.

In multicellular species such as animals and plants, the program not only controls cellular features but also determines the arrangement of cells in the organisms. In this section, we will examine some of the general issues associated with the development of multicellular species. Later sections will focus on specific examples of development in animals and plants.

Developmental Biologists Have Chosen a Few Model Organisms to Study Development

The development of even a simple multicellular organism involves many types of changes in form and function. For this reason, the research community has focused its efforts on only a few model organisms with the expectation that knowledge of the developmental programs in these model organisms will also apply to many other species. With regard to animal development, the two organisms that have been the most extensively investigated are two invertebrate species: the fruit fly *Drosophila melanogaster* and the nematode worm *Caenorhabditis elegans* (**Figure 19.1a,b**). *Drosophila* has been studied for a variety of reasons. First, researchers have exposed this organism to mutagens and identified many mutant organisms with altered developmental pathways. The techniques for generating and analyzing mutants in this organism are more advanced than in any other animal. Second, in all of its life stages, *Drosophila* is large enough and has distinct morphological features to easily identify the effects of mutations, yet it is small enough to determine where particular genes are expressed at critical stages of development. *C. elegans* is used by developmental geneticists for its simplicity. The adult organism is a small transparent worm composed of only about a thousand somatic cells. Starting with a fertilized egg, the pattern of cell division and the fate of each cell within the embryo are completely known.

Embryologists have also studied the morphological features of development in many vertebrate species. Historically, amphibians and birds have been studied extensively, because their eggs are rather large and easy to manipulate. From a morphological point of view, the developmental stages of the chicken (*Gallus gallus*) and the African clawed frog (*Xenopus laevis*) have been described in great detail. More recently, several vertebrate species have been the subject of genetic studies of development. These include the house mouse (*Mus musculus*) and the small aquarium zebrafish (*Brachydanio rerio*) (**Figure 19.1c,d**).

In the study of plant development, the model organism for genetic analysis is thale cress (*Arabidopsis thaliana*), more commonly called *Arabidopsis* by researchers (**Figure 19.1e**). *Arabidopsis* is a small annual weed that belongs to the wild mustard family. It occurs naturally throughout temperate regions of the world. *Arabidopsis* has a short generation time of about two months and a small genome size of 14×10^7 bp, which is similar to *Drosophila* and *C. elegans*. A flowering *Arabidopsis* plant is small enough to be grown in the laboratory and produces a large number of seeds.

Both Animals and Plants Develop by Pattern Formation

Development in plants and animals produces a body plan or pattern. At the cellular level, the body pattern is due to the arrangement of cells and their differentiation. The coordination of these events leads to the formation of a body with a particular pattern, a process called **pattern formation**. The end result of animal development is usually the formation of an adult body pattern organized along three axes: the **dorsoventral axis**, the **anteroposterior axis**, and the **right-left axis** (**Figure 19.2a**). In addition, many animal bodies are then **segmented** into separate sections containing specific body parts such as wings or legs.

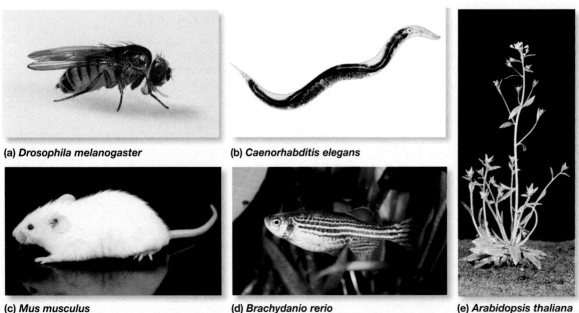

(a) *Drosophila melanogaster*

(b) *Caenorhabditis elegans*

(c) *Mus musculus*

(d) *Brachydanio rerio*

(e) *Arabidopsis thaliana*

Figure 19.1 Model organisms used to study the genetics of development.

By comparison, the body pattern of plants is quite different, being formed along a **root-shoot axis** and in a **radial pattern** (**Figure 19.2b**). The root and shoot axes are determined at the first division of the fertilized egg, and growth occurs in a radial pattern around those axes. As we'll see later, the identification of mutant alleles that disrupt development has permitted great insight into the genes controlling pattern formation.

Pattern Formation Depends on Positional Information

Before we examine how genes affect pattern formation, let's consider a central concept in developmental biology—**positional information**. For an organism to develop the correct morphological features or pattern, each cell of the body must become the appropriate cell type based on its position relative to other cells. Each cell receives positional information that provides it with instructions on where to go and what type of cell to become. Later in this chapter, we will examine how the expression of genes and the function of gene products provide this information.

A cell may respond to positional information in one of four ways: cell division, cell migration, cell differentiation, and cell death (**Figure 19.3**). First, positional information may stimulate a cell to divide. Second, positional information in animals may cause the migration of a cell or group of cells in a particular direction from one region of the embryo to another. Third, it may cause a cell to differentiate into a specific cell type such as a nerve cell. Finally, positional information may promote cell death. This process, known as **apoptosis**, is a necessary event during normal development.

As an example of how the coordination of these four processes is required for pattern formation, let's consider the development of the vertebrate limb. **Figure 19.4a** shows the growth and development of a human arm during the embryonic stage. Cell division with accompanying cell growth increases the size of the limb (**Figure 19.4b**). Cell migration is also important in this process. For example, embryonic cells that eventually form muscles in the arm and hand must migrate long distances to reach their correct location within the limb.

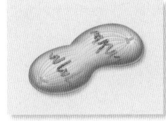

(a) Cell division

(b) Cell migration

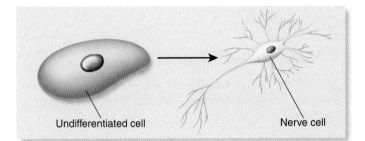

Undifferentiated cell   Nerve cell

(c) Cell differentiation

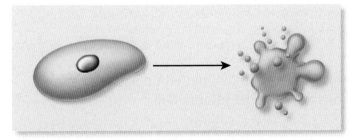

(d) Cell death (apoptosis)

Figure 19.3   Four types of cellular responses to positional information in animals.

Biological inquiry: Which of these four responses do you expect to be more prevalent in the early stages of development, and which would become more prevalent in later stages?

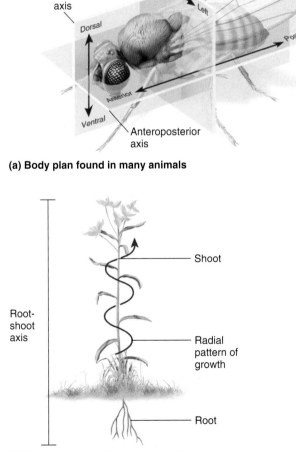

Right-left axis

Right

Dorso-ventral axis

Dorsal

Left

Posterior

Anterior

Ventral

Anteroposterior axis

(a) Body plan found in many animals

Shoot

Root-shoot axis

Radial pattern of growth

Root

(b) Body plan found in many seed-bearing plants

Figure 19.2   Body plan axes in animals and plants.

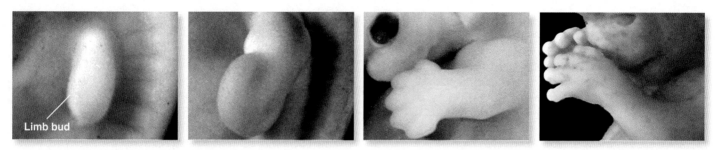

(a) Limb development in a human embryo

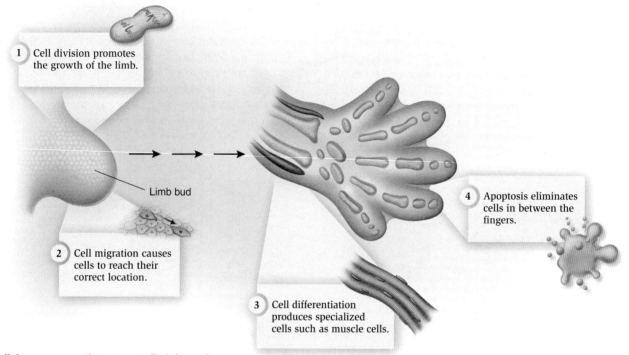

1. Cell division promotes the growth of the limb.

2. Cell migration causes cells to reach their correct location.

3. Cell differentiation produces specialized cells such as muscle cells.

4. Apoptosis eliminates cells in between the fingers.

Limb bud

(b) Four cellular processes that promote limb formation

Figure 19.4 **Limb development in humans.** **(a)** Photographs of limb development in human embryos. The limb begins as a protrusion called a limb bud that eventually forms an arm and hand. **(b)** The development of a human limb from an embryonic limb bud.

As development proceeds, cell differentiation produces the various tissues that will eventually be found in the fully developed limb. Some cells will become nerve cells, others will be muscle cells, and still others will form the outer layer of skin. Finally, apoptosis is important in the formation of fingers. If apoptosis did not occur, a human hand would have webbed fingers.

Morphogens and Cell-to-Cell Contacts Convey Positional Information

How does positional information lead to the development of a body plan? Two main molecular mechanisms are used to communicate positional information. One of these mechanisms involves molecules called morphogens. **Morphogens** impart positional information and promote developmental changes at the cellular level. A morphogen influences the fate of a cell by promoting cell division, cell migration, cell differentiation, or apoptosis. A key feature of morphogens is that they act in a concentration-dependent manner. At a high concentration, a morphogen will restrict a cell into a particular developmental pathway, while at a lower concentration it will not. There is often a critical **threshold concentration** above which the morphogen will exert its effects but below which it is ineffective.

Morphogens typically are distributed asymmetrically along a concentration gradient. This may happen at one of two times during early development. First, morphogenic gradients may be established in the oocyte or egg cell precursor (**Figure 19.5a**). Second, a morphogenic gradient can be established in the embryo by secretion and transport (**Figure 19.5b**). A certain cell or group of cells may synthesize and secrete a morphogen at a specific stage of development. After secretion, the morphogen is transported to neighboring cells. The concentration of the morphogen is usually highest near the cells that secrete it. The morphogen may then influence the developmental fate of cells that are exposed to it. The process by which a cell or group of cells governs the developmental fate of neighboring cells is known as **induction**.

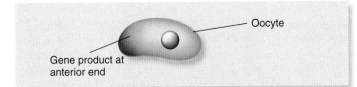

(a) Asymmetric distribution of morphogens in the oocyte

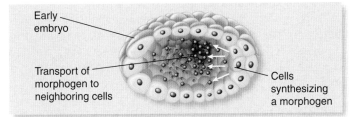

(b) Induction: Asymmetric synthesis and extracellular distribution of a morphogen

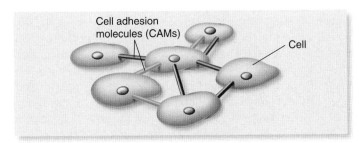

(c) Cell adhesion: Cell-to-cell contact conveys positional information

Figure 19.5 Molecular mechanisms that convey positional information. Asymmetric distribution of a morphogen in the (a) oocyte or (b) embryo. (c) Positional information may also be obtained by cell-to-cell contact.

Another mechanism used to convey positional information involves **cell adhesion** (**Figure 19.5c**). Each cell makes its own collection of surface receptors that enable it to adhere to other cells. Such receptors, which are known as **cell adhesion molecules (CAMs)**, are described in Chapter 10. The positioning of a cell within a multicellular organism is strongly influenced by the combination of contacts it makes with other cells and with the extracellular matrix.

The phenomenon of cell adhesion, and its role in multicellular development, was first recognized by H. V. Wilson in 1907. He took multicellular sponges and separated them into individual cells. Remarkably, the cells actively migrated until they adhered to one another to form a new sponge, complete with the chambers and canals that characterize a sponge's internal structure! When sponge cells from different species were mixed, they sorted themselves properly, adhering only to cells of the same species. Overall, these results indicate that cells possess specific CAMs, which are critical in cell-to-cell recognition. Cell adhesion plays an important role in governing the position that a cell will adopt during development. Together, the actions of morphogens and CAMs convey positional information to the cells in a developing organism.

A Hierarchy of Transcription Factors Plays a Key Role in Orchestrating a Program of Development in Plants and Animals

The formation of a body, both in plants and animals, occurs in a series of organizational phases. As an overview of this process, let's consider four general phases of pattern development in an animal (**Figure 19.6**). The first phase is to organize the body along major axes. The anteroposterior axis determines the organization from head to tail, the dorsoventral axis governs the structure from front (ventral) to back (dorsal), and the right-left axis provides organization from side to side. During the second phase, the body becomes organized into smaller regions that will eventually contain organs and other structures such as legs. In insects, these regions form well-defined segments. In mammals, some segmentation of the body is apparent during embryonic development but defined boundaries are lost as the embryo proceeds to the fetal and adult stages. After segmentation, the third phase occurs when the cells within the segments organize themselves in ways that will produce particular body parts. Finally, during the fourth phase, the cells themselves change their morphologies and become differentiated. This final phase of development produces an organism with many types of tissues, organs, and other body parts with specialized functions.

Certain genes are expressed at a specific phase of development in a particular cell type, a phenomenon called **differential gene regulation**. Developmental geneticists have discovered a parallel between the expression of specific transcription factors and the four major phases of animal development. Many morphogens, particularly those that act at an early phase of development, are transcription factors. Such transcription factors regulate the expression of genes in a way that controls the formation of the body axes. Next, these early transcription factors cause the expression of other transcription factors that direct the segmented body plan. After the body plan has been segmented, a third category of transcription factors controls what structures will be made within each segment. Finally, a fourth category of transcription factors controls the differentiation of each cell type. Overall, as diagrammed in Figure 19.6, development is largely determined by a hierarchy of transcription factors that controls a program of developmental changes.

When considering the events of animal development, the plan progresses from large to small. First, the axes of the entire animal are determined. Next, the entire animal is divided into regions or segments. Then, the organization within each region leads to the formation of tissues, organs, and other body structures. And finally, the cells within each region become differentiated. As these phases occur, the transcription factors controlling each phase regulate the production of proteins that convey positional information to cells, which alters their behavior and characteristics. In the early phases, this positional information mostly promotes cell division and cell migration. During the last phase, cell differentiation becomes a common event, and some cells die via apoptosis.

Hierarchy of transcription factors

Posterior
Right
Dorsal (ventral is underneath)
Anterior
Left

1 **Phase 1:**
Transcription factors determine the formation of the body axes and control the transcription factors of phase 2.

Evidence of segmentation

2 **Phase 2:**
Transcription factors cause the embryo to become subdivided into regions that have properties of individual segments. They also control transcription factors of phase 3.

Head forming

Limbs forming

3 **Phase 3:**
Transcription factors cause each segment and groups of segments to develop specific body parts and control transcription factors of phase 4.

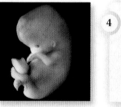

4 **Phase 4:**
Transcription factors cause cells to differentiate into specific cell types such as skin, nerve, and muscle cells.

Figure 19.6   Pattern development in a human embryo. As shown here, pattern formation in animals occurs in four phases that are controlled by a hierarchy of transcription factors. The example shown here involves human development, although research suggests that pattern formation in all complex animals follows a similar plan. The ideas in this scenario are based largely on analogies between pattern formation in mammals and *Drosophila*. Many of the transcription factors that are likely to control the early phases of pattern formation in mammals have yet to be identified. Also, note that the phases of development are overlapping. For example, cell differentiation begins to occur as the cells are adopting their correct locations.

19.2 Development in Animals

In the previous section, we considered the general features of development as they apply to animals and plants. In this section, we will begin by examining the steps of *Drosophila* development, and then focus our attention on embryonic development.

During this stage, the overall body plan is determined. We will see how the differential expression of particular genes and the localization of positional information within the embryo control the developmental processes of pattern formation and segmentation. The roles of genes in the organization of mammalian embryos are not as well understood as they are in *Drosophila*. Even so, the analysis of the genomes of mammals and many other species has revealed many interesting parallels in the developmental program of all animals.

This section will end with an examination of cell differentiation. This process is better understood in mammals than in *Drosophila* because researchers have been studying mammalian cells in the laboratory for many decades. To explore cell differentiation, we will consider mammals as our primary example.

Embryonic Development Determines the Pattern of Structures in the Adult: The Development of *Drosophila*

Figure 19.7 illustrates the general sequence of events in *Drosophila* development. Let's examine these steps before we consider the differential gene expression that causes them to happen. The oocyte is critical to establishing the pattern of development that will ultimately produce an adult organism. It is an elongated cell that contains positional information: As noted in Figure 19.7a, the oocyte already has anterior and posterior ends that correspond to those found in the adult (compare Figure 19.7a and f). After fertilization takes place, the zygote develops into a **blastoderm**. First, the zygote goes through a series of nuclear divisions that are not accompanied by cytoplasmic division, producing many free nuclei. Initially, these nuclei are scattered throughout the yolk (Figure 19.7b), but eventually they migrate to the periphery, producing a stage called a syncytial blastoderm (Figure 19.7c). After the nuclei have lined up along the cell membrane, individual cells are formed as portions of the membrane surround each nucleus to create a cellular blastoderm (Figure 19.7d).

After blastoderm formation is complete, some dramatic changes occur during **gastrulation**, which involves a well-ordered rearrangement of cells in the embryo. During this process, some cells migrate into the interior. The result is three cellular layers called the ectoderm, endoderm, and mesoderm, producing a **gastrula** (Figure 19.7e). In general, the ectoderm remains on the outside of the gastrula, the endoderm is on the inside, and the mesoderm is wedged in the middle. These cell layers are important because each of them leads to the development of specific cell types at a later stage. For example, the mesoderm produces muscle cells.

A key process in *Drosophila* embryonic development is the creation of a segmented body pattern. The embryo is subdivided into visible units. In *Drosophila* the segments can be grouped into three general areas, the head, the thorax, and the abdomen. Figure 19.7f shows the segmented pattern of a *Drosophila* embryo about 10 hours after fertilization. Later in this section, we will explore how the coordination of gene expression underlies the formation of these segments.

An embryo then develops into a **larva**, which is a free-living organism that is morphologically very different from the adult (Figure 19.7g). *Drosophila* undergoes three larval stages. After the third larval stage, the organism becomes a **pupa** (Figure 19.7h) and proceeds through a process known as **metamorphosis**, in which it changes into a mature adult. Each segment in the adult develops its own characteristic structures; for example, the wings are on a thoracic segment. The adult fly then emerges from the pupal case (Figure 19.7i). From beginning to end, this process takes about 10 days.

Phase 1 Pattern Development: Maternal Effect Genes Promote the Formation of the Main Body Axes

The first phase in *Drosophila* embryonic pattern development is the establishment of the body axes, which occurs before the embryo becomes segmented (see Figure 19.6). The morphogens necessary to establish these axes are distributed prior to fertilization. In most invertebrates and some vertebrates, certain morphogens, which are important in early developmental stages, are deposited asymmetrically within the egg as it develops. Later, after the egg has been fertilized and development begins, these morphogens will initiate developmental programs that govern the formation of the body axes of the embryo. The products of several different genes ensure that proper development will occur.

As an example of one morphogen that plays a role in axis formation, let's consider the product of a gene in *Drosophila* called *bicoid*. Its name is derived from the observation that a mutation that inactivates the gene results in a larva with two posterior ends (**Figure 19.8**). During normal egg development, the *bicoid* gene product accumulates in the anterior region of the oocyte, causing the development of the anterior end of the embryo.

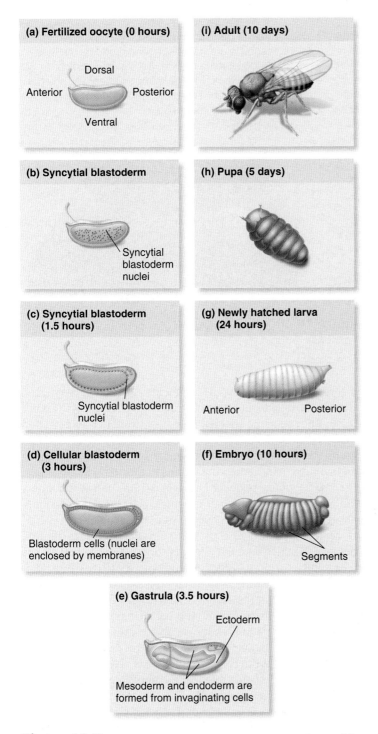

(a) Fertilized oocyte (0 hours)

Dorsal

Anterior — Posterior

Ventral

(b) Syncytial blastoderm

Syncytial blastoderm nuclei

(c) Syncytial blastoderm (1.5 hours)

Syncytial blastoderm nuclei

(d) Cellular blastoderm (3 hours)

Blastoderm cells (nuclei are enclosed by membranes)

(e) Gastrula (3.5 hours)

Ectoderm

Mesoderm and endoderm are formed from invaginating cells

(i) Adult (10 days)

(h) Pupa (5 days)

(g) Newly hatched larva (24 hours)

Anterior   Posterior

(f) Embryo (10 hours)

Segments

Figure 19.7   Developmental stages of the fruit fly *Drosophila*.

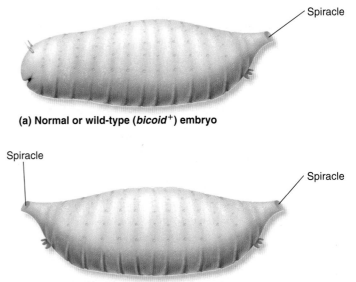

Spiracle

(a) Normal or wild-type (*bicoid*⁺) embryo

Spiracle

Spiracle

(b) Mutant (*bicoid*⁻) embryo

Figure 19.8   The *bicoid* mutation in *Drosophila*. (a) A normal *bicoid*⁺ embryo. (b) A *bicoid*⁻ embryo, in which both ends of the larva develop posterior structures. For example, both ends develop a spiracle, which normally is found only at the posterior end.

Biological inquiry: What would you expect the phenotype to be of a larva in which the bicoid gene was expressed in both the anterior region and the posterior region of the oocyte?

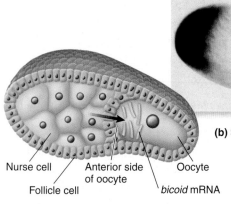

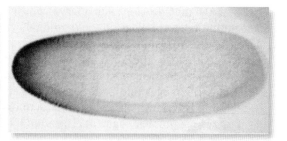

(b) Staining of *bicoid* mRNA in oocyte

(c) Staining of Bicoid protein in an embryonic blastoderm

Nurse cell   Anterior side   Oocyte
of oocyte

Follicle cell   *bicoid* mRNA

(a) Transport of maternal effect gene products (*bicoid* mRNA) into the oocyte

Figure 19.9 Asymmetric localization of gene products during egg development in *Drosophila*. **(a)** The nurse cells transport maternal effect gene products such as *bicoid* mRNA into the anterior end of the developing oocyte. **(b)** Staining of *bicoid* mRNA in an oocyte prior to fertilization. The *bicoid* mRNA is trapped near the anterior end. **(c)** Staining of Bicoid protein after fertilization. The Bicoid protein forms a gradient, with its highest concentration near the anterior end.

How does the *bicoid* gene product accumulate in the anterior region of the oocyte? The answer involves specialized nurse cells that are found next to the oocyte in the ovaries of female flies. As discussed in Chapter 17, nurse cells supply the products (for example, mRNA) of maternal effect genes. These genes cause an unusual pattern of inheritance called **maternal effect** (refer back to Figure 17.15). In *Drosophila*, the nurse cells are located toward the anterior end of the oocyte rather than the posterior end. The *bicoid* gene is transcribed in the nurse cells, and *bicoid* mRNA is then transported into the anterior end of the oocyte and trapped there (**Figure 19.9a**). Prior to fertilization, the *bicoid* mRNA is highly concentrated near the anterior side of the oocyte (**Figure 19.9b**). After fertilization, the *bicoid* mRNA is translated and a gradient of Bicoid protein is established across the zygote (**Figure 19.9c**). This gradient starts a progression of developmental events that will provide the positional information that causes the end of the zygote with a high Bicoid protein concentration to become the anterior end of the embryo.

The Bicoid protein is a morphogen that functions as a transcription factor to activate particular genes at specific times. The ability of Bicoid to activate a given gene is tuned exquisitely to its concentration. Due to its asymmetric distribution, the Bicoid protein will activate genes only in certain regions of the embryo. For example, a high concentration of Bicoid stimulates the expression of a gene called *hunchback* (that also encodes a transcription factor) in the anterior half of the embryo, but its concentration is too low in the posterior half to activate the *hunchback* gene. The ability of Bicoid to activate genes in certain regions but not others plays a role in the second phase of pattern formation, which is segmentation.

The Study of *Drosophila* Mutants Has Identified Genes That Control the Development of Segments and Their Characteristics

As described earlier in Figure 19.6, the second phase of pattern formation is the development of segments. The normal *Drosophila*

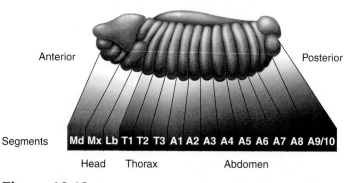

Anterior   Posterior

Segments | Md Mx Lb T1 T2 T3 A1 A2 A3 A4 A5 A6 A7 A8 A9/10 |

Head   Thorax   Abdomen

Figure 19.10 The organization of segments in the *Drosophila* embryo.

embryo is subdivided into 15 segments: three head segments, three thoracic segments, and nine abdominal segments (**Figure 19.10**). The segments are identified by a two-letter abbreviation or a letter and number corresponding to the area of the segment and its sequential order. Each segment of the embryo will give rise to unique morphological features in the adult. For example, the second thoracic segment (T2) will become part of the thorax with a pair of legs and a pair of wings, and the eighth abdominal segment (A8) will become a part of the abdomen.

Mutations that alter *Drosophila* development have allowed scientists to understand the normal process. **Figure 19.11** shows an example of a normal fly and one with mutations in a complex of genes called the *bithorax* complex. In a normal fly, two wings are found on the second thoracic segment, and two halteres, which together function as a balancing organ that resembles a pair of miniature wings, are found on the third thoracic segment. In this mutant fly, the third thoracic segment has the characteristics of the second, so the fly has no halteres and four wings. The term *bithorax* refers to the duplicated characteristics of the second thoracic segment.

Edward Lewis, a pioneer in the genetic study of development, became interested in the bithorax phenotype and began

Figure 19.11 The *bithorax* mutation in *Drosophila*. (a) A normal fly has two wings on the second thoracic segment, and two halteres on the third thoracic segment. (b) This mutant fly contains mutations in a complex of genes called the *bithorax* complex. In this fly, the third thoracic segment has the same characteristics as the second thoracic segment, thereby producing a fly with four wings instead of two.

(a)

(b)

Normal Mutant	Normal Mutant	Normal Mutant
Krüppel	*Even-skipped*	*Gooseberry*
(a) Gap gene	**(b) Pair-rule gene**	**(c) Segment-polarity gene**

Figure 19.12 Phenotypic effects in *Drosophila* larvae that have mutations in segmentation genes. In each pair, the larva on the *left* is normal. The *right* side shows a larva with a defect in **(a)** a gap gene called *Krüppel*, **(b)** a pair-rule gene called *Even-skipped*, and **(c)** a segment-polarity gene called *Gooseberry*.

investigating it in 1946. He discovered that the mutant chromosomal region actually contains a complex of three genes that play a role in the third phase of development, which will be described later in this chapter. During the 1960s and 1970s, as the study of molecular genetics blossomed, it became clear that the genomes of multicellular organisms contain groups of genes that initiate a program of development.

In the 1970s, Christiane Nüsslein-Volhard and Eric Wieschaus undertook a systematic search for *Drosophila* mutants with disrupted development. In particular, they focused their search on genes that alter the segmentation pattern of the *Drosophila* embryo and larva, called **segmentation genes**. Their pioneering efforts identified most of the genes required for the embryo to develop a segmented pattern. Based on the characteristics of abnormal larva, they identified three classes of segmentation genes that they termed gap genes, pair-rule genes, and segment-polarity genes (**Figure 19.12**). When a mutation inactivates a **gap gene**, several adjacent segments are missing in the larva (Figure 19.12a).

A defect in a **pair-rule gene** may cause alternating segments or parts of segments to be deleted (Figure 19.12b). Finally, **segment-polarity gene** mutations cause portions of segments to be missing either an anterior or a posterior region and for adjacent regions to become mirror images of each other (Figure 19.12c).

An important advantage of isolating mutant fruit flies as well as other mutant organisms is that a mutation allows researchers to eventually identify, at the molecular level, both the mutant allele and normal allele. Using molecular techniques similar to those described in Chapter 20, the DNA from mutant and normal strains can be compared to determine which gene has become altered to cause the observed changes in development. Once the gene has been identified, the expression of the gene can be studied throughout development. For example, after the *bicoid* gene was identified at the molecular level, researchers could study when and where the Bicoid protein was made, as described earlier in Figure 19.9. As discussed next, this approach also made it possible to understand how gene expression leads to pattern formation.

Phase 2 Pattern Development: Segmentation Genes Act Sequentially to Divide the *Drosophila* Embryo into Segments

The study of segmentation genes has provided information regarding the process by which segments are formed. To create a segment, a group of genes acts sequentially to govern the fate of a given region of the body. A simplified scheme of gene expression that leads to a segmented pattern in the *Drosophila* embryo

is shown in **Figure 19.13**. Many more genes are actually involved in this process.

In general, maternal effect genes, which promote phase 1 pattern development, activate gap genes; this activation is seen as broad bands of gap gene expression in the embryo (Figure 19.13, step 2). The gap genes and maternal effect genes then activate the pair-rule genes in alternating stripes in the embryo (Figure 19.13, step 3). Once the pair-rule genes are activated, their gene products then regulate the segment-polarity genes. As you follow the progression from maternal effect genes to segment-polarity genes, notice that a body pattern is emerging in the embryo that matches the segmentation pattern found in the larva and adult animal. As you can see in step 4 of Figure 19.13, the expression of a segment-polarity gene corresponds to portions of segments in the adult fly. To appreciate this phenomenon, notice that the embryo at this stage is curled up and folded back on itself. If you imagine that the embryo was stretched out linearly, the 15 stripes seen in this embryo correspond to portions of the 15 segments of an adult fly.

Phase 3 Pattern Development: Homeotic Genes Control the Development of Segment Characteristics

Thus far, we have considered how the *Drosophila* embryo becomes organized along axes and then into a segmented body pattern. During the third phase of pattern development, each segment begins to develop its own unique characteristics (see Figure 19.6).

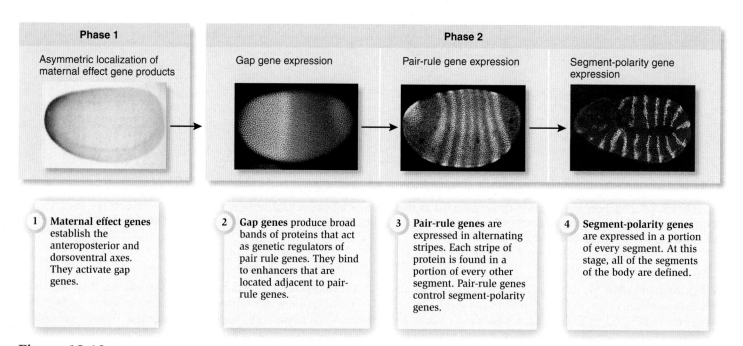

Phase 1	**Phase 2**		
Asymmetric localization of maternal effect gene products	Gap gene expression	Pair-rule gene expression	Segment-polarity gene expression

1 **Maternal effect genes** establish the anteroposterior and dorsoventral axes. They activate gap genes.

2 **Gap genes** produce broad bands of proteins that act as genetic regulators of pair rule genes. They bind to enhancers that are located adjacent to pair-rule genes.

3 **Pair-rule genes** are expressed in alternating stripes. Each stripe of protein is found in a portion of every other segment. Pair-rule genes control segment-polarity genes.

4 **Segment-polarity genes** are expressed in a portion of every segment. At this stage, all of the segments of the body are defined.

Figure 19.13 Overview of segmentation in *Drosophila*.   The *left* micrograph shows an embryonic blastoderm, and the other three depict the progression during the first few hours of embryonic development. The micrographs show the expression of protein products of a maternal effect gene (step 1) or segmentation genes (steps 2–4). In step 1, the protein is stained brown and is found in the *left* side of the embryonic blastoderm. In step 2, three different proteins encoded by gap genes are stained green, yellow, and red. In step 3, a protein encoded by a pair-rule gene is stained in light blue. In step 4, a protein encoded by a segment-polarity gene is stained pink. When comparing steps 3 and 4, the embryo has undergone a 180° turn, folding back on itself.

Geneticists use the term **cell fate** to describe the ultimate morphological features that a cell or group of cells will adopt. For example, as mentioned previously, the fate of cells in segment T2 is to develop into a thoracic segment containing two legs and two wings. In *Drosophila*, the cells in each segment of the body have their fate determined at a very early stage of embryonic development, long before the morphological features become apparent.

Our understanding of developmental fate has been greatly aided by the identification of mutant genes that alter cell fates. In animals, the first mutant of this type was described by the German entomologist G. Kraatz in 1876. He observed a sawfly (*Climbex axillaris*) in which part of an antenna was replaced with a leg. During the late 19th century, the English zoologist William Bateson collected many of these types of observations and published them in 1894 in a book entitled *Materials for the Study of Variation Treated with Especial Regard to Discontinuity in the Origin of Species*. In this book, Bateson coined the term **homeotic** to describe changes in which one body part is replaced by another. We now know these are caused by mutant alleles of what we call **homeotic genes**.

The *bithorax* mutation discussed earlier that was studied by Edward Lewis is an example of a homeotic mutation (see Figure 19.11). Lewis also studied strains of *Drosophila* with other homeotic mutations. From these studies, he eventually discovered that each homeotic gene controls the fate of a particular region of the body. *Drosophila* contains two clusters of homeotic genes called the *bithorax* complex and the *Antennapedia* complex (**Figure 19.14**). The *Antennapedia* complex contains

five genes, designated *lab*, *pb*, *Dfd*, *Scr*, and *Antp*. The *bithorax* complex has three genes, *Ubx*, *abd-A*, and *Abd-B*. Both of these complexes are located on the same chromosome, but a long stretch of DNA separates them. Homeotic genes are controlled by many genes that are initially expressed during embryonic development, including segmentation genes.

Interestingly, the order of homeotic genes along the chromosome correlates with their expression along the anteroposterior axis of the body. This phenomenon is called the **colinearity rule**. For example, *lab* (for labial) is expressed in the anterior segment and governs the formation of mouth structures. The *Antp* (for antennapedia) gene is expressed in the thoracic region during embryonic development and controls the formation of thoracic structures. Transcription of the *Abd-B* (for abdominal B) gene occurs in the posterior region of the embryo and promotes the formation of the posteriormost abdominal segments.

The role of homeotic genes in determining the identity of particular segments has been revealed by mutations that alter their function. For example, a mutation in the *Antp* gene has been identified in which the gene is incorrectly expressed in an anterior segment (**Figure 19.15**). A fly with this mutation has the bizarre trait in which it develops legs where antennae are normally found!

Homeotic genes encode homeotic proteins that function as transcription factors. The coding sequence of homeotic genes contains a 180-bp sequence known as a **homeobox** (**Figure 19.16a**). This sequence was first discovered in the *Antp* and *Ubx* genes, and it has since been found in many *Drosophila* homeotic genes. The homeobox is also found in other genes affecting pattern development, such as *bicoid*, which is not a homeotic gene. The homeobox encodes a region of the protein called a **homeodomain**, which functions in binding to the DNA (**Figure 19.16b**). The arrangement of α helices in the homeodomain promotes the binding of the protein to the DNA.

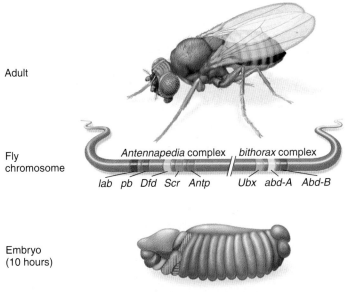

Adult

Fly chromosome

Antennapedia complex | *bithorax* complex

lab pb Dfd Scr Antp   *Ubx abd-A Abd-B*

Embryo (10 hours)

Figure 19.14 **Expression pattern of homeotic genes in *Drosophila*.** The order of homeotic genes, *labial* (*lab*), *proboscipedia* (*pb*), *deformed* (*Dfd*), *sex combs reduced* (*Scr*), *antennapedia* (*Antp*), *ultrabithorax* (*Ubx*), *abdominal A* (*abd-A*), and *abdominal B* (*Abd-B*), correlates with their spatial order of expression in the embryo.

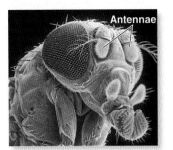
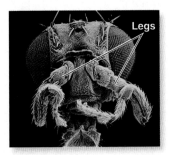

(a) Normal fly   **(b) Mutant fly**

Figure 19.15 **The *Antennapedia* mutation in *Drosophila*.** (a) A normal fly with antennae. (b) This fly has a mutation in which the *Antp* gene is expressed in the embryonic segment that normally gives rise to antennae. The abnormal expression of *Antp* causes this region to have legs rather than antennae.

Biological inquiry: What phenotype would you expect if the Antp *gene was expressed in the last abdominal segment?*

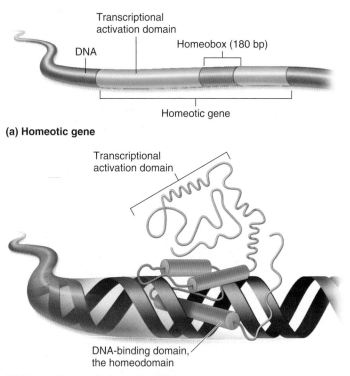

(a) Homeotic gene

(b) Homeotic protein bound to DNA

Figure 19.16 Molecular features of homeotic genes and proteins. **(a)** A homeotic gene (shown in blue) contains a 180-bp sequence called the homeobox (shown in dark blue). **(b)** Homeotic genes encode transcription factors. The homeobox encodes a region of the protein called a homeodomain, which functions in binding to the DNA at a regulatory site such as an enhancer that controls the expression of a gene.

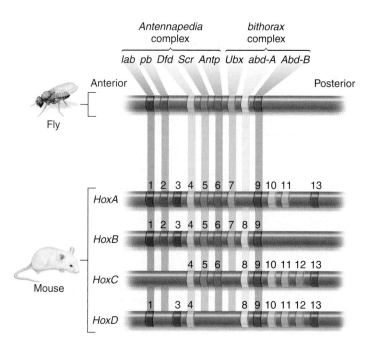

Figure 19.17 **A comparison of homeotic genes in *Drosophila* and the mouse.** The mouse and other mammals have four gene complexes, *HoxA–D*, that correspond to certain homeotic genes found in *Drosophila*. Thirteen different types of homeotic genes are found in the mouse, although each *Hox* complex does not contain all 13 genes. In this drawing, homologous genes are aligned in columns. For example, *lab* is the homologue to *HoxA-1*, *HoxB-1*, and *HoxD-1*; *Ubx* is homologous to *HoxA-7* and *HoxB-7*.

The primary function of homeotic proteins is to activate the transcription of specific genes that promote developmental changes in the animal. The homeodomain binds to enhancer sequences, which are described in Chapter 13. These enhancers are found in the vicinity of specific genes that control development. Most homeotic proteins also contain a transcriptional activation domain (Figure 19.16). After the homeodomain binds an enhancer, the transcriptional activation domain of the homeotic protein activates RNA polymerase to begin transcription. Some homeotic proteins also function as repressors of transcription.

GENOMES & PROTEOMES

A Homologous Group of Homeotic Genes Is Found in All Animals

Because most vertebrates have long generation times and produce relatively few offspring, it is not practical for researchers to screen large numbers of their embryos or offspring in search of mutant phenotypes with developmental defects. As an alternative,

the most successful way of identifying genes that affect vertebrate development has been the use of molecular techniques to identify vertebrate genes that are homologous to those that control development in simpler organisms such as *Drosophila*. Homologous genes are evolutionarily derived from the same ancestral gene and have similar DNA sequences. For this reason, a DNA strand from one gene (a probe) will bind to a complementary DNA strand of a homologous gene. As described in Chapter 20, this phenomenon, termed hybridization, can be used to identify homologous genes among different species.

The general approach of using *Drosophila* genes as probes to identify homologous vertebrate genes has been quite successful. Researchers have found complexes of homeotic genes in vertebrate species that bear striking similarities to those in the fruit fly. In the mouse, these groups of adjacent homeotic genes are called ***Hox* complexes**. The mouse has four *Hox* complexes, designated *HoxA*, *HoxB*, *HoxC*, and *HoxD*. Thirty-eight genes are found in the four complexes, which represent 13 different gene types. You can see from **Figure 19.17** that several of the homeotic genes in fruit flies and mammals are evolutionarily related. Among the first six types of *Hox* genes, five of them are homologous to genes found in the *Antennapedia* complex of

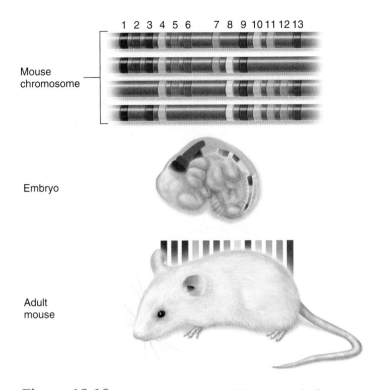

Mouse chromosome

Embryo

Adult mouse

Figure 19.18 Expression pattern of *Hox* genes in the mouse. A schematic illustration of *Hox* gene expression in the embryo and the corresponding regions in the adult. The order of *Hox* gene expression, from anterior to posterior, parallels the order of genes along the chromosome.

Drosophila. Among the last seven (genes numbered 7–13), three are homologous to the genes of the *bithorax* complex.

Like the *Antennapedia* and *bithorax* complexes in *Drosophila*, the arrangement of *Hox* genes along the mouse chromosome follows the colinearity rule, reflecting their pattern of expression from the anterior to the posterior end (**Figure 19.18**). Mutations in these mouse genes affect the formation of structures along this axis. These results indicate that the *Hox* genes play a role in determining the fates of regions along the anteroposterior axis. Nevertheless, additional research will be necessary to understand the individual role that each of the 38 *Hox* genes plays during embryonic development.

Hox genes are present in all animals. Sponges, which are a primitive type of animal, have a single *Hox* gene. The study of *Hox* genes in many different animal species has shown that the *Hox* cluster, with its colinear expression and its role of determining the anteroposterior axis, originated very early in the evolution of animals. At the level of genetics, fundamental similarities are observed in the ways that animals, such as worms, fruit flies, and mammals, undergo embryonic development. Researchers have suggested there is a "universal body plan" for animal development. A portion of the genome of all animals is devoted to the execution of this plan. The biological diversity that we see among animals is due to variation from this common plan.

Phase 4 Pattern Development: Stem Cells Can Divide and Differentiate into Specialized Cell Types

Thus far we have focused our attention on patterns of gene expression that occur during the early stages of development. These genes control the basic body plan of the organism. During the fourth phase of pattern development, the emphasis shifts to cell differentiation (see Figure 19.6). But how do cells differentiate into the necessary cell types? For example, how does an undifferentiated mesodermal cell differentiate into a specialized muscle cell, or how does an ectodermal cell differentiate into a nerve cell?

Although invertebrates have been instrumental in our understanding of pattern formation in animals, cell differentiation has been studied more extensively in mammals. One reason is because researchers have been able to grow mammalian cells in the laboratory for many decades. The availability of laboratory-grown cells makes it much easier to analyze the process of cell differentiation.

By studying mammalian cells in the laboratory, geneticists have determined that the profound morphological differences between two different types of differentiated cells, such as muscle and nerve cells, arise through gene regulation. Though nerve and muscle cells contain the same set of genes, they regulate the expression of their genes in very different ways. Certain genes that are transcriptionally active in muscle cells are inactive in nerve cells, and vice versa. Therefore, nerve and muscle cells express different proteins that affect the morphological and physiological characteristics of the respective cells in distinct ways. In this manner, differential gene expression underlies cell differentiation.

To understand the process of cell differentiation in a multicellular organism, we need to consider the special properties of **stem cells**, which divide and supply the cells that construct the bodies of all animals and plants. Stem cells have two common characteristics. First, they have the capacity to divide, and second, their daughter cells can differentiate into one or more specialized cell types. The two daughter cells that are produced from the division of a stem cell may have different fates (**Figure 19.19**). One of the cells may remain an undifferentiated stem cell, while the other daughter cell can differentiate into a specialized cell type. With this asymmetric division/differentiation pattern, the population of stem cells remains constant, yet stem cells provide a population of specialized cells. For example, in mammals, this mechanism is needed to replenish cells that have a finite life span such as skin cells and red blood cells.

In mammals, stem cells are commonly categorized according to their developmental stage and their ability to differentiate (**Figure 19.20**). The ultimate stem cell is the fertilized egg, which, via multiple cellular divisions, can give rise to an entire organism. A fertilized egg is considered to be **totipotent** because it can produce all of the cell types in the adult organism. The early mammalian embryo contains **embryonic stem cells (ES cells)**, which are initially found in the inner cell mass of the blastocyst.

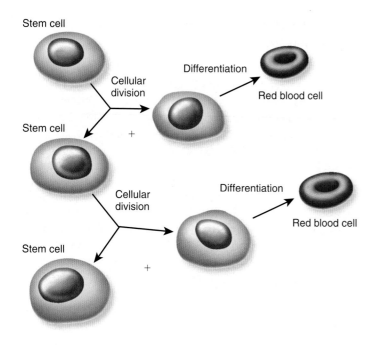

Figure 19.19 **Growth pattern of stem cells.** When a stem cell divides, one of the two daughter cells may remain a stem cell, while the other cell can differentiate into a specialized cell type, such as the red blood cells shown here.

Biological inquiry: What are the two key features of stem cells?

The blastocyst stage of embryonic development occurs prior to uterine implantation. Embryonic stem cells are **pluripotent**, which means they can also differentiate into almost every cell type of the body. However, a single embryonic stem cell has lost the ability to produce an entire, intact individual. At the early fetal stage of development, the cells that later give rise to sperm or eggs cells, known as the **embryonic germ cells (EG cells)**, also are pluripotent.

During the embryonic and fetal stages of development, multipotent and unipotent stem cells are produced that give rise to the differentiated cells within the tissues and organs of an animal's body. Adults also have multipotent and unipotent stem cells. A **multipotent** stem cell can differentiate into several cell types, but far fewer than an embryonic stem cell can differentiate into. For example, hematopoietic stem cells (HSCs) found in the bone marrow can supply cells that populate two different tissues, namely, the blood and lymphoid tissues (**Figure 19.21**). Furthermore, each of these tissues contains several cell types. Multipotent hematopoietic stem cells can follow a pathway in which cell division produces a myeloid progenitor cell, which then differentiates into various cells of the blood and immune systems. Alternatively, an HSC can follow a path in which it becomes a lymphoid progenitor cell that differentiates into lymph cell types. Other stem cells found in the adult seem to be **unipotent**, or only able to produce daughter cells that differentiate into one cell type. For example, stem cells in the testis produce daughter cells that only differentiate into sperm.

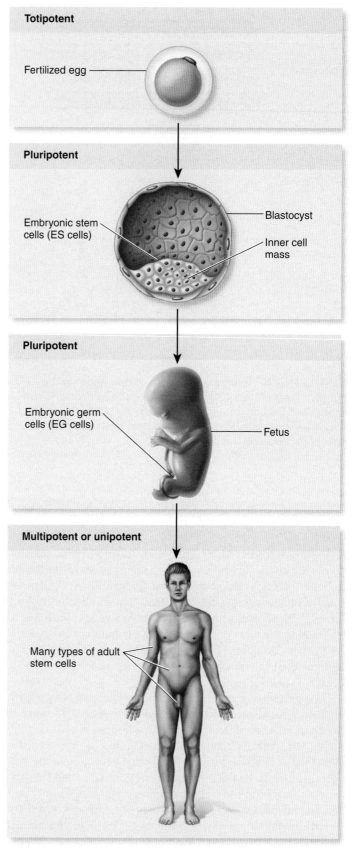

Figure 19.20 Occurrence of stem cells at different stages of mammalian development.

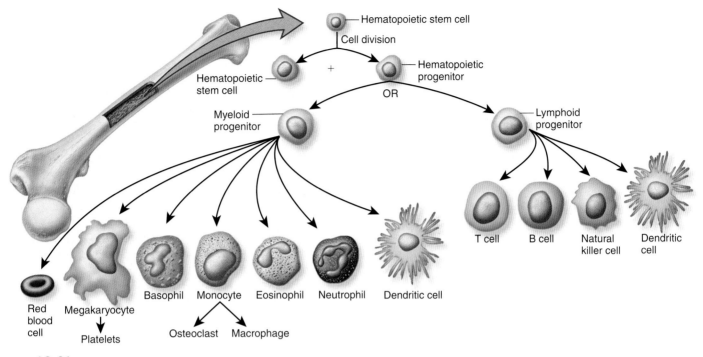

Figure 19.21 Fates of hematopoietic stem cells.

A compelling medical reason why researchers are interested in stem cells is the potential to treat human diseases or injuries that cause cell and tissue damage. This application has already become a reality in certain cases. For example, bone marrow transplants are used to treat patients with certain forms of cancer. When bone marrow from a healthy person is injected into the body of a patient who has had their immune system wiped out via radiation, the stem cells within the transplanted marrow have the ability to proliferate and differentiate into various types of blood cells within the body of the patient.

Renewed interest in the use of stem cells in the potential treatment of many other diseases has been fostered by studies in 1998 that showed researchers can obtain ES cells from blastocysts and EG cells from aborted fetuses and successfully propagate them in the laboratory. As mentioned, ES and EG cells are pluripotent. Therefore, they have the capacity to produce many different kinds of tissue. Embryonic cells could potentially be used to treat a wide variety of diseases associated with cell and tissue damage (**Table 19.1**). Much progress has been made in testing the use of stem cells in animal models. However, more research will need to be done before the use of stem cells to treat such diseases in humans is realized.

From an ethical perspective, the primary issue that raises debate is the source of stem cells for research and potential treatments. Most ES cells have been derived from human embryos that were produced from *in vitro* fertilization, and that were not used by the couples. Most EG cells are obtained from aborted fetuses, either those that were spontaneously aborted or those in which the decision to abort was not related to donating the fetal tissue to research. Some feel that it is morally wrong to use such tissue in research and/or the treatment of disease. Furthermore, some people fear that this technology could lead to intentional abortions for the sole purpose of obtaining fetal tissues for

Table 19.1	Some Potential Uses of Stem Cells to Treat Diseases
Cell/tissue type	**Disease treatment**
Nerve	Implantation of cells into the brain to treat Parkinson disease
	Treatment of injuries such as those to the spinal cord
Skin	Treatment of burns and skin disorders
Cardiac	Repair of heart damage associated with heart attacks
Cartilage	Repair of joints damaged by injury or arthritis
Bone	Repair of damaged bone or replacement with new bone
Liver	Repair or replacement of liver tissue damaged by injury or disease
Skeletal muscle	Repair or replacement of damaged muscle

transplantation. Alternatively, others feel that the embryos and fetuses that have been the sources of ES and EG cells were not going to become living individuals, and therefore it is beneficial to study these cells and to use them in a positive way to treat human diseases and injury. It is not clear whether these two opposing viewpoints can reach a common ground. Many governments are enacting laws that limit or prohibit the use of embryos or fetuses to obtain stem cells, yet permit the use of stem cell lines that are already available in research laboratories. In the U.S., for example, federal law prohibits the use of government funding for research projects that involve the destruction of embryos to obtain stem cells. However, government-sponsored research can be done on stem cell lines that were created prior to this legislation.

FEATURE INVESTIGATION

Davis, Weintraub, and Lasser Identified Genes Encoding Transcription Factors That Promote Muscle Cell Differentiation

A key question regarding the study of stem cells is, What causes a stem cell to differentiate into a particular cell type? Though the answer is not understood in great detail, researchers have discovered that certain proteins function as "master transcription factors" that cause a cell to differentiate into a specific cell type. The investigation described here was one of the first studies to reveal this phenomenon.

In 1987, Robert Davis, Harold Weintraub, and Andrew Lasser conducted a study to identify genes that promote muscle cell differentiation. The initial strategy for their experiments was to identify genes that are expressed only in differentiating muscle cells, not in nonmuscle cells. Though methods of gene cloning are described in Chapter 20, let's briefly consider these scientists'

cloning methods so we can understand their approach. They began with two different laboratory cell lines that could differentiate into muscle cells. From these two cell lines, they identified about 10,000 different genes that were transcribed into mRNA. Next, they compared the expressed genes in these two muscle cell lines with genes that were expressed in a nonmuscle cell line. Davis, Weintraub, and Lasser's comparison revealed 26 genes that were expressed only in the two muscle cell lines but not in the nonmuscle cell line. To narrow their search further, they compared these 26 genes with other nonmuscle cell lines they had available. Among the 26, only 3 of them were not expressed in other nonmuscle cell lines. Based on these criteria, they were left with three different cloned genes, which they termed *MyoA*, *MyoD*, and *MyoH*.

In the experiment shown in **Figure 19.22**, the scientists' goal was to determine if any of these three cloned genes could cause nonmuscle cells to differentiate into muscle cells. Using

Figure 19.22   Davis, Weintraub, and Lasser and the promotion of muscle cell differentiation in fibroblasts by the expression of *MyoD*.

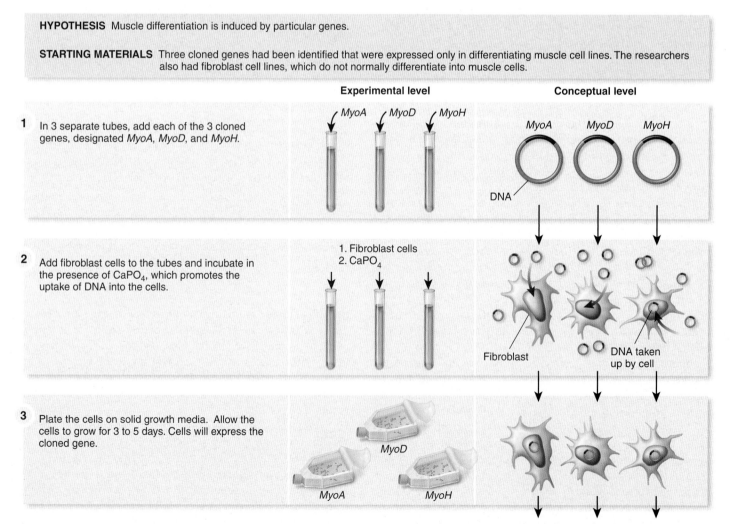

HYPOTHESIS Muscle differentiation is induced by particular genes.

STARTING MATERIALS Three cloned genes had been identified that were expressed only in differentiating muscle cell lines. The researchers also had fibroblast cell lines, which do not normally differentiate into muscle cells.

Experimental level	Conceptual level

1 In 3 separate tubes, add each of the 3 cloned genes, designated *MyoA*, *MyoD*, and *MyoH*.

MyoA *MyoD* *MyoH*

DNA

2 Add fibroblast cells to the tubes and incubate in the presence of CaPO₄, which promotes the uptake of DNA into the cells.

1. Fibroblast cells
2. CaPO₄

Fibroblast

DNA taken up by cell

3 Plate the cells on solid growth media. Allow the cells to grow for 3 to 5 days. Cells will express the cloned gene.

MyoD

MyoA   *MyoH*

4 Examine the cells under a microscope to determine if they have the morphology of differentiating muscle cells.

5 Also, determine if the cells are synthesizing myosin, which is a protein that is abundantly made in muscle cells. This is done by adding a labeled antibody that recognizes myosin and determining the amounts of antibody that bind.

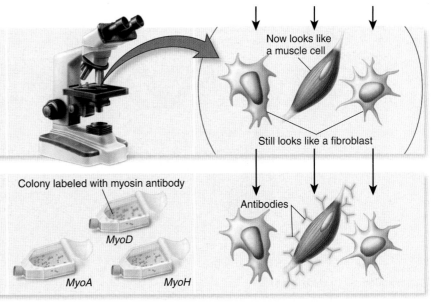

Now looks like a muscle cell

Still looks like a fibroblast

Colony labeled with myosin antibody

MyoD

MyoA   MyoH

Antibodies

6 **THE DATA**

Results from step 4:

DNA added	Microscopic morphology of cells
MyoA	Fibroblasts
MyoD	Muscle cells
MyoH	Fibroblasts

Results from step 5:

DNA added	Colonies labeled with antibody that binds to myosin?
MyoA	No
MyoD	Yes
MyoH	No

molecular techniques described in Chapter 20, the coding sequence of each cloned gene was placed next to an active promoter, and then the genes were introduced into fibroblasts, which are a type of cell that normally differentiates into osteoblasts (bone cells), chrondrocytes (cartilage cells), adipocytes (fat cells), and smooth muscle cells, but *in vivo* never differentiates into skeletal muscle cells. However, when the cloned *MyoD* gene was expressed in fibroblast cells in a laboratory, the fibroblasts differentiated into skeletal muscle cells. These cells contained large amounts of myosin, which is a protein that is expressed in muscle cells. The other two cloned genes (*MyoA* and *MyoH*) did not cause muscle cell differentiation or promote myosin production.

Since this initial discovery, researchers have found that *MyoD* belongs to a small group of genes termed **myogenic bHLH genes** that initiate muscle development. Myogenic bHLH genes encode transcription factors that contain a basic domain and a helix-loop-helix domain (bHLH). They are found in all vertebrates, and they have been identified in several invertebrates, such as *Drosophila* and *C. elegans*. In all cases, myogenic bHLH genes are activated during skeletal muscle cell development.

At the molecular level, two key features enable myogenic bHLH proteins to promote muscle cell differentiation. First, the basic domain binds specifically to an enhancer DNA sequence that is adjacent to genes that are expressed only in muscle cells. Second, the protein contains a transcriptional activation domain that stimulates the ability of RNA polymerase to initiate transcription. Therefore, bHLH proteins can bind to enhancers and activate the expression of many different muscle cell–specific genes. When the encoded proteins are synthesized, they change the characteristics of an undifferentiated cell into those of a highly specialized skeletal muscle cell.

The function of myogenic bHLH proteins relies on their interaction with other cellular proteins. Myogenic bHLH proteins can form a dimer with a protein called Id (for inhibitor of differentiation) or with an E protein, which is also expressed in cells that should differentiate into muscle cells. The helix-loop-helix domain is necessary for dimer formation. The Id protein is produced during early stages of development and prevents myogenic bHLH proteins from promoting muscle differentiation too soon. When a dimer forms between a myogenic bHLH protein and Id protein, the dimer cannot bind to DNA because the Id protein lacks a basic domain (**Figure 19.23a**). At later stages of development, the amount of Id protein falls, and myogenic bHLH proteins can then combine with E proteins to induce muscle differentiation. When a dimer forms between a myogenic bHLH protein and an E protein, the dimer binds to the DNA and activates gene expression (**Figure 19.23b**).

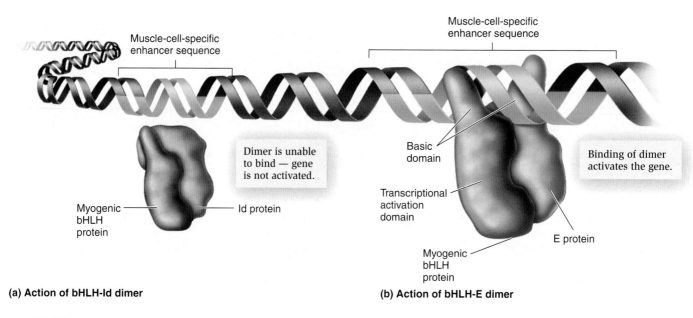

Muscle-cell-specific
enhancer sequence

Muscle-cell-specific
enhancer sequence

Dimer is unable
to bind — gene
is not activated.

Basic
domain

Binding of dimer
activates the gene.

Transcriptional
activation
domain

Myogenic
bHLH
protein

Id protein

E protein

Myogenic
bHLH
protein

(a) Action of bHLH-Id dimer

(b) Action of bHLH-E dimer

Figure 19.23   Regulation of muscle-cell-specific genes by myogenic bHLH proteins.   (a) When a myogenic bHLH protein forms a dimer with an Id protein, it cannot bind to the DNA and therefore does not activate gene transcription. **(b)** A dimer formed from a myogenic bHLH protein and an E protein can bind to a muscle-cell-specific enhancer sequence and activate gene expression. This leads to muscle cell differentiation.

19.3 Development in Plants

As discussed at the beginning of the chapter, the morphology of more complex plants has two key features (see Figure 19.2b). The first is the root-shoot axis. Most plant growth occurs via cell division near the tips of the shoots and the bottoms of the roots. Second, this growth occurs in a well-defined radial pattern. For example, early in *Arabidopsis* growth, a rosette of leaves is produced from leaf buds that emanate in a spiral pattern directly from the main shoot (**Figure 19.24**). Later, the shoot generates branches that produce leaf buds as they grow. Overall, the radial pattern in which a plant shoot generates buds is an important mechanism that determines much of the general morphology of the plant.

At the cellular level too, plant development differs markedly from animal development. For example, cell migration does not occur during plant development. In addition, the development of a plant does not rely on morphogens that are deposited asymmetrically in the oocyte, as in complex animals. In plants, an entirely new individual can be regenerated from many types of somatic cells. Certain somatic cells of plants are totipotent, with the ability to produce an entire individual.

In spite of these apparent differences, the underlying molecular mechanisms of pattern formation in plants still share some similarities with those in animals. Like animals, plants use the mechanism of differential gene expression to coordinate the

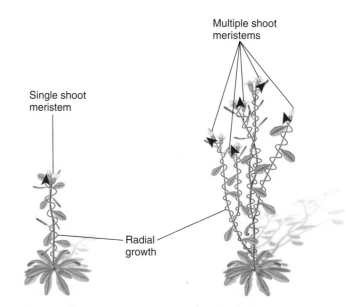

Multiple shoot
meristems

Single shoot
meristem

Radial
growth

Figure 19.24   Radial pattern of shoot growth in plants. Early in development, as shown here in *Arabidopsis*, a single shoot promotes the formation of early leaves on the plant. Later, buds will form from this main shoot that will go on to form branches. The buds that produce the branches and the leaves that form on the branches are also produced in a radial manner.

development of a body plan. Like their animal counterparts, a plant's developmental program relies on the use of transcription factors, determining when and how much gene products are made. In this section, we will consider pattern formation in plants, and study how transcription factors play a key role in plant development.

Plant Development Occurs from Meristems That Are Formed in the Embryo

Before we discuss genes that control plant development, let's examine pattern formation in the early development of the plant embryo. **Figure 19.25** illustrates a common sequence of events that takes place in the embryonic development of seed plants such as *Arabidopsis*. After fertilization, the first cellular division is asymmetrical and produces a smaller apical cell and a larger basal cell (Figure 19.25a). The apical cell will give rise to most of the embryo and later develop into the shoot of the plant. In *Arabidopsis*, the basal cell will give rise to the root, along with a structure called the suspensor, which will become tissue that is required for seed formation (Figure 19.25b).

At the heart stage, which is composed of only about 100 cells, the basic organization of the plant has been established (Figure 19.25c). Plants have organized groups of actively dividing stem cells called **meristems**. As discussed earlier in this chapter, stem cells retain the ability to divide and their daughter cells differentiate into multiple cell types. The meristem produces offshoots of proliferating and differentiating cells. The **root meristem** gives rise only to the root, while the **shoot meristem** produces all aerial parts of the plant, which include the stem as well as lateral structures such as leaves and flowers.

The heart stage then progresses to the formation of a seedling that has two cotyledons, which are embryonic leaves that store nutrients for the developing embryo and seedling. In the seedling shown in Figure 19.25d, you can see three main regions. The **apical region** produces the leaves and flowers of the plant. The **central region** creates the stem. The radial pattern of cells in the central region causes the radial growth observed in plants. Finally, the **basal region** produces the roots. Each of these three regions develops differently, as indicated by their unique cell division patterns and distinct morphologies.

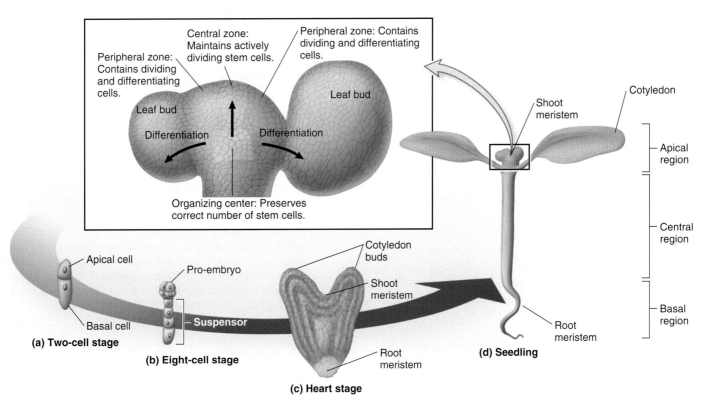

Figure 19.25 **Developmental steps in the formation of a plant embryo.** (a) The two-cell stage consists of the apical cell and basal cell. (b) The eight-cell stage consists of a pro-embryo and a suspensor. The suspensor gives rise to extra embryonic tissue that is needed for seed formation. (c) At the heart stage, all of the plant tissues have begun to form. Note that the shoot meristem is located between the future cotyledons, while the root meristem is on the opposite side. (d) A seedling showing basal, central, and apical regions. The inset shows the organization of the shoot meristem. Note: The steps shown in this figure occur during seed formation, and the embryo would be enclosed within a seed.

Table 19.2	Examples of *Arabidopsis* Apical-Basal-Patterning Genes
Region: *Gene*	**Description**
Apical:	
Gurke	Mutations that cause minor loss of function in the *Gurke* gene produce seedlings with highly reduced or no cotyledons. Complete loss of function eliminates the entire shoot.
Aintegumenta	Encodes a transcription factor that is also expressed in the peripheral zone. Its expression maintains the growth of lateral buds.
Central:	
Scarecrow	Encodes a transcription factor that plays a role in the asymmetric division that produces the radial pattern of growth in the stem. Note: The Scarecrow protein also affects cell division patterns in roots and plays a role in sensing gravity.
Basal:	
Monopterous	Encodes a transcription factor. When this gene is defective, the plant embryo cannot initiate the formation of root structures, but root structures can be formed post-embryonically under the correct growth conditions. This gene seems to be required for organizing root formation in the embryo, but is not required for root formation *per se*.

As you can see in the inset to Figure 19.25d, the shoot meristem is organized into three areas called the organizing center, the central zone, and the peripheral zone. The **organizing center** ensures the proper organization of the meristem and preserves the correct number of actively dividing stem cells. The **central zone** (not to be confused with the central region) is an area where undifferentiated stem cells are always maintained. The **peripheral zone** contains dividing cells that will eventually differentiate into plant structures. For example, the peripheral zone may form a bud that will produce a leaf or flower.

By analyzing mutants that disrupt the developmental process, researchers have discovered that the apical, central, and basal regions of a growing plant express different sets of genes. G. Jürgens and his colleagues began a search to identify a category of genes, known as the **apical-basal-patterning genes**, that are important in early stages of plant development. A few examples are described in **Table 19.2**. Defects in apical-basal-patterning genes cause dramatic effects in one of these three regions. For example, the *gurke* gene is necessary for apical development. When it is defective, the embryo lacks apical structures.

Plant Homeotic Genes Control Flower Development

Although William Bateson coined the term homeotic to describe such mutations in animals, the first known homeotic genes were described in plants. Naturalists in ancient Greece and Rome, for example, recorded their observations of double flowers in which stamens were replaced by petals. In current research, geneticists are studying these types of mutations to better understand devel-

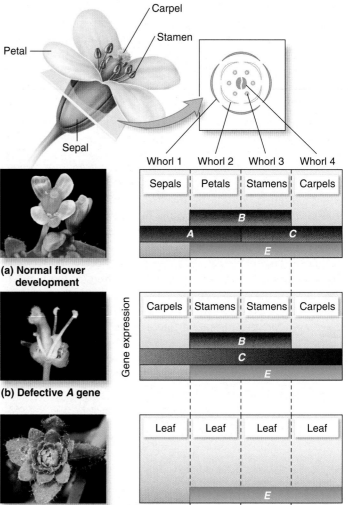

Figure 19.26 Normal and mutant homeotic gene action in *Arabidopsis.* (a) A normal flower composed of four concentric whorls of structures: sepals, petals, stamens, and carpels. To the right is the ABC model of homeotic gene action in *Arabidopsis*. This is a revised model based on the recent identification of *E* genes. (b) A homeotic mutant defective in gene *A* in which the sepals have been transformed into carpels, and the petals have been transformed into stamens. (c) A triple mutant defective in the *A*, *B*, and *C* genes, producing a flower with all leaves.

Biological inquiry: What pattern would you expect if the B gene was expressed in whorls 2, 3, and 4?

opmental pathways in plants. Many homeotic mutations affecting flower development have been identified in *Arabidopsis* and also in the snapdragon (*Antirrhinum majus*).

A normal *Arabidopsis* flower is composed of four concentric whorls of structures (**Figure 19.26a**). The first, outer whorl contains four **sepals**, which protect the flower bud before it opens. The second whorl is composed of four **petals**, and the third whorl contains six stamens. The **stamens** are structures that make the male gametophyte, pollen. Finally, the fourth, innermost whorl contains two carpels that are fused together. The **carpel** produces the female gametophyte.

By analyzing the effects of many different homeotic mutations in *Arabidopsis*, Elliot Meyerowitz and his colleagues in 1994 proposed the ABC model for flower development. In this model, three classes of genes, called *A*, *B*, and *C*, govern the formation of sepals, petals, stamens, and carpels. More recently, a fourth category of genes called the *E* genes was found to also be required for this process. In *Arabidopsis*, Figure 19.26a illustrates how these genes affect normal flower development. In whorl 1, gene *A* product is made. This promotes sepal formation. In whorl 2, *A*, *B*, and *E* gene products are made, which promotes petal formation. In whorl 3, the expression of genes *B*, *C*, and *E* causes stamens to be made. Finally, in whorl 4, the products of *C* and *E* genes promote carpel formation.

Now let's consider what happens in certain homeotic mutants that undergo transformations of particular whorls. According to the original ABC model, genes *A* and *C* repress each other's expression, and gene *B* functions independently. In a mutant defective in gene *A* expression, gene *C* will also be expressed in whorls 1 and 2. This produces a carpel-stamen-stamen-carpel arrangement in which the sepals have been transformed into carpels and the petals into stamens (**Figure 19.26b**). When gene *B* is defective, a flower cannot make petals or stamens. Therefore, a gene *B* defect yields a flower with a sepal-sepal-carpel-carpel

arrangement. When gene *C* is defective, gene *A* is expressed in all four whorls. This results in a sepal-petal-petal-sepal pattern. If the expression of *E* genes is defective, the flower consists entirely of sepals.

Overall, the genes described in Figure 19.26 promote a pattern of development that leads to sepal, petal, stamen, or carpel structures. But what happens if genes *A*, *B*, and *C* are all defective? This produces a flower that is composed entirely of leaves (**Figure 19.26c**). These results indicate that the leaf structure is the default pathway and that the *A*, *B*, *C*, and *E* genes cause development to deviate from a leaf structure in order to make something else. In this regard, the sepals, petals, stamens, and carpels can be viewed as modified leaves. With astonishing insight, Johann Goethe, a German poet, novelist, playwright, and natural philosopher, originally proposed this idea over 200 years ago!

Like the *Drosophila* homeotic genes, plant homeotic genes are part of a hierarchy of gene regulation. Genes that are expressed within the flower bud produce proteins that activate the expression of these homeotic genes. Once they are transcriptionally activated, the homeotic genes then regulate the expression of other genes, the products of which promote the formation of sepals, petals, stamens, or carpels.

CHAPTER SUMMARY

- Development refers to a series of changes in the state of a cell, tissue, organ, or organism.

19.1 General Themes in Development

- A cell that is determined has a particular cell fate. A cell that is differentiated has a specialized morphology and function.

- *Drosophila*, *C. elegans*, the mouse, zebrafish, and *Arabidopsis* are model organisms studied by developmental geneticists. (Figure 19.1)

- The process that gives rise to a plant or animal with a particular body structure is called pattern formation. (Figure 19.2)

- Four responses to positional information are cell division, cell migration, cell differentiation, and apoptosis. (Figures 19.3, 19.4)

- Morphogens and direct contacts between adjacent cells are two ways that cells obtain positional information. (Figure 19.5)

- Transcription factors control the program of development in animals and plants. In animals, this program occurs in four phases. (Figure 19.6)

19.2 Development in Animals

- Embryonic development in *Drosophila* occurs in a series of steps, starting with a zygote, then an embryo, larvae, pupa, and an adult. The basic body plan is established in the embryo. (Figure 19.7)

- Maternal effect genes control the formation of body axes. (Figures 19.6, phase 1, 19.8, 19.9)

- *Drosophila* is divided into many segments. The identification of *Drosophila* mutants has aided in our understanding of segmentation. (Figures 19.10, 19.11, 19.12)

- The sequential expression of three categories of segmentation genes divides the embryo into segments. (Figure 19.13)

- Homeotic genes control the development of structures in a particular segment or group of segments. (Figures 19.14, 19.15, 19.16)

- Invertebrates, such as *Drosophila*, and vertebrates, such as the mouse, both have a homologous set of homeotic genes. In mammals, these are called the *Hox* genes. (Figure 19.17, 19.18)

- Stem cells have the ability to divide and to differentiate. (Figure 19.19)

- In mammals, a fertilized egg is totipotent, certain embryonic and fetal cells are pluripotent, while stem cells in the adult are multipotent or unipotent. (Figures 19.20, 19.21)

- Stem cells have the potential to treat a variety of human disorders. (Table 19.1)

- Certain "master transcription factors" control the cell differentiation process. An example is *MyoD*, which Davis, Weintraub, and Lasser showed experimentally can cause a cell to differentiate into a muscle cell. (Figures 19.22, 19.23)

19.3 Development in Plants

- Plants grow in a radial pattern along a root and shoot axis. (Figure 19.24)

- Plant meristems contain dividing cells that promote the development of plant structures such as roots, stems, leaves, and flowers. (Figure 19.25)

- Several types of genes have been identified in plants that influence pattern development. (Table 19.2)

- Four classes of homeotic genes in plants, *A*, *B*, *C*, and *E*, control flower formation. (Figure 19.26)

TEST YOURSELF

1. The process whereby a cell's morphology and function has changed is called
   a. determination.
   b. cell fate.
   c. differentiation.
   d. genetic engineering.
   e. both a and c.

2. Pattern formation in plants is along the _____ axis.
   a. dorsoventral
   b. anteroposterior
   c. right-left
   d. root-shoot
   e. all of the above

3. Positional information is important in determining the destiny of a cell in a multicellular organism. Cells respond to positional information by
   a. dividing.
   b. migrating.
   c. differentiating.
   d. undergoing apoptosis.
   e. all of the above.

4. Morphogens are
   a. molecules that disrupt normal development.
   b. molecules that convey positional information.
   c. mutagenic agents that cause apoptosis.
   d. receptors that allow cells to adhere to the extracellular matrix.
   e. both a and c.

5. What group of molecules plays a key role in controlling the program of developmental changes?
   a. motor proteins
   b. transporters
   c. transcription factors
   d. restriction endonucleases
   e. cyclins

6. Using the following list of events, determine the proper sequence for the events of animal development:
   1. Formation of tissues, organs, and other body structures in each segment.
   2. Axes of the entire animal are determined.
   3. Cells become differentiated.
   4. The entire animal is divided into segments.

   a. 2, 3, 4, 1
   b. 1, 2, 4, 3
   c. 2, 4, 3, 1
   d. 3, 2, 4, 1
   e. 2, 4, 1, 3

7. The homeotic genes in *Drosophila*
   a. determine the structural and functional characteristics of different segments of the developing fly.
   b. encode motor proteins that transport morphogens throughout the embryo.
   c. are dispersed apparently randomly throughout the genome.
   d. are expressed in similar levels in all parts of the developing embryo.
   e. both a and c.

8. Which of the following genes do not play a role in the process whereby segments are formed in the fruit fly embryo?
   a. homeotic genes
   b. gap genes
   c. pair-rule genes
   d. segment-polarity genes
   e. All of the above play a role in segmentation.

9. A type of stem cell that can give rise to any type of cell of an adult organism but cannot produce an entire, intact individual is called
   a. totipotent.
   b. pluripotent.
   c. multipotent.
   d. unipotent.
   e. antipotent.

10. During plant development, the leaves and the flowers of the plant are derived from
    a. the central region.
    b. the basal region.
    c. the suspensor.
    d. the apical region.
    e. both a and d.

CONCEPTUAL QUESTIONS

1. Define development.

2. Explain the hierarchy of gene expression that controls segmentation in *Drosophila*.

3. What are two characteristics of the proteins that are encoded by homeotic genes that give clues to their function?

EXPERIMENTAL QUESTIONS

1. What was the goal of the research conducted by Davis, Weintraub, and Lasser?

2. How did Davis, Weintraub, and Lasser's research identify the candidate genes for muscle differentiation?

3. Once the researchers identified the candidate genes for muscle differentiation, how did they test the effect of each gene on cell differentiation? What were the results of the study?

COLLABORATIVE QUESTIONS

1. Discuss four types of cellular responses that cells can exhibit in response to positional information.

2. Discuss meristems in plants.

---

www.brookerbiology.com
This website includes answers to the Biological Inquiry questions found in the figure legends and all end-of-chapter questions.

20
GENETIC TECHNOLOGY

Copycat, the first cloned pet. In 2002, the cat shown here called Copycat or CC (for carbon copy) was produced by cloning, a procedure described later in this chapter.

The past several decades have witnessed amazing technical advances in the ability of biologists to study life at the molecular level. **Recombinant DNA technology** is the use of laboratory techniques to isolate and manipulate fragments of DNA. Such technology produces **recombinant DNA**, which is any DNA molecule that has been manipulated so that it contains DNA from two or more sources. In the early 1970s, the first successes in making recombinant DNA molecules were accomplished independently by two groups at Stanford University: David Jackson, Robert Symons, and Paul Berg; and Peter Lobban and A. Dale Kaiser. Both groups were able to isolate and purify pieces of DNA in a test tube, and then covalently link together two or more DNA fragments. Shortly thereafter, these recombinant DNA molecules were introduced into living cells. Once inside a host cell, the recombinant molecules are replicated to produce many identical copies, a process known as cloning. When copies contain a gene of interest, the technique is called **gene cloning**. Recombinant DNA technology and cloning have enabled geneticists to probe relationships between gene sequences and phenotypic consequences and have been fundamental to our understanding of gene structure and function.

Later in this chapter we will consider the topic of genomics, which involves the molecular analysis of the entire genome of a species. In recent years, molecular techniques have progressed to the point where researchers can study the structure and function of many genes as large integrated units. For example, the expression of all genes in a genome can be analyzed under different conditions, such as in normal versus cancer cells. This information can help us to understand how changes in gene expression can cause uncontrolled cell growth.

In the last section of this chapter, we will explore the topic of **biotechnology**, which is the use of living organisms or the products of living organisms for human benefit. We will learn that **genetic engineering**, the direct manipulation of genes for practical purposes, is playing an ever-increasing role in the creation of strains of microorganisms, plants, and animals that have characteristics that are useful to people. These include bacteria that make human insulin, plants that are resistant to herbicides, and farm animals that make human medicines.

20.1 Gene Cloning

Molecular biologists want to understand how the molecules within living cells contribute to cell structure and function. Because proteins are the workhorses of cells, many researchers focus their attention on the structure and function of proteins, or the genes that encode them. In any given laboratory, it is common to study just one or perhaps a few different genes or proteins. At the molecular level, this poses a daunting task. Any given cell can express thousands of different proteins, making the study of any single gene or protein akin to finding a needle in a haystack. To overcome this truly formidable obstacle, researchers frequently take the approach of cloning the genes that encode their proteins of interest.

As mentioned, the term gene cloning refers to procedures that lead to the formation of many copies of a particular gene. **Figure 20.1** provides an overview of the steps and goals of gene cloning. The process is usually done with one of two goals in mind.

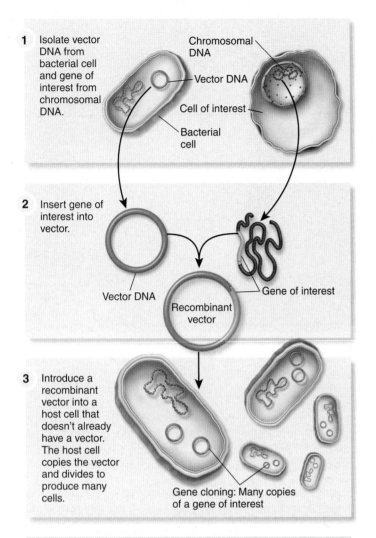

1 Isolate vector DNA from bacterial cell and gene of interest from chromosomal DNA.

Chromosomal DNA

Vector DNA

Cell of interest

Bacterial cell

2 Insert gene of interest into vector.

Vector DNA

Recombinant vector

Gene of interest

3 Introduce a recombinant vector into a host cell that doesn't already have a vector. The host cell copies the vector and divides to produce many cells.

Gene cloning: Many copies of a gene of interest

4 Gene cloning is done to achieve one of two main goals:

Producing large amounts of DNA of a specific gene	Expressing the cloned gene to produce the encoded protein
Examples	*Examples*
• Cloned genes provide enough DNA for DNA sequencing. The sequence of a gene can help us understand how a gene works and identify mutations that cause diseases.	• Large amounts of the protein can be purified to study its structure and function.
• Cloned DNA can be used as a probe to identify the same gene or similar genes in other organisms.	• Cloned genes can be introduced into bacteria or livestock to make pharmaceutical products such as insulin.
	• Cloned genes can be introduced into plants and animals to alter their traits.
	• Cloned genes can be used to treat diseases—a clinical approach called gene therapy.

Figure 20.1 **Gene cloning.** The process of gene cloning produces copies of a gene or copies of its protein product, which have many uses.

One goal would be that a researcher or clinician wants many copies of the gene, perhaps to study the DNA directly or to use the DNA as a tool. Alternatively, the goal may be to obtain lots of the gene product—mRNA or protein. For example, biochemists use gene cloning to obtain large amounts of proteins to study their structure and function. In modern molecular biology, the many uses for gene cloning are remarkable. Gene cloning has provided the foundation for critical technical advances in a variety of disciplines including molecular biology, genetics, cell biology, biochemistry, and medicine. In this section, we will examine the procedures that are used to copy genes.

Step 1: Vector DNA and Chromosomal DNA Are the Starting Materials to Clone a Gene

In the first step of gene cloning, a key material is a type of DNA known as a **vector** (Figure 20.1). The role of vector DNA is to act as a carrier of the DNA segment that is to be cloned. (The term vector comes from a Latin term meaning "carrier.") In cloning experiments, a vector may carry a small segment of chromosomal DNA, perhaps only a single gene. By comparison, a chromosome carries many more genes, perhaps a few thousand. When a vector is introduced into a living cell, it can replicate, and so the DNA that it carries is also replicated. This produces many identical copies of the inserted gene.

The vectors commonly used in gene cloning experiments were derived originally from two natural sources, plasmid or viral. Some vectors are **plasmids**, which are small circular pieces of DNA. As discussed in Chapter 18, plasmids are found naturally in many strains of bacteria and also occasionally in eukaryotic cells. Commercially available plasmids have been genetically engineered for effective use in cloning experiments. They contain unique sites into which geneticists can easily insert pieces of DNA. An alternative type of vector used in cloning experiments is a **viral vector**. Viruses can infect living cells and propagate themselves by taking control of the host cell's metabolic machinery. When a chromosomal gene is inserted into a viral vector, the gene will be replicated whenever the viral DNA is replicated. Therefore, viruses can be used as vectors to carry other pieces of DNA.

Another material that is needed to clone a gene is the gene itself, which we will call the "gene of interest." If a scientist wants to clone a particular gene, the source of the gene is the chromosomal DNA that carries the gene. To prepare chromosomal DNA, an experimenter first obtains some cells from the organism of interest. The preparation of chromosomal DNA involves breaking open cells and extracting and purifying the DNA using biochemical techniques such as chromatography and centrifugation.

Step 2: Cutting Chromosomal and Vector DNA into Pieces and Linking Them Together Produces Recombinant Vectors

The second step in a gene cloning experiment is the insertion of chromosomal DNA into a plasmid or viral vector (Figure 20.1).

Table 20.1 — Examples of Restriction Enzymes Used in Gene Cloning

Restriction enzyme*	Bacterial source	Sequence recognized†
*Bam*HI	*Bacillus amyloliquefaciens* H	5′–GGATCC–3′ 3′–CCTAGG–5′
*Eco*RI	*Escherichia coli* RY13	5′–GAATTC–3′ 3′–CTTAAG–5′
*Sac*I	*Streptomyces achromonogenes*	5′–GAGCTC–3′ 3′–CTCGAG–5′

* Restriction enzymes are named according to the species in which they are found. The first three letters are italicized because they indicate the genus and species names. Since a species may produce more than one restriction enzyme, the enzymes are designated I, II, III, etc., to indicate the order in which they were discovered in a given species.

† The arrows show the locations in the upper and lower DNA strands where the restriction enzymes cleave the DNA backbone.

DNA fragments must be cut and pasted to produce recombinant vectors. To cut DNA, researchers use enzymes known as **restriction enzymes** or restriction endonucleases. These enzymes, which were discovered by Werner Arber, Hamilton Smith, and Daniel Nathans in the 1960s and 1970s, are made naturally by many different species of bacteria. Restriction enzymes protect bacterial cells from invasion by foreign DNA, particularly that of bacteriophages. Currently, several hundred different restriction enzymes from various bacterial species have been identified and are available commercially to molecular biologists.

The restriction enzymes used in cloning experiments bind to a specific base sequence and then cleave the DNA backbone at two defined locations, one in each strand. The sequences recognized by restriction enzymes are called **restriction sites**. Most restriction enzymes recognize sequences that are palindromic, which means that the sequence is identical when read in the opposite direction in the complementary strand (**Table 20.1**). For example, the sequence recognized by the restriction enzyme *Eco*RI is 5′–GAATTC–3′ in the top strand. Read in the opposite direction in the bottom strand, this sequence is also 5′–GAATTC–3′.

Figure 20.2 shows the action of a restriction enzyme and the insertion of a gene into a vector. This vector also carries the *amp*^R and *lac*Z genes, whose functions will be discussed later. Certain restriction enzymes are useful in cloning because they digest DNA into fragments with **sticky ends**. This means that these DNA fragments have single-stranded ends that will hydrogen-bond to each other due to their complementary sequences.

1. Cut vector and chromosomal DNA with *Eco*RI, a restriction enzyme that recognizes the sequence **GAATTC** and cuts at the arrows. **CTTAAG**

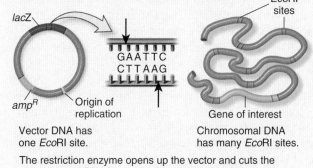

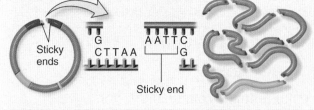

2. Allow sticky ends to hydrogen-bond with each other due to complementary sequences.

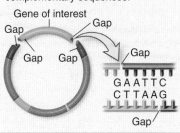

In this example, a fragment of DNA carrying the gene of interest has hydrogen-bonded to the vector. Four gaps are found where covalent bonds in the DNA backbone are missing.

3. Add DNA ligase to close the gaps by catalyzing the formation of covalent bonds in the DNA backbone.

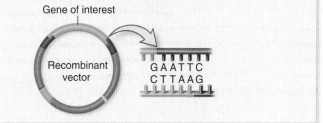

Figure 20.2 Step 2 of gene cloning: The actions of a restriction enzyme and DNA ligase to produce a recombinant vector. The restriction enzyme binds to a specific sequence in both the chromosomal and vector DNA. It then cleaves the DNA backbones, producing DNA fragments. The complementary single-stranded ends of the DNA fragments hydrogen-bond with each other. The enzyme DNA ligase then covalently links these fragments, producing a recombinant vector.

Biological inquiry: In the experiment shown in this figure, has the gene of interest been cloned?

The hydrogen bonding between the sticky ends of DNA fragments promotes temporary interactions between two DNA fragments. However, this interaction is not stable, because it involves only a few hydrogen bonds between complementary bases. To establish a permanent connection between two DNA fragments, the sugar-phosphate backbones within the DNA strands must be covalently linked together. This linkage is catalyzed by DNA ligase (Figure 20.2, step 3).

In some cases, the two ends of the vector will simply ligate back together, restoring it to its original circular structure; this forms what is called a recircularized vector. In other cases, a fragment of chromosomal DNA may become ligated to both ends of the vector. When this happens, a segment of chromosomal DNA has been inserted into the vector. The result is a vector containing a piece of chromosomal DNA, which is called a **recombinant vector** or a hybrid vector. We then have a recombinant vector that is ready to be cloned. A recombinant vector may contain the gene of interest or it may contain a different piece of chromosomal DNA.

Step 3: Putting Recombinant Vectors into Host Cells and Allowing Those Cells to Propagate Achieves Gene Cloning

The procedure we have considered in Figure 20.2 seeks to insert a chromosomal gene into a plasmid vector. The vector already carries an antibiotic-resistance gene called the *amp^R* gene. Such a gene is called a **selectable marker** because the presence of the antibiotic selects for the growth of cells expressing the *amp^R* gene. The *amp^R* gene encodes an enzyme known as β-lactamase that degrades the antibiotic ampicillin, which normally kills bacteria. Bacteria containing the *amp^R* gene can grow on media containing ampicillin because they can degrade it. In a cloning experiment where the *amp^R* gene is found within the plasmid, the growth of cells in the presence of ampicillin identifies bacteria that contain the plasmid.

The third step in gene cloning is the actual cloning of the gene of interest. In this step, the goal is for the recombinant vector carrying the desired gene to be taken up by bacterial cells treated with agents that render them permeable to DNA molecules (**Figure 20.3**). This procedure is called transformation, when a plasmid vector is used, or transfection, when a viral vector is introduced into a cell. During transformation, some cells take up a single plasmid while most cells fail to take up a plasmid. The bacteria are then streaked on petri plates containing a bacterial growth medium and ampicillin. In the experiment shown here, the bacterial cells were originally sensitive to ampicillin. Bacteria that have not taken up a plasmid will be killed by the antibiotic. In contrast, any bacterium that has taken up a plasmid carrying the *amp^R* gene will grow and divide many times to form a bacterial colony containing tens of millions of cells. Because each cell in a single colony is derived from the same original cell that took up a single plasmid, all cells within a colony contain the same type of plasmid DNA.

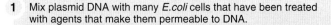

1 Mix plasmid DNA with many *E. coli* cells that have been treated with agents that make them permeable to DNA.

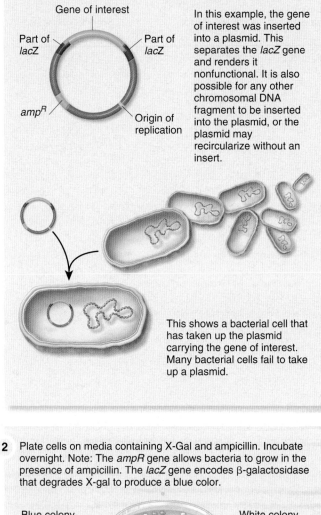

In this example, the gene of interest was inserted into a plasmid. This separates the *lacZ* gene and renders it nonfunctional. It is also possible for any other chromosomal DNA fragment to be inserted into the plasmid, or the plasmid may recircularize without an insert.

This shows a bacterial cell that has taken up the plasmid carrying the gene of interest. Many bacterial cells fail to take up a plasmid.

2 Plate cells on media containing X-Gal and ampicillin. Incubate overnight. Note: The *ampR* gene allows bacteria to grow in the presence of ampicillin. The *lacZ* gene encodes β-galactosidase that degrades X-gal to produce a blue color.

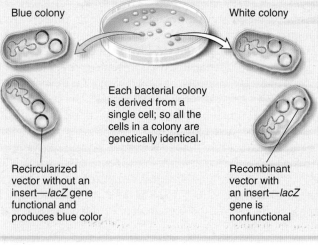

Blue colony

White colony

Each bacterial colony is derived from a single cell; so all the cells in a colony are genetically identical.

Recircularized vector without an insert—*lacZ* gene functional and produces blue color

Recombinant vector with an insert—*lacZ* gene is nonfunctional

Figure 20.3 Step 3 of gene cloning: The cloning of a recombinant vector. For cloning to occur, a recombinant vector is introduced into a host cell, which copies the vector and divides to produce many cells. This produces many copies of the gene of interest.

In the experiment shown in Figure 20.3, the experimenter can distinguish bacterial colonies that contain a recombinant vector from those containing a recircularized vector—and therefore no chromosomal DNA. In a recombinant vector, a piece of chromosomal DNA carrying a desired gene has been inserted into a region of the vector that contains the *lacZ* gene, which encodes the enzyme β-galactosidase. The insertion of chromosomal DNA into the vector disrupts the *lacZ* gene. By comparison, a recircularized vector has a functional *lacZ* gene. The functionality of *lacZ* can be determined by providing the growth medium with a colorless compound, X-Gal, which is cleaved by β-galactosidase into a blue dye. Bacteria grown in the presence of X-Gal will form blue colonies if they have a functional β-galactosidase enzyme, and white colonies if they do not. In this experiment, therefore, bacterial colonies containing recircularized vectors will form blue colonies, while colonies containing recombinant vectors will be white.

The net result of gene cloning is an enormous number of gene copies within a huge number of recombinant vectors. During transformation, a single bacterial cell usually takes up a single recombinant vector. Two subsequent events lead to the production of many copies of the cloned gene. First, because the vector has a highly active origin of replication, the bacterial host cell produces many copies of the recombinant vector per cell. Second, the bacterial cells divide approximately every 20 minutes. Following overnight growth, a population of many millions of bacteria will be obtained from a single cell. Each of these bacterial cells will contain many copies of the cloned gene. For example, a bacterial colony may comprise 10 million cells, with each cell containing 50 copies of the recombinant vector. Therefore, this bacterial colony would have 500 million copies of the cloned gene!

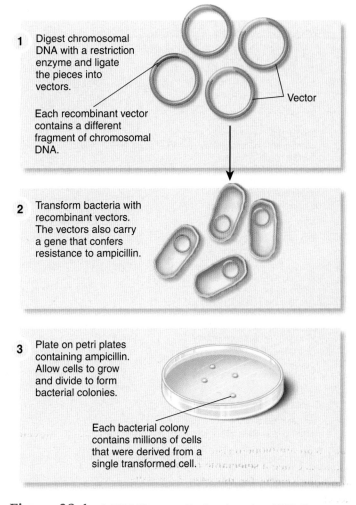

1   Digest chromosomal DNA with a restriction enzyme and ligate the pieces into vectors.

Each recombinant vector contains a different fragment of chromosomal DNA.

Vector

2   Transform bacteria with recombinant vectors. The vectors also carry a gene that confers resistance to ampicillin.

3   Plate on petri plates containing ampicillin. Allow cells to grow and divide to form bacterial colonies.

Each bacterial colony contains millions of cells that were derived from a single transformed cell.

Figure 20.4   **A DNA library.**   Each colony in a DNA library contains a vector with a different piece of chromosomal DNA.

A Cloned Gene of Interest Can Be Identified in a DNA Library with the Use of Colony Hybridization

In our previous discussion of gene cloning, we focused on the insertion of a particular gene into a vector, and its introduction into a host cell. But how do we know that a vector contains a gene of interest? In a typical cloning experiment, such as the one described in Figures 20.2 and 20.3, the treatment of chromosomal DNA with restriction enzymes actually yields tens of thousands of different DNA fragments. Therefore, after the DNA fragments are ligated individually to vectors, a researcher has a collection of many recombinant vectors, with each vector containing a particular fragment of chromosomal DNA. This collection of vectors is known as a **DNA library** (**Figure 20.4**). Researchers make DNA libraries using the methods shown in Figures 20.2 and 20.3 and then use those libraries to obtain clones of genes in which they are interested.

Two types of DNA libraries are commonly made. The library is called a **genomic library** when the inserts are derived from chromosomal DNA. Alternatively, researchers can isolate mRNA

and use the enzyme reverse transcriptase, which is described in Chapter 18, to make DNA molecules using mRNA as a starting material. Such DNA is called **complementary DNA** or **cDNA**. A **cDNA library** is a collection of recombinant vectors that have cDNA inserts. From a research perspective, an important advantage of cDNA is that it lacks introns. Because introns can be quite large, it is much simpler for researchers to insert cDNAs into vectors rather than work from chromosomal DNA if they want to focus their attention on the coding sequence of a gene.

In many cloning experiments, the ultimate goal is to clone a specific gene. After making a DNA library, how is it possible to identify those rare colonies that happen to contain the gene of interest? One method that can be used is **colony hybridization**, in which a researcher uses a probe to identify colonies that contain a desired gene. The master plate shown at the top of **Figure 20.5** has many bacterial colonies, each composed of bacterial cells containing a recombinant vector with a different piece of chromosomal DNA. In this example, the goal is to identify a colony that contains the gene of interest.

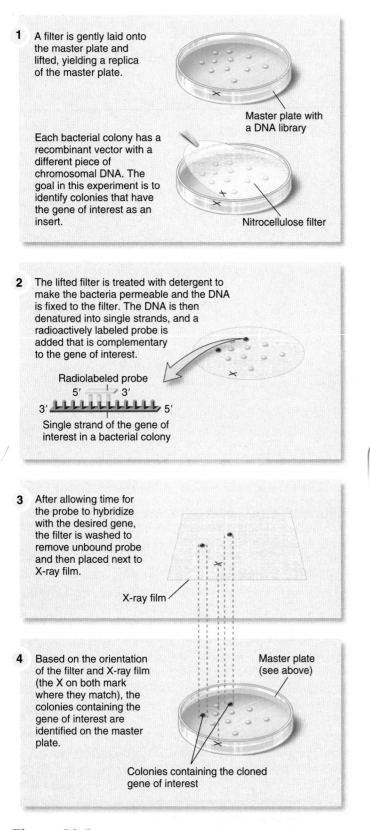

1 A filter is gently laid onto the master plate and lifted, yielding a replica of the master plate.

Master plate with a DNA library

Each bacterial colony has a recombinant vector with a different piece of chromosomal DNA. The goal in this experiment is to identify colonies that have the gene of interest as an insert.

Nitrocellulose filter

2 The lifted filter is treated with detergent to make the bacteria permeable and the DNA is fixed to the filter. The DNA is then denatured into single strands, and a radioactively labeled probe is added that is complementary to the gene of interest.

Radiolabeled probe
5' 3'
3' 5'
Single strand of the gene of interest in a bacterial colony

3 After allowing time for the probe to hybridize with the desired gene, the filter is washed to remove unbound probe and then placed next to X-ray film.

X-ray film

4 Based on the orientation of the filter and X-ray film (the X on both mark where they match), the colonies containing the gene of interest are identified on the master plate.

Master plate (see above)

Colonies containing the cloned gene of interest

Figure 20.5 **Identifying a cloned gene by colony hybridization.** This method can be used to identify colonies containing recombinant vectors produced by a gene cloning experiment.

A nitrocellulose filter is laid gently onto the master plate containing many bacterial colonies. When the filter is lifted, some cells from each colony are attached to it. In this way, the filter paper contains a replica of the colonies on the master plate.

The cells attached to the filter are exposed to a detergent, which makes them permeable, and the DNA inside the cells is fixed to the filter. The DNA is denatured into single strands and the filter is then submerged in a solution containing a radiolabeled DNA probe. In this case, the probe is single-stranded DNA that is complementary to the gene of interest. A probe that is 15 to 20 nucleotides long is usually of sufficient length to specifically bind to a gene. The probe is given time to hybridize to the DNA on the filter. If a bacterial colony contains the desired gene, the probe will hybridize to the DNA in this colony. Most bacterial colonies are not expected to contain this gene. The unbound probe is then washed away, and the filter is placed next to X-ray film. If the DNA within a bacterial colony did hybridize to the probe, a dark spot will appear on the developed film in the corresponding location. Therefore, a dark spot identifies a bacterial colony that contains the desired gene. Following the identification of labeled colonies, the researcher can go back to the master plate to select and grow bacteria containing the cloned gene. The growth of cells will produce many copies of the cloned gene.

GENOMES & PROTEOMES

Blotting Methods Can Identify Gene Families

As we have just seen, blotting methods such as colony hybridization can allow us to detect a gene within a DNA library. Another blotting method, **Southern blotting**, can detect the presence of a particular gene within a mixture of many chromosomal DNA fragments separated on a gel. This method, developed by Edwin Southern in 1975, can be used to determine the number of copies of a gene within the genome of an organism. For example, Southern blotting has revealed that rRNA genes are found in multiple copies, whereas structural genes are usually unique. In the study of human genetic diseases, Southern blotting can also detect small gene deletions. Another common use of Southern blotting is to identify gene families, in which two or more genes are derived from the same ancestral gene. The members of a gene family are homologous, having similar but not identical DNA sequences. Southern blotting can distinguish the homologous members of a gene family within a single species or identify homologous genes among different species.

Before we discuss the Southern blotting method, we first need to examine **gel electrophoresis**, a technique that is used to separate macromolecules on a gel. In the example shown in **Figure 20.6**, gel electrophoresis was used to separate different fragments of chromosomal DNA based on their masses. It consists of a flat, semisolid gel called a slab gel with depressions at

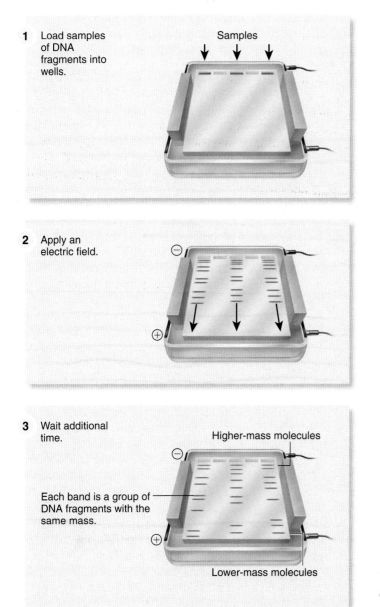

1 Load samples of DNA fragments into wells.

Samples

2 Apply an electric field.

3 Wait additional time.

Higher-mass molecules

Each band is a group of DNA fragments with the same mass.

Lower-mass molecules

Figure 20.6 **Separation of molecules by gel electrophoresis.** In this example, samples containing many fragments of DNA are loaded into wells at the top of the gel and then subjected to an electric field that causes the fragments to move toward the bottom of the gel. This separates the fragments according to their masses, with the smaller DNA fragments near the bottom of the gel.

Biological inquiry: One DNA fragment contains 600 bp and another has 1,300 bp. Following electrophoresis, which would be closer to the bottom of a gel?

the top called wells where samples are added. An electric field is applied to the gel, causing charged molecules to migrate from the top of the gel toward the bottom—a process called electrophoresis. DNA is negatively charged and moves toward the positive end of the gel, which is at the bottom in this figure.

Smaller DNA fragments move more quickly through the gel polymer and therefore are located closer to the bottom of the gel compared to larger ones. After the slab gel has run, the fragments are separated into bands within the gel. The fragments in each band can then be stained with a dye for identification. Alternatively, in a Southern blotting experiment, specific bands in a gel are identified with a probe, as described next.

To conduct a Southern blotting experiment, a strand of DNA from a gene of interest is labeled (for example, radiolabeled) *in vitro*, and then the labeled strand is used as a probe to detect the presence of the gene within a mixture of many DNA fragments obtained from chromosomal DNA. The basis for a Southern blotting experiment, as with other hybridization experiments, is that two DNA fragments will bind to each other only if they have complementary sequences. In such an experiment, a labeled strand of the probe will pair specifically with a complementary DNA strand from a chromosomal fragment.

To begin a Southern blotting experiment, the chromosomal DNA is isolated and digested with a restriction enzyme (**Figure 20.7**). Because the restriction enzyme cuts the chromosomal DNA at many different sites within the chromosomes, this step produces thousands of DNA pieces of different sizes. The chromosomal pieces are loaded onto a gel that separates them according to their masses. The DNA pieces within the gel are denatured into single-stranded molecules and then transferred from the gel to a nitrocellulose filter, a process called blotting. At this point, the filter contains many unlabeled DNA strands that have been separated according to mass.

The next step is to determine if any of these unlabeled strands contain sequences complementary to a probe. The filter, which has the unlabeled chromosomal DNA attached to it, is submerged in a solution containing the radiolabeled probe. If the labeled probe and a strand of chromosomal DNA are complementary, they will hydrogen bond to each other. Any unbound radiolabeled probe is then washed away, and the filter is exposed to X-ray film. Locations where the labeled probe has bound appear as dark bands on the developed X-ray film.

Important variables in the Southern blotting procedure are temperature and the ionic strength of the hybridization and wash steps. If these steps are done at very high temperatures and/or at low salt concentrations, the probe and chromosomal DNA must be very complementary—nearly a perfect match—to hybridize. This condition is called high stringency and is used to detect a close match between the probe and a chromosomal DNA fragment. If the temperature is lower and/or the ionic strength is higher, DNA sequences that are similar but not necessarily identical may hybridize to the probe. This is called low stringency and can be used to detect homologous genes. In the results shown in Figure 20.7, conditions of high stringency reveal that the probe, which is a strand of DNA from a gene of interest, binds to a single chromosomal DNA fragment, indicating that the gene is found in a single copy in the genome. At low stringency, however, two other bands are detected. These results suggest that this gene is a member of a gene family composed of three distinct members.

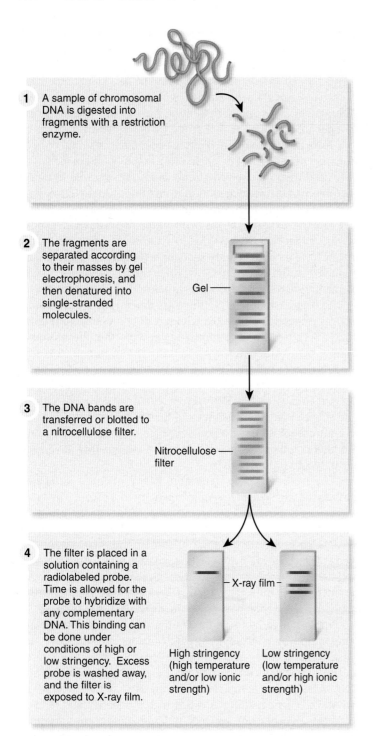

1. A sample of chromosomal DNA is digested into fragments with a restriction enzyme.

2. The fragments are separated according to their masses by gel electrophoresis, and then denatured into single-stranded molecules.

Gel

3. The DNA bands are transferred or blotted to a nitrocellulose filter.

Nitrocellulose filter

4. The filter is placed in a solution containing a radiolabeled probe. Time is allowed for the probe to hybridize with any complementary DNA. This binding can be done under conditions of high or low stringency. Excess probe is washed away, and the filter is exposed to X-ray film.

X-ray film

High stringency (high temperature and/or low ionic strength)

Low stringency (low temperature and/or high ionic strength)

Figure 20.7 **The technique of Southern blotting.** This technique is used to identify genes that are complementary to a labeled DNA probe from a known gene. Under conditions of low stringency, the probe may bind to genes that are similar but not identical to the probe, thereby identifying homologous genes that constitute a gene family.

Biological inquiry: The globin gene family is described in Chapter 21 (look ahead to Figure 21.8). Let's suppose you started with human chromosomal DNA and a probe that is complementary to the β-globin gene. How many bands would you see under conditions of high stringency? How many under low stringency?

Polymerase Chain Reaction (PCR) Can Also Be Used to Make Many Copies of DNA

Thus far we have examined cloning methods in which the gene of interest was inserted into a vector and then introduced into a host cell. Another way to copy DNA without the aid of vectors and host cells is a technique called **polymerase chain reaction (PCR)** that was developed by Kary Mullis in 1985 (**Figure 20.8**). The goal of PCR is to make many copies of DNA in a defined region, perhaps encompassing a gene or part of a gene. Several reagents are required for synthesis of the DNA. These include a high concentration of two primers that are complementary to sequences at the ends of the DNA region to be amplified, deoxynucleoside triphosphates (dNTPs), and a heat-stable form of DNA polymerase called *Taq* **polymerase** (isolated from the bacterium *Thermus aquaticus*, which lives in hot springs). A heat-stable form of DNA polymerase is necessary because PCR involves heating steps that would inactivate most other natural forms of DNA polymerase.

To make copies, a sample of chromosomal DNA, called the double-stranded template DNA, is denatured into single-stranded molecules by heat treatment and then the primers bind to the DNA as the temperature is lowered. The binding of the primers to the specific sites in the template DNA is called annealing. Once the primers have annealed, *Taq* polymerase will catalyze the synthesis of complementary DNA strands, thereby doubling the amount of DNA in the region that is flanked by the primers. The sequential process of denaturation—annealing—synthesis is then repeated to double the amount of template DNA many times over. This method is called a chain reaction because the products of each previous step are used as reactants in subsequent steps.

A thermal reactor that automates the timing of each cycle, known as a thermocycler, is used to carry out PCR. The experimenter mixes the DNA sample, an excess amount of primers, *Taq* polymerase, and deoxynucleotides together in a single tube. Then the experimenter places the tube in a thermocycler, and sets the machine to operate within a defined temperature range and number of cycles. During every cycle, the thermocycler increases the temperature to denature the DNA strands and then lowers the temperature to allow annealing and DNA synthesis to occur. After a few minutes, the cycle is repeated by increasing and then lowering the temperature. A typical PCR run is likely to involve 20 to 30 cycles of replication and takes a few hours to complete. The PCR technique can amplify the amount of DNA by a staggering amount. After 30 cycles of amplification, a DNA sample will increase 2^{30}-fold, which is approximately a billionfold!

The PCR reaction shown in Figure 20.8 seeks to amplify a particular DNA segment. The sequences of the PCR primers are complementary to two specific sequences within the DNA. Therefore, the two primers bind to these sites and the intervening region is replicated. To conduct this type of PCR experiment, the researcher must have prior knowledge about the sequence of the DNA in order to design primers that are complementary to the ends of the DNA sequence. When specific primers can be constructed, PCR can amplify a specific region of DNA from a

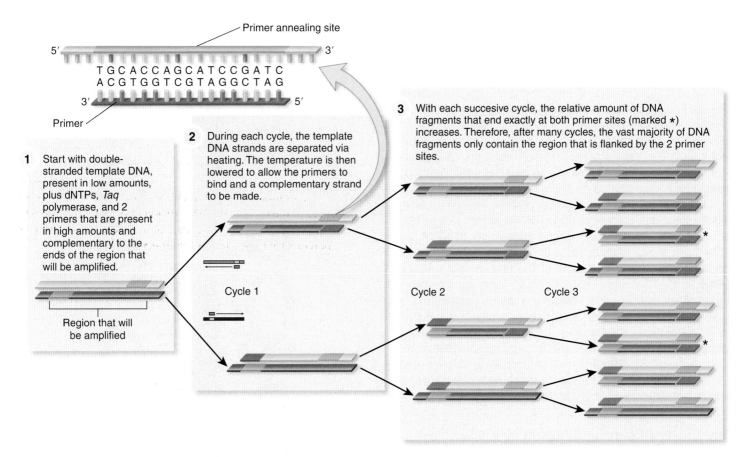

Figure 20.8   Polymerase chain reaction (PCR).   During each PCR cycle, short DNA sequences that are complementary to the ends of the targeted DNA sequence bind to the DNA and act as primers for the synthesis of this DNA region (see small illustrations next to arrows in step 2). The primers used in actual PCR experiments are usually 15 to 20 nucleotides in length. The region between the two primers is typically hundreds of nucleotides in length, not just several nucleotides as shown here. The net result of PCR is the synthesis of many copies of DNA in the region that is flanked by the two primers.

complex mixture of DNA or from a small amount of DNA. For example, PCR is routinely used in crime investigations to amplify DNA from small samples of cells that may have been left behind at a crime scene.

20.2 Genomics and Proteomics

As discussed throughout Unit III, the genome is the total genetic composition of an organism. As genetic technology has progressed over the past few decades, researchers have gained an increasing ability to analyze the composition of genomes as a whole unit. The term **genomics** refers to the molecular analysis of the entire genome of a species. The two main types of genomic research are structural and functional. **Structural genomics** is aimed at the direct analysis of the DNA itself. Segments of chromosomes are cloned and analyzed in progressively smaller pieces, the locations of which are known on the intact chromosomes. This is the mapping phase of structural genomics. The mapping of a genome ultimately progresses to the determination of the complete DNA sequence, which provides the most detailed description available of an organism's genome at the

molecular level. By comparison, **functional genomics** is aimed at studying the expression of a genome. For example, functional genomics can be used to analyze which genes are turned on or off in particular cell types, such as nerve and muscle cells. In this section, we will consider a few of the methods that are used in structural and functional genomics.

A long-term goal of researchers is to determine the roles of all cellular proteins, as well as how those proteins interact to produce the characteristics of particular cell types, and the traits of complete organisms. As discussed at the end of this section, **proteomics** involves techniques that are used to identify and study groups of proteins.

BAC Cloning Vectors Are Used to Make Contigs of Chromosomes to Map a Genome

A goal of structural genomics is to clone and analyze the entire genome of a species. For large eukaryotic genomes, cloning an entire genome is much easier when a cloning vector can accept very large chromosomal DNA inserts. In general, most plasmid and viral vectors can accommodate inserts only a few thousand to perhaps tens of thousands of nucleotides in length.

If a plasmid or viral vector has a DNA insert that is too large, it will have difficulty with DNA replication and is likely to suffer deletions in the insert. By comparison, a type of cloning vector known as a **bacterial artificial chromosome (BAC)** can reliably contain much larger inserted DNA fragments. BACs are derived from large plasmids called F factors (see Chapter 18). They can typically contain inserts up to 500,000 bp. BACs are used in genomic research in the same way as other types of vectors. Similarly, YACs, yeast artificial chromosomes, are used as vectors in yeast. An insert in a YAC can be several hundred thousand to perhaps 2 million nucleotides in length.

The term **mapping** refers to the process of determining the relative locations of genes or other DNA segments along a chromosome. After many large fragments of chromosomal DNA have been inserted into BACs or YACs, the first step of mapping is to determine the relative locations of the inserted chromosomal pieces as they would occur in an intact chromosome. This is called physical mapping. To obtain a complete physical map of a chromosome, researchers need a series of clones that contain overlapping pieces of chromosomal DNA. Such a collection of clones, known as a **contig**, contains a contiguous region of a chromosome that is found as overlapping regions within a group of vectors (**Figure 20.9**). These overlapping regions allow researchers to identify their order along the chromosome.

The Dideoxy Chain-Termination Method Is Used to Determine the Base Sequence of DNA

Once researchers have cloned DNA into vectors and obtained a physical map, the next phase of genomic research is **DNA sequencing**, which is a method to determine the base sequence of DNA. Scientists can learn a great deal of information about the function of a gene if its nucleotide sequence is known. For example, the investigation of genetic sequences has been vital in our understanding of the genetic basis of human diseases.

During the 1970s, two methods for DNA sequencing were devised. One method, developed by Alan Maxam and Walter Gilbert, involves the base-specific cleavage of DNA. Another method, developed by Frederick Sanger and colleagues, is known as the **dideoxy chain-termination method** or more simply, **dideoxy sequencing**. Because it has become the more popular method of DNA sequencing, we will consider the dideoxy method here.

The dideoxy procedure of DNA sequencing is based on our knowledge of DNA replication. As described in Chapter 11, DNA polymerase connects adjacent deoxynucleotides by catalyzing a covalent linkage between the 5′–phosphate on one nucleotide and the 3′—OH group on the previous nucleotide. Chemists, however, can synthesize nucleotides, called dideoxynucleoside triphosphates (ddNTPs), that are missing the —OH group at the 3′ position (**Figure 20.10,** step 1). Note: The prefix "dideoxy-" refers to the fact that there are two (di) removed (de) oxygens (oxy) compared with ribose, which has —OH groups at both the 2′ and 3′ positions. Sanger reasoned that if a dideoxynucleotide is added to a growing DNA strand, the strand can no longer grow because the 3′—OH group is missing, which results in chain termination.

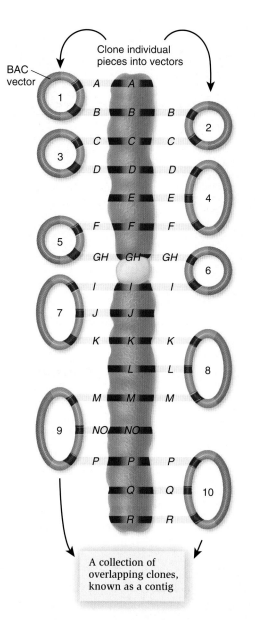

Figure 20.9 A contig. As shown here, a contig is a collection of clones that have overlapping pieces of DNA from a particular chromosome. The numbers denote the order of the members of the contig. The chromosome is labeled with letters that denote the locations of particular genes. The members of the contig have overlapping regions that have the same genes, which allows you to order them.

Biological inquiry: What does it mean when we say that two members of a contig have overlapping regions?

Before describing the steps of this DNA sequencing protocol, we need to become acquainted with the DNA segments that are used in a sequencing experiment. The segment of DNA to be sequenced must be obtained in large amounts by using the gene cloning or PCR techniques that were described earlier in this chapter. In Figure 20.10, the segment of DNA to be sequenced was inserted into a vector next to a primer-annealing site, the site where the primer will bind.

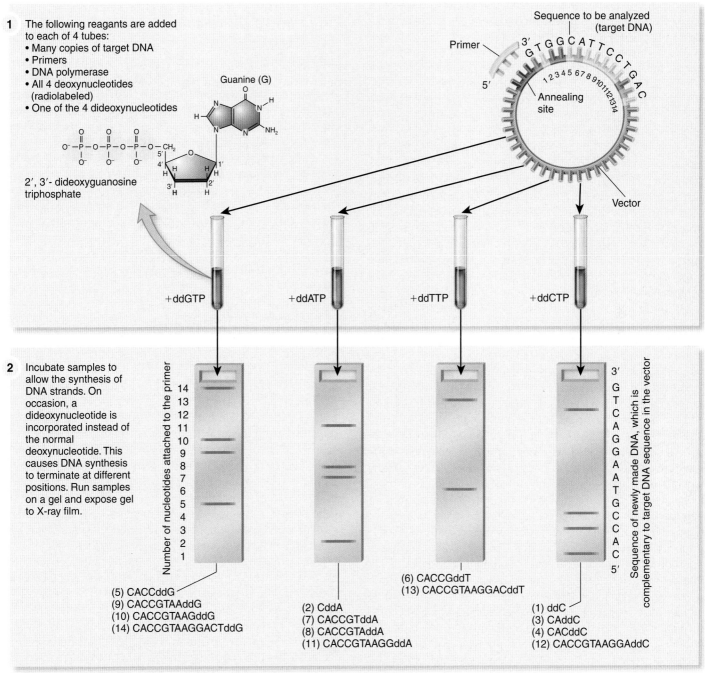

1 The following reagants are added to each of 4 tubes:
- Many copies of target DNA
- Primers
- DNA polymerase
- All 4 deoxynucleotides (radiolabeled)
- One of the 4 dideoxynucleotides

Guanine (G)

2′, 3′- dideoxyguanosine triphosphate

Sequence to be analyzed (target DNA)

Primer

Annealing site

Vector

+ddGTP   +ddATP   +ddTTP   +ddCTP

2 Incubate samples to allow the synthesis of DNA strands. On occasion, a dideoxynucleotide is incorporated instead of the normal deoxynucleotide. This causes DNA synthesis to terminate at different positions. Run samples on a gel and expose gel to X-ray film.

Number of nucleotides attached to the primer

14
13
12
11
10
9
8
7
6
5
4
3
2
1

3′
G
T
C
A
G
G
A
A
T
G
C
C
A
C
C
5′

Sequence of newly made DNA, which is complementary to target DNA sequence in the vector

(5) CACCddG
(9) CACCGTAAddG
(10) CACCGTAAGddG
(14) CACCGTAAGGACTddG

(2) CddA
(7) CACCGTddA
(8) CACCGTAddA
(11) CACCGTAAGGddA

(6) CACCGddT
(13) CACCGTAAGGACddT

(1) ddC
(3) CAddC
(4) CACddC
(12) CACCGTAAGGAddC

(a) The procedure used in traditional dideoxy sequencing

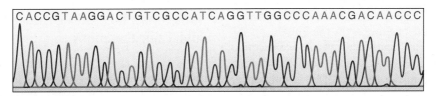

CACCGTAAGGACTGTCGCCATCAGGTTGGCCCAAACGACAACCC

(b) Output from automated dideoxy sequencing

Figure 20.10 **DNA sequencing by the dideoxy method.** **(a)** Traditional dideoxy sequencing used a mixture of radiolabeled deoxy and nonlabeled dideoxynucleoside triphosphates. Following electrophoresis, the bands were detected by autoradiography. Step 1 shows the structure of dideoxyguanosine triphosphate, abbreviated ddGTP. It has a hydrogen, shown in red, instead of a hydroxyl group. **(b)** Automated sequencing uses a fluorescence detector that measures the four kinds of dideoxynucleotides as they emerge from the gel.

Let's now examine the steps involved in DNA sequencing. The DNA containing the sequence to be analyzed is actually double stranded. Figure 20.10a, step 1, shows the DNA after it has been denatured into a single strand by heat treatment. Many copies of this single-stranded DNA are placed into four tubes and mixed with primers that bind to the primer-annealing site. All four types of regular deoxynucleotides and DNA polymerase are also added to each tube. Finally, each of the four tubes has a low concentration of just one of the four possible dideoxynucleotides (ddGTP, ddATP, ddTTP, or ddCTP). The tubes are then incubated to allow DNA polymerase to make strands complementary to one strand of the target DNA sequence. However, the dideoxynucleotides will occasionally cause DNA synthesis to terminate early. For example, let's consider the third tube, which contains ddTTP. Synthesis of new DNA strands will occasionally stop at the sixth or thirteenth position after the annealing site if a ddTTP, instead of a dTTP, is incorporated into the growing DNA strand. The target DNA has a complementary A at the sixth and thirteenth positions.

Within the four tubes, mixtures of DNA strands of different lengths are made. These DNA strands are separated according to their lengths by subjecting them to gel electrophoresis. The shorter strands move to the bottom of the gel more quickly than do the longer strands. To detect the newly made DNA strands, the deoxynucleotides that were added to each reaction are radiolabeled. This enables the strands to be visualized as bands when the gel is exposed to X-ray film. In Figure 20.10a, the DNA strands in the four tubes were run in separate lanes on a gel. Because we know which dideoxynucleotide was added to each tube, we also know which base is at the very end of each DNA strand separated on this gel, because dideoxynucleotides cause chain termination. Therefore, we can read the DNA sequence by reading which base is at the end of every DNA strand and matching this sequence with the length of the strand.

Dideoxy sequencing can now be done much more quickly with automated sequencing. Instead of having four separate tubes with a single type of dideoxynucleotide in each tube, automated sequencing uses one tube containing all four dideoxynucleotides. Each dideoxynucleotide (ddGTP, ddATP, ddTTP, and ddCTP) has a different-colored fluorescent label attached. After incubating the template DNA with deoxynucleotides, the four types of fluorescent dideoxynucleotides, and DNA polymerase, the sample is then loaded into a single lane of a gel and the fragments are separated by electrophoresis. Electrophoresis is continued until each band emerges from the bottom of the gel, where a laser excites the fluorescent dye. A fluorescence detector records the amount of fluorescence emission at four wavelengths, corresponding to the four dyes. An example of a printout from a fluorescence detector is shown in Figure 20.10b. The peaks of fluorescence correspond to the DNA sequence that is complementary to the target DNA. The intensity of the fluorescent peaks is not always the same because dideoxynucleotides get incorporated at certain sites more readily than at other sites.

A Microarray Can Identify Which Genes Are Transcribed by a Cell

Let's now turn our attention to functional genomics. Researchers have developed an exciting new technology, called **DNA microarrays** (or gene chips), that is used to monitor the expression of thousands of genes simultaneously. A DNA microarray is a small silica, glass, or plastic slide that is dotted with many different sequences of single-stranded DNA, each corresponding to a short sequence within a known gene. Each spot contains multiple copies of a specific DNA sequence. For example, one spot in a microarray may correspond to a sequence within the β-globin gene, while another spot could correspond to a different gene, such as a gene that encodes an iron transporter. A single slide contains tens of thousands of different spots in an area the size of a postage stamp. The DNA sequences of each spot is known. These microarrays are typically produced using spotting technologies that are quite similar to the way that an inkjet printer works.

A DNA microarray is used as a hybridization tool. In the experiment shown in **Figure 20.11**, mRNA was isolated from a sample of cells and then used to make fluorescently labeled cDNA. The labeled cDNAs were then incubated with a DNA microarray. The DNA in the microarray is single stranded and corresponds to the sense strand—the strand that has a sequence like mRNA. Those cDNAs that are complementary to the DNAs in the microarray will hybridize and thereby remain bound to the microarray. The array is then washed and placed in a microscope equipped with a computer that scans the fluorescence intensity for each spot. If the fluorescence intensity in a spot is high, there was a large amount of cDNA in the sample that hybridized to the DNA at this location. For example, if the β-globin gene was expressed in the cells being tested, a large amount of cDNA for this gene would be made, and the fluorescence intensity for that spot would be high. Because the DNA sequence of each spot is already known, a fluorescent spot identifies cDNAs that are complementary to those DNA sequences. Furthermore, because the cDNA was generated from mRNA, this technique identifies genes that have been transcribed in a particular cell type under a given set of conditions. Thus far, the most common use of microarrays is to study gene expression patterns. In addition, the technology of DNA microarrays has found many other important uses (**Table 20.2**).

Proteomics Identifies Cellular Proteins and Their Functions

Thus far we have considered ways to characterize the genome of a given species and to study its function. Because most genes encode proteins, a logical next step is to examine the functional roles of the proteins that a species can make. As previously mentioned, this technical approach is called proteomics, and the entire collection of a species' proteins is its proteome. As we move into the 21st century, a key challenge facing molecular biologists is the study of proteomes. Much like genomic research, this study

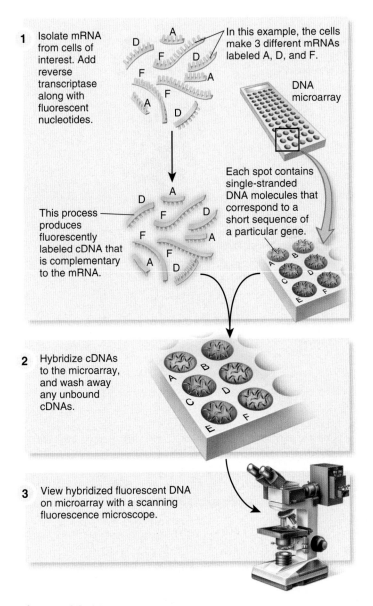

1 Isolate mRNA from cells of interest. Add reverse transcriptase along with fluorescent nucleotides.

In this example, the cells make 3 different mRNAs labeled A, D, and F.

DNA microarray

This process produces fluorescently labeled cDNA that is complementary to the mRNA.

Each spot contains single-stranded DNA molecules that correspond to a short sequence of a particular gene.

2 Hybridize cDNAs to the microarray, and wash away any unbound cDNAs.

3 View hybridized fluorescent DNA on microarray with a scanning fluorescence microscope.

Figure 20.11   Identifying transcribed genes within a DNA microarray.   In this simplified example, only three cDNAs specifically hybridize to spots on the microarray. Those genes were expressed in the cells from which the mRNA was isolated. In an actual experiment, there are typically hundreds or thousands of different cDNAs and tens of thousands of different spots on the array.

Table 20.2	Applications of DNA Microarrays
Application	**Description**
Cell-specific gene expression	A comparison of microarray data using cDNAs derived from mRNA of different cell types can identify genes that are expressed in a cell-specific manner.
Gene regulation	Environmental conditions play an important role in gene regulation. A comparison of microarray data using cDNA derived from mRNA isolated from cells exposed to two different environmental conditions may reveal genes that are induced under one set of conditions and repressed under another set of conditions.
Elucidation of metabolic pathways	Genes that encode proteins that participate in a common metabolic pathway are often expressed in a parallel manner and can be revealed from a microarray analysis.
Tumor profiling	Different types of cancer cells exhibit striking differences in their profiles of gene expression, which can be revealed by a DNA microarray analysis. This approach is gaining widespread use to classify tumors that are sometimes morphologically indistinguishable.
Genetic variation	A mutant allele may not hybridize to a spot on a microarray as well as a wild-type allele. Therefore, microarrays are gaining widespread use as a tool to detect genetic variation. This application has been used to identify disease-causing alleles in humans, and to elucidate mutations that contribute to quantitative traits in plants and other species.
Microbial strain identification	Microarrays can distinguish between closely related bacterial species and subspecies.

One major goal of researchers in the field of proteomics is to identify and functionally characterize all the proteins that a cell type will make. Because cells produce thousands of different proteins, this is a daunting task. Nevertheless, along with genomic research, the past decade has seen important advances in our ability to isolate and identify cellular proteins.

An important technique in the field of proteomics is **two-dimensional (2D) gel electrophoresis**, which is a technique that can separate hundreds or even thousands of different proteins within a cell extract (**Figure 20.12**). To begin this procedure, an experimenter loads a sample of cellular proteins onto the top of a gel that is shaped like a tube. This tube gel separates proteins according to their net charge at a given pH. A protein migrates to the point in the gel where its net charge is zero, a process termed isoelectric focusing. After the tube gel has run, the experimenter places it on top of a slab gel that contains a detergent called sodium dodecyl sulfate (SDS). The SDS is negatively charged and coats each protein uniformly. During this process, the proteins lose their three-dimensional structure or become denatured. The proteins in the slab gel are separated according to their molecular masses. The end result is a collection of spots, each spot corresponding to a unique cellular protein.

will require the collective contributions of many research scientists, as well as improvements in technologies that are aimed at explaining the complexities of the proteome.

Due to gene regulation, any given cell will produce only a subset of the proteins that are found in the proteome of a species. For example, the human genome has approximately 20,000 to 25,000 different genes, yet a human red blood cell makes less than half that number of different proteins. The subset of proteins that a cell makes depends primarily on what type of cell it is, its stage of development, and its environmental conditions.

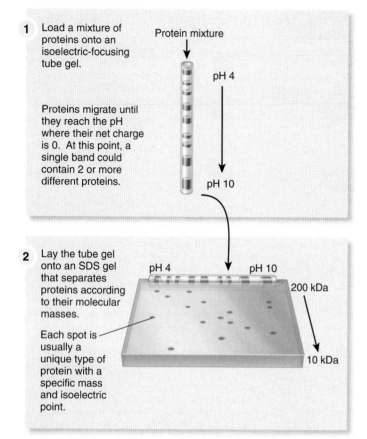

1   Load a mixture of proteins onto an isoelectric-focusing tube gel.

Protein mixture

pH 4

Proteins migrate until they reach the pH where their net charge is 0. At this point, a single band could contain 2 or more different proteins.

pH 10

2   Lay the tube gel onto an SDS gel that separates proteins according to their molecular masses.

pH 4   pH 10

200 kDa

Each spot is usually a unique type of protein with a specific mass and isoelectric point.

10 kDa

Figure 20.12   Two-dimensional gel electrophoresis. The technique involves two electrophoresis steps. First, a mixture of proteins is separated on an isoelectric focusing gel that has the shape of a tube. Proteins migrate to the point where their net charge is zero. This tube gel is placed into a long well on top of an SDS-polyacrylamide gel. This second gel separates the proteins according to their masses. In this diagram, only a few spots are seen, but an actual experiment would involve a mixture of hundreds or thousands of different proteins.

Two-dimensional gel electrophoresis is used as a method to separate a mixture of cellular proteins. The next step is to identify those proteins, based on the concept that each protein has its own unique amino acid sequence. To identify a protein, a spot on a 2D gel can be cut out of the gel to obtain a tiny amount of the protein within the spot. In essence, the 2D gel procedure purifies a small amount of the cellular protein of interest. A protein in a given spot can then be identified via a technique called **mass spectrometry** that allows researchers to determine the amino acid sequence of short regions within a protein. The technique determines the mass of the peptide fragments produced by digesting a purified protein with an enzyme called a protease that cuts the protein into small peptide fragments. The peptides are mixed with an organic acid and dried onto a metal slide, and then the sample is struck with a laser. This causes the peptides to temporarily vaporize in the form of an ionized gas, in which the peptide contains one or more positive charges. The charged gaseous peptides are then accelerated by an electric field toward a detector. The time of flight to the detector is determined

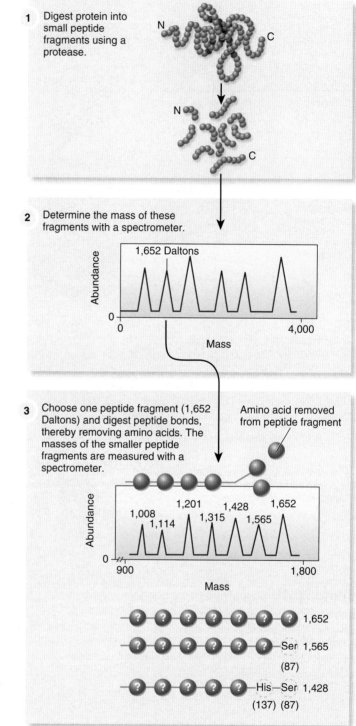

1   Digest protein into small peptide fragments using a protease.

N   C

N

C

2   Determine the mass of these fragments with a spectrometer.

1,652 Daltons

Abundance

0
 0   4,000

Mass

3   Choose one peptide fragment (1,652 Daltons) and digest peptide bonds, thereby removing amino acids. The masses of the smaller peptide fragments are measured with a spectrometer.

Amino acid removed from peptide fragment

Abundance

1,201   1,428   1,652
1,008   1,315   1,565
    1,114

0
 900   1,800

Mass

? ? ? ? ? ? ?   1,652

? ? ? ? ? ? Ser 1,565
                (87)

? ? ? ? ? His─Ser 1,428
          (137) (87)

Figure 20.13   Tandem mass spectrometry. This technique is used to determine the amino acid sequence of a peptide.

by their mass and net charge. A measurement of the flight time provides an extremely accurate way to measure the mass of a peptide.

To determine the amino acid sequence of a given peptide, two mass spectrometers are used, a method called tandem mass spectrometry (**Figure 20.13**). The first spectrometer measures the mass of a given peptide, and then this same peptide is analyzed by a second spectrometer after the peptide has been digested into smaller fragments. The differences in the masses of the peaks in the second spectrum reveal the amino acid sequence of the peptide, because the masses of all 20 amino acids are known. For example, suppose that a peptide had a mass of 1,652 Daltons. If one amino acid at the end were removed, and the smaller peptide had a mass that was 87 Daltons less (that is, 1,565 Daltons), this would indicate that a serine is at one end of the peptide because the mass of serine within a polypeptide chain is 87 Daltons. If two amino acids were removed at one end and the mass was 224 Daltons less, this would correspond to the removal of one serine (87 Daltons) and one histidine (137 Daltons). Thus, from these measurements, we would conclude that the amino acid sequence from one end of the peptide was serine-histidine.

After a researcher has obtained a few short peptide sequences from a given protein, genomic information can readily predict the entire amino acid sequence of the protein. For example, if a peptide had the sequence serine-histidine-leucine-asparagine-serine-asparagine, a researcher could determine the possible codon sequences that could encode such a peptide. More than one sequence is possible due to the degeneracy of the genetic code. Using computer software described in Chapter 21, the codon sequences would be used to scan the entire genomic sequence if it has been determined in a genome sequencing project. For example, the DNA sequence of the entire human genome has already been determined. Computer software can locate a match between a predicted codon sequence and a specific gene within the human genome. In this way, mass spectrometry makes it possible to identify the gene that encodes the entire protein. The gene sequence, in turn, can be used to predict the amino acid sequence of the whole protein.

20.3 Biotechnology

As mentioned at the beginning of this chapter, biotechnology is defined as technologies that involve the use of living organisms, or products from living organisms, as a way to benefit humans. Biotechnology is not a new topic. Its use began about 12,000 years ago when humans began to domesticate animals and plants for the production of food. Since that time, many species of microorganisms, plants, and animals have become routinely used by people. More recently, the term biotechnology has become associated with molecular genetics. Since the 1970s, molecular genetic tools have provided new ways to make use of living organisms to benefit humans.

In this section, we will consider genetic techniques to alter the characteristics of microorganisms, plants, and animals. Examples include bacteria that can degrade pollutants in the environment and livestock that make human hormones. We will also examine several topics that you often hear about in the news. We will explore the technical side of mammalian cloning and also touch upon the ethical issues. Another newsworthy topic is DNA fingerprinting, a common method used to analyze the DNA from an individual. Like conventional fingerprinting,

this technique can be used for identification purposes, as in a crime investigation. Finally, we will learn how certain inherited diseases may be treated via gene therapy. In this relatively new approach, cloned genes are introduced into individuals with genetic diseases in an attempt to compensate for mutant genes.

Many Important Medicines Are Produced by Recombinant Microorganisms

In 1982, the U.S. Food and Drug Administration approved the sale of human insulin made by recombinant bacteria. In healthy individuals, insulin is produced by the beta cells of the pancreas. Insulin functions to regulate several physiological processes, particularly the uptake of glucose into fat and muscle cells. Persons with insulin-dependent diabetes cannot synthesize an adequate amount of insulin due to a defect in their beta cells. Today, people with this disorder are usually treated with human insulin that is made by genetically engineered bacteria. Prior to 1982, insulin was isolated from pancreases removed from cattle. Unfortunately, in some cases, diabetic individuals became allergic to cow insulin. These allergic patients had to use expensive combinations of insulin from human cadavers and other animals. Now, of course, they can use human insulin made by recombinant bacteria.

As shown in **Figure 20.14**, insulin is a hormone composed of two polypeptides, called the A and B chains. To make this hormone using bacteria, the coding sequence of the A or B chains is inserted into a plasmid vector next to the coding sequence of the E. coli protein, β-galactosidase. After transforming bacteria with such vectors, the cells produce many copies of a fusion protein comprising β-galactosidase and the A or B chain. This step is necessary because the A and B chains are rapidly degraded when expressed in bacterial cells by themselves; the fusion proteins, however, are not. These fusion proteins are then extracted from the bacterial cells and treated with a chemical, cyanogen bromide (CNBr), that cleaves after methionine to separate β-galactosidase from the A or B chain. The A and B chains are purified and mixed together under conditions in which they will fold and associate with each other to form a functional insulin hormone molecule.

Microorganisms Can Reduce Environmental Pollutants

Bioremediation is the use of microorganisms to decrease pollutants in the environment. As its name suggests, it is a biological remedy for pollution. During bioremediation, enzymes produced by a microorganism modify a toxic pollutant by altering or transforming its structure, an event called biotransformation. In many cases, biotransformation results in biodegradation, in which the toxic pollutant is degraded, yielding less complex, nontoxic metabolites.

Since the early 1900s, people have intentionally used microorganisms in the treatment and degradation of sewage. More recently, the field of bioremediation has expanded into the treatment of hazardous and refractory chemical wastes (that is,

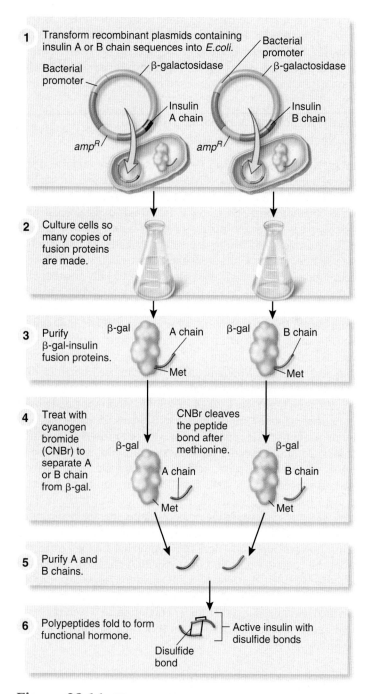

1 Transform recombinant plasmids containing insulin A or B chain sequences into *E.coli.*

Bacterial promoter
β-galactosidase
Insulin A chain
amp^R

Bacterial promoter
β-galactosidase
Insulin B chain
amp^R

2 Culture cells so many copies of fusion proteins are made.

3 Purify β-gal-insulin fusion proteins.

β-gal   A chain   Met

β-gal   B chain   Met

4 Treat with cyanogen bromide (CNBr) to separate A or B chain from β-gal.

CNBr cleaves the peptide bond after methionine.

β-gal   A chain   Met

β-gal   B chain   Met

5 Purify A and B chains.

6 Polypeptides fold to form functional hormone.

Disulfide bond

Active insulin with disulfide bonds

Figure 20.14   The use of bacteria to make human insulin.

encode enzymes involved in bioremediation. The characterization of the relevant genes greatly enhances our understanding of how microbes can modify toxic pollutants. In addition, recombinant strains created in the laboratory can be more efficient at degrading certain types of pollutants.

In 1980, in a landmark case (*Diamond v. Chakrabarty*), the U.S. Supreme Court ruled that a live, recombinant microorganism is patentable as a "manufacture or composition of matter." The first recombinant microorganism to be patented was an "oil-eating" bacterium, which contained a laboratory-constructed plasmid. This strain can oxidize the hydrocarbons commonly found in petroleum. It grew faster on crude oil than did any of the natural strains that were tested. However, it has not been a commercial success because this recombinant strain metabolizes only a limited number of toxic compounds. Unfortunately, the strain did not degrade many higher-molecular-weight compounds that tend to persist in the environment.

Currently, bioremediation should be considered a developing industry. In the future, recombinant microorganisms may provide an effective way to decrease the levels of toxic chemicals within our environment. However, this approach will require careful studies to demonstrate that recombinant organisms are effective at reducing pollutants and safe when released into the environment.

Gene Replacements and Knockouts in Mice Can Be Used to Understand Gene Function and Human Disease

Let's now turn our attention to the genetic engineering of animals. Researchers can introduce a cloned gene into a fertilized egg or embryonic cells and create animals that carry the cloned gene. The term **transgenic** is used to describe an organism that carries genes that were introduced using molecular techniques such as gene cloning. Transgenics are also called **genetically modified organisms (GMOs)**. In some cases, the cloned gene will recombine with the normal gene on a chromosome, a phenomenon called **gene replacement**. For eukaryotic species that are diploid, only one of the two copies is initially replaced. In other words, the initial gene replacement creates a heterozygote carrying one normal copy of the gene and one copy that has been replaced with a cloned gene. Heterozygotes can be crossed to each other to obtain homozygotes, which carry both copies of the cloned gene. If the cloned gene carries a mutation that inactivates its function, such a homozygote is said to have undergone a **gene knockout**. The inactive cloned gene has replaced both copies of the normal gene. In this way, the function of the normal gene has been "knocked out." Gene replacements and gene knockouts have become powerful tools for understanding gene function.

A particularly exciting avenue of gene replacement research is its application in the study of human disease. As an example, let's consider the disease cystic fibrosis (CF), which is one of the most common and severe inherited human disorders. In humans, the defective gene that causes CF has been identified. Likewise, the homologous gene in mice was later identified. Using the technique of gene replacement, researchers produced

chemicals that are difficult to degrade), usually associated with chemical and industrial activity. These pollutants include heavy metals, petroleum hydrocarbons, and halogenated organic compounds such as those with chlorine atoms, as well as pesticides, herbicides, and organic solvents. Many new applications that use microorganisms to degrade these pollutants are being tested. The field of bioremediation has been fostered by better knowledge of how pollutants are degraded by microorganisms, the identification of new and useful strains of microbes, and the ability to enhance bioremediation through genetic engineering. Molecular genetic technology is critical to identifying genes that mice that are homozygous for the same type of mutation that is found in humans with CF. Such mice exhibit disease symptoms resembling those found in humans, namely, respiratory and digestive abnormalities. Therefore, these mice can be used as model organisms to study this human disease. Furthermore, these mice models have been used to test the effects of various therapies in the treatment of the disease.

Biotechnology Holds Promise in Producing Transgenic Livestock

The technology of creating transgenic mice has been extended to other animals, and much research is under way to develop transgenic species of livestock including fish, sheep, pigs, goats, and cattle. A novel avenue of research involves the production of medically important proteins in the mammary glands of live-stock. This approach is sometimes called **molecular pharming**. (The word pharming refers to the use of farm animals to make pharmaceuticals.) Several human proteins have been success-fully produced in the milk of domestic livestock such as goats, sheep, and cattle. These include Factor IX to treat a certain type of hemophilia, tissue plasminogen activator to dissolve blood clots, and α-1-antitrypsin for the treatment of emphysema.

Compared with the production of proteins in bacteria, one advantage of molecular pharming is that certain proteins are more likely to function properly when expressed in mammals. This may be due to the post-translational modification of pro-teins that occurs in mammals but not in bacteria. In addition, certain proteins may be degraded rapidly or folded improperly when expressed in bacteria. Furthermore, the yield of recombi-nant proteins in milk can be quite large. Each dairy cow, for example, produces about 10,000 liters of milk per year. In most cases, a transgenic cow can produce approximately 1 g/L of the transgenic protein in its milk.

To introduce a human gene into a mammal so that the en-coded protein will be secreted into its milk, the strategy is to clone the gene next to the promoter of a gene that is specifically expressed in mammary cells. As we learned in Chapter 13, gene regulation promotes the expression of genes in certain cell types and/or under certain conditions. Researchers have identified several genes that are transcribed only in mammary cells. One example is β-lactoglobulin, a protein that is specifically found in milk. To express a human hormone into a domestic animal's milk, the promoter for one of these milk-specific genes is linked to the coding sequence of a human hormone gene and inserted into a plasmid vector (**Figure 20.15**). In addition to the pro-moter, a short signal sequence may also be necessary so that the protein hormone will be secreted from the mammary cells and into the milk. Such a vector is then injected into an oocyte, such as a sheep oocyte, where it integrates into the genome. The egg is then fertilized by exposure to sperm, and then implanted into the uterus of a female sheep. The resulting offspring carries the cloned gene. If the offspring is a female, the protein encoded by the human gene will be expressed within the mammary gland and secreted into the milk. The milk can then be obtained from the animal, and the human protein isolated and purified.

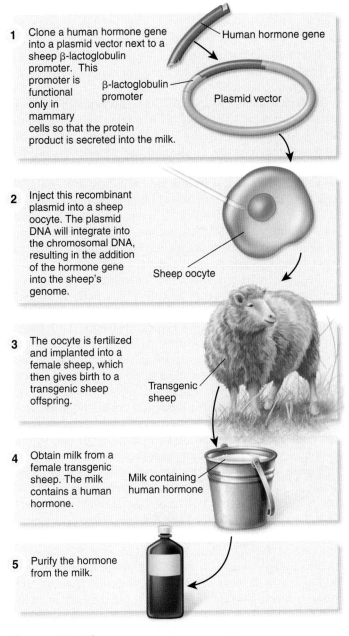

1 Clone a human hormone gene into a plasmid vector next to a sheep β-lactoglobulin promoter. This promoter is functional only in mammary cells so that the protein product is secreted into the milk.

Human hormone gene

β-lactoglobulin promoter

Plasmid vector

2 Inject this recombinant plasmid into a sheep oocyte. The plasmid DNA will integrate into the chromosomal DNA, resulting in the addition of the hormone gene into the sheep's genome.

Sheep oocyte

3 The oocyte is fertilized and implanted into a female sheep, which then gives birth to a transgenic sheep offspring.

Transgenic sheep

4 Obtain milk from a female transgenic sheep. The milk contains a human hormone.

Milk containing human hormone

5 Purify the hormone from the milk.

Figure 20.15 Molecular pharming.   This procedure is used for expressing human genes in a domestic animal's milk. The β-lactoglobulin promoter is normally expressed in mammary cells, whereas the human hormone gene is not. To express the human hormone gene in milk, the promoter from the β-lactoglobulin gene is linked to the coding sequence of the hormone gene. This causes the human hormone to be expressed in the milk of the animal, from which it can be purified.

Agrobacterium tumefaciens Can Be Used to Make Transgenic Plants

As we have just seen, the introduction of cloned genes into oocytes or embryonic cells can produce transgenic animals. The production of transgenic plants is somewhat easier because plant cells are totipotent, which means that an entire organism can be regenerated from somatic cells. Therefore, a transgenic plant can be made by the introduction of cloned genes into somatic tissue, such as the tissue of a leaf. After the cells of a leaf have become transgenic, an entire plant can be regenerated by treating the leaf with plant growth hormones that cause it to form roots and shoots.

Molecular biologists can use the bacterium *Agrobacterium tumefaciens*, which naturally infects plant cells and causes tumors, to produce transgenic plants. A plasmid from the bacterium is known as the **Ti plasmid**, for tumor-inducing plasmid. This plasmid contains a region called the T DNA (for transferred DNA) that is transferred from the bacterium to the plant cell. The T DNA from the Ti plasmid becomes integrated into the chromosomal DNA of the plant cell. After this occurs, genes within the T DNA that encode plant growth hormones cause uncontrolled plant cell growth and produce a crown gall tumor, a bulbous growth on the plant.

Researchers have modified the Ti plasmid to use it as a vector to introduce cloned genes into plant cells. In the modified Ti plasmid, the genes that cause a crown gall tumor to form have been removed. Fortunately for genetic engineers, the T DNA is still taken up into plant cells and integrated into the plant chromosomal DNA. In addition, a selectable marker gene has been inserted into the T DNA to allow selection of plant cells that have taken up the T DNA. A gene that provides resistance to the antibiotic kanamycin, called *Kan^R*, is a commonly used selectable marker. Finally, the Ti plasmid used in cloning experiments has been modified to contain unique restriction sites for the convenient insertion of any gene of interest.

Figure 20.16 shows the general strategy for producing transgenic plants via T DNA–mediated gene transfer. A desired gene is inserted into the T DNA of a genetically engineered Ti plasmid and then transformed into *A. tumefaciens*. Plant cells are exposed to the transformed *A. tumefaciens*. After allowing time for infection, the plant cells are grown on a solid medium that contains kanamycin and carbenicillin. Carbenicillin kills *A. tumefaciens*, and kanamycin kills any plant cells that have not taken up the T DNA. Therefore, the only surviving cells are those plant cells that have integrated the T DNA into their genome. Because the T DNA also contains the cloned gene of interest, the selected plant cells are expected to have received this cloned gene as well. The cells are then transferred to a medium that contains the plant growth hormones necessary for the regeneration of entire plants.

Transgenic plants are approved for human consumption. Their production has become routine practice for many agriculturally important plant species, including alfalfa, corn, cotton, soybean, tobacco, and tomato. Transgenic plants can be given characteristics that are agriculturally useful, such as those that improve plant quality and protection. In terms of quality, gene additions have been made to improve the nutritional value of some plants, such as making the canola grain produce more oil, and making the Brazil nut contain more of the amino acid methionine. Frequently, transgenic research has sought to produce plant strains resistant to insects, disease, and herbicides. For example, transgenic plants highly tolerant of particular herbicides have been made. The Monsanto Company has produced

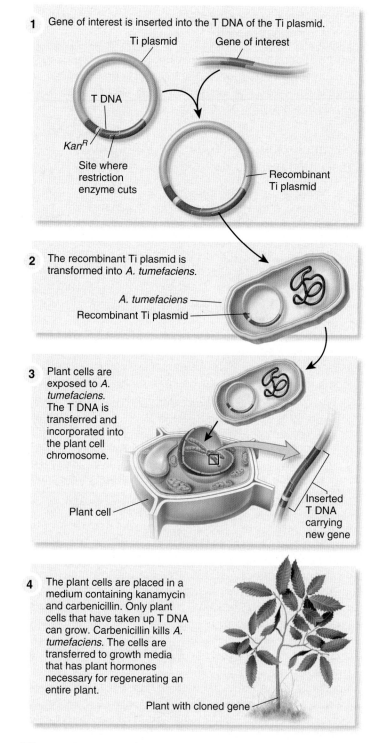

1 Gene of interest is inserted into the T DNA of the Ti plasmid.

Ti plasmid   Gene of interest

T DNA

Kan^R

Site where restriction enzyme cuts

Recombinant Ti plasmid

2 The recombinant Ti plasmid is transformed into *A. tumefaciens*.

A. tumefaciens

Recombinant Ti plasmid

3 Plant cells are exposed to *A. tumefaciens*. The T DNA is transferred and incorporated into the plant cell chromosome.

Plant cell

Inserted T DNA carrying new gene

4 The plant cells are placed in a medium containing kanamycin and carbenicillin. Only plant cells that have taken up T DNA can grow. Carbenicillin kills *A. tumefaciens*. The cells are transferred to growth media that has plant hormones necessary for regenerating an entire plant.

Plant with cloned gene

Figure 20.16 Using the Ti plasmid and *Agrobacterium tumefaciens* to make transgenic plants.

transgenic plant strains tolerant of glyphosate, the active agent in the herbicide Roundup™. Compared with nontransgenics, these plants grow quite well in the presence of glyphosate-containing herbicides (**Figure 20.17**). Another important approach is to make plant strains that are disease resistant. In many cases, virus-resistant plants have been developed by introducing a gene that encodes a viral coat protein. When the plant

Figure 20.17 Transgenic plants that are resistant to glyphosate. This field of soybean plants has been treated with the herbicide glyphosate. The plants in the back rows have been genetically engineered to contain a herbicide-resistance gene, so they are resistant to killing by glyphosate. By comparison, the dead or stunted plants in the front row with the orange stick do not contain this gene.

cells express the viral coat protein, they become resistant to infection by that pathogenic virus.

Research into transgenic plants has also led to the development of new products such as biodegradable plastics and medical products. For example, transgenic plants have been modified to produce vaccines against many human and animal diseases, including hepatitis B, cholera, and malaria. Transgenic plants have been made that produce a variety of medicines, such as human epidermal growth factor for wound repair and human interferon to fight viral diseases and cancer.

Researchers Have Succeeded in Cloning Mammals from Somatic Cells

We now turn our attention to cloning as a way to genetically manipulate plants and animals. The term cloning has many different meanings. At the beginning of this chapter we discussed gene cloning, which involves methods that produce many copies of a gene or its protein product. The cloning of an entire organism is a different matter. By accident, this happens in nature. Identical twins are genetic clones that began from the same fertilized egg. Similarly, researchers can take mammalian embryos at an early stage of development (for example, the two-cell to eight-cell stage), separate the cells, implant them into the uterus of a female, and obtain multiple births of genetically identical individuals.

In the case of plants, the cloning of new individuals is relatively easy. Plants can be cloned from somatic cells. In most cases, it is easy to take a cutting from a plant, expose it to growth hormones, and obtain a separate plant that is genetically identical to the original. However, this approach has not been possible with mammals. For several decades, scientists believed that chromosomes within the somatic cells of mammals had incurred irreversible genetic changes that render them unsuitable for cloning. However, this hypothesis has proven to be incorrect.

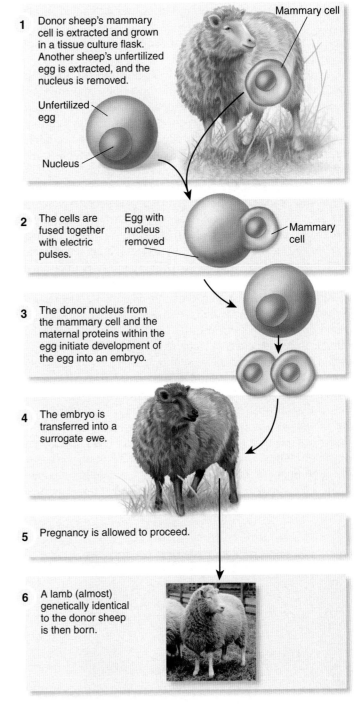

Figure 20.18 Protocol for the successful cloning of sheep. In this protocol, the genetic material from a somatic cell is used to make a cloned mammal, in this case the sheep Dolly.

In 1996, Ian Wilmut and his colleagues at the Roslin Institute in Edinburgh, Scotland, created clones of sheep using the genetic material from somatic cells. As you may have heard, they named the first cloned lamb Dolly.

Figure 20.18 illustrates how Dolly was created. The researchers removed mammary cells from an adult female sheep and grew them in the laboratory. The researchers then extracted the nucleus from a sheep oocyte and fused a diploid mammary cell with the enucleated oocyte cell. Fusion was promoted by electric pulses. After fusion, the zygote was implanted into the uterus of an adult sheep. One hundred and forty-eight days later, Dolly was born.

Dolly was (almost) genetically identical to the sheep that donated a mammary cell to create her. Dolly and the donor sheep were (almost) genetically identical in the same way that identical twins are. They carry the same set of genes and look remarkably similar. However, Dolly and her somatic cell donor may have some minor genetic differences due to possible differences in their mitochondrial DNA and may exhibit some phenotypic differences due to maternal effect genes.

Mammalian cloning is still at an early stage of development. Nevertheless, creating Dolly was a breakthrough that showed that it is technically possible. In recent years, cloning using somatic cells has been achieved in several mammalian species, including sheep, cows, mice, goats, pigs, and cats. In 2002, the first pet was cloned, which was named Copycat (see chapter-opening photo). The cloning of mammals provides the potential for many practical applications. With regard to livestock, cloning would enable farmers to use the somatic cells from their best animals to create genetically homogeneous herds. This could be advantageous in terms of agricultural yield, although such a genetically homogeneous herd may be more susceptible to certain diseases.

Aside from the practical uses of cloning agricultural species, however, people have become greatly concerned with the possibility of human cloning. This prospect has raised serious ethical questions. Some people feel that it is morally wrong and threatens the basic fabric of parenthood and family. Others feel that it is a modern technology that offers a new avenue for reproduction, one that could be offered to infertile couples, for example. Human cloning is a complex subject with many more viewpoints than these two. In the public sector, the sentiment toward human cloning has been generally negative. Indeed, many countries have issued a complete ban on human cloning, while others permit limited research in this area. In the future, our society will have to wrestle with the legal and ethical aspects of cloning as it applies not only to animals but also to people.

DNA Fingerprinting Is Used for Identification and Relationship Testing

DNA fingerprinting, also known as DNA profiling, is a technology that identifies particular individuals using properties of their DNA. Like the human fingerprint, the DNA of each individual is a distinctive characteristic that provides a means of identification. When subjected to DNA fingerprinting, the chromosomal DNA produces a series of bands on a gel (**Figure 20.19a**). The unique pattern of these bands is a distinguishing feature of each individual.

In the past two decades, the technique of DNA fingerprinting has become automated, much like the automation that changed the procedure of DNA sequencing described earlier in this chapter. DNA fingerprinting is now done using PCR, which amplifies **short tandem repeat sequences (STRs)**. STRs are found in multiple sites in the genome of humans and other species, and they vary in length among different individuals. Using primers, the STRs from a sample of DNA are amplified by PCR and then separated by gel electrophoresis according to their molecular masses. Like automated DNA sequencing, the amplified STR fragments are fluorescently labeled. A laser excites the fluorescent molecule within an STR, and a detector records the amount of fluorescence emission for each STR. The DNA finger-

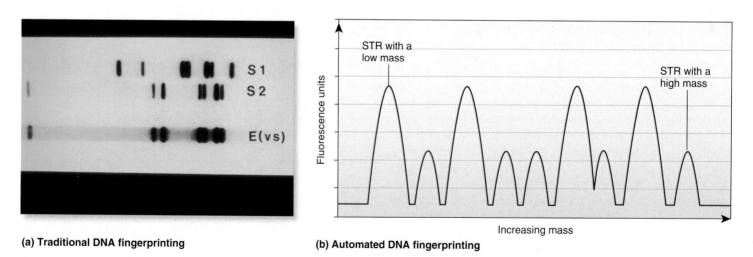

(a) Traditional DNA fingerprinting

(b) Automated DNA fingerprinting

Figure 20.19 DNA fingerprinting. (a) Chromosomal DNA from two different individuals was subjected to traditional DNA fingerprinting. Their DNA appears as a series of bands on a gel. The dissimilarities in the patterns of these bands distinguish different individuals, much as the differences in physical fingerprint patterns can be used for identification. DNA evidence at a crime scene (E) matches suspect 2 (S2) but not suspect 1 (S1). (b) Automated DNA fingerprinting compares fluorescently labeled DNA sequences called short tandem repeat sequences (STRs), which differ in length among various individuals. A printout from the fluorescence detector is shown here.

print yields a series of peaks, each peak having a characteristic molecular mass (**Figure 20.19b**). In this automated approach, the pattern of peaks is an individual's DNA fingerprint.

Within the past decade or so, the uses of DNA fingerprinting have expanded in many ways. DNA fingerprinting has gained acceptance as a precise method of identification. In medicine, it is used to identify different species of bacteria and fungi, and it can even distinguish among closely related strains of the same species. This is useful so that clinicians can treat patients with the appropriate antibiotic or fungicide.

A second common use is forensics—providing evidence in a criminal case. DNA fingerprinting can be used as evidence that an individual was at a crime scene. Forensic DNA was first used in the U.S. court system in 1986. When a sample taken from a crime scene matches the DNA fingerprint of an individual, the probability that a match could occur simply by chance can be calculated. Each STR size is given a probability score based on its observed frequency within a reference human population (Caucasian, Asian, etc.). An automated DNA fingerprint contains many peaks, and the probability scores for each peak are multiplied together to arrive at the likelihood that a particular pattern of peaks would be observed. For example, if a DNA fingerprint contains 20 fluorescent peaks, and the probability of an individual having each peak is 1/4, then the likelihood of having that pattern would be $(1/4)^{20}$, or roughly 1 in 1 trillion. Therefore, a match between two samples is rarely a matter of random chance.

Another important application of DNA fingerprinting is relationship testing. Persons who are related genetically will have some peaks in common. The number they share depends on the closeness of their genetic relationship. For example, offspring are expected to receive half of their peaks from one parent and half from the other. Therefore, DNA fingerprinting can be used as evidence in paternity cases.

FEATURE INVESTIGATION

Blaese and Colleagues Performed the First Gene Therapy to Treat ADA Deficiency

Gene therapy is the introduction of cloned genes into living cells in an attempt to cure disease. It represents a new, potentially powerful means of treating a wide variety of illnesses. Many current research efforts in gene therapy are aimed at alleviating inherited human diseases. More than 4,000 human genetic diseases involve a single gene abnormality—common examples include cystic fibrosis and sickle-cell anemia. Many inherited diseases have been investigated as potential targets for gene therapy. These include metabolic diseases, such as phenylketonuria, and blood disorders, such as hemophilia and severe combined immunodeficiency. In addition, gene therapies have also been aimed at treating diseases, such as cancer and cardiovascular disease, that may occur later in life. Some researchers are even pursuing research that will use gene therapy to combat infectious diseases such as AIDS. Human gene therapy is still at an early stage of development. Relatively few patients have been successfully treated with gene therapy in spite of a large research effort. In addition, experimental gene therapy in humans has been associated with adverse reactions. In 1999, a patient even died from a reaction to a gene therapy treatment.

Adenosine deaminase (ADA) deficiency was the first inherited disease treated with gene therapy. When present, the adenosine deaminase enzyme deaminates the nucleoside deoxyadenosine. This is an important step in the proper metabolism of nucleosides. If both copies of the *ADA* gene are defective, however, deoxyadenosine will accumulate within the cells of the individual. At high concentration, deoxyadenosine is particularly toxic to lymphocytes in the immune system, namely, T cells and B cells. In affected individuals, the destruction of T and B cells leads to a severe combined immunodeficiency disease (SCID). If left untreated, SCID is typically fatal at an early age (generally 1 to 2 years old), because the compromised immune system of these individuals cannot fight infections.

Three approaches are used to treat ADA deficiency. In some cases, a bone marrow transplant is received from a compatible donor. A second method is to treat SCID patients with purified ADA enzyme that is coupled to polyethylene glycol (PEG). This PEG-ADA is taken up by lymphocytes and can correct the ADA deficiency. Unfortunately, these two approaches are not always available and/or successful. A third, more recent approach is to treat ADA patients with gene therapy.

On September 14, 1990, the first human gene therapy was approved for a young girl suffering from ADA deficiency. This work was carried out by a large team of researchers including R. Michael Blaese, Kenneth Culver, W. French Anderson, and several collaborators at the National Institutes of Health (NIH). Prior to this clinical trial, the normal gene for ADA had been cloned into a retroviral vector. The retroviral vector also contained mutations that prevented it from causing a viral disease, yet it still enabled the virus to infect human cells. The general aim of this therapy was to remove lymphocytes from the blood of the girl with SCID, introduce the normal *ADA* gene into the cells, and then return them to her bloodstream.

Figure 20.20 outlines the protocol for the experimental treatment. The researchers removed lymphocytes from the girl and cultured them in a laboratory. The lymphocytes were then transfected with a recombinant retrovirus that contained the normal *ADA* gene. During the life cycle of a retrovirus, the retroviral genetic material is inserted into the host cell's DNA.

Figure 20.20   The first human gene therapy for adenosine deaminase (ADA) deficiency by Blaese and colleagues.

HYPOTHESIS   Infecting lymphocytes with a retrovirus containing the *ADA* normal gene will correct the inherited deficiency of the mutant *ADA* gene in patients with ADA deficiency.

STARTING MATERIALS   A retrovirus with the normal *ADA* gene.

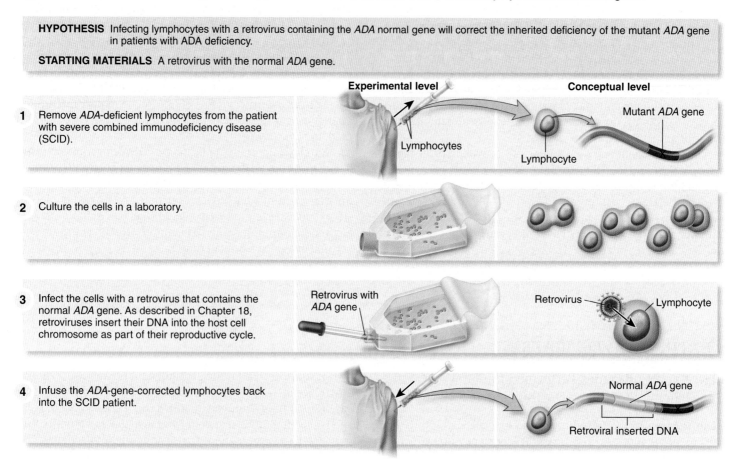

Experimental level   Conceptual level

1   Remove *ADA*-deficient lymphocytes from the patient with severe combined immunodeficiency disease (SCID).

Lymphocytes

Lymphocyte   Mutant *ADA* gene

2   Culture the cells in a laboratory.

3   Infect the cells with a retrovirus that contains the normal *ADA* gene. As described in Chapter 18, retroviruses insert their DNA into the host cell chromosome as part of their reproductive cycle.

Retrovirus with *ADA* gene

Retrovirus   Lymphocyte

4   Infuse the *ADA*-gene-corrected lymphocytes back into the SCID patient.

Normal *ADA* gene

Retroviral inserted DNA

5   **THE DATA**

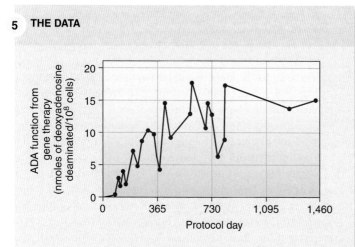

Therefore, because this retrovirus contained the normal *ADA* gene, this gene also was inserted into the chromosomal DNA of the girl's lymphocytes. After this occurred in the laboratory, the cells were reintroduced back into the patient.

In this clinical trial, two U.S. patients were enrolled, and a third patient was later treated in Japan. The data in Figure 20.20 show the results of this trial for one of the three patients. Over the course of two years, this patient was given 11 infusions of lymphocytes that had been corrected with the normal *ADA* gene. Even after four years, this patient's lymphocytes were still making ADA. Therefore, these results suggest that this first gene therapy trial may offer some benefit. However, because the patients were also treated with PEG-ADA, the researchers could not determine whether gene transfer into T cells was of significant clinical benefit.

CHAPTER SUMMARY

20.1 Gene Cloning

- Recombinant DNA technology is the use of laboratory techniques to isolate and manipulate fragments of DNA.

- Gene cloning is the process of making multiple copies of a gene. The procedure is used to obtain large amounts of the DNA that encodes a particular gene, or to obtain large amounts of the gene product. (Figure 20.1)

- Plasmid and viral vectors are used in gene cloning. To obtain recombinant DNA, a vector and chromosomal DNA are cut with restriction enzymes. The DNA fragments bind to each other via their sticky ends, and the pieces are linked together via DNA ligase. (Table 20.1, Figure 20.2)

- When a recombinant vector is introduced into a bacterial cell, the cell replicates the vector and also divides to produce many cells. This achieves gene cloning. (Figure 20.3)

- A collection of recombinant vectors, each with a particular piece of chromosomal DNA, is introduced into bacterial cells to create a DNA library. If the DNA inserts are derived from cDNA, which is made from mRNA, this is a cDNA library. (Figure 20.4)

- Colony hybridization is a method that uses a labeled probe that recognizes a specific gene to identify that gene in a DNA library. (Figure 20.5)

- Gel electrophoresis is used to separate macromolecules by using an electric field that causes them to pass through a gel matrix. Gel electrophoresis typically separates molecules according to their masses. (Figure 20.6)

- In a Southern blotting experiment, a labeled probe, which is a strand of DNA from a specific gene, is used to identify that gene in a mixture of many chromosomal DNA fragments. (Figure 20.7)

- Polymerase chain reaction (PCR) is another technique to make many copies of a gene. Primers are used that flank the region of DNA to be amplified. (Figure 20.8)

20.2 Genomics and Proteomics

- Genomics is the study of genomes as whole units, while proteomics is the study of large groups of proteins made by a particular cell or organism.

- For structural genomics, large fragments of chromosomal DNA are cloned into vectors such as BACs. One goal of structural genomics is to make a contig, which is a collection of clones that cover a contiguous region of a chromosome. This is a type of mapping— determining the relative locations of genes or other DNA segments along a chromosome. (Figure 20.9)

- The dideoxy method of DNA sequencing uses dideoxynucleotides to determine the base sequence of a segment of DNA. (Figure 20.10)

- An important technique in functional genomics is a microarray that contains a group of spots, each with a specific type of DNA. It is used as a hybridization tool to study the expression of groups of genes. (Figure 20.11, Table 20.2)

- Two-dimensional gel electrophoresis is used to separate complex mixtures of proteins, such as all the proteins that are made by a particular cell type. The first dimension separates proteins according to their isoelectric points and the second dimension separates according to mass. (Figure 20.12)

- Tandem mass spectrometry is a method to determine short amino acid sequences within proteins. These sequences are then used to identify the protein, provided the genome sequence encoding the protein is already known. (Figure 20.13)

20.3 Biotechnology

- Microorganisms can be genetically engineered to produce human products such as insulin. (Figure 20.14)

- Microorganisms are also used to reduce pollutants in the environment, a phenomenon called bioremediation.

- Transgenic organisms, also called genetically modified organisms, are made via gene replacement or gene addition.

- Transgenic livestock can be genetically engineered to produce human hormones in their milk. (Figure 20.15)

- The Ti plasmid in *Agrobacterium tumefaciens* has been extensively used to produce transgenic plants, such as plants that are resistant to herbicides. (Figures 20.16, 20.17)

- The cloning of mammals can be achieved by fusing a somatic cell with an egg that has had its nucleus removed. The possibility of human cloning raises serious ethical questions. (Figure 20.18)

- DNA fingerprinting is a method of identification based on the occurrence of segments of DNA in the genomes of all individuals, such as STRs, that are variable in length among different individuals. (Figure 20.19)

- Gene therapy is a method to treat human diseases by the introduction of cloned genes into cells. The first gene therapy involved a disease called severe combined immunodeficiency syndrome (SCID). (Figure 20.20)

TEST YOURSELF

1. Vectors used to clone genes were derived originally from
   a. proteins.
   b. plasmids.
   c. viruses.
   d. all of the above.
   e. b and c only.

2. Restriction enzymes
   a. are used to cut DNA into pieces for gene cloning.
   b. are naturally produced by bacteria cells to prevent viral infection.
   c. produce sticky ends on DNA fragments.
   d. all of the above.
   e. a and c only.

3. A DNA library produced by isolating mRNA from a cell and using reverse transcriptase to make DNA molecules is called a _____ library.
   a. genomic
   b. mRNA
   c. proteonomic
   d. cDNA
   e. chromosomal

4. Researchers can identify the colonies that contain the vector with the gene of interest by
   a. screening the different colonies using a probe that is complementary to the gene of interest.
   b. keeping records of the particular colonies that were supposedly inoculated with the particular probe.
   c. using PCR to determine the gene sequence of the DNA in the different colonies.
   d. using DNA fingerprinting techniques to identify the particular gene of interest.
   e. none of the above.

5. A method used to detect a particular DNA sequence within a mixture of many DNA fragments is
   a. PCR.
   b. colony hybridization.
   c. DNA fingerprinting.
   d. DNA sequencing.
   e. Southern blotting.

6. Why is *Taq* polymerase used in PCR rather than other DNA polymerases?
   a. *Taq* polymerase is a synthetic enzyme that produces DNA strands at a faster rate than natural polymerases.
   b. *Taq* polymerase is a heat-stable form of DNA polymerase that can function after exposure to the high temperatures that are necessary for PCR.
   c. *Taq* polymerase is easier to isolate than other DNA polymerases.
   d. *Taq* polymerase is the DNA polymerase commonly produced by most eukaryotic cells.
   e. All of the above.

7. The method of determining the base sequence of DNA is
   a. PCR.
   b. gene cloning.
   c. DNA fingerprinting.
   d. DNA sequencing.
   e. gene mapping.

8. During bioremediation, microorganisms are used to
   a. clone genes from eukaryotic organisms.
   b. introduce correct genes into individuals with genetic diseases.
   c. decrease pollutants in the environment.
   d. produce useful products such as insulin.
   e. all of the above.

9. Organisms that carry genes that were introduced using molecular techniques are called
   a. transgenics.
   b. clones.
   c. mutants.
   d. genetically modified organisms.
   e. both a and d.

10. DNA fingerprinting is used
   a. to provide a means of precise identification of an organism, such as the identification of specific strains of bacteria.
   b. as a forensics tool to provide evidence in a criminal case.
   c. to determine genetic relationships between individuals.
   d. to determine the identity of an individual.
   e. all of the above.

CONCEPTUAL QUESTIONS

1. Define recombinant DNA technology and recombinant DNA.

2. Explain how using one restriction enzyme to cut both a plasmid and a gene of interest will allow the gene to be inserted into the plasmid.

3. Explain how gel electrophoresis separates DNA fragments.

EXPERIMENTAL QUESTIONS

1. What is gene therapy? What is ADA deficiency?

2. In the investigation of Figure 20.20, how did the researchers treat ADA deficiency?

3. How successful was the gene therapy for ADA deficiency?

COLLABORATIVE QUESTIONS

1. Discuss the use of microorganisms for bioremediation.

2. Discuss the process of molecular pharming.

www.brookerbiology.com

This website includes answers to the Biological Inquiry questions found in the figure legends and all end-of-chapter questions.

21

GENOMES, PROTEOMES, AND BIOINFORMATICS

CHAPTER OUTLINE

Genomes and computer technology. The amount of data derived from the analyses of genomes is so staggering in size and complexity that researchers have turned to the use of computers to unravel the amazing information that genomes contain.

The unifying theme of biology is evolution. The genome of every living species is the product of over 3.5 billion years of evolution. We can understand the unity of modern organisms by realizing that all species evolved from an interrelated group of ancestors. Throughout this textbook, each chapter touches upon "Genomes & Proteomes" as a way to see the evolutionary connections among all forms of life, and also to understand how the genetic material produces the form and function of living organisms. By now, you may feel familiar with the concept of a **genome**, which is the complete genetic makeup of a cell or organism. The genome of each species is critical to its existence in several ways, including:

- The genome stores information in the form of genes, which provide a blueprint to produce the characteristics of organisms.

- The genome is copied and transmitted from generation to generation.

- The accumulation of genetic changes over the course of many generations produces the evolutionary changes that alter species and create new species.

An extension of genome analysis is the study of **proteomes**, which refers to the entire complement of proteins that a cell or organism can make. The function of most genes is to encode proteins, which are the key participants in maintaining cell structure and carrying out most cell functions. Analyzing the proteome of a single species and comparing the proteomes of different species allow biologists to understand the structure and function of cells, multicellular organisms, and the interactions between organisms and their environment.

The terms genomes and proteomes are sometimes confused with the terms genomics and proteomics. When we speak of genomes and proteomes, we are talking about characteristics of living organisms. By comparison, genomics and proteomics are experimental approaches to studying genomes and proteomes, respectively. An exciting advance in biology has been the ability to analyze the DNA sequence of genomes and the expression of groups of genes. This technology is genomics. Likewise, other tools have been invented to study large groups of proteins; this technology is proteomics. The techniques of genomics and proteomics are discussed in Chapter 20.

In the first two sections of this chapter, our focus will be primarily on biology, not on experimental techniques. We will consider the sizes and compositions of the genomes and proteomes of prokaryotic and eukaryotic species. As you will learn, genomes and proteomes are full of surprises. For example, did you know that most of your DNA has no known function? From a molecular perspective, genomes and proteomes contain extensive and intriguing information, which has challenged researchers to study them using computer technology. In the last section of this chapter, we will consider how the field of **bioinformatics**, which uses computers to study biological information, has been critical in the study of genomes and proteomes.

21.1 Genome Sizes and Composition

The past decade has seen remarkable advances in our overall understanding of the entire genomes of several species. As genetic technology has progressed, researchers have gained an increasing ability to analyze the composition of genomes as a whole unit.

For many species, we now know their complete DNA sequence, which provides the most detailed description available of an organism's genome at the molecular level. In this section, we will survey the sizes of genomes in prokaryotic and eukaryotic species and examine their compositions. Genomes consist not only of genes but also of other types of noncoding sequences. For example, the genomes of all species have repetitive DNA— short repeated sequences. We will learn how certain types of repetitive DNA sequences are formed by a process called transposition. We will also examine how the duplication of genes can lead to families of related genes.

Prokaryotic Genomes Often Contain a Circular Chromosome with a Few Thousand Genes and Little Repetitive DNA

Geneticists have made great progress in the study of bacterial and archaeal genomes. Some of the key features of prokaryotic chromosomes are described in Chapter 18 (refer back to Figure 18.2). Researchers are interested in the genomes of prokaryotic species for three main reasons. First, bacteria cause many different diseases that affect people, as well as plants and animals. Studying the genomes of bacteria reveals important clues about the process of infection and also may help us find ways to combat bacterial infection. A second reason for studying prokaryotic genomes is that the information we learn about these tiny creatures often applies to more complex organisms. For example, basic genetic mechanisms such as DNA replication and gene regulation were first understood in *Escherichia coli*. That knowledge provided a critical foundation to understand how these processes work in humans and other eukaryotic species. A third reason is evolution. The origin of the first eukaryotic cell probably involved a union between an archaeal and a bacterial cell. The study of prokaryotic genomes helps us understand how all modern species evolved.

Because they are both relatively small and of great interest to us, the entire genomes of many prokaryotic species have been sequenced and analyzed. The chromosomes of prokaryotes are usually in the range of several hundred thousand to a few million base pairs in length. Genomic researchers refer to 1 million base pairs as 1 megabase pair, abbreviated Mbp or Mb. Most prokaryotes that have been studied contain a single type of chromosome, though multiple copies may be present in a single cell. However, some prokaryotes are known to have different chromosomes. For example, *Vibrio cholerae*, the bacterium that causes the diarrheal disease known as cholera, has two different chromosomes in each cell, one 2.9 Mb and the other 1.1 Mb. Bacterial chromosomes are often circular. For example, the two chromosomes in *V. cholerae* are circular, as is the single type of chromosome found in *E. coli*. However, linear chromosomes are found in some prokaryotic species, such as *Borrelia burgdorferi*, which is the bacterium that causes Lyme disease, the most common tick-borne disease in the U.S. Certain bacterial species may even contain both linear and circular chromosomes. *Agrobacterium tumefaciens*, which infects plants

Table 21.1 | Examples of Prokaryotic Genomes That Have Been Sequenced*

Species	Genome size (Mb)†	Number of genes‡	Description
Methanobacterium thermoautotrophicum	1.7	1,869	An archaeon that produces methane.
Haemophilus influenzae	1.8	1,743	One of several different bacterial species that causes respiratory illness and meningitis.
Sulfolobus solfataricus	3.0	3,032	An archaeon that metabolizes sulfur.
Lactobacillus plantarum	3.3	3,052	A type of lactic acid bacterium used in the production of cheese and yogurt.
Mycobacterium tuberculosis	4.4	4,033	The bacterium that causes the respiratory disease tuberculosis.
Escherichia coli	4.6	4,289	A common intestinal bacterium; certain strains can cause human illness.
Bacillus anthracis	5.2	5,439	The bacterium that causes the disease anthrax.

* Prokaryotic species often exist in different strains that may differ slightly in their genome size and number of genes. The data are from common strains of the indicated species. The species shown in this table have only one type of chromosome.

† Mb equals 1 million base pairs.

‡ The number of genes is an estimate based on the analysis of genome sequences.

and causes a disorder called crown gall disease, has one linear chromosome (2.1 Mb) and one circular chromosome (3.0 Mb).

Table 21.1 compares the sequenced genomes from several prokaryotic species. They range in size from 1.7 to 5.2 Mb. The total number of genes is correlated with the total genome size. Roughly 1,000 genes are found for every Mb of DNA. Compared to eukaryotic genomes described later, prokaryotic genomes are less complex. They lack centromeres and telomeres, and they have a single origin of replication. Also, compared to their eukaryotic counterparts, prokaryotic chromosomes have relatively little repetitive DNA.

In addition to one or more chromosomes, prokaryotes often have plasmids, usually circular pieces of DNA that exist independently of the bacterial chromosome. Plasmids are typically small, in the range of a few thousand to tens of thousands of base pairs in length, though some can be quite large, even hundreds of thousands of base pairs. The various functions of plasmids are described in Chapter 18. Their use as vectors in genetic engineering is discussed in Chapter 20.

FEATURE INVESTIGATION

Venter, Smith, and Colleagues Sequenced the First Complete Genome, That of *Haemophilus influenzae*

The first genome to be entirely sequenced was that of the bacterium *Haemophilus influenzae*. This bacterium causes a variety of diseases in humans, including respiratory illnesses and bacterial meningitis. *H. influenzae* has a relatively small genome size, approximately 1.8 Mb.

Scientists can follow different strategies when tackling a genome-sequencing project. One strategy, which has been used for larger eukaryotic genomes, requires extensive mapping. This means that the genome is cut into large pieces of DNA whose locations are known within a given chromosome. These large pieces are then cut into smaller and smaller pieces, whose relative locations are known within the larger pieces. Once small DNA pieces have been mapped throughout the whole genome, their DNA sequences are determined by the procedure known as dideoxynucleotide DNA sequencing, which is described in Chapter 20.

An alternative strategy for sequencing an entire genome is called **shotgun DNA sequencing**. In this approach, researchers use the technique of dideoxy sequencing to randomly sequence many DNA fragments from the genome. As a matter of chance, some of the fragments are overlapping, which allows researchers to order them as they are found in the intact chromosome (refer back to Figure 20.9). The advantage of shotgun DNA sequencing is that it does not require extensive genetic mapping, which can be time-consuming. A disadvantage is that researchers will waste time sequencing the same region of DNA more than once.

To obtain a complete sequence of a genome with the shotgun approach, how do researchers decide how many fragments to sequence? We can calculate the probability that a base will not be sequenced using this approach with the following equation:

$$P = e^{-m}$$

where P is the probability that a base will be left unsequenced, e is the base of the natural logarithm ($e = 2.72$), and m is the number of bases sequenced divided by the total genome size. For example, in the case of *H. influenzae*, with a genome size of 1.8 Mb, if researchers sequenced 9.0 Mb, $m = 5$ (i.e., 9.0 Mb divided by 1.8 Mb):

$$P = e^{-m} = e^{-5} = 0.0067, \text{ or } 0.67\%.$$

This means that if we randomly sequence 9.0 Mb, which is five times the length of a single genome, we are likely to miss only 0.67% of the genome. With a genome size of 1.8 Mb, we would miss about 12,060 nucleotides out of approximately 1.8 million. Such missed sequences are typically on small DNA fragments that—as a matter of random chance—did not happen to be sequenced. The missing links in the genome can be sequenced later using mapping methods.

The general protocol conducted by Craig Venter, Hamilton Smith, and colleagues in this discovery-based investigation is described in **Figure 21.1**. The researchers isolated chromosomal DNA from *H. influenzae* and broke the DNA into small fragments, approximately 2,000 bp in length. These fragments were randomly cloned into vectors that allow the DNA to be propagated in *E. coli*. Each *E. coli* clone carried a vector with a different piece of DNA from *H. influenzae*. As discussed in Chapter 20, this collection of clones is called a DNA library. The researchers then subjected many of these clones to the procedure of DNA sequencing. They sequenced a total of approximately 10.8 Mb of DNA.

Figure 21.1 Determination of the complete genome sequence of *Haemophilus influenzae* by Venter, Smith, and colleagues.

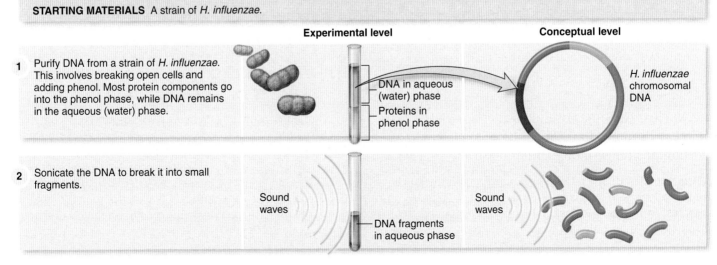

GOAL The goal is to obtain the entire genome sequence of *Haemophilus influenzae*. This information will reveal its genome size and also which genes the organism has.

STARTING MATERIALS A strain of *H. influenzae*.

Experimental level   **Conceptual level**

1. Purify DNA from a strain of *H. influenzae*. This involves breaking open cells and adding phenol. Most protein components go into the phenol phase, while DNA remains in the aqueous (water) phase.

DNA in aqueous (water) phase
Proteins in phenol phase

H. influenzae chromosomal DNA

2. Sonicate the DNA to break it into small fragments.

Sound waves

Sound waves

DNA fragments in aqueous phase

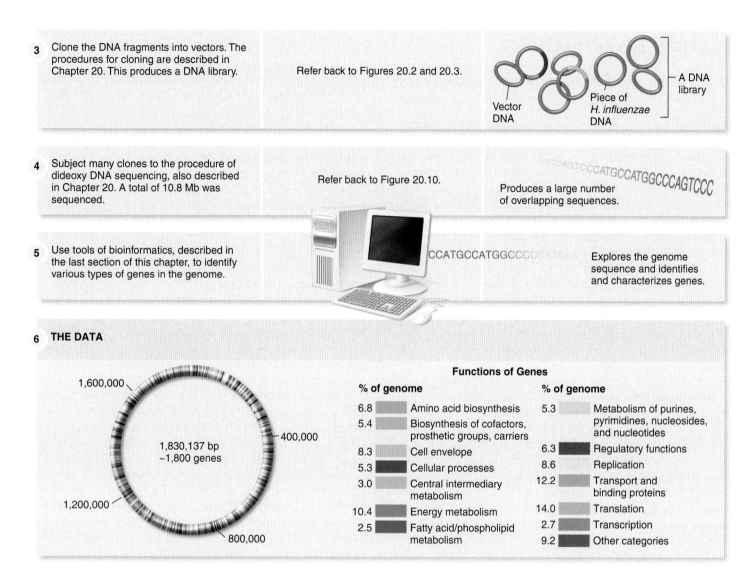

3 Clone the DNA fragments into vectors. The procedures for cloning are described in Chapter 20. This produces a DNA library.

Refer back to Figures 20.2 and 20.3.

Vector DNA

Piece of *H. influenzae* DNA

A DNA library

4 Subject many clones to the procedure of dideoxy DNA sequencing, also described in Chapter 20. A total of 10.8 Mb was sequenced.

Refer back to Figure 20.10.

Produces a large number of overlapping sequences.

CCAGTCCCATGCCATGGCCCAGTCCC

5 Use tools of bioinformatics, described in the last section of this chapter, to identify various types of genes in the genome.

Explores the genome sequence and identifies and characterizes genes.

CCATGCCATGGCCCC

6 THE DATA

1,600,000

400,000

1,830,137 bp ~1,800 genes

1,200,000

800,000

Functions of Genes

% of genome			% of genome		
6.8		Amino acid biosynthesis	5.3		Metabolism of purines, pyrimidines, nucleosides, and nucleotides
5.4		Biosynthesis of cofactors, prosthetic groups, carriers			
8.3		Cell envelope	6.3		Regulatory functions
5.3		Cellular processes	8.6		Replication
3.0		Central intermediary metabolism	12.2		Transport and binding proteins
10.4		Energy metabolism	14.0		Translation
2.5		Fatty acid/phospholipid metabolism	2.7		Transcription
			9.2		Other categories

The outcome of this genome-sequencing project was a very long DNA sequence. In 1995, Venter, Smith, and colleagues published the entire DNA sequence of *H. influenzae*. The researchers then analyzed the genome sequence by computer to obtain information about the properties of the genome. Questions that they asked included, How many genes does the genome contain? and What are the likely functions of those genes? Later in this chapter, we will learn how scientists can answer such questions with the use of computers. The data in Figure 21.1 summarize the results that the researchers obtained. The *H. influenzae* genome is composed of 1,830,137 bp of DNA. The computer analysis predicted 1,743 genes. Based on their similarities to known genes in other species, the researchers also predicted the functions of nearly two-thirds of these genes. The diagram shown in the data of Figure 21.1 places genes in various categories based on their predicted function. These results gave the first complete "genome picture" of a living organism!

The Nuclear Genomes of Eukaryotes Are Sets of Linear Chromosomes That Vary Greatly in Size and Composition Among Different Species

As we learned in Chapter 15, the genome found in the nucleus of eukaryotic species is usually found in sets of linear chromosomes. In humans, for example, one set contains 23 chromosomes—22 autosomes and one sex chromosome, X or Y.

In addition, certain organelles in eukaryotic cells contain a small amount of their own DNA. These include the mitochondrion, which plays a role in ATP synthesis, and the chloroplast found in plants and algae, which carries out photosynthesis. The DNA found in these organelles is referred to as extranuclear DNA to distinguish it from the DNA that is found in the cell nucleus. In this chapter, we will focus on the nuclear genome of eukaryotes.

Table 21.2	Examples of Eukaryotic Nuclear Genomes That Have Been Sequenced		
Species	**Nuclear genome size (Mb)**	**Number of genes**	**Description**
Saccharomyces cerevisiae (baker's yeast)	12.1	~6,200	One of the simplest eukaryotic species; it has been extensively studied by researchers to understand eukaryotic cell biology and other molecular mechanisms.
Caenorhabditis elegans (nematode worm)	100	~19,000	A model organism used to study animal development.
Drosophila melanogaster (fruit fly)	180	~14,000	A model organism used to study many genetic phenomena, including development.
Arabidopsis thaliana (thale cress)	120	~26,000	A model organism studied by plant biologists.
Oryza sativa (rice)	460	~40,000–50,000	A cereal grain with a relatively small genome; it is very important worldwide as a food crop.
Homo sapiens (humans)	2,900	~20,000–25,000	The sequencing of the human genome will help to elucidate our inherited traits and may aid in the identification and treatment of diseases.

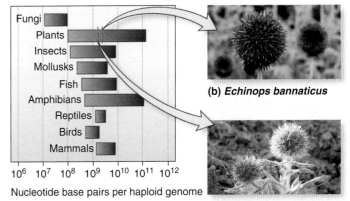

(b) *Echinops bannaticus*

(a) Genome size   **(c)** *Echinops nanus*

Figure 21.2   Genome sizes among selected groups of eukaryotes.   (a) Genome sizes among various groups of eukaryotes are shown on a log scale. As an example for comparison, two closely related species of globe thistle are pictured, *Echinops bannaticus* in (b) and *Echinops nanus* in (c). These species have similar characteristics, but *E. bannaticus* has nearly double the amount of DNA that *E. nanus* does due to the accumulation of repetitive DNA sequences.

Biological inquiry: What are two reasons why the groups of species shown in (a) have variation in their total amount of DNA?

In the past decade or so, the DNA sequence of entire nuclear genomes has been determined for several eukaryotic species (**Table 21.2**). The genomes of simpler eukaryotes, such as yeast, carry a few thousand different genes, while the genomes of more complex eukaryotes contain tens of thousands of genes. Motivation to sequence these genomes comes from four main sources. First, the work of researchers can greatly benefit from genome sequences that allow them to identify and characterize the genes of model organisms. This has been the impetus for genome projects involving baker's yeast (*Saccharomyces cerevisiae*), the fruit fly (*Drosophila melanogaster*), a nematode worm (*Caenorhabditis elegans*), the simple plant called thale cress (*Arabidopsis thaliana*), and the mouse (*Mus musculus*). A second reason for genome sequencing is to gather more information in order to identify and treat human diseases, which is an important aim for sequencing the human genome. Researchers hope that the DNA sequence of the human genome will help to identify genes in which mutation plays a role in disease. Third, the genomes of agriculturally important species have also been

the subject of genome-sequencing projects. An understanding of a species' genome may help us to develop new strains of livestock and plant species that have improved traits from an agricultural perspective. Fourth and finally, evolutionary biologists are interested in genome sequences as a way to establish evolutionary relationships.

When we speak of genome size, this means the total amount of DNA, often measured in megabase pairs. Genome size is not the same as the number of genes. The relative sizes of nuclear genomes vary dramatically among different eukaryotic species (**Figure 21.2a**). In general, increases in the amount of DNA are correlated with increases in cell size, cell complexity, and body complexity. For example, yeast have smaller genomes than animals. However, major variations in genome sizes are observed within organisms that are similar in form and function. For example, the total amount of DNA found within different species of amphibians can vary over 100-fold. The DNA content of closely related species can also vary. **Figure 21.2b,c** compares two closely related species of the plant called the globe thistle, *Echinops bannaticus* and *Echinops nanus*. These species have similar numbers of chromosomes, but *E. bannaticus* has nearly double the amount of DNA. The larger genome of *E. bannaticus* is not likely to contain twice as many genes. Rather, the genome composition includes many **repetitive sequences**, which are short DNA sequences that are present in many copies. Repetitive sequences are often abundant in eukaryotic species.

Repetitive sequences fall into two broad categories, moderately and highly repetitive. **Moderately repetitive sequences** are found a few hundred to several thousand times in the genome.

In some cases, these sequences are multiple copies of the same gene. For example, the genes that encode ribosomal RNA (rRNA) are found in many copies. The cell needs a large amount of rRNA for its cellular ribosomes. This is accomplished by having and expressing multiple copies of the genes that encode rRNA. In addition, other types of functionally important sequences can be moderately repetitive. For example, multiple copies of origins of replication are found in eukaryotic chromosomes. Other moderately repetitive sequences may play a role in the regulation of gene transcription and translation.

Highly repetitive sequences are those that are found tens of thousands or even millions of times throughout the genome. Each copy of a highly repetitive sequence is relatively short, ranging from a few nucleotides to several hundred nucleotides in length. Most of these sequences have no known function, and whether they benefit the organism is a matter of debate. A widely studied example is the *Alu* family of sequences found in humans and other primates. The *Alu* sequence is approximately 300 bp long. This sequence derives its name from the observation that it contains a site for cleavage by a restriction endonuclease known as *Alu*I. The *Alu* sequence is present in about 1 million copies in the human genome. It represents about 10% of the total human DNA and occurs approximately every 5,000–6,000 bases. Evolutionary studies suggest that the *Alu* sequence arose 65 million years ago from a section of a single ancestral gene known as the 7SL RNA gene. Remarkably, over the course of 65 million years, the *Alu* sequence has been copied and inserted into the human genome so often that it now appears more than 1 million times. The mechanism for the proliferation of *Alu* sequences will be described later.

Some highly repetitive sequences, like the *Alu* family, are interspersed throughout the genome. However, other highly repetitive sequences are clustered together in a tandem array in which a very short nucleotide sequence is repeated many times in a row. In *Drosophila*, for example, 19% of the chromosomal DNA is highly repetitive DNA found in tandem arrays. An example is shown here:

A A T A T A A T A T A A T A T A A T A T A A T A T A A T
T T A T A T T A T A T T A T A T T A T A T T A T A T A T T A

In this particular tandem array, two related sequences, AATAT and AATATAT, are repeated many times. Highly repetitive sequences, which contain tandem arrays of short sequences, are commonly found in centromeric regions of chromosomes and can be quite long, sometimes more than 1 million bp in length!

Figure 21.3 shows the composition of the relative classes of DNA sequences that are found in the nuclear genome of humans. Surprisingly, the coding regions of genes make up only about 2% of our genome! The other 98% is composed of noncoding sequences. Though we often think of genomes as being the repository of sequences that code for proteins, most eukaryotic genomes are largely composed of other types of sequences. Intron DNA is the second most common category at 24%. Unique noncoding DNA, whose function is largely unknown, constitutes 15%. Repetitive DNA makes up 59% of the DNA in the genome. Much of the repetitive DNA is derived from transposable elements, which are described next.

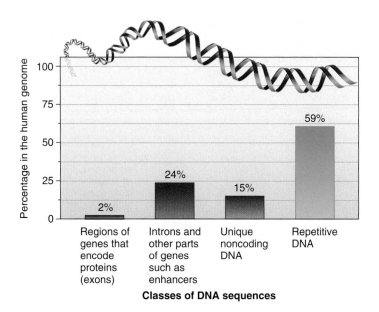

Figure 21.3   The composition of DNA sequences that are found in the nuclear genome of humans.   Note that only about 2% of our genome codes for proteins; most of our genome is made up of repetitive sequences.

Transposable Elements Can Move from One Chromosomal Location to Another

As we have seen, genomes are composed of several types of DNA sequences. Such sequences include not just the coding sequences of genes but also other sequences, including introns and repetitive sequences. Much of the moderately repetitive and highly repetitive DNA that is found in genomes is derived from a process called **transposition**, in which a short segment of DNA moves within a cell from its original site to a new site in the genome. The DNA segments that transpose themselves are known as **transposable elements (TEs)**. They range from a few hundred to several thousand base pairs in length. TEs have sometimes been referred to as "jumping genes," because they are inherently mobile. Barbara McClintock first identified transposable elements in the late 1940s from her studies with corn plants (**Figure 21.4**). She identified a segment of DNA that could move into and out of a gene that affected the color of corn kernels, producing a speckled appearance. Since that time, biologists have discovered many different types of TEs in prokaryotes, protists, fungi, plants, and animals—transposable elements have been found in all species examined.

Though Barbara McClintock identified TEs in corn in the late 1940s, her work was met with great skepticism because many researchers had trouble believing that DNA segments could be mobile. The advent of molecular technology in the 1960s and 1970s allowed scientists to understand more about the characteristics of TEs that enable their movement. Most notably, research involving bacterial TEs eventually progressed to a molecular understanding of the transposition process. In 1983, over 30 years after her initial discovery, McClintock was awarded the Nobel Prize.

(a)　　　　**(b)**

Figure 21.4 **Barbara McClintock, who discovered transposable elements.** As shown in part **(b)**, when a transposable element is found within a pigment gene in corn, its frequent movement causes the kernel color to be speckled.

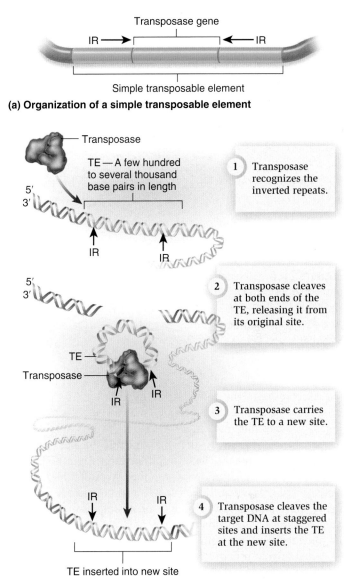

(a) Organization of a simple transposable element

Transposase gene

IR ⟶　⟵ IR

Simple transposable element

Transposase

TE—A few hundred to several thousand base pairs in length

5′
3′

IR　　　IR

1　Transposase recognizes the inverted repeats.

5′
3′

2　Transposase cleaves at both ends of the TE, releasing it from its original site.

TE →
Transposase —

IR　　IR

3　Transposase carries the TE to a new site.

IR　　IR

4　Transposase cleaves the target DNA at staggered sites and inserts the TE at the new site.

TE inserted into new site

(b) Cut-and-paste mechanism of transposition

Figure 21.5 **Simple transposable elements and their mechanism of transposition.** **(a)** Simple transposable elements (TEs) contain inverted repeat (IR) sequences at each end and a gene that encodes transposase in the middle. **(b)** Simple transposition occurs by a cut-and-paste mechanism.

As researchers have studied TEs from many species, they have found that DNA sequences within transposable elements are organized in several different ways, and they can move by different molecular mechanisms. Both ends of many TEs have inverted repeats (IRs), DNA sequences that are identical (or very similar) but run in opposite directions (**Figure 21.5a**), such as the following:

　　5′—CTGACTCTT—3′　and　5′—AAGAGTCAG—3′
　　3′—GACTGAGAA—5′　　　　3′—TTCTCAGTC—5′

Depending on the particular transposable element, inverted repeats range from 9 to 40 base pairs in length. In addition, TEs may contain a central region that encodes **transposase**, an enzyme that facilitates transposition.

As shown in **Figure 21.5b**, one type of transposition occurs by a cut-and-paste mechanism. Transposase first recognizes the inverted repeats in the TE and then removes the sequence from its original site. Next, the transposase/TE complex moves to a new location, and transposase then inserts the sequence into the chromosome. Transposition may occur when a cell is in the process of DNA replication. If a transposable element is removed from a site that has already replicated, and is inserted into a chromosomal site that has not yet replicated, the TE will increase in number after DNA replication is complete. This is one way for TEs to become more prevalent in a genome.

Another category of transposable elements moves via an RNA intermediate. This form of transposition is very common but is found only in eukaryotic species. These types of elements are known as **retroelements** or **retrotransposons**. By comparison, TEs that move via DNA are called **transposons**. The *Alu* sequence in the human genome is an example of a retroelement. Some retroelements contain genes that encode the enzymes reverse transcriptase and integrase, which are needed in the transposition process (**Figure 21.6a**). Retroelements may also contain repeated sequences called terminal repeats at each end that facilitate their recognition. The mechanism of retroelement

movement is shown in **Figure 21.6b**. First, the element is transcribed into RNA by the enzyme RNA polymerase. Reverse transcriptase uses this RNA as a template to synthesize a double-stranded DNA molecule. The ends of the double-stranded DNA are then recognized by integrase, which catalyzes the insertion of the DNA into the host chromosomal DNA. The integration of retroelements can occur at many locations within the genome. Furthermore, because a single retroelement can be copied into many RNA transcripts, retroelements may accumulate rapidly within a genome. This explains how the *Alu* element in the human genome was able to proliferate and constitute 10% of our genome.

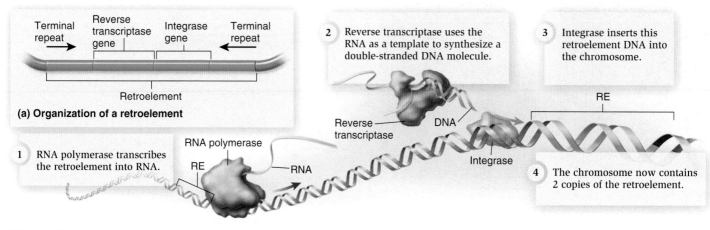

(a) Organization of a retroelement

Terminal repeat | Reverse transcriptase gene | Integrase gene | Terminal repeat

Retroelement

2 Reverse transcriptase uses the RNA as a template to synthesize a double-stranded DNA molecule.

3 Integrase inserts this retroelement DNA into the chromosome.

Reverse transcriptase   DNA

RE

1 RNA polymerase transcribes the retroelement into RNA.

RNA polymerase

RE   RNA

Integrase

4 The chromosome now contains 2 copies of the retroelement.

(b) Mechanism of movement of a retroelement

Figure 21.6 **Retroelements and their mechanism of transposition.** Retroelements are found only in eukaryotic species. (a) Some retroelements contain terminal repeats and may contain genes that encode the enzymes reverse transcriptase and integrase, which are needed in the transposition process. (b) The process that adds a copy of a retroelement into a host chromosome.

Biological inquiry: Based on their mechanism of movement, which type of TEs do you think would proliferate more rapidly in a genome, simple TEs (see Figure 21.5b) or retroelements?

The biological significance of TEs remains a matter of debate. According to the **selfish DNA hypothesis**, TEs exist because they have the characteristics that allow them to insert themselves into the host cell DNA. In other words, they resemble parasites in the sense that they inhabit the host without offering any advantage. They can proliferate within the host as long as they do not harm the host to the extent that they significantly disrupt survival. However, TEs can do harm. For example, if they jump into the middle of an important gene and thereby disrupt its function, this may have a negative impact on the phenotype of an organism.

Alternatively, other biologists have argued that TEs may provide benefits to a given species. For example, bacterial TEs often carry an antibiotic-resistance gene that provides the organism with a survival advantage. In addition, TEs may cause greater genetic variability by promoting chromosomal rearrangements. As discussed next, such rearrangements can cause a misaligned crossover during meiosis, and promote the formation of a gene family.

Gene Duplications Provide Additional Material for Genome Evolution, Sometimes Leading to the Formation of Gene Families

Let's now turn our attention to a way that the number of genes in a genome can increase. Small chromosomal duplications are important because they provide raw material for the addition of more genes into a species' genome. Such duplications can create two or more copies of the same gene (**Figure 21.7a**). Over the course of many generations, each version of the gene accumulates different mutations, producing **homologous genes**—two or more similar genes that are derived from a single ancestral gene. During evolution, this type of event can occur several times, creating a family of many similar genes called a **gene family**. Homologous genes within a single species are also called **paralogs**.

A mechanism that produces gene duplications is a misaligned crossover (**Figure 21.7b**). In this example, two homologous chromosomes have paired with each other during meiosis, but the homologs are misaligned. If a crossover occurs, this produces one chromosome with a gene duplication, one with a deletion, and two normal chromosomes. Each of these chromosomes will be segregated into different haploid cells. If a haploid cell carrying the chromosome with the gene duplication participates in fertilization with another gamete, an offspring with a gene duplication is produced. In this way, gene duplications can form and be transmitted to future generations. The presence of multiple copies of the same transposable element in a genome can foster this process, because the chromosomes may misalign while attempting to align TEs that are at different locations in the same chromosome.

A well-studied example of gene duplications is the globin gene family found in animals. The globin genes encode polypeptides that are subunits of proteins that function in oxygen binding. Hemoglobin, which is made in red blood cells, carries oxygen throughout the body. In humans, the globin gene family is composed of 14 homologous genes that were originally derived from a single ancestral globin gene (**Figure 21.8**). According to an evolutionary analysis, the ancestral globin gene duplicated between 500 and 600 million years ago. Since that time, additional duplication events and chromosomal rearrangements have occurred to produce the current number of 14 genes on three different human chromosomes. Four of these are pseudogenes, which are genes that have been produced by gene duplication but have accumulated mutations that make them nonfunctional so they are not transcribed into RNA.

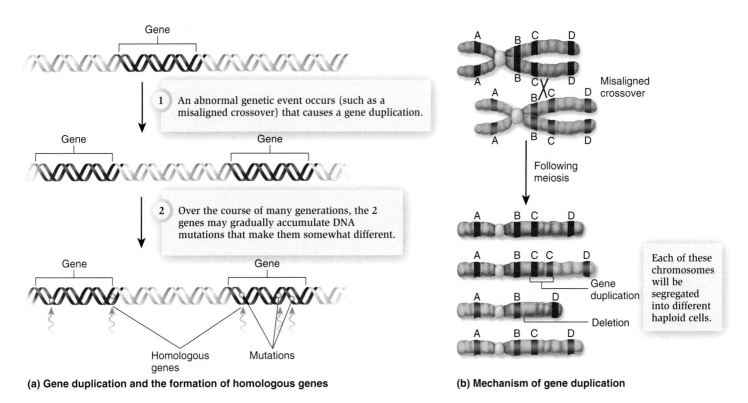

(a) Gene duplication and the formation of homologous genes

(b) Mechanism of gene duplication

Figure 21.7 **Gene duplication and the evolution of homologous genes.** **(a)** A gene duplication produces two copies of the same gene. Over time, these copies accumulate different random mutations, which results in homologous genes with similar but not identical DNA sequences. **(b)** Mechanism of gene duplication. If two homologous chromosomes misalign during meiosis, a crossover will produce a chromosome with a gene duplication.

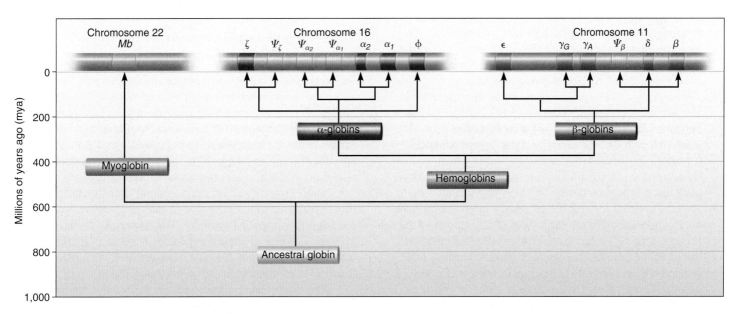

Figure 21.8 **The evolution of the globin gene family in humans.** The globin gene family evolved from a single ancestral globin gene. The first gene duplication produced two genes that accumulated mutations and became the genes encoding myoglobin and the family of hemoglobins. The modern myoglobin gene is found on chromosome 22. An ancestral hemoglobin gene duplicated to produce the α- and β-globins. Further duplications of ancestral α-globin and β-globin genes produced several homologous genes on chromosomes 16 and 11, respectively. The four genes shown in gray are nonfunctional pseudogenes.

Biological inquiry: What is the advantage of a gene family?

Gene families have been important in the evolution of complex traits. Even though all of the globin polypeptides are subunits of proteins that play a role in oxygen binding, the accumulation of different mutations in the various family members has created globins that are more specialized in their function. For example, myoglobin is better at binding and storing oxygen in muscle cells, whereas the hemoglobins are better at binding and transporting oxygen via the red blood cells. Also, different globin genes are expressed during different stages of human development. For instance, the epsilon (ε)- and zeta (ζ)-globin genes are expressed very early in embryonic life, while the gamma (γ)-globin genes are maximally expressed during the second and third trimesters of gestation. Following birth, the γ-globin genes are turned off and the β-globin gene is turned on. These differences in the expression of the globin genes reflect the differences in the oxygen transport needs of humans during the embryonic, fetal, and postpartum stages of life.

The Human Genome Project Has Stimulated Genomic Research

Before ending our discussion of genomes, let's consider what is now the largest genome project in history, the **Human Genome Project**. Scientists had been discussing how to undertake this project since the mid-1980s. In 1988, the National Institutes of Health in Bethesda, Maryland, established an Office of Human Genome Research with James Watson as its first director. The Human Genome Project, which officially began on October 1, 1990, was a 13-year effort coordinated by the U.S. Department of Energy and the National Institutes of Health. From its outset, the Human Genome Project had the following goals:

1. *To identify all human genes.* This involved mapping the locations of genes throughout the entire genome. The data from the Human Genome Project suggest that humans have about 20,000 to 25,000 different genes.
2. *To obtain the DNA sequence of the entire human genome.* The first draft of a nearly completed DNA sequence was published in February 2001, and a second draft was published in 2003. The entire genome is approximately 2.9 billion nucleotides in length. If the entire human genome were typed in a textbook like this, with about 3,000 letters per page, it would be nearly 1 million pages long!
3. *To develop technology for the generation and management of human genome information.* Some of the efforts of the Human Genome Project have involved improvements in molecular genetic technology such as gene cloning, DNA sequencing, and so forth. The Human Genome Project has also developed computer tools to allow scientists to easily access up-to-date information from the project, and analytical tools to interpret genomic information.
4. *To analyze the genomes of model organisms.* These include *E. coli, S. cerevisiae, D. melanogaster, C. elegans, A. thaliana,* and *M. musculus.*
5. *To develop programs focused on understanding and addressing the ethical, legal, and social implications of the results obtained from the Human Genome Project.*

An important question is: Who should have access to genetic information? Should employers, insurance companies, law enforcement agencies, and schools know our genetic makeup? The answer is complex and will require discussion among many groups.

A great benefit expected from the characterization of the human genome is the ability to identify our genes. A complete DNA sequence has made it profoundly easier for researchers to locate such genes. Furthermore, the DNA sequence of the human genome provides researchers with insight into the types of proteins encoded by these genes.

21.2 Proteomes

Thus far in this chapter, we have considered the genome characteristics of many different species, including humans. Because most genes encode proteins, a logical next step is to examine the functional roles of the proteins that a species can make. As mentioned, the entire collection of proteins that a cell or organism produces is called a proteome. As we move into the 21st century, a key challenge facing molecular biologists is the study of proteomes. Much like the study of genomes, this will require the collective contributions of many research scientists, as well as improvements in technologies that are aimed at explaining the complexities of the proteome. In this section, we will begin by considering the functional categories of proteins and examining their relative abundance in the proteome. Then, we will explore the molecular mechanisms that cause an organism's proteome to be much larger than its genome.

The Proteome Is a Diverse Array of Proteins with Many Kinds of Functions

The genomes of simple, unicellular organisms such as bacteria and yeast contain thousands of structural genes, while the genomes of complex, multicellular organisms contain tens of thousands. Such genome sizes can produce proteomes with thousands of different proteins. To bring some order to this large amount of complex information, researchers often organize proteins into different categories based on their functions.

Table 21.3 describes some general categories of protein function and provides examples of each type. Many approaches are used to categorize proteins. The categories listed in Table 21.3 are just one of the more general ways to evaluate protein function. For example, the data of Figure 21.1 categorize protein function in a different, more complicated way.

Dividing proteins into categories by function allows scientists to compare the amount of each type of protein, the protein abundance. The relative abundance of proteins can be viewed at two levels. First, we can consider abundance in the genome: the numbers of genes in the genome that encode a particular type or category of protein. For example, if a cell has genes that encode 10,000 different types of proteins, and 1,500 of these are different types of transporters, we would say that 15% of the genome is composed of transporters. However, such an analysis

Table 21.3 Categories of Proteins Found in the Proteome

Function	Examples
Metabolic enzymes	Hexokinase: Phosphorylates glucose during the first step in glycolysis. Glycogen synthetase: Uses glucose to synthesize a large carbohydrate known as glycogen.
Structural proteins	Tubulin: Forms cytoskeletal structures known as microtubules. Collagen: Found abundantly in the extracellular matrix of animals.
Motor proteins	Myosin: Involved in muscle cell contraction. Kinesin: Involved in the movement of chromosomes during cell division.
Cell-signaling proteins	Insulin: Influences target cell metabolism and growth. Insulin receptor: Recognizes insulin and initiates a cellular response.
Transport proteins	Lactose permease: Transports lactose across the bacterial cell membrane. Hemoglobin: Found in red blood cells, and transports oxygen throughout the body.
Gene expression and regulation proteins	Transcription factors: Regulate the expression of genes. Ribosomal proteins: Make up the structure of ribosomes, which are needed for the synthesis of new proteins.
Protective proteins	Antibodies: Fight viral and bacterial infections in vertebrate species. Antiviral proteins: Prevent viruses from binding to cells or successfully infecting cells in plants and animals.

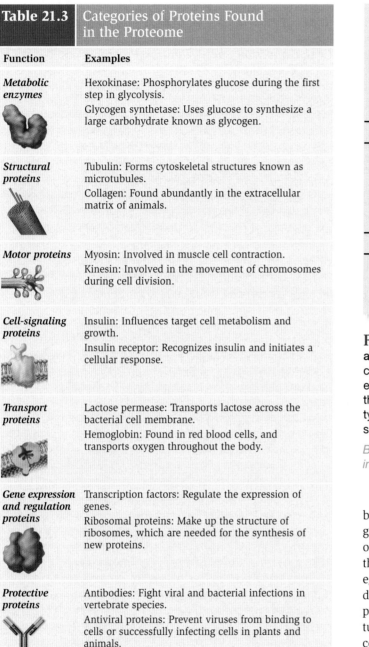

Figure 21.9 **A comparison of the proteomes in human liver and skeletal muscle cells.** Because all cells of the human body carry the same genome, the percentages of proteins that are encoded in the genome are the same in each cell type. However, the relative amounts of proteins that are made in different cell types can be vastly different, as is the case between liver and skeletal muscle cells.

Biological inquiry: What genetic process explains the differences in protein abundance in liver cells versus muscle cells?

ignores the phenomenon that genes are expressed at different levels—in other words, different types of proteins are made in different amounts. Therefore, a second way to view protein abundance is to consider abundance in the cell: the amount of a given protein or protein category that is actually made by a living cell. For example, less than 1% of human genes encode proteins, such as collagen, that are found in the extracellular matrix. Even so, these genes are highly expressed in certain cells, so that a large amount of this type of protein is made compared to other types.

Figure 21.9 is a general comparison of protein abundance in two cell types in humans—liver and muscle cells. The liver plays a key role in metabolism, while muscles are involved in

bodily movements. Both liver and muscle cells have the same genes. Therefore, at the level of the genome, the percentages of the different protein categories are identical. However, at the cellular level, the relative abundance of certain protein categories is quite different. Liver cells make a large number of different enzymes that play a role in the metabolism of fats, proteins, and carbohydrates. By comparison, their level of structural and motor proteins is relatively small. In contrast, muscle cells have fairly low levels of enzymes, but they have a high percentage of structural and motor proteins.

The following discussion describes each category of protein, its relative abundance in both the genome and the cell, and trends in abundance that are observed among different groups of species and cell types.

Metabolic Enzymes Metabolic enzymes, which accelerate chemical reactions within the cell, are a key category of proteins. Some enzymes assist in the breakdown of molecules or macromolecules into smaller units. These are known as catabolic enzymes and are important in generating cellular energy. In contrast, anabolic enzymes function in the synthesis of molecules and macromolecules. For example, glycogen synthetase is required for the synthesis of glycogen from glucose building blocks. In terms of abundance in the genome, typically 20–30% of a cell's genes encode metabolic enzymes.

However, many of these genes are expressed at relatively low levels, so that the abundance of metabolic enzymes in most cell types may be less than 20–30% of the total protein.

Structural Proteins   Structural proteins are involved in providing shape and form to cells and organisms. Examples include cytoskeletal proteins, such as actin or tubulin, and proteins of the extracellular matrix, such as collagen. Structural proteins are important in all species, but they tend to be more abundant in multicellular organisms such as plants and animals. In terms of abundance in the genome, only about 5% of a eukaryotic cell's genes encode structural proteins. However, certain structural proteins are expressed at very high levels, so these proteins can be much more abundant than 5% of the proteins made by particular cells. For example, in mammals, roughly 25% of the total protein in the body is a type of structural protein called collagen.

Motor Proteins   Motor proteins use energy to facilitate intracellular movements and the movements of whole cells. Examples are dynein and myosin. These motor proteins provide the power to move eukaryotic flagella and cause muscles to contract, respectively. Overall, genes that encode motor proteins constitute a small percentage (that is, less than 2%) of the genome of a cell or organism. However, in certain cell types, motor proteins can be abundant. For example, myosin is very abundant in muscle cells. In skeletal or heart muscle cells, usually 25–40% of the total protein is myosin.

Cell-Signaling Proteins   Many different types of proteins are needed so that cells can respond to environmental signals and send signals to each other. Proteins involved in cell signaling include hormones, hormone receptors, and intracellular proteins that form signal cascades, allowing a cell to respond to a signaling molecule. Cell signaling is needed by all species to respond to environmental changes. Among multicellular organisms, such as plants and animals, cell-to-cell signaling is more complex, so that the cells of the body can coordinate their activities. In terms of abundance in the human genome, roughly 12% of our genes encode proteins that are devoted to cell signaling. However, many proteins involved with cell signaling are only needed in small amounts, so their abundance in cells is considerably less than 12%.

Transport Proteins   A diverse array of membrane proteins are involved in the transport of ions and molecules across membranes. Among all species, typically 10–15% of the genome is made up of genes that encode proteins involved in transport. The protein abundance in the cell is usually less than 10–15%, though certain specialized cells, such as nerve cells, have a greater abundance of transport proteins in their membranes. Most transport proteins are involved in the transmembrane transport of solutes. These include proteins such as sugar transporters, ion pumps, and ion channels. Other proteins are involved in intercellular transport. For example, hemoglobin plays a role in the transport of oxygen throughout the bodies of vertebrates.

Gene Expression and Regulation Proteins   For proteins to be made, genes must be expressed. This process involves transcription, mRNA modification, and translation. In addition, genes are regulated so that their encoded product will be expressed at the correct times and in the correct amounts. Genes that encode proteins involved in gene expression and regulation constitute a large percentage of the genomes of all organisms, perhaps 25–30%. In terms of abundance in the cell, well over 25% of the different types of proteins made by cells are devoted to gene expression and regulation. Among more complex eukaryotes, the percentage tends to be higher. This increase can be attributed largely to the complexities of gene regulation and mRNA modification. In complex plants and animals, a substantial percentage of the genome is made up of genes that encode transcription factor proteins that regulate gene expression. For example, in both *Arabidopsis* and humans, roughly 5% of all genes encode proteins that function as transcription factors.

Protective Proteins   All species make proteins that help them to survive environmental stress. These include proteins that help organisms withstand high temperatures and other protective proteins that prevent damage caused by infectious agents. Mammals, for example, produce proteins called antibodies that attack infectious agents such as viruses and bacteria and thereby help to ward off infection. Plants also produce a diverse array of proteins that prevent viral invasion.

From the perspective of abundance, this category of proteins is somewhat difficult to calculate. At the genome level, the percentage of genes that encode proteins involved in protection would be relatively small, less than 2%. However, in vertebrates, genes that encode antibodies undergo a specialized type of rearrangement in the B cells of the immune system (see Chapter 53). With regard to the whole organism, antibodies are by far the most diverse of all proteins; a single person can make millions of different types of antibodies, each with a slightly different amino acid sequence. Even so, each B cell makes only one type of antibody. The amount of protein made by a given cell can vary greatly depending on the cell's environment. For example, in the absence of an infection, the amounts of antiviral proteins that are synthesized by animal and plant cells are relatively small. In contrast, when confronted with a pathogen, the amounts of protective proteins produced by these cells greatly increase.

Proteomes Are Larger Than Genomes

From the sequencing and analysis of genomes, researchers can identify all or nearly all of the genes that a given species has. For example, the human genome is predicted to contain between 20,000 and 25,000 different genes that encode proteins. Even so, humans can make many more than 25,000 different types of proteins. The larger size of the proteome relative to the genome is primarily due to two types of cellular processes, as described next.

First, changes in pre-mRNA structure may ultimately affect the resulting amino acid sequence of a protein. The most important alteration that occurs commonly in eukaryotic species is **alternative splicing**, which is described in Chapter 13. For many genes, a single pre-mRNA can be spliced into more than one version (**Figure 21.10a**). The splicing is often cell specific, or it may be related to environmental conditions. Alternative splicing is widespread, particularly among more complex eukaryotes (refer back to Table 13.1). It can lead to the production of several or perhaps dozens of different polypeptide sequences from the same pre-mRNA. This greatly increases the number of potential proteins in the proteome.

A second process that greatly diversifies the composition of a proteome is the phenomenon of **post-translational covalent modification** (**Figure 21.10b**). Such modifications can be permanent or reversible. Permanent modifications are primarily involved with the assembly and construction of functional proteins. These alterations include proteolytic processing (the cleavage of a polypeptide to a smaller unit), disulfide bond formation, and the attachment of prosthetic groups, sugars, or lipids. In contrast, reversible modifications, such as phosphorylation, acetylation, and methylation, often transiently affect the function of a protein. The covalent bonds are formed and then later broken by cellular enzymes. Because a given type of protein may be subjected to several different types of modifications, this can greatly increase the forms of a particular protein that are found in a cell at any given time.

21.3 Bioinformatics

In the two previous sections, we have learned that the number of genes in a genome and the number of proteins that are made by a given cell type are extremely large. In the 1960s and 1970s, when the tools of molecular biology first became available, researchers tended to focus on the study of just one or a few genes and proteins at a time. While this is a useful approach, scientists came to realize that certain properties of life arise by complex interactions involving the expression of many genes and the functioning of many different proteins. This is the field of systems biology. Such an awareness challenged researchers to invent new tools to study many genes and many proteins at the same time. These tools involved experimental procedures—genomics and proteomics—that allowed researchers to study complex interactions. To manage the huge amounts of data produced by these studies, researchers turned to the use of computers.

As a very general definition, bioinformatics describes any use of computers to handle biological information. Scientists primarily use computers to store and analyze data. We often think of bioinformatics in the context of analyzing genetic data. Even so, bioinformatics can also be applied to information from various sources, such as patient statistics and scientific literature.

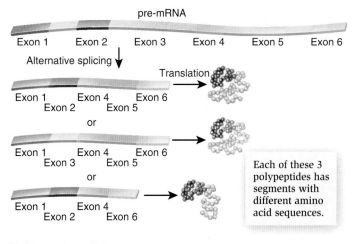

(a) Alternative splicing

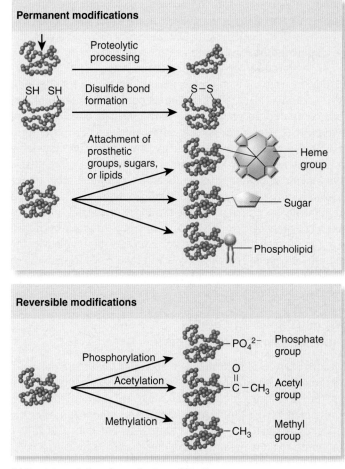

(b) Post-translational covalent modification

Figure 21.10 Cellular mechanisms that increase protein diversity. (a) During alternative splicing, the pattern of exons that remains in a mature mRNA can be different, creating multiple types of transcripts from the same gene. (b) In post-translational covalent modification, after a protein is made, it can be modified in a variety of ways, some of which are permanent and some reversible.

This rapidly developing branch of biology is highly interdisciplinary, using principles from mathematics, statistics, biology, chemistry, and physics.

But why do we need bioinformatics? Simply put, the main issues are size and speed. Earlier in this chapter, we learned that the human genome has been sequenced and that it is approximately 2.9 billion base pairs long. A single person, or even a group of talented mathematicians, cannot, in a reasonable length of time, analyze such an enormous amount of data. Instead, the data are put into computers and then scientists devise computational procedures to study it. The ability of computers to analyze data at a rate of millions or even billions of operations per second has made it possible to solve biological problems that were thought intractable a few decades ago. Biologists have made many important discoveries using a bioinformatic approach. For example, we can compare the entire genome sequences of different organisms, which contain millions or even billions of base pairs, and identify genes that are homologous in two or more different species. We can also compare genes in the same species and discover how mutations in particular genes are correlated with the development of diseases such as cancer and diabetes.

In this section, we will consider the branch of bioinformatics that focuses on using molecular information to study biology. This area, also called **computational molecular biology**, uses computers to characterize the molecular components of living things. Molecular genetic data, which comes in the form of a DNA, RNA, or protein sequence, are particularly amenable to computer analysis. Such studies usually rely on three basic components: a computer (or hardware), a computer program (or software), and some type of data. In genetic research, the data are typically a particular genetic sequence or several sequences that a researcher or clinician wants to study. In this section, we will first survey the fundamental concepts that underlie the analysis of genetic sequences. We will then consider how these methods are used to provide knowledge about how biology works at the molecular level.

Sequence Files Are Stored and Analyzed by Computers

The first step in bioinformatics is to collect and store data in a computer. As an example, let's consider a gene sequence as a type of data. The gene sequence must first be determined experimentally using the technique of DNA sequencing. After the sequence is obtained, the next step is to put that data into a computer. To store data in a computer, a scientist creates a computer data file and enters the data into the file. For short genetic sequences, less than a few hundred nucleotides in length, this may be done using a keyboard to enter (that is, type) the DNA sequence into the file. For very long genetic sequences, however, like those obtained in genome-sequencing projects, keyboarding data would be a tedious and error-prone process. Typically, genetic sequence data are entered into a computer file by laboratory instruments that can read experimental data—such as data from a DNA-sequencing experiment—and enter the sequence directly into a computer.

Genetic sequence data in a computer data file can then be investigated in many different ways, corresponding to the many questions a researcher might ask about the sequence and its functional significance, including:

1. Does a sequence contain a gene?
2. Does a gene sequence contain a mutation that might cause a disease?
3. Where are functional sequences such as promoters, regulatory sites, and splice sites located within a particular gene?
4. From the sequence of a structural gene, what is the amino acid sequence of the polypeptide encoded by that gene?
5. Is a genetic sequence similar to any other known sequences?
6. What is the evolutionary relationship between two or more genetic sequences?

To answer these and many other questions, computer programs have been written to analyze genetic sequences in particular ways.

As an example, let's consider a computer program aimed at translating a DNA sequence into an amino acid sequence. **Figure 21.11** shows a short computer data file of a DNA sequence that is presumed to be part of the coding sequence of a structural gene. In this figure, only the coding strand of DNA is shown. A computer program can analyze this sequence and print out the possible amino acid sequences that this DNA sequence would encode. The program relies on the triplet genetic code (refer back to Table 12.1). In the example shown in Figure 21.11, the computer program shows the results for all three possible reading frames, beginning at nucleotide 1, 2, or 3, respectively. In a newly obtained DNA sequence, a researcher would not know the proper reading frame, so the computer program provides all three. If you look at the results, reading frames 1 and 2 include several stop codons, while reading frame 3 does not. From these results, reading frame 3 is likely to be the correct one. Also, for a new DNA sequence, a researcher may not know which DNA strand is the coding strand. Therefore, the sequence of the other DNA strand, which is not shown in this figure, would also be analyzed by this computer program.

The Scientific Community Has Collected Computer Data Files and Stored Them in Large Computer Databases

Over the past several decades, the amount of genetic information generated by researchers and clinicians has become enormous. The Human Genome Project, for example, has produced more data than any other undertaking in the history of biology. With these advances, scientists have realized that another critical use of computers is to store the staggering amount of data produced from genetic research.

When a large number of computer data files are collected, stored in a single location, and organized for rapid search and retrieval, this collection is called a **database**. The files within databases are often annotated, which means they contain a concise description of each gene sequence, the name of the organism from which the sequence was obtained, and the function of the encoded protein, if it is known. The file may also

Computer DNA sequence file

```
5′ ATGTCCACGC  GGTCCTGGAA  AACCCAGGCT  TGGGCAGGAA  ACTCTCTGAC
   TTTGGACAGG  AAACAAGCTA  TATTGAAGAC  AACTGCAATC  AAAATGGTGC
   CATATCACTG  ATCTTCTCAC  TCAAAGAAGA  AGTTGGTGCA  TTGGCCAAAG
   TATTGCGCTT  ATTTGAGGAG  AATGATGTAA  ACCTGACCCA  CATTGAATCT
   AGACCTTCTC  GTTTAAAGAA  AGATGAGTAT  GAATTTTTCA  CCCATTTGGA
   TAAACGTAGC  CTGCCTGCTC  TGACAAACAT  CATCAAGATC  TTGAGGCATG
   ACATTGGTGC  CACTGTCCAT  GAGCTTTCAC  GAGATAAGAA  GAAAGACACA
   GTGCCCTGGT  TTCCCAAG 3′
```

Run a computer program that translates this DNA sequence into an amino acid sequence in all 3 reading frames.

Possible amino acid sequences

5′ ➤ 3′ Frame 1

Met Ser Thr Arg Ser Trp Lys Thr Gln Ala Trp Ala Gly Asn Ser Leu Thr Leu Asp Arg Lys Gln Ala Ile Leu Lys Thr Thr Ala Ile Lys Met Val Pro Tyr His **STOP** Ser Ser His Ser Lys Lys Lys Leu Val His Trp Pro Lys Tyr Cys Ala Tyr Leu Arg Arg Met Met **STOP** Thr **STOP** Pro Thr Leu Asn Leu Asp Leu Leu Val **STOP** Arg Lys Met Ser Met Asn Phe Ser Pro Ile Trp Ile Asn Val Ala Cys Leu Leu **STOP** Gln Thr Ser Ser Arg Ser **STOP** Gly Met Thr Leu Val Pro Leu Ser Met Ser Phe His Glu Ile Arg Arg Lys Thr Gln Cys Pro Gly Ser Gln

5′ ➤ 3′ Frame 2

Cys Pro Arg Gly Pro Gly Lys Pro Arg Leu Gly Gln Glu Thr Leu **STOP** Leu Trp Thr Gly Asn Lys Leu Tyr **STOP** Arg Gln Leu Gln Ser Lys Trp Cys His Ile Thr Asp Leu Leu Thr Gln Arg Arg Ser Trp Cys Ile Gly Gln Ser Ile Ala Leu Ile **STOP** Gly Glu **STOP** Cys Lys Pro Asp Pro His **STOP** Ile **STOP** Thr Phe Ser Phe Lys Glu Arg **STOP** Val **STOP** Ile Phe His Pro Phe Gly **STOP** Thr **STOP** Pro Ala Cys Ser Asp Lys His His Gln Asp Leu Glu Ala **STOP** His Trp Cys His Cys Pro **STOP** Ala Phe Thr Arg **STOP** Glu Glu Arg His Ser Ala Leu Val Pro Lys

5′ ➤ 3′ Frame 3

Val His Ala Val Leu Glu Asn Pro Gly Leu Gly Arg Lys Leu Ser Asp Phe Gly Gln Glu Thr Ser Tyr Ile Glu Asp Asn Cys Asn Gln Asn Gly Ala Ile Ser Leu Ile Phe Ser Leu Lys Glu Glu Val Gly Ala Leu Ala Lys Val Leu Arg Leu Phe Glu Glu Asn Asp Val Asn Leu Thr His Ile Glu Ser Arg Pro Ser Arg Leu Lys Lys Asp Glu Tyr Glu Phe Phe Thr His Leu Asp Lys Arg Ser Leu Pro Ala Leu Thr Asn Ile Ile Lys Ile Leu Arg His Asp Ile Gly Ala Thr Val His Glu Leu Ser Arg Asp Lys Lys Lys Asp Thr Val Pro Trp Phe Pro

Figure 21.11   **The use of a computer program to translate a DNA sequence into an amino acid sequence.**   The top part of this figure shows the sequence of a segment of the coding strand of a structural gene (artificially divided into groups of 10 nucleotide bases for ease of reading). A computer program translates the DNA sequence into an amino acid sequence based on the genetic code. The program produces three different amino acid sequences, as shown at the bottom of the figure. In this example, reading frame 3 is likely to be the correct reading frame because it does not contain any stop codons.

describe other features of significance and provide a published reference that contains the sequence.

The research community has collected genetic information from thousands of research labs and created several large databases. **Table 21.4** describes some of the major genetic databases in use worldwide. These databases enable researchers to access and compare genetic sequences that are obtained by many laboratories. Later in this chapter, we will learn how researchers can use databases to analyze genetic sequences.

The databases described in Table 21.4 collect genetic information from many different species. Scientists have also created more specialized databases, called **genome databases**, that focus on the genetic characteristics of a single species. Genome databases have been created for most model organisms such as bacteria (*E. coli*), yeast (*S. cerevisiae*), worms (*C. elegans*), fruit flies (*D. melanogaster*), plants (*A. thaliana*), mice (*M. musculus*), and humans (*H. sapiens*). The primary aim of genome databases is to organize the information from sequencing and mapping projects for a single species. Genome databases identify the known genes within an organism and describe their map locations in the genome. In addition, a genome database may provide information concerning gene alleles, bibliographic information, a directory of researchers who study the species, and other pertinent information.

Table 21.4	Examples of Major Computer Databases
Type	**Description**
Nucleotide sequence	DNA sequence data are collected into three internationally collaborating databases: GenBank (a U.S. database), EMBL (European Molecular Biology Laboratory Nucleotide Sequence Database), and DDBJ (DNA Databank of Japan). These databases receive sequence and sequence annotation data from genome projects, sequencing centers, individual scientists, and patent offices. These databases are accessed via the Internet and on CD-ROM.
Amino acid sequence	Amino acid sequence data are collected into a few international databases, including Swissprot (Swiss protein database), PIR (Protein Information Resource), Genpept (translated peptide sequences from the GenBank database), and TrEMBL (Translated sequences from the EMBL database).
Three-dimensional structure	PDB (Protein Data Bank) collects the three-dimensional structures of biological macromolecules with an emphasis on protein structure. These are primarily structures that have been determined by X-ray crystallography and nuclear magnetic resonance (NMR), but some models are included in the database. These structures are stored in files that can be viewed on a computer screen.

From: Persson, Bengt (2000) Bioinformatics in protein analysis. EXS 88, 215–231.

Computer Programs Can Identify Homologous Sequences

Let's now turn our attention to the use of computer technology to identify genes that are evolutionarily related. Organisms that are closely related evolutionarily tend to have genes with similar DNA sequences. As an example, let's consider the gene that encodes β-globin. As discussed earlier, β-globin is a polypeptide found in hemoglobin, which carries oxygen in red blood cells. The β-globin gene is found in humans and other vertebrates. **Figure 21.12a** compares a short region of this gene from the laboratory mouse (*Mus musculus*) and rat (*Rattus norvegicus*). As you can see, the gene sequences are similar but not identical. In this 60-nucleotide sequence, five differences are observed. The reason for the sequence similarity is that the genes are derived from the same ancestral gene. This idea is shown schematically in **Figure 21.12b**. An ancestral gene was found in a rodent species that was a common ancestor to both mice and rats. During evolution, this ancestral species diverged into different species, which eventually gave rise to several modern rodent species, including mice and rats. Following divergence, the β-globin genes accumulated distinct mutations that produced somewhat different base sequences for this gene. Therefore, in mice and rats, the β-globin genes have homologous sequences—their sequences are similar because they are derived from the same ancestral gene, but they are not identical because each species has accumulated a few different random mutations. Homologous genes in different species are also called **orthologs**.

Analyzing genes that are homologous to each other helps biologists understand the evolutionary relationships among modern species. This topic is considered in Units IV and V. Later in this section, we will also see how the study of homology may provide important clues about gene function.

But how do researchers, with the aid of computers, determine if two genes are homologous to each other? To evaluate the similarity between two sequences, a matrix can be constructed. In a general way, **Figure 21.13** illustrates the use of a simple dot matrix. In Figure 21.13a, the word BIOLOGY is compared with itself. Each point in the grid corresponds to one position of each sequence. The matrix allows all such pairs to be compared simultaneously. Dots are placed where the same letter occurs at the two corresponding positions. Sequences that are alike produce a diagonal line on the matrix. In contrast, Figure 21.13b compares two similar but different sequences: BIOLOGY and ECOLOGY. This comparison produces only a partial diagonal line. Overall, the key observation is that regions of similarity are distinguished by the occurrence of many dots along a diagonal line within the matrix. This same concept holds true when genetic sequences are compared with each other.

To relate homologous genes in different species, researchers must compare relatively long DNA sequences. For such long sequences, a simple dot matrix approach is not adequate. Instead, dynamic computer programming methods are used to identify similarities between genetic sequences. This approach was

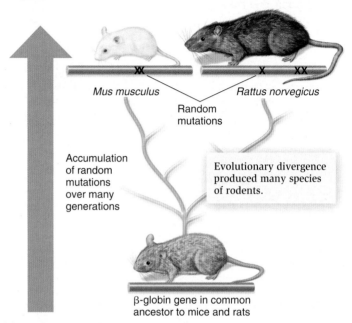

(a) A comparison of one DNA strand of the mouse and rat β-globin genes

(b) The formation of homologous β-globin genes during evolution of mice and rats

Figure 21.12   Structure and formation of the homologous β-globin genes in mice and rats.   (a) A comparison of a short region of the gene that encodes β-globin in laboratory mice (*Mus musculus*) and rats (*Rattus norvegicus*). Only one DNA strand is shown. Bases that are identical between the two sequences are connected by a vertical line. The β-globin genes are similar because they are derived from the same ancestral gene, but they are not identical because each species has accumulated a few different random mutations since the divergence of their latest common ancestor. (b) The formation of these homologous β-globin genes during evolution. An ancestral β-globin gene was found in a rodent species that was a common ancestor to both mice and rats. This ancestral species later diverged into different species, which gave rise to modern rodent species, such as mice and rats. During this process, the β-globin genes accumulated different random mutations, causing the DNA sequences of these two homologous genes to be slightly different from each other.

proposed originally by Saul Needleman and Christian Wunsch in 1970. Dynamic programming methods are theoretically similar to a dot matrix, but they involve mathematical operations that are beyond the scope of this textbook. In their original work, Needleman and Wunsch demonstrated that whale myoglobin and human β-globin genes have similar sequences.

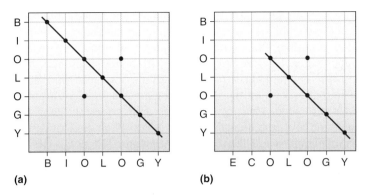

Figure 21.13   **The use of a simple dot matrix.**   In these comparisons, a diagonal line indicates sequence similarity. **(a)** The word BIOLOGY is compared with itself. Dots are placed where the same letter occurs at the two corresponding positions. Notice the diagonal line that is formed. **(b)** Two similar but different sequences, BIOLOGY and ECOLOGY are compared with each other. Notice that only a partial line is formed by this comparison.

A Database Can Be Searched to Identify Homologous Sequences and Thereby Identify Gene Function

Because they are derived from the same ancestral gene, homologous genes usually carry out similar or identical functions. In many cases, the first indication of the function of a newly determined gene sequence is through homology to known sequences in a database. An example is the gene that is altered in cystic fibrosis patients. After this gene was identified in humans, bioinformatic methods revealed that it is homologous to several genes found in other species. Moreover, a few of the homologous genes were already known to encode proteins that function in the transport of ions and small molecules across the plasma membrane. This observation provided an important clue that cystic fibrosis involves a defect in membrane transport.

The ability of computer programs to identify homology between genetic sequences provides a powerful tool for predicting the function of genetic sequences. In 1990, Stephen Altschul, David Lipman, and their colleagues developed an approach called a **b**asic **l**ocal **a**lignment **s**earch **t**ool, or **BLAST**. The BLAST program has been described by many biologists as the single most important tool in computational molecular biology. This computer program can start with a particular genetic sequence—either a nucleotide or an amino acid sequence—and then locate homologous sequences within a large database.

As an example of how the BLAST program works, let's consider the human enzyme phenylalanine hydroxylase, which functions in the metabolism of phenylalanine, an amino acid. Recessive mutations in the gene that encodes this enzyme are responsible for the disease called phenylketonuria (PKU). The computational experiment shown in **Table 21.5** started with the amino acid sequence of this protein and used the BLAST program to search the Swissprot database, which contains hundreds

Table 21.5	Results from a BLAST Program Comparing Human Phenylalanine Hydroxylase with Database Sequences		
Match	**Percentage of identical* amino acids**	**Species**	**Function of of sequence†**
1	100	Human (*Homo sapiens*)	Phenylalanine hydroxylase
2	99	Orangutan (*Pongo pygmaeus*)	Phenylalanine hydroxylase
3	95	Mouse (*Mus musculus*)	Phenylalanine hydroxylase
4	95	Rat (*Rattus norvegicus*)	Phenylalanine hydroxylase
5	89	Chicken (*Gallus gallus*)	Phenylalanine hydroxylase
6	82	Pipid frog (*Xenopus tropicalis*)	Phenylalanine hydroxylase
7	82	Green pufferfish (*Tetradon nigroviridis*)	Phenylalanine hydroxylase
8	82	Zebrafish (*Danio rerio*)	Phenylalanine hydroxylase
9	80	Japanese pufferfish (*Takifugu rubripes*)	Phenylalanine hydroxylase
10	75	Fruit fly (*Drosophila melanogaster*)	Phenylalanine hydroxylase

* The number indicates the percentage of amino acids that are identical with the amino acid sequence of human phenylalanine hydroxylase.

† In some cases, the function of the sequence was determined by biochemical assay. In other cases, the function was inferred due to the high degree of sequence similarity with other species.

of thousands of different protein sequences. The BLAST program can determine which sequences in the Swissprot database are the closest matches to the amino acid sequence of human phenylalanine hydroxylase. Table 21.5 shows the results—the 10 best matches to human phenylalanine hydroxylase that were identified by the program. Because this enzyme is found in nearly all eukaryotic species, the program identified phenylalanine hydroxylase from many different species. The column to the right of the match number shows the percentage of amino acids that are identical between the species indicated and the human sequence. Because the human phenylalanine hydroxylase sequence is already in the Swissprot database, the closest match of human phenylalanine hydroxylase is to itself. The next nine sequences are in order of similarity. The next most similar sequence is from the orangutan, a close relative of humans. This is followed by two mammals, the mouse and rat, and then five vertebrates that are not mammals. The tenth best match is from *Drosophila*, an invertebrate.

You can see two trends in Table 21.5. First, the order of the matches follows the evolutionary relatedness of the various species to humans. The similarity between any two sequences is related to the time that has passed since they diverged from a common ancestor. Among the species listed in this table, humans are most similar to themselves, followed by the orangutan, other mammals, other vertebrates, and finally invertebrates. A second trend you may have noticed is that several of the matches involve species that are important from a research, medical, or agricultural perspective. Currently, our databases are biased toward organisms that are of interest to humans, particularly model organisms such as mice and *Drosophila*. Over the next several decades, the sequencing of genomes from many different species will tend to lessen this bias.

Overall, Table 21.5 is an example of the remarkable computational abilities of current computer technology. In less than a minute, the amino acid sequence of human phenylalanine hydroxylase can be compared with hundreds of thousands of different sequences to yield the data shown in this table! The main power of the BLAST program is its use with newly identified sequences, in which a researcher does not know the function of a gene or an encoded protein. When the BLAST program identifies a match to a sequence whose function is already known, it is likely that the newly identified sequence has an identical or similar function.

CHAPTER SUMMARY

21.1 Genome Sizes and Composition

- The genome is the complete genetic makeup of a cell or organism.
- Prokaryotic genomes are typically a single circular chromosome that has several hundred thousand to a few million base pairs of DNA. Such genomes usually have a few thousand different genes. (Table 21.1)
- Venter, Smith, and colleagues used a shotgun DNA-sequencing strategy to determine the first sequence of a prokaryotic genome, that of *Haemophilus influenzae*. (Figure 21.1)
- The nuclear genomes of eukaryotic species are composed of sets of linear chromosomes with a total length of several million to billions of base pairs. They typically contain several thousand to tens of thousands of genes. (Table 21.2)
- Genomes sizes vary among eukaryotic species. In many cases, this variation is due to the accumulation of noncoding regions of DNA, particularly repetitive DNA sequences. (Figures 21.2, 21.3)
- Transposable elements are segments of DNA that can move from one site to another, a process called transposition. (Figure 21.4)
- The enzyme transposase mediates the movement of transposable elements by a cut-and-paste mechanism. (Figure 21.5)
- Retroelements move to new sites in the genome via RNA intermediates. (Figure 21.6)
- Gene duplication may occur by a misaligned crossover. This is one mechanism that creates gene families, which are composed of two or more homologous genes. The members of a gene family often have similar yet specialized functions. (Figures 21.7, 21.8)
- The Human Genome Project undertook the DNA sequencing of the entire human genome.

21.2 Proteomes

- A proteome is the collection of proteins that a given cell makes.
- Proteins are often placed into broad categories based on their functions. These include metabolic enzymes, structural proteins, motor proteins, cell-signaling proteins, transport proteins, proteins involved with gene expression and regulation, and those involved with protection. (Table 21.3)
- Protein abundance can refer to their relative abundance in the genome or in the cell. (Figure 21.9)
- Protein diversity can increase via alternative splicing and post-translational modifications. (Figure 21.10)

21.3 Bioinformatics

- Bioinformatics involves the use of computers to analyze biological information, particularly genetic data such as DNA and protein sequences.
- Genetic information is stored in data files that can be analyzed using computer programs. (Figure 21.11)
- The research community has collected genetic information and created several large databases. (Table 21.4)
- Homologous genes are derived from the same ancestral genes and have accumulated random mutations that make their sequences slightly different. (Figure 21.12)
- A simple dot matrix illustrates the approach of identifying regions of similarity between two sequences. (Figure 21.13)
- Computer programs, such as BLAST, can identify homologous genes that are found in a database. (Table 21.5)

TEST YOURSELF

1. The entire collection of proteins produced by a cell or organism is
   a. a genome.
   b. bioinformatics.
   c. a proteome.
   d. a gene family.
   e. a protein family.

2. Important reasons for studying the genomes of prokaryotes include all of the following except:
   a. It may provide information that helps us understand how prokaryotes infect other organisms.
   b. It may provide a basic understanding of cellular processes that allow us to determine eukaryotic cellular function.
   c. It may provide the means to understand evolutionary processes.
   d. It will reveal the approximate number of genes that an organism has in its genome.
   e. All of the above are important reasons.

3. The enzyme that allows short segments of DNA to move within a cell from one location in the genome to another is
   a. transposase.
   b. DNA polymerase.
   c. protease.
   d. restriction endonuclease.
   e. reverse transcriptase.

4. A gene family includes
   a. a specific gene found in several different species that has a similar DNA sequence.
   b. all of the genes on the same chromosome.
   c. two or more homologous genes found within a single species.
   d. genes that code for structural proteins.
   e. both a and c.

5. Which of the following is not a goal of the Human Genome Project?
   a. Identify all human genes.
   b. Sequence the entire human genome.
   c. Address the legal and ethical implications resulting from the project.
   d. Develop programs to manage the information gathered from the project.
   e. Clone a human.

6. Bioinformatics is
   a. the analysis of DNA by molecular techniques.
   b. the use of computers to analyze and store biological information.
   c. a collection of gene sequences from a single individual.
   d. cloning.
   e. all of the above.

7. Using bioinformatics, evolutionary relationships between species can be characterized by identifying and analyzing
   a. phenotypes of selected organisms.
   b. homologous DNA sequences.
   c. fossils of ancestral species.
   d. all of the above.
   e. a and b only.

8. The BLAST program is a tool for
   a. inserting many DNA fragments into a cell at the same time.
   b. translating a DNA sequence into an amino acid sequence.
   c. identifying homology between a selected sequence and genetic sequences in large databases.
   d. all of the above.
   e. both b and c.

9. Repetitive sequences
   a. are short DNA sequences that are found many times throughout the genome.
   b. may be multiple copies of the same gene found in the genome.
   c. are more common in eukaryotes.
   d. all of the above.
   e. a and c only.

10. Proteins that provide movement of material inside the cell and movement of the entire cell are
    a. enzymes.
    b. structural proteins.
    c. motor proteins.
    d. transport proteins.
    e. antibodies.

CONCEPTUAL QUESTIONS

1. Define bioinformatics.

2. Explain how homologous genes may have arisen.

3. What are seven general categories of proteins found in the proteome?

EXPERIMENTAL QUESTIONS

1. What was the goal of the experiment conducted by Venter, Smith, and their colleagues?

2. How does shotgun DNA sequencing differ from procedures that involve mapping? What are an advantage and a disadvantage of the shotgun DNA sequencing approach?

3. What were the results of the study described in Figure 21.1?

COLLABORATIVE QUESTIONS

1. Compare the genomes of prokaryotic and eukaryotic organisms.

2. Discuss the concept and importance of transposable elements.

www.brookerbiology.com

This website includes answers to the Biological Inquiry questions found in the figure legends and all end-of-chapter questions.

GLOSSARY

A

A band   A wide, dark band produced by the orderly parallel arrangement of the thick filaments in the middle of each sarcomere.

abiotic   The term used to describe interactions between organisms and their nonliving environment.

aboral   Refers to the region opposite the mouth.

abortion   Procedures or circumstances that cause the death of an embryo or fetus after implantation.

abscisic acid   One of several plant hormones that help a plant to cope with environmental stress.

absorption   The process in which digested nutrients are transported from the digestive cavity into an animal's circulatory system.

absorption spectrum   A diagram that depicts the wavelengths of electromagnetic radiation that are absorbed by a pigment.

absorptive nutrition   The process whereby an organism secretes enzymes into food substrates, breaking down complex organic molecules into small organic molecules that are absorbed as food.

absorptive state   One of two alternating phases in the utilization of nutrients; occurs when ingested nutrients enter the blood from the gastrointestinal tract. The other phase is the postabsorptive state.

acclimatization   The process of fine-tuning an animal's adaptive mechanisms to a changing environment.

accommodation   The process in which contraction and relaxation of the ciliary muscles adjust the lens according to the angle at which light enters the eye.

acid   A molecule that releases hydrogen ions in solution.

acid hydrolase   A hydrolytic enzyme found in lysosomes that functions at acidic pH and uses a molecule of water to break a covalent bond.

acid rain   Precipitation with a pH of less than 5.6; acid rain results from the burning of fossil fuels, which releases sulfur dioxide and nitrogen oxide into the atmosphere, which in turn react with oxygen in the air to form sulfuric acid and nitric acid, and falls to the surface in rain or snow.

acidic   A solution that has a pH below 7.

acoelomate   An animal that lacks a body cavity entirely.

acquired antibiotic resistance   The common phenomenon of a previously susceptible strain becoming resistant to a specific antibiotic.

acquired immune deficiency syndrome (AIDS)   A disease caused by the human immunodeficiency virus (HIV) that leads to a defect in the immune system of infected individuals.

acrocentric   A chromosome in which the centromere is near one end.

acromegaly   A condition in which a person's GH levels remain elevated after growth has ceased, and the continued excess GH causes many bones, like those of the hands and feet, to thicken and enlarge.

acrosomal reaction   An event in fertilization in which the binding of a sperm cell to proteins located in the egg cell plasma membrane triggers a series of events producing the fast block to polyspermy and the entry of the sperm cell's nucleus into the egg cell.

acrosome   A special structure at the tip of a sperm's head that contains proteolytic enzymes that help break down the plasma membrane of the ovum at fertilization.

actin   A cytoskeletal protein.

actin filament   A thin type of protein filament composed of actin proteins that forms part of the cytoskeleton and supports the plasma membrane and plays a key role in cell strength, shape, and movement.

action potential   The movement of an electrical impulse along the plasma membrane, which occurs in animal nerve axons and some plant cells.

action spectrum   The rate of photosynthesis plotted as a function of different wavelengths of light.

activation energy   An initial input of energy in a chemical reaction that allows the molecules to get close enough to cause a rearrangement of bonds.

activator   A transcription factor that binds to DNA and increases the rate of transcription.

active immunity   The acquired response to exposure to any type of antigen.

active site   The location in an enzyme where the chemical reaction takes place.

active transport   The transport of a solute across a membrane against its gradient—that is, from a region of low concentration to higher concentration. In the case of ions, active transport is against an electrochemical gradient.

adaptation   The processes and structures by which organisms adjust to short-term or long-term changes in their environment.

adaptive radiation   The process whereby a single ancestral species evolves into a wide array of descendant species that differ greatly in their habitat, form, or behavior.

adenine (A)   A purine base found in DNA and RNA.

adenosine triphosphate (ATP)   A nucleotide that is a common energy source of all cells.

adenylyl cyclase   An enzyme in the plasma membrane that synthesizes cAMP from ATP.

adiabatic cooling   The process in which increasing elevation leads to a decrease in air temperature.

adventitious root   A root that is produced on the surfaces of stems (and sometimes leaves) of vascular plants; also, roots that develop at the bases of stem cuttings.

aerenchyma   Spongy plant tissue with large air spaces.

aerobic   Refers to a process that occurs in the presence of oxygen; a form of metabolism that requires oxygen.

aerobic respiration   During this type of respiration, O_2 is consumed and CO_2 is released.

aerotolerant anaerobe   A microorganism that does not use oxygen but is not poisoned by it either.

afferent arterioles   Blood vessels that provide a pathway for blood into a tissue or organ. For example, afferent arterioles in the kidney that supplies each glomerulus with blood.

aflatoxins   Fungal toxins that cause liver cancer and are a major health concern worldwide.

age structure   The relative numbers of individuals of each defined age group in a population.

age-specific fertility rate   The rate of offspring production for females of a certain age; used to help calculate how a population grows.

AIDS   Acquired immune deficiency syndrome. AIDS reduces the body's immunity by killing helper T cells.

air sac   A component of the avian respiratory system; air sacs—not lungs—expand when a bird inhales and shrink when it exhales, and they do not participate in gas exchange.

akinete   A thick-walled cell used to survive unfavorable conditions in a dormant state.

aldosterone   A steroid hormone made by the adrenal glands.

algae   A term that applies to about 10 phyla of protists that include both photosynthetic and nonphotosynthetic species.

alimentary canal   The single elongated tube of a digestive system with an opening at either end, through which food and eventually wastes pass from one end to the other.

alkaline   A solution that has a pH above 7.

alkaloids   A group of structurally related secondary metabolites that all contain nitrogen and usually have a cyclic, ringlike structure.

allantois   An extraembryonic membrane in the amniotic egg that serves as a disposal sac for metabolic wastes.

allee effect   The phenomenon that some individuals will fail to mate successfully purely by chance, for example, because of the failure to find a mate.

allele   A variant form of a gene.

allele frequency   The number of copies of an allele divided by the total number of alleles in a population.

allelochemical   A powerful plant chemical, often a root exudates, that kills other plant species.

allelopathy   The suppressed growth of one species due to the release of toxic chemicals by another species.

allodiploid   An alloploid that has only one set of chromosomes from two different species.

allometric growth   The pattern whereby different parts of the body grow at different rates with respect to each other.

allopatric   The term used to describe species occurring in different geographic areas.

allopatric speciation   A form of speciation that occurs when a population becomes geographically isolated from other populations and evolves into one or more new species.

alloploid   An organism that contains at least one set of chromosomes from two or more different species.

allopolyploid   An organism that contains two or more complete sets of chromosomes from two or more different species.

allosteric site   A site where a molecule can bind noncovalently and affect the function of the active site. The binding of a molecule to an allosteric site causes a conformational change in the enzyme that inhibits its catalytic function.

allotetraploid   A type of allopolyploid that contains two complete sets of chromosomes from two species for a total of four sets.

alternation of generations   The phenomenon that occurs in plants (and some protists) in which the life cycle alternates between multicellular diploid organisms, called sporophytes, and multicellular haploid organisms, called gametophytes.

alternative splicing   The splicing of pre-mRNA in more than one way to create two or more different polypeptides.

altruism   Behavior that appears to benefit others at a cost to oneself.

alveolus   A saclike structure of the lungs where gas exchange occurs.

Alzheimer's disease (AD)   The leading worldwide cause of dementia; characterized by a loss of memory and intellectual and emotional function.

amensalism   One-sided competition, where the interaction is detrimental to one species but not to the other.

Ames test   A test that helps ascertain whether or not an agent is a mutagen by using a strain of a bacterium, *Salmonella typhimurium.*

amino acid   The building block of proteins. Amino acids have a common structure in which a carbon atom, called the α-carbon, is linked to an amino group (NH_2) and a carboxyl group (COOH). The α-carbon also is linked to a hydrogen atom and a particular side chain.

amino terminus   *See* N-terminus.

aminoacyl site (A site)   One of the three sites for tRNA binding to the ribosome; the others are the peptidyl site (P site) and the exit site (E site). The A site is the site where incoming tRNA molecules bind to the mRNA (except for the initiator tRNA).

aminoacyl tRNA   *See* charged tRNA.

aminoacyl-tRNA synthetase   An enzyme that catalyzes the attachment of amino acids to tRNA molecules.

ammonia (NH_3)   One of the most highly toxic of the nitrogenous wastes because it disrupts pH, ion electrochemical gradients, and many chemical reactions that involve oxidations and reductions; typically produced in many aquatic species.

ammonification   The conversion of organic nitrogen to NH_3 and NH_4^+.

amnion   An innermost extraembryonic membrane in the amniotic egg; it protects the developing embryo in a fluid-filled sac called the amniotic cavity.

amniotes   A group of tetrapods with amniotic eggs that includes turtles, lizards, snakes, crocodiles, birds, and mammals.

amniotic egg   The structure that contains the developing embryo and the four separate extraembryonic membranes that it produces: the amnion, the yolk sac, the allantois, and the chorion.

amoeba   A protist cell that moves by pseudo-podia, which involves extending cytoplasm into filaments or lobes.

amoebocyte   A mobile cell within a sponge's mesohyl that absorbs food from choanocytes, digests it, and carries the nutrients to other cells.

amphibian   A tetrapod that has successfully invaded the land but must return to the water to reproduce.

amphipathic   In molecules, meaning they have a hydrophobic (water-fearing) region and a hydrophilic (water-loving) region.

ampulla   A muscular sac at the base of each tube foot of a echinoderm that stores water.

amygdala   An area of the brain known to be critical for understanding and remembering emotional situations.

amylase   A digestive enzyme involved in the digestion of starch.

anabolic reaction   A metabolic pathway that promotes the synthesis of larger molecules from smaller precursor molecules.

anabolism   The synthesis of cellular molecules and macromolecules, which usually requires an input of energy.

anaerobic   Refers to a process that occurs in the absence of oxygen; a form of metabolism that does not require oxygen.

anaerobic respiration   The breakdown of organic molecules in the absence of oxygen.

anagenesis   The pattern of speciation in which a single species is transformed into a different species over the course of many generations.

analogous structure   A trait that is the result of convergent evolution; structures have arisen

independently, two or more times, because species have occupied similar types of environments on the Earth.

anaphase   The phase of mitosis during which the sister chromatids separate from each other and move to opposite poles; the poles themselves also move farther apart.

anatomy   The study of the morphology of living organisms, such as plants and animals.

anchoring junctions   A junction found in between animal cells that attaches cells to each other and to the ECM.

androgens   Steroid hormones produced by the male testes that affect most aspects of male reproduction.

anemia   A condition characterized by lower than normal levels of hemoglobin, which reduces the amount of oxygen that can be stored in the blood.

aneuploidy   An alteration in the number of particular chromosomes so that the total number of chromosomes is not an exact multiple of a set.

angina pectoris   Chest pain during exertion due to the tissues being deprived of oxygen.

angiosperm   A flowering plant; the term means "enclosed seed," which reflects the presence of seeds within fruits.

animal cap assay   A type of experiment used extensively to identify factors (proteins) secreted by embryonic cells that induce cells in the animal pole to differentiate into mesoderm.

animal pole   In triploblast organisms, the pole of the egg where less yolk and more cytoplasm are concentrated.

Animalia   One of the four traditional eukaryotic kingdoms of the domain Eukarya.

anion   An ion that has a net negative charge.

annual   A plant that dies after producing seed during its first year of life.

anther   A cluster of microsporangia in a flower that produces pollen and then opens to release it.

antheridia   Round or elongate gametangia that produce plant sperm.

antagonist   A muscle or group of muscles that produces oppositely directed movements at a joint.

antenna complex   *See* light-harvesting complex.

anterior   Refers to the end of an animal where the head is found.

anteroposterior axis   One of the three axes along which the adult body pattern is organized; the others are the dorsoventral axis and the right-left axis.

anthropoid   A member of a class of primates that includes the monkeys and the hominoids; species are larger-brained and diurnal.

antibiotic   A chemical, usually made by microorganisms, that inhibits the growth of certain microorganisms.

antibody   A protein secreted by plasma cells that is part of the immune response; antibodies travel all over the body to reach antigens identical to those that stimulated their production, and then they combine with these antigens and guide an attack that eliminates the antigens or the cells bearing them.

anticodon   A three-nucleotide sequence in tRNA that is complementary to a codon in mRNA.

antidiuretic hormone (ADH)   A hormone secreted by the posterior pituitary gland that acts on kidney cells to decrease urine production.

antigen   Any foreign molecule that the host does not recognize as self and that triggers a specific immune response.

antigen-presenting cells (APCs)   Cells bearing fragments of antigen, called antigenic determinants or epitopes, complexed with the cell's MHC proteins.

antiparallel   An arrangement in DNA where one strand runs in the 5′ to 3′ direction while the other strand is oriented in the 3′ to 5′ direction.

antiporter   A type of transporter that binds two or more ions or molecules and transports them in opposite directions.

apical-basal-patterning genes   A category of genes that are important in early stages of plant development during which the apical and basal axes are formed.

apical-basal polarity   An architectural feature of plants in which they display an upper, apical pole and a lower, basal pole; shoot apical meristem occurs at the apical pole, and the root apical meristem occurs at the basal pole.

apical constriction   A cellular process during gastrulation that is crucial to development; a reduction in the diameter of the actin rings connected to the adherens junctions causes the cells to elongate toward their basal end.

apical region   The region of a plant that projects upwards, usually from the soil, and produces the leaves and flowers.

apomixis   A natural asexual reproductive process in which plant fruits and seeds are produced in the absence of fertilization.

apoplast   The continuum of water-soaked cell walls and intercellular spaces in a plant.

apoplastic transport   The movement of solutes through cell walls and the spaces between cells.

apoptosis   Programmed cell death.

aposematic coloration   Warning coloration that advertises an organism's unpalatable taste.

appendix   A finger-like projection in the gastrointestinal tract of animals having no known essential function but that may at one time have been an important part of the body's defense mechanisms.

aquaporin   A three-dimensional cell pore that allows water to diffuse through the membrane.

aqueous humor   A thin liquid in the anterior cavity behind the cornea of the vertebrate eye.

aqueous solution   A solution made with water.

aquifer   An underground water supply.

arbuscular mycorrhizae   Symbiotic associations between glomalean fungi and the roots of vascular plants.

Archaea   One of the three domains of life; the other two are Bacteria and Eukarya.

archaea   When not capitalized refers to a cell or species within the domain Archaea.

archegonia   Flask-shaped plant gametangia that enclose an egg cell.

archenteron   A cavity formed by the embryo during gastrulation that will become the organism's digestive tract.

area hypothesis   The proposal that larger areas contain more species than smaller areas because they can support larger populations and a greater range of habitats.

arteriole   A single-celled layer of endothelium surrounded by one or two layers of smooth muscle and connective tissue that delivers blood to the capillaries and distributes blood to regions of the body in proportion to metabolic demands.

artery   A blood vessel that carries blood away from the heart.

artificial selection   *See* selective breeding.

asci   Fungal sporangia shaped like sacs that produce and release sexual ascospores.

ascocarp   The type of fruiting body produced by ascomycete fungi.

ascomycetes   A phylum of fungi that produce sexual spores in saclike asci located at the surfaces of fruiting bodies known as ascocarps.

ascospore   The type of sexual spore produced by the ascomycete fungi.

aseptate   The condition of not being partitioned into smaller cells; usually refers to fungal cells.

asexual reproduction   A reproductive strategy that occurs when offspring are produced from a single parent, without the fusion of gametes from two parents. The offspring are therefore clones of the parent.

assimilation   In the case of nitrogen, the process by which plants and animals incorporate the ammonia and NO_3^- formed through nitrogen fixation and nitrification.

associative learning   A change in behavior due to an association between a stimulus and a response.

assort   The process of distributing.

asthma   A disease in which the smooth muscles around the bronchioles contract more than usual, increasing resistance to airflow.

AT/GC rule   Refers to the phenomenon that an A in one DNA strand always hydrogen bonds with a T in the opposite strand, while a G in one strand bonds with a C.

atherosclerosis   The condition in which large plaques may occlude (block) the lumen of an artery.

atmospheric (barometric) pressure   The pressure exerted by the gases in air on the body surfaces of animals.

atom   The smallest functional unit of matter that forms all chemical substances and cannot be further broken down into other substances by ordinary chemical or physical means.

atomic mass   An atom's mass relative to the mass of other atoms. By convention, the most common form of carbon, which has six protons and six neutrons, is assigned an atomic mass of exactly 12.

atomic nucleus   The center of an atom.

atomic number   The number of protons in an atom.

ATP synthase   An enzyme that utilizes the energy stored in an H^+ electrochemical gradient for the synthesis of ATP via chemiosmosis.

ATP-dependent chromatin remodeling enzyme   An enzyme that catalyzes a loosening in the compaction of chromatin; this loosening facilitates the ability of RNA polymerase to recognize and transcribe a gene.

ATP-driven pump   A common category of pump found in all living cells; this transporter has a binding site for ATP and hydrolyzes ATP to actively transport solutes against a gradient.

atrial natriuretic peptide (ANP)   A peptide secreted from the atria of the heart whenever blood levels of sodium increase; ANP causes a natriuresis by decreasing sodium reabsorption in the kidney tubules.

atrioventricular (AV) node   Specialized cardiac cells that sit near the junction of the atria and ventricles and conduct the electrical events from the atria to the ventricles.

atrioventricular (AV) valve   A one-way valve into the ventricles of the vertebrate heart through which blood moves from the atria.

atrium   In the heart, a single filling chamber to collect blood from the tissues.

atrophy   A reduction in the size of a structure, such as a muscle.

atropine   A potent toxin derived from the deadly nightshade plant.

audition   The sense of hearing.

autoimmune disease   In humans and many other vertebrates, the situation in which the body's normal state of immune tolerance breaks down, with the result that antibody-mediated or T-cell mediated attacks are directed against the body's own cells and tissues.

autonomic nervous system   The division of the peripheral nervous system that regulates homeostasis and organ function; also called the visceral nervous system.

autophagosome   Cellular material enclosed in a double membrane, produced by the process of autophagy.

autophagy   Meaning "the eating of one's self." A process whereby cellular material, such as a worn-out organelle, becomes enclosed in a double membrane and degraded.

autosomes   All of the chromosomes found in the cell nucleus of eukaryotes except for the sex chromosomes.

autotomy   In echinoderms, the ability to intentionally detach a body part, such as a limb, that will later regenerate.

autotroph   An organism that has metabolic pathways that directly harness energy from either inorganic molecules or light.

auxin   One of several types of hormones considered to be the "master" plant hormones because they influence plant structure, development, and behavior in many ways, often working with other hormones.

auxin efflux carrier One of several types of PIN proteins, which transport auxin out of plant cells.

auxin influx carrier A plasma membrane protein that transports auxin into plant cells.

avirulence gene (*Avr* gene) A gene that encodes a virulence-enhancing elicitor, which causes plant disease.

Avogadro's number As first described by Italian physicist Amedeo Avogadro, 1 mole of any element contains the same number of particles—6.022×10^{23}.

axillary bud A bud that occurs in the axil, the upper angle where a twig or leaf joins the stem.

axon An extension of the plasma membrane that is involved in sending signals from a neuron to neighboring cells.

axon hillock The part of the axon closest to the cell body.

axoneme The internal structure of eukaryotic cilia and flagella consisting of microtubules, the motor protein dynein, and linking proteins.

B

B cell A type of lymphocyte responsible for specific immunity.

bacilli Rods; one of the five major shapes of prokaryotic cells.

backbone The linear arrangement of phosphates and sugar molecules in a DNA or RNA strand.

Bacteria One of the three domains of life; the other two are Archaea and Eukarya.

bacteria When not capitalized, refers to a cell or species within the domain Bacteria.

bacterial artificial chromosome (BAC) A cloning vector in bacteria that can contain large DNA inserts.

bacterial colony A clone of genetically identical cells formed from a single cell.

bacteriophage A virus that infects bacteria.

bacteroid A modified bacterial cell of the type known as rhizobia present in mature root nodules of some plants.

balanced polymorphism The phenomenon in which two or more alleles are kept in balance, and therefore are maintained in a population over the course of many generations.

balancing selection A type of natural selection that maintains genetic diversity in a population.

balloon angioplasty A common treatment to restore blood flow through a blood vessel. A thin tube with a tiny, inflatable balloon at its tip is threaded through the artery to the diseased area; inflating the balloon compresses the plaque against the arterial wall, widening the lumen.

baroreceptor A pressure-sensitive region within the walls of certain arteries that contains the endings of nerve cells; these regions help to maintain blood pressure in the normal range for an animal.

Barr body A highly condensed X chromosome.

basal body A site at the base of flagella or cilia from which microtubules grow. Basal bodies are anchored on the cytosolic side of the plasma membrane.

basal metabolic rate (BMR) The metabolic rate of an animal under resting conditions, in good health, and not under stress of any kind.

basal nuclei Clusters of neuronal cell bodies that surround the thalamus on both sides and lie beneath the cerebral cortex; involved in planning and learning movements and also function via a complex circuitry to initiate or inhibit movements.

basal region The region of a plant that produces the roots.

basal transcription A low level of transcription resulting from the core promoter.

basal transcription apparatus For a eukaryotic structural gene, refers to the complex of GTFs, RNA polymerase II, and a DNA sequence containing a TATA box.

base 1. A component of nucleotides that is a single or double ring of carbon and nitrogen atoms. 2. A molecule that when dissolved in water lowers the H^+ concentration.

base pairs The structure in which two bases in opposite strands of DNA hydrogen bond with each other.

base substitution A mutation that involves the substitution of a single base in the DNA for another base.

basic local alignment search tool (BLAST) A computer program that can identify homologous genes that are found in a database.

basidiocarp The type of fruiting body produced by basidiomycete fungi.

basidiomycetes A group of fungi whose sexual spores are produced on the surfaces of club-shaped structures (basidia).

basidiospore A sexual spore of the basidiomycete fungi.

basophil A cell that secretes the anticlotting factor heparin at the site of an infection, which helps the circulation flush out the infected site; basophils also secrete histamine, which attracts infection-fighting cells and proteins.

Batesian mimicry The mimicry of an unpalatable species (the model) by a palatable one (the mimic).

behavior The observable response of organisms to external or internal stimuli.

behavioral ecology A subdiscipline of organismal ecology that focuses on how the behavior of an individual organism contributes to its survival and reproductive success, which in turn eventually affects the population density of the species.

benign tumor A precancerous condition.

bidirectional replication In DNA replication, the two DNA strands unwind, and DNA replication proceeds outward from the origin in opposite directions.

biennial A plant that does not reproduce during the first year of life but may reproduce within the following year.

bilateral Refers to living organisms, such as animals, or parts of living organisms, such as leaves, that can be divided along a vertical plane at the midline to create two halves.

bile A substance produced by the liver that contains bicarbonate ions, cholesterol, phospholipids, a number of organic wastes, and a group of substances collectively termed bile salts.

bile salts A group of substances produced in the liver that solubilize dietary fat and increase its accessibility to digestive enzymes.

binary fission The process of cell division in bacteria and archaea in which the cells divide into two cells.

binocular vision A type of vision in animals with both eyes located at the front of the head; the overlapping images coming into both eyes are processed together in the brain to form one perception.

binomial A two-part description used by biologists to provide each species with a unique scientific name. For example, the scientific name of the jaguar is *Panthera onca*. The first part is the genus and the second part is the specific epithet or species descriptor.

binomial nomenclature The standard method for naming species. Each species has a genus name and species epithet.

biochemistry The study of the chemistry of living organisms.

biodiversity Biological diversity, including genetic diversity, species diversity, and ecosystem diversity.

biodiversity crisis The idea that there is currently an elevated loss of species on Earth, far beyond the normal extinction rate of species.

biofilm An aggregation of microorganisms that secrete adhesive mucilage, thereby gluing themselves to surfaces.

biogeochemical cycle The continuous movement of nutrients such as phosphorus, carbon, and nitrogen from the physical environment to organisms and back.

biogeography The study of the geographic distribution of extinct and modern species.

bioinformatics A field of study that uses computers to study biological information.

biological evolution The phenomenon that populations of organisms change over the course of many generations. As a result, some organisms become more successful at survival and reproduction.

biological species concept A species is a group of individuals whose members have the potential to interbreed with one another in nature to produce viable, fertile offspring but cannot successfully interbreed with members of other species.

biology The study of life.

bioluminescence A phenomenon that results from chemical reactions that give off light rather than heat.

biomagnification The increase in the concentration of a substance in living organisms with each increase in trophic level in a food web.

biomass A quantitative estimate of the total mass of living matter in a given area, usually measured in grams or kilograms per square meter.

biome A major type of habitat characterized by distinctive plant and animal life.

bioremediation The use of living organisms, usually microbes or plants, to detoxify polluted habitats such as dump sites or oil spills.

biosphere The regions on the surface of the Earth and in the atmosphere where living organisms exist.

biosynthetic reaction Also called an anabolic reaction; a chemical reaction to make larger molecules and macromolecules.

biotechnology The use of living organisms or the products of living organisms for human benefit.

biotic The term used to describe interactions among organisms.

biparental inheritance A pattern in which both the male and female gametes contribute particular genes to the offspring.

bipedal Having the ability to walk on two feet.

bipolar cells Cells in the eye that make synapses with the photoreceptors and relay responses to the ganglion cells.

bivalent Homologous pairs of sister chromatids associated with each other, lying side by side.

blastocyst The mammalian counterpart of a blastula.

blastoderm An early stage of embryonic development in animals that is composed of a mass of cells with an internal cavity.

blastomeres The two half-size daughter cells produced by each cell division during cleavage.

blastopore A small opening created when a band of tissue invaginates during gastrulation. It forms the primary opening of the archenteron to the outside.

blastula An animal embryo at the stage when it forms an outer epithelial layer and an inner cavity.

blending inheritance An early hypothesis of inheritance that stated that the seeds that dictate hereditary traits blend together from generation to generation, and the blended traits are then passed to the next generation.

blood A fluid connective tissue consisting of cells and (in mammals) cell fragments suspended in a solution of water containing dissolved nutrients, proteins, gases, and other molecules. Blood has four components: plasma, leukocytes, erythrocytes, and thrombocytes or platelets.

blood-doping An example of hormone misuse in which the number of red blood cells in the circulation is boosted to increase the oxygen-carrying capacity of the blood.

blood pressure The force exerted by blood on the walls of blood vessels; blood pressure is responsible for moving blood through the vessels.

blood vessels A system of hollow tubes within the body through which blood travels.

body mass index (BMI) A method of assessing body fat and health risk that involves calculating the ratio of weight compared to height; weight in kilograms is divided by the square of the height in meters.

bottleneck effect A form of genetic drift in which a population size is dramatically reduced and then rebounds. Genetic drift is common when the population size is small.

Bowman's capsule A sac at the beginning of the tubular component of a nephron in the mammalian kidney.

brain The structure in the head of animals that controls sensory and motor functions of the entire body.

brassinosteroid One of several plant hormones that help a plant to cope with environmental stress.

bronchiole A thin-walled, small tube that can dilate or constrict to prevent foreign particles from reaching delicate lung tissue.

bronchodilator A compound that binds to the muscles around bronchioles and causes them to relax and widen.

bronchus A tube branching from the trachea and leading to a lung.

brown fat A specialized tissue in small mammals such as hibernating bats, small rodents living in cold environments, and many newborn mammals including humans that can help to generate heat and maintain body temperature.

brush border The combination of villi and microvilli in the small intestine, which increase the surface area about 600-fold over that of a flat-surfaced tube having the same length and diameter.

bryophytes Liverworts, mosses, and hornworts, the modern non-vascular land plants.

buccal pumping A form of breathing in which animals open their mouths to let in air and then close and raise the floor of the mouth, creating a positive pressure and pumping water or air across the gills or into the lungs; found in fish and amphibians.

bud A miniature plant shoot having a dormant shoot apical meristem.

budding A form of asexual reproduction in which a portion of the parent organism pinches off to form a complete new individual.

buffer A compound that acts to minimize pH fluctuations in the fluids of living organisms. Buffer systems can raise or lower pH as needed.

bulbourethral glands Paired accessory glands that secrete an alkaline mucus that protects sperm by neutralizing the acidity in the urethra.

bulk flow The mass movement of liquid in a plant caused by pressure, gravity, or both.

C

C$_3$ plant A plant that can only incorporate CO_2 into organic molecules via RuBP to make 3PG, a three-carbon molecule.

C$_4$ plant A plant that uses PEP carboxylase to initially fix CO_2 into a 4-carbon molecule and later uses rubisco to fix CO_2 into simple sugars.

cadherin A cell adhesion molecule found in animal cells that promotes cell-to-cell adhesion.

callose A carbohydrate that plays crucial roles in plant development, and plugging wounds in plant phloem.

calorie The amount of heat required to raise the temperature of 1 gram of water 1 degree Celsius.

Calvin cycle The cycle that includes carbon fixation, reduction and carbohydrate production, and regeneration of ribulose bisphosphate (RuBP). During this process, ATP is used as a source of energy and NADPH is used as a source of high-energy electrons so that CO_2 can be incorporated into carbohydrate.

calyx The sepals that form the outermost whorl of a flower.

CAM plants C$_4$ plants that take up carbon dioxide at night.

Cambrian explosion An event during the Cambrian period (543 to 490 mya) in which there was an abrupt increase (on a geological scale) in the diversity of animal species.

cancer A disease caused by gene mutations that lead to uncontrolled cell growth.

canopy The uppermost layer of tree foliage.

CAP site One of two regulatory sites near the *lac* promoter; this site is a DNA sequence recognized by the catabolite activator protein (CAP).

capillary A thin-walled vessel that is the site of gas and nutrient exchange between the blood and interstitial fluid.

capping A 7-methylguanosine covalently attached at the 5′ end of mature mRNAs of eukaryotes.

capsid A protein coat enclosing a virus's genome.

capsule A very thick, gelatinous glycocalyx produced by certain strains of bacteria that invade animals' bodies that may help them avoid being destroyed by the animal's immune (defense) system.

carapace The hard protective covering of a crustacean.

carbohydrate An organic molecule with the general formula, $C(H_2O)$; a carbon-containing compound that is hydrated (that is, contains water).

carbon fixation In this process, inorganic CO_2 is incorporated into an organic molecule such as a carbohydrate.

carboxyl terminus *See* C-terminus.

carcinogen An agent that increases the likelihood of developing cancer, usually a mutagen.

carcinoma A cancer of epithelial cells.

cardiac cycle The events that produce a single heartbeat, which can be divided into two phases, diastole and systole.

cardiac muscle A type of muscle tissue found only in hearts in which physical and electrical connections between individual cells enable many of the cells to contract simultaneously.

cardiac output (CO) The amount of blood the heart pumps per unit time, usually expressed in units of L/min.

cardiovascular disease   Diseases affecting the heart and blood vessels.

cardiovascular system   A system containing three components: blood or hemolymph, blood vessels, and one or more hearts.

carnivore   An animal that consumes animal flesh or fluids.

carotenoid   A type of pigment found in chloroplasts that imparts a color that ranges from yellow to orange to red.

carpel   A flower shoot organ that produces ovules that contain female gametophytes.

carrier   *See* transporter.

carrying capacity (K)   The upper boundary for a population size.

Casparian strips   Suberin ribbons on the walls of endodermal cells of plant roots; prevent apoplastic transport of ions into vascular tissues.

caspase   An enzyme that is activated during apoptosis.

catabolic reaction   The breakdown of a molecule into smaller components, usually releasing energy.

catabolism   A metabolic pathway that results in the breakdown of larger molecules into smaller molecules. Such reactions are often exergonic.

catabolite activator protein (CAP)   An activator protein also known as the cAMP receptor protein (CRP). CAP is needed for activation of the *lac* operon.

catabolite repression   In bacteria, a process whereby transcriptional regulation is influenced by the presence of glucose.

catabolized   Broken down.

catalase   An enzyme within peroxisomes that breaks down hydrogen peroxide to water and oxygen gas.

catalyst   An agent that speeds up the rate of a chemical reaction without being consumed during the reaction.

cataract   An accumulation of protein in the lens of the eye; causes blurring, poor night vision.

cation   An ion that has a net positive charge.

cation exchange   With regard to soil, the process in which hydrogen ions are able to replace mineral cations on the surfaces of humus or clay particles.

cDNA library   A collection of recombinant vectors that have cDNA inserts.

cecum   The first portion of the large intestine in humans and other similarly sized animals and mammals.

celiac sprue   A common genetic disorder in humans that results in a loss of intestinal surface area due to an allergic sensitivity to the wheat protein gluten.

cell   The simplest unit of a living organism.

cell adhesion   The phenomenon in which cells adhere to each other. Cell adhesion provides one way to convey positional information between neighboring cells.

cell adhesion molecule (CAM)   A membrane protein found in animal cells that promotes cell adhesion.

cell biology   The study of individual cells and their interactions with each other.

cell body   A part of a neuron that contains the cell nucleus and other organelles.

cell coat   Also called the glycocalyx, the carbohydrate-rich zone on the surface of certain animal cells that shields the cell from mechanical and physical damage.

cell communication   The process through which cells can detect and respond to signals in their extracellular environment. In multicellular organisms, cell communication is also needed to coordinate cellular activities within the whole organism.

cell cycle   The series of phases eukaryotic cells progress through to divide.

cell differentiation   Refers to the phenomenon in which cells become specialized into particular cell types.

cell doctrine   *See* cell theory.

cell fate   The ultimate morphological features that a cell or group of cells will adopt.

cell junctions   Specialized structures that adhere cells to each other and to the ECM.

cell-mediated immunity   A type of specific immunity in which cytotoxic T cells directly attack and destroy infected body cells, cancer cells, or transplanted cells.

cell nucleus   The membrane-bounded area of a eukaryotic cell in which the genetic material is found.

cell plate   In plant cells, a structure that forms a cell wall between the two daughter cells.

cell signaling   A vital function of the plasma membrane that involves cells sensing changes in their environment and interacting with each other.

cell surface receptor   A receptor found in the plasma membrane that enables a cell to respond to different kinds of signaling molecules.

cell theory   A theory that states that all organisms are made of cells. Cells come from pre-existing cells by cell division.

cell wall   A relatively rigid, porous structure that supports and protects the plasma membrane and cytoplasm of prokaryotic, plant, fungal, and certain protist cells.

cellular differentiation   The process by which different cells within a developing organism acquire specialized forms and functions, due to the expression of cell-specific genes.

cellular respiration   A process by which living cells obtain energy from organic molecules.

cellular response   Adaptation at the cellular level that often times involves a cell responding to signals in its environment.

cellulose   The main macromolecule of the primary cell wall of plants and green algae; a polymer made of repeating molecules of glucose attached end to end.

centimorgan (cM)   *See* map unit (mu).

central cell   In the female gametophyte of a flowering plant, a large cell that contains two nuclei; after double fertilization it forms the first cell of the nutritive endosperm tissue.

central dogma   Refers to the steps of gene expression at the molecular level. DNA is transcribed into mRNA and mRNA is translated into a polypeptide.

central region   The region of a plant apical meristem that produces stem tissue.

central vacuole   An organelle that often occupies 80% or more of the cell volume of plant cells and stores a large amount of water, enzymes, and inorganic ions.

central zone   The area of a plant apical meristem where undifferentiated stem cells are maintained.

centrioles   A pair of structures within the centrosome of animal cells. Most plant cells and many protists lack centrioles.

centromere   The region where the two sister chromatids are tightly associated; the centromere binds to the kinetochore.

centrosome   A single structure often near the cell nucleus of eukaryotic cells that forms a nucleating site for the growth of microtubules.

cephalization   The localization of sensory structures at the anterior end of the body of animals.

cephalothorax   The fused head and thorax structure in species of the class Arachnida and Crustacea.

cerebellum   The part of the hindbrain, along with the pons, responsible for monitoring and coordinating body movements.

cerebral cortex   The surface layer of gray matter that covers the cerebrum of the brain.

cerebral ganglia   In flatworms, a paired structure that receives input from photoreceptors in eyespots and sensory cells.

cerebrospinal fluid   Fluid that surrounds the exterior of the brain and spinal cord and absorbs physical shocks to the brain resulting from sudden movements or blows to the head.

cerebrum   A group of structures in the forebrain that are responsible for the higher functions of conscious thought, planning, and emotion in vertebrates.

cervix   A fibrous structure at the end of the female vagina that forms the opening to the uterus.

channel   A transmembrane protein that forms an open passageway for the direct diffusion of ions or molecules across a membrane.

character   A visible characteristic, such as the appearance of seeds, pods, flowers, and stems.

character displacement   The tendency for two species to diverge in morphology and thus resource use because of competition.

charophyceans   The lineage of freshwater green algae that is most closely related to the land plants.

charged tRNA   A tRNA with its attached amino acid; also called aminoacyl tRNA.

checkpoint   One of three critical regulatory points found in the cell cycle of eukaryotic cells. At these checkpoints, a variety of proteins act as sensors to determine if a cell is in the proper condition to divide.

checkpoint protein   A protein that senses if a cell is in the proper condition to divide and prevents a cell from progressing through the cell cycle if it is not.

chemical element Each specific type of atom—nitrogen, hydrogen, oxygen, and so on.

chemical energy The potential energy contained within covalent bonds in molecules.

chemical equilibrium In a chemical reaction, occurs when the rate of formation of products equals the rate of formation of reactants.

chemical mutagen A chemical that causes mutations.

chemical reaction A reaction that occurs when one or more substances are changed into other substances. This can happen when two or more elements or compounds combine with each other to form a new compound, when one compound breaks down into two or more molecules, or when electrons are added to or taken away from an atom.

chemical synapse A synapse in which a chemical called a neurotransmitter is released from the nerve terminal and acts as a signal from the presynaptic to the postsynaptic cell.

chemiosmosis A process for making ATP in which energy stored in an ion electrochemical gradient is used to make ATP from ADP and P_i.

chemoautotroph An organism able to use energy obtained by chemical modifications of inorganic compounds to synthesize organic compounds.

chemoorganotroph An organism that must obtain organic molecules both for energy and as a carbon source.

chemoreceptor Specialized cells located in the vertebrate aorta, carotid arteries, and brainstem that detect the circulating levels of hydrogen ions and the partial pressures of carbon dioxide and oxygen, and relay that information through nerves or interneurons to the respiratory centers.

chiasma The connection at a crossover site of two chromosomes.

chimeric gene A gene formed from the fusion of two gene fragments to each other.

chitin A tough, nitrogen-containing polysaccharide that forms the external skeleton of many insects and the cell walls of fungi.

chlorophyll A green pigment found in photosynthetic plants, algae, and bacteria.

chlorophyll *a* Type of chlorophyll pigment found in cyanobacteria, and photosynthetic algae and plants.

chlorophyll *b* Type of chlorophyll pigment found in green algae and plants.

chloroplast genome The chromosome found in chloroplasts.

chlorosis The yellowing of plant leaves caused by various types of mineral deficiency.

choanocyte A specialized cell of sponges that functions to trap and eat small particles.

chondrichthyans Members of the class Chondrichthyes, including sharks, skates, and rays.

chordate An organism with a spinal cord.

chorion An extraembryonic membrane in the amniotic egg that, along with the allantois,

exchanges gases between the embryo and the surrounding air.

chorionic gonadotropin (CG) An LH-like hormone made by the placenta that maintains the corpus luteum.

chromatin Refers to the biochemical composition of chromosomes, which contain DNA and many types of proteins.

chromosome territory A distinct, nonoverlapping area where each chromosome is located within the cell nucleus of eukaryotic cells.

chromosome theory of inheritance An explanation of how the steps of meiosis account for the inheritance patterns observed by Mendel.

chromosome A unit of genetic material composed of DNA and associated proteins. Each cell has a characteristic number of chromosomes. Eukaryotes have chromosomes in their cell nuclei and in plastids and mitochondria.

chylomicrons Large fat droplets coated with amphipathic proteins that perform an emulsifying function similar to that of bile salts; they are formed in intestinal epithelial cells from absorbed fats in the diet.

chyme A solution of food particles in the stomach that contains water, salts, molecular fragments of proteins, nucleic acids, polysaccharides, droplets of fat, and various other small molecules.

chymotrypsin A protease involved in the breakdown of proteins in the small intestine.

chytrids Simple, early-diverging lineages of fungi; commonly found in aquatic habitats and moist soil, where they produce flagellate reproductive cells.

ciliate A protist that moves by means of cilia, which are tiny hairlike extensions on the outsides of cells.

cilium (plural, **cilia**) A cell appendage that functions like flagella to facilitate cell movement; cilia are shorter and more numerous on cells than are flagella.

circulatory system A system that transports necessary materials to all cells of an animal's body, and transports waste products away from cells. The three basic types of circulatory systems are gastrovascular cavities, open systems, and closed systems.

***cis*-acting element** *See cis*-effect.

***cis*-effect** A DNA segment that must be adjacent to the gene(s) that it regulates. The *lac* operator site is an example of a cis-acting element.

cisternae Flattened, fluid-filled tubules within the cell.

citric acid cycle A cycle that results in the breakdown of carbohydrates to carbon dioxide; also known as the Krebs cycle.

clade *See monophyletic group.*

cladistic approach An approach that reconstructs a phylogenetic tree by comparing primitive and shared derived characters.

cladogenesis A pattern of speciation in which a species is divided into two or more species.

cladogram A phylogenetic tree based on a cladistic approach.

clamp connection In basidiomaycete fungi, a structure that helps distribute nuclei during cell division.

clasper An extension of the pelvic fin of a chondrichthyan, used by the male to transfer sperm to the female.

class A subdivision of a phylum.

classical conditioning A type of associative learning in which an involuntary response comes to be associated positively or negatively with a stimulus that did not originally elicit the response.

cleavage A succession of rapid cell divisions with no significant growth that produces a hollow sphere of cells called a blastula.

cleavage furrow In animal cells, an area that constricts like a drawstring to separate the cells.

climate The prevailing weather pattern in a given region.

climax community A distinct end point of succession.

clitoris Located at the anterior part of the labia minora, erectile tissue that becomes engorged with blood during sexual arousal and is very sensitive to sexual stimulation.

clonal deletion A process that explains why individuals normally lack active lymphocytes that respond to self components; T cells with receptors capable of binding self proteins are destroyed by apoptosis.

clonal inactivation A process that explains why individuals normally lack active lymphocytes that respond to self components; the process occurs outside the thymus and causes potentially self-reacting T cells to become nonresponsive.

clonal selection The process by which certain cells are selected to proliferate as clones.

cloning Methods that produce many copies of something. These can be copies of genes, copies of genetically identical cells, or copies of genetically identical organisms. The cloning of mammals can be achieved by fusing a somatic cell with an egg that has had its nucleus removed. Plants can be cloned simply by removing cells, and growing them in particular mixtures of hormones.

closed circulatory system A circulatory system in which blood flows throughout an animal entirely within a series of vessels and is kept separate from the interstitial fluid.

closed conformation Tightly packed chromatin that cannot be transcribed into RNA.

clumped The most common pattern of dispersion within a population, in which individuals are gathered in small groups.

cnidocil On the surface of a cnidocyte, a hairlike trigger that detects stimuli.

cnidocyte A characteristic feature of cnidarians; a stinging cell that functions in defense or the capture of prey.

coacervates Droplets that form spontaneously from the association of charged polymers such as proteins, carbohydrates, or nucleic acids.

coat protein   A protein that surrounds a membrane vesicle and facilitates vesicle formation.

cocci   Spheres; one of the five major shapes of prokaryotic cells.

cochlea   A coiled structure containing the hair cells and other structures that generate the responses that travel via the auditory nerve to the brain.

coding strand   The DNA strand opposite to the template (or noncoding strand).

codominance   The phenomenon in which a single individual expresses both alleles.

codon   A sequence of three nucleotide bases that specifies a particular amino acid or a stop codon; codons function during translation.

coefficient of relatedness   The probability that any two individuals will share a copy of a particular gene is a quantity, r.

coelom   A fluid-filled body cavity in an animal.

coelomate   An animal with a true coelom.

coenzyme   An organic molecule that participates in the chemical reaction but is left unchanged after the reaction is completed.

coevolution   The process by which two or more species of organisms influence each other's evolutionary pathway.

cofactor   Usually an inorganic ion that temporarily binds to the surface of an enzyme and promotes a chemical reaction.

cognitive learning   The ability to solve problems with conscious thought and without direct environmental feedback.

cohesion-tension theory   The explanation for long-distance water transport as the combined effect of the cohesive forces of water and evaporative tension.

cohort   A group of organisms of the same age.

coleoptile   A protective sheath that encloses the first bud of an epicotyl in a mature monocot embryo.

coleorhiza   A protective envelope that encloses the monocot hypocotyl.

colinearity rule   The phenomenon whereby the order of homeotic genes along the chromosome correlates with their expression along the anteroposterior axis of the body.

collagen   A protein secreted from animal cells that forms large fibers in the extracellular matrix.

collecting duct   A tubule in the human vertebrate kidney that collects urine from nephrons.

collenchyma cells   Flexible cells that make up collenchyma tissue.

collenchyma tissue   A tissue that provides support to plant organs.

colligative property   Properties of water that depend on the amounts of dissolved substances. For example, the colligative properties of water cause certain solutes to function as antifreeze in certain organisms, and thereby lower the freezing point.

colloid   A gel-like substance in the follicles of the thyroid gland.

colon   A part of the large intestine consisting of three relatively straight segments—the ascending, transverse, and descending portions. The terminal portion of the descending colon is S-shaped, forming the sigmoid colon, which empties into the rectum.

colony hybridization   A method that uses a labeled probe that recognizes a specific gene to identify that gene in a DNA library.

combinatorial control   The phenomenon whereby a combination of many factors determines the expression of any given gene.

commensalism   An interaction that benefits one species and leaves the other unaffected.

communication   The use of specially designed visual, chemical, auditory or tactile signals to modify the behavior of others.

community   An assemblage of many populations that live in the same place at the same time.

community ecology   The study of how populations of species interact and form functional communities.

compartmentalization   A characteristic of eukaryotic cells that is defined by many organelles that separate the cell into different regions. Cellular compartmentalization allows a cell to carry out specialized chemical reactions in different places.

competent   The term used to describe bacterial strains that have the ability to take up DNA from the environment.

competition   An interaction that affects two or more species negatively, as they compete over food or other resources.

competitive exclusion hypothesis   The proposal that two species with the same resource requirements cannot occupy the same niche.

competitive inhibitor   A molecule that binds to the active site of an enzyme and inhibits the ability of the substrate to bind.

complement   The family of plasma proteins that provides a means for extracellular killing of microbes without prior phagocytosis.

complementary DNA (cDNA)   DNA molecules that are made from mRNA as a starting material.

complementary   In DNA, you can predict the sequence in one DNA strand if you know the sequence in the opposite strand according to the AT/GC rule.

complete flower   A flower that possesses all four types of flower organs.

complete metamorphosis   A dramatic change in body form in the majority of insects, from larva to a very different looking adult.

compound   A molecule composed of two or more different elements.

compound eyes   Image-forming eyes in arthropods and some annelids consisting of several hundred to several thousand light detectors called ommatidia.

computational molecular biology   An area of study that uses computers to characterize the molecular components of living things.

concentration   The amount of a solute dissolved in a unit volume of solution.

condensation reaction   *See* dehydration reaction.

conditioned response   The response that is created by a new, conditioned stimulus.

conditioned stimulus   A new stimulus that is delivered at the same time as the old stimulus, and that over time, is sufficient to elicit the same response.

condom   A sheathlike membrane worn over the penis that collects the ejaculate; in addition to their contraceptive function, condoms significantly reduce the risk of STDs such as HIV infection, syphilis, gonorrhea, chlamydia, and herpes.

conduction   The process in which the body surface loses or gains heat through direct contact with cooler or warmer substances.

cone pigments   The several types of visual pigments in cones.

cones   1. Photoreceptors found in the vertebrate eye; they are less sensitive to low levels of light but can detect color. Cones are used in daylight by most diurnal vertebrate species and by some insects. 2. The reproductive structures of conifer plants.

congestive heart failure   The condition resulting from the failure of the heart to pump blood normally; this results in fluid build-up in the lungs (congestion).

conidia   A type of asexual reproductive cell produced by many fungi.

Conifers   A phylum of gymnosperm plants, Coniferophyta.

conjugation   A type of genetic transfer between bacteria that involves a direct physical interaction between two bacterial cells.

connective tissue   Clusters of cells that connect, anchor, and support the structures of an animal's body; derived from mesenchyme and include blood, adipose (fat-storing) tissue, bone, cartilage, loose connective tissue, and dense connective tissue.

connexon   A channel that forms gap junctions consisting of six connexin proteins in one cell aligned with six connexin proteins in an adjacent cell.

conservation biology   The study that uses principles and knowledge from molecular biology, genetics, and ecology to protect the biological diversity of life at all levels.

conservative mechanism   In this incorrect model for DNA replication, both parental strands of DNA remain together following DNA replication. The original arrangement of parental strands is completely conserved, while the two newly made daughter strands are also together following replication.

constant regions   In immunology, the amino acid sequences of the Fc domains, which are identical for all immunoglobulins of a given class.

constitutive gene   An unregulated gene that has essentially constant levels of expression in all conditions over time.

contig   A series of clones that contain overlapping pieces of chromosomal DNA.

continental drift The phenomenon whereby, over the course of billions of years, the major landmasses, known as the continents, have shifted their positions, changed their shapes, and in some cases have become separated from each other.

continuous trait A trait that shows continuous variation over a range of phenotypes.

contraception The use of birth control procedures to prevent fertilization or implantation of a fertilized egg.

contractile vacuole A small, membrane enclosed, water-filled space that eliminates excess liquid from the cells of certain protists.

contrast In microscopy, relative differences in the lightness, darkness, or color between adjacent regions in a sample. Contrast improves the ability to discern adjacent objects.

control sample The sample in an experiment that is treated just like an experimental sample except that it is not subjected to one particular variable. For example, the control and experiment samples may be treated identically except that the temperature may vary for the experimental sample.

convection The transfer of heat by the movement of air or water next to the body.

convergent evolution The process whereby two different species from different lineages show similar characteristics because they occupy similar environments.

convergent extension A cellular process during gastrulation that is crucial to development; two rows of cells merge to form a single elongated layer.

convergent trait *See* analogous structure.

copulation The process of sperm being deposited within the reproductive tract of the female.

coral reef A type of aquatic biome found in warm, marine environments.

core promoter For a eukaryotic structural gene, refers to the transcriptional start site and TATA box.

corepressor A small effector molecule that binds to a repressor protein to inhibit transcription.

Coriolis effect The effect of the Earth's rotation on the surface flow of wind.

cork cambium A secondary meristem in a plant that produces cork tissue.

cornea A thin, clear layer on the front of the vertebrate eye.

corolla The petals of a flower, which occur in the whorl to the inside of the calyx and the outside of the stamens.

corona The ciliated crown of members of the phylum Rotifera.

coronary artery An artery that carries oxygen and nutrients to the heart muscle.

coronary artery bypass A common treatment to restore blood flow through a blood vessel. A small piece of healthy blood vessel is removed from one part of the body and surgically grafted onto the coronary circulation in such a way that blood bypasses the diseased artery.

coronary artery disease A condition that occurs when plaques form in the coronary vessels.

corpus callosum The major tract that connects the two hemispheres of the cerebrum.

corpus luteum A structure that is responsible for secreting hormones that stimulate the development of the uterus needed for sustaining the embryo in the event of a pregnancy. If pregnancy does not occur, the corpus luteum degenerates.

correlation A meaningful relationship between two variables.

cortex The area of a plant stem or root beneath the epidermis that is largely composed of parenchyma tissue.

cortical reaction An event in fertilization in which calcium and IP$_3$ produce additional barriers to more than one sperm cell binding to and uniting with an egg, a process called the slow block to polyspermy.

cotranslational sorting The sorting process in which the synthesis of certain eukaryotic proteins begins in the cytosol and then halts temporarily until the ribosome has become bound to the ER membrane. After this occurs, translation resumes and the polypeptide is synthesized into the ER lumen or ER membrane.

cotransporter *See* symporter.

cotyledon An embryonic seed leaf.

countercurrent exchange mechanism An arrangement of water and blood flow in which water enters a fish's mouth and flows between the lamellae of the gills in the opposite direction to blood flowing through the lamellar capillaries.

countercurrent heat exchange A method of regulating heat loss to the environment; many animals conserve heat by returning it to the body's core and keeping the core much warmer than the extremities.

covalent bond A chemical bond that occurs when atoms share pairs of electrons.

CpG island A cluster of CpG sites. CpG refers to the nucleotides of C and G in DNA that are connected by a phosphodiester linkage.

cranial nerve A nerve in the peripheral nervous system that is directly connected to the brain; cranial nerves are located in the head and transmit incoming and outgoing information between the peripheral nervous system and the brain.

craniate A chordate that has a brain encased in a skull and possesses a neural crest.

cranium A protective bony or cartilaginous housing that encases the brain of a craniate.

crenation The process of cell shrinkage that occurs if animal cells are placed in a hypertonic medium—water exits the cells via osmosis and equalizes solute concentrations on both sides of the membrane.

critical period A limited time period of development in which many animals develop species-specific patterns of behavior.

crop A storage organ that is a dilation of the lower esophagus; found in most birds and many invertebrates, including insects and some worms.

cross-bridge cycle During muscle contraction, the sequence of events that occurs between the time when a cross-bridge binds to a thin filament and when it is set to repeat the process.

cross-bridge A region of myosin molecules that extend from the surface of the thick filaments toward the thin filaments in a skeletal muscle.

cross-fertilization The fusion of gametes formed by different individuals.

crossing over The exchange of genetic material between homologous chromosomes, which allows for increased variation in the genetic information each parent may pass to the offspring.

cross-pollination The process in which a stigma receives pollen from a different plant of the same species.

cryptic coloration The blending of an organism with the background color of its habitat; also known as camouflage.

cryptochrome A type of blue-light receptor in plants and protists.

CT scan Computerized tomography, which is a technique for examining the structure and activity level of the brain without anesthesia or surgery. An X-ray beam and a series of detectors rotate around the head, producing slices of images that are reconstructed into three-dimensional images based on differences in the density of brain tissue.

C-terminus The location of the last amino acid in a polypeptide.

cupula A gelatinous structure within the lateral line organ that helps an organism to detect changes in water movement.

cuticle A coating of wax and cutin that helps to reduce water loss from plant surfaces. Also, a nonliving covering that serves to both support and protect an animal.

cutin A polyester polymer produced at the surfaces of plants; helps to prevent attack by pathogens.

cycads A phylum of gymnosperm plants, Cycadophyta.

cyclic adenosine monophosphate (cAMP) A second messenger molecule.

cyclic AMP (cAMP) Cyclic adenosine monophosphate; a small effector molecule that is produced from ATP via an enzyme known as adenylyl cyclase.

cyclic electron flow *See* cyclic photophosphorylation.

cyclic photophosphorylation A pattern of electron flow in the thylakoid membrane that is cyclic and generates ATP alone.

cyclin A protein responsible for advancing a cell through the phases of the cell cycle by binding to a cyclin-dependent kinase.

cyclin-dependent kinase (cdk) A protein responsible for advancing a cell through the phases of the cell cycle. Its function is dependent on the binding of a cyclin.

cyst A one-to-few celled structure that often has a thick, protective wall and can remain dormant through periods of unfavorable climate or low food availability.

cytogenetics The field of genetics that involves the microscopic examination of chromosomes.

cytokines A family of proteins that function in both nonspecific and specific immune defenses by providing a chemical communication network that synchronizes the components of the immune response.

cytokinesis The division of the cytoplasm to produce two distinct daughter cells.

cytokinin A type of plant hormone; promotes cell division in addition to other effects.

cytoplasm The region of the cell that is contained within the plasma membrane.

cytoplasmic inheritance *See* extranuclear inheritance.

cytoplasmic streaming A phenomenon in which the cytoplasm circulates throughout the cell to distribute resources efficiently in large cells, such as algal or plant cells.

cytosine (C) A pyrimidine base found in DNA and RNA.

cytoskeleton In eukaryotes, a network of three different types of protein filaments called microtubules, intermediate filaments, and actin filaments.

cytosol The region of a eukaryotic cell that is inside the plasma membrane and outside the organelles.

cytotoxic T cell A type of lymphocyte that travels to the location of its target, binds to the target by combining with an antigen on it, and directly kills the target via secreted chemicals.

D

dalton (Da) One-twelfth the mass of a carbon atom, or about the mass of a proton or a hydrogen atom.

Darwinian fitness The relative likelihood that a genotype will contribute to the gene pool of the next generation as compared with other genotypes.

database A large number of computer data files that are collected, stored in a single location, and organized for rapid search and retrieval.

daughter strand The newly made strand in DNA replication.

day-neutral plant A plant that flowers regardless of the night length, as long as day length meets the minimal requirements for plant growth.

deafness Hearing loss, usually caused by damage to the hair cells within the cochlea, although some cases result from functional problems in brain areas that process sound or in nerves that carry information to the brain from the hair cells.

decomposer *See* detritivore.

defecation The expulsion of feces that occurs through the final portion of the digestive canal, the anus; contractions of the rectum and relaxation of associated sphincter muscles expel the feces.

defensive mutualism A mutually beneficial interaction often involving an animal defending a plant or a herbivore in return for food or shelter.

deficiency The term used to describe when a segment of chromosomal material is missing.

deforestation The conversion of forested areas by humans to nonforested land.

degenerate In the genetic code, this means that more that one codon can specify the same amino acid.

dehydration A reduction in the amount of water in the body.

dehydration reaction A reaction that involves the removal of a water molecule, and the formation of a covalent bond between two separate molecules.

delayed implantation A reproductive cycle in which a fertilized egg reaches the uterus but does not implant until later, when environmental conditions are more favorable for the newly produced young.

delayed ovulation A reproductive cycle in which the ovarian cycle in females is halted before ovulation, and sperm are stored and nourished in the female's uterus over the winter. Upon arousal from hibernation in the spring, the female ovulates one or more eggs, which are fertilized by the stored sperm.

deletion The term used to describe a missing region of a chromosome.

demographic transition The shift in birth and death rates accompanying human societal development.

demography The study of birth rates, death rates, age distributions, and the sizes of populations.

dendrite A type of extension or projection that arises from the cell body; chemical and electrical messages from other neurons are received by the dendrites, and electrical signals move toward the cell body.

dendritic cell A type of cell derived from bone marrow stem cells that plays an important role in nonspecific immunity; these cells are scattered throughout most tissues, where they perform various macrophage functions.

denitrification The reduction of nitrate to gaseous nitrogen.

density In the context of populations, the numbers of organisms in a given unit area.

density-dependent factor A mortality factor whose influence varies with the density of the population.

density-independent factor A mortality factor whose influence is not affected by changes in population size or density.

deoxynucleoside triphosphates Individual nucleotides with three phosphate groups.

deoxyribonucleic acid (DNA) One of two classes of nucleic acids; the other is ribonucleic acid (RNA). A DNA molecule consists of two strands of nucleotides coiled around each other to form a double helix, held together by hydrogen bonds according to the AT/GC rule.

deoxyribose A five-carbon sugar found in DNA.

depolarization The change in the membrane potential that occurs when the cell becomes less polarized, that is, less negative relative to the surrounding fluid.

dermal tissue The covering on various parts of a plant.

descent with modification Darwin's theory that existing life-forms on our planet are the product of the modification of pre-existing life-forms.

desertification The overstocking of land with domestic animals that can greatly reduce grass coverage through overgrazing, turning the area more desert-like.

determinate cleavage A characteristic of protostome development in which the fate of each embryonic cell is determined very early.

determined The term used to describe a cell that is destined to differentiate into a particular cell type.

detritivore A consumer that gets its energy from the remains and waste products of organisms.

detritus Unconsumed plants that die and decompose, along with the dead remains of animals and animal waste products.

deuterostome An animal exhibiting radial cleavage, indeterminate cleavage and where the blastopore becomes the anus; includes echinoderms and vertebrates.

development In biology, a series of changes in the state of a cell, tissue, organ, or organism; the underlying process that gives rise to the structure and function of living organisms.

developmental genetics A field of study aimed at understanding how gene expression controls the process of development.

diaphragm A large muscle that subdivides the thoracic cavity from the abdomen in mammals.

diastole The first phase of the cardiac cycle, in which the ventricles fill with blood coming from the atria through the open AV valves.

diazotroph A bacterium that fixes nitrogen.

dideoxy chain-termination method The most common method of DNA sequencing that utilizes dideoxynucleotides as a reagent.

dideoxy sequencing *See* dideoxy chain-termination method.

differential gene regulation The phenomenon in which the expression of genes is altered. Differential gene expression allows cells to adapt to environmental conditions, change during development, and differentiate into particular cell types.

differentiated The term used to describe the actual alteration of a cell's morphology and physiology.

diffusion For dissolved substances, occurs when a solute moves from a region of high concentration to a region of lower concentration.

digestion The process of breaking down nutrients in food into smaller molecules that can be directly used by cells.

digestive system In a vertebrate, this system consists of the alimentary canal plus several associated structures, not all of which are found in all vertebrates: the tongue, teeth, salivary glands, liver, gallbladder, and pancreas.

dihybrid   An offspring that is a hybrid with respect to two traits.

dikaryotic   The occurrence of two genetically distinct nuclei in the cells of fungal hyphae after mating has occurred.

dinosaur   A term meaning "terrible lizard" used to describe some of the extinct fossil reptiles.

dioecious   The term to describe plants that produce staminate and carpellate flowers on separate plants.

diploblastic   Having two distinct germ layers—ectoderm and endoderm but not mesoderm.

diploid cell   A cell that carries two sets of chromosomes.

diploid-dominant species   Species in which the diploid organism is the prevalent organism in the life cycle. Animals are an example.

direct calorimetry   A method of determining basal metabolic rate that involves quantifying the amount of heat generated by the animal.

direct repair   Refers to a DNA repair system in which an enzyme finds an incorrect structure in the DNA and directly converts it back to the correct structure.

directional selection   A pattern of natural selection that favors individuals at one extreme of a phenotypic distribution that have greater reproductive success in a particular environment.

directionality   In a DNA or RNA strand, refers to the orientation of the sugar molecules within that strand. Can be 5′ to 3′ or 3′ to 5′.

disaccharide   A carbohydrate composed of two monosaccharides.

discontinuous trait   A trait with clearly defined phenotypic variants.

discovery science   *See* discovery-based science.

discovery-based science   The collection and analysis of data without the need for a preconceived hypothesis. Also called discovery science.

dispersion   A pattern of spacing in which individuals in a population are clustered together or spread out to varying degrees.

dispersive mechanism   In this incorrect model for DNA replication, segments of parental DNA and newly made DNA are interspersed in both strands following the replication process.

dispersive mutualism   A mutually beneficial interaction often involving plants and pollinators that disperse their pollen, and plants and fruit eaters that disperse the plant's seeds.

disruptive selection   A pattern of natural selection that favors the survival of two or more different genotypes that produce different phenotypes.

dissociation constant   An equilibrium constant between a ligand and a protein, such as a receptor or an enzyme.

distal convoluted tubule   A structure in the tubule of the nephron through which fluid flows into one of the many collecting ducts in the kidney.

diversity-stability hypothesis   The proposal that species-rich communities are more stable than those with fewer species.

DNA (deoxyribonucleic acid)   The genetic material that provides a blueprint for the organization, development, and function of living things.

DNA fingerprinting   A technology that identifies particular individuals using properties of their DNA.

DNA helicase   An enzyme that uses ATP and separates DNA strands.

DNA library   A collection of vectors each containing a particular fragment of chromosomal DNA or cDNA.

DNA ligase   An enzyme that catalyzes the formation of a covalent bond between nucleotides in adjacent DNA fragments to complete the replication process in the lagging strand.

DNA methylase   An enzyme that attaches methyl groups to bases in DNA.

DNA methylation   A process in which methyl groups are attached to bases in DNA. This usually inhibits gene transcription by preventing the binding of activator proteins or by promoting the compaction of chromatin.

DNA microarray   A technology used to monitor the expression of thousands of genes simultaneously.

DNA polymerase   An enzyme responsible for covalently linking nucleotides together to form DNA strands.

DNA primase   An enzyme that synthesizes a primer for DNA replication.

DNA repair systems   One of several systems to reverse DNA damage before a permanent mutation can occur.

DNA replication   The mechanism by which DNA can be copied.

DNA sequencing   A method to determine the base sequence of DNA.

DNA topoisomerase   An enzyme that affects the level of DNA supercoiling.

DNase   An enzyme that digests DNA.

domain   1. A defined region of a protein with a distinct structure and function. 2. One of the three major categories of life: Bacteria, Archaea, and Eukarya.

dominant   A term that describes the displayed trait in a heterozygote.

domestication   A process that involves artificial selection of plants or animals for traits desirable to humans.

dominant species   A species that has a large effect in a community because of its high abundance or high biomass.

dormancy   A phase of metabolic slowdown in a plant.

dorsal   Refers to the upper side of an animal.

dorsoventral axis   One of the three axes along which the adult body pattern is organized; the others are the anteroposterior axis and the right-left axis.

dosage compensation   The phenonomen that gene dosage is compensated between males and females. In mammals, the inactivation of one X chromosome in the female reduces the number of expressed copies (doses) of X-linked genes from two to one.

double bond   A bond that occurs when the atoms of a molecule share two pairs of electrons.

double fertilization   In angiosperms, the process in which two different fertilization events occur, producing both a zygote and a nutritive endosperm tissue.

double helix   Two strands of DNA hydrogen-bonded with each other. In a DNA double helix, two DNA strands are twisted together to form a structure that resembles a spiral staircase.

Down syndrome   A human disorder caused by the inheritance of three copies of chromosome 21.

duplication   The term used to describe when a section of a chromosome occurs two or more times.

dynamic instability   The oscillation of a single microtubule between growing and shortening phases; important in many cellular activities including the sorting of chromosomes during cell division.

E

ecdysis   The periodic shedding and re-formation of the exoskeleton.

ecdysone   A steroid hormone synthesized and secreted by the prothoracic glands of certain invertebrates such as arthropods; in response to this hormone, a larva molts and begins a new growth period until it must shed its skin again in response to another episode of ecdysone secretion.

Ecdysozoa   A clade of moulting animals that encompasses primarily the arthropods and nematodes.

echolocation   The phenomenon in which certain species generate high-frequency sound waves in order to determine the distance and location of an object.

ecological footprint   The amount of productive land needed to support each person on Earth.

ecological niche   The unique set of habitat resources that a species requires, as well as its influence on the environment and other species.

ecological species concept   A species concept that considers a species within its native environment. Each species occupies its own ecological niche.

ecology   The study of interactions among organisms and between organisms and their environments.

ecosystem   The biotic community of organisms in an area as well as the abiotic environment affecting that community.

ecosystem engineer   A species that creates, modifies, and maintains habitats.

ecosystems ecology   The study of the flow of energy and cycling of nutrients among organisms within a community and between organisms and the environment.

ectoderm   The outer most layer of cells formed during gastrulation that covers the surface of the embryo and differentiates into the epidermis and nervous system.

ectomycorrhizae Beneficial interactions between temperate forest trees and soil fungi whose hyphae coat tree-root surfaces and grow into the spaces between root cells.

ectoparasite A parasite that lives on the outside of the host's body.

ectotherm An animal whose body temperature changes with the environmental temperature.

edge effect A special physical condition that exists at the boundary or "edge" of an area.

effective population size The number of individuals that contribute genes to future populations, often smaller than the actual population size.

effector A molecule that directly influences cellular responses.

effector cell A component of the immune response; these cells carry out the attack response.

egg Also, egg cell. The female gamete.

ejaculation The movement of semen through the urethra by contraction of muscles at the base of the penis.

ejaculatory duct The structure within the male penis through which semen is released.

elastin A protein that makes up elastic fibers in the extracellular matrix of animals.

electrical synapse A synapse that directly passes electric current from the presynaptic to the postsynaptic cell via gap junctions.

electrocardiogram (ECG or EKG) A record of the electrical impulses generated during the cardiac cycle.

electrochemical gradient The combined effect of both an electrical and chemical gradient; determines the direction that an ion will move.

electrogenic pump A pump that generates an electrical gradient across a membrane.

electromagnetic receptor A sensory receptor that detects radiation within a wide range of the electromagnetic spectrum, including visible, ultraviolet, and infrared light, as well as electrical and magnetic fields in some animals.

electromagnetic spectrum All possible wavelengths of electromagnetic radiation, from relatively short wavelengths (gamma rays) to much longer wavelengths (radio waves).

electron A negatively charged particle found in orbitals around the nucleus. For atoms, the number of protons is equal to the number of electrons.

electron microscope A microscope that uses an electron beam for illumination.

electron transport chain A group of protein complexes and small organic molecules embedded in the inner mitochondrial membrane. The components accept and donate electrons to each other in a linear manner. The movement of electrons produces an H^+ electrochemical gradient.

electronegativity A measure of an atom's ability to attract electrons to its outer shell from another atom.

elicitor A compound produced by bacterial and fungal pathogens that promotes virulence.

elongation factor In translation, a protein that is needed for the growth of a polypeptide.

elongation stage The second step in transcription or translation where RNA strands or polypeptides are made, respectively.

embryo The early stages of development in a multicellular organism during which the organization of the organism is largely formed.

embryogenesis The process by which embryos develop from single-celled zygotes by mitotic divisions.

embryonic development The process by which a fertilized egg is transformed into an organism with distinct physiological systems and body parts.

embryonic germ cell (EG cell) At the early fetal stage of development, the cells that later give rise to sperm or eggs cells. These cells are pluripotent.

embryonic stem cell (ES cell) A cell in the early mammalian embryo that can differentiate into almost every cell type of the body.

embryophyte A synonym for the land plants.

emerging virus A newly arising virus.

empirical thought Thought that relies on observation to form an idea or hypothesis, rather than trying to understand life from a nonphysical or spiritual point of view.

emulsification A process during digestion that disrupts the large lipid droplets into many tiny droplets, thereby increasing their total surface area and exposure to lipase action.

enantiomer A type of stereoisomer that exists as a mirror image of another molecule.

endangered species Those species that are in danger of extinction throughout all or a significant portion of their range.

endemic The term to describe organisms that are naturally found only in a particular location.

endergonic Refers to a reaction that has a positive free energy change and does not proceed spontaneously.

endocrine disruptor A chemical found in water and soil exposed to pollution runoff that has a molecular structure that in some cases resembles estrogen sufficiently to bind to estrogen receptors.

endocrine gland A structure that contains epithelial cells that secrete hormone molecules into the bloodstream, where they circulate throughout the body.

endocrine system All the endocrine glands and other organs with hormone-secreting cells.

endocytic pathway A pathway to take substances into the cell; the reverse of the secretory pathway.

endocytosis A process in which the plasma membrane invaginates, or folds inward, to form a vesicle that brings substances into the cell.

endoderm The inner most layer of cells formed during gastrulation that lines the gut and gives rise to many internal organs.

endomembrane system A network of membranes that includes the nuclear envelope, which encloses the nucleus, and the

endoplasmic reticulum, Golgi apparatus, lysosomes, vacuoles, and plasma membrane.

endomycorrhizae Partnerships between plants and fungi in which the fungal hyphae grow into the spaces between root cell walls and plasma membranes.

endoparasite A parasite that lives inside the host's body.

endophyte A mutualistic fungus that lives compatibly within the tissues of various types of plants.

endoplasmic reticulum (ER) A convoluted network of membranes in the cell's cytoplasm that forms flattened, fluid-filled tubules or cisternae.

endoskeleton An internal skeleton covered by soft tissue; composed of calcareous plates overlaid by a thin skin in echinoderms and composed of a bony skeleton overlaid by muscles in invertebrates.

endosperm A nutritive tissue that increases the efficiency with which food is stored and used in the seeds of flowering plants.

endospore A cell with a tough coat that is produced inside the cells of certain bacteria and then released when the enclosing cell dies and breaks down.

endosporic gametophyte A plant gametophyte that grows within the confines of microspore and megaspore walls.

endosymbiosis A symbiotic relationship in which the smaller species—the symbiont— lives inside the larger species.

endosymbiosis theory A theory that mitochondria and chloroplasts originated from bacteria that took up residence within a primordial eukaryotic cell.

endosymbiotic Describes a relationship in which one organism lives inside the other.

endothelium A single-celled inner layer of a blood vessel, which forms a smooth lining in contact with the blood.

endotherm An animal that generates its own internal heat.

endothermic A term to describe that ability of an organism to generate and retain body heat through its metabolism.

energy The ability to promote change.

energy expenditure The amount of energy an animal uses in a given period of time to power all of its metabolic requirements.

energy flow The movement of energy through an ecosystem.

energy intermediate A molecule such as ATP and NADH that is directly used to drive endergonic reactions in cells.

enhancement effect The phenomenon whereby maximal activation of the pigments in photosystems I and II is achieved when organisms are exposed to two wavelengths of light.

enhancer A response element in eukaryotes that increases the rate of transcription.

enterocoelous In deuterostomes, a pattern of development in which a layer of mesoderm cells forms outpockets that bud off from the developing gut to form the coelom.

enthalpy (H) The total energy of a system.

entomology The study of insects.

entropy The degree of disorder of a system.

environmental science The application of ecology to real-world problems.

enzyme A protein responsible for speeding up a chemical reaction in a cell.

enzyme-linked receptor A receptor found in all living species that typically has two important domains: an extracellular domain, which binds a signaling molecule, and an intracellular domain, which has a catalytic function.

enzyme-substrate complex The binding between an enzyme and substrate.

eosinophil A type of phagocyte found in large numbers lining the mucosal surfaces lining the gastrointestinal, respiratory, and urinary tracts, where they fight off parasitic invasions.

epicotyl The portion of an embryonic plant stem with two tiny leaves in a first bud that is located above the point of attachment of the cotyledons.

epidermis A layer of dermal tissue that helps protect a plant from damage.

epididymis A coiled, tubular structure located on the surface of the testis; it is here that the sperm complete their differentiation by becoming motile and gaining the capacity to fertilize ova.

epigenetic inheritance An inheritance pattern in which modification of a gene or chromosome during egg formation, sperm formation, or early stages of embryo growth alters gene expression in a way that is fixed during an individual's lifetime.

epinephrine A hormone secreted by the adrenal glands; also known as adrenaline.

episome A plasmid that can integrate into the bacterial chromosome.

epistasis A gene interaction in which the alleles of one gene mask the expression of the alleles of another gene.

epithelial placodes Regions of slightly thickened epithelial cells.

epithelial tissue A sheet of densely packed cells that covers the body or individual organs or lines the walls of various cavities inside the body.

epitopes Antigenic determinants; the peptide fragments of the antigen that are complexed to the MHC proteins and presented to the helper T cell.

equilibrium 1. In a chemical reaction, occurs when the rate of the forward reaction is balanced by the rate of the reverse reaction. 2. In a population, the situation in which the population size stays the same.

equilibrium model of island biogeography A model to explain the process of succession on new islands that states that the number of species on an island tends toward an equilibrium number that is determined by the balance between immigration rates and extinction rates.

equilibrium potential In membrane physiology, the membrane potential at which the flow of an ion is at equilibrium—no net movement in either direction.

ER lumen A single compartment enclosed by the ER membrane.

ER signal sequence A sorting signal in a polypeptide usually located near the amino terminus that is recognized by SRP (signal recognition particle) and directs the polypeptide to the ER membrane.

erythrocyte *See* red blood cell.

erythropoietin (EPO) A hormone made by the liver and kidneys in response to any situation where additional blood cells are required, such as when animals lose blood following an injury; when abused, as in blood-doping, the concentration of red blood cells reaches such high levels that the blood becomes much thicker than normal.

esophagus The tubular structure that forms a pathway from the throat to the stomach.

essential amino acids Those amino acids that are required in the diet of particular organisms. In humans, they include isoleucine, leucine, lysine, methionine, phenylalanine, threonine, tryptophan, and valine.

essential fatty acid A polyunsaturated fatty acid such as linoleic acid that cannot be synthesized by animal cells and must therefore be consumed in the diet.

essential nutrient A compound that cannot be synthesized from any ingested or stored precursor molecule and so must be obtained in the diet in its complete form.

estradiol The major estrogen in many animals, including humans.

estrogens Steroid hormones produced by the female ovaries that affect most aspects of reproduction.

ethology Scientific studies of animal behavior.

ethylene A plant hormone that is particularly important in coordinating plant developmental and stress responses.

euchromatin The less condensed regions of a chromosome; areas that are capable of gene transcription.

eudicots One of the two largest lineages of flowering plants; the embryo possesses two seed leaves.

Eukarya One of the three domains of life; the other two are Bacteria and Archaea.

eukaryote Cells or organisms from the domain Eukarya. The distinguishing feature is cell compartmentalization, which includes a cell nucleus. "Eukaryote" means "true nucleus," and includes protists, fungi, plants, and animals.

eukaryotic Refers to organisms having cells with internal compartments that serve various functions; includes all members of the domain Eukarya.

Eumetazoa A subgroup of animals having more than one type of tissue and, for the most part, different types of organs.

euphotic zone A fairly narrow zone close to the surface of an aquatic environment, where light is sufficient to allow photosynthesis to exceed respiration.

euphyll A leaf with branched veins.

euphyllophytes The clade that includes pteridophytes and seed plants.

euploid An organism that has a chromosome number that is a multiple of a chromosome set ($1n$, $2n$, $3n$, etc.).

eusociality The phenomenon whereby sterile castes evolve in social insects in which the vast majority of females, known as workers, rarely reproduce themselves but instead help one reproductive female (the queen) to raise offspring.

Eustachian tube A connection from the middle ear to the pharynx, which maintains the pressure in the middle ear at atmospheric pressure.

eustele A ring of vascular tissue arranged around a central pith of nonvascular tissue; typical of progymnosperms, gymnosperms, and angiosperms.

eutherian A placental mammal and member of the subclass Eutheria.

eutrophic Waters that contain relatively high levels of nutrients such as phosphate or nitrogen and typically exhibit high levels of primary productivity and low levels of biodiversity.

eutrophication The process by which elevated nutrient levels lead to an overgrowth of algae or aquatic plants and the subsequent depletion of water oxygen levels.

evaporation The transformation of water from the liquid to the gaseous state.

evapotranspiration rate The rate at which water moves into the atmosphere through the processes of evaporation from the soil and transpiration of plants.

evolutionarily conserved The term used to describe DNA sequences that are very similar or identical between different species.

evolutionary developmental biology (evo-devo) A field of biology that compares the development of different organisms in an attempt to understand ancestral relationships between organisms and the developmental mechanisms that bring about evolutionary change.

evolutionary species concept A species is derived from a single lineage that is distinct from other lineages and has its own evolutionary tendencies and historical fate.

excitable cell The term used to describe neurons and muscle cells, because they have the capacity to generate electrical signals.

excitation-contraction coupling The sequence of events by which an action potential in the plasma membrane of a muscle fiber leads to cross-bridge activity.

excitatory postsynaptic potential (EPSP) The response from an excitatory neurotransmitter depolarizing the postsynaptic membrane; the depolarization brings the membrane potential closer to the threshold potential that would trigger an action potential.

excurrent siphon A tunicate structure used to expel water.

exercise   Any physical activity that increases an animal's metabolic rate.

exergonic   Refers to reactions that release free energy and occur spontaneously.

exit site (E site)   A site for the tRNA binding in the ribosome; the other two are the peptidyl site (P site) and the aminoacyl site (A site). This is the site where the uncharged tRNA exits.

exocytosis   A process in which material inside the cell is packaged into vesicles and excreted into the extracellular medium.

exon shuffling   A form of mutation in which exons are inserted into genes and thereby create proteins with additional functional domains.

exon   A portion of RNA that is found in the mature RNA molecule after splicing is finished.

exoskeleton   An external skeleton that surrounds and protects most of the body surface of animals such as insects.

exotic species   Species moved from a native location to another location, usually by humans.

expansin   A protein that occurs in the plant cell wall and fosters cell enlargement.

experimental sample   The sample in an experiment that is subjected to some type of variation that does not occur for the control sample.

exponential growth   J-shaped, rapid population growth that occurs when the per capita growth rate remains above zero.

extensor   A muscle that straightens a limb.

external fertilization   Fertilization that occurs in aquatic environments, when eggs and sperm are released into the water in close enough proximity for fertilization to occur.

extinction   The end of the existence of a species or group of species.

extinction vortex   A downward spiral toward extinction from which a species cannot naturally recover.

extracellular fluid   The fluid in an organism's body that is outside of the cells.

extracellular matrix (ECM)   A network of material that is secreted from cells and forms a complex meshwork outside of cells. The ECM provides strength, support, and organization.

extranuclear inheritance   In eukaryotes, the transmission of genes that are located outside the cell nucleus.

extremophile   An organism that occurs primarily in extreme habitats.

eye   The visual organ that detects light and sends signals to the brain.

eyecup   In planaria, a primitive eye that detects light and its direction.

F

F factor   A segment of DNA called a fertility factor that plays a role in bacterial conjugation.

F_1 generation   The first filial generation in a genetic cross.

F_2 generation   The second filial generation in a genetic cross.

facilitated diffusion   A method of passive transport that involves the aid of a transport protein.

facilitation   A mechanism for succession in which a species facilitates or makes the environment more suitable for subsequent species.

facultative aerobe   A microorganism that can use oxygen in aerobic respiration, obtain energy via anaerobic fermentation, or use inorganic chemical reactions to obtain energy.

facultative mutualism   An interaction that is beneficial but not essential to the survival and reproduction of either species.

family   A subdivision of an order.

fast block to polyspermy   A depolarization of the egg that blocks other sperm from binding to the egg membrane proteins.

fast fiber   A muscle fiber containing myosin with high ATPase activity.

fast-glycolytic fiber   A skeletal muscle fiber that has high myosin activity but cannot make as much ATP as oxidative fibers, because its source of ATP is glycolysis; best suited for rapid, intense actions.

fast-oxidative fiber   A skeletal muscle fiber that has high myosin activity and can make large amounts of ATP; used for long-term activities.

fat   A molecule formed by bonding glycerol to three fatty acids.

fate mapping   A technique in which a small population of cells within an embryo is specifically labeled with a harmless dye, and the fate of these labeled cells is followed to a later stage of embryonic development.

feedback inhibition   A form of regulation in which the product of a metabolic pathway inhibits an enzyme that acts early in the pathway, thus preventing the over accumulation of the product.

feedforward regulation   The process by which an animal's body begins preparing for a change in some variable before it even occurs.

female-enforced monogamy hypothesis   The suggestion that males are monogamous because females stop their male partners from being polygynous.

female gametophyte   In plants, a haploid generation that produces one or more eggs, but does not produce sperm cells.

fermentation   The breakdown of organic molecules to produce energy without any net oxidation of an organic molecule.

ferrell cell   The middle cell in the three-cell circulation of wind in each hemisphere.

fertilization   The union of two gametes, such as an egg cell with a sperm cell, to create a zygote.

fertilizer   A soil addition that enhances plant growth by providing essential elements.

fetus   The maturing embryo, after the eighth week of gestation in humans.

fibrous root system   The root system of monocots, which consists of multiple adventitious roots that grow from the stem base.

fibrin   A protein that forms a meshwork of threadlike fibers that wrap around and between platelets and blood cells, enlarging and thickening a blood clot.

fidelity   Refers to the high level of accuracy in DNA replication.

filament   The elongate portion of a flower's stamen; contains vascular tissue that delivers nutrients from parental sporophytes to anthers.

filtrate   In the process of filtration, the material that passes through the filter and enters the excretory organ for either further processing or excretion.

filtration   The passive removal of water and small solutes from the blood.

first law of thermodynamics   The first law states that energy cannot be created or destroyed; it is also called the law of conservation of energy.

5′ cap   The 7-methylguanosine cap structure found on mRNA in eukaryotes.

fixed action pattern (FAP)   An animal behavior that, once initiated, will continue until completed.

flagellate   A protist that uses one or more flagella to move in water or cause water motions useful in feeding.

flagellum (plural, flagella)   A relatively long cell appendage that facilitates cellular movement or the movement of extracellular fluids.

flagship species   A single large or instantly recognizable species.

flame cell   A cell that exists primarily to maintain osmotic balance between an organism's body and surrounding fluids; present in flatworms.

flatus   Intestinal gas, which is a mixture of nitrogen and carbon dioxide, with small amounts of hydrogen, methane, and hydrogen sulfide.

flavonoid   A type of phenolic secondary metabolite that provides plants with protection from UV damage or colors organs such as flower petals.

flexor   A muscle that bends a limb at a joint.

florigen   The hypothesized flowering hormone, now identified as the mRNA that produces FT (flowering time) protein in the shoot apex.

flower   A reproductive shoot; a stem branch that produces reproductive organs instead of leaves.

flowering plants   The angiosperms, which produce ovules within the protective ovaries of flowers; when ovules develop into seeds, angiosperm ovaries develop into fruits, which function in seed dispersal.

flow-through system   A form of ventilation in fish in which water moves unidirectionally such that the gills are constantly in contact with fresh, oxygenated water.

fluid-feeder   An animal that licks or sucks fluid from plants or animals and does not need teeth except to puncture an animal's skin.

fluidity   A quality of biomembranes that means that individual molecules remain in close association yet have the ability to move laterally or rotationally within the plane of the membrane. Membranes are semi-fluid.

fluid-mosaic model The accepted model of the plasma membrane; its basic framework is the semi-fluid phospholipid bilayer with a mosaic of proteins. Carbohydrates may be attached to lipids or proteins.

follicle A structure within the ovary where each ovum undergoes growth and development before it is ovulated.

follicle-stimulating hormone (FSH) A gonadotropin that stimulates follicle development.

food chain A linear depiction of energy flow between organisms, with each organism feeding on and deriving energy from the preceding organism.

food vacuole *See* phagocytic vacuole.

food web A complex model of interconnected food chains in which there are multiple links between species.

food-induced thermogenesis A rise in metabolic rate for a few hours after eating that produces heat.

foot In mollusks, a muscular structure usually used for movement.

forebrain One of three major divisions of the vertebrate brain; the other two divisions are the midbrain and hindbrain.

fossil Recognizable remains of past life on Earth.

fossil fuel A fuel formed in the Earth from protist, plant or animal remains, such as coal, petroleum, and natural gas.

founder effect A small group of individuals separates from a larger population and establishes a colony in a new location; genetic drift is common due to the small population size.

fovea A small area on the retina directly behind the lens that is responsible for the sharpness with which we and other animals see in daylight.

frameshift mutation A mutation that involves the addition or deletion of nucleotides that are not in multiples of three nucleotides.

free energy (G) In living organisms, the usable energy, that is, the amount of available energy that can be used to do work.

free radical A molecule containing an atom with a single, unpaired electron in its outer shell. A free radical is unstable and interacts with other molecules by "stealing" electrons from their atoms.

frequency In regard to sound, the number of complete wavelengths that occur in 1 second, measured in number of waves per second, or Hertz (Hz).

frugivore An herbivore that is adapted primarily to feed on fruits.

fruit A structure that develops from flower organs, encloses seeds, and fosters seed dispersal in the environment.

fruiting bodies The visible fungal reproductive structures; composed of densely packed hyphae that typically grow out of the substrate.

functional genomics Genomic methods aimed at studying the expression of a genome.

functional group A group of atoms with chemical features that are functionally important. Each functional group exhibits the same properties in all molecules in which it occurs.

fundamental niche The optimal range in which a particular species best functions.

Fungi One of the four traditional eukaryotic kingdoms of the domain Eukarya.

fungus-like protist A heterotrophic organism that often resembles true fungi in having threadlike, filamentous bodies and absorbing nutrients from its environment.

G

G_0 A stage in which cells exit the cell cycle and postpone making the decision to divide.

G_1 The first gap phase of the cell cycle.

G_2 The second gap phase of the cell cycle.

G banding A staining procedure for chromosomes that produces an alternating pattern of G bands that is unique for each type of chromosome.

G protein An intracellular protein that binds guanosine triphosphate (GTP) and guanosine diphosphate (GDP) and participates in intracellular signaling pathways.

gallbladder A small sac underneath the liver that is a storage site for bile that allows the release of large amounts of bile to be precisely timed to the consumption of fats.

gametangia Specialized structures produced by many land plants in which developing gametes are protected by a jacket of tissue.

gamete A cell that is involved with sexual reproduction, such as a sperm or egg cell.

gametic life cycle In this type of life cycle, all cells except the gametes are diploid, and gametes are produced by meiosis.

gametogenesis The formation of gametes.

gametophyte In plants and many multicellular protists, the haploid stage that produces gametes by mitosis.

ganglia Groups of neuronal cell bodies that perform basic functions of integrating inputs from sense organs and controlling motor outputs, usually in the peripheral nervous system.

ganglion cells Cells that send their axons out of the eye into the optic nerve.

gap gene A type of segmentation gene; when a mutation inactivates a gap gene, several adjacent segments are missing in the larva.

gap junction A type of junction between animal cells that provides a passageway for intercellular transport.

gas exchange The process of moving oxygen and carbon dioxide in opposite directions between cells and blood, and between blood and the environment.

gas vesicle A cytoplasmic structure used by cyanobacteria and some other bacteria that live in aquatic habitats to adjust their buoyancy.

gastrovascular cavity A body cavity with a single opening to the outside; it functions as both a digestive system and circulatory system.

gastrula An embryo that is the result of gastrulation, which has three cellular layers called the ectoderm, endoderm, and mesoderm.

gastrulation A process in which an area in the blastula invaginates and folds inward, creating different embryonic cell layers called germ layers.

gated channel A channel that can open to allow the diffusion of solutes and close to prohibit diffusion.

gel electrophoresis A technique used to separate macromolecules by using an electric field that causes them to pass through a gel matrix.

gene A unit of heredity that contributes to the characteristics or traits of an organism. At the molecular level, a gene is composed of organized sequences of DNA.

gene amplification An increase in the copy number of a gene.

gene cloning The process of making multiple copies of a gene.

gene expression Gene function both at the level of traits and at the molecular level.

gene family A group of homologous genes within a single species.

gene flow Occurs when individuals migrate between different populations and cause changes in the genetic composition of the resulting populations.

gene interaction A situation in which a single trait is controlled by two or more genes.

gene knockout An organism in which both copies of a functional gene have been replaced with nonfunctional copies. Experimentally, this can occur via gene replacement.

gene mutation A relatively small change in DNA structure that alters a particular gene.

gene pool All of the genes in a population.

gene regulation The ability of cells to control their level of gene expression.

gene replacement The phenomenon in which a cloned gene recombines with the normal gene on a chromosome and replaces it.

gene silencing The ability of one gene to silence the effect of another via small RNA molecules called microRNAs.

gene therapy The introduction of cloned genes into living cells in an attempt to cure disease.

general transcription factors (GTFs) Five different proteins that play a role in initiating transcription at the core promoter of structural genes in eukaryotes.

generative cell In seed plants, the male gametophyte cell that divides to produce sperm cells.

genetic code A code that specifies the relationship between the sequence of nucleotides in the codons found in mRNA and the sequence of amino acids in a polypeptide.

genetic drift The random change in a population's allele frequencies from one generation to the next that is attributable to chance. It occurs more quickly in small populations.

genetic engineering The direct manipulation of genes for practical purposes.

genetic linkage map A diagram that describes the linear arrangement of genes that are linked to each other along the same chromosome.

genetic mosaic An individual with somatic regions that are genetically different from each other.

genetic transfer The process by which genetic material is transferred from one bacterial cell to another.

genetically modified organisms (GMOs) *See* transgenic.

genome The complete genetic composition of a cell or a species.

genome database A specialized database that focuses on the genetic characteristics of a single species.

genomic imprinting A phenomenon in which a segment of DNA is imprinted, or marked, in a way that affects gene expression throughout the life of the individual who inherits that DNA.

genomic library A type of DNA library in which the inserts are derived from chromosomal DNA.

genomics Techniques that are used in the molecular analysis of the entire genome of a species.

genotype The genetic composition of an individual.

genotype frequency The number of individuals with a given genotype divided by the total number of individuals.

genus In taxonomy, a subdivision of a family.

geological timescale A time line of the Earth's history from its origin about 4.55 billion years ago to the present.

germ layer An embryonic cell layer such as ectoderm, mesoderm, or endoderm.

germ line Cells that give rise to gametes such as egg and sperm cells.

germ plasm Cytoplasmic determinants that help define and specify the primordial germ cells in the gastrula stage.

gestation *See* pregnancy.

giant axon A very large axon in certain species such as squid that facilitates high-speed nerve conduction and rapid responses to stimuli.

gibberellic acid A type of gibberellin.

gibberellin A plant hormone that stimulates both cell division and cell elongation.

gills Specialized filamentous organs in aquatic animals that aid in obtaining oxygen and eliminating carbon dioxide.

ginkgos A phylum of gymnosperms; Ginkgophyta.

gizzard In the stomach of a bird, the muscular structure with a rough inner lining capable of grinding food into smaller fragments.

glaucoma A condition in which drainage of aqueous humor in the eye becomes blocked and the pressure inside the eye increases as the fluid level rises. If untreated, damages cells in the retina and leads to irreversible loss of vision.

glia Cells that surround the neurons; a major class of cells in nervous systems that perform various functions.

global warming A gradual elevation of the Earth's surface temperature caused by the greenhouse effect.

glomerulus A cluster of interconnected, fenestrated capillaries in the renal corpuscle of the kidney; the site of filtration in the kidneys.

glucocorticoid A steroid hormone that regulates glucose balance and helps prepare the body for stress situations.

gluconeogenesis A mechanism for maintaining blood glucose levels; enzymes in the liver convert noncarbohydrate precursors into glucose, which are then secreted into the blood.

glucose sparing A metabolic adjustment that reserves the glucose produced by the liver for use by the nervous system.

glycocalyx 1. An outer viscous covering surrounding a bacterium. The glycocalyx, which is secreted by the bacterium, traps water and helps protect bacteria from drying out. 2. A carbohydrate covering that is found outside of animal cells.

glycogen A polysaccharide found in animal cells and sometimes called animal starch.

glycogenolysis A mechanism for maintaining blood glucose levels; stored glycogen can be broken back down into molecules of glucose by hydrolysis.

glycolipid A lipid that has carbohydrate attached to it.

glycolysis A metabolic pathway that breaks down glucose to pyruvate.

glycolytic fiber A skeletal muscle fiber that has few mitochondria but possesses both a high concentration of glycolytic enzymes and large stores of glycogen.

glycoprotein A protein that has carbohydrate attached to it.

glycosaminoglycan (GAG) The most abundant type of polysaccharide in the extracellular matrix (ECM) of animals, consisting of repeating disaccharide units that give a gel-like character to the ECM of animals.

glycosylation The attachment of carbohydrate to a protein or lipid, producing a glycoprotein or glycolipid.

glyoxysome A specialized organelle within plant seeds that contains enzymes needed to convert fats to sugars.

gnathostomes All vertebrate species that possess jaws.

gnetophytes A phylum of gymnosperms; Gnetophyta.

Golgi apparatus A stack of flattened, membrane-bounded compartments that performs three overlapping functions: secretion, processing, and protein sorting.

gonadotropins Hormones secreted by the anterior pituitary gland that are the same in both sexes; gonadotropins influence the ability of the testes and ovaries to produce the sex steroids.

gonads The testes in males and the ovaries in females, where the gametes are formed.

G-protein-coupled receptors (GPCRs) A common type of receptor found in the cells of eukaryotic species that interacts with G proteins to initiate a cellular response.

gradualism A concept that suggests that species evolve continuously over long spans of time.

grain The characteristic single-seeded fruit of cereal grasses such as rice, corn, barley, and wheat.

granum A structure composed of stacked tubules within the thylakoid membrane of chloroplasts.

gravitropism Plant growth in response to the force of gravity.

gray matter Brain tissue that consists of neuronal cell bodies, dendrites, and some unmyelinated axons.

grazer An herbivore that feeds almost constantly on grasses.

greenhouse effect The process in which short-wave solar radiation passes through the atmosphere to warm the Earth but is radiated back to space as long-wave infrared radiation. Much of this radiation is absorbed by atmospheric gases and reradiated back to Earth's surface, causing its temperature to rise.

groove In the DNA double helix, an indentation where the atoms of the bases make contact with the surrounding water.

gross primary production (GPP) The measure of biomass production by photosynthetic organisms; equivalent to the carbon fixed during photosynthesis.

ground meristem A type of primary plant tissue meristem that gives rise to ground tissue.

ground tissue Most of the body of a plant, which has a variety of functions, including photosynthesis, storage of carbohydrates, and support. Ground tissue can be subdivided into three types: parenchyma, collenchyma, and sclerenchyma.

group selection The premise that natural selection produces outcomes beneficial for the whole group or species rather than for individuals.

growth An increase in weight or size.

growth factors A group of proteins in animals that stimulate certain cells to grow and divide.

growth hormone (GH) A hormone produced in vertebrates by the anterior pituitary gland; GH acts on the liver to produce insulin-like growth factor-1 (IGF-1).

guanine (G) A purine base found in DNA and RNA.

guard cell A specialized plant cell that allows epidermal pores (stomata) to close when conditions are too dry, and open under moist conditions, allowing the entry of CO_2 needed for photosynthesis.

gustation The sense of taste.

gut The gastrointestinal (GI) tract.

gymnosperm A plant that produces seeds that are exposed rather than enclosed in fruits.

gynoecium The aggregate of carpels that forms the innermost whorl of flower organs.

H

H zone   A narrow, light region in the center of the A band of the sarcomere that corresponds to the space between the two sets of thin filaments in each sarcomere.

H⁺ electrochemical gradient   A transmembrane gradient for H⁺ composed of both a membrane potential and a concentration difference for H⁺ across the membrane.

habituation   The form of nonassociative learning in which an organism learns to ignore a repeated stimulus.

Hadley cell   The most prominent of the three cells in the three-cell circulation of wind in each hemisphere.

hair cell   A mechanoreceptor that is a specialized epithelial cell with deformable stereocilia.

half-life   1. In the case of organic molecules in a cell, refers to the time it takes for 50% of the molecules to be broken down. 2. In the case of radioisotopes, the time it takes for half the molecules to decay and emit radiation.

halophyte   A plant that can tolerate higher than normal salt concentrations in the cell sap, and thus can occupy coastal salt marshes or saline deserts.

Hamilton's rule   The proposal that an altruistic gene will be favored by natural selection when r > C/B where r is the coefficient of relatedness of the donor (the altruist) to the recipient, B is the benefit received by the recipients of the altruism, and C is the cost incurred by the donor.

haplodiploidy   A genetic system in which females develop from fertilized eggs and are diploid but males develop from unfertilized eggs and are haploid.

haploid   Containing one set of chromosomes; or 1*n*.

haploid-dominant species   Species in which the haploid organism is the prevalent organisms in the life cycle. Examples include fungi and some protists.

Hardy-Weinberg equation   An equation ($p^2 + 2pq + q^2 = 1$) that relates allele and genotype frequencies; the equation predicts an equilibrium if the population size is very large, mating is random, the populations do not migrate, no natural selection occurs, and no new mutations are formed.

heart attack   *See* myocardial infarction (MI).

heart   A muscular structure that pumps blood through blood vessels.

heat of fusion   The amount of heat energy that must be withdrawn or released from a substance to cause it to change from the liquid to the solid state.

heat of vaporization   The heat required to vaporize 1 mole of any substance at its boiling point under standard pressure.

heat shock protein   A protein that helps to protect other proteins from heat damage and refold them to their functional state.

heavy chain   A part of an immunoglobulin molecule.

helper T cell   A type of lymphocyte that assists in the activation and function of B cells and cytotoxic T cells.

hematocrit   The volume of blood that is composed of red blood cells, usually between 40 and 65% among vertebrates.

hemiparasite   A parasitic organism that generally photosynthesizes, but lacks a root system to draw water and thus depends on its hosts for that function.

hemispheres   The two halves of the cerebrum.

hemizygous   The term used to describe the single copy of an X-linked gene in a male.

hemocyanin   A copper-containing pigment that gives the blood or hemolymph a bluish tint.

hemodialysis   A medical procedure used to artificially perform the kidneys' excretory filtration and cleansing functions.

hemoglobin   An oxygen-binding protein found within the cytosol of red blood cells.

hemolymph   Blood and interstitial fluid combined in one fluid compartment in many invertebrates.

hemorrhage   A loss of blood from a ruptured blood vessel.

herbaceous plant   A plant that produces little or no wood and is composed mostly of primary vascular tissues.

herbivore   An animal that eats only plants.

herbivory   Refers to herbivores feeding on plants.

hermaphrodite   An individual that can produce both sperm and eggs.

hermaphroditism   A form of sexual reproduction in which individuals have both male and female reproductive systems.

heterochromatin   The highly compacted regions of chromosomes; in general, these regions are transcriptionally inactive because of their tight conformation.

heterocyst   A specialized cell of some cyanobacteria in which nitrogen fixation occurs.

heterodimer   The structure that results when two different proteins come together.

heterokaryon   In fungi, a mycelium having nuclei of two or more genetic types.

heterospory   In plants, the formation of two different types of spores, microspores and megaspores; microspores produce male gametophytes and megaspores produce female gametophytes.

heterotherm   An animal that has a body temperature that varies with the environment.

heterotroph   Organisms that cannot produce their own organic food and thus must obtain organic food from other organisms.

heterotrophic   Requiring organic food from the environment.

heterozygote advantage   A phenomenon in which a heterozygote has a higher Darwinian fitness compared to the corresponding homozygotes.

heterozygous   An individual with two different alleles of the same gene.

hibernation   The state of torpor in an animal over months.

high affinity   Refers to the binding of an ion or molecule to a protein very tightly. The substance will bind at a very low concentration.

highly repetitive sequence   A DNA sequence found tens of thousands or even millions of times throughout a genome.

hindbrain   One of three major divisions of the vertebrate brain; the other two divisions are the midbrain and forebrain.

hippocampus   The area of the brain whose main function appears to be establishing memories for spatial locations, facts, and sequences of events; composed of several layers of cells that are connected together in a circuit.

histone acetyltransferase   An enzyme that attaches acetyl groups to histone proteins.

histone code hypothesis   Refers to the pattern of histone modification recognized by particular proteins. The pattern of covalent modifications of amino terminus tails provides binding sites for proteins that subsequently affect the degree of chromatin compaction.

histones   A group of proteins involved in the formation of nucleosomes.

holoblastic cleavage   A complete type of cell cleavage in certain animals in which the entire zygote is bisected into two equal-sized blastomeres.

holoparasite   A parasitic organism that lacks chlorophyll and is totally dependent on the host plant for its water and nutrients.

homeobox   A 180-bp sequence within the coding sequence of homeotic genes.

homeodomain   A region of a protein that functions in binding to the DNA.

homeostasis   The process whereby living organisms regulate their cells and bodies to maintain relatively stable internal conditions.

homeotherm   An animal that maintains its body temperature within a narrow range.

homeotic   A term that describes changes in which one body part is replaced by another.

homeotic gene   A gene that controls the developmental fate of particular segments or regions of an animal's body.

hominoid   A gibbon, gorilla, orangutan, chimpanzee, or human.

hominin   Either an extinct or modern form of humans.

homodimer   The structure that results when two identical proteins come together.

homologous genes   Genes derived from the same ancestral gene that have accumulated random mutations that make their sequences slightly different.

homologous structures   Structures that are similar to each other because they are derived from the same ancestral structure.

homologue   A member of a pair of chromosomes in a diploid organism that are evolutionarily related.

homology   A fundamental similarity that occurs due to descent from a common ancestor.

homozygous   An individual with two identical copies of an allele.

horizontal gene transfer The transfer of genes between different species.

hormone A chemical messenger that is produced in a gland or other structure and acts on distant target cells in one or more parts of the body.

hornworts A phylum of bryophytes; Anthocerophyta.

host The prey organism in a parasitic association.

host cell A cell that is infected by a virus, fungus, or a bacterium.

host plant resistance The ability of plants to prevent herbivory.

host range The number of species and cell types that a virus or bacterium can infect.

hot spot A human-impacted geographic area with a large number of endemic species. To qualify as a hot spot, a region must contain at least 1,500 species of vascular plants as endemics and have lost at least 70% of its original habitat.

Hox **complex** A group of adjacent homeotic genes in vertebrates that controls the formation of structures along the anteroposterior axis.

Hox **genes** A class of genes involved in pattern formation in early embryos.

Human Genome Project The largest genome project in history, which was a 13-year effort coordinated by the U.S. Department of Energy and the National Institutes of Health. The goals of the project were to identify all human genes, to sequence the entire human genome, to address the legal and ethical implications resulting from the project, and to develop programs to manage the information gathered from the project.

human immunodeficiency virus (HIV) A retrovirus that is the causative agent of acquired immune deficiency syndrome (AIDS).

humoral immunity A type of specific immunity in which plasma cells secrete antibodies that bind to antigens.

humus A collective term for the organic constituents of soils.

hybrid zone An area where two populations can interbreed.

hybridization A situation in which two individuals with different characteristics are mated or crossed to each other; the offspring are referred to as hybrids.

hydrocarbon Molecules with predominantly hydrogen-carbon bonds.

hydrogen bond Electrostatic attraction between a hydrogen atom of a polar molecule and an electronegative atom of another polar molecule.

hydrolysis The process in which reactions utilize water to break apart other molecules.

hydrophilic "Water-loving"—generally, ions and molecules that contain polar covalent bonds will dissolve in water and are said to be hydrophilic.

hydrophobic "Water-fearing"—molecules that are not attracted to water molecules. Such molecules are composed predominantly of carbon and hydrogen and are relatively insoluble in water. Because carbon-carbon and carbon-hydrogen bonds are nonpolar, the atoms in such compounds are electrically neutral.

hydrostatic skeleton A fluid-filled body cavity surrounded by muscles that gives support and shape to the body of organisms.

hydroxide ion An anion with the formula, OH^-.

hypermutation A process that primarily involves numerous C to T point mutations that is crucial to enabling lymphocytes to produce a diverse array of immunoglobulins capable of recognizing many different antigens.

hyperpolarization The change in the membrane potential that occurs when the cell becomes more polarized.

hypersensitive response (HR) A plant's local defensive response to pathogen attack.

hyperthyroidism A hyperactive thyroid gland.

hypertonic When the solute concentration inside the cell is higher relative to the outside of the cell.

hypha A microscopic, branched filament of the body of a fungus.

hypocotyl The portion of an embryonic plant stem located below the point of attachment of the cotyledons.

hypothalamus A gland located below the thalamus at the floor of the forebrain; it controls functions of the gastrointestinal and reproductive systems, and it regulates many basic behaviors such as eating and drinking.

hypothesis In biology, a proposed explanation for a natural phenomenon based on previous observations or experimental studies.

hypothesis testing Also known as the scientific method, a strategy for testing the validity of a hypothesis.

hypothyroidism An underactive thyroid gland.

hypotonic When the solute concentration outside the cell is lower relative to the inside of the cell.

I

I band A light band that lies between the A bands of two adjacent sarcomeres.

immune system The cells and organs within an animal's body that contribute to immune defenses.

immune tolerance The process by which the body distinguishes between self and nonself components.

immunity The ability of an animal to ward off internal threats, including the invasion of potentially harmful microorganisms such as bacteria, the presence of foreign molecules such as the products of microorganisms, and the presence of abnormal cells such as cancer cells.

immunization *See* vaccination.

immunoglobulin A Y-shaped protein with two heavy chains and two light chains that provide immunity to foreign substances; antibodies are a type of immunoglobulin.

immunological memory The immune system's ability to produce a secondary immune response.

imperfect flower A flower that lacks either stamens or carpels.

implantation The first event of pregnancy, when the blastocyst embeds within the uterine endometrium.

imprinted In genetics, a marked segment of DNA.

imprinting 1. The development of a species-specific pattern of behavior. A form of learning, with a large innate component, within a limited time period. 2. In genetics, the marking of DNA that occurs differently between males and females.

in vitro Literally, "in glass." An approach to studying a process in living cells that involves isolating and purifying cellular components, outside the cell.

in vivo Meaning, "in life." An approach to studying a process in living cells.

inactivation gate A string of amino acids that juts out from a channel protein into the cytosol.

inborn error of metabolism A genetic defect in the ability to metabolize certain compounds.

inbreeding Mating among genetically related relatives.

inbreeding depression The phenomenon whereby inbreeding produces homozygotes that are less fit, thereby decreasing the reproductive success of a population.

inclusive fitness The term used to designate the total number of copies of genes passed on through one's relatives, as well as one's own reproductive output.

incomplete dominance The phenomenon in which a heterozygote that carries two different alleles exhibits a phenotype that is intermediate between the corresponding homozygous individuals.

incomplete flower A flower that lacks one or more of the four flower organ types.

incomplete metamorphosis A gradual change in body form in some insects from different nymphal stages, called instars, into a similar looking adult.

incurrent siphon A tunicate structure used to draw water through the mouth.

indeterminate cleavage A characteristic of deuterostome development in which each cell produced by early cleavage retains the ability to develop into a complete embryo.

indeterminate growth Growth in which plant shoot apical meristems continuously produce new stem tissues and leaves as long as conditions remain favorable.

indicator species A species whose status provides information on the overall health of an ecosystem.

indirect calorimetry A method of determining basal metabolic rate in which the rate at which an animal uses oxygen is measured.

individual selection The proposal that adaptive traits generally are selected for because they benefit the survival and reproduction of the individual rather than the group.

individualistic model A view of the nature of a community that considers it to be an assemblage of species coexisting primarily because of similarities in their physiological requirements and tolerances.

induced fit   Occurs when a substrate(s) binds to an enzyme and the enzyme undergoes a conformational change that causes the substrate(s)s to bind more tightly to the enzyme.

induced mutation   A mutation brought about by environmental agents that enter the cell and then alter the structure of DNA.

inducer   In transcription, a small effector molecule that increases the rate of transcription.

inducible operon   In this type of operon, the presence of a small effector molecule causes transcription to occur.

induction   1. In development, the process by which a cell or group of cells governs the developmental fate of neighboring cells. 2. In molecular genetics, refers to the process by which transcription has been turned on by the presence of a small effector molecule.

industrial nitrogen fixation   The human activity of producing nitrogen fertilizers.

infertility   The inability to produce viable offspring.

inflammation   An innate local response to infection or injury characterized by local redness, swelling, heat, and pain.

inflorescence   A cluster of flowers on a plant.

infundibular stalk   The structure that is physically connected to a multilobed endocrine gland sitting directly below the hypothalamus, called the pituitary gland.

ingroup   A monophyletic group in a cladogram of interest.

inheritance of acquired characteristics   Jean-Baptiste Lamarck's hypothesis that species change over the course of many generations by adapting to new environments. He thought behavioral changes modified traits, and he hypothesized that such modified traits were inherited by offspring.

inhibition   A mechanism for succession in which space is all-important, and order of colonization determines subsequent community structure.

inhibitory postsynaptic potential (IPSP)   The response from an inhibitory neurotransmitter hyperpolarizing the postsynaptic membrane, which reduces the likelihood of an action potential.

initiation stage   In transcription or translation, the first step that initiates the process.

initiator tRNA   A specific tRNA that recognizes the start codon AUG in mRNA and binds to it.

innate   The term used to describe behaviors that seem to be genetically programmed.

inner bark   The thin layer of phloem that conducts most of the sugar transport in a woody stem.

inner ear   One of the three main compartments of the mammalian ear; composed of the bony cochlea and the vestibular system, which plays a role in balance.

inorganic chemistry   The study of the nature of atoms and molecules, with the exception of those that contain rings or chains of carbon.

instar   A stage of growth in an insect with incomplete metamorphosis.

insulin-like growth factor-1 (IGF-1)   A hormone that in mammals stimulates the elongation of bones, especially during puberty, when mammals become reproductively mature.

integral membrane protein   A protein that cannot be released from the membrane unless it is dissolved with an organic solvent or detergent—in other words, you would have to disrupt the integrity of the membrane to remove it.

integrase   An enzyme, sometimes encoded by viruses, that catalyzes the integration of the viral genome into a host-cell chromosome.

integrin   A cell-surface receptor protein found in animal cells that connects cells and the extracellular matrix.

integument   In plants, a modified leaf that encloses the megasporangium to form an ovule.

intercostal muscles   Muscles that surround and connect the ribs in the chest.

interference competition   Competition in which individuals interact directly with one another by physical force or intimidation.

interferon   A protein that generally inhibits viral replication inside host cells.

intermediate-disturbance hypothesis   The proposal that moderately disturbed communities are more diverse than undisturbed or highly disturbed communities.

intermediate filament   A type of protein filament within the cytoskeleton that helps maintain cell shape and rigidity.

internal fertilization   Fertilization that occurs in terrestrial animals, in which sperm are deposited within the reproductive tract of the female during the act called copulation.

interneuron   A type of neuron that forms interconnections between other neurons.

internode   The region of stem on a plant between adjacent nodes.

interphase   The G_1, S, and G_2 phases of the cell cycle. It is phase of the cell cycle during which the chromosomes are decondensed and found in the nucleus.

intersexual selection   Sexual selection between members of the opposite sex.

interspecies hybrid   The offspring resulting from two species mating.

interspecific competition   The term used to describe competition between individuals of different species.

interstitial   Refers to the fluid between cells.

intertidal zone   The area where the land meets the sea, which is alternately submerged and exposed by the daily cycle of tides.

intracellular fluid   The fluid inside cells.

intranuclear spindle   A spindle that forms within an intact nuclear envelope during nuclear division.

intrasexual selection   Sexual selection between members of the same sex.

intraspecific competition   The term used to describe competition between individuals of the same species.

intrinsic rate of increase   The situation in which conditions are optimal for a population, and the per capita growth rate is at its maximum rate.

introduced species   A species moved by humans from a native location to another location.

intron   Intervening DNA sequences that are found in between the coding sequences of genes.

invasive   The term used to describe introduced species that spread on their own, often outcompeting native species for space and resources.

invasive cell   A cell that can invade healthy tissues.

inversion   A change in the direction of the genetic material along a single chromosome.

invertebrate   An animal that lacks vertebrae.

iodine-deficient goiter   An overgrown gland that is incapable of making thyroid hormone.

ion   At atom or molecule that gains or loses one or more electrons and acquires a net electric charge.

ion electrochemical gradient   A dual gradient for an ion that is composed of both an electrical gradient and chemical gradient for that ion.

ionic bond   The bond that occurs when a cation binds to an anion.

ionotropic receptor   A ligand-gated ion channel that opens in response to binding of a neurotransmitter molecule.

iris   The circle of pigmented smooth muscle that is responsible for eye color.

iron regulatory element (IRE)   A response element within the ferritin mRNA to which the iron regulatory protein binds.

iron regulatory protein (IRP)   An RNA-binding protein that regulates the translation of the mRNA that encodes ferritin.

islets of Langerhans   Spherical clusters of endocrine cells that are scattered in great numbers throughout the endocrine pancreas.

isomers   Two structures with an identical molecular formula but different structures and characteristics.

isotonic   When the solute concentrations on both sides of the plasma membrane are equal.

isotope   An element that exists in multiple forms that differ in the number of neutrons they contain.

iteroparity   The pattern of repeated reproduction at intervals throughout the life cycle.

J

joint   The juncture where two or more bones of a vertebrate endoskeleton come together.

K

K/T event   An ancient cataclysm that involved at least one large meteorite or comet that crashed into the Earth near the present-day Yucatan Peninsula in Mexico about 65 million years ago.

karyogamy   The process of nuclear fusion.

karyotype   A photographic representation of the chromosomes that reveals how many chromosomes are found within an actively dividing cell.

K_D   The dissociation constant between a ligand and its receptor.

ketones   Small compounds generated from carbohydrates, fatty acids, or amino acids. Ketones are made in the liver and released into the blood to provide an important energy source during prolonged fasting for many tissues, including the brain.

keystone species   A species within a community that has a role out of proportion to its abundance.

kilocalorie (kcal)   One thousand calories; the common unit of measurement when measuring biological activities.

kin selection   Selection for behavior that lowers an individual's own fitness but enhances the reproductive success of a relative.

kinesis   A movement in response to a stimulus, but one that is not directed toward or away from the source of the stimulus.

kinetic energy   Energy associated with movement.

kinetic skull   A defining characteristic of the class Lepidosauria, in which the joints between various parts of the skull are extremely mobile.

kinetochore   A group of proteins necessary for sorting each chromosome that binds to the centromere.

kingdom   A taxonomic group that contains one or more phyla.

knowledge   The awareness and understanding of information.

Koch's postulates   A series of steps used to determine whether a particular organism causes a specific disease.

K-selected species   A type of life history strategy, where species have a low rate of per capita population growth but good competitive ability.

L

labia majora   In the female genitalia, large outer folds that surround the external opening of the reproductive tract.

labia minora   In the female genitalia, smaller, inner folds near the external opening of the reproductive tract.

labor   A three-stage process that includes (1) dilation and thinning of the cervix to allow passage of the fetus out of the uterus; (2) the movement of the fetus through the cervix and the vagina and out into the world; and (3) the contraction of blood vessels within the placenta and umbilical cord, blocking further blood flow, making the newborn independent from the mother; the placenta detaches from the uterine wall and is delivered a few minutes after the birth of the baby.

lac operon   An operon in the genome of *E. coli* that contains the genes for the enzymes that allow it to metabolize lactose.

lac repressor   A repressor protein that regulates the lac operon.

lactation   In most mammals, a period after birth in which the young are nurtured by milk produced by the mother.

lacteal   A vessel in the center of each intestinal villus; lipids are absorbed by the lacteals, which eventually empty into the circulatory system.

lagging strand   A strand of DNA made as a series of small Okazaki fragments that are eventually connected to each other to form a continuous strand. The synthesis of these DNA fragments occurs in the direction away from the replication fork.

lamellae   Platelike structures in the internal gills of fish that branch from structures called filaments.

landscape ecology   The study of the influence of large-scale spatial patterns of land use or habitat type on ecological processes.

larva   A free-living organism that is morphologically very different from the embryo and adult.

larynx   The area beyond the throat where the vocal cords lie.

latent   The term used to describe when a prophage or provirus remains inactive for a long time.

lateral line system   A sensory system in fish and some toads that allows them to detect changes in their environment; hair cells detect changes in water currents brought about by waves, nearby moving objects, and low-frequency sounds traveling through the water.

law of independent assortment   The alleles of different genes assort independently of each other during gamete formation.

law of segregation   The phenomenon that the two copies of a gene segregate from each other during gamete formation and transmission from parent to offspring.

leaching   The dissolution and removal of inorganic ions as water percolates through materials such as soil.

leading strand   A DNA strand made in the same direction that the replication fork is moving. The strand is synthesized as one long continuous molecule.

leaf   A flattened plant organ that emerges from stems and functions in photosynthesis or other ways.

leaf abscission   The process by which a leaf drops after the formation of an abscission zone at the point where a leaf petiole connects with the stem. The abscission zone consists of an inner protective layer of cork cells whose tough walls help to prevent pathogen attack and dehydration, and an outer layer of cells having thin walls that break easily.

leaf primordia   Small bumps that occur at the sides of a shoot apical meristem and develop into young leaves.

leaflet   1. Half of a phospholipid bilayer. 2. A portion of a compound leaf.

learning   The ability of an animal to make modifications to a behavior based on previous experience.

legume   A member of the pea (bean) family. Also their distinctive fruits, dry pods that develop from one carpel and open down both sides when seeds are mature.

lek   A designated communal courting area.

lens   1. A structure of the eye that focuses light. 2. The glass components of a light microscope or electromagnetic parts of an electron microscope that allow the production of magnified images of microscopic structures.

lentic   Referring to a freshwater habitat characterized by standing water.

leptin   A molecule produced by adipose cells in proportion to fat mass; controls appetite and metabolic rate.

leukocyte   A white blood cell; involved in nonspecific immunity.

lichens   Mutualistic partnerships of particular fungi and certain photosynthetic green algae or cyanobacteria, and sometimes both to form a body distinctive from that of either partner alone.

Liebig's law of the minimum   Species biomass or abundance is limited by the scarcest factor.

life cycle   The sequence of events that characterize the steps of development of the individuals of a given species.

leukocyte   A white blood cell.

life table   A table that provides data on the number of individuals alive in particular age classes.

ligand   An ion or molecule that binds to a protein, such as an enzyme or a receptor.

ligand-gated channel   A channel controlled by the noncovalent binding of small molecules—called ligands—such as hormones or neurotransmitters.

ligand-gated ion channels   A type of cell surface receptor that binds a ligand and functions as an ion channel. Ligand binding either opens or closes a channel.

light chain   A part of an immunoglobulin molecule.

light-harvesting complex   A component of photosystem II and photosystem I composed of several dozen pigment molecules that are anchored to proteins. The role of these complexes is to absorb photons of light.

light microscope   A microscope that utilizes light for illumination.

light reactions   One of two stages in the process of photosynthesis. During the light reactions, photosystem II and photosystem I absorb light energy and produce O_2, ATP, and NADPH.

lignin   A tough polymer that adds strength and decay resistance to cell walls of tracheids, vessel elements, and other cells of plants.

lignophytes   Modern and fossil seed plants, and seedless ancestors that produced wood.

limbic system   The system primarily involved in the formation and expression of emotions, and also plays a role in learning, memory, and the perception of smells; includes certain areas of the telencephalon and parts of the diencephalon.

limiting factor The factor that is most scarce in relation to need.

lineage The genetic relationship between an individual or group of individuals and its ancestors. A series of ancestors in a population that shows a progression of changes.

linkage group A group of genes that usually stay together during meiosis.

linkage The phenomenon of two genes that are close together on the same chromosome tending to be transmitted as a unit.

lipase The major digestive fat-digesting enzyme from the pancreas.

lipid A molecule composed predominantly of hydrogen and carbon atoms. Lipids are nonpolar and therefore very insoluble in water. They include fats, phospholipids, and steroids.

lipid anchor A way for proteins to associate with the plasma membrane; involves the covalent attachment of a lipid to an amino acid side chain within a protein.

lipid exchange protein A protein that extracts a lipid from one membrane, diffuses through the cell, and inserts the lipid into another membrane.

lipopolysaccharides Lipids having covalently-bound carbohydrates. Major components of the thin, outer envelope that encloses the cell walls of Gram-negative bacteria.

liposome A vesicle surrounded by a lipid bilayer.

liver An organ in vertebrates that performs diverse metabolic functions and is the site of bile production.

liverworts A phylum of bryophytes; formally called Hepatophyta.

lobe fins The Actinistia (coelacanths) and Dipnoi (lungfish) and tetrapods; also called Sarcopterygii.

lobe-finned fish Fish in which the fins are part of the body, and they are supported by skeletal extensions of the pectoral and pelvic areas that are moved by muscles residing in the fins.

locomotion The movement of an animal from place to place.

locus The physical location of a gene on a chromosome.

logistic equation $dN / dt = rN (K - N) / K$, where dN / dt is the rate of population change, r is the per capita rate of population growth, N is the population size, and $(K - N) / K$ represents the proportion of unused resources remaining.

logistic growth The S-shaped pattern in which the growth of a population typically slows down as it approaches carrying capacity.

long-day plant A plant that flowers in spring or early summer, when the night period is shorter (and thus the day length is longer) than a defined period.

long-term potentiation (LTP) The long-lasting strengthening of the connection between neurons.

loop domain A chromosomal segment that is folded into loops by the attachment to proteins; a method of compacting chromosomes.

loop of Henle A sharp, hairpin-like loop in the tubule of the nephron of the kidney consisting of a descending limb coming from the proximal tubule and an ascending limb leading to the distal convoluted tubule.

lophophore A horseshoe-shaped crown of tentacles used for feeding.

Lophotrochozoa A clade that encompasses the annelids, mollusks, and several other phyla; they are distinguished by two morphological features—the lophophore, a crown of tentacles used for feeding, and the trochophore larva, a distinct larval stage.

lotic Referring to a freshwater habitat characterized by running water.

lumen The internal space of an organelle.

lung A structure used to bring oxygen into the circulatory system and remove carbon dioxide.

lungfish The Dipnoi; fish with primitive lungs which live in oxygen-poor freshwater swamps and ponds.

luteinizing hormone (LH) A gonadotropin.

lycophyll A relatively small leaf having a single unbranched vein; the type of leaf produced by lycophytes.

lycophytes Members of a phylum of vascular land plants whose leaves are lycophylls; Lycopodiophyta.

lymphatic system A system of vessels along with a group of organs and tissues where most leukocytes reside. The lymphatic vessels collect excess interstitial fluid and return it to the blood.

lymphocytes A type of leukocyte that is responsible for specific immunity; the two types are B cells and T cells.

Lyon hypothesis *See* X inactivation.

lysogenic cycle The growth cycle of a bacteriophage consisting of integration, prophage replication, and excision.

lysogeny Latency in bacteriophages.

lysosome A small organelle found in animal cells that contains acid hydrolases that degrade macromolecules.

lytic cycle The growth cycle of a bacteriophage in which the production and release of new viruses lyses the host cell.

M

M line A narrow, dark band in the center of the H zone that corresponds to proteins that link together the central regions of adjacent thick filaments.

M phase The sequential events of mitosis and cytokinesis.

macroevolution Evolutionary changes that create new species and groups of species.

macromolecule Many molecules bonded together to form a polymer. Carbohydrates, proteins, and nucleic acids (for example, DNA

and RNA) are important macromolecules found in living organisms.

macronutrient An element required by plants in amounts of at least 1 g/kg of plant dry matter.

macroparasite A parasite that lives in the host but releases infective juvenile stages outside the host's body.

macrophage A type of phagocyte capable of engulfing viruses and bacteria; strategically located where it will encounter invaders.

macular degeneration A condition in which photoreceptor cells in and around the fovea of the retina are lost; associated with loss of sharpness and color vision.

madreporite A sieve-like plate on the surface of an echinoderm where water enters the water vascular system.

magnification The ratio between the size of an image produced by a microscope and its actual size.

major groove A groove that spirals around the DNA double helix. The major groove provides a location where a protein can bind to a particular sequence of bases and affect the expression of a gene.

major histocompatibility complex (MHC) A gene family that encodes the plasma membrane self proteins that must be complexed with the antigen in order for T-cell recognition to occur.

malabsorption Any interference with the secretion of bile or the action of bile salts in the intestine that decreases the absorption of fats, including fat-soluble vitamins.

male assistance hypothesis A hypothesis to explain the existence of monogamy that maintains that males remain with females to help them rear their offspring.

male gametophyte A haploid plant life cycle phase that produces sperm.

malignant tumor A growth of cells that has progressed to the cancerous stage.

Malpighian tubules Delicate projections from the digestive tract of insects and some other taxa that protrude into the hemolymph and function as an excretory organ.

mammal A vertebrate that is a member of the class Mammalia that nourishes its young with milk secreted by mammary glands. Another distinguishing feature is hair.

mammary gland A gland in female members of mammal species that secretes milk.

manganese cluster A site where the oxidation of water occurs in photosystem II.

mantle A fold of skin draped over the visceral mass of a mollusk that secretes a shell in those species that form shells.

mantle cavity The chamber in a mollusk mantle that houses delicate gills.

many eyes hypothesis The idea that increased group size decreases predators' success because of increased predator detection ability.

map distance The distance between genes along chromosomes which is calculated as the number of recombinant offspring divided by the total number of offspring times 100.

mapping   The process of determining the relative locations of genes or other DNA segments along a chromosome.

map unit (mu)   A unit of distance equivalent to a 1% recombination frequency.

mark-recapture technique   The capture and tagging of animals so they can be released and recaptured, allowing an estimate of population size.

marsupial   A member of a group of seven mammalian orders and about 280 species found in the subclass Metatheria.

mass extinction   When many species become extinct at the same time.

mass-specific BMR   The amount of energy expended per gram of body mass.

mass spectrometry   A method to determine the masses of molecules such as short peptide fragments within proteins. Tandem mass spectrometry can be used to determine the amino acid sequences of proteins.

mast cell   A type of cell derived from bone marrow stem cells that plays an important role in nonspecific immunity; these cells are found throughout connective tissues, and secrete many locally acting molecules, notably histamine.

mastax   The circular muscular pharynx in the mouth of rotifers.

masting   The synchronous production of many progeny by all individuals in a population to satiate predators and thereby allow some progeny to survive.

mate-guarding hypothesis   A hypothesis to explain the existence of monogamy that theorizes that males stay with a female to prevent her from being fertilized by other males.

maternal effect gene   A gene that follows a maternal effect inheritance pattern.

maternal effect   An inheritance pattern in which the genotype of the mother determines the phenotype of her offspring.

maternal inheritance   A phenomenon in which offspring inherit particular genes only from the female parent (through the egg).

maturation promoting factor (MPF)   The factor that causes oocytes to progress (or mature) from G_2 to M phase.

mature mRNA   In eukaryotes, transcription produces a longer RNA, pre-mRNA, which undergoes certain processing events before it exits the nucleus; mature mRNA is the final functional product.

mean fitness of the population   The average reproductive success of members of a population.

mechanosensitive channel   A channel that is sensitive to changes in membrane tension.

mechanoreceptor   A sensory receptor that transduces mechanical energy such as pressure, touch, stretch, movement, and sound.

mediator   A large protein complex that plays a role in initiating transcription at the core promoter of structural genes in eukaryotes.

medulla oblongata   The part of the hindbrain that coordinates many basic reflexes and bodily functions, such as breathing, that maintain the normal homeostatic processes of the animal.

megadiversity country   Those countries with the greatest numbers of species; used in targeting areas for conservation.

megaspore   In seed plants and some seedless plants, a large spore that produces a female gametophyte within the spore wall.

meiosis   The process by which haploid cells are produced from a cell that was originally diploid.

meiosis I   The first division of meiosis when the homologues are separated into different cells.

meiosis II   The second division of meiosis in which sister chromatids are separated to different cells.

meiotic nondisjunction   An occurrence during meiosis I or meiosis II that produces haploid cells that have too many or too few chromosomes.

Meissner's corpuscles   Structures that sense touch and light pressure and that lie just beneath the skin surface.

membrane attack complex (MAC)   A multiunit protein formed by the activation of complement proteins; the complex creates water channels in the microbial plasma membrane and causes the microbe to swell and burst.

membrane potential   The difference between the electric charges inside and outside the cell; also called a potential difference (or voltage).

membrane transport   The movement of ions or molecules across a cell membrane.

memory   The retention of information over time.

memory cells   A component of the immune response; these cells remain poised to recognize the antigen if it returns in the future.

Mendelian inheritance   The inheritance patterns of genes that segregate and assort independently.

meninges   A protective structure in the central nervous system consisting of three layers of sheathlike membranes.

meningitis   A potentially life-threatening infectious disease in which the meninges become inflamed.

menopause   The event during which a woman permanently stops having ovarian cycles.

menstrual cycle   Also called the uterine cycle; the cyclical changes in the lining of the uterus that occur in parallel with the ovarian cycle.

menstruation   A period of bleeding at the beginning of the menstrual cycle.

meristem   In plants, an organized tissue that includes actively dividing cells and a reservoir of stem cells.

meroblastic cleavage   An incomplete type of cell cleavage, in which only the region of the egg containing cytoplasm at the animal pole undergoes cell division.

merozygote   A strain of bacteria containing an F′ factor.

mesoderm   A layer of cells formed during gastrulation that develops between the ectoderm and endoderm; gives rise to

skeleton, muscles and much of the circulatory system.

mesoglea   A gelatinous substance that connects the two germ layers in the Radiata.

mesohyl   A gelatinous, protein-rich matrix in between the choanocytes and the epithelial cells of a sponge.

mesophyll   The internal tissue of a plant leaf whose cells carry out photosynthesis.

messenger   A molecule that transmits messages from many types of activated sensors to effector molecules.

messenger RNA (mRNA)   RNA that contains the information to specify a polypeptide with a particular amino acid sequence; its job is to carry information from the DNA to the ribosome.

metabolic cycle   A biochemical cycle in which particular molecules enter while others leave; the process is cyclical because it involves a series of organic molecules that are regenerated with each turn of the cycle.

metabolic enzyme   A protein that accelerates chemical reactions within the cell.

metabolic pathway   In living cells, a series of chemical reactions; each step is catalyzed by a specific enzyme.

metabolic rate   The total energy expenditure of an organism per unit of time.

metabolism   The sum total of all chemical reactions that occur within an organism. Also, a specific set of chemical reactions occurring at the cellular level.

metabotropic receptor   A G-protein-coupled receptor that is coupled to an intracellular signaling pathway that initiates changes in a postsynaptic cell.

metacentric   A chromosome in which the centromere is near the middle.

metagenomics   A field of study that seeks to identify and analyze the collective microbial genomes contained in a community of organisms, including, for microbial genomes, those that are not easily cultured in the laboratory.

metamerism   The division of the body into identical subunits called segments.

metamorphosis   The process in which a pupal or juvenile organism changes into a mature adult with very different characteristics.

metanephridia   Excretory organs found in a variety of invertebrates; a type of tubular nephridium.

metanephridial system   The filtration system used by annelids to cleanse the blood.

metaphase   The phase of mitosis during which the chromosomes are aligned along the metaphase plate.

metaphase plate   A plane halfway between the poles on which the sister chromatids align during metaphase.

metastatic cell   A cancer cell that can migrate to other parts of the body.

Metazoa   The collective term for animals.

methanogens   Several groups of anaerobic archaea that convert CO_2, methyl groups, or acetate to methane and release it from their cells.

methanotroph An aerobic bacterium that consumes methane.

methyl-CpG-binding protein A protein that binds methylated sequences.

methyl-directed mismatch repair A DNA repair system that involves the participation of several proteins that detect the mismatch and specifically remove a segment from the newly made strand.

micelle The sphere formed by long amphipathic molecules when mixed with water. In animals, micelles aid in the absorption of poorly soluble products during digestion; they consist of bile salts, phospholipids, fatty acids, and monoglycerides clustered together.

microclimate Local variations of the climate within a given area.

microevolution The term used to describe changes in a population's gene pool from generation to generation.

microfilament *See* actin filament.

micronutrient An element required by plants in amounts at or less than 0.1 g/kg of plant dry mass.

microparasite A parasite that multiplies within its host, usually within the cells.

micropyle A small opening in the integument of a seed plant ovule through which pollen tubes grow.

microRNAs (miRNAs) Small RNA molecules, typically 22 nucleotides in length, that silence the expression of specific mRNAs, either by inhibiting translation or by promoting the degradation of mRNAs.

microscope A magnification tool that enables researchers to study the structure and function of cells.

microsphere A small water-filled vesicle surrounded by a macromolecular boundary.

microspore In seed plants and some seedless plants, a relatively small spore that produces a male gametophyte within the spore wall.

microtubule A type of hollow protein filament composed of tubulin proteins that is part of the cytoskeleton and is important for cell organization, shape, and movement.

microtubule-organizing center *See* centrosome.

microvilli Small projections in the surface membranes of epithelial cells in the small intestine.

midbrain One of three major divisions of the vertebrate brain; the other two divisions are the hindbrain and forebrain.

middle ear One of the three main compartments of the mammalian ear; contains three small bones called ossicles that connect the eardrum with the oval window.

middle lamella A layer composed primarily of carbohydrate that cements adjacent plant cell walls together.

migration Long-range seasonal movement among animals in order to feed or breed.

mimicry The resemblance of an organism (the mimic) to another organism (the model).

mineral An inorganic ion required by a living organism. Minerals are used to build skeletons, maintain the balance of salts and water in the body, provide a source of electric current across plasma membranes, and provide a mechanism for exocytosis and muscle contraction, among other functions.

mineralization The general process by which phosphorus, nitrogen, CO_2, and other minerals are released from organic compounds.

mineralocorticoid A steroid hormone such as aldosterone that regulates the balance of certain minerals in the body.

minor groove A smaller groove that spirals around the DNA double helix.

missense mutation A base substitution that changes a single amino acid in a polypeptide sequence.

mitochondrial genome The chromosome found in mitochondria.

mitochondrial matrix A compartment inside the inner membrane of a mitochondrion.

mitochondrion Literally, "thread granule." An organelle found in eukaryotic cells that supplies most of the cell's ATP.

mitosis In eukaryotes, the process in which nuclear division results in two nuclei; each daughter cell receives the same complement of chromosomes.

mitotic cell division A process whereby a eukaryotic cell divides to produce two new cells that are genetically identical to the original cell.

mitotic spindle *See* mitotic spindle apparatus.

mitotic spindle apparatus The structure responsible for organizing and sorting the chromosomes during mitosis.

mixotroph An organism that is able to use autotrophy as well as phagotrophy or osmotrophy to obtain organic nutrients.

moderately repetitive sequence A DNA sequence found a few hundred to several thousand times in a genome.

modern synthesis of evolution Within a given population of interbreeding organisms, natural variation exists that is caused by random changes in the genetic material. Such genetic changes may affect the phenotype of an individual in a positive, negative, or neutral way. If a genetic change promotes an individual's reproductive success, natural selection may increase the prevalence of that trait in future generations.

molar An adjective to describe the number of moles of a solute dissolved in 1 L of water.

molarity The number of moles of a solute dissolved in 1 L of water.

mole The amount of any substance that contains the same number of particles as there are atoms in exactly 12 g of carbon: 12 grams of carbon equals 1 mole, while 1 g of hydrogen equals 1 mole.

molecular biology A field of study spawned largely by genetic technology that looks at the structure and function of the molecules of life.

molecular clock A clock on which to measure evolutionary time.

molecular evolution The molecular changes in genetic material that underlie the phenotypic changes associated with evolution.

molecular formula A representation of a molecule that consists of the chemical symbols for all of the atoms present and subscripts that tell you how many of those atoms are present.

molecular homologies Similarities that indicate that living species evolved from a common ancestor or interrelated group of common ancestors.

molecular machine A machine that is measured in nanometers, which has moving parts and does useful work.

molecular mass The sum of the atomic masses of all the atoms in a molecule.

molecular pharming An avenue of research that involves the production of medically important proteins in agricultural crops or animals.

molecular recognition The process whereby surfaces on various protein subunits recognize each other in a very specific way, causing them to bind to each other and promote the assembly process.

molecular systematics A field of study that involves the analysis of genetic data, such as DNA sequences, to identify and study genetic homology and construct phylogenetic trees.

molecule Formed from two or more atoms that are connected by chemical bonds.

monocots One of the two largest lineages of flowering plants; the embryo produces a single seed leaf.

monocular vision A type of vision in animals with eyes on the sides of the head; the animal sees a wide area at one time, though depth perception is reduced.

monocyte A type of phagocyte that circulates in the blood for only a few days, after which it takes up permanent residence as a macrophage in different organs.

monoecious The term to describe plants that produce carpellate and staminate flowers on the same plant.

monogamy A mating system in which one male mates with one female, and most individuals have mates.

monohybrid The F_1 offspring, also called single-trait hybrids, of true-breeding parents that differ with regard to a single trait.

monomorphic gene A gene that exists predominantly as a single allele in a population.

monophagous The term used to define parasites that feed on one species or two or three closely related hosts.

monophyletic group A group of species, a taxon, consisting of the most recent common ancestor and all of its descendants.

monosaccharide A simple sugar.

monosomic An aneuploid organism that has one too few chromosomes.

monotreme One of three species in the mammalian order Monotremata, which are found in Australia and New Guinea: the duck-billed platypus and two species of echidna.

morphogen A molecule that imparts positional information and promotes developmental changes at the cellular level.

morphogenetic field A group of embryonic cells that ultimately produce a specific body structure.

morula The mammal embryo in species having undergone holoblastic cleavage.

mosaic An individual in which some cells throughout the body show genetic differences. For example, in female mammals, about half of the somatic cells will express one X-linked allele, while the rest of the somatic cells will express the other allele.

mosses A phylum of bryophytes; Bryophyta.

motif The structure that occurs when a domain or portion of a domain has a very similar structure in many different proteins.

motor neuron A neuron that sends signals away from the central nervous system and elicits some type of response.

motor protein A category of cellular proteins that uses ATP as a source of energy to promote movement; consists of three domains called the head, hinge, and tail.

movement corridor Thin strips of habitat that may permit the movement of individuals between larger habitat patches.

mucigel A gooey plant substance that lubricates roots, aiding in their passage through the soil; helps in water and mineral absorption, prevents root drying, and provides an environment hospitable to beneficial microbes.

Müllerian mimicry A type of mimicry in which many noxious species converge to look the same, thus reinforcing the basic distasteful design.

multicellular Consisting of more than one cell, with cells attached to each other; cells able to communicate with each other by chemical signaling, and some cells able to specialize.

multimeric protein A protein with more than one polypeptide chain; also said to have a quarternary structure.

multiple alleles Refers the occurrence of a gene that exists as three or more alleles in a population.

multiple sclerosis (MS) A disease in which the patient's own body, for reasons that are unknown, attacks and destroys myelin as if it were a foreign substance. Eventually, these repeated attacks leave scarred (sclerotic) areas of tissue in the nervous system and impair the function of myelinated neurons that control movement, speech, memory, and emotion.

multipotent A term used to describe a stem cell that can differentiate into several cell types, but far fewer than pluripotent cells.

muscle A grouping of muscle tissue bound together by a succession of connective tissue layers.

muscle tissue Clusters of cells that are specialized to contract, generating the mechanical forces that produce body movement, exert pressure on a fluid-filled cavity, or decrease the diameter of a tube.

mutagen An agent known to cause mutation.

mutant allele An allele that has been altered by mutation.

mutation A heritable change in the genetic material.

mutualism An interaction in which both species benefit.

mycelium A fungal body composed of microscopic branched filaments known as hyphae.

mycorrhizae Associations between the hyphae of certain fungi and the roots of most plants.

myelin sheath In the nervous system, an insulating layer made up of specialized glial cells wrapped around the axons.

myocardial infarction (MI) The death of cardiac muscle cells, which can occur if a region of the heart is deprived of blood for an extended time.

myofibrils Cylindrical bundles within muscle fibers, each of which contains thick and thin filaments.

myogenic bHLH genes A small group of genes that initiates muscle development in animals.

myogenic heart A heart in which the signaling mechanism that initiates contraction resides within the cardiac muscle itself.

myoglobin An oxygen-binding protein that increases the availability of oxygen in the muscle fiber by providing an intracellular reservoir of oxygen.

myosin A motor protein in muscle.

N

nacre The smooth, iridescent lining of the shells of oysters, mussels, abalone, and other mollusks.

NAD⁺ Nicotinamide adenine dinucleotide; a dinucleotide that functions as an energy intermediate molecule. It combines with two electrons and H⁺ to form NADH.

NADP⁺ Nicotinamide adenine dinucleotide phosphate; a dinucleotide that functions as an energy intermediate molecule in chloroplasts. It combines with two electrons and H⁺ to form NADPH.

natural killer (NK) cells A type of leukocyte that is part of the body's nonspecific defenses because it recognizes general features on the surface of cancer cells or any virus-infected cells.

natural selection The process that culls out those individuals that are less likely to survive and reproduce in a particular environment, while allowing other individuals with traits that confer greater reproductive success to increase in numbers.

nauplius The first larval stage in a crustacean.

navigation A mechanism of migration that involves the ability not only to follow a compass bearing but also to set or adjust it.

nectar A sugar-rich substance produced by many flowers that serves as a food reward for pollinators; in some plants, nectar is produced by other plant parts as a reward for insects that protect the plant.

negative control Transcriptional regulation by repressor proteins.

negative feedback loop A system in which a change in the variable being regulated brings about responses that move the variable in the opposite direction.

negative frequency-dependent selection A pattern of natural selection in which the fitness of a genotype decreases when its frequency becomes higher; the result is a balanced polymorphism.

negative pressure filling The mechanism by which reptiles, birds, and mammals ventilate their lungs.

nekton Free-swimming animals in the open ocean that can swim against the currents to locate food.

nematocyst In a cnidarian, a powerful capsule with an inverted coiled and barbed thread that functions to immobilize small prey so they can be passed to the mouth and ingested.

neocortex The layer of the brain that evolved most recently in mammals.

nephron One of several million single-cell-thick tubules that are the functional units of the kidney.

Nernst equation The formula that gives the equilibrium potential for an ion at any given concentration gradient.

nerve A structure found in the peripheral nervous system that is composed of multiple neurons and makes contact with structures outside the central nervous system and transmits signals that enter or leave the CNS.

nerve cord In more complex invertebrates, a structure that extends from the anterior end of the animal to the tail.

nerve impulse A way that neurons communicate, involving changes in the amount of electric charge across a cell's plasma membrane.

nerve net Interconnected neurons with no central control organ.

nervous system Groups of cells that sense internal and environmental changes and transmit signals that enable an animal to respond in an appropriate way.

nervous tissue Clusters of cells that initiate and conduct electrical signals from one part of an animal's body to another part.

net primary production (NPP) Gross primary production minus the energy lost in plant cellular respiration.

net reproductive rate The population growth rate per generation.

neural crest A cell lineage that gives rise to all neurons and supporting cells of the peripheral nervous system in vertebrates; in addition, it gives rise to melanocytes and to cells that form facial cartilage and parts of the adrenal gland.

neural tube In chordates, a structure formed from ectoderm located dorsal to the notochord; all neurons and their supporting cells in the central nervous system originate from neural precursor cells derived from the neural tube.

neurogenesis   The production of new neurons by cell division.

neurogenic heart   A heart that will not beat unless it receives regular electrical impulses from the nervous system.

neurohormone   A hypothalamic releasing hormone made in and secreted by neurons whose cell bodies are in the hypothalamus.

neuromodulator   Short chains of amino acids that can alter or modulate the response of the postsynaptic neuron to other neurotransmitters.

neuromuscular junction   The junction between a motor neuron's axon and a muscle fiber.

neuron   Another name for a nerve cell. A highly specialized cell that communicates with another cell of its kind and with other types of cells by electrical or chemical signals.

neuroscience   The scientific study of nervous systems.

neurotransmitter   A small signaling molecule that is synthesized and stored in nerve cells.

neurulation   The embryological process responsible for initiating central nervous system formation.

neutral mutation   A mutation that does not affect the function of the encoded protein.

neutral theory of evolution   States that most genetic variation is due to the accumulation of neutral mutations that have attained high frequencies in a population via genetic drift.

neutral variation   Variation that does not favor any particular genotype.

neutralism   The phenomenon in which two species occur together but in fact do not interact in any measurable way.

neutron   A neutral particle found in the center of the atom.

neutrophil   A type of phagocyte and the most abundant type of leukocyte; neutrophils engulf bacteria by endocytosis.

niche   The physical distribution and ecological role of an organism.

nitrification   The conversion by soil bacteria of NH_3 or NH_4^+ to nitrate (NO_3^-), a form of nitrogen commonly used by plants.

nitrogen fixation   A specialized metabolic process in which certain prokaryotes use the enzyme nitrogenase to convert inert atmospheric nitrogen gas into ammonia; also, the industrial process by which humans produce ammonia fertilizer from nitrogen gas.

nitrogenase   An enzyme used in the biological process of fixing nitrogen.

nitrogen-limitation hypothesis   The proposal that organisms select food based on its nitrogen content.

nitrogenous wastes   Molecules that include nitrogen from amino groups; these wastes are toxic at high concentrations and must be eliminated from the body.

nociceptor   A sensory receptor that responds to extreme heat, cold, and pressure, as well as to certain molecules such as acids; also known as a pain receptor.

nocturnal enuresis   Bed-wetting.

Nod factor   Nodulation factor; a substance produced by nitrogen-fixing bacteria in response to flavonoids secreted from the roots of potential host plants; the Nod factors bind to receptors in plant root membranes, starting a process that allows the bacteria to invade roots.

node   The region of a plant stem from which one or more leaves, branches, or buds emerge.

nodes of Ranvier   Exposed areas in the axons of myelinated neurons that contain many voltage-gated Na^+ channels.

nodule   A small swelling on a plant root that contains nitrogen-fixing bacteria.

nodulin   One of several plant proteins that foster root nodule development.

noncoding strand   *See* template.

noncompetitive inhibitor   A molecule that binds to an enzyme at a location that is outside the active site and inhibits the enzyme's function.

noncyclic electron flow   The combined action of photosystem II and photosystem I in which electrons flow in a linear manner to produce NADPH.

non-Darwinian evolution   Also "survival of the luckiest" to contrast it with Darwin's "survival of the fittest" theory; the idea that much of the modern variation in gene sequences is explained by neutral variation rather than adaptive variation.

nondisjunction   An event in which the chromosomes do not sort properly during cell division.

nonparental type   *See* recombinant.

nonpolar molecule   A molecule composed predominantly of nonpolar bonds.

nonrandom mating   The phenomenon that individuals choose their mates based on their genotypes or phenotypes.

nonrecombinant   An offspring whose combination of traits has not changed from the parental generation.

nonsense codon   *See* stop codon.

nonsense mutation   A mutation that changes a normal codon into a stop codon; this causes translation to be terminated earlier than expected, producing a truncated polypeptide.

nonspecific (innate) immunity   The body's defenses that are present at birth and act against foreign materials in much the same way regardless of the specific identity of the invading material; includes the body's external barriers, plus a set of cellular and chemical defenses that oppose substances that breach those barriers.

nonvascular plant   A plant that does not produce lignified vascular tissue, such as a modern bryophyte or extinct pretracheophyte polysporangiophytes.

norepinephrine   A neurotransmitter; also known as noradrenaline.

norm of reaction   A description of how a trait may change depending on the environmental conditions.

notochord   A single flexible rod that lies between the digestive tract and the nerve cord in a chordate.

N-terminus   The location of the first amino acid in a polypeptide.

nuclear envelope   A double-membrane structure that encloses the cell's nucleus.

nuclear genome   The chromosomes found in the nucleus of the cell.

nuclear lamina   A collection of filamentous proteins that line the inner nuclear membrane.

nuclear matrix   A filamentous network of proteins that is found inside the nucleus and lines the inner nuclear membrane. The nuclear matrix serves to organize the chromosomes.

nuclear pore   A passageway for the movement of molecules and macromolecules into and out of the nucleus; formed where the inner and outer nuclear membranes make contact with each other.

nucleic acid   An organic molecule composed of nucleotides. The two types of nucleic acids are deoxyribonucleic acid (DNA) and ribonucleic acid (RNA).

nucleoid   A region of a bacterial cell where the genetic material (DNA) is located.

nucleolus   A prominent region in the nucleus of nondividing cells where ribosome assembly occurs.

nucleosome   A structural unit of eukaryotic chromosomes composed of an octamer of histones (eight histone proteins) wrapped with DNA.

nucleotide   An organic molecule having three components: a phosphate group, a five-carbon sugar (either ribose or deoxyribose), and a single or double ring of carbon and nitrogen atoms known as a base.

nucleotide excision repair (NER)   A common type of DNA repair system that removes (excises) a region of the DNA where the damage occurs. This system can fix many different types of DNA damage, including UV-induced damage, chemically modified bases, missing bases, and various types of cross-links.

nucleus (plural, **nuclei**)   1. In cell biology, an organelle found in eukaryotic cells that contains most of the cell's genetic material. The primary function involves the protection, organization, and expression of the genetic material. 2. In chemistry, the region of an atom that contains protons and neutrons. 3. In neurobiology, a group of neuronal cell bodies in the brain that are devoted to a particular function.

nutrient   Any substance taken up by a living organism that is needed for survival, growth, development, repair, or reproduction.

O

obese   According to current National Institutes of Health guidelines, a person having a BMI of 30 or more.

obligatory mutualism   An interaction in which neither species can live without the other.

ocelli   Photosensitive organs.

octet rule   The phenomenon that atoms are most stable when their outer shell is full. For many, but not all, types of atoms, their outer shell is full when they contain eight electrons, an octet.

Okazaki fragments   Short segments of DNA synthesized in the lagging strand during DNA replication.

olfaction   The sense of smell.

oligotrophic   The term used to describe a young lake that starts off clear and with little plant life.

ommatidium   An independent visual unit in the eye of insects that functions as a separate photoreceptor capable of forming an independent image.

omnivore   An animal that has the ability to eat and survive on both plant and animal products.

oncogene   A type of mutant gene derived from a protooncogene; an oncogene is overactive, thus contributing to uncontrolled cell growth and promoting cancer.

one gene–one polypeptide theory   The concept that one structural gene codes for one polypeptide.

oogenesis   Gametogenesis, which results in the production of egg cells.

oogonia   In animals, diploid germ cells that give rise to the female gametes, the eggs.

open circulatory system   A circulatory system in which hemolymph, which is not different than the interstitial fluid, flows throughout the body and is not confined to special vessels.

open complex   Also called the transcription bubble; a small bubble-like structure between two DNA strands that occurs during transcription.

open conformation   Loosely packed chromatin that can be transcribed into RNA.

operant conditioning   A form of behavior modification; a type of associative learning in which an animal's behavior is reinforced by a consequence, either a reward or a punishment.

operator   A DNA sequence in bacteria that is recognized by activator or repressor proteins that regulate the level of gene transcription.

operculum   A protective flap that covers the gills of a bony fish.

operon   An arrangement of two or more genes in bacteria that are under the transcriptional control of a single promoter.

opsin   A protein that is a component of visual pigments.

opportunistic   A term used to describe animals that have a strong preference for one type of food but can adjust their diet if the need arises.

opposable thumb   A thumb that can be placed opposite the fingers of the same hand, which gives animals a precision grip that enables the manipulation of small objects.

optic nerve   A structure of the eye that carries the electrical signals to the brain.

optimal foraging   The concept that in a given circumstance, an animal seeks to obtain the most energy possible with the least expenditure of energy.

optimality theory   The theory that predicts that an animal should behave in a way that maximizes the benefits of a behavior minus its costs.

oral   Refers to the region of an animal where the mouth is located; refers to the top side of a radial animal.

orbital   The region surrounding the nucleus of an atom where the probablility is high of finding a particular electron.

order   In taxomony, a subdivision of a class.

organ   Two or more types of tissue combined to perform a common function. For example, the heart is composed of several types of tissues, including muscle, nervous, and connective tissue.

organ system   Different organs that work together to perform an overall function in an organism.

organelle   A subcellular structure or membrane-bounded compartment with its own unique structure and function.

organelle genome   In eukaryotes, the genetic material found in mitochondria and plastids.

organic chemistry   The study of carbon-containing molecules.

organic farming   The production of crops without the use of commercial inorganic fertilizers, growth substances, and pesticides.

organic molecule   A carbon-containing molecule, so named because they are found in living organisms.

organism   A living thing that maintains an internal order that is separated from the environment.

organismal ecology   The investigation of how adaptations and choices by individuals affect their reproduction and survival.

organismic model   A view of the nature of a community that considers it to be equivalent to a superorganism; individuals, populations, and communities have a relationship to each other that resembles the associations found between cells, tissues, and organs.

organizing center   A group of cells that ensures the proper organization of the meristem and preserves the correct number of actively dividing stem cells.

organogenesis   The developmental stage during which cells and tissues form organs.

orientation   A mechanism of migration in which animals have the ability to follow a compass bearing and travel in a straight line.

origin of replication   A site within a chromosome that serves as a starting point for DNA replication.

ortholog   A homologous gene in different species.

osculum   A large opening at the top of a sponge.

osmoconformer   An animal whose osmolarity conforms to that of its environment.

osmolarity   The solute concentration of a solution of water, expressed as milliOsmoles/liter (mOsm/L).

osmoregulator   An animal that maintains constant stable internal salt concentrations and osmolarities, even when living in water with very different osmolarities than its body fluids.

osmosis   The movement of water across membranes to balance solute concentrations. Water diffuses from a solution that is hypotonic

(lower solute concentration) into a solution that is hypertonic (higher solute concentration).

osmotic adjustment   The process by which a plant increases the solute concentration of its cytosol.

osmotic pressure   The hydrostatic pressure required to stop the net flow of water across a membrane due to osmosis.

osmotroph   An organism that relies on osmotrophy (uptake of small organic molecules via osmosis) as a form of nutrition.

ostia   Small openings in the heart of an arthropod, through which hemolymph re-enters the heart.

ostracoderms   An umbrella term for several classes of primitive, heavily armored fish, now extinct, that lacked a jaw.

otoliths   Granules of calcium carbonate found in the gelatinous substance that embeds hair cells in the ear.

outer bark   Protective layers of mostly dead cork cells that cover the outside of woody stems and roots.

outer ear   One of the three main compartments of the mammalian ear; consists of the external ear, or pinna, and the auditory canal.

outer segment   The highly convoluted plasma membranes found in the rods and cones of the eye.

outgroup   A species or group of species that is most closely related to an ingroup.

ovaries   In animals, the female gonads where eggs are formed.

overweight   According to current National Institutes of Health guidelines, a person having a BMI of 25 or more.

oviduct   A thin tube with undulating fimbriae (fingers) that extend out to the ovary; also called the fallopian tube.

oviparity   Development of the embryo within an egg outside the mother.

ovoviparous   The term used to describe an organism that retains eggs inside the body, where the young hatch.

ovoviviparity   Development of the embryo involving aspects of both viviparity and oviparity; eggs covered with a thin shell are produced and hatch inside the mother's body, but the offspring receive no nourishment from the mother.

ovulation   The release of the ovum from the ovary.

ovule   In a plant, a megaspore-producing megasporangium and enclosing tissues known as integuments.

ovum   *See* egg.

oxidation   A process that involves the removal of electrons; occurs during the breakdown of small organic molecules.

oxidative fiber   A skeletal muscle fiber that contains numerous mitochondria and has a high capacity for oxidative phosphorylation.

oxidative phosphorylation   A process during which NADH and $FADH_2$ are oxidized to make more ATP via the phosphorylation of ADP.

oxytocin A hormone secreted by the posterior pituitary gland that stimulates contractions of the smooth muscles in the uterus of a pregnant mammal, which facilitates the birth process; after birth, it is important in milk secretion.

P

P generation The parental generation in a genetic cross.

P protein Phloem protein; the proteinaceous material produced in sieve tube elements of plant phloem as a response to wounding.

pacemaker See sinoatrial (SA) node.

Pacinian corpuscles Structures located deep beneath the surface of the skin that respond to deep pressure or vibration.

paedomorphosis The retention of juvenile traits in an adult organism.

pair-rule gene A type of segmentation gene; a defect in this gene may cause alternating segments or parts of segments to be deleted.

paleontologist A scientist who studies fossils.

palisade parenchyma Photosynthetic ground tissue in a plant that consists of closely packed, elongate cells adapted to absorb sunlight efficiently.

pancreas An elongated gland located behind the stomach that secretes digestive enzymes and a fluid rich in bicarbonate ions.

pangenesis An idea proposed by the ancient Greek physician Hippocrates that suggested that "seeds" produced by all parts of the body are collected and transmitted to offspring at the time of conception, and that these seeds cause offspring to resemble their parents.

parabronchi A series of parallel air tubes that make up the lungs and are the regions of gas exchange in birds.

paracrine Refers to a type of cellular communication in which molecules are released into the interstitial fluid and act on nearby cells.

paralogs Homologous genes within a single species.

paraphyletic taxon A group that contains a common ancestor and some, but not all, of its descendants.

parapodia Fleshy, footlike structures in the polychaetes that are pushed into the substrate to provide traction during movement.

parasite A predatory organism that feeds off another organism but does not normally kill it outright.

parasitism An association in which one organism feeds off another, but does not normally kill it outright.

parasympathetic division The division of the autonomic nervous system that is involved in maintaining and restoring body functions.

parathyroid hormone (PTH) A hormone that acts on bone to stimulate the activity of cells that dissolve the mineral part of bone.

Parazoa A subgroup of animals that are not generally thought to possess specialized tissue types or organs, although they may have several distinct types of cells; the one phylum in this group is the Porifera (sponges).

parenchyma cell A type of plant cell that is thin-walled and alive at maturity.

parenchyma tissue Plant tissue that is composed of parenchyma cells.

parental strand The original strand in DNA replication.

parental type See nonrecombinant.

parthenogenesis An asexual process in which an offspring develops from an unfertilized egg.

partial pressure The individual pressure of each gas in the air; the sum of these pressures is known as atmospheric pressure.

particulate inheritance The idea that the determinants of hereditary traits are transmitted intact from one generation to the next.

parturition The birth of an organism.

passive diffusion Diffusion that occurs through a membrane without the aid of a transport protein.

passive immunity A type of acquired immunity that confers protection against disease through the direct transfer of antibodies from one individual to another.

passive transport The diffusion of a solute across a membrane in a process that is energetically favorable and does not require an input of energy.

paternal inheritance A pattern in which only the male gamete contributes particular genes to the offspring.

pathogen A microorganism that causes disease symptoms in its host.

pattern formation The process that gives rise to a plant or animal with a particular body structure.

pedal glands Glands in the foot of a rotifer that secrete a sticky substance that aids in attachment to the substrate.

pedicel A narrow, waistlike point of attachment in a spider or insect body.

pedicellariae The spines and jawlike pincers that cover the skeleton of an echinoderm and deter the settling of animals such as barnacles.

pedigree analysis An examination of the inheritance of human traits in families.

pedipalps In spiders, a pair of appendages that have various sensory, predatory, or reproductive functions.

peduncle The tip of a flower stalk.

pelagic zone The open ocean, where water depth averages 4,000 m and nutrient concentrations are typically low.

penis A male external accessory sex organ found in many animals that is involved in copulation.

pentadactyl limb A limb ending in five digits.

PEP carboxylase An enzyme in C_4 plants that adds CO_2 to phosphoenolpyruvate (PEP) to produce the four-carbon compound oxaloacetate.

pepsin An active enzyme in the stomach that begins the digestion of protein.

peptide bond The covalent bond that links together amino acids in a polypeptide.

peptidoglycan A polymer composed of protein and carbohydrate that is an important component of the cell walls of most bacteria.

peptidyl site (P site) One of the three sites for tRNA binding to the ribosome; the others are the aminoacyl site (A site) and the exit site (E site) The polypeptide is usually in the P site.

peptidyl transfer reaction As a peptide bond is formed, the polypeptide is removed from the tRNA in the P site and transferred to the amino acid at the A site.

per capita growth rate The per capita birth rate minus the per capita death rate; the rate that determines how populations grow over any time period.

perception An awareness of the sensations that are experienced.

perennial A plant that lives for more than two years, often producing seeds each year after it reaches reproductive maturity.

perfect flower A flower that has both stamens and carpels.

perianth The term that refers to flower petals and sepals collectively.

pericarp The wall of a plant's fruit.

pericycle A cylinder of plant tissue having cell division (meristematic) capacity that encloses the root vascular tissue.

peripheral membrane protein A protein that is noncovalently bound to regions of integral membrane proteins that project out from the membrane, or they are noncovalently bound to the polar head groups of phospholipids.

peripheral zone The area of a plant that contains dividing cells that will eventually differentiate into plant structures.

peristalsis Rhythmic, spontaneous waves of muscle contraction that begin near the mouth in the esophagus and move toward the stomach.

peritubular capillaries Capillaries near the junction of the cortex and medulla in the nephron of the kidney.

permafrost A layer of permanently frozen soil.

peroxisome A relatively small organelle found in all eukaryotic cells that catalyzes detoxifying reactions.

petal A flower organ that usually serves to attract insects or other animals for pollen transport.

phage See bacteriophage.

phagocyte A cell capable of phagocytosis; phagocytes provide nonspecific defense against pathogens that enter the body.

phagocytic vacuole A vacuole that is formed by the process of phagocytosis.

phagocytosis A form of endocytosis that involves the formation of a membrane vesicle called a phagosome, or phagocytic vacuole, that engulfs a large particle such as a bacterium.

phagotroph An organism that specializes in phagotrophy (particle feeding) by means of phagocytosis as a form of nutrition.

phagotrophy The use of phagocytosis as a feeding mechanism.

pharyngeal slit A filter-feeding device in primitive chordates.

pharynx The area at the back of the throat where inhaled air from the mouth and nose converges.

phenolic Refers to compounds that contain a cyclic ring of carbon with three double bonds, known as a benzene ring, that is covalently linked to a single hydroxyl group.

phenotype The characteristics of an organism that are the result of the expression of its genes.

phenotypic plasticity The phenomenon in which individual members of the same plant species that experience different environmental conditions may display considerable variation in structure or behavior.

pheromone A powerful chemical attractant used to manipulate the behavior of others.

phloem A specialized conducting tissue at the center of a plant's stem.

phloem loading The process of conveying sugars to sieve-tube elements for long-distance transport.

phoresy A form of commensalism in which individuals of one species use individuals of a second species for transportation.

phosphodiester linkage Refers to a double linkage (two phosphoester bonds) that holds together adjacent nucleotides in DNA and RNA strands.

phosphodiesterase An enzyme that breaks down cAMP into AMP.

phospholipid A class of lipids that are similar in structure to triglycerides, but the third hydroxyl group of glycerol is linked to a phosphate group instead of a fatty acid. They are a key component of biological membranes.

phospholipid bilayer The basic framework of the cellular membrane, consisting of two layers of lipids.

photoautotroph An organism that uses the energy from light to make organic molecules from inorganic sources.

photoheterotroph An organism that is able to use light energy to generate ATP, but which must take in organic compounds from the environment.

photon A massless particle traveling in a wavelike pattern and moving at the speed of light.

photoperiodism A plant's ability to measure and respond to night length, and indirectly, day length, as a way of detecting seasonal change.

photoreceptor An electromagnetic receptor that responds to visible light energy.

photorespiration The metabolic process occurring in C_3 plants that results when the enzyme rubisco combines with oxygen instead of carbon dioxide and produces only one molecule of PGA instead of two PGA, thereby reducing photosynthetic efficiency.

photosynthesis The process whereby light energy is captured by plant, algal, or cyanobacterial cells and used to synthesize organic molecules from CO_2 and H_2O (or H_2S).

photosystem I (PSI) One of two distinct complexes of proteins and pigment molecules

in the thylakoid membrane that absorbs light.

photosystem II (PSII) The complex of proteins and pigment molecules in the thylakoid membrane that generates oxygen from water.

phototropin The main blue-light sensor involved in phototropism.

phototropism The tendency of a plant to grow toward a light source.

phragmoplast A plant organelle involved in the construction of cell plate that produces an intervening cell wall between two dividing plant cells.

phyla The subdivisions of a kingdom.

phylogenetic species concept The members of a single species are identified by having a unique combination of characteristics.

phylogenetic tree A diagram that describes a phylogeny; such a tree is a hypothesis of the evolutionary relationships among various species, based on the information available to and gathered by systematists.

physical mutagen A physical agent, such as UV light, that causes mutations.

physiological ecology A subdiscipline of organismal ecology that investigates how organisms are physiologically adapted to their environment and how the environment impacts the distribution of species.

physiology The study of the functions of cells and body parts of living organisms.

phytochrome A red- and far-red-light receptor.

phytoplankton Microscopic algae and cyanobacteria that float or actively move through water.

phytoremediation The process of removing harmful metals from soils by growing hyperaccumulator plants on metal-contaminated soils, then harvesting and burning the plants to ashes for disposal and/or metal recovery.

pigment A molecule that can absorb light energy.

pili (singular, **pilus**) Threadlike surface appendages that allow prokaryotes to attach to each other during mating, or to move across surfaces.

piloting A mechanism of migration in which an animal moves from one familiar landmark to the next.

pinocytosis A form of endocytosis that involves the formation of membrane vesicles from the plasma membrane as a way for cells to internalize the extracellular fluid.

pistil A flower structure that may consist of a single carpel or multiple, fused carpels, and is differentiated into stigma, style, and ovary.

pit A small cavity in a plant cell wall where secondary wall materials such as lignin are absent.

pitch The tone of a sound wave.

pituitary dwarfism A condition in which a person's anterior pituitary gland fails to make adequate amounts of GH during childhood, so the concentrations of GH and IGF-1 in the blood will be lower than normal and growth is stunted.

pituitary giant A person who has a tumor of the GH-secreting cells of the anterior pituitary gland and thus produces excess GH during childhood and adulthood; the person can grow very tall.

pituitary gland A multilobed endocrine gland sitting directly below the hypothalamus.

placenta A structure through which humans and other eutherian mammals retain and nourish their young within the uterus via the transfer of nutrients and gases.

placental transfer tissue In plants, a nutritive tissue that aids in the transfer of nutrients from maternal parent to embryo.

plant A member of the kingdom Plantae.

plant tissue culture A laboratory process to produce thousands of identical plants having the same desirable characteristics.

Plantae A eukaryotic kingdom of the domain Eukarya; includes multicellular organisms having cellulose-rich cell walls and plastids, and which are adapted in many ways to terrestrial habitats (or if aquatic, derived from terrestrial ancestors).

plaque 1. A deposit of lipids, fibrous tissue, and smooth muscle cells that may develop in blood vessels. 2. A bacterial biofilm that may form on the surfaces of teeth.

plasma The fluid part of blood.

plasma cell A cell that synthesizes and secretes antibodies.

plasma membrane The biomembrane that separates the internal contents of a cell from its external environment.

plasmid A small circular piece of DNA found naturally in many strains of bacteria and also occasionally in eukaryotic cells.

plasmodesma (plural, **plasmodesmata**) A membrane-lined, ER-containing channel that connects the cytoplasm of adjacent plant cells.

plasmogamy The fusion of the cytoplasm between two gametes.

plasmolysis The shrinkage of algal or plant cytoplasm that occurs when water leaves the cell by osmosis, with the result that the plasma membrane no longer presses on the cell wall.

plastid A general name given to organelles found in plant and algal cells, which are bound by two membranes and contain DNA and large amounts of chlorophyll (chloroplasts), carotenoids (chromoplasts), or starch (amyloplasts).

platelets Cell fragments in the blood of mammals that play a crucial role in the formation of blood clots.

pleiotropy The phenomenon in which a mutation in a single gene can have multiple effects on an individual's phenotype.

pleural sac A double layer of sheathlike moist membranes that encases each lung.

pluripotent Refers to the ability of embryonic stem cells to differentiate into almost every cell type of the body.

point mutation A mutation that affects only a single base pair within the DNA or that involves the addition or deletion of a single base pair to a DNA sequence.

polar cell The highest latitude cell in the three-cell circulation of wind in each hemisphere.

polar covalent bond The bond that forms when two atoms with different electronegativities form a covalent bond and the shared electrons are closer to the atom of higher electronegativity rather than the atom of lower electronegativity. The distribution of electrons around the atoms creates a polarity, or difference in electric charge, across the molecule.

polar molecule A molecule containing significant numbers of polar bonds.

polar transport The process whereby auxin primarily flows downward in shoots and into roots.

polarized 1. In cell biology, refers to cells that have different sides, such as the apical and basal sides of epithelial cells. 2. In neuroscience, refers to the electrical gradient across the membrane. A neuron with a large electrical gradient across its plasma membrane is said to be highly polarized.

pole A structure of the spindle apparatus defined by each centrosome.

pole cells The primordial germ cells that are the first cells to form at the posterior end of the embryo in certain species such as *Drosophila*.

pollen Tiny male gametophytes enclosed by sporopollenin-containing microspore walls.

pollen coat A layer of material that covers the sporopollenin-rich pollen wall.

pollen grain An immature male gametophyte.

pollen tube A mature male gametophyte consisting of a germinated pollen grain with a long, thin pollen tube that carries haploid sperm cells.

pollen wall A tough, sporopollenin wall at the surface of a pollen grain.

pollination The process in which pollen grains are transported to an angiosperm flower or a gymnosperm cone by means of wind or animal pollinators.

pollination syndromes The pattern of coevolved traits between particular types of flowers and specific pollinators.

pollinator An animal that carries pollen between angiosperm flowers (or cones of gymnosperms).

poly A tail A string of adenine nucleotides at the 3′ end of most mature mRNAs in eukaryotes.

polyandry A form of mating in which one female mates with several males.

polycistronic mRNA An mRNA that contains the coding sequences for two or more structural genes.

polycythemia A condition of increased hemoglobin due to increased hematocrit.

polygenic A trait in which several or many genes contribute to the outcome of the trait.

polygyny A form of mating in which one male mates with more than one female in a single breeding season, but females mate only with one male.

polyketides A group of secondary metabolites that are produced by bacteria, fungi, plants, insects, dinoflagellates, mollusks, and sponges.

polymer A large molecule formed by linking together many smaller molecules called monomers.

polymerase chain reaction (PCR) A technique to make many copies of a gene in vitro; primers are used that flank the region of DNA to be amplified.

polymorphic gene A gene that commonly exists as two or more alleles in a population.

polymorphism The phenomenon that many traits or genes may display variation within a population.

polypeptide A linear sequence of amino acids; the term denotes structure.

polyphagous The term used to define parasites that feed on many host species.

polyphyletic taxon A group that consists of members of several evolutionary lines and does not include the most recent common ancestor of the included lineages.

polyploid An organism that has three or more sets of chromosomes.

polysaccharide Many monosaccharides linked together to form long polymers.

pons The part of the hindbrain, along with the cerebellum, responsible for monitoring and coordinating body movements.

population A group of individuals of the same species that can interbreed with one another.

population ecology The study of how populations grow and what promotes and limits growth.

population genetics The study of genes and genotypes in a population.

portal vein A vein that not only collects blood from capillaries—like all veins—but also forms another set of capillaries, as opposed to returning the blood directly to the heart like other veins.

positional information Molecules that are provided to a cell that allow it to determine its position relative to other cells.

positive control Transcriptional regulation by activator proteins.

positive feedback loop The acceleration of a process, leading to what is sometimes called an explosive system.

positive pressure filling The method by which amphibians ventilate their lungs—the animals gulp air and force it under pressure into the lungs, as if inflating a balloon.

postabsorptive state One of two alternating phases in the utilization of nutrients; occurs when the gastrointestinal tract is empty of nutrients and the body's own stores must supply energy. The other phase is the absorptive state.

posterior Refers to the rear (tail-end) of an animal.

postsynaptic cell The cell that receives the electrical or chemical signal sent from a neuron.

post-translational covalent modification A process of changing the structure of a protein, usually by covalently attaching functional groups.

post-translational sorting The uptake of proteins into the nucleus, mitochondria, chloroplasts, and peroxisomes that occurs after the protein is completely made (that is, completely translated).

postzygotic isolating mechanism A mechanism that prevents interbreeding by blocking the development of a viable and fertile individual after fertilization has taken place.

potential energy The energy that a substance possesses due to its structure or location.

power stroke The process in which an energized myosin cross-bridge binds to actin and triggers the movement of the bound cross-bridge.

prebiotic soup The medium formed by the slow accumulation of molecules in the early oceans over a long period of time prior to the existence of life.

predation An interaction where the action of the predator results in the death of the prey.

predator An animal that kills live prey.

pregnancy The time during which a developing embryo and fetus grows within the uterus of the mother. The period of pregnancy is also known as gestation.

preinitiation complex The structure of the completed assembly of GTFs and RNA polymerase II at the TATA box prior to transcription of eukaryotic structural genes.

pre-mRNA The RNA transcript prior to any processing.

presynaptic cell The cell that sends the electrical or chemical signal from a neuron to another cell.

prezygotic isolating mechanism A mechanism that stops interbreeding by preventing the formation of a zygote.

primary active transport A type of transport that involves pumps that directly use energy and generate a solute gradient.

primary cell wall In plants, a cell wall that is synthesized first between the two newly made daughter cells.

primary consumer An organism that obtains its food by eating primary producers; also called an herbivore.

primary electron acceptor The molecule to which a high-energy electron from an excited pigment molecule such as P680* is transferred.

primary endosymbiosis The process by which a eukaryotic host cell acquires prokaryotic endosymbionts or plastids or mitochondria derived from a prokaryotic endosymbiont.

primary immune response The response to an initial exposure to an antigen.

primary meristem A meristematic tissue that increases plant length and produces new organs.

primary metabolism The synthesis and breakdown of molecules and macromolecules that are found in all forms of life and are essential for cell structure and function.

primary oocytes   In animals, the first stage of producing female gametes by meiosis.

primary plastid   A plastid that arose from a prokaryote as the result of endosymbiosis.

primary producer   An autotroph, which typically harvests light energy from the sun; located at the base of the food chain.

primary spermatocytes   In animals, the first stage of producing sperm by meiosis.

primary structure   The linear sequence of amino acids of a polypeptide. One of four levels of protein structure.

primary succession   Succession on newly exposed sites that were not previously occupied by soil and vegetation.

primary tissue   Plant tissue generated as the result of primary growth at apical meristems.

primary vascular tissue   Plant tissue composed of primary xylem and phloem.

primer   A short segment of RNA, typically 10 to 12 nucleotides in length, that is needed to begin DNA replication.

principle of parsimony   The preferred hypothesis is the one that is the simplest.

principle of species individuality   A view of the nature of a community; each species is distributed according to its physiological needs and population dynamics; most communities intergrade continuously and competition does not create distinct vegetational zones.

probability   The chance that an event will have a particular outcome.

proboscis   The coiled tongue of a butterfly or moth, which can be uncoiled, enabling it to drink nectar from flowers.

procambium   A type of primary plant tissue meristem that produces vascular tissue.

producer   An organism that synthesizes the organic compounds used by other organisms for food.

product   During a chemical reaction, the reactants are converted to products.

product rule   The probability that two or more independent events will occur is equal to the product of their individual probabilities.

production efficiency   The percentage of energy assimilated by an organism that becomes incorporated into new biomass.

productivity hypothesis   The proposal that greater production by plants results in greater overall species richness.

progesterone   A hormone secreted by the female ovaries that plays a key role in pregnancy.

progymnosperms   An extinct group of plants having wood but not seeds, that evolved before the gymnosperms.

prokaryote   One of the two categories into which all forms of life can be placed, based on cell structure; the other is eukaryote. Prokaryotic cells lack a nucleus having an envelope with pores; includes bacteria and archaea.

prometaphase   A phase of mitosis during which the mitotic spindle is completely formed.

promoter   The site in the DNA where transcription begins.

proofreading   The ability of DNA polymerase to identify a mismatched nucleotide and remove it from the daughter strand.

prophage   Refers to the DNA of a phage that has become integrated into the bacterial chromosome.

prophase   A phase of mitosis during which the chromosomes condense and the nuclear membrane fragments.

proplastid   Unspecialized structures that form plastids.

prosimian   A member of a class of primates that includes the smaller species such as bush babies, lemurs, pottos, and tarsiers.

prostate gland   A structure that secretes a thin fluid that protects sperm once they are deposited within the female reproductive tract.

prosthetic group   Small molecules that are permanently attached to the surface of an enzyme and aid in catalysis.

protease   An enzyme that cuts proteins into smaller polypeptides.

proteasome   A molecular machine that is the primary pathway for protein degradation in archaea and eukaryotic cells.

protein   A functional unit composed of one or more polypeptides. Each polypeptide is composed of a linear sequence of amino acids.

protein kinase   An enzyme that transfers phosphate groups from ATP to a protein.

protein phosphatase   An enzyme responsible for removing phosphate groups from proteins.

protein-protein interactions   Specific interactions between proteins that can carry out cellular processes that occur as a series of steps or build larger structures that provide organization to the cell.

proteoglycan   A glycosaminoglycan in the extracellular matrix linked to a core protein.

proteolysis   A processing event within the cell in which enzymes called proteases cut proteins into smaller polypeptides.

proteome   All of the types and relative amounts of proteins that are made in a particular cell at a particular time and under specific conditions. The proteome largely determines a cell's structure and function.

proteomics   Techniques used to identify and study groups of proteins.

protist   A eukaryotic organism that is not a member of the animal, plant, or fungal kingdoms; lives in moist habitats, and is typically microscopic in size.

Protista   In traditional classification systems, a eukaryotic kingdom of the domain Eukarya.

protobiont   The term used to describe the first nonliving structures that evolved into living cells.

protoderm   A type of primary plant tissue meristem that generates the outermost dermal tissue.

proton   A positively charged particle found in the nucleus of the atom. The number of

protons in an atom is called the atomic number and defines each type of element.

protonephridia   The simplest filtration mechanism for cleansing the blood; used in flatworms.

proton-motive force   *See* H$^+$ electrochemical gradient.

proto-oncogene   A normal gene that, if mutated, can become an oncogene.

protostome   An animal that exhibits spiral determinate cleavage, and where the blastopore becomes the mouth; includes mollusks, annelid worms, and arthropods.

provirus   Refers to viral DNA that has become incorporated into a eukaryotic chromosome.

protozoa   A term commonly used to describe diverse heterotrophic protists.

proventriculus   The glandular portion of the stomach of a bird.

proximal convoluted tubule   The segment of the tubule of the nephron in the kidney that drains Bowman's capsule.

proximate cause   A specific genetic and physiological mechanism of behavior.

pseudocoelom   A coelom that is not completely lined by tissue derived from mesoderm.

pseudocoelomate   An animal with a pseudocoelom.

pteridophytes   A phylum of of vascular plants having euphylls, but not seeds; Pteridophyta.

pulmocutaneous circulation   The routing of blood from the heart through different vessels to the gas exchange organs (lungs and skin) of frogs and some other amphibians.

pulmonary circulation   The pumping of blood from the right side of the heart to the lungs to release carbon dioxide and pick up oxygen from the atmosphere.

pulse-chase experiment   A procedure in which researchers administer a pulse of radioactively labeled materials to cells so that they make radioactive products. This is followed by the addition of nonlabeled materials called a chase.

pump   A transporter that directly couples its conformational changes to an energy source, such as ATP hydrolysis.

punctuated equilibrium   A concept that suggests that the tempo of evolution is more sporadic than gradual. Species rapidly evolve into new species followed by long periods of equilibrium with little evolutionary change.

Punnett square   A common method for predicting the outcome of simple genetic crosses.

pupa   The organism resulting after the larval stages in insects that gives rise to an adult organism via metamorphosis.

pupil   A small opening in the eye of a vertebrate that transmits different patterns of light emitted from images in the animal's field of view.

purine   The bases adenine (A) and guanine (G), with double (fused) rings of nitrogen and carbon atoms.

pyramid of biomass   A measure of efficiency in which the organisms at each trophic level are weighed.

pyramid of numbers An expression of trophic-level transfer efficiency, in which the number of individuals decreases at each trophic level, with a huge number of individuals at the base and fewer individuals at the top.

pyramid of production A measure of efficiency in which rates of production are shown rather than standing crop; the laws of thermodynamics ensure that the highest amounts of free energy are found at the lowest trophic levels.

pyrimidine The bases cytosine (C), thymine (T), and uracil (U) with a single ring.

Q

quantitative trait *See* continuous trait.

quaternary structure The association of two or more polypeptides to form a protein. One of four levels of protein structure.

quorum sensing A mechanism by which prokaryotic cells are able communicate when they reach a critical population size.

R

radial cleavage A mechanism of development in which the cleavage planes are either parallel or perpendicular to the vertical axis of the egg.

radial loop domain A loop of chromatin, often 25,000 to 200,000 base pairs in size, that is anchored to the nuclear matrix.

radial pattern A characteristic of the body pattern of plants.

radial symmetry 1. In plants, an architectural feature of plants in which plant embryos display a cylindrical shape, which is retained in the stems and roots of seedlings and mature plants. In addition, new leaves or flower parts are produced in circular whorls, or spirals, around shoot tips. 2. In animals, an architectural feature in which the body can be divided into symmetrical halves by many different longitudinal planes along a central axis.

Radiata Radially symmetric animals, which means they can be divided equally by a longitudinal plane passing through the central axis; includes cnidarians and ctenophores.

radiation The emission of electromagnetic waves by the surfaces of objects.

radicle An embryonic root, which extends from the plant hypocotyl.

radioisotope An isotope found in nature that is inherently unstable and does not exist for long periods of time. Such isotopes emit energy called radiation in the form of subatomic particles.

radioisotope dating A common way to estimate the age of a fossil by analyzing the elemental isotopes within the accompanying rock.

radula A unique, protrusible, tonguelike organ in a mollusk that has many teeth and is used to eat plants, scrape food particles off rocks, or bore into shells of other species and tear flesh.

rain shadow An area where precipitation is noticeably less, such as on the side of the mountain sheltered from air movement.

ram ventilation A mechanism used by fish to ventilate their gills; fish swim or face upstream with their mouths open, allowing water to enter into their buccal cavity and from there across their gills.

random A pattern of dispersion within a population, in which individuals do not appear to be specially positioned relative to anyone else.

random genetic drift A change in allele frequencies due to random sampling error.

random sampling error The deviation between the observed and expected outcomes.

rate-limiting step The slowest step in a pathway.

ray-finned fish The Actinopterygii, which includes all bony fish except the coelacanths and lungfish.

reabsorption In the production of urine, the process in which useful solutes in the filtrate are recaptured and transported back into the body fluids of an animal.

reactant A substance that participates in a chemical reaction and becomes changed by that reaction.

reaction mechanism In the case of the Na^+/K^+-ATPase, a molecular roadmap of the steps that direct the pumping of ions across the plasma membrane.

reading frame Refers to the idea that codons are read from the start codon in groups of three bases each.

receptacle The enlarged region at the tip of a flower peduncle to which flower parts are attached.

receptor A cellular protein that recognizes a signaling molecule.

receptor-mediated endocytosis A common form of endocytosis in which a receptor is specific for a given cargo.

receptor potential The membrane potential in a sensory receptor cell.

recessive A term that describes a trait that is masked by the presence of a dominant trait in a heterozygote.

reciprocal translocation The process in which two different types of chromosomes exchange pieces, thereby producing two abnormal chromosomes carrying translocations.

recombinant An offspring that has a different combination of traits from the parental generation.

recombinant DNA Any DNA molecule that has been manipulated so that it contains DNA from two or more sources.

recombinant DNA technology The use of laboratory techniques to isolate and manipulate fragments of DNA.

recombinant vector A vector containing a piece of chromosomal DNA.

recombination frequency The frequency of crossing over between two genes.

red blood cell A cell that serves the critical function of transporting oxygen throughout the body; also known as an erythrocyte.

redox reaction A type of reaction in which the electron that is removed during the oxidation of an atom or molecule must be transferred to another atom or molecule, which becomes reduced; short for a reduction-oxidation reaction.

reduction A process that involves the addition of electrons to an atom or molecule.

reductionism An approach that involves reducing complex systems to simpler components as a way to understand how the system works. In biology, reductionists study the parts of a cell or organism as individual units.

redundancy hypothesis An extension of the rivet hypothesis. In this model, also called the passenger hypothesis, most species are more like passengers on a plane—they take up space but do not add to the airworthiness. The species are said to be redundant because they could simply be eliminated or replaced by others with no loss in function.

reflex arc A simple circuit that allows an organism to respond rapidly to inputs from sensory neurons and consists of only a few neurons.

refractory The term used to describe a cell that is unresponsive to another stimulus.

regeneration A form of asexual reproduction in which a complete organism forms from small fragments of its body.

regulatory gene A gene whose function is to regulate the expression of other genes.

regulatory sequence In the regulation of transcription, a sequence that functions as a site for genetic regulatory proteins. Regulatory sequences control whether a gene is turned on or off.

regulatory transcription factor A protein that binds to DNA in the vicinity of a promoter and affects the rate of transcription of one or more nearby genes.

relative abundance The frequency of occurrence of species in a community.

relative water content (RWC) The property often used to gauge the water content of a plant organ or entire plant; RWC integrates the water potential of all cells within an organ or plant and is thus a measure of relative turgidity.

release factor A protein that recognizes the three stop codons in the termination stage of translation and promotes the termination of translation.

renal corpuscle A filtering component in the nephron of the kidney.

repetitive sequence Short DNA sequences that are present in many copies in a genome.

replica plating A technique in which a replica of bacterial colonies is transferred to a new petri plate.

replication 1. The performing of experiments several or many times. 2. The copying of DNA strands.

replication fork The area where two DNA strands have separated and new strands are being synthesized.

repressible operon In this type of operon, a small effector molecule inhibits transcription.

repressor A transcription factor that binds to DNA and inhibits transcription.

reproduction The process by which organisms produce offspring.

reproductive isolating mechanisms The mechanisms that prevent interbreeding between different species.

reproductive isolation Mechanisms that prevent one species from successfully interbreeding with other species.

resistance (R) The tendency of blood vessels to slow down the flow of blood through their lumens.

resistance gene (R gene) A plant gene that has evolved as part of a defense system in response to pathogen attack.

resolution In microscopy, the ability to observe two adjacent objects as distinct from one another; a measure of the clarity of an image.

resonance energy transfer The process by which the energy (not the electron itself) can be transferred to adjacent pigment molecules.

resource competition Competition in which organisms compete indirectly through via the consumption of a limited resource, with each obtaining as much as it can.

resource partitioning The differentiation of niches, both in space and time, that enables similar species to coexist in a community.

respiratory centers Several regions of the brainstem in vertebrates that initiate expansion of the lungs.

respiratory chain *See* electron transport chain.

respiratory distress syndrome of the newborn The situation in which a human baby is born prematurely, before sufficient surfactant is produced, and many alveoli quickly collapse.

respiratory pigment A large protein that contains one or more metal atoms that bind to oxygen.

respiratory system All components of the body that contribute to the exchange of gas between the external environment and the blood; in mammals, includes the nose, mouth, airways, lungs, and muscles and connective tissues that encase these structures within the thoracic (chest) cavity.

response elements DNA sequences that are recognized by regulatory transcription factors and regulate the expression of genes.

response regulator In bacteria, a protein that interacts with a sensor kinase and regulates the expression of many genes.

resting potential The difference in charges across the plasma membrane in an unstimulated cell.

restoration ecology The full or partial repair or replacement of biological habitats and/or their populations that have been damaged.

restriction enzyme An enzyme that recognizes particular DNA sequences and cleaves the DNA backbone at two sites.

restriction point A point in the cell cycle in which a cell has become committed to divide.

restriction sites The base sequences recognized by restriction enzymes.

retina A sheetlike layer of photoreceptors at the back of the eye.

retinal A derivative of vitamin A that is capable of absorbing light energy.

retroelement A type of transposable element that moves via an RNA intermediate.

retrotransposon *See* retroelement.

retrovirus An RNA virus that utilizes reverse transcription to produce viral DNA that can be integrated into the host cell genome.

reverse transcriptase A viral enzyme that catalyzes the synthesis of viral DNA starting with viral RNA as a template.

rhizobia The collective term for proteobacteria involved in the nitrogen-fixation symbioses with plants that are important in nature and to agriculture.

rhizomorphs Fungal mycelia that have the shape of roots.

rhodopsin The visual pigment in rods.

ribonucleic acid (RNA) One of two classes of nucleic acids; the other is deoxyribonucleic acid (DNA). RNA consists of a single strand of nucleotides.

ribose A five-carbon sugar found in RNA.

ribosomal initiation factor In the initiation stage of translation, a protein that facilitates the interactions between mRNA, the first tRNA, and the ribosomal subunits.

ribosomal RNA (rRNA) An RNA that forms part of ribosomes, which provide the site where translation occurs.

ribosome A structure composed of proteins and rRNA that provides the site where protein synthesis occurs.

ribozyme A biological catalyst that is an RNA molecule.

right-left axis In bilateral animals, one of the three axes along which the adult body pattern is organized; the others are the dorsoventral axis and the anteroposterior axis.

ring canal A structure in the water vascular system of echinoderms.

rivet hypothesis An alternative to the diversity-stability hypothesis. In this model, species are like the rivets on an airplane, with each species playing a small but critical role in keeping the plane (the ecosystem) airborne. The loss of a rivet weakens the plane and causes it to lose a little airworthiness. The loss of a few rivets could probably be tolerated, while the loss of more rivets would prove critical to the airplane's function.

RNA *See* ribonucleic acid.

RNA-induced silencing complex (RISC) The complex that mediates RNA interference via microRNAs.

RNA interference (RNAi) Refers to a type of mRNA silencing; miRNA interferes with the proper expression of an mRNA.

RNA polymerase The enzyme that synthesizes strands of RNA during gene transcription.

RNA processing A step in gene expression between transcription and translation; the RNA transcript, termed pre-mRNA, is modified in ways that make it a functionally active mRNA.

RNA world A hypothetical period on primitive Earth when both the information needed for life and the enzymatic activity of living cells were contained solely in RNA molecules.

RNase An enzyme that digests RNA.

rods Photoreceptors found in the vertebrate eye; they are very sensitive to low-intensity light but do not readily discriminate different colors. Rods are utilized mostly at night, and they send signals to the brain that generate a black-and-white visual image.

"Roid" rage Extreme aggressive behavior brought about by androgen administration.

root A plant organ that provides anchorage in the soil and also fosters efficient uptake of water and minerals.

root apical meristem (RAM) The region where new root tissues of plants are produced.

root cap A protective covering on the root tip that is produced by the root apical meristem of a plant.

root cortex A region of ground parenchyma located between the epidermis and vascular tissue of mature plant roots.

root hair A long, thin root epidermal cell that functions to absorb water and minerals, usually from soil.

root meristem The collection of cells at the root tip that generate all of the tissues of a plant root.

root pressure A process whereby plants are able to refill embolized vessels.

root system The collection of roots and root branches produced by root apical meristems.

rough endoplasmic reticulum (rough ER) The part of the ER that is studded with ribosomes; this region plays a key role in the initial synthesis and sorting of proteins that are destined for the ER, Golgi apparatus, lysosomes, vacuoles, plasma membrane, or outside of the cell.

r-selected species A type of life history strategy, where species have a high rate of per capita population growth but poor competitive ability.

rubisco The enzyme that that catalyzes the first step in the Calvin cycle in which CO_2 is incorporated into an organic molecule.

S

S The synthesis phase of the cell cycle.

saltatory conduction A type of conduction in which sodium ions move into the cell and the charge moves rapidly through the cytosol to the next node, where the action potential continues.

sarcoma A tumor of connective tissue such as bone or cartilage.

sarcomere One compete unit of the repeating pattern of thick and thin filaments within a myofibril.

sarcoplasmic reticulum A cellular organelle that provides a muscle fiber's source of the cytosolic calcium involved in muscle contraction.

satiety A feeling of fullness.

satiety signal A response to eating that removes the sensation of hunger and sets the time period before hunger returns again.

saturated fatty acid A fatty acid in which all the carbons are linked by single covalent bonds.

scaffold An area in metaphase chromosomes formed from proteins that holds the radial loops in place.

scanning electron microscopy (SEM) A type of microscopy that utilizes an electron beam to produce an image of the three-dimensional surface of biological samples.

scavenger An animal that eats the remains of dead animals.

schizocoelous In protostomes, a pattern of development in which a solid mass of mesoderm cells splits to form the cavity that becomes the coelom.

Schwann cells The glial cells that form myelin on axons that travel outside the brain and spinal cord.

science In biology, the observation, identification, experimental investigation, and theoretical explanation of natural phenomena.

scientific method A series of steps to test the validity of a hypothesis. The experimentation often involves a comparison between control and experimental samples.

sclera The white of the vertebrate eye; a strong outer sheet that in the front is continuous with a thin, clear layer known as the cornea.

sclereid Star- or stone-shaped plant cells having tough, lignified cell walls.

sclerenchyma tissue Rigid plant tissue composed of tough-walled fibers and sclereids.

scurvy A potentially fatal vitamin C deficiency in humans characterized by weakness, bleeding gums, and tooth loss.

seaweed A multicellular protist that occurs in marine habitats and may display a relatively large and complex body.

second law of thermodynamics The second law states that the transfer of energy or the transformation of energy from one form to another increases the entropy, or degree of disorder of a system.

second messengers Small molecules or ions that relay signals inside the cell.

secondary active transport A type of transport that involves the utilization of a pre-existing gradient to drive the active transport of another solute.

secondary cell wall A plant cell wall that is synthesized and deposited between the plasma membrane and the primary cell wall after a plant cell matures and has stopped increasing in size.

secondary consumer An organism that eats primary consumers; also called a carnivore.

secondary endosymbiosis A process that occurs when a eukaryotic host cell acquires a eukaryotic endosymbiont having a primary plastid.

secondary immune response An immediate and heightened production of additional specific antibodies against a particular antigen that elicited a primary immune response.

secondary metabolism Involves the synthesis of chemicals that are not essential for cell structure and growth and are usually not required for cell survival, but are advantageous to the organism.

secondary metabolite Molecules that are produced by secondary metabolism.

secondary oocyte In animals, the large egg cell that is the result of meiosis I.

secondary phloem The inner bark of a woody plant.

secondary plastid A plastid that has originated by the endosymbiotic incorporation into a host eukaryote of a eukaryotic cell having a primary plastid.

secondary production The measure of production of heterotrophs and decomposers.

secondary spermatocytes The haploid cells produced in the primary spermatocyte by meiosis I.

secondary structure The bending or twisting of proteins into helices or β sheets. One of four levels of protein structure.

secondary succession Succession on a site that has already supported life but that has undergone a disturbance.

secretion In the production of urine, the process in which some solutes are actively transported from the interstitial fluid surrounding the epithelial cells of the tubules of an excretory organ, into the tubule lumens.

secretory pathway A pathway for the movement of larger substances, such as proteins and carbohydrates, out of the cell.

secretory vesicle A membrane vesicle carrying different types of materials that later fuses with the cell's plasma membrane to release the contents extracellularly.

seed A reproductive structure produced by flowering plants and other seed plants, usually as the result of sexual reproduction.

seed coat A hard and tough covering that develops from the ovule's integument and protects the plant's embryo.

seed plant The informal name for gymnosperms and angiosperms.

segment 1. The portion of the eye cell that contains the cell nucleus and other cytoplasmic organelles. 2. A body part in animals that is repeated many times in a row along an anteroposterior axis.

segmentation gene A gene that controls the segmentation pattern of an animal embryo.

segment-polarity gene A type of segmentation gene; a mutation in this gene causes portions of segments to be missing either an anterior or a posterior region and for adjacent regions to become mirror images of each other.

segregate To separate, as in chromosomes during mitosis.

selectable marker A gene whose presence can allow organisms (such as bacteria) to grow under a certain set of conditions. For example, an antibiotic resistance gene is a selectable marker that allows bacteria to grow in the presence of the antibiotic.

selective breeding Programs and procedures designed to modify traits in domesticated species.

selectively permeable The phenomenon that membranes allow the passage of certain ions or molecules but not others.

self-fertilization Fertilization that involves the union of a female gamete and male gamete from the same individual.

self-incompatibility (SI) Rejection of pollen that is genetically too similar to the pistil of a plant.

selfish DNA hypothesis The hypothesis that transposable elements exist because they have the characteristics that allow them to insert themselves into the host cell DNA but do not provide any advantage.

self-pollination The process in which pollen from the anthers of a flower is transferred to the stigma of the same flower, or between flowers of the same plant.

self-splicing The phenomenon that RNA itself can catalyze the removal of its own intron(s).

semelparity A reproductive pattern in which organisms produce all of their offspring in a single reproductive event.

semen A mixture containing fluid and sperm that is released during ejaculation.

semicircular canals Structures of the ear that can detect circular motions of the head.

semiconservative mechanism In this model for DNA replication, the double-stranded DNA is half conserved following the replication process such that the new double-stranded DNA contains one parental strand and one daughter strand.

semifluid A quality of motion within biomembranes; considered two-dimensional, which means that movement occurs within the plane of the membrane.

semilunar valves A one-way valve into the systemic and pulmonary arteries through which blood is pumped.

seminal vesicles Paired accessory glands that secrete fructose, the main nutrient for sperm, into the urethra to mix with the sperm.

seminiferous tubule A tightly packed structure in the testis, where spermatogenesis takes place.

sense A system that consists of sensory cells that respond to a specific type of chemical or physical stimulus and send signals to the central nervous system, where the signals are received and interpreted.

senescent Cells that have doubled many times and have reached a point where they have lost the capacity to divide any further.

sensor kinase An enzyme-linked receptor that recognizes a signal found in the environment and also has the ability to hydrolyze ATP and phosphorylate itself.

sensory neuron A neuron that detects or senses information from the outside world, such as light, sound, touch, and heat; sensory neurons also detect internal body conditions such as blood pressure and body temperature.

sensory receptor A specialized cell whose function it is to receive sensory inputs.

sensory transduction The process by which incoming stimuli are converted into neural signals.

sepal A flower organ that often functions to protect the unopened flower bud.

septum (plural, **septa**) A cross wall; examples include the cross walls that divide the hyphae of most fungi into many small cells, and the structure that separates the old and new chambers of a nautilus.

sere Each phase of succession in a community; also called a seral stage.

setae Chitinous bristles in the integument of many invertebrates.

sex chromosomes A distinctive pair of chromosomes that are different in males and females.

sex linked Refers to genes that are found on one sex chromosome but not on the other.

sex pili Hair-like structures made by bacterial F^+ cells that bind specifically to F^- cells.

sex-influenced inheritance The phenomenon in which an allele is dominant in one sex but recessive in the other.

sexual reproduction A process that requires a fertilization event in which two gametes unite to create a cell called a zygote.

sexual selection A type of natural selection that is directed at certain traits of sexually reproducing species that make it more likely for individuals to find or choose a mate and/or engage in successful mating.

Shannon diversity index (H_S) A means of measuring the diversity of a community; $H_S = -\Sigma p_i \ln p_i$.

shared derived character A trait that is shared by a group of organisms but not by a distant common ancestor.

shared primitive character A trait shared with a distant ancestor.

shattering The process by which ears of wild grain crops break apart and disperse seeds.

shell A tough, protective covering that is impermeable to water and prevents the embryo from drying out.

shivering thermogenesis Rapid muscle contractions in an animal, without any locomotion, in order to raise body temperature.

shoot apical meristem (SAM) The region of rapidly dividing plant cells at plant shoot apices.

shoot meristem The tissue that produces all aerial parts of the plant, which include the stem as well as lateral structures such as leaves and flowers.

shoot system The collection of plant organs produced by shoot apical meristems.

short-day plant A plant that flowers only when the night length is longer than a defined period. Such night lengths occur in late summer, fall, or winter, when days are short.

short tandem repeat sequences (STRs) Sequences found in multiple sites in the genome of humans and other species, and they vary in length among different individuals.

shotgun DNA sequencing A strategy for sequencing an entire genome by randomly sequencing many different DNA fragments.

sickle-cell anemia A disease due to a genetic mutation in a hemoglobin gene in which sickle-shaped red blood cells are less able to move smoothly through capillaries and can block blood flow, resulting in severe pain and cell death of the surrounding tissue.

sieve elements Thin-walled living cells that are arranged end to end in a plant to form pipelines.

sieve plate The perforated end wall of a mature sieve-tube element.

sieve plate pore One of many perforations in a plant's sieve plate.

sieve-tube elements A component of the phloem tissues of flowering plants; these structures are arranged end to end to form transport pipes.

sigma factor A protein that plays a key role in bacterial promoter recognition and recruits RNA polymerase to the promoter.

sign stimulus A factor that initiates a fixed-action pattern of behavior.

signal recognition particle (SRP) A protein/RNA complex that recognizes the ER signal sequence and pauses translation; then, SRP binds to a receptor in the ER membrane, which docks the ribosome over a translocation channel.

signal transduction pathway A group of proteins that convert an initial signal to a different signal inside the cell.

signal An agent that can influence the properties of cells.

silencer A response element that prevents transcription of a given gene when its expression is not needed.

silent mutation A gene mutation that does not alter the amino acid sequence of the polypeptide, even though the nucleotide sequence has changed.

simple Mendelian inheritance The inheritance pattern of traits affected by a single gene that is found in two variants, one of which is completely dominant over the other.

simple translocation A single piece of chromosome that is attached to another chromosome.

single-factor cross A cross in which an experimenter follows the variants of only one trait.

single-strand binding protein A protein that binds to both of the single strands of parental DNA and prevents them from re-forming a double helix.

sinoatrial (SA) node A collection of modified cardiac cells that spontaneously and rhythmically generates action potentials that spread across the entire atria; also known as the pacemaker of the heart.

sister chromatids The two duplicated chromatids that are still joined to each other after DNA replication.

skeletal muscle A type of muscle tissue that is attached to bones in vertebrates and to the exoskeleton of invertebrates.

skeleton A structure or structures that serve one or more functions related to support, protection, and locomotion.

sliding filament mechanism The way in which a muscle fiber shortens during muscle contraction.

SLOSS debate In conservation biology, the debate over whether it is preferable to protect one single, large reserve or several smaller ones (SLOSS stands for single large or several small).

slow block to polyspermy Events that produce barriers to more sperm penetrating an already fertilized egg.

slow fiber A skeletal muscle fiber with a low rate of myosin ATP hydrolysis.

slow-oxidative fiber A skeletal muscle fiber that has a low rate of myosin ATP hydrolysis but has the ability to make large amounts of ATP; used for prolonged, regular activity.

small effector molecule With regard to transcription, refers to a molecule that exerts its effects by binding to a regulatory transcription factor and causing a conformational change in the protein.

small intestine A tube that leads from the stomach to the large intestine where nearly all digestion of food and absorption of food nutrients and water occur.

smooth endoplasmic reticulum (smooth ER) The part of the ER that is not studded with ribosomes. This region is continuous with the rough ER and functions in diverse metabolic processes such as detoxification, carbohydrate metabolism, accumulation of calcium ions, and synthesis and modification of lipids.

smooth muscle A type of muscle tissue that surrounds hollow tubes and cavities inside the body's organs, such that their contraction can propel the contents of those organs.

soil horizon Layers of soil, ranging from topsoil to bedrock.

solute A substance dissolved in a liquid.

solute potential (S) The osmotic potential; an element in the water potential equation.

solution A liquid that contains one or more dissolved solutes.

solvent The liquid in which a solute is dissolved.

soma See cell body.

somatic cell The type of cell that constitutes all cells of the body excluding the germ-line cells. Examples include skin cells and muscle cells.

somatic embryogenesis The production of plant embryos from body (somatic) cells.

somatic nervous system The division of the peripheral nervous system that senses the external environmental conditions and controls skeletal muscles.

somites Blocklike structures of mesoderm that are formed during neurulation.

soredia   Small clumps of hyphae surrounding a few algal cells that can disperse in wind currents; an asexual reproductive structure produced by lichens.

sorting signal   A short amino acid sequence in a protein's structure that directs the protein to its correct location.

source pool   The pool of species on the mainland that is available to colonize an island.

Southern blotting   A method in which a labeled probe, which is a strand of DNA from a specific gene, is used to identify that gene in a mixture of many chromosomal DNA fragments.

speciation   The formation of new species.

species   A group of related organisms that share a distinctive form in nature.

species-area effect   The relationship between the amount of available area and the number of species present.

species concepts   Different approaches for distinguishing species, including the phylogenetic, biological, evolutionary, and ecological species concepts.

species interactions   A part of the study of population ecology which focuses on interactions such as predation, competition, parasitism, mutualism and commensalism.

species richness   The numbers of species in a community.

specific (acquired) immunity   An immunity defense that develops only after the body is exposed to foreign substances.

Spemann's organizer   An extremely important morphogenetic field in the early gastrula; the organizer secretes morphogens responsible for inducing the formation of a new embryonic axis.

sperm   Refers to a "male" gamete that is generally smaller than the female gamete (egg); the male gamete is often motile or in many species of plants transported to the egg via a pollen tube.

sperm storage   A method of synchronizing the production of offspring with favorable environmental conditions; female animals store and nourish sperm in their reproductive tract for long periods of time.

spermatids   In animals, the haploid cells produced by the secondary spermatocytes by meiosis II; these cells eventually differentiate into sperm cells.

spermatogenesis   The formation of sperm.

spermatogonia   In animals, diploid germ cells that give rise to the male gametes, the spermatozoa.

spermatophytes   All of the living and fossil seed plant phyla.

spinal cord   In vertebrates and simpler chordates, the structure that connects the brain to all areas of the body and constitutes the central nervous system.

spinal nerve   A nerve that connects the peripheral nervous system and the spinal cord.

spinneret   A spider's abdominal silk gland; also found in the mouths of caterpillars.

spiracle   A pore to the trachea that is found in the bodies of insects.

spiral cleavage   A mechanism of development in which the planes of cell cleavage are oblique to the axis of the embryo.

spirilli   Rigid, spiral-shaped prokaryotic cells.

spirochaetes   Flexible, spiral-shaped prokaryotic cells.

spliceosome   A complex of several subunits known as snRNPs that removes introns from eukaryotic pre-mRNA.

splicing   The process whereby introns are removed from RNA and the remaining exons are connected to each other.

spongin   A tough protein that lends skeletal support to a sponge.

spongocoel   A central cavity in the body of a sponge.

spongy parenchyma   Photosynthetic tissue of the plant leaf mesophyll that contains cells separated by abundant air spaces.

spontaneous mutation   A mutation resulting from abnormalities in biological processes.

sporangia   Structures that produce and disperse the spores of plants, fungi, or protists.

spore   Single-celled reproductive structure that is dispersed into the environment and is able to grow into new fungal hyphae, plant gametophytes, or protists if they find suitable habitats.

sporic life cycle   *See* alternation of generations.

sporophyte   The diploid generation of plants or multicellular protists that have a sporic life cycle; this generation produces haploid spores by the process of meiosis.

sporopollenin   The tough material that composes much of the walls of plant spores and helps to prevent cellular damage during transport in air.

stabilizing selection   A pattern of natural selection that favors the survival of individuals with intermediate phenotypes.

stamen   A flower structure that makes the male gametophyte, pollen.

standard metabolic rate (SMR)   A method for measuring the BMR of ectotherms at a standard temperature for each species—one that approximates the average temperature that a species normally encounters.

standing crop   The total biomass in an ecosystem at any one point in time.

starch   A polysaccharide produced by the cells of plants and some algal protists.

start codon   A three-base sequence—usually AUG—that specifies the first amino acid in a polypeptide.

statocyst   An organ of equilibrium found in many animal species.

statoliths   1. Tiny granules of sand or other dense objects that aid in equilibrium in many vertebrates. 2. In plants, a starch-heavy plastid that allows both roots and shoots to detect gravity.

stem   A plant organ that produces buds, leaves, branches and reproductive structures.

stem cell   A cell that divides and supplies the cells that construct the bodies of all animals and plants.

stereoisomers   Isomers with identical bonding relationships, but the spatial positioning of the atoms differs in the two isomers.

sternum   The breastbone.

steroid   A lipid molecule with a chemical structure containing four interconnected rings of carbon atoms.

steroid receptor   A transcription factor that recognizes a steroid hormone and usually functions as a transcriptional activator.

steroid   A lipid containing four interconnected rings of carbon atoms; function as hormones in animals and plants.

sticky ends   Single-stranded ends of DNA fragments that will hydrogen-bond to each other due to their complementary sequences.

stigma   1. A topmost portion of the pistil, which receives and recognizes pollen of the appropriate species or genotype. 2. In many protists, a red-colored cellular structure.

stomach   A saclike organ that most likely evolved as a means of storing food; it partially digests some of the macromolecules in food and regulates the rate at which the contents empty into the small intestine.

stomata   Surface pores on plant surfaces that can be closed to retain water or open to allow the entry of CO_2 needed for photosynthesis and the exit of oxygen and water vapor.

stop codon   One of three three-base sequences—UAA, UAG, and UGA—that signals the end of translation.

strain   Within a given species, a lineage that has genetic differences compared to another strain.

strand   A structure of DNA (or RNA) formed by the covalent linkage of nucleotides in a linear manner.

stretch receptor   A type of mechanoreceptor found widely in organs and muscle tissues that can be distended.

striated muscle   Skeletal and cardiac muscle with a series of light and dark bands perpendicular to the muscle's long axis.

stroke   The condition that occurs when blood flow to part of the brain is disrupted.

stroke volume (SV)   The amount of blood ejected with each beat, or stroke, of the heart.

stroma   The fluid-filled region of the chloroplast between the thylakoid membrane and the inner membrane.

stromatolite   A layered calcium carbonate structure in an aquatic environment produced by cyanobacteria.

strong acid   An acid that completely ionizes in solution.

structural gene   Refers to most genes, which produce an mRNA molecule that contains the information to specify a polypeptide with a particular amino acid sequence.

structural genomics   Genomic methods aimed at the direct analysis of the DNA itself.

structural isomers Isomers that contain the same atoms but in different bonding relationships.

style The elongate pistil structure through which the pollen tube of a flower grows.

stylet A sharp, piercing organ in the mouth of nematodes and some insects.

submetacentric A chromosome in which the centromere is off center.

subsidence zones Areas of high pressure that are the sites of the world's tropical deserts, because the subsiding air is relatively dry, having released all of its moisture over the equator.

substrate-level phosphorylation A method of synthesizing ATP that occurs when an enzyme directly transfers a phosphate from one molecule to a different molecule.

substrates The reactant molecules and/or ions that bind to an enzyme at the active site and participate in a chemical reaction.

succession The gradual and continuous change in species composition and community structure over time.

sugar sink The plant tissues or organs in which more sugar is consumed than is produced by photosynthesis.

sugar source The plant tissues or organs that produce more sugar than they consume in respiration.

sum rule The probability that one of two or more mutually exclusive outcomes will occur is the sum of the probabilities of the possible outcomes.

supercoiling A method of compacting chromosomes; the phenomenon of forming additional coils around the long, thin DNA molecule.

supergroup A proposed way to organize eukaryotes into monophyletic groups.

surface area/volume (SA/V) ratio The ratio between a structure's surface area and the volume in which the structure is contained.

surfactant A mixture of proteins and amphipathic lipids produced in certain alveolar cells that acts to reduce surface tension in lungs.

survivorship curve A graphical plot of the numbers of surviving individuals at each age.

suspension feeder An aquatic animal that sifts water, filtering out the organic matter and expelling the rest.

suspensor A short chain of cells at the base of an early angiosperm embryo that provides anchorage and nutrients.

swim bladder A gas-filled, balloon-like structure that helps a fish to remain buoyant in the water even when it is completely stationary.

swimmeret An abdominal appendage in a crustacean that provides movement.

symbiosis An intimate association between two or more organisms.

symbiotic Describes a relationship in which two or more different species live in direct contact with each other.

sympathetic division The division of the autonomic system that is responsible for rapidly activating systems that provide immediate energy to the body in response to danger or stress.

sympatric The term used to describe species occurring in the same geographic area.

sympatric speciation A form of speciation that occurs when members of a species that initially occupy the same habitat within the same range diverge into two or more different species.

symplast All of a plant's protoplasts (the cell contents without the cell walls) and plasmodesmata.

symplastic transport The movement of a substance from the cytosol of one cell to the cytosol of an adjacent cell via membrane-lined channels called plasmodesmata.

symplesiomorphy *See* shared primitive character.

symporter A type of transporter that binds two or more ions or molecules and transports them in the same direction.

synapomorphy *See* shared derived character.

synapse A junction where a nerve terminal meets a target neuron, muscle cell, or gland and can communicate with other cells.

synapsis The process of forming a bivalent.

synaptic cleft The extracellular space between two neurons.

synaptic integration The summation of EPSPs generated at one time, which can bring the membrane potential to the threshold potential at the axon hillock for action potential firing.

synaptic plasticity A change in synapses that occurs as a result of learning.

synaptic signaling A specialized form of paracrine signaling that occurs in the nervous system of animals.

synergids In the female gametophyte of a flowering plant, the two cells adjacent to the egg cell that help to import nutrients from maternal sporophyte tissues.

systematics The study of biological diversity and evolutionary relationships among organisms, both extinct and modern.

systemic acquired resistance (SAR) A whole-plant defensive response to pathogenic microorganisms.

systemic circulation The pumping of blood from the left side of the heart to the body to drop off oxygen and nutrients and pick up carbon dioxide and wastes. The blood then returns to the right side of the heart.

systemic hypertension An arterial blood pressure above normal, which in humans ranges from systolic/diastolic pressures of about 90/60 to 120/80 mmHg; often called hypertension or high blood pressure.

systems biology A field of study in which researchers study living organisms in terms of their underlying network structure—groups of structural and functional connections—rather than their individual molecular components.

systole The second phase of the cardiac cycle, in which the ventricles contract and eject the blood through the open semilunar valves.

T

T cell A type of lymphocyte that directly kills infected, mutated, or transplanted cells.

tagmata The fusion of body segments into functional units.

tapetum lucidum A reflective layer of tissue located beneath the photoreceptors at the back of the eye of certain animals.

taproot system The root system of eudicots, which has one main root with many branch roots.

***Taq* polymerase** A heat-stable form of DNA polymerase; one of several reagents required for synthesis of DNA via polymerase chain reaction (PCR).

TATA box One of three features found in most promoters; the others are the transcriptional start site and response elements.

taxis A directed type of response either toward or away from a stimulus.

taxon A group of species that are evolutionarily related to each other. In taxonomy, each species is placed into several taxons that form a hierarchy from large (domain) to small (genus).

taxonomy The field of biology that is concerned with the theory, practice, and rules of classifying living and extinct organisms and viruses.

telocentric A chromosome in which the centromere is at the end.

telomerase An enzyme that catalyzes the replication of the telomere.

telomere A region at the ends of eukaryotic chromosomes where a specialized form of DNA replication occurs.

telophase The phase of mitosis during which the chromosomes decondense and the nuclear membrane re-forms.

temperate phage A bacteriophage that may spend some of its time in the lysogenic cycle.

template The DNA strand that is used as a template for RNA synthesis or DNA replication.

termination codon *See* stop codon.

termination stage The final stage of transcription or translation in which the process ends.

terminator A sequence that specifies the end of transcription.

territory A fixed area in which an individual or group excludes other members of its own species, and sometimes other species, by aggressive behavior or territory marking.

tertiary consumer An organism that feeds on secondary consumers.

tertiary endosymbiosis The acquisition by eukaryotic protist host cells of plastids from cells that possess secondary plastids.

tertiary plastid A plastid acquired by the incorporation into a host cell of an endosymbiont having a secondary plastid.

tertiary structure The three-dimensional shape of a single polypeptide. One of four levels of protein structure.

testcross A cross in which an individual with a dominant phenotype is mated with a homozygous recessive individual.

testes The male gonads of certain animals, where sperm are produced.

testosterone The primary androgen in vertebrates.

tetrad *See* bivalent.

tetraploid   An organism or cell that has four sets of chromosomes.

tetrapod   A vertebrate animal having four legs or leglike appendages.

thalamus   In vertebrates, a gland that plays a major role in relaying sensory information to appropriate parts of the cerebrum and, in turn, sending outputs from the cerebrum to other parts of the brain. It receives input from all sensory systems and is involved in the perception of pain and the degree of mental arousal in the cortex.

theory   In biology, a broad explanation of some aspect of the natural world that is substantiated by a large body of evidence. Biological theories incorporate observations, hypothesis testing, and the laws of other disciplines such as chemistry and physics. A theory makes valid predictions.

thermodynamics   The study of energy interconversions.

thermoreceptor   A sensory receptor that responds to cold and heat.

theropods   A group of bipedal saurischian dinosaurs.

thick filament   A skeletal muscle structure composed almost entirely of the motor protein myosin.

thigmotropism   Touch responses in plants.

thin filament   A skeletal muscle structure that contains the cytoskeletal protein actin, as well as two other proteins—troponin and tropomyosin—that play important roles in regulating contraction.

30-nm fiber   Nucleosome units organized into a more compact structure that is 30 nm in diameter.

thoracic breathing   Breathing in which coordinated contractions of muscles expand the rib cage, creating a negative pressure to suck air in and then forcing it out later; found in amniotes.

threatened species   Those species that are likely to become endangered in the future.

threshold concentration   The concentration above which a morphogen will exert its effects but below which it is ineffective.

thrifty genes   Genes that boosted our ancestors' ability to store fat from each feast in order to sustain them through the next famine.

thrombocytes   Intact cells in the blood of vertebrates other than mammals that play a crucial role in the formation of blood clots.

thylakoid   A flattened, plate-like membranous region found in cyanobacterial cells and the chloroplasts of photosynthetic protists and plants; the location of the light reactions of photosynthesis.

thylakoid lumen   The fluid-filled compartment within the thylakoid.

thylakoid membrane   A membrane within the chloroplast that forms many flattened, fluid-filled tubules that enclose a single, convoluted compartment. It is the site where the light-dependent reactions of photosynthesis occurs.

thymine (T)   A pyrimidine base found in DNA.

thymine dimer   One type of pyrimidine dimer; a site where two adjacent thymine bases become covalently cross-linked to each other.

thyroglobulin   A protein found in the colloid of the thyroid gland.

thyroxine   A thyroid hormone that contains iodine and helps regulate metabolic rate.

Ti plasmid   Tumor-inducing plasmid found in *Agrobacterium tumefaciens;* it is used as a cloning vector to transfer genes into plant cells.

tidal ventilation   A type of breathing in which the lungs are inflated with air, and then the chest muscles and diaphragm relax and recoil back to their original positions as an animal exhales. During exhalation, air leaves via the same route that it entered during inhalation, and no new oxygen is delivered to the airways at that time.

tidal volume   The volume of air that is normally breathed in and out at rest.

tight junction   A type of junction that forms a tight seal between adjacent epithelial cells and thereby prevents molecules from leaking between cells; also called an occluding junction.

tissue   The association of many cells of the same type, for example, muscle tissue.

tolerance   A mechanism for succession in which any species can start the succession, but the eventual climax community is reached in a somewhat orderly fashion; early species neither facilitate nor inhibit subsequent colonists.

tonoplast   The membrane of the central vacuole in a plant or algal cell.

torpor   The strategy in the smallest endotherms of lowering internal body temperature to just a few degrees above that of the environment in order to conserve energy.

torus   The nonporous, flexible central region of a conifer pit that functions like a valve.

total fertility rate   The average number of live births a female has during her lifetime.

total peripheral resistance (TPR)   The sum of all the resistance in all the arterioles.

totipotent   The ability of a fertilized egg to produce all of the cell types in the adult organism; also the ability of unspecialized plant cells to regenerate an adult plant.

trachea   1. A sturdy tube arising from the spiracles of an insect's body involved in respiration. 2. The name of the tube leading to the lungs of air-breathing vertebrates.

tracheal system   In insects, a series of finely branched air tubes called tracheae that lead into the body from pores called spiracles.

tracheid   A type of dead, lignified plant cell that conducts water, along with dissolved minerals and certain organic compounds.

tracheophytes   A term used to describe the vascular plants.

tract   A parallel bundle of myelinated axons.

traffic signal   *See* sorting signal.

trait   *See* character.

transcription   The use of a gene sequence to make a copy of RNA: transcription occurs in three stages called initiation, elongation, and termination.

transcription factor   A protein that influences the ability of RNA polymerase to transcribe genes.

transcriptional start site   The site in a promoter where transcription begins.

transduction   A type of genetic transfer between bacteria in which a virus infects a bacterial cell and then transfers some of that cell's DNA to another bacterium.

***trans*-effect**   In both prokaryotes and eukaryotes, a form of genetic regulation that can occur even though two DNA segments are not physically adjacent. The action of the *lac* repressor on the *lac* operon is a trans-effect.

transfer RNA (tRNA)   An RNA that carries amino acids and is used to translate mRNA into polypeptides.

transformation   A type of genetic transfer between bacteria in which a segment of DNA from the environment is taken up by a competent cell and incorporated into the bacterial chromosome.

transgenic   The term used to describe an organism that carries genes that were introduced using molecular techniques such as gene cloning.

transition state   In a chemical reaction, a state in which the original bonds have stretched to their limit; once this state is reached, the reaction can proceed to the formation of products.

transitional form   An organism that provides a link between earlier and later forms in evolution.

translation   The process of synthesizing a specific polypeptide on a ribosome; a nucleotide sequence in mRNA is used to make an amino acid sequence of a polypeptide. The process of translation occurs in three stages: initiation, elongation, and termination.

translocation   A phenomenon that occurs when one segment of a chromosome becomes attached to a different chromosome.

transmembrane gradient   The phenomenon that the concentration of a solute is higher on one side of a membrane than on the other.

transmembrane protein   A protein that has one or more regions that are physically embedded in the hydrophobic region of the cell's phospholipid bilayer.

transmembrane segment   A region of a membrane protein that is a stretch of hydrophobic amino acids that spans or traverses the membrane from one leaflet to the other.

transmission electron microscopy (TEM)   A type of microscopy in which a beam of electrons is differentially transmitted through a biological sample to form an image on a photographic plate or screen.

transpiration   The evaporative loss of water from plant surfaces into sun-heated air.

transport protein   Proteins embedded within the phospholipid bilayer that allow plasma membranes to be selectively permeable by providing a passageway for the movement of some but not all substances across the membrane.

transporter A membrane protein that binds a solute and undergoes a conformational change to allow the movement of the solute across a membrane.

transposable element (TE) A segment of DNA that can move from one site to another.

transposase An enzyme that facilitates transposition.

transposition The process in which a short segment of DNA moves within a cell from its original site to a new site in the genome.

transposon A transposable element that moves via DNA that is removed from one site and inserted into a new site.

transverse tubules (T-tubules) Invaginations of the plasma membrane that open to the extracellular fluid and conduct action potentials from the outer surface of a muscle fiber to the myofibrils.

trichome A projection from the epidermal tissue of a plant that offers protection from excessive light, ultraviolet radiation, extreme air temperature, or attack by herbivores.

trichromatic color vision The ability to distinguish blue, green, and red colors.

triiodothyronine A thyroid hormone that contains iodine and helps regulate metabolic rate.

triplet A group of three bases that function as a codon.

triploblastic Having three distinct germ layers—endoderm, ectoderm, and mesoderm.

triploid An organism or cell that has three sets of chromosomes.

trisomic An aneuploid organism that has one too many chromosomes.

trochophore larva A distinct larval stage of many invertebrate phyla.

trophic level Each feeding level in a food chain.

trophic-level transfer efficiency The amount of energy at one trophic level that is acquired by the trophic level above and incorporated into biomass.

tropomyosin A protein that plays an important role in regulating muscle contraction.

troponin A protein that plays an important role in regulating muscle contraction.

trp operon An operon of *E. coli* that encodes enzymes required to make the amino acid tryptophan, a building block of cellular proteins.

true-breeding line A strain that continues to exhibit the same trait after several generations of self-fertilization or inbreeding.

trypsin A protease involved in the breakdown of proteins in the small intestine.

t-snare A protein in a target membrane that recognizes a v-snare in a membrane vesicle.

tube cell In a plant, the cell that stores proteins and lipids that may be used during later stages of male gametophyte development, after pollen has germinated. This cell forms the pollen tube that after pollination grows to reach the female gametophyte.

tube feet Echinoderm structures that function in movement, gas exchange, and feeding.

tumor An overgrowth of cells that serves no useful purpose.

tumor-suppressor gene A gene that when normal (that is, not mutant) encodes a protein that prevents cancer; however, when a mutation eliminates its function, cancer may occur.

tunic A nonliving structure that encloses a tunicate, made of a protein and a cellulose-like material called tunicin.

turgid The term used to describe a plant cell whose cytosol is so full of water that the plasma membrane presses right up against the cell wall; as a result, turgid cells are firm or swollen.

turgor pressure *See* osmotic pressure.

two-component regulatory system A signaling system found in bacteria and plants composed of an enzyme-linked receptor called a sensor kinase and a response regulator, which is usually a protein that regulates the expression of many genes.

two-dimensional (2D) gel electrophoresis A technique to separate many different proteins that utilizes an isoelectric-focusing tube gel in the first dimension, and an SDS-slab gel in the second dimension.

two-factor cross Crosses in which the inheritance of two different traits are followed.

type 1 diabetes mellitus (T1DM) A disease in which the pancreas does not produce sufficient insulin; as a result, extracellular glucose cannot cross plasma membranes, and so glucose accumulates to very high concentrations in the blood.

type 2 diabetes mellitus (T2DM) A disease in which the pancreas functions normally, but the cells of the body lose much of their ability to respond to insulin.

U

ubiquitin A small protein in eukaryotic cells that directs unwanted proteins to a proteasome by its covalent attachment.

ultimate cause The reason a particular behavior evolved, in terms of its effect on reproductive success.

umbrella species A species whose habitat requirements are so large that protecting them would protect many other species existing in the same habitat.

unconditioned response A response that is already elicited by an unconditioned stimulus.

unconditioned stimulus A factor that originally elicits a response.

uniform A pattern of dispersion within a population, in which individuals maintain a certain minimum distance between themselves to produce an evenly spaced distribution.

uniporter A type of transporter that binds a single molecule or ion and transports it across the membrane.

unipotent A term used to describe a stem cell found in the adult that can only produce daughter cells that differentiate into one cell type.

unsaturated The quality of a lipid when a double bond is formed.

unsaturated fatty acid A fatty acid that contains one or more $C = C$ double bonds.

upwelling In the ocean, a phenomenon that carries mineral nutrients from the bottom waters to the surface.

uracil (U) A pyrimidine base found in RNA.

urea A nitrogenous waste commonly produced in many terrestrial species.

uremia A condition characterized by the presence of nitrogenous wastes, such as urea, in the blood; typically results from kidney disease.

ureter A structure in the kidney through which urine flows from the kidney into the urinary bladder.

urethra The structure through which urine is eliminated from the body.

uric acid A nitrogenous waste commonly produced in many terrestrial species.

urinary bladder The structure that collects urine before it is eliminated.

urinary system The structures which collectively act to filter blood or hemolymph and excrete wastes while recapturing useful compounds. In humans, it includes the two kidneys, two ureters, the urinary bladder, and the urethra.

uterine cycle The cyclical changes that occur in the uterus in parallel with the ovarian cycle in a female mammal.

uterus A large, conical-shaped structure that consists of an inner lining of glandular and secretory cells called the endometrium, and a thick muscular layer called the myometrium; the uterus is specialized for carrying the developing fetus.

V

vaccination The injection into the body of small quantities of living or dead microbes, small quantities of toxins, or harmless antigenic molecules derived from a microorganism or its toxin, resulting in a primary immune response including the production of memory cells. Subsequent natural exposure to the immunizing antigen results in a response that can prevent or reduce the severity of disease.

vacuole Specialized compartments found in eukaryotic cells that function in storage, the regulation of cell volume, and degradation.

vagina A tubular, smooth muscle structure in the female into which sperm are deposited.

vaginal diaphragm A barrier method of preventing fertilization in which the device is placed in the upper part of the vagina just prior to intercourse and blocks movement of sperm to the cervix.

valence electron An electron in the outer shell that is available to combine with other atoms. Such electrons allow atoms to form chemical bonds with each other.

variable region A domain within an immunoglobulin that serves as the antigen-binding site.

vas deferens   A muscular tube through which sperm leave the epididymis.

vasa recta capillaries   Capillaries in the medulla in the nephron of the kidney.

vascular bundle   Primary plant vascular tissues that occur in a group.

vascular cambium   A meristematic tissue of plants that produces both wood and inner bark.

vascular plant   A plant that can transport water, sugar, and salts throughout the plant body via xylem and phloem tissues.

vascular tissue   Plant tissue that makes up the vascular system, which conducts materials within the plant body and also provides support.

vasectomy   A surgical procedure in men that severs the vas deferens, thereby preventing the release of sperm at ejaculation.

vasoconstriction   A decrease in blood vessel radius; an important mechanism for directing blood flow away from specific regions of the body.

vasodilation   An increase in blood vessel radius; an important mechanism for directing blood flow to specific regions of the body.

vasotocin   A peptide that is responsible for regulating salt and water balance in the blood of nonmammalian vertebrates.

vector   A type of DNA that acts as a carrier of a DNA segment that is to be cloned.

vegetal pole   In triploblast organisms, the pole of the egg where the yolk is most concentrated.

vegetative growth   The production of new tissues by the shoot apical meristem and root apical meristem during seedling development and growth of mature plants.

vein   In animals, a blood vessel that returns blood to the heart.

veliger   A free-swimming larva that has a rudimentary foot, shell, and mantle.

ventral   Refers to the lower side of an animal.

ventricle   In the heart, a chamber to pump blood out of the heart.

venule   In animals, a small, thin-walled extension of a capillary that empties into larger vessels called veins that return blood to the heart for another trip around the circulation.

vernalization   The induction of flowering by cold treatment.

vertebrae   A bony or cartilaginous column of interlocking structures that provides support and also protects the nerve cord, which lies within its tubelike structure.

vertebrate   An organism with a backbone.

vertical evolution   A process that involves genetic changes in a series of ancestors, which form a lineage.

vesicle   A small membrane-enclosed sac within a cell.

vessel   In a plant, a pipeline-like file of dead, water-conducting vessel elements.

vessel element   A type of plant cell that conducts water, along with dissolved minerals and certain organic compounds.

vestibular system   The organ of balance in vertebrates, located in the inner ear next to the cochlea.

vestigial structure   An anatomical feature that has no apparent function but resembles a structure of a presumed ancestor.

vibrios   Comma-shaped prokaryotic cells.

villi   Finger-like projections extending from the luminal surface into the lumen of the small intestine; these are specializations that aid in digestion and absorption.

viral envelope   A structure enclosing the capsid that consists of a membrane derived from the plasma membrane of the host cell and embedded with virally encoded spike glycoproteins.

viral genome   The genetic material in a virus.

viral reproductive cycle   When a virus infects a cell, the series of steps that results in the production of new viruses.

viral vector   A type of vector used in cloning experiments that is derived from a virus.

viroid   An RNA particle that infects plant cells.

virulent phage   A phage that follows only the lytic cycle.

virus   A small infectious particle that consists of nucleic acid enclosed in a protein coat.

visceral mass   In mollusks, a structure that rests atop the foot and contains the internal organs.

vitamin   An important organic nutrient that serves as a coenzyme for metabolic and biosynthetic reactions.

vitamin D   A molecule that regulates calcium levels in the blood through an action on intestinal transport of calcium ions.

vitreous humor   A thick liquid in the large posterior cavity of the vertebrate eye, which helps maintain the shape of the eye.

viviparity   The process in which the embryo develops within the mother.

viviparous   The term used to describe an organism whose eggs develop within the uterus, receiving nourishment from the mother via a placenta.

volt   A unit of measurement of the difference in charge (the electrical force) such as between the interior and exterior of the cell.

voltage-gated channel   A channel that opens and closes in response to changes in the amount of electric charge across the membrane.

v-snare   A protein incorporated into the vesicle membrane during vesicle formation that is recognized by a t-snare in a target membrane.

W

water potential   The potential energy of water.

water vascular system   A network of canals powered by hydraulic power, that is, by water pressure generated by the contraction of muscles that enables the extension and contraction of the tube feet, allowing echinoderms to move slowly.

wavelength   The distance from the peak of one sound or light wave to the next.

waxy cuticle   A protective, water-proof layer of wax and polyester present on most surfaces of vascular plant sporophytes.

weak acid   An acid that only partially ionizes in solution.

weathering   The physical and chemical breakdown of rock.

white blood cell   A cell that develops from the inner parts (the marrow) of certain bones in vertebrates; all white blood cells (known as leukocytes) perform vital functions that defend the body against infection and disease.

white matter   Brain tissue that consists of myelinated axons that are bundled together in large numbers to form tracts.

whorls   In a flower, four concentric rings of sepals, petals, stamens, and carpels.

wild-type allele   One or more prevalent alleles in a population.

wood   A secondary plant tissue composed of numerous pipelike arrays of dead, empty, water-conducting cells whose walls are strengthened by an exceptionally tough secondary metabolite known as lignin.

X

X inactivation   The phenomenon in which one X chromosome in the somatic cells of female mammals is inactivated, meaning that its genes are not expressed.

X inactivation center (Xic)   A short region on the X chromosome known to play a critical role in X inactivation.

xenoestrogen   A synthetic compound that exerts estrogen-like actions or, in some cases, inhibits the actions of the body's own estrogen; the consequences are dramatic on fertility and development of embryos and fetuses.

X-linked gene   A gene found on the X chromosome but not on the Y.

X-linked inheritance   The pattern displayed by pairs of dominant and recessive alleles located on X chromosomes.

X-ray crystallography   A technique in which researchers must purify a molecule such as a protein or protein complex and expose it to conditions that cause the proteins molecules to associate with each other in an ordered array. In other words, the proteins form a crystal. When a crystal is exposed to X-rays, the resulting pattern can be analyzed mathematically to determine the three-dimensional structure of the crystal's components.

xylem   A plant vascular tissue that conducts water, minerals, and organic compounds.

xylem loading   The process by which root xylem parenchyma cells transport ions and water across their membranes into the xylem apoplast, which includes the vessel elements and tracheids.

Y

yeast   A fungus that can occur as a single cell and that reproduces by budding.

yolk sac   An extraembryonic membrane in the amniotic egg that encloses a stockpile of nutrients, in the form of yolk, for the developing embryo.

Z

Z line   A network of proteins that anchor thin filaments in a sarcomere.

Z scheme   According to this scheme, an electron absorbs light energy twice, and it loses some of that energy as it flows along the electron transport chain in the thylakoid membrane. The energy diagram of this process occurs in a zigzag pattern.

zero population growth   The situation in which no changes in population size occur.

zoecium   A nonliving case that houses a bryozoan.

zone of elongation   The area above the root apical meristem of a plant where cells extend by water uptake, thereby dramatically increasing root length.

zone of maturation   The area above the zone of elongation in a plant where root cell differentiation and tissue specialization occur.

zooplankton   Aquatic organisms including minute animals consisting of some worms, copepods, tiny jellyfish, and the small larvae of invertebrates and fish that graze on the phytoplankton.

zygomycete   A type of fungus that produces distinctive, large zygospores as the result of sexual reproduction.

zygospore   A dark-pigmented, thick-walled spore that matures within the zygosporangium of zygomycete fungi.

zygote   A diploid cell formed by the fusion of two gametes that divides and develops into an embryo, and eventually into an adult organism.

zygotic life cycle   In this type of life cycle, haploid cells transform into gametes.

zooplankton   Open-ocean organisms including minute animals consisting of some worms, copepods, tiny jellyfish, and the small larvae of invertebrates and fish that graze on the phytoplankton.

PHOTO CREDITS

Page numbers followed by *t* and *f* indicate tables and figures, respectively.

Q